P9-CBL-426

■ A Practical Guide to the Marine Animals of Northeastern North America

A Practical Guide to the Marine Animals of Northeastern North America

Leland W. Pollock

Rutgers University Press
New Brunswick, New Jersey, and London

This book is dedicated to the pioneering marine scientists and artists of the Northeast upon whose foundational works this volume is based.

Library of Congress Cataloging-in-Publication Data

Pollock, Leland W., 1943–
A practical guide to the marine animals of northeastern North America / Leland W. Pollock.
p. cm.
Includes bibliographical references (p.) and index.
ISBN 0-8135-2398-2 (alk. paper). — ISBN 0-8135-2399-0 (pbk. : alk. paper)
1. Marine animals—Atlantic Coast (North America)—Identification. I. Title.
QL 157.A84P65 1997
591.77'0974—DC21 96-39284
CIP

Printed on partially recycled paper.

Manufactured in the United States of America.

■ Contents

■ Acknowledgments

Although this survey has been compiled by an individual, it is, in fact, largely a synthesis of the work of many, who, over the years, have contributed their expert observations to the marine biological literature. In addition to my own observations, I have relied on some of the same distinguishing characters that have appeared in key after key, either because they are obvious points of difference or because time-honored use has proven them valuable. Similarly, for biological and ecological notes on each species, I have drawn from my experience and have used sources cited in each section. In addition, I have found valuable descriptive material in the works of Miner (1950) and Gosner (1971) on marine invertebrates of the Northeast, Hayward and Ryland (1990) for northern animals, and Lippson and Lippson (1997), Ruppert and Fox (1988), and Sterrer (1986) for Virginian and Carolinian fauna and in the many monographic sources listed by faunal group in the bibliography.

Most figures are taken from older literature. Although our understanding of animal construction and relationships has improved importantly over the years, we have seldom been able to match the artistic skill employed in the past. An intentional and significant contribution made by the present volume is the opportunity to breathe a second life into these masterful figures. Generous permissions granted for the use of those works protected by copyright laws are gratefully acknowledged. In particular, I would like to mention the following sources. The Putnam Publishing Group granted permission to use figures of invertebrates from R. W. Miner's *Field Book of Seashore Life* (1950). Their status as government documents make available illustrations of fishes from Bigelow and Schroeder (1953) and Hildebrand and Schroeder (1927) and decapod crustacea from Williams (1965). The author and the National Museum of Natural History permitted the use of figures of polychaete worms from Pettibone (1963). The Director of Communications of the Marine Biological Laboratory at Woods Hole, Massachusetts allowed me to reprint invertebrate figures from Smith (1964). The Canadian Museum of Nature in Toronto made available figures of amphipod crustaceans from Bousfield (1973), and the University of Toronto Press granted permission to reproduce hydroid illustrations from Fraser (1944). For their generosity in sharing these works, I am profoundly grateful. Finally, a number of original figures have been drawn by the author and by Eric Brothers of the American Museum of Natural History. My use of all figures, including those from journals and from the older literature, is cited in the Illustration Credits.

A survey of this breadth requires input from experts on individual faunal groups. I deeply appreciate the time and attention spent by the following scientists who reviewed and commented on specific sections of the manual: Dr. Dale R. Calder, Royal Ontario Museum (hydroids); Dr. Robert Bullock, University of Rhode Island (chitons); Dr. Paula M. Mikkelsen and Ms. E. A. Peiche, Delaware Museum of Natural History (gastropods); Dr. Alan Kuzirian, Marine Biological Laboratory, Woods Hole and Dr. Paula M. Mikkelsen, Delaware Museum of Natural History (nudibranchs); Dr. Paula

M. Mikkelsen, Delaware Museum of Natural History and Dr. Kenneth Thomas, University of Rhode Island (bivalve mollusks); Dr. Marian H. Pettibone, National Museum of Natural History (polychaete annelids); Dr. Edward L. Bousfield, Researcher Emeritus, Canadian Museum of Nature, Ottawa, Canada (amphipods); Dr. Christopher Boyko, University of Rhode Island and Dr. Gerhard Pohle, Atlantic Reference Centre, St. Andrews, New Brunswick (decapod crustacea); Dr. John Dearborn, University of Maine, and Ms. Suzette Smith, Drew University (echinoderms); Dr. Howard Evans, Cornell University (fishes); and Mr. Eric Brothers, American Museum of Natural History (mammals). In addition, Dr. Arthur Borror, University of New Hampshire; Dr. David Campbell, Rider University and Shoals Marine Laboratory; Dr. Judith Grassle, Rutgers University; and Dr. Paula Mikkelsen, Delaware Museum of Natural History generously offered valuable comments on introductory material and other sections of the manual. Their suggestions and corrections have greatly improved the accuracy and utility of this material. In acknowledging their help, however, I hasten to point out that the responsibility for all errors and omissions from these pages is mine alone.

I am most grateful for assistance in the production of this volume. Two Faculty Research Grants from Drew University provided financial support. Ms. Josepha Cook of the Drew University library staff helped with access to research materials. Ms. Teresa Carson adjusted, polished, and clarified the text through her skills as copy editor, and Ms. Ellen C. Dawson astonished me by her masterly handling of the difficult typesetting and layout of this unusual project. I am especially grateful to Ms. Karen Reeds, recently of the Rutgers University Press, for her enthusiastic encouragement and guidance through the early stages of its production, and to Ms. Marilyn Campbell, managing editor, for her capable guidance in bringing it to completion.

Above all, I acknowledge my debt to dozens of curious, energetic, and always stimulating students who have sustained me and nourished this evolving project over the past twenty-five years. As fellow learners, we have stumbled through foggy, low tides at dawn, waded just over the tops of our hip boots, and then spent endless hours in the laboratory, enduring the hard work of analysis, but exulting together in the rewards of discovery. I have been fortunate to have been able to work with an exceptional group of colleagues and especially students both at Drew University and at the Shoals Marine Laboratory in Maine.

Finally, as I have become increasingly consumed by the drive to complete this project, I have come to understand why so many books are dedicated to the patient understanding of the author's spouse and children. I hereby acknowledge that debt of love owed to my wife, Sylvia, and my children, Joshua and Jessica.

Leland W. Pollock
Madison, New Jersey

■ A Practical Guide to the Marine Animals of Northeastern North America

■ Introduction

> We need another and a wiser and perhaps a more mystical concept of animals. Remote from universal nature, and living by complicated artifice, man in civilization surveys the creature through the glass of his knowledge and sees thereby a feather magnified and the whole image in distortion. We patronize them for their incompleteness, for their tragic fate of having taken a form so far below ourselves. And therein we err, and greatly err. For the animals shall not be measured by man. In a world older and more complete than ours[,] they move finished and complete, gifted with extensions of the senses we shall never hear. They are not brethren, they are not underlings; they are other nations, caught with ourselves in the net of life and time, fellow prisoners of the splendour and travail of the earth.
>
> —Henry Beston, *The Outermost House*

It is both interesting and important to know the identity of the organisms that surround us. Ecologists, environmentalists, and other scientists work from an outright obligation to be precisely descriptive about the creatures they study in order to communicate their findings to others. Conversely, awareness of organismal diversity is prerequisite to understanding the reports of others in the literature. Periodic review of regional biodiversity monitors environmental well-being. The presence of indicator species with narrow tolerances for key environmental conditions demonstrates that those circumstances are present in a particular setting. In some cases, comparatively inconspicuous species have been shown to play keystone roles in controlling the composition and stability of the communities in which they reside. But beyond these practical motives, for many of us, professionals and amateurs alike, there simply is satisfaction in extending a general awareness of our surroundings to a more detailed and specific acquaintance with its inhabitants.

Increasingly, attention to organismal science is being de-emphasized in college and university curricula and by research funding sources. Shifting priorities—often driven by a narrowly focused redirection of funds toward other, conspicuously advancing subdisciplines—threaten to break long chains of knowledge generated by our predecessors in fields such as marine invertebrate biology, ornithology, ichthyology, mammalogy, and plant taxonomy. Obviously areas currently in vogue are significant: for example, application of techniques of molecular analysis are making major, exciting contributions to all of the fields just mentioned. Much is still to be learned at macro levels, however, and without a thorough grasp of the whole organism as a context, information regarding the fine details of its structure or function are of limited value. We should not allow our pursuit of knowledge in all these fields to be cast in an either-or situation. Questions of structure and function at the macro level are every bit as interesting and worthy as those asked at the micro level, and plenty of young scientists would be delighted to pursue them. This manual is an attempt to link another generation with marine interests to the wealth of information that exists regarding the identity of marine animals of northeastern North America. The most engaging setting in which to nurture an appreciation for these fascinating organisms is holding them alive in one's hands. Much of the value in attempting to identify such specimens is found in the process that requires careful observation in order to answer the analytical questions asked. Whether determining the animal's name is ultimately important to the task, one will have come to know its type by being forced to examine the way it is put together.

Initial reviewers of a regional fauna face the enormous task of building identification aids from direct observations of specimens in hand. Those of us arriving later in the process have the advantage of combining our own experiences with the collected observations of pioneering predecessors in the form of literature and identified collections. Personal fieldwork suggests which species are most apt to be encountered and what sort of characteristics are most effective for nonexperts to use. The available works of others offer time-tested key features and detailed descriptions that can be used for species confirmation. This guide represents an opportunity to compile practical field experience with the rich historical legacy of marine science in the Northeast.

For 30 years, Smith's (1964) *Keys to Marine Invertebrates of the Woods Hole Region* has served as the standard resource for the identification of the northeastern marine invertebrates for serious students of marine biology. Unfortunately, this work is limited in regional scope, has become dated in its taxonomy, and is now out of print. Carriker (1996) reviewed the enormous group effort of assembling an inclusive set of regional keys in the National Marine Fisheries Series, *Marine Flora and Fauna of the Eastern United States.* Although this project eventually will provide the definitive treatment required by professionals, 25 years after its onset, many major groups remain to be covered. The keys offered here have more modest goals: that is, to provide reasonably comprehensive coverage of the broad northeastern region in a manner accessible to students and other nonexperts.

The geographic range covered in this manual includes regions that biogeographers term the cold-water, boreal Acadian Faunal Subprovince, from Cape Cod north into the maritime provinces of Canada, and the warmer water Virginian Faunal Subprovince, from Cape Cod south to Cape Hatteras (FIG. 0.1). A few species are indicated as being more typical of faunal subprovinces bordering those treated here—the Carolinian Faunal Subprovince to the south and the Polar Faunal Subprovince to the north. Vertically, only macrofaunal animals (i.e., those visible to the unaided eye) occurring from the supralittoral splash zone, above average high-tidal water levels, to the margin of the offshore continental shelf (i.e., about 200–300 m depth) are included. The northern half of this coverage is characterized mainly by hard substrata, rocky shores, and a glacially influenced continental shelf, whereas the southern half is dominated by particulate sediments, with marshes, beaches, flats, and barrier islands along shorelines and vast plains of sandy-muddy sediments subtidally. Combined, this northeastern coastline offers a rich variety of marine habitats and supports a diverse biota.

Identification aids to the fauna from this geographic region include elementary descriptions for the casual observer and technical manuals suitable for experts alone. This coverage misses those wishing more than the common name of the most obvious marine organisms but who lack the background, time, or patience to wade through complex terminology and difficult observational procedures. There is a need for a treatment that is reasonably comprehensive and provides species-level identifications without relying on out-of-print or obsolete supports, but at the same time is not so steeped in zoological jargon as to require an advanced degree to use it or so expensive as to lie beyond the reach of all but the most serious student.

Achieving these objectives simultaneously requires compromises. Organisms selected for inclusion here represent those aquatic, marine animals that are large enough to be visible to the unaided eye and most frequently encountered in the northeastern United States and maritime Canada. Reluctantly, this rather artificially excludes important terrestrial contributers to marine ecosystems such as the insects, arachnids, and birds. Fortunately, excellent identification sources already exist for these animals. I have composed the species list for inclusion largely from regional checklists (e.g., Smith 1964; Wass 1972; Watling and Maurer 1973; Larsen, et al. 1977; Borror 1995; and many others listed in the Bibliography) and from other guidebooks and monographs (e.g, Hildebrand and Schroeder 1927; Bigelow and Schroeder 1953; Pettibone 1963; Gosner 1971, 1979; Bousfield 1973; Brinkhurst et al. 1975; McClane 1978; Meinkoth 1981; Robins and Ray 1986; Ruppert and Fox 1988; Scott and Scott 1988; Abbott 1993;

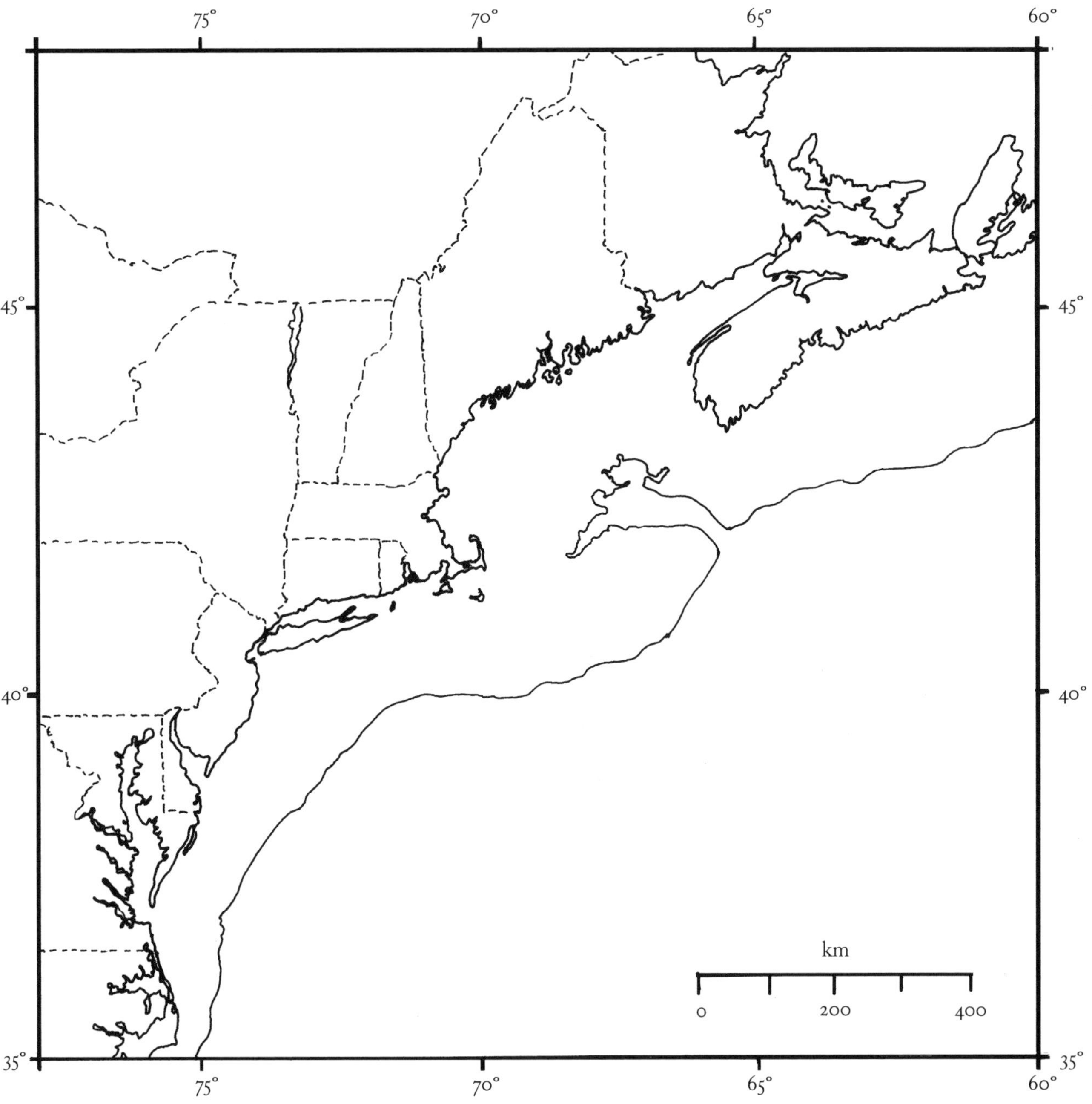

FIG. 0.1

Lippson and Lippson 1997; and others found in the Bibliography). Many more species are excluded from these pages than are included in them, for example, more than 1000 species are recorded from the coast of Maine alone (Center for Natural Areas 1976).

In attempting to make this manual more accessible, I have simplified terminology and avoided features that are difficult to observe. (Most characters used here should be observable using a good hand lens, although more sophisticated optical equipment—i.e., a low-power microscope—is certainly helpful.) The keys provided are superficial in that they are based on only enough morphological features to distinguish among the particular species included here, but they do not necessarily use the characters defining the species in a formal taxonomic sense. Therefore, some identifications "to species" may be misleading. It is possible that a less common species not included here might match the characters used in these keys. Variability in features often leaves even experts in disagreement regarding the identity of individual organisms. In short, positive identification of an organism is not achieved simply by correctly answering the few related questions in a key and arriving at a name. This is especially true if the questions asked have been simplified, the characteristics examined are limited to only the most obvious, and only the most common organisms have been included. Still, careful use of this identification aid can lead to the most likely identity and additional references suggested can be used to confirm it.

0.1 Keys

Traditionally, aids to the identification of marine animals have included two basic formats. In field guides, organisms are arranged in systematic order and then are described individually, usually with accompanying plates. The principal drawback is that, without some advance knowledge of the identity of the organism in hand, the search for a compatible match can be time-consuming. Further, this experience yields little more than a name.

The alternative, the key, involves a stepwise analytic procedure to assist in identification. The most common type of key is the *dichotomous key,* which presents paired statements representing alternatives regarding a distinctive characteristic and asks the user to select the option that most closely corresponds to the organism in hand. Each option ends with an instruction to proceed to another set of paired statements. The process is repeated. If all the statements can be and have been answered correctly using the specimen under study, the process ultimately leads to the name of the organism.

Unfortunately, useful as this sort of key can be, dichotomous keys possess several inherent drawbacks. Successful application of a dichotomous key requires that all questions be answered in the sequence in which they are presented in the key. If the specimen in question is immature or damaged, there is a good chance that the sequence of questions cannot be followed to completion. If the specimen is male and the shape of the female's tail is a question early on in the sequence, the organism cannot be identified. If the question calls for a character difficult to observe because of size, timing (e.g., developmental characteristics or those available only seasonally), or location (e.g., internal features), the sequence cannot be followed. In addition, at the end of the identification process, one knows only the name of the known species that also happens to follow the sequence of statements traced using the unknown specimen. Should the statements not be specific enough, a new or related species of organism close to the one listed in the key could easily be misidentified. Lastly, if a single new species is encountered, the entire dichotomous key describing the group to which the new species belongs must be rewritten.

An alternative type of key, the *tabular key,* is used here. Although its format is simple, it avoids several of the problems just described. For an unknown organism, the user is presented with a series of choices regarding distinctive characters, as before. Following each character is a list of several possible choices, each represented by a symbol. After recording the symbol representing the choice that best corresponds to the specimen, the user proceeds to the next character. The process is repeated until a choice has been made for each character and a symbolic "sentence" has been generated, forming a description of the unknown specimen. This symbolic sentence is then compared to sentences describing known species in a table that follows the choices. In most cases, a match will be found, indicating the likely identity of the unknown specimen.

This technique has several advantages. Each character is presented for each organism, and each organism is compared to all others near it in taxonomic placement. If, for some reason, a choice cannot be made for a particular character (e.g., because of damage, immaturity, etc.), a symbolic sentence missing only that character can still be produced, allowing a best-fit attempt at identification. Because the organism has been described by a series of choices, each one independent of the others, it is less likely that choices will be made in biased prejudgment of the likely identity of the unknown animal. A descriptive sentence is produced in this process and similar sentences are available to describe all the known species in the larger group to which the unknown belongs. Therefore the user has a better means for broader comparison with related species. In other words, the user has more information than simply a name. Finally, if additional species or ones not included in a particular key are discovered, the entire key need not be rewritten. A new descriptive sentence and species designation are simply added to the table.

FIG. 0.2

To illustrate the method for using a tabular key and to clarify the format for choices and illustrations as used here, the user should attempt to identify an unknown bolt or screw such as the object illustrated in figure 0.2.

Turn to the appropriate key, such as the one that follows, and, based on the specimen in hand, record the choice for each character. Numbers on figures accompanying the key match question numbers and guide the user as to where to look for features asked. Where appropriate, an asterisk (*) marks the condition illustrated from among the choices offered. In some cases, several alternatives are illustrated and labeled. Terms requiring explanation are defined when used (or in immediately preceding introductory sections) rather than in a remotely located glossary. Many, but not all, species are illustrated by figures. Most animals are oriented with the anterior end upward or to the left. Exceptions are noted in captions. An annotated listing follows, describing more specific characters and ecological information. Linkage to additional information is available through parenthetic reference to a page number in an appropriate source. In this example, the source, the Random House Dictionary, is coded as R.

Key to Certain Bolts and Screws

Reference: (R) Random House Dictionary of the English Language, 2d ed., unabridged, 1987.

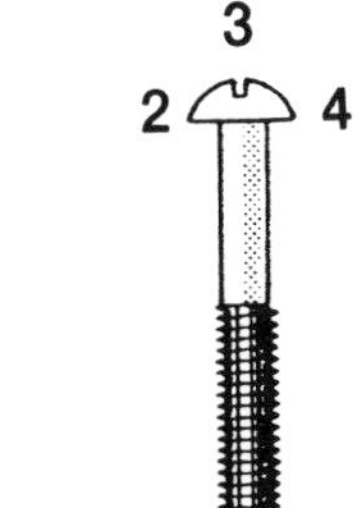

FIG. 0.3

1. Shape of tip (FIG. 0.3)
   - A. *Pointed* sharply — po
   - B. *Blunt* — bl*
2. Slots on head (FIG. 0.3)
   - A. *Single* slot cut across head — si*
   - B. Two slots or *crossed* appearance — cr
   - C. Slots *absent* — x
3. Shape of head, seen from above (FIG. 0.3)
   - A. Head *round* in outline — ro*
   - B. Head *square* in outline — sq
   - C. Head *hexagonal* (six-sided) in outline — he
4. Shape of head, seen from the side (from the top to the beginning of threads) (FIG. 0.3)
   - A. *Domed* top — do*
   - B. *Squared* top — sq
   - C. *Triangular* top — tr

Tip	Slots	Head (top)	Head (side)	Type of Bolt or Screw
po	si	ro	do	Roundhead screw
po	cr	ro	tr	Flathead Phillips screw
po	x	sq,he	sq	Lag screw
bl	si	ro	do	Roundhead stove bolt
bl	si	ro	tr	Flathead stove bolt
bl	x	ro	do	Carriage bolt
bl	x	sq	sq	Machine bolt

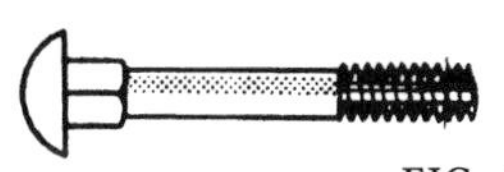
FIG. 0.4

Carriage bolt. Roundheaded bolt; threaded along part of its shank; fastened with a nut. (R319). (FIG. 0.4)

FIG. 0.5

Flathead Phillips screw. Sharply pointed screw; head with crossed slots sunken into surface; requires special screwdriver to turn. (R1454). (FIG. 0.5)

FIG. 0.6

Flathead stove bolt. Like a machine screw except that threads are coarser; requires a nut. (R1878). (FIG. 0.6)

FIG. 0.7

Lag screw. Pointed fastener with square or hexagonal head; driven by a wrench. (R1076). (FIG. 0.7)

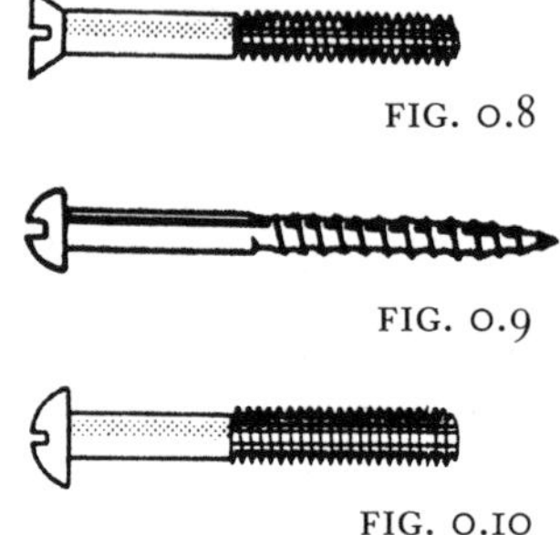
FIG. 0.8
FIG. 0.9
FIG. 0.10

Machine bolt (screw). Threaded shank; usually small diameter; fastened by nut. (RI151). (FIG. 0.8)

Roundhead screw. Pointed fastener with domed head that remains raised above surface upon tightening. (RI722). (FIG. 0.9)

Roundhead stove bolt. Similar to flathead stove bolt except for domed head. (RI878). (FIG. 0.10)

For the object illustrated, the appropriate choices generate the following symbolic sentence: po cr ro tr. Comparison to the sentences for known types of bolts or screws shows the object to be a flathead Phillips screw described on page 1454 of the Random House Dictionary. This can be a straightforward exercise using new, well-made fasteners such as those shown, but older hardware can be bent and rusted with stripped threads and broken points, making it difficult to identify clear matches with key choices offered. Similarly, individual variations among animals can present difficulties in finding a good fit among the ideal characters listed. Sometimes identifications will have to be made with reservations. Identifications made using any key must be considered tentative until they have been verified using more detailed descriptions and illustrations in specialized literature.

In an attempt to limit the number of characters described, the key often leads from the more general to the more specific in a series of steps. For example, if the group to which the wormlike specimen belongs is uncertain, begin with the key to Groups of Marine Invertebrates in section 1.1 or the key to wormlike organisms in section 3.1. Once it is clear that this hypothetical animal belongs to the Phylum Annelida, Class Polychaeta, for example, proceed to the overall key, Polychaeta in section 14.3. From this key, move to a subgroup within the Polychaeta, and then to the particular family of these worms. Finally, after the fourth or fifth set of characters has been described, the specimen can be identified. As familiarity with these organisms increases, however, one can skip directly to the known taxonomic level. Before long one can easily identify a polychaete scale worm as a member of the family Polynoidae and turn directly to the last step only for identification. The only assumption made initially is the ability to distinguish the classes of marine vertebrate organisms from the invertebrates and from one another. Keys to the marine vertebrates, including sharks, rays, skates, fishes, reptiles, and mammals begin in chapter 17 in the subsection Subphylum Vertebrata. Finally, I suggest the robin theory approach to identifications, which assumes that the unknown specimen is the most common possible member of the group of organisms to which it belongs, unless it is obviously something less common. Thus, because of the relative abundance of robins, all birds are considered robins until the details of their size, shape, and coloration prove otherwise. For each group, a smiling face symbol, ☺, identifies the "robins," that is, the most common forms.

These keys are designed to shortcut the identification process. In all instances, identifications made here should be considered tentative until reference to the specialized literature can be used for confirmation. For those engaged in critical marine work, such confirmation may be absolutely required, and these keys should help to narrow the range of possibilities needing such follow-up. For others, the "most likely" identifications offered here may suffice.

0.2 Errors: Sources and Remedies

Of many possible explanations for a failure to identify a specimen, three are most frequently encountered. Often juvenile or immature individuals have not yet grown into their full adult complement of body parts, coloration, or size. As a result, although it is possible to rule out matches in which the specimen's characters exceed those described, instances in which the specimen falls short of the mark should be viewed with greater caution. A bivalve that exceeds 25 mm in length as an adult, may only measure 5 mm after its settlement from the plankton. Second, damaged specimens may offer regenerating, truncated, or missing features. Amphipod crustaceans without antennae or body fragments of polychaete worms can prove either misleading or impossible to identify. To aid in identification, protect delicate specimens in the field by isolating them in small containers, and look for signs of breakage and injury in the laboratory. Finally, some species are not common enough to have been included in this manual.

Be flexible in interpreting features described, but be prepared to accept that the specimen may not be covered and that you must consult more detailed works for help.

Other sources of error inevitably reside in a work of this sort. Taxonomic changes will threaten to make this treatment progressively more outdated from the moment of its publication. Furthermore, I have included hundreds of species, each described by a list of morphological and ecological details. Despite care in production, testing, and proofreading, errors are certain to be present. Finally, choices that seem clear to me may not always seem that way to others.

Fortunately, modern communications permit corrections and adjustments to be offered for these sorts of problems as they become apparent, avoiding the need for users to wait for subsequent editions for updates. As I learn of corrections or improvements, I will maintain a cumulative notice of such items as part of my Internet Web site. This will be accessible through the World Wide Web by way of my name as a member of the Biology Department of Drew University (http://www.drew.edu/cla/depts/biol). In this spirit, I welcome your suggestions, comments, and corrections that can be sent to me by way of the internet (lpollock@drew.edu) or by hard copy to the Biology Department, Drew University, Madison, NJ 07940.

0.3 Notations Used in These Listings

Scientific names follow Parker (1982) for higher taxa and Gosner (1979) for species, with updates as appropriate. For cases in which a newer taxonomic designation has supplanted an older but well-known name, the older name (preceded by an equals sign) is given in parentheses. In many cases, the common name(s) is given in lowercase lettering to emphasize subordinate status of common names compared to the universal Latin name. In recent years, committees of experts on several groups, working under the auspices of the American Fisheries Society, have prepared lists of recommended common names (see Cairns et al. 1991, Cnidaria; Robbins et al., 1991, Fishes; Turgeon et al., 1988, Mollusca; Williams et al., 1989, Decapod Crustacea). Where available, their official common names (designated by a dagger, †) are offered for purposes of standardization. Although I emphatically urge the use of these official names, widely used alternative common names are given in parentheses to help users to find connections among the common names of the past and to make transition to official common names hereafter. When higher taxonomic categories are listed, they are preceded by the following identifying abbreviations: Ph, phylum, SP, subphylum, Cl , class, Or, order, and IO, infraorder. If the final grouping of species is not based on taxonomic family, the family name (without identifying abbreviation, but recognizable by its ending of "-idae") appears in parentheses. The final component of an official species name is the identity of its describer and the date of its publication. To save space in the text, the authority and date for each species accompanies its listing in the index.

The common name is followed by useful notes on appearance and biology. A series of five coded entries is used to describe geographic distribution, habitat preference, depth distribution, trophic category, and maximal size. Within any of these categories, hyphens will be used to separate instances in which more than one code applies to the species in question. An uncertain listing is indicated by a question mark; probable designations are listed by coding followed by a question mark. Codes are listed in the tables that follow.

Biogeographic Subprovince

Code	Description
Po	Polar, south to Newfoundland
Ac	Acadian, north side of Cape Cod to Newfoundland
Vi	Virginian, south side of Cape Cod to Cape Hatteras
Ca	Carolinian, Cape Hatteras through Florida
Bo	Both Acadian & Virginian, range covers both subprovinces

A plus symbol (+) follows the subprovince indication when the animal's distribution extends somewhat northward of the subprovince limit; a minus symbol (–) indicates a distribution extending south of the subprovince boundary. Many of these designations are drawn from the data of Gosner (1971).

Habitat Preference: Substrate

Code	Description	Code	Description
Ha	hard substrate; rocky	SM	salt marsh
Gr	gravel, cobble	SG	seagrass beds
Sa	sand	Al	epiphytic on surfaces of algae
Mu	mud	Co	commensal, parasitic
Pe	pelagic, open water	Ot	other

Habitat Preference: Depth

Code	Description	Code	Description
Es	estuarine, brackish water	De	deeper than 100 ft
Li	littoral, intertidal	Ps	pelagic (shallow to 100 ft)
Su	sublittoral to 100 ft	Pd	pelagic (water deeper than 100 ft)

A minus symbol (–) added to a depth symbol indicates distribution that extends just below the lower limit for that habitat. A plus symbol (+) indicates distribution extending above the habitat limit. Many of these designations are based on the data of Gosner (1971).

Trophic Category

Code	Description	Code	Description
pr	predator	sc	scavenger
gr	grazer or herbivore	om	omnivore
ff	filter or suspension feeder	pa	parasite
df	deposit feeder		

Body Length

For body length, the animal's longest dimension is measured. In some cases, the largest recorded size is given but is followed, in brackets, by a more "typical" large size. Measurements are given in both metric and English terms. For example, for a species listed at "23 cm (9 in). [14 cm (5.5 in).]", a record length of 23 cm or 9 inches has been found, but 14 cm or 5.5 inches is a more typical upper limit for length.

References

Finally, with a name for an organism in hand, it is essential to confirm the identification by referring to a more complete description. In most cases, parenthetic references (e.g., G79) indicate more complete descriptions of the organism. The letter refers to the author of the reference volume and numbers to appropriate pages therein. As a standard, most references include listings related to the two most comprehensive and readily available field guides for the area. These citations begin with G representing Kenneth L. Gosner's *A Field Guide to the Atlantic Seashore* (1979) or NM for Norman A. Meinkoth's *The Audubon Society Field Guide to North American Seashore Creatures* (1981). Appropriate additional reference sources are listed in the introductions to many of the animal groups. Other useful treatments may be found, arranged by faunal group, in the Bibliography.

A smiling face symbol (☺) will appear if the animal is among the "robins" or most common species of its type (see discussion of "robin theory" above).

0.4 Fieldwork and Habitat Types

Northeastern North America offers access to an exceptional diversity of marine habitats. Although each habitat supports its own particular array of animal species and each requires site-specific study techniques, some general background and tips on making observations in the marine world may be useful initially.

Tides

Snorkeling and scuba techniques make intertidal areas accessible even during flood-tidal conditions, but intertidal observations during ebb-tidal exposure are easiest for most of us. A good strategy for intertidal observation is to approach these settings during maximal low-tidal exposure, a task requiring some elementary understanding of tidal behavior and some field trip planning in consultation with tables for local tides.

Tidal rise and fall is controlled by predictable, large-scale celestial relationships (among the earth, the sun, and the moon), by varied but regionally constant features of geography (the shape of the local coastline and ocean floor), and finally by transient weather conditions (onshore vs. offshore winds, changes in atmospheric pressure, storms, etc.). Detailed description of the interplay of these factors in the creation of tides goes beyond the scope of this manual, but may be found in any good elementary text on oceanography. For now, suffice it to say, the result of these influences in our region is the production of a semidiurnal (twice daily) rise and fall of seawater in a pattern that places successive high tides or successive low tides approximately 12 hours and 25 minutes apart, making daytime low tides occur about 50 minutes later each day. Because both the sun and the moon exert an influence over tidal rise and fall, the positions of the sun and moon relative to the earth determine the extent of tidal change on a particular day. The tidal influence of the moon, however, is more than twice that of the sun because the moon is much closer to the earth. The moon orbits the earth every 28 days, causing the earth, moon, and sun to align with one another at two points (e.g., at day 0 and day 14) at which lunar and solar influences become additive, producing tides known as *spring tides.* At these times, the degree of tidal change should be maximal: the highest high tides and the lowest low tides. During the intervening period of the lunar orbit, however (e.g., at days 7 and 21), the moon is located at right angles to the sun. Here lunar and solar influences conflict, producing minimal net tidal rise and fall, forming *neap tides* of low high tides and high low tides.

Unless local storms or other atmospheric influences modify these effects, tide tables, based on celestial events and regional geography, can be used to predict the times at which high and low tides are expected each day and the extent of their expected rise and fall, all marked relative to the average low-tidal level at 0.0 feet. Official tidal tables (formerly produced by the Coast and Geodetic Survey arm of the U.S. Commerce Department, and now available from various commercial sources) provide a calendar of predicted times and expected heights for high and low tides for a series of key reference locations along the coastline. But local geography contributes an important, peculiar, but predictable influence over both times and heights of tides. Available in the original Commerce Department tide tables is a listing of intervening localities, each including correction factors for both time and heights to be applied to calendar listings for the nearest reference site. For example, corrections for tides at sites along the northern side of Cape Cod are referenced to Boston, Massachusetts listings, whereas those along the southern side refer to Newport, Rhode Island listings. Because most table listings are given in Eastern Standard Time (EST), it may be necessary to include a correction for daylight savings time. Local weather may modify predicted conditions considerably. Observation trips are best at low tide, but actual low tide may occur somewhat later within a complex network of intertidal channels, for example, at a marsh, than at the open coast upon which the tidal tables were calculated. Ask for local advice on this sort of adjustment.

General Procedures and Safety

A typical intertidal study trip involves arriving at the observation site about 1 hour before the predicted low tide. Generally, marine organisms find low-tidal exposure to contrasting atmospheric conditions stressful. Consequently, only specifically adapted organisms extend their distribution high up into the intertidal zone. As a result, intertidal species richness follows a gradient from maximal richness toward the low-tide line to minimal species numbers farther up the intertidal. Again, as a general principle, the upper limit to the distribution of many intertidal organisms appears to be determined by their tolerance to physical conditions (desiccation, temperature extremes, etc.), whereas their lower limits tend to be established by biological interactions such as competition for suitable substrate or predation upon them. Thus, maximal low-tidal exposure should provide optimal conditions for intertidal observation. Spring tides produce maximal intertidal exposure at low tides (i.e., the lowest low tides). As noted, all tidal heights are listed in tidal tables relative to the mean low-tidal level, reported as 0.0 ft in tide tables. Spring low tides are the predicted tides with the largest negative numbers (e.g., a low tide predicted to be at −1.2 should fall to 1.2 feet below an average low tide, whereas one predicted at 0.5 will fail to reach the average low-tide level by half a foot).

Upon arriving at the site, go immediately to the falling water line and keep following it as it recedes for the next hour or so. This technique insures observation of maximal intertidal exposure, including the greatest intertidal diversity at the lowest possible water level during that low-tide series. As the water begins to rise after the low tide is the time to visit higher intertidal sites for observations. Many an eager observer has become stalled upon arrival by some interesting event at midtide level, only to discover that the time of lowest tide has passed, and the water level has started to rise.

The intertidal zone can be a dangerous place for human explorers. The next section provides site-specific tips, but some overall rules for safety are listed here.

- Always observe in pairs or groups. Sustaining a fall, suffering an injury, or getting stuck in an intertidal setting could mean serious trouble.
- Approach the area cautiously until you get a feel for the substrate and your ability to negotiate it safely.
- In areas where waves or boat wakes could catch you unaware, remember never to turn your back to the sea.

Conservation, Humane Treatment, and Laboratory Work

Many marine communities are challenged by rigorous conditions that have both natural and human causes. The process of learning more about these fascinating creatures should include efforts to avoid stressing or threatening them further. Tread as lightly as possible in marine habitats, disrupting things minimally. Make as many field observations as possible to avoid the need to remove specimens from their natural setting. Although many animals can survive careful handling if they are returned to their natural setting, some cannot.

Accurate study and limited access time often requires that some specimens be removed from the field for closer observation. A few carefully selected and treated individuals should suffice for this purpose. Do not overcollect by bringing in more organisms than you need or will have time to handle properly. Cover collected specimens with fresh seawater in a bucket or, in the case of small or delicate subjects, in small plastic (not glass) jars of seawater. Shield the bucket from strong light and temperature exposure. Use portable air pumps or oxygen-releasing tablets (available from many pet supply or sporting goods stores).

During study in the laboratory or at home, maintain specimens in cool saltwater and dimly lit, well-oxygenated conditions. Use a properly adjusted aquarium and air stones to reduce stress. During examination, limit the time specimens spend in a bowl of seawater under the lights of a dissecting microscope to minimize exposure to bright light, elevated temperature, and stagnant conditions. Try to work quickly with material removed from the field, and if specimens have not been exposed to damaging or toxic chemicals during handling, when possible, return them *to the original locality* when you are done. Even individuals that do not ultimately survive this handling will contribute natural fodder to the trophic structure of the community. *Do not,* however, return specimens from one geographic location to a different one. Inadvertent species introductions occur frequently enough, often with disastrous results for the local biota. Well-intentioned but inappropriate release of animals can create serious problems for residents.

A good magnifying lens is often adequate for closer viewing, although a dissecting microscope will vastly improve visibility of smaller organisms. Dissecting microscopes can be purchased from scientific supply houses for as little as $200–300. Questions or entire sections requiring the use of dissecting or more powerful compound microscopes will be noted. A collection of probes, pins, and glass or plastic containers is also needed. Fine-tipped, jeweler's forceps along with dental probes (sometimes available as retired items from your dentist) are especially handy. A good, concentrated light source (e.g., a small tensor lamp), is essential. One with adjustable orientation has a strong advantage because overhead lighting works well for some specimens, transmitted light from below is more effective on others, and often, lighting directly from the side will highlight and add contrast to fine features.

In each key, I have listed first characteristics that are based on intact, living specimens. In a some cases, temporary immobilization in an anesthetizing agent may be helpful. Brief exposure to such agents and prompt return to fresh seawater usually does not permanently injure specimens. In some cases, however, accurate identification may require access to internal features requiring the sacrifice of the animal. Overanesthetization (e.g., 20–30 minutes in 7% solution of magnesium chloride for many invertebrates, or 20 minutes in 0.5% MS–222 for fishes) is recommended in these cases. Naturally, this practice should be minimized. Animal groups respond differently to the several media usually used for this purpose. Faunal group-specific procedures recommended for anesthetizing, fixing, and preserving specimens are offered in the Appendix.

Temporary wet mounts may be made of subjects requiring closer scrutiny using a compound microscope. Place the specimen in a small drop of seawater on a glass slide. Just touch each corner of a coverslip to a lump of modeling clay to create tiny pedestals that should prevent the specimen from becoming squashed when the coverslip is added. Be prepared to add another drop of seawater between the coverslip and slide if the preparation begins to dry out. Add enough gentle pressure on the coverslip using a probe or pin to restrict the animal's movements, but not so much as to damage it. Sometimes a coverslip-sized square of clear plastic wrap will accomplish the same objective and is easier to remove later on.

Habitats

Hard Substrata

Along erosional shorelines, wave action rinses away any loose material, leaving hard intertidal surfaces behind. These settings force biota to occupy rock faces where they must endure biological pressure from predation and the physical rigors of rhythmic tidal exposure and storm waves. The result is a stark interplay of factors arrayed along a vertical, intertidal gradient that creates striking patterns of distribution by zones especially by the primary space occupiers that are attached directly to the substratum. Rocky intertidal settings are among the clearest for the study of plant and animal distribution and community ecology.

Tips on observing the intertidal rocky shore must begin with safety precautions. Footing on slippery, algal-coated rocks or exposure to unexpected surges of water can be especially treacherous. A fall in the rocky shore can place you abruptly on sharp, hard surfaces often coated with razor-sharp barnacle plates. Extreme caution is advised, especially initially until you get a feel for what movements the habitat will and will not allow. Proper apparel includes gripping footwear and clothing that can withstand getting very dirty and wet. Dress for optimal scrambling ability and abrasion protection, not to keep dry or clean. Avoid using hip boots or waders because a slip into the water can cause such boots to fill with water and weigh you down as you attempt to scramble back to safety.

Make notes on distribution patterns in the field. Seaweeds visually dominate surfaces exposed to light, typically with small holdfast attachments to connect the plant to a solid base. Invertebrates tend to occupy any available surface including all rock faces, exposed to light or not. They also live *epiphytically* on plant surfaces and *epizoically* on one another. *Filter feeders* draw water currents through a filtering device to extract and sort edible organic matter (e.g., plankton or bits of floating detritus) from the particles they collect. *Suspension feeders* accomplish the same end more passively by screening particles from externally produced water flow. They are often found on the sides and undersurfaces of stones, which are settings representing a compromise between optimal water flow and protection from predation or atmospheric contrasts. Invertebrate *predators* (animal eaters), *grazers* (plant eaters), *omnivores* (eaters of plant and anmal material) and *scavengers* (feeders on dead material) are minimally active during low-tidal exposure, hiding in nooks and crannies within the rock complex. They become more mobile as they scour the microlandscape for food when flooding waters return. As can be seen by the saturated and sometimes overlapping occupation of all available surfaces here, attachment space is frequently a limited resource at intertidal rocky shores. Be sure to return to the rocky intertidal during a flood tide (preferably with snorkeling or scuba gear) to discover that this setting looks completely different when covered with water. Our image of the area at low tide is of dense layers of seaweeds draped over rock surfaces. But most seaweeds are somewhat buoyant so that, with support from flooding seawater, the scene created is more forestlike, with rock surfaces exposed for easy faunal mobility among the upright stipes and blades of seaweeds. The low-tidal period may be convenient for observers, but it is not as pleasant for marine organisms, many of which must hunker down, waiting for the return of seawater to resume more activity.

"Rock-rolling," to expose the diversity of animal life normally hidden from superficial view, is often a fruitful study technique in this habitat. (A pair of inexpensive gardening gloves may help here). The undersurface of an otherwise undisturbed, lower intertidal rock can offer an "invertebrate text book"-style review of animal diversity. Sizes are small, however, so most specimens will be of modest dimension (e.g., some less than 6 mm (¼ in) in length). A paint scraper, dive knife, or pair of forceps can be useful for removing representative specimens for further study. These organisms live on their particular rock surface for good reasons: if the rolled rock is left upside-down, all the organisms on both sides may die. Return any rolled rock to its original position before moving on.

Subtidal, hard-substrate settings are best accessed by direct observation, especially using scuba techniques. The topographically varied and resilient nature of hard surfaces makes them difficult to sample using remote devices such as dredges or grab-samplers, which scrape across or dig into the substrate.

Soft Sediments

The speed of moving water is related to the size and weight of particles it can transport. Fast-water currents resuspend and carry away particles ranging from the finest, lightest grain sizes to those of coarse dimensions, leaving only the heaviest ones behind to form the substrate. As water flow slows, the heavier particles that it is transporting will settle out. Water flow must virtually cease before the finest particles drop out and accumulate as fine sedimentary deposits. Therefore, the particle size of sediments reflects the water movements the region typically experiences. Habitats formed in particulate or depositional conditions range from steep, coarse, cobble beaches to fine-grained sand or mud flats.

Sediments include particles of *inorganic material* eroded from rocks as well as *organic detritus* or debris formed from broken bits of animal or plant origin. Organic detritus serves as nourishing substrate for bacterial and fungal decomposers. Detritus and its accompanying decomposer community members serve as fodder for the many deposit-feeding invertebrates. *Selective deposit feeders,* including many polychaete worms and crabs, sort out edible organic particles from the substrate before ingesting it. *Nonselective deposit feeders,* such as acorn worms and some polychaetes, merely ingest volumes of the unprocessed sediment, relying on digestion to extract useful organics from unusable inorganic matter on its way through the gut. Nonselective deposit feeders produce volumes of rejected matter as feces that accumulate as *castings* or coiled threads of sand, mucus, and some unharvested organic matter. Such castings can be used as visual evidence for the presence of these animals hidden in sediments beneath. Castings along with smaller but richer fecal pellets produced by more selective types, still contain useful unharvested organic matter and bacteria. They become a rich food source for other consumers. *Coprophagy,* or the ingestion of fecal material, provides for a repetitive working and reworking of sediments until most usable organic matter has been processed.

Visually, particulate settings differ completely from rocky ones. Plants do not survive well amid moving, abrasive sedimentary particles. Lacking the protection of plant cover, surface-dwelling animals, the *epifauna,* are scarce. But particulate habitats afford inhabitants the opportunity to live as *infauna,* to penetrate the substrate where they are protected from surface predators (e.g., birds in intertidal settings or fish), buffered from stressful atmospheric exposure during low tides and surrounded by organic detrital food sources. Perhaps the greatest difficulty facing such infauna is that restricted water flow through narrow spaces between stacked sedimentary particles can lead to a depletion of oxygen. Chemical reactions and biological functions such as respiration use up the oxygen present. Because diffusion from overlying oxygenated water is very slow, and circulation of surface waters though the sediments can be poor, stagnation results and can lead to anaerobic conditions that many animals find intolerable. Anaerobic conditions also are responsible for the "rotten egg," (hydrogen sulfide) smell of mud flats.

As mentioned above, particulate substrates range from coarse to fine. Water circulation, and thus the maintenance of favorable levels of oxygen, follows this gradient from least oxygen stress in coarse sediments (which restrict flow minimally) to the most serious oxygen deficiency in the finest mud deposits. Ease of burrowing into the sediment also follows this gradient, but in reverse. It is easy for animals to penetrate fine muddy deposits but difficult to penetrate coarser particles. It is much easier to press your finger into mud as opposed to a coarse sandy beach. Thus infaunal organisms face lives of compromise. In fine sand or mud flats, penetration is easy but the organisms must maintain connection to overlying waters (burrows, long siphons, periodic return to the surface, etc.) to meet oxygen needs. Coarse-sediment dwellers have reduced oxygen problems but either must possess an effective digging capability or be small and slender enough to penetrate interstitial spaces without needing to move the particles themselves. Organisms known as *meiofauna* are important members of this latter group, although they occur in other habitats as well. They include representatives of nearly all invertebrate phyla but are too small (about 200–2000 μm in body length) to be included in this guide.

Study techniques specific to particulate settings involve gathering a volume of substrate and isolating organisms, usually by sifting or screening. A shovelful of intertidal material or a grab sample from the subtidal will include all the biota from this volume of sediment. The size of the mesh used to separate biota from substrate is determined by the requirements of the study. The sample is transferred to a high-sided sorting screen or sifting box. The screen is held with its undersurface just below the water surface and is agitated from side to side to encourage particulates to fall through the screen mesh while larger biota are retained by it. With most sediment removed, the screen contents can be examined carefully and a spoon or forceps can be used to transfer individuals and/or collected debris to a storage container for closer sorting and separation later on.

Weather conditions will probably dictate whether hip boots or shorts and sneakers are more appropriate apparel. Sinking deeply into soft sediments is an occupational hazard in these habitats. Careful negotiation with hip boots or waders may keep the observer drier and warmer, but in decent weather, simply charging in with old sneakers is usually the best choice. Working barefoot in these habitats is not advisable because sharp shell fragments and miscellaneous human debris are apt to be lurking below the sedimentary surface.

Collection of subtidal sediment samples can be done by scuba or by the use of a grab-sampler. Descriptions of the variety and operation of grab-samplers can be found in textbooks on marine ecology, such as Nybakken (1988). Coyer and Witman (1990) offer valuable suggestions on research methods using scuba.

It is valuable to retain a representative, undisturbed sample of sediment for subsequent grain-size analysis. Such samples can be thoroughly dried, weighed, and passed through a series of sieves of progressively smaller mesh sizes. The sediment collected by each sieve-mesh interval is reweighed and its contribution to the sample expressed as a percentage of the total sample weight. These data are then used to calculate descriptive statistics regarding the sample.

Open-Water Habitats

The water column forms a medium through which many organisms pass and in which some reside more or less permanently. Open-water or pelagic animals are divided into two groups on the basis of their mobility through the water. *Plankton,* including *phytoplankton* (microscopic plants) and *zooplankton* (such as jellyfish, copepods, and other crustaceans) may possess short-range locomotory powers, but they are basically swept along at the mercy of water mass movements. The *nekton,* including fish and squid, are stronger swimmers capable of more directly controlling their own location within the water column.

Many bottom-dwelling or *benthic* invertebrates produce planktonic larvae that serve an important role as dispersing agents for otherwise bottom-bound species. Such organisms that spend only part of their lives in the plankton community are termed *meroplanktonic* in contrast to *holoplanktonic* types found permanently in the plankton. Both types are best collected by towing a *plankton net* slowly through the water. Conical plankton nets are made of fabrics manufactured at specific mesh sizes. As the net is towed, water and the tiniest objects pass through the mesh openings, and planktonic organisms and debris are held within. Retained objects accumulate toward the pointed end of the net in a *plankton bucket* which, at the end of the towing period (often 5 or 10 minutes), can be removed or drained into a storage jar. Holding a jar of freshly caught plankton up to the light will usually reveal a confusion of tiny, darting, pulsing life-forms, just visible to the naked eye. A key to zooplankton is provided here for those wishing to come closer to identifying the plankton. This varied and exciting material is much more interesting to work with when it is fresh and alive in a shallow container under a dissecting microscope. If you must preserve some or all of the sample, 5% buffered formalin or Steedman's solution is traditionally used. See Appendix I for further suggestions and recipes for preservation.

Many larger invertebrate members of the open-water community, such as jellyfishes, comb jellies, siphonophores, and salps, are gelatinous, which increases buoyancy and reduces their tendency to sink, makes them transparent so that they avoid predation, and renders them efficient small-mesh feeding nets that capture tiny members of the plankton community. A key to Gelatinous Organisms (2.1) helps in separating these groups. Animals of this sort are best observed in the field because they tend to be easily damaged in net collections and do not survive well in aquaria. Underwater photography is an optimal way to "capture" these animals.

Nektonic organisms may be observed in nature, captured by various forms of commercial and sport fishing, or encountered incidentally as dead specimens left stranded on the intertidal zone by storms or receding tides. Identification aids to the vertebrate nekton include keys of fishes, reptiles, and mammals.

Salt Marshes and Grass Beds

Habitats dominated by higher vascular (i.e., nonalgal) plant forms such as grasses, include upper-intertidal salt marshes and submerged sea-grass beds. In both cases, the abundance of rapidly growing plant material, often enriched by a coating of epiphytic microalgae, provides rich organic productivity as well as a spatially complex substratum capable of supporting a diversity of epiphytic plant and animal types. These habitats are among the most productive settings in nature.

Salt marshes in northeastern North America are dominated by two species of emergent terrestrial grasses capable of invading marine conditions to the extent of allowing their roots and bases to be covered briefly by seawater. *Spartina alterniflora,* tall salt marsh grass, is the more tolerant of marine conditions and presses from the land into the upper reaches of the intertidal toward the average height reached by neap high tides. Above the mean high-tide line, however, *S. patens,* salt marsh hay, dominates. Although it is less tolerant of frequent seawater cover, it is the better competitor in higher and drier conditions. Salt marsh grasses replace most of their aboveground biomass each growing season. With plants reaching 2.4 meters (8 ft) in height, this represents an enormous amount of organic detritus entering the tidal creeks and nearby lower intertidal flats associated with the marsh. Filter and deposit feeders abound here as do predators and scavengers, including many species of crabs, fishes, and birds. Because few animals harvest the marsh grasses directly, tidal channels and their banks are often the best places to seine for fish and look for invertebrates feeding on more edible detritus derived from the plants of the high marsh. Salt marshes also attract a variety of terrestrial animals such as insects, spiders, reptiles, birds, and mammals.

Seagrass beds in the Northeast are formed by *Zostera marina,* eel grass, and occur below the low water line in shallow subtidal areas where particulate sediments permit them to root. The intertwined blades and root systems of dense growths of eel grass help to stabilize surrounding sediments and support a rich community of epifaunal and infaunal animal life. Passing a beach seine through such areas will gather various shrimp, gastropods and fishes. A shovelful of sediment from within the grass bed will often be rich in burrowing infaunal invertebrates. It is important, however, to minimize disturbance to grassbeds because they are very slow to recover.

Chapter 1 ■ Groups of Marine Invertebrates

1.1

Typically keys are written under the assumption that the user is already familiar with most major groups, for example, the phyla and classes, of organisms. Because this is often not the case, these keys begin from the real starting point: a totally unknown organism in the hand. Surprisingly, this sort of general key is about the most difficult to construct. The options presented are broad categories. Do not allow the illustrations provided to limit your interpretation of their ranges. Be generous with your answers to questions in this key and be prepared to explore more than one possible outcome from its use. (Specimens are assumed to be live or at least fresh and intact.)

1. Texture of body
   - A. *Soft*-bodied — so
   - B. *Firm* but without obvious solid covering; does not droop much when removed from water — fi
   - C. Hard, solid *shell(s)* or shelly material; animal itself is inside shell(s) and is either firm or soft — sh
   - D. Hard, solid *exoskeleton*: thin, tough exterior like a crab, scales, or some hydroids, but not like clam shell — ex
   - E. Hard-bodied, surface densely covered by large or small *spines* — sp
2. Basic shape (choose the body shape that best describes the organism)
   - A. *Flattened* or laterally compressed—ranges from thin and disclike to somewhat more inflated and clamlike; *not* firmly attached to substrate — fl
   - B. Thin or *sheet*like; firmly attached to substrate — sh
   - C. Roughly *cylindrical* — cy
   - D. Generally *slender* with flattened ventral surface, with or without a shell (includes many snails) — sl
   - E. *Dome*- or cap-shaped or spherical — do
   - F. Irregularly *lobed* or lumpy — lo
   - G. *Arborescent*: slender stalk attached at base, with or without branches — ar
   - H. *Star*-shaped — st

3. Mobility
   A. *Firmly attached* to hard objects (some organisms are firmly attached when alive but are found "loose" on the shore when broken free or dead) fa
   B. Firmly attached to other animals (i.e., *ectoparasitic*) pa
   C. Freely moving: *crawling* or gliding on surface; easily dislodged cr
   D. Freely moving: *burrowing* into substrate bu
   E. Freely moving: *swimming* in water column sw

4. Special constructional features (dissecting microscope)
   A. *Jointed* legs jo
   B. More than four *tentacles* (often tentacles are withdrawn when the animal is disturbed; allow specimens to relax in seawater to evaluate this feature) te
   C. *Two siphons* (siphon is a mound or tube with an opening at its peak) 2s
   D. Primary features of surface are scattered *holes* ho
   E. These distinctive features *absent* x

5. Segmentation
   A. Body and/or legs *segmented* se
   B. Obvious segmentation of body or legs *absent* x

1.2

Texture	Shape	Mobility	Features	Segments	Group (Ph: phylum; Cl: class; Or: order; Fa: family)	
so	ar	fa,bu	te,x	se,x	Various cnidarians (Ph: Cnidaria, includes linkage to Ph: Ectoprocta & Ph: Entoprocta), 8.3	(FIG. 1.1)
so	ar	fa	2s	x	Tunicate, sea squirt, (Ph: Chordata, Cl: Ascidiacea), 17.4	(FIG. 1.2)
so	sl	sw	te,x	x	Sea butterflies (Ph: Mollusca, Cl: Gastropoda, Fa: Clionidae), 13.70	(FIG. 1.3)
so	sl	cr,bu	te,x	se	Polychaetes (Ph: Annelida, Cl: Polychaeta), 14.3	(FIG. 1.4)
so	sl	cr	x	x	Body entirely smooth:	
					Flatworms (Ph: Platyhelminthes, Cl: Turbellaria), 10.1	(FIG. 1.5)
					OR Body with bumps, papillae, or other projections: Aeolid and ridge-backed nudibranchs (Ph: Mollusca, Cl: Gastropoda, Or: Nudibranchia & Saccoglossa), 13.76	(FIG. 1.6)
so	cy	fa	ho	x	Sponges (Ph: Porifera), 7.1	(FIG. 1.7)
so	cy	fa	x	x	Egg masses, 6.1	
so	cy	fa	te	x	Miscellaneous worm-shaped organisms, 3.1	(FIG. 1.8)
so	cy	cr,bu	te,x	se	Miscellaneous worm-shaped organisms, 3.1; either Polychaeta or Oligochaeta (Ph: Annelida)	(FIG. 1.8)
so	cy	cr,bu	x	x	Ph: Nemertea, ribbon worms, 11.1	(FIG. 1.9)
so	cy	cr,sw	te	x	Swimming, cylindrical, soft-bodied animals with tentacles), 1.3	
so	cy	sw	x	se	Polychaetes (Ph: Annelida, Cl: Polychaeta), 14.3	(FIG. 1.10)
so	cy	sw	te	se	Siphonophores (Ph: Cnidaria, Cl: Hydrozoa, Or: Siphonophora), 8.38	(FIG. 1.11)
so	cy	pa	jo,te	se,x	Parasites, 4.1	
so	cy	pa	x	se	Leeches (Ph: Annelida, Cl: Hirudinea), 14.88	(FIG. 1.12)
so	lo	fa	x	x	Colonial tunicates, sea squirts (Ph: Chordata, Cl: Ascidiacea), 17.4	(FIG. 1.13)
so	do	fa	te,x	x	Anemone (Ph: Cnidaria, Cl: Anthozoa), 8.42	(FIG. 1.14)
so	do	cr	te,x	x	Dorid nudibranch (Ph: Mollusca, Cl: Gastropoda, Or: Nudibranchia), 13.76	(FIG. 1.15)
so	do,fl	sw	te,x	x	Gelatinous Animals, 2.1	
so	fl,sl	cr	x	x	Flatworms (Ph: Platyhelminthes, Cl: Turbellaria), 10.1	(FIG. 1.16)
so	fl	fa	x	x	Colonial tunicates, sea squirts (Ph: Chordata, Cl: Ascidiacea), 17.4 OR Egg masses, 6.1	(FIG. 1.17)
so	fl	pa	x	se	Leeches (Ph: Annelida, Cl: Hirudinea), 14.88	(FIG. 1.18)
so	sh	fa	ho	x	Sponges (Ph: Porifera), 7.1	(FIG. 1.19)
so	sh	fa	x	x	Soft, sheetlike, firmly attached animals, 1.5	
fi	ar	fa	te	se,x	Various cnidarians (Ph: Cnidaria), 8.1	(FIG. 1.20)
fi	ar,cy	fa	ho	x	Sponges (Ph: Porifera), 7.1	(FIG. 1.21)
fi	cy	fa	2s	x	Solitary tunicates, sea squirts (Ph: Chordata, Cl: Ascidiacea), 7.4	(FIG. 1.22)

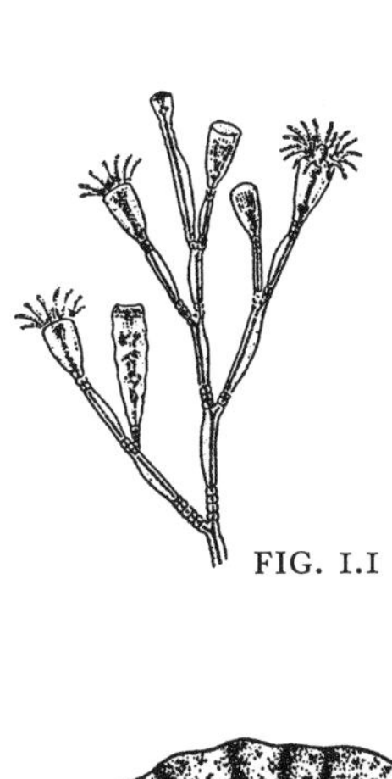
FIG. 1.1

FIG. 1.2

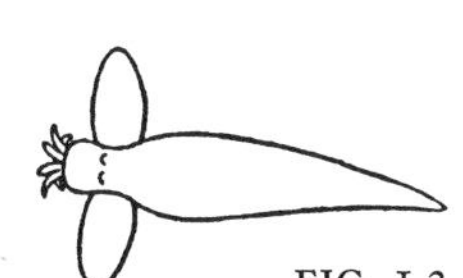
FIG. 1.3

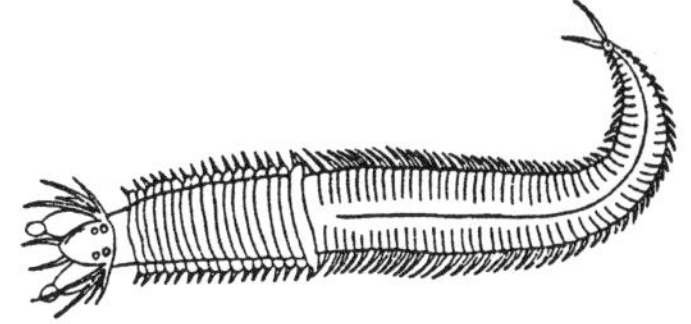

FIG. 1.4

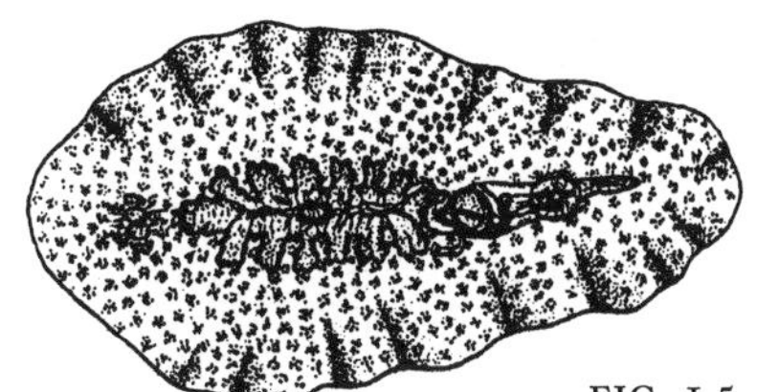
FIG. 1.5

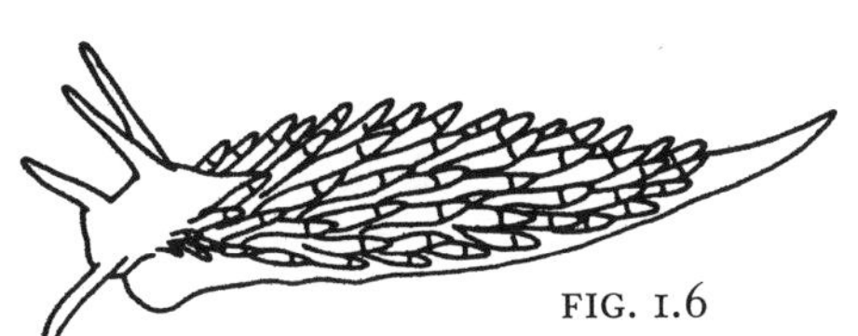
FIG. 1.6

FIG. 1.7

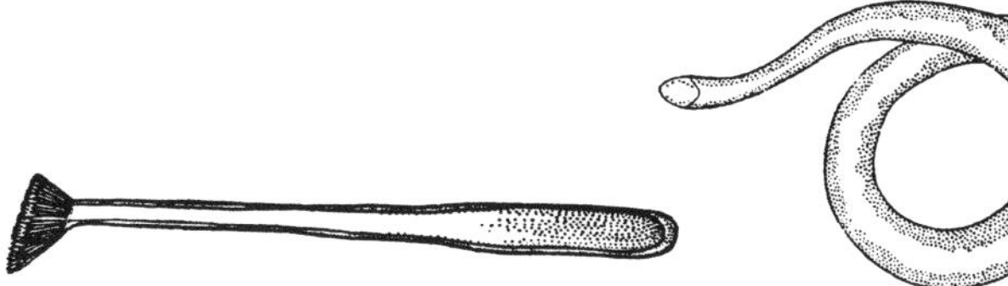
FIG. 1.8

FIG. 1.9

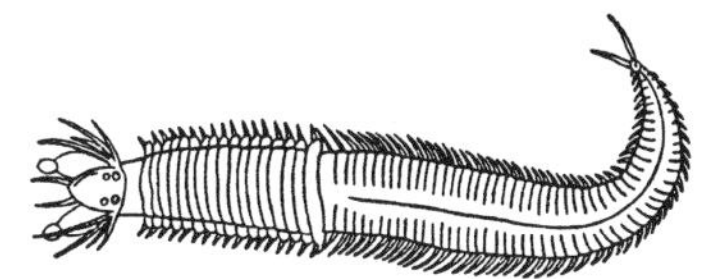
FIG. 1.10

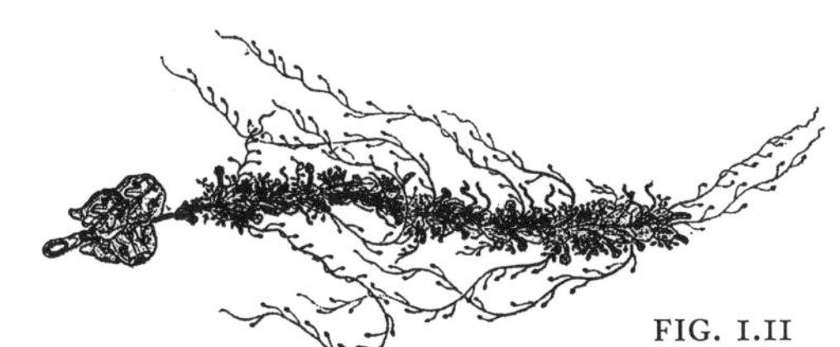
FIG. 1.11

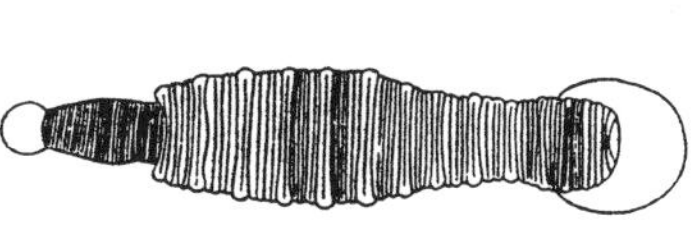
FIG. 1.12

FIG. 1.13

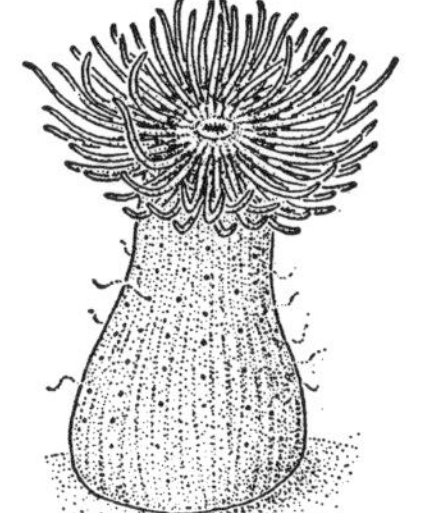

FIG. 1.14

FIG. 1.15

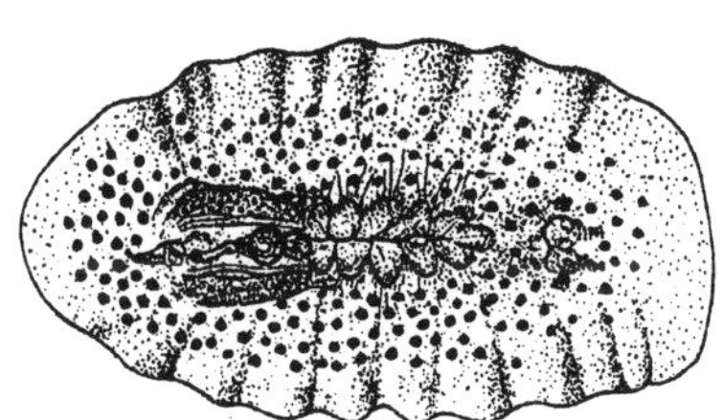
FIG. 1.16

FIG. 1.17

FIG. 1.18

FIG. 1.19

FIG. 1.20

FIG. 1.21

FIG. 1.22

Texture	Shape	Mobility	Features	Segments	Group (Ph: phylum; Cl: class; Or: order; Fa: family)	
fi	cy	bu	x	x	Miscellaneous worm-shaped organisms, 3.1	
fi	cy	sw	te	x	Squid (Ph: Mollusca, Cl: Cephalopoda), 13.161	(FIG. 1.23)
fi	cy	sw	x	x	*Nectonema agile,* horsehair worm (Ph: Nemertea)	(FIG. 1.24)
fi	cy,do	fa	x	x	Colonial tunicates, sea squirts (Ph: Chordata, Cl: Ascidiacea), 17.4 OR Egg masses, 6.1	(FIG. 1.25)
fi	sl	fa	te	x	Sea cucumber (Ph: Echinodermata, Cl: Holothuroidea, 16.12	(FIG. 1.26)
fi	sl	cr,bu	te,x	se	Polychaete (Ph: Annelida, Cl: Polychaeta), 14.3	(FIG. 1.27)
fi	do,lo,sh	fa	ho	x	Sponges (Ph: Porifera), 7.1	(FIG. 1.28)
fi	do,lo	fa	2s	x	Solitary tunicates, sea squirts (Ph: Chordata, Cl: Ascidiacea), 17.4	(FIG. 1.29)
fi	lo,fl	fa,sw	x	x	Colonial tunicates, sea squirts (Ph: Chordata, Cl: Ascidiacea), 17.4	(FIG. 1.25)
fi	fl,cy	bu,sw	te	se	*Branchiostoma caribaeum,* Caribbean lancelet (Ph: Chordata)	(FIG. 1.30)
fi	sh	fa	te	x,se	Encrusting ectoprocts (Ph: Ectoprocta), 12.1	(FIG. 1.31)
fi	sh	fa	ho	x	Sponges (Ph: Porifera), 7.1	(FIG. 1.32)
fi	sh	fa	2s	x	Solitary tunicates, sea squirts (Ph: Chordata, Cl: Ascidiacea), 17.4	(FIG. 1.33)
fi	sh	fa	x	x	Colonial tunicates, sea squirts (Ph: Chordata, Cl: Ascidiacea), 17.4	(FIG. 1.34)
fi	st	cr,bu,fa	x	x	Sea stars or starfish, 1.7	(FIG. 1.35)
sh	ar	fa	te	se,x	Gooseneck barnacles (Ph: Arthropoda, Cl: Cirripedia), 15.18	(FIG. 1.37)
sh	ar	fa	te	s	Sea whip (Ph: Cnidaria, Cl: Anthozoa), 8.42	
sh	ar	fa	x	se	*Corallina officinalis,* a coralline red alga (Division: Rhodophyta). On rocks, shells, seaweeds. Not an animal; not included here.	(FIG. 1.36)
sh	do,sl,fl	cr,bu	x	x	Shelled, motile animals, 1.9	
sh	do	fa	jo	se	Gooseneck barnacles (Ph: Arthropoda, Cl: Cirripedia), 15.18	(FIG. 1.37)
sh	do	fa	x	x	Snails or chitons (Ph: Mollusca), 13.1	
sh	cy	fa	te	se,x	Serpulid polychaetes (tube is not segmented, but worm inside is) (Ph: Annelida, Cl: Polychaeta), 14.64	(FIG. 1.38)
sh	cy	bu	te	x	Tooth shells (Ph: Mollusca, Cl: Scaphopoda), 13.74	(FIG. 1.39)
sh	cy	ru	x	x	Cylindrical mollusks, 1.11	
sh	fl	fa	te	se,x	Gooseneck barnacles (Ph: Arthropoda, Cl: Cirripedia), 15.18	(FIG. 1.37)
sh	fl	fa	x	x	Attached, flattened, shelled animals, 1.13	
sh	fl	bu	2s	x	Bivalve mussels, clams, etc. (Ph: Mollusca, Cl: Bivalvia), 13.92	(FIG. 1.40)
sh	sh	fa	te	x	Encrusting ectoprocts (Ph: Ectoprocta), 12.1	(FIG. 1.41)
ex	ar	fa	te,x	se,x	Attached, arborescent colonies, 1.15	
ex	cy	cr,bu,sw	jo	se	Crustaceans (Ph: Arthropoda, SP: Crustacea), 15.4	(FIG. 1.42)
ex	cy,sl	cr	te	se	Polychaete worm (see FIG. 1.27), 15.4	
ex	cy,do	fa	x	x	Egg masses, 6.1	
ex	cy	sw	x	x	Rotifers (Ph: Rotifera)	
ex	do	cr	jo	se	Crabs (Ph: Arthropoda, SP: Crustacea, Cl: Decapoda), 15.104	(FIG. 1.43)
ex	do,lo	fa	2s	x	Solitary tunicates, sea squirts (Ph: Chordata, Cl: Ascidiacea), 17.4	(FIG. 1.44)
ex	fl	cr,bu,sw	jo	se	Crustaceans (Ph: Arthropoda, SP: Crustacea), 15.4	(FIG. 1.45)
ex	sh	fa	te	se,x	Encrusting ectoprocts (Ph: Ectoprocta), 12.1	(FIG. 1.41)
sp	st	cr,bu,fa	x	x	Sea stars or starfish, 1.7	(FIG. 1.46)
sp	do	fa,cr	x	x	Sea urchins (Ph: Echinodermata, Cl: Echinoidea), 16.14	(FIG. 1.47)
sp	fl	bu,cr	x	x	Sand dollar (Ph: Echinodermata, Cl: Echinoidea), 16.14	(FIG. 1.48)
sp	lo,do,cy	fa	ho	x	Sponges (Ph: Porifera), 7.1	(FIG. 1.49)

FIG. 1.23

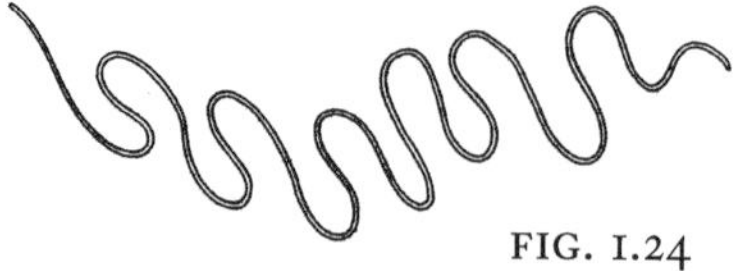

FIG. 1.24

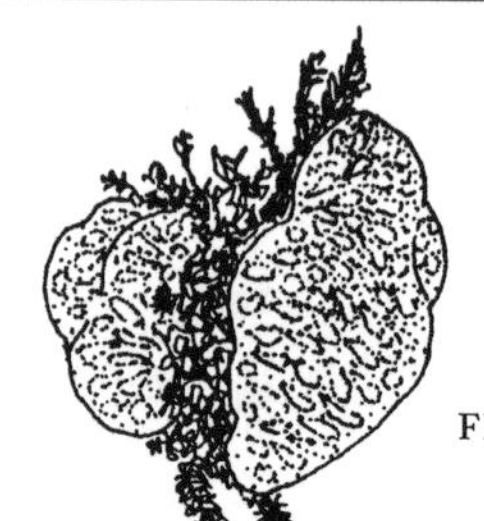

FIG. 1.25

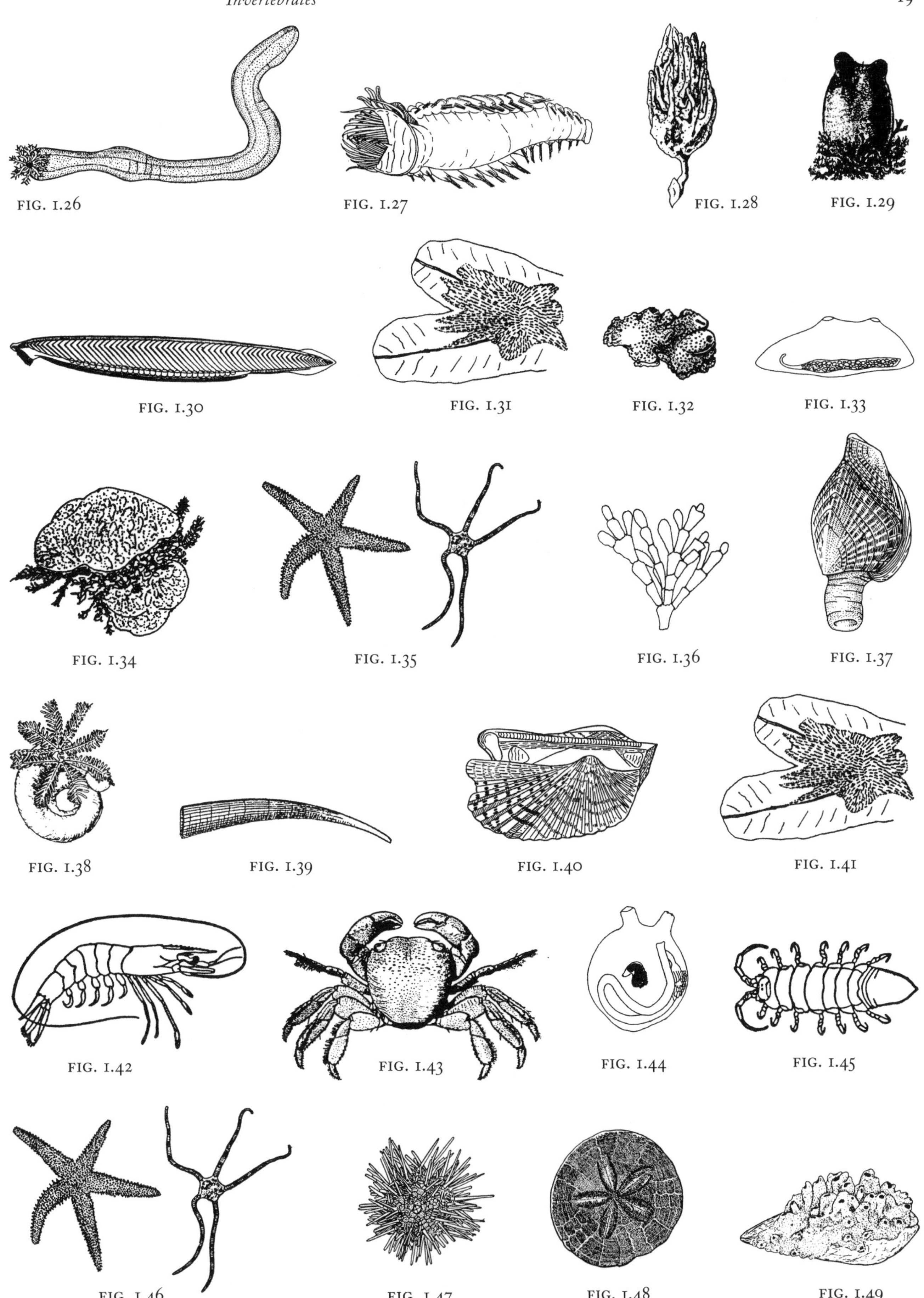
FIG. 1.26 FIG. 1.27 FIG. 1.28 FIG. 1.29

FIG. 1.30 FIG. 1.31 FIG. 1.32 FIG. 1.33

FIG. 1.34 FIG. 1.35 FIG. 1.36 FIG. 1.37

FIG. 1.38 FIG. 1.39 FIG. 1.40 FIG. 1.41

FIG. 1.42 FIG. 1.43 FIG. 1.44 FIG. 1.45

FIG. 1.46 FIG. 1.47 FIG. 1.48 FIG. 1.49

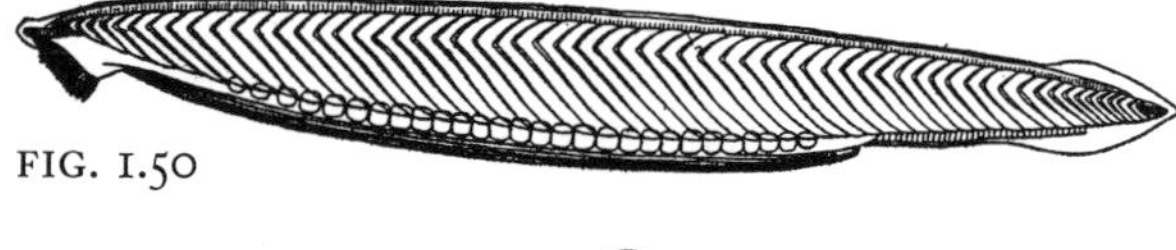

FIG. 1.50

FIG. 1.51

Branchiostoma caribaeum, Caribbean lancelet (Ph: Chordata, SP: Cephalochrodata). Fishlike; burrows in shallow sand; head out for filter feeding. Ca+,Sa,Su,ff,51 mm (2 in), (NM745). (FIG. 1.50)

Nectonema agile, horsehair worm (Ph: Nematomorpha). Long, thin; exterior with tiny hairs. Nocturnal, planktonic. Bo,Pe,Ps,?,20.3 cm (8 in), (GI08). (FIG. 1.51)

Rotifers (Ph: Rotifera). Largest rotifers visible to unaided eye. Brackish water. No key provided; refer to Pennak (1989).

1.3 Swimming, Cylindrical, Soft-Bodied Animals with Tentacles (from 1.2)

1. Body length
   - A. Animal *large,* greater than 3.8 cm (1.5 in) in length — la
   - B. Animal *small,* less than 3.8 cm (1.5 in) in length — sm
2. Suckers on tentacles
   - A. Large, *sucker*-bearing tentacles — su
   - B. One large pair and three short pairs of tentacles, suckers *absent* — x

1.4

Length	Suckers	Group	
la	su	Cephalopods (Ph: Mollusca, Cl: Cephalopoda), 13.161	(FIG. 1.23)
sm	x	Sea butterflies (Ph: Mollusca, Cl: Gastropoda, Or: Gymnosomata, Fa: Clionidae), 13.70; try *Clione limacina,* naked sea butterfly	(FIG. 1.3)

1.5 Soft, Sheetlike Animals Firmly Attached to Substrata (from 1.2)

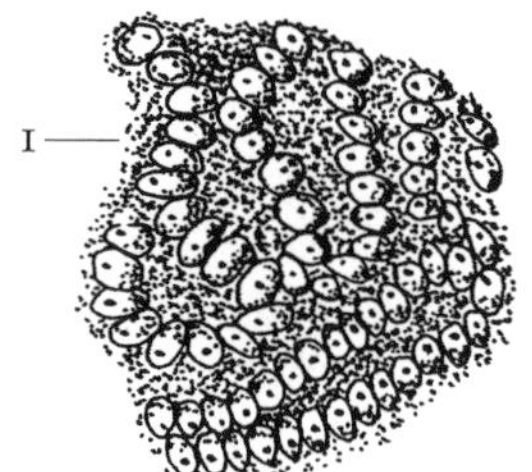

FIG. 1.52

1. Overall composition (dissecting microscope) (FIG. 1.52)
   - A. A collection of adjacent, fused, but still identifiable *individual* units comprise the sheet — in*
   - B. Sheet without identifiable individuals but with scattered *pores* — po
   - C. Sheet with *no* identifiable units — x
2. Presence of tentacles (provide plenty of time for animals to relax enough to reveal these, if present)
   - A. *Tentacles* present — te
   - B. Tentacles *absent* — x

1.6

Composition	Tentacles	Group	
in,po	x	Colonial tunicates, sea squirts (Ph: Chordata, Cl: Ascidiacea), 17.4	(FIG. 1.52)
in,po	te	Encrusting ectoprocts or bryozoa (Ph: Ectoprocta), 12.3	(FIG. 1.31)
x	x	*Halisarca* sp., slimy sponge	

Halisarca sp. (Halisarcidae). Uniform orange-brown, slippery coating on hard surfaces. No spicules. Ac,Ha–Al,Li–,ff,15.2 cm (6 in).

1.7 Sea Stars or Starfish (from 1.2)

1. Shape of arms or rays
   - A. Arms *fat,* abut one another around the central body disk — fa
   - B. Arms *slender* with spaces between bases as they join the central body disk — sl

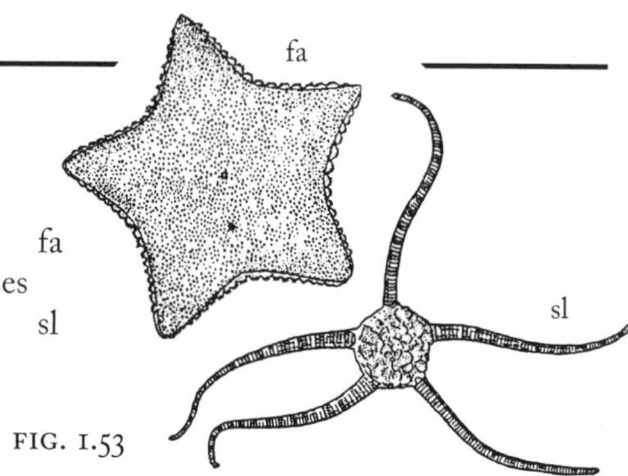

FIG. 1.53

1.8

Arms	Group	
fa	Sea stars (Ph: Echinodermata, Cl: Asteroidea), 16.2	(FIG. 1.53)
sl	Brittle stars (Ph: Echinodermata, Cl: Ophiuroidea) 16.8	(FIG. 1.53)

1.9 Shelled, Motile Animals (from 1.2)

1. Number of pieces (valves) comprising shelled portion (FIG. 1.54)
   - A. Shell a *single* piece, may be cap-shaped, spiraled, or tubular — 1
   - B. Shell comprising *two pieces* or valves — 2
   - C. Shell comprising a row of *eight* platelike valves — 8
2. Shape of ventral, muscular foot
   - A. Foot is *flattened* ventrally — fl
   - B. Foot is *wedge,* hatchet, or finger-shaped — we

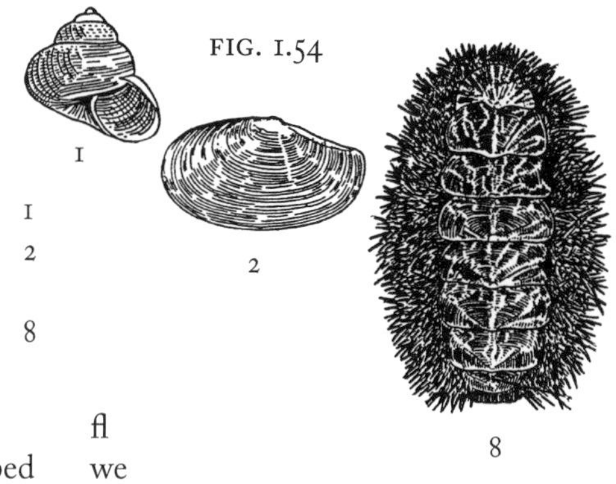

FIG. 1.54

1 2 8

1.10

Valves	Foot	Group	
1	fl	Snails, whelks, etc. (Ph: Mollusca, Cl: Gastropoda), 13.5	(FIG. 1.54)
2	we	Clams, mussels, etc. (Ph: Mollusca, Cl: Bivalvia), 13.92	(FIG. 1.54)
8	fl	Chitons (Ph: Mollusca, Cl: Polyplacophora), 13.3	(FIG. 1.54)

1.11 Cylindrical Mollusks (from 1.2)

1. Shell shape
   - A. Shell *spiraled* — sp
   - B. Shell *not* spiraled — x
2. Number of valves or pieces comprising shell: Provide *number* of valves to shell (1 or 2) — —
3. Burrows into wood (FIG. 1.55)
   - A. Animal found in tunnels within *wood* — wo*
   - B. Animal found on or in *non*wood substrata — x

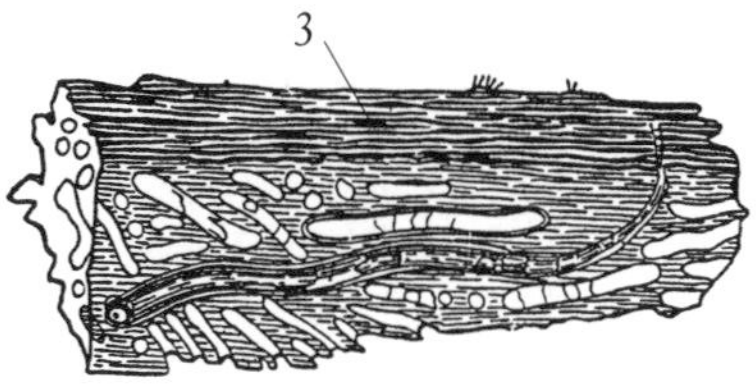

3

FIG. 1.55

1.12

Spiral	Valves	Wood	Group	
sp	1	x	Cylindrical snails (Ph: Mollusca, Cl: Gastropoda), 13.5	(FIG. 1.56)
x	1	x	Tooth shells (Ph: Mollusca, Cl: Scaphopoda), 13.74	(FIG. 1.57)
x	2	wo	Ship worms (Ph: Mollusca, Cl: Bivalvia), 3.7	(FIG. 1.55)
x	2	x	Bivalve clams, etc. (Ph: Mollusca, Cl: Bivalvia), 13.92	(FIG. 1.58)

FIG. 1.56

FIG. 1.57

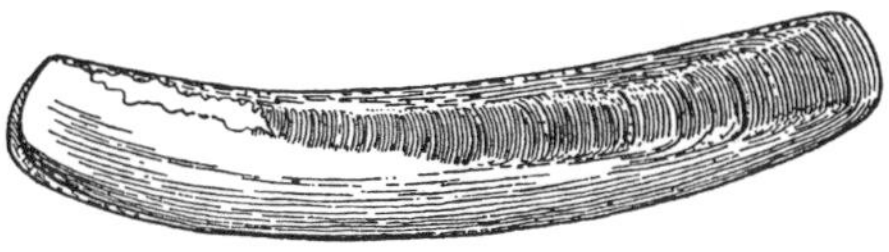

FIG. 1.58

1.13 Attached, Flattened, Shelled Animals (from 1.2)

1. Attachment to substrate
   - A. Shelled portion attached by way of an exposed, cylindrical, muscular *stalk* — st
   - B. Shell attached to substrate by exposed *byssus* threads or tough fibers — by
   - C. Shelled portion attached *directly* to substrate — di
2. Number of pieces (valves) to shell
   - A. Shell formed of *two* pieces or valves — 2
   - B. Shell comprised of *more than two* pieces or valves — >2

1.14

Attach	Valves	Group	
st	2	*Terebratulina septentrionalis,* northern lamp shell	
st	>2	Gooseneck barnacle (Ph: Arthropoda, Cl: Cirripedia), 15.16	(FIG. 1.37)
by,di	2	Mollusca, Bivalvia, 13.92	

Terebratulina septentrionalis, northern lamp shell (Ph: Brachiopoda). Most common living member of this once plentiful phylum. Despite appearances, not a bivalve mollusk. Ac–,Ha,Su–De,ff,3.3 cm (1.3 in), (GI21,NM723). (FIG. 1.59. Lateral view; dorsal view.)

FIG. 1.59

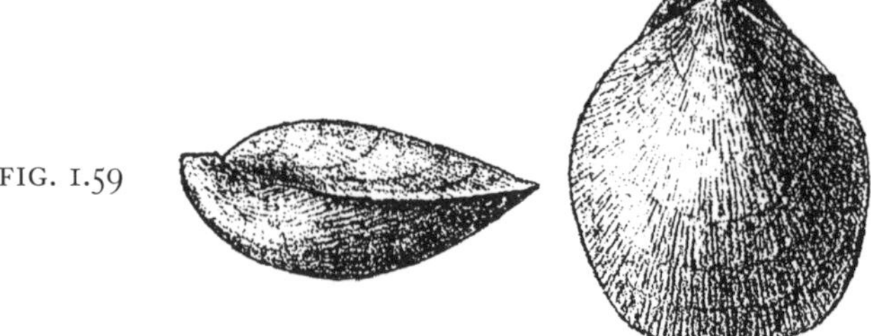

1.15 Attached, Arborescent, Firm-Bodied Colonies (from 1.2)

Animals fitting this general description are small, colonial forms in which many individuals produce plantlike clusters attached to solid substrates. Close examination (a dissecting microscope is best) reveals that tentacled individuals or zooids extend from stalklike bases to access food items in the surrounding water column. Obviously features offered here cannot be used for dead, decomposed specimens, although such remains may be identifiable. Try each of the groups listed as endpoints in the key below. Despite superficial similarities in appearance, arborescent colonies fall into two rather distantly related phyla, the Cnidaria and the Ectoprocta.

1. Prominence of tentacular cilia as a feeding technique (observe the behavior of individuals of an undisturbed, living colony in seawater) (dissecting microscope)
   - A. Tentacular *cilia* produce visible water currents in the vicinity of tentacles — ci
   - B. Tentacles are extended but no water currents are visible (i.e., tentacular cilia are *absent*) — x
2. Appearance of tentacle surfaces (dissecting microscope)
   - A. Tentacles are *bumpy* or nonuniform in appearance (bumps correspond to clusters of nematocysts or stinging cells) — bu
   - B. Tentacles appear *smooth* or uniform in construction — sm

1.16

Cilia	Tentacles	Group
ci	sm	Erect Ectoprocta, 12.11
x	bu	Hydroids (Ph: Cnidaria, Cl: Hydrozoa), 8.3

Chapter 2 ■ Gelatinous Organisms

2.1

Several types of invertebrates that spend time in the water column are gelatinous to increase buoyancy, transparent to avoid detection, and distasteful to discourage predation. Certain attached organisms have found gelatinous construction useful as well. This key should help identify major groups of gelatinous organisms. Because these creatures are often comparatively fragile, thorough observations in the field and especially careful handling are important.

1. Pattern of movement
   - A. Motion occurs in *pulses* — pu
   - B. Motion is *continuous* — co
   - C. Organism *attached,* that is, does not change location — at
   - D. Organism appears to be intact but free-floating, that is, there is *no* movement — x
2. Method of locomotion
   - A. Unified contractive pulses involve the *entire* organism — en
   - B. *Grouped* contractive pulses involve several small saclike components — gr
   - C. Beating of *cilia,* that is, animal does not change shape — ci
   - D. Locomotion is *absent* — x
3. Special structures for feeding (FIG. 2.1)
   - A. *Marginal* tentacles form peripheral fringe or clumps — ma
   - B. Single *pair* of *tentacles* (sometimes branched) extend from the body — 2t*
   - C. Two open-ended *siphons* serve as incurrent and excurrent canals for water flow — si
   - D. Distinct tentacles and siphons *absent* — x

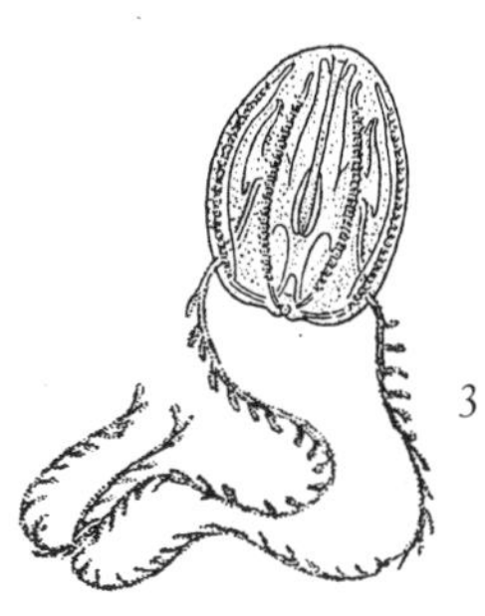

FIG. 2.1

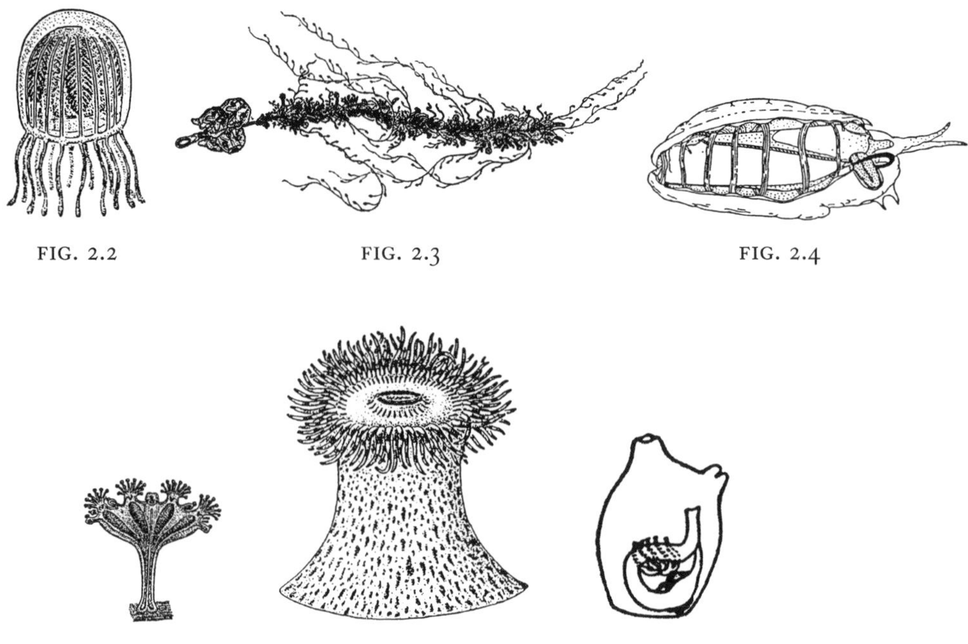

FIG. 2.2 FIG. 2.3 FIG. 2.4

FIG. 2.5 FIG. 2.6 FIG. 2.7

2.2

Pattern	Method	Tentacles	Group (Ph: Phylum; Cl: Class; Or: Order)	
pu	en	ma	Ph: Cnidaria, Cl: Hydrozoa and Scyphozoa, jellyfishes, 8.1	(FIG. 2.2)
co	gr	ma	Ph: Cnidaria, Cl: Hydrozoa, siphonophores, 8.38	(FIG. 2.3)
co	ci	2t,x	Ph: Ctenophora, comb jellies, 9.1	(FIG. 2.1)
co	en	x	Ph: Chordata, Cl: Thaliacea, salps, 17.14	(FIG. 2.4)
at	x	ma	Tentacles in ball-shaped clusters around the margin: Ph: Cnidaria, Cl: Scyphozoa, Or: Stauromedusae, 8.40 OR Tentacles uniformly distributed around the margin: Ph: Cnidaria, Cl: Anthozoa, Anemones, 8.42	(FIG. 2.5) (FIG. 2.6)
at	x	si	Ph: Chordata, Cl: Ascidiacea, sea squirts, 17.4	(FIG. 2.7)
at,x	x	x	Egg masses, 6.1	

Chapter 3 ■ Miscellaneous Worm-Shaped Organisms

3.1 A wormlike body plan has proved successful for a variety of only remotely related invertebrates. Use this key to distinguish among "worms," which, in fact, may range from anemones to hemichordates. The animal should be relaxed in fresh seawater, perhaps by using an anesthetizing agent (see the Appendix for recommended procedures).

1. Shape of body
   - A. Body divided into many (>3) *segments* marked by superficial annulations — se
   - B. Body divided into *three* distinct regions or sections — 3
   - C. Body divided into *two* distinct regions (the first is a retractile, anterior proboscis) — 2
   - D. Body not segmented but *flattened* dorsoventrally — fl
   - E. Body *worm*like but lacking these distinctive features — wo
2. Dominant appendages on anterior end (usually visible only when worm is thoroughly relaxed) (dissecting microscope helpful)
   - A. Cylindrical or bulbous *proboscis; retractable* into main body section — pr
   - B. Cylindrical or bulbous *proboscis; not* retractable — px
   - C. Hard *shell* valves — sh
   - D. Single *threadlike* process (*may* be long and sticky) — th
   - E. Several *finger*like, unbranched tentacles — fi
   - F. *Branched*, treelike appendages — br
   - G. *Sucker* — su
   - H. Clusters of bristles or *setae* or hairs — se
   - I. Specific appendages or processes *absent* on head — x
3. Overall length of body
   - A. *Less than 1 inch* — <1
   - B. *More than 1 inch* — >1
4. General rigidity or body firmness (remove animal from seawater to test this)
   - A. Body *firm*, rather rigid — fi
   - B. Body *soft*, droops — so

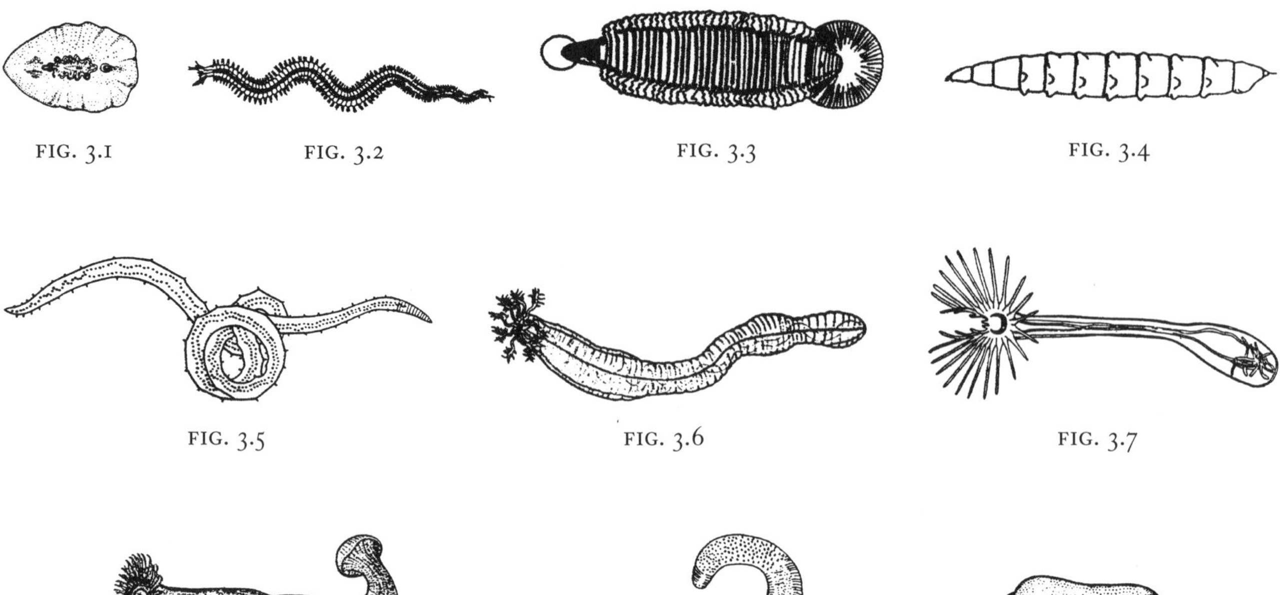

FIG. 3.1 FIG. 3.2 FIG. 3.3 FIG. 3.4

FIG. 3.5 FIG. 3.6 FIG. 3.7

FIG. 3.8 FIG. 3.9 FIG. 3.10

3.2

Shape	Anterior End	Length	Soft/Firm	Group or Organism	
fl	x,fi	<1	so	Flatworms (Ph: Platyhelminthes), 10.1	(FIG. 3.1)
se	th,fi,br	<1,>1	so	Segmented, soft-bodied worms, 3.3	(FIG. 3.2)
se	fi	>1	fi	*Branchiostoma caribaeum,* Caribbean lancelet	
se	se	>1	so	Polychaete worms (Ph: Annelida, Cl: Polychaeta), 14.3	(FIG. 3.2)
se	su	<1,>1	so	Leech (Ph: Annelida, Cl: Hirudinea), 14.88	(FIG. 3.3)
se	x	<1	fi	Ph: Arthropoda, Cl: Insecta. No key provided. Larvae (especially dipteran "maggots" occasionally in littoral zone or tidal pools. Refer to a key to freshwater invertebrates (e.g., Pennak 1989).	(FIG. 3.4)
se	x	<1,>1	so	Tips of some abdominal setae (last half of body) with curved and forked tips covered by tiny chitinous envelope (i.e., hooded hooks); some distinction between anterior thorax and more posterior abdominal region: Ph: Annelida, Cl: Polychaeta, Fa: Capitellidae, 14.24 OR No abdominal setae with hooded tips; body not regionalized: Ph: Annelida, Cl; Oligochaeta, 14.83	(FIG. 3.5)
wo	br	>1	so	Burrowing sea cucumbers (Ph: Echinodermata, Cl: Holothuroidea), 16.12	(FIG. 3.6)
wo	fi	<1	fi,so	Single pair of antennae, terminal segment with adhesive lobes: *Polygordius* sp. (Ph: Annelida, Cl: Archiannelida) OR U-shaped tentacle cluster: *Phoronis architecta* (Ph: Phoronida)	(FIG. 3.7)
wo	fi	>1	so	Soft-bodied, nonsegmented worms with tentacles, 3.5	(FIG. 3.8)
wo	sh	<1,>1	so	Shipworms (Ph: Mollusca, Cl: Bivalvia), 3.7	
wo	pr,fi,x	>1	fi	Peanut worms (Sipuncula), 3.9	(FIG. 3.9)
wo	x,th	<1,>1	fi	Roundworms (Ph:Nematoda). No key. Thrash but never change length or width. Many species possible.	
wo	x	<1	so	*Crystallophrisson* sp. (Ph: Mollusca; Cl: Aplacophora)	
2	pr,fi,x	>1	fi	Peanut worms (Ph: Sipuncula), 3.9	(FIG. 3.9)
2	px	>1	so	Echiuroid worms (Ph: Echiuroidea), 3.11	(FIG. 3.10)
3	x	>1	so	Three-sectioned worms, 13.13	

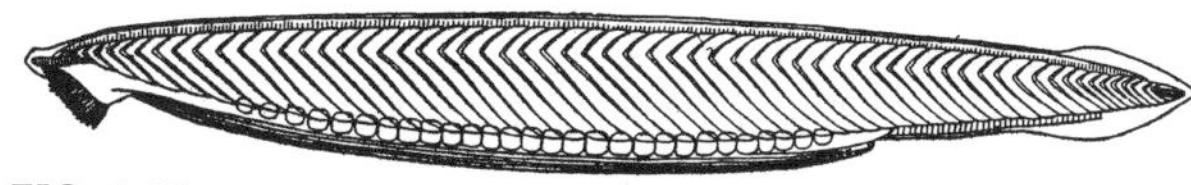
FIG. 3.11

Branchiostoma caribaeum, Caribbean lancelet (Ph: Chordata, SP: Cephalochordata). Fishlike; burrows in shallow sand; head out for filter feeding. Ca+,Sa,Su,ff,51 mm (2 in), (NM745). (FIG. 3.11)

Crystallophrisson sp. (Ph: Mollusca; Cl: Aplacophora). Small, mud-dwelling worms with fuzzy appearance from calcareous spicules; two small feather-shaped gills terminally. Other species are possible. Ac,Mu,Su,df,2.5 cm (1 in), (GI22).

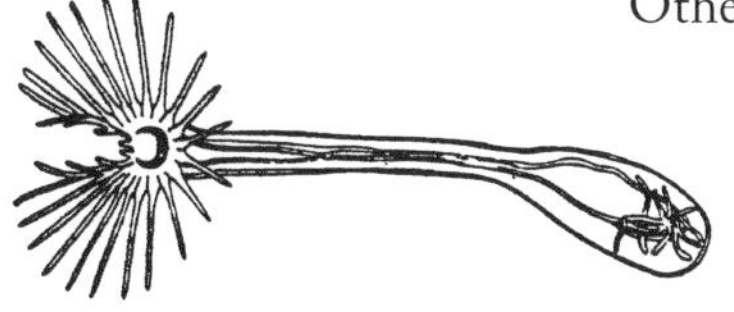
FIG. 3.12

Phoronis architecta (Ph:Phoronida). Small, slender, flesh-colored, nonsegmented worms without setae, form clusters of sand-covered tubules in subtidal waters and fine sand. Feed using U-shaped lophophore of ciliated tentacles. Ca+,Sa–Mu,Su,ff,5.1 cm (2 in), (GI21). (FIG. 3.12)

Polygordius spp. (Ph: Annelida, Cl: Archiannelida or Polychaeta). Tiny, slender worms that lack segmentation and setae; ciliated groove runs ventrally along the body. Taxonomic placement of "archiannelids" is unclear. Vi,Sa,Su,df,15 mm (0.6 in).

3.3 Segmented, Soft-Bodied Worms (from 3.2)

1. Features of lateral margins of trunk segments
   - A. *Parapodia* or lateral flaps present — pa
   - B. If no distinct parapodia, with groups of needlelike *setae* or bristles present — se
   - C. Both parapodia and regularly spaced setal clumps *absent* — x

3.4

Parapodia	Group or Organism	
pa	Polychaetes, (Ph: Annelida, Cl: Polychaeta), very common, 14.3	
se	Tips of some abdominal setae (last half of body) with curved and forked tips covered by tiny chitinous envelope (i.e., hooded hooks); some distinction between anterior thorax and more posterior abdominal region (dissecting microscope): Ph: Annelida, Cl: Polychaeta, Capitellidae, 14.24 OR No abdominal setae with hooded tips; body not regionalized: Ph: Annelida, Cl: Oligochaeta, 14.83	(FIG. 3.5)
x	*Priapulus caudatus,* priapulid, uncommon	

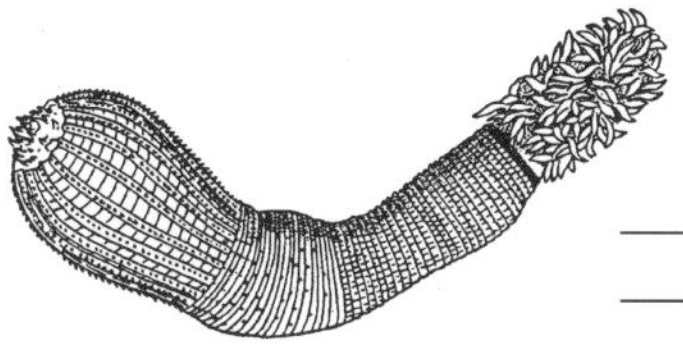

FIG. 3.13

Priapulus caudatus, priapulid (Ph: Priapulida). Stout worm; anterior section is retractable into body; cluster of tentaclelike respiratory structures extend posteriorly. Ac,Mu,Su,df,7.6 cm (3 in), (GI09,NM445). (FIG. 3.13)

3.5 Nonsegmented, Soft-Bodied Worms with Simple Tentacles (from 3.2)

1. Arrangement of tentacles (dissecting microscope)
   - A. Tentacles in *U-shaped* grouping — u
   - B. Tentacles generally in rings or clusters, but *not* U-shaped grouping — x
2. Dwelling tubes
   - A. Lives in *slender, permanent* sand-covered tubes — sp
   - B. Lives in *non*permanent burrows in sediment — x

3.6

Tentacles	Tubes	Group or Organism
u	sp	Ph: Phoronida: *Phoronis architecta*
x	x	Burrowing anemone or hydroid (Ph: Cnidaria), 8.1

FIG. 3.14

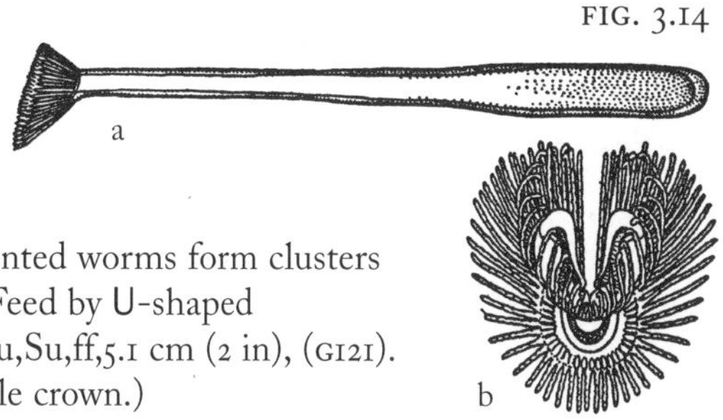

a

b

Phoronis architecta. Small, slender, nonsegmented worms form clusters of sand-covered tubules in subtidal waters. Feed by U-shaped lophophore of ciliated tentacles. Ca+,Sa–Mu,Su,ff,5.1 cm (2 in), (GI21). (FIG. 3.14. a, animal; b, lophophore or tentacle crown.)

3.7 Shipworms (Ph: Mollusca, Cl: Bivalvia) (from 3.2)

Shipworms burrow into wood and live in shell-lined galleries (there are many species in this difficult group; consult more detailed references for identification [Turner 1966]. Only the two most frequently listed types are included here.)

sp

se

FIG. 3.15

1. Shape of calcareous "pallets" located at the posterior end of the worm; pallets are used to plug the comparatively tiny opening to the outside of the penetrated wood (FIG. 3.15)
   - A. Pallet is a *single piece,* deep funnel — sp
   - B. Pallet is *segmented,* looks like a stack of nested funnels — se

3.8

Pallet	Organism
sp	*Teredo navalis,* common shipworm
se	*Bankia gouldii*

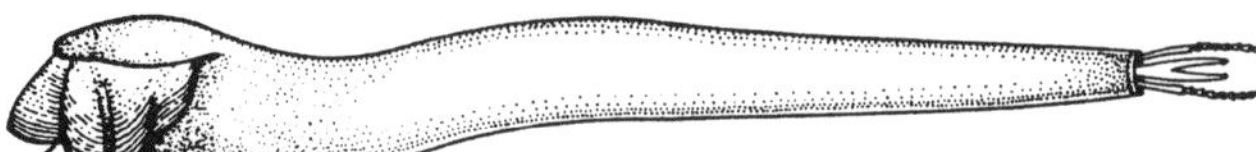

FIG. 3.16

Bankia gouldii, Vi,Ot,Ps,ot,7.6 cm (3 in), (GI59,NM576). (FIG. 3.16)

Teredo navalis, common shipworm. Bo,Ot,Ps,ot,7.6 cm (3 in), (GI59,NM576).

3.9 Peanut Worms (Ph: Sipuncula) (from 3.2)

Body is divided into a trunk and an evertable/retractable "introvert" or proboscis, with the mouth and often tentacles at its tip. One regional species occupies empty gastropod shells or worm tubes, whereas the others are typically burrowers. Positive identifications often require dissection of the specimen (see Cutler 1977).

1. Ratio of fully extended introvert length compared to trunk length (allow animal to relax thoroughly in fresh seawater; see the Appendix): Provide estimate of *ratio* — __:__
2. Habitat
   - A. Worm within abandoned *snail* shells or worm tubes — sn
   - B. Worm burrows in sediment; *not* within snail shell — x
3. Body length
   - A. Worm small, *less than 2.5 cm (1 in)* in length — <1
   - B. Worm larger, *greater than 2.5 cm (1 in)* in length — >1

3.10

Introvert:Trunk	Shell	Size	Organism
1:1	sn	<1	*Phascolion strombi*
1:3	x	>1	*Phascolopsis gouldii,* Gould peanut worm
1:2	x	<1	*Golfingia eremita*

Golfingia eremita (Golfingiidae). Fat worm with thick body wall; skin rough, with ridges and papillae. Ac,Sa–Mu,De,df,2.5 cm (1 in).

Phascolion strombi (Golfingiidae). In abandoned gastropod shells; muddy tube plugs aperture. Typically animal is seen as its cylindrical proboscis slowly extends from its muddy tube and then contracts back again. Bo,Mu,Li–Su,df,2.5 cm (1 in), (GI98), ☺. (FIG. 3.17)

Phascolopsis gouldii, Gould peanut worm (Sipunculidae). Light colored; in mud or sand. Ac–,Sa–Mu,Li–Su,df,30.5 cm (12 in), (GI98,NM447), ☺. (FIG. 3.18)

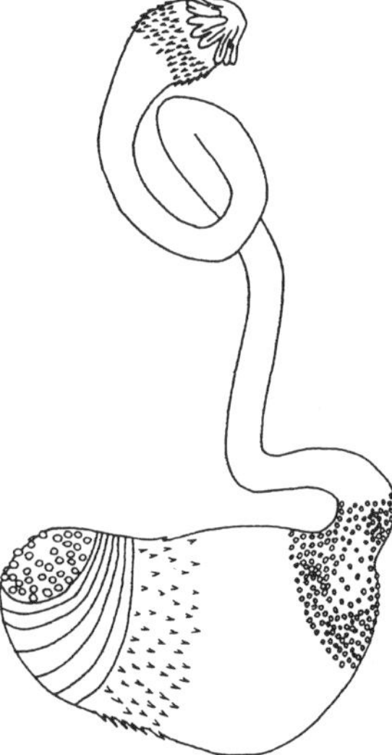

FIG. 3.17

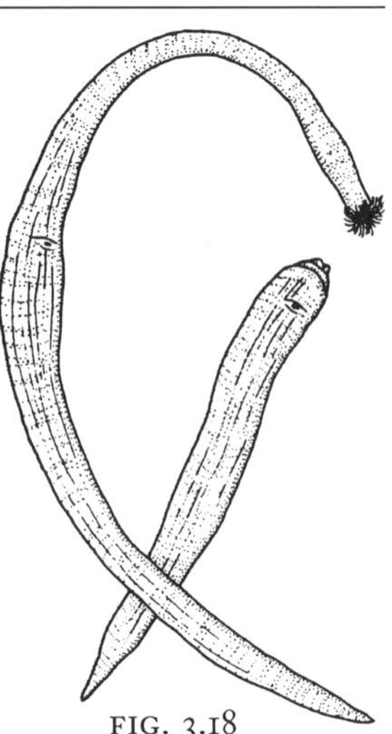

FIG. 3.18

3.11 Echiuroid Worms (Ph: Echiuroidea) (from 3.2)

Uncommon, cylindrical, soft-bodied, nonsegmented worms with a large, nonretractable proboscis anteriorly; burrows in sand or mud.

1. Size
   - A. Worms large, *greater than 3.7 cm (1.5 in)* >1.5
   - B. Worms small, *less than 3.7 cm (1.5 in)* <1.5

3.12

Size	Organism
>1.5	*Echiurus echiurus*
<1.5	*Thalassema* spp.

FIG. 3.19

Echiurus echiurus (Echiuridae). Scoop-shaped proboscis; rare. Ac,Mu,Su,df,30.5 cm (12 in), (G199). (FIG. 3.19)

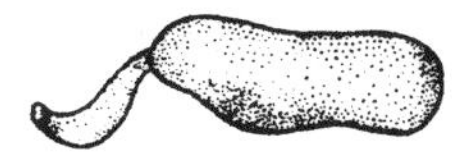

FIG. 3.20

Thalassema spp. (Echiuridae). Several species of small worms (0.3–1 in) from subtidal mud. Bo,Mu,Su,df,2.5 cm (1 in), (G199,NM451). (FIG. 3.20)

3.13 Three-Sectioned Worms (from 3.2)

An odd pairing of coincidentally shaped wormlike animals.

1. Appearance of middle section
   - A. Middle section a comparative *short* orange *collar* sc*
   - B. Middle section *large, bumpy* surfaced, with *annulations* lba
2. Appearance of posterior section
   - A. Posterior section, a *long,* fragile *trunk* lt*
   - B. Posterior section a cluster of short, *tentaclelike* structures te

3.14

Middle	Posterior	Organism
sc	lt	*Saccoglossus kowalewskii,* Kowalewsky acorn worm
lba	te	*Priapulus caudatus,* priapulid

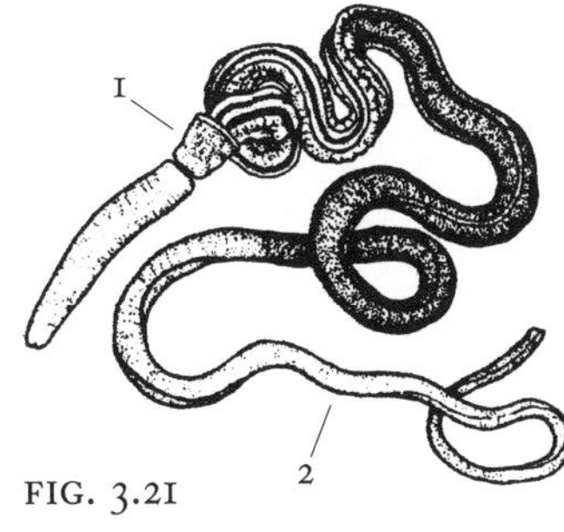

FIG. 3.21

Priapulus caudatus, priapulid (Ph: Priapulida, Priapulidae). Small, light-colored, mud-dwelling worm. Bulbous, anterior body section can be retracted entirely into the trunk section. Rare. Ac,Mu,Su,df,7.6 cm (3 in), (G109,NM445). (FIG. 3.13)

Saccoglossus kowalewskii, Kowalewsky's hemichordate or acorn worm (Ph: Hemichordata, Harrimaniidae). Cream colored proboscis; orange collar, green-brown, fragile trunk. Infaunal, nonselective deposit feeder leaves coiled casting at sand-mud surface. Vi+,Sa–Mu,Li–,df,15.2cm (6 in), (G265), ☺. (FIG. 3.21)

Chapter 4 ■ Ectoparasites and Commensals

4.1 The outcome of interactions occurring between any two populations of organisms can be beneficial, detrimental, or neutral with respect to each of the parties involved. When one population gains at the expense of another, the relationship falls along a continuum from predation (in which the gains for one, the predator, and the loss for the other, the prey, are substantial) to parasitism (in which parasite gains and host losses are much more modest). Evolutionarily refined parasites tend to maintain long-term, generally low-impact relations with their much larger hosts. The successful parasite meets its needs without seriously debilitating the host on which it depends. To extend this interactive spectrum one step further, parasitic relationships blend into *commensalism* if the commensal gains, but the host neither loses nor benefits from their association.

Parasitic or commensalistic/parasitic associations are not uncommon among marine animals. Treatment here is limited to the most frequently observed, external parasites, termed *ectoparasites,* and the commensals (that is, animals found closely associated others, but which have not been demonstrated to be harmful to the host). Coverage of the myriad of internal parasites found in marine organisms lies beyond the scope of this work. The following treatment follows a different format. Because encounters with ectoparasites occur in association with their more conspicuous hosts, the first step in making an identification should be to refer to lists of parasites known to be linked to a particular host. In many cases you will be referred to species listings located within their particular phylum or group. Annotated coverage appears at the end of this chapter for parasites/commensals not treated elsewhere.

4.2 Phylum Porifera

There are many invertebrates living within channels and tissues of sponges. These are fundamentally free-living organisms and should be identified using keys provided for each phylum.

4.3 Phylum Cnidaria, Class Scyphozoa

Host	Parasite/Commensal
Aurelia aurita, moon jelly†	*Hyperia galba,* big-eye amphipod, 15.35
Cyanea capillata, lion mane†	*H. medusarum,* big-eye amphipod, 15.35 *Peachia parasitica,* parasitic anemone, 8.43

4.4 Phylum Ctenophora

Host	Parasite/Commensal
Bolinopsis infundibulum, northern comb jelly	*Hyperoche tauriformis,* big-eye amphipod, 15.35
Mnemiopsis leidyi, sea walnut	*Edwardsia leidyi,* parasitic anemone, 8.42

4.5 Phylum Mollusca, Class Bivalvia

Host	Parasite/Commensal
Anomia simplex, jingle†	*Pinnotheres maculatus,* squatter pea crab†, 15.109 *P. ostreum,* oyster pea crab†, 15.109
Argopecten irradians, bay scallop†	*P. maculatus,* squatter pea crab†, 15.109
Crassostrea virginica, eastern oyster†	*Cliona* spp., sulfur or boring sponges, 7.1 *Macrobdella grossa,* leech ribbon worm, 11.2 *Pinnotheres ostreum,* oyster pea crab†, 15.109 *Polydora websteri,* oyster mudworm, 14.71
Geukensia demissus, ribbed mussel†	*Pinnotheres ostreum,* oyster pea crab†, 15.109
Laevicardium mortoni, Morton eggcockle†	*P. maculatus,* squatter pea crab†, 15.109
Modiolus modiolus, northern horsemussel†	*P. maculatus,* squatter pea crab†, 15.109
Mya arenaria, softshell†	*P. maculatus,* squatter pea crab†, 15.109
Mytilus edulis, blue mussel†	*P. maculatus,* squatter pea crab†, 15.109 *P. ostreum,* oyster pea crab†, 15.109
Solemya spp., awningclams	*Listriella* spp., amphipod, 15.72 *Pinnixa sayana,* 15.109

4.6 Phylum Annelida, Class Polychaeta

Host	Parasite/Commensal
Amphitrite ornata, ornate spaghetti worm	*Pinnixa chaetopterana,* tube pea crab, 15.109 *Listriella barnardi,* amphipod, 15.72
Arenicola cristata, lugworm	*P. cylindrica,* 15.109 *P. sayana* (maybe), 15.109
Chaetopterus variopedatus, parchment worm	*Pinnotheres ostreum,* oyster pea crab, 15.109 *Pinnixa chaetopterana,* tube pea crab, 15.109 *Polyonyx gibbesi,* eastern tube crab, 15.109
Clymenella torquata, bambooworm	*Listriella clymenellae,* amphipod, 15.72
Diopatra cuprea, plumed worm	*Arabella iricolor,* opal worm; young dwell in tubes, 14.23
Enoplobranchus sanguineus	*Microphthalmus aberrans,* hesionid polychaete, 14.37
Neoamphitrite johnstoni, Johnston ornate terrebellid	*Lepidametria commensalis,* commensal scaleworm, 14.59

4.7 Phylum Arthropoda, Subphylum Chelicerata, Class Merostomata

Host	Parasite/Commensal
Limulus polyphemus, horseshoe crab, 15.1, 15.105	On gills: *Bdelloura candida,* limulus "leech", 10.2 *Syncoelidium pellucidum,* 10.2 On carapace: *Chelonibia patula,* barnacle, 15.19 *Crepidula plana,* eastern white slippersnail, 13.26

4.8 Phylum Arthropoda, Subphylum Crustacea

Host	Parasite/Commensal
Callianassa setimanus	*Pinnixa chaetopterana,* tube pea crab, 15.109
Callinectes sapidus, blue crab	*Carcinonemertes carcinophila,* crab nemertean, on gills and egg masses, 11.2 *Chelonibia patula,* barnacle, 15.19
Carcinus meanas, green crab	*Sacculina carcini,* parasite barnacle
Paleomonetes spp., grass shrimp	*Probopyrus pandalicola,* shrimp parasitic isopod, especially in gill chamber, 15.25
Pagurus spp., hermit crabs	*Crepidula plana,* eastern white slippersnail, 13.26 *Lepidonotus sublevis,* commensal 12-scaled polychaete worm, 14.59 *Polydora commensalis,* hermit crab worm, 14.71
Upogebia affinis, coastal mudshrimp	*Pinnixa retinens,* 15.109 *P. sayana,* 15.109

Sacculina carcini, parasitic barnacle (Ph: Arthropoda, SP: Crustacea, Cl: Cirripedia, Or: Rhizocephala, Sacculinidae). Parasite's body ramifies throughout host's tissues, but produces external, cream to brown reproductive sack under crab's abdomen. (Do not confuse this with normal and common, orange to brownish egg mass in same location.)

4.9 Phylum Echinodermata

Host	Parasite/Commensal
Asterias rubens, northern sea star	Caprellid amphipods, especially *Aeginina spinosa,* 15.87 and *Caprella unica,* 15.87 *Pinnotheres maculatus,* squatter pea crab, 15.109
Echinarachnius parma, sand dollar; and *Mellita quinquiesperforata,* keyhole dollar	*Dissodactylus mellitae,* sand-dollar pea crab, 15.109

4.10 Phylum Chordata, Subphylum Urochordata

Host	Parasite/Commensal
Bostrichobranchis pilularis, tunicate	*Pinnotheres maculatus,* squatter pea crab, 15.109
Salpa spp. and *Thalia* spp., salps	*Phronima sedentaria,* big-eye amphipod, juvenile stages, 15.35

4.11 Phylum Chordata, Subphylum Vertebrata, Classes Chondryichthes and Osteichthyes

Too many fishes are known to be hosts to parasites and commensals to list them individually here. Leeches, copepods, and isopods are among the common external parasites of fishes. Leeches (Ph: Annelida, Cl: Hirudinea) attach to surfaces by large anterior and posterior suckers (see 14.88). They lack lateral appendages, although some possess pairs of conspicuous gills along their sides. Bopyrid and cymothuroid isopods (Ph: Arthropoda, SP: Crustacea, Cl: Malacostraca, Or: Isopoda) (see 15.24) may be found in the mouth cavity or on gills, especially of bluefish, striped bass, menhaden, and white and silver perch. Although some parasitic copepods retain recognizable features of their class (Ph: Arthropoda, SP: Crustacea, Cl: Copepoda), many are so highly modified that their relationship is not immediately apparent. Small, attached parasites on the fins, body surface, mouth cavity, or gills of fishes may be identified using other keys to parasitic copepods (Cressey 1978; Ho, J.-S. 1971, 1977, 1978).

Chapter 5 ■ Zooplankton

5.1 Zooplankton include heterotrophic organisms living freely suspended in the water column (also see discussion of open-water habitats, 0.4). The locomotory power of some zooplankton is limited, and they are transported primarily by mass movements of their surroundings. Detailed treatment of the thousands of different species that occur within the rich plankton communities of the Northeast lies beyond the scope of this manual. To encourage interest in this fascinating and critical segment of the regional biota, however, this key provides crude separation of the most common, larger types, often to class, order, family, or, infrequently, to genus.

Zooplankton can be divided into groups by taxonomic, life-historical, or size criteria. *Protozooplankton* include individual or colonial members of the Kingdom Protista. They contrast with the *metazooplankton* which belong to the Kingdom Metazoa or Animalia, the multicellular animals. Organisms that spend their entire lives as members of either of these planktonic groups are termed *holoplankton.* Many other invertebrates produce larval stages which are *meroplanktonic,* living as plankton for only a portion of their lives. This strategy is especially common in the case of benthic (i.e., bottom-dwelling) invertebrates that use the planktonic larval stage provides to disperse to new locations.

Planktonic organisms can also be characterized by body length. The smallest plankton are designated picoplankton, nannoplankton, and microplankton—categories that include organisms from bacterial size to those measuring 0.2 mm (0.008 in). in length. Mindful of our criterion of including animals large enough to see with the unaided eye, this treatment is limited to the mesoplankton, 0.2–20 mm (0.008–0.8 in) long, and macroplankton, 2–20 cm (0.8–8 in). Many larger macroplankton and any common megaplankton, larger than 20 cm (8 in) in length, are treated in their proper taxonomic grouping elsewhere in this manual. Some of these may be found among Gelatinous Organisms (2.1).

If possible, it is best to work initially with living plankton so that locomotory styles may be observed. Unfortunately, once collected, these sensitive animals do not survive well, and although refrigerating samples may prolong their lives, samples that cannot be reviewed within an hour or two of collection must be preserved. Add enough formaldehyde solution to some or all of the sample to bring its final sample concentration to 4% (4 parts formalin [i.e.,

40% formaldehyde solution] to 96 parts water). Animals of this size range must be examined using the magnifying power of a dissecting microscope at least. See instructions for preparing temporary wet mounts (0.7) and recipies for preservation in the Appendix.

1. Description of the most conspicuous appendages or body adornments
   - A. Two or more *jointed* (segmented), *slender* appendages — js
   - B. Two or more *non*jointed, *slender* appendages (including tentacles) — xs
   - C. A single, slender *tail* — ta
   - D. *Fins* — fi
   - E. Ciliated *lobes* — lo
   - F. *Setae* or bristles in the absence of jointed appendages — se
   - G. Appendages *absent;* animals various shapes, basically smooth outline — x
2. Description of locomotion (ignore this question for preserved material)
   - A. Moves by *pulsed contractions* of all or part of body — pc
   - B. Uses one or more pairs of *appendages* or setae to move — ap
   - C. Uses *ciliary currents* to move (usually smooth, more or less continuous movement) — cc
   - D. *Thrashes* a tail or all of its body for movement — th
   - E. *No* self-generated movement (observe for a reasonable period of time) — x

5.2

Appendages	Locomotion	Group
js	ap	Crustacea, 5.3
xs,ta,fi	th	Thrashing swimmers, 5.15
xs	cc	Lacks shelllike covering: Larvae with elongated, ciliated appendages, 5.17 OR With shell-like covering: Larvae with ciliated lobes, 5.19
lo	cc	Larvae with ciliated lobes, 5.19
se	ap	Nectochaeta larva of polychaete worms
x	cc	Nonlobed, ciliated larvae, 5.23
x	pc	Gelatinous animals, 2.1
x	x	Miscellaneous: eggs, cast-off body parts, etc. Most are unidentifiable using this source. Spherical eggs with bandlike, segmented larvae are apt to be fish eggs. Exoskeletal ecdyses or cast-off, molted parts of crustaceans are common in nearshore samples; often these include transparent, outline shapes of copepods or the thoracic legs of barnacles

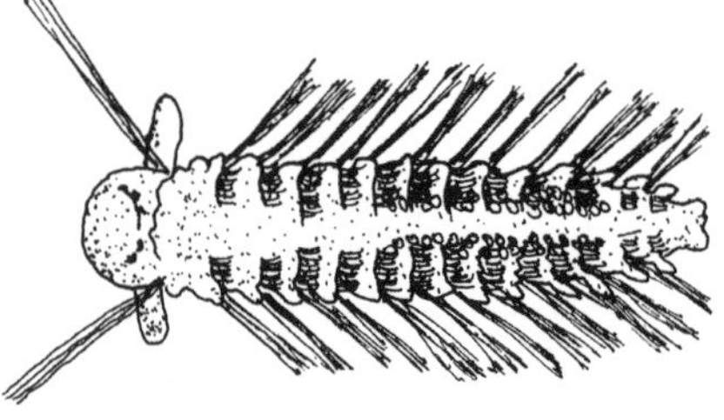

FIG. 5.1

Nectochaeta larva of polychaete worms (Ph: Annelida, Cl: Polychaeta). In this stage, paddle-shaped swimming setae have replaced ciliary bands typical of early trochophore (5.20) stages. Eventually, simple parapodia develop as well. (FIG. 5.1)

5.3 Crustacea (Ph: Arthropoda, SP: Crustacea) (from 5.2, 5.5, 5.11)

Crustacea are major components in most plankton samples and include both holoplanktonic forms, such as copepods and cladocera and meroplanktonic, larval stages. Most crustaceans have a segmented body divided into three regions: the head, thorax, and abdomen. Typically, the head and thorax are continuous and, in many cases, are covered by a complete or partial exoskeletal plate, the *carapace.* The more slender abdomen tends to be distinctly set off from the thorax. A terminal lobe, the *telson,* is often found posterior to the last abdominal segment. Crustaceans possess many pairs of appendages, the number and morphology of which can be useful in their separation. There are two pairs of antennae (first antennae are *antennae;* second antennae are *antennules*). Thoracic legs may have pincer tips (*chelate*) or not. The most terminal pairs of abdominal legs are called *uropods.*

The keys that follow help the reader find the most likely crustacean group from among those most commonly encountered in the plankton.

1 3, 4 2

FIG. 5.2

1. Carapace covers at least part of thorax (FIG. 5.2)
   - A. *Single*-piece *carapace* covers one-third to *one-half of thorax* (i.e., some thoracic segments are visible — 1c½t
   - B. *Single*-piece *carapace* covers all or nearly all of the *thorax* — 1ct*
   - C. *Bivalved carapace* (open along one seam) covers entire *head* and thorax, clamlike — 2ch
   - D. *Bivalved carapace* covers thorax but *not* the head — 2cx
   - E. Carapace *absent*, segmentation of body visible right up to the head — x
2. Number of pairs of bristle-bearing appendages (FIG. 5.2)
   - A. Only *three* pairs of bristle-bearing appendages — 3
   - B. *More than three* pairs of bristle-bearing appendages — >3*
3. Eye construction
   - A. *Stalked* eyes (occupy tips of slender shafts) present — st*
   - B. Eyes *sessile* (attached directly to head) — se
   - C. Eyes *absent* — x
4. Eye size (FIG. 5.2)
   - A. Eyes *large* — la*
   - B. Eyes *small* — sm
   - C. Eyes *absent* — x

5.4

Carapace	Bristle Appendage	Eye Type	Eye Size	Group
1ct	3	se	sm	Nauplius larvae, 5.5
1ct	>3	st	la	Larger crustacean larvae, 5.7
1c½t	>3	se	sm	Class Copepoda, 5.11
2ch	>3	se,x	sm,x	Both ends rounded; major appendages are anteriorly placed: Ostracoda. No key provided. (FIG. 5.3) OR Anterior rounded, posterior pointed; major appendages are posteriorly placed: Cypris larvae of barnacles (SP: Crustacea, Cl: Cirripedia) (FIG. 5.4)
2cx	>3	se	la	Or: Cladocera, 5.13
x	>3	se	la	Hyperiid Amphipoda, 5.34

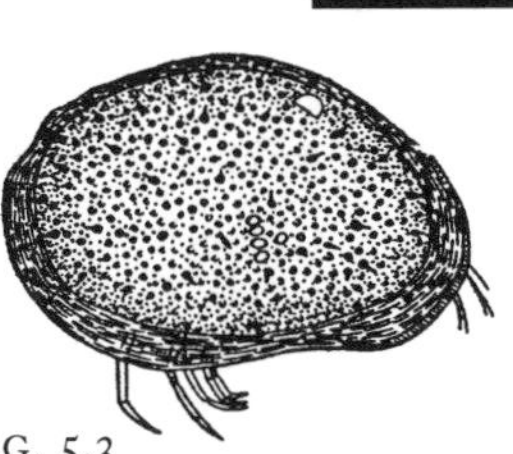

FIG. 5.3

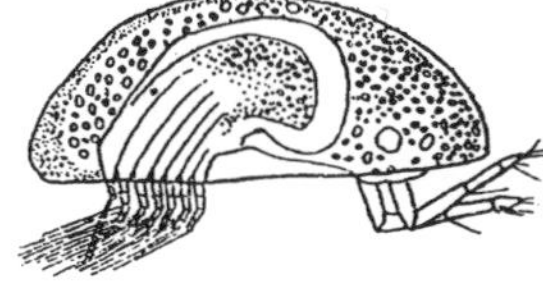

FIG. 5.4

5.5

Nauplius Larvae (from 5.4)

The nauplius (first larval stage) of most Crustacea possesses only three pairs of appendages, corresponding to antennae 1, antennae 2, and mandibles. The triangular shape of barnacle nauplii distinguishes them from the rest. Subsequent larval stages of Crustacea are treated in 5.7 (Larger Crustacean Larvae), below.

1. Nauplius shape
   - A. Larvae *triangular* with distinct lateral points along anterior margin — tr
   - B. Larvae various shapes but anterior marginal points *absent* — x

5.6

Shape	Type
tr	Barnacle nauplius (SC: Cirripedia) (FIG. 5.5)
x	Nauplius larva of other Crustacea. Based on typical abundance, copepods or possibly crabs are the best guess. (FIG. 5.6)

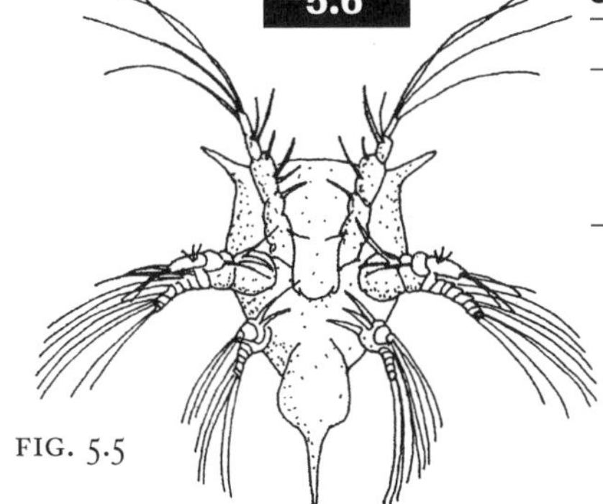

FIG. 5.5

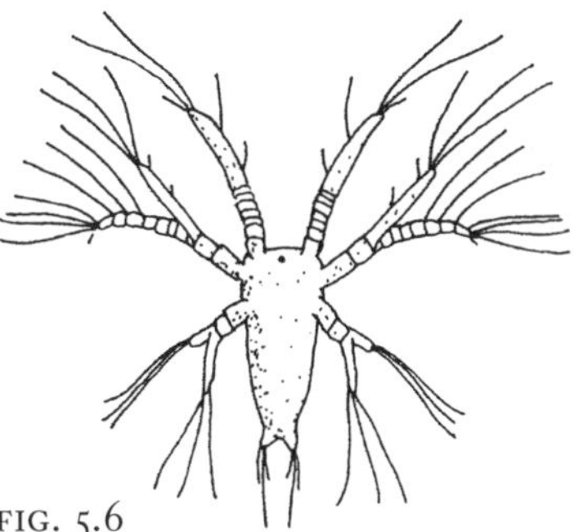

FIG. 5.6

5.7 Larger Crustacean Larvae (from 5.4)

Crustaceans grow by means of a sequence of molts during which they shed the existing exoskeleton and replace it with a new one. In many instances, dramatic changes in appearance accompany such molts, producing a series of species-specific, life-history stages. Because of the dozens of species, each displaying several distinctive stages, identification of particular crustacean larvae is not attempted here. Since common crustacean groups pass through a roughly comparable series of larval stages, a coarse breakdown by stage and faunal group is practical.

FIG. 5.7

1. Presence of chelate legs (legs with pincer endings) (FIG. 5.7)
   - A. At least one pair of appendages is *chelate* — ch*
   - B. All appendages end simply; chelate appendages are *absent* — x
2. Number of pairs of thoracic appendages: Provide *number* of pairs of limbs (note: does not include appendages of the head; i.e., two pairs of antennae plus one pair of mandibles) (FIG. 5.7) — —
3. Presence of uropods (terminal legs) flanking the telson (FIG. 5.7)
   - A. Uropods *present* — ur*
   - B. Uropods *absent* — x

5.8

Chela	No. Legs	Uropods	Larval Type	
x	3	x	Nauplius larva, 5.5	
x	4–7	x	Protozoea larva of Mysidacea, Euphausiacea (FIG. 5.8), or Shrimp (FIG. 5.9); most other groups pass through this stage in the egg.	(FIG. 5.8) (FIG. 5.9)
x	8	ur,x	Zoea larva, 5.9	
ch	8	ur	Postlarva: Wide, rounded thorax; small abdomen (i.e., crablike): Megalops larva of crabs (Or: Decapoda; adults, 15.104) OR Slender, laterally compressed carapace; well-developed abdomen (i.e., shrimplike): Mastigopus larva of shrimp (Or: Decapoda; adults, 15.92)	(FIG. 5.10) (FIG. 5.11)

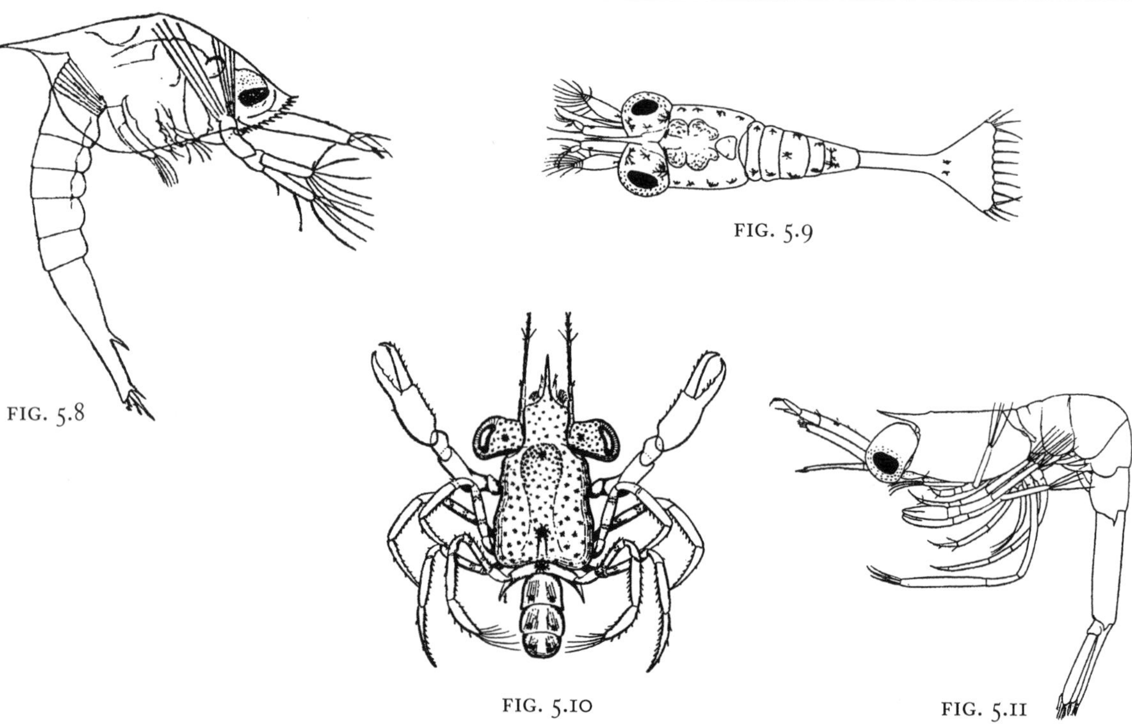
FIG. 5.8
FIG. 5.9
FIG. 5.10
FIG. 5.11

5.9 Zoea Larvae (from 5.8)

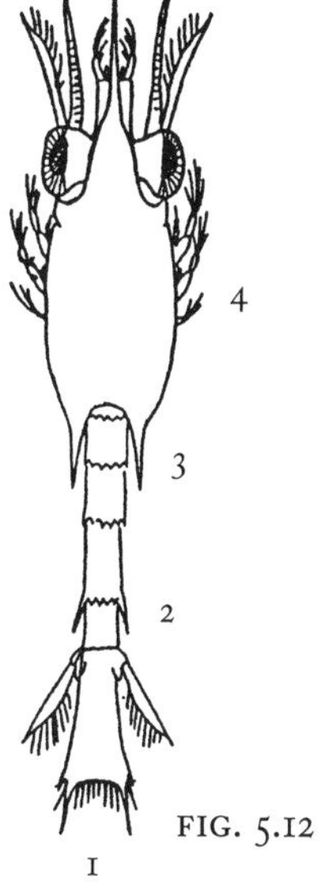

FIG. 5.12

Zoea of many crustacean species may be found in plankton samples. This key is limited to the most typical representatives. Exceptions may be found. The zoea larvae of mysid shrimp (Or: Mysidacea; adults, 15.90) can be distinguished from all other groups by a carapace attached only to the first three thoracic segments and by uropods bearing basal statocysts (spherical bodies involved in balance perception).

1. General shape of the telson (FIG. 5.12)
   - A. Telson *expands* posteriorly (i.e., is narrowest where it attaches) ex*
   - B. Telson is roughly *rectangular*, or triangular re
2. Presence of spines on at least some abdominal segments (FIG. 5.12)
   - A. Abdominal *spines* present sp*
   - B. Abdominal segments *lack* spines x
3. Presence of points along the posterior margin of the carapace (FIG. 5.12)
   - A. Posterior margin of carapace with at least one *point* po*
   - B. Posterior margin of carapace *lacks* points x
4. Presence of obvious dorsal spine midway along carapace (FIG. 5.12)
   - A. Dorsal *spine* present sp
   - B. Dorsal spine *absent* x*

5.10

Telson	Abdominal Spines	Carapace Points	Dorsal Spine	Group	
re	x	po,x	x	Euphausiacea, krill (Or: Euphausiacea; adults, 15.88). FIG. 5.13 a, early zoea, dorsal; b, later zoea, lateral.	(FIG. 5.13)
re	sp,x	x	x	Penaeid shrimp (Or: Decapoda; Penaeidae; adults, 15.93).	
ex	x	po	x	Caridean shrimp (Or: Decapoda; IO: Caridea; adults, 15.92).	
ex	sp	x	x	Lobster zoea (Or: Decapoda; adults, 15.7).	(FIG. 5.14)
ex	sp	po	x	Pagurid crabs (Or: Decapoda; IO: Anomura; adults, 15.119).	(FIG. 5.15)
ex	sp	po	sp	Brachyuran crabs (Or: Decapoda; IO: Brachyura; adults, 15.104).	(FIG. 5.16)

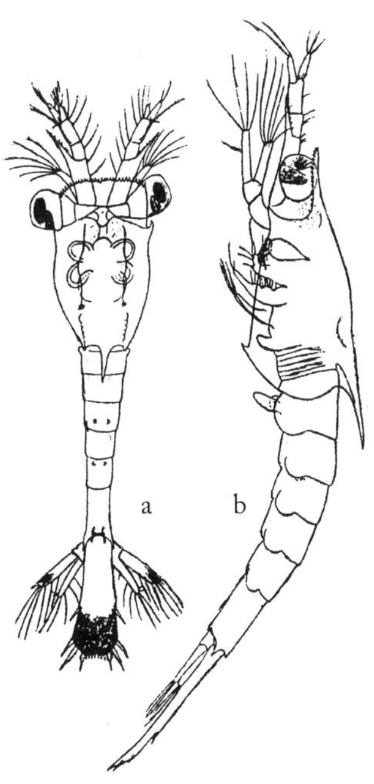

FIG. 5.13

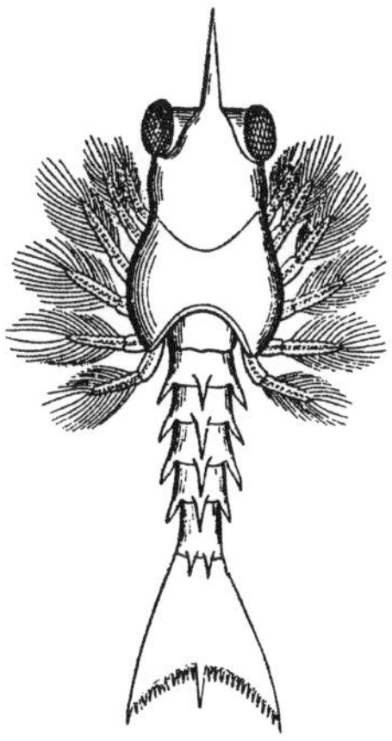

FIG. 5.14

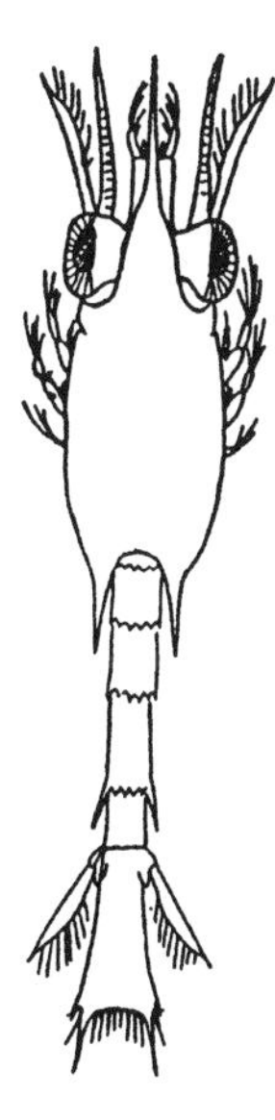
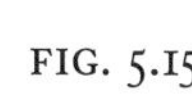
FIG. 5.15

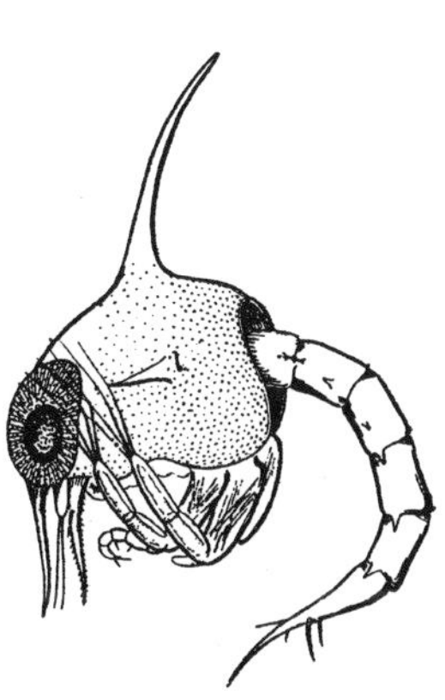
FIG. 5.16

5.11

Copepods (Ph: Arthropoda; SP: Crustacea, Cl: Copepoda) (from 5.4)

Juvenile and adult copepods are usually abundant in plankton samples. Most free-living members of the class are included in three orders: Cyclopoidea and Calanoidea tend to be pelagic holoplankters; the Harpacticoidea are more common in and on bottom sediments and are not treated further here. The body of a copepod is divided into a combined head plus thorax portion (6 segments) and a more slender, posterior abdominal section (2–5 segments). A single-piece carapace covers the head and one or more anterior thoracic segments. Calanoids possess antennules (second antennae) longer than the thorax and tend to exhibit a distinct difference between the width of the ovaloid anterior trunk region and the slender abdomen. Cyclopoids have shorter antennules and a more gradually tapered body shape. Males are usually smaller and less numerous than females.

. This key should help to distinguish among the most common genera, although it is far from a comprehensive treatment of the diversity of the cyclopoids and especially of the more abundant and diverse calanoids. Genera included here are in the order Calanoidea unless noted otherwise.

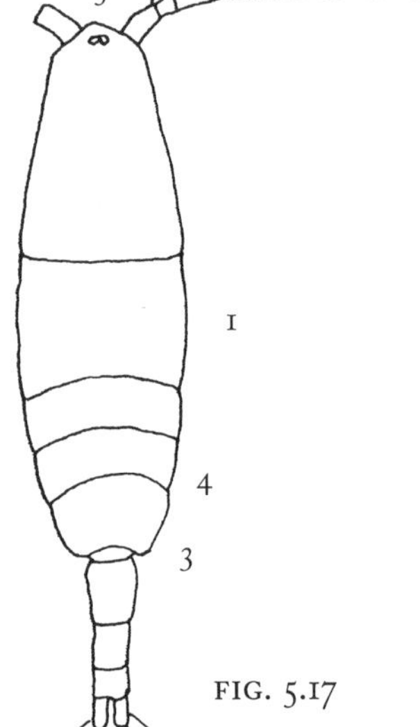

FIG. 5.17

1. The number of distinct segments counting from the anterior end (the carapace counts as one) to the start of the more slender abdomen (dorsal view): Provide the *number* of segmented areas (FIG. 5.17) —
2. Select the body shape that best describes the specimen (FIG. 5.17)
   - A. Body is basically *cylindrical* with a rounded head cy*
   - B. Body is basically cylindrical, with a distinctly *triangular* head tr
   - C. Body is *bulb* shaped or tapered gradually bu
3. Presence of a pair of rudimentary appendages on the first segment in the more slender abdominal section of the body (technically, this segment is part of the thorax in cyclopoid copepods) (FIG. 5.17)
   - A. A pair of tiny *appendages* present ap
   - B. Tiny appendages *absent* from first segment x*
4. Shape of posterior margin of the thorax as it joins the slender abdomen (FIG. 5.17)
   - A. Posterior margin of the thorax *flared*, forming lateral points fl
   - B. Posterior margin of the thorax *rounded*, not flared ro*
5. Number of eyes visible dorsally (FIG. 5.17)
   - A. Provide *number* of eyes visible dorsally —*
   - B. Eyes *absent* x

5.12

Thoracic Segments	Shape	Tiny Appendages	Thorax Margin	Eyes	Genus
6	tr	x	fl	4	*Anomalocera* sp.
6	cy	x	fl,ro	2	*Centropages* sp.
6	cy	x	fl	1	*Eurytemora* sp.
6	cy	x	ro	1	*Calanus* sp.
4	cy	x	ro	1	*Paracalanus* sp.
4	cy	x	ro	x	*Pseudocalanus* sp.
5	cy	x	fl	x	*Metridia* sp.
5	cy	x	ro	1,2	*Acartia* sp.
5	bu	x	ro	1	*Temora* sp.
5	bu	ap	ro	1	*Oithona* sp.
5	bu	ap	ro	x	*Oncaea* sp.

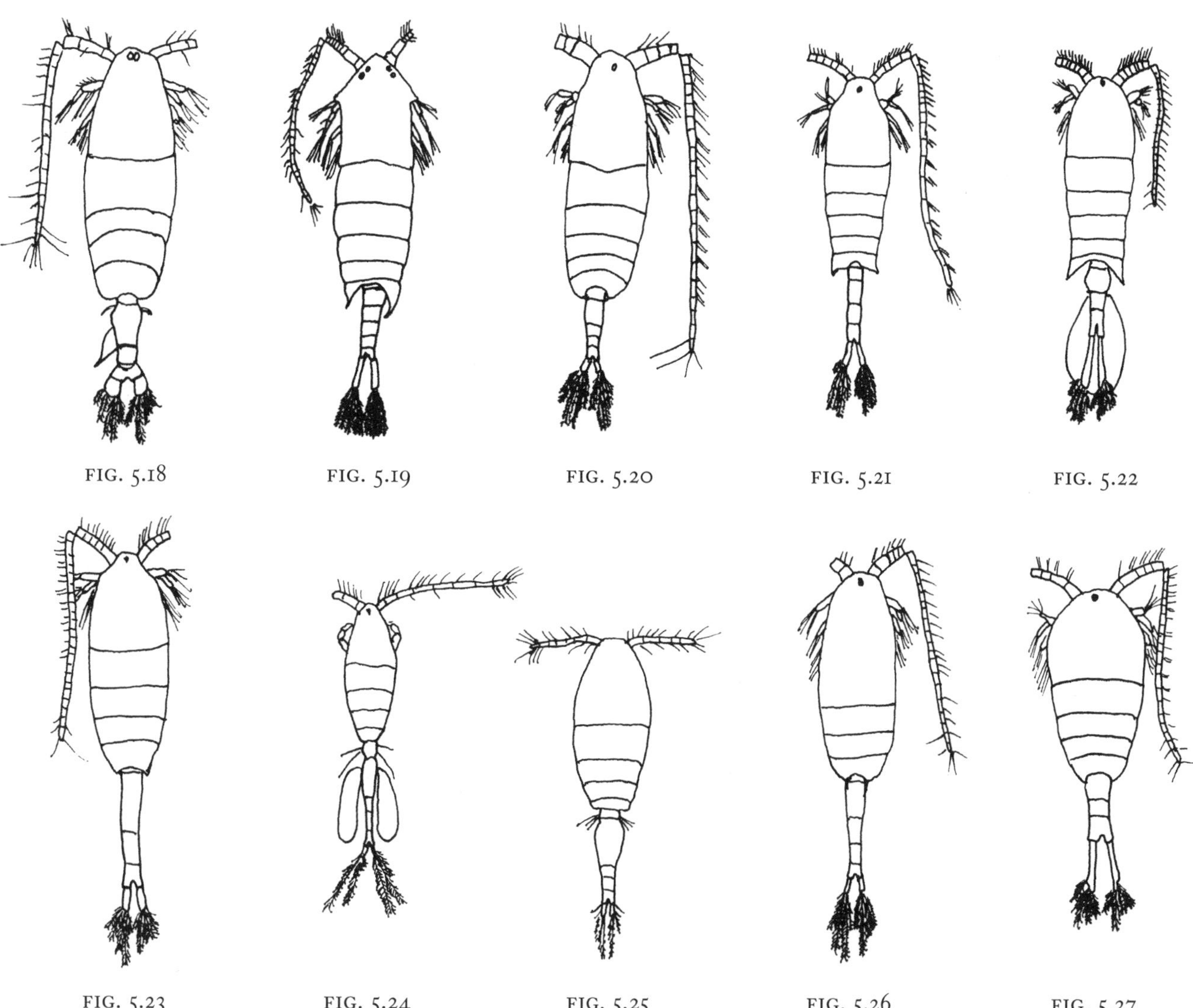

FIG. 5.18 FIG. 5.19 FIG. 5.20 FIG. 5.21 FIG. 5.22

FIG. 5.23 FIG. 5.24 FIG. 5.25 FIG. 5.26 FIG. 5.27

Acartia spp. (Acartiidae). First segment following carapace is much longer than subsequent segments; caudal setae are unequal in length. Bo,Pe,Es–Ps,om,1.6 mm (<0.1 in). (FIG. 5.18)

Anomalocera spp. (Pontellidae). Posterior margin of thorax forms asymmetrical points. Bo,Pe,Ps,om,?. (FIG. 5.19)

Calanus spp. (Calanidae). Bo,Pe,Ps,om,6.5 mm (0.3 in). (FIG. 5.20)

Centropages spp. (Centropagidae). Bo,Pe,Es–Ps,om,?. (FIG. 5.21)

Eurytemora spp. (Temoridae). Slender. Vi,Pe,Es–Ps,om,1.6 mm (<0.1 in). (FIG. 5.22)

Metridia spp. (Metrididae). Ac,Pe,Ps,om,4.5 mm (0.2 in). (FIG. 5.23)

Oithona spp. (Or:Cyclopoidea,Oithonidae). The abdominal segment following the segment with tiny appendages is shorter than the next three segments together. Bo,Pe,Es–Ps,om,1 mm (<0.1 in). (FIG. 5.24)

Oncaea spp. (Or:Cyclopoidea,Oncaeidae) The abdominal segment following the segment with tiny appendages is clearly longer than the next three segments together. Bo,Pe,Ps,om,1 mm (<0.1 in). (FIG. 5.25)

Paracalanus spp. (Paracalanidae). Females reddish; males yellowish. Bo,Pe,Es–Ps,om,1 mm (<0.1 in).

Pseudocalanus spp. (Pseudocalanidae). Reddish; females may bear ovisacs. Bo,Pe,Es–Ps,ff,1.6 mm (<0.1 in). (FIG. 5.26)

Temora spp. (Temoridae). Northern. Ac,Pe,Es–Ps,ff-om,1.4 mm (<0.1 in). (FIG. 5.27)

5.13 Cladocera (Ph: Arthropoda, SP: Crustacea, Cl: Branchiopoda, Or: Cladocera, Fa: Polyphemidae) (from 5.4)

Water fleas, well known as members of the freshwater plankton, are common also in marine samples. A large, pigmented, compound eye, embedded within a rounded head, is a good initial characteristic to look for. The two most common genera differ in trunk shape.

1. Shape of the trunk region
   - A. Trunk basically *conical,* with pointed tip — co
   - B. Trunk basically *rounded* — ro

5.14

Abdomen	Genus
co	*Evadne* spp.
ro	*Podon* spp.

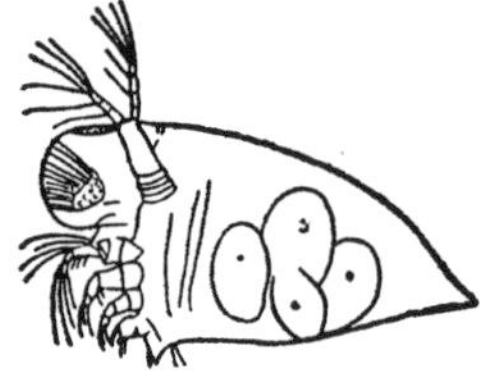

FIG. 5.28

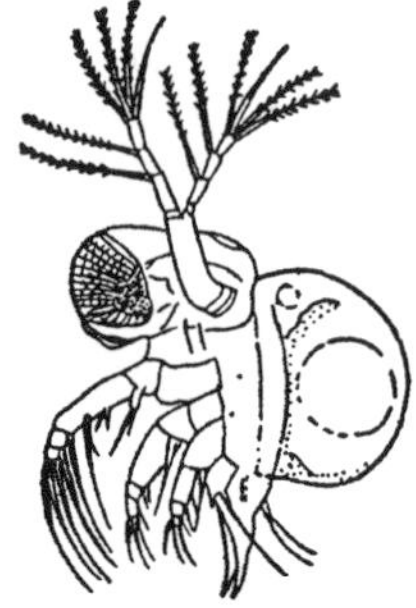

FIG. 5.29

Evadne spp. Female often carries several, dark-eyed young in a brood chamber within its broadly conical carapace. Bo,Pe,Ps–Pd,ff,1.5 mm (<0.1 in), (G211). (Fig 5.28)

Podon spp. Bo,Pe,Ps–Pd,ff,1.5 mm (<0.1 in), (G211). (FIG. 5.29)

5.15 Thrashing Swimmers (from 5.2)

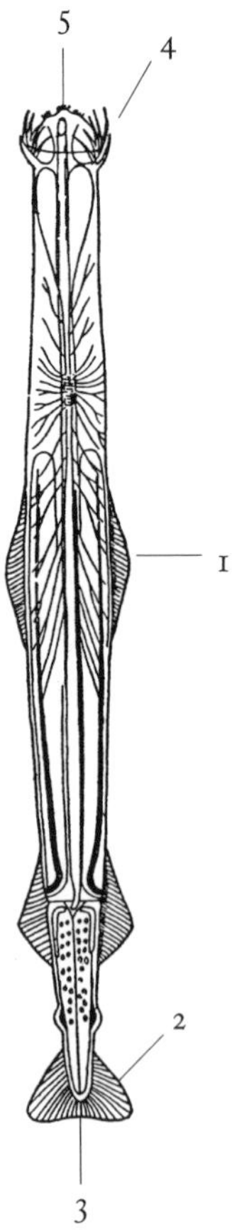

This artificial assemblage of unrelated organisms is unified functionally by their use of thrashing movements or undulations of parts or all of the body for individual locomotion.

1. Fins along the body (FIG. 5.30)
   - A. At least some distinct *fins* (flattened folds of body tissue) present along the body (does not include a slender tail only) — fi*
   - B. Fins *absent* along the body — x
2. Major locomotory appendage(s) (FIG. 5.30)
   - A. A tail that connects at the *midpoint* of the trunk — mp
   - B. A tail that connects at the *posterior* of the trunk — po*
   - C. One to four lateral flaps or *fins* (must be more than simply a flattened tail) — fi
   - D. More than four lateral flaps (*parapodia*) along sides of trunk — pa
3. Tip of tail (FIG. 5.30)
   - A. Tail is *forked* — fo
   - B. Tail is *squared* — sq*
   - C. Tail is *pointed* — po
   - D. Tail is *rounded* — ro
   - E. Tail *not* visible (e.g., encased in a shell) — x
4. Deployment of hard parts (e.g., shell, bristles, or setae) (FIG. 5.30)
   - A. Row of strong, hooked, anteriorly directed *bristles* on head for feeding — br*
   - B. *Paired* tufts of hard bristles or setae along body — pa
   - C. Encased in a *shell* — sh
   - D. Such bristles or shell *absent* — x
5. Presence of eyes (FIG. 5.30)
   - A. Distinct *eyes* present, even if small — ey*
   - B. Eyes apparently *absent* — x

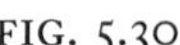

FIG. 5.30

5.16

Fins	Locomotory Appendages	Tail Tip	Bristles	Eyes	Type
fi	po,fi	po,ro,sq	x	ey	Fish larva (Ph: Chordata, Cl: Osteichthys)
fi	po,fi	sq	br	ey	Arrow worm (Ph: Chaetognatha)
fi	fi	po	x	x	Pteropoda—naked: sea butterfly, *Clione limacina*
x	fi	x	sh	x	Pteropoda—shelled
x	po	po,ro	x	ey,x	One or two suckers located at tip and toward base of anterior section of body: Cercaria larva of parasitic fluke OR Two or more small adhesive papillae, all located at tip of anterior section of body, or, no suckers or papillae visible: Tadpole larva of sea squirts or tunicates
x	pa	po,ro	pa	ey	Polychaete worm larvae
x	pa	po	x	ey	*Tomopteris* spp., planktonic polychaete, 14.82
x	mp	po,ro	x	x	Larvacean: *Oikopleura* spp.
x	mp	fo	x	x	Larvacean: *Fritillaria* spp.

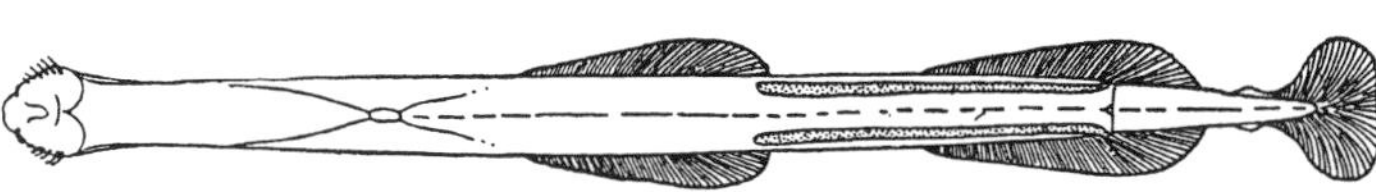

FIG. 5.31

Arrow worm (Ph: Chaetognatha, Sagittidae). Slender, transparent, planktonic predators of copepods and fish larvae. Members of the genus, *Sagitta,* are most likely to be seen. Bo,Pe,Ps,pr,17.8 mm (0.7 in), (GI20). (FIG. 5.31)

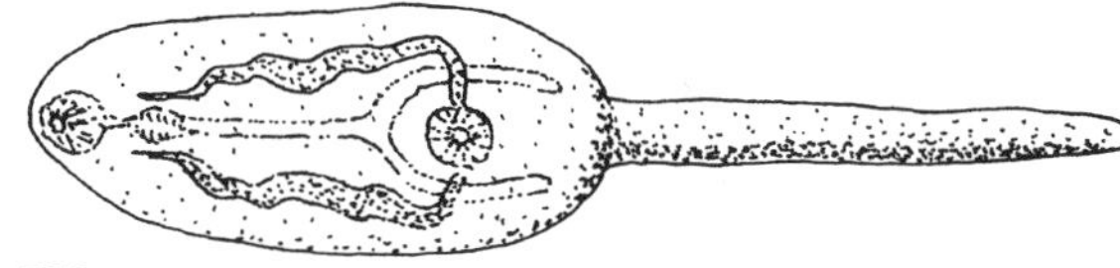

FIG. 5.32

Cercaria larva of parasitic fluke (Ph: Platyhelminthes, Cl: Trematoda). Cercariae are a free-living stage in the life cycle of trematode flukes. They are sometimes found in the water column in transit between their gastropod intermediate host and their next target host. (FIG. 5.32)

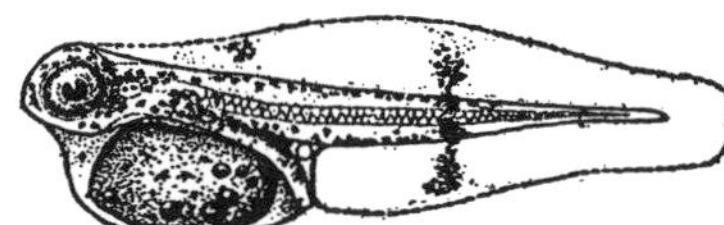

FIG. 5.33

Fish larva (Ph: Chordata, Cl: Osteichthyes). Usually recognizable because of their shape, complex eyes, and often by the presence of stellate (star-shaped) pigment spots. Many types are possible (e.g., Elliott and Jimenez 1981). (FIG. 5.33)

Larvacean (Ph: Chordata, SP: Urochordata, Cl: Appendicularia): Larvaceans produce delicate, mucous chambers that often break apart during the collection procedure. Bo,Pe,Ps,ff,<7.5 mm (0.3 in).

Fritillaria spp. (Fritillaridae). Trunk slender; tail short. Attaches to the outside surface of its mucous housing. (FIG. 5.34)

Oikopleura spp. (Oikopleuridae). Trunk ovaloid, tail long. Uses tail to create water flow through the interior of its 40 mm (1.5 in) diameter mucous housing. (FIG. 5.35)

Advanced polychaete worm larva (Ph: Annelida, Cl: Polychaeta). (FIG. 5.36)

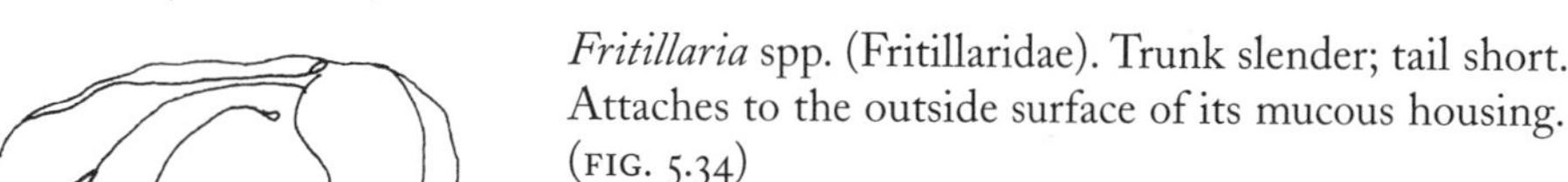

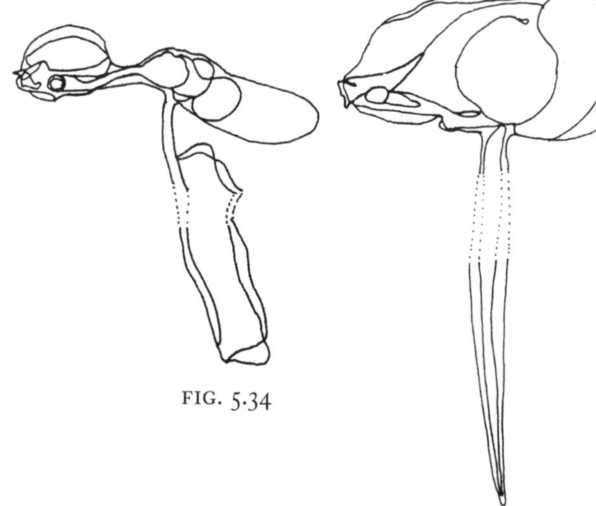

FIG. 5.34

FIG. 5.35

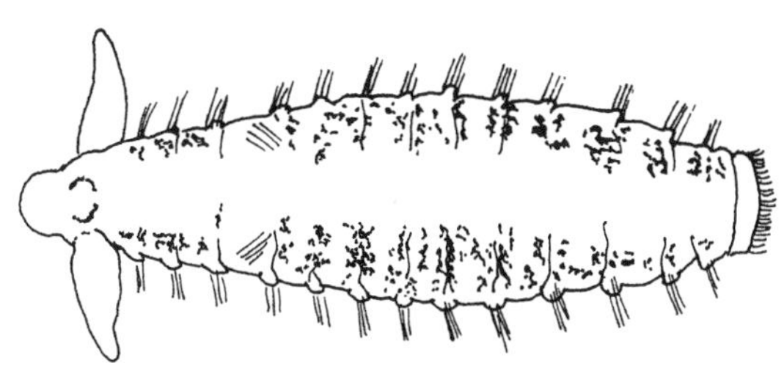

FIG. 5.36

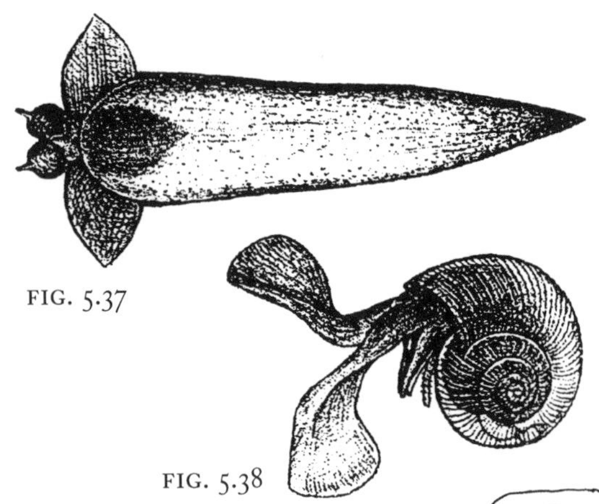

FIG. 5.37

FIG. 5.38

Pteropods and sea butterflies (Ph: Mollusca, Cl: Gastropoda)

Clione limacina, naked sea butterfly (Or: Gymnosomata, Clionidae). Gray-yellow-pinkish. Sluglike pteropod with parapodial "wings" for swimming. Bo,Pe,Ps,pr,25 mm (1 in), (G137), 13.70. (FIG. 5.37)

Shelled pteropods (Or: Thecosomata). Swim by flapping foot lobes. Bo,Pe,Ps,ff,8 mm (<0.3), (G137), 13.71. (FIG. 5.38)

Tadpole larva of sea squirts or tunicates (Ph: Chordata, SP: Urochordata, Cl: Ascidiacea). Small size and thrashing tails make tadpole larvae superficially resemble large sperm cells, the larvae of fish, or larvaceans. (FIG. 5.39)

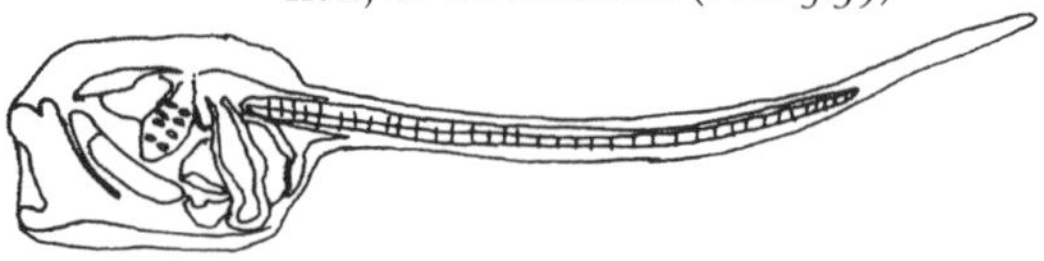

FIG. 5.39

5.17 Larvae with Elongated, Ciliated Appendages (from 5.2)

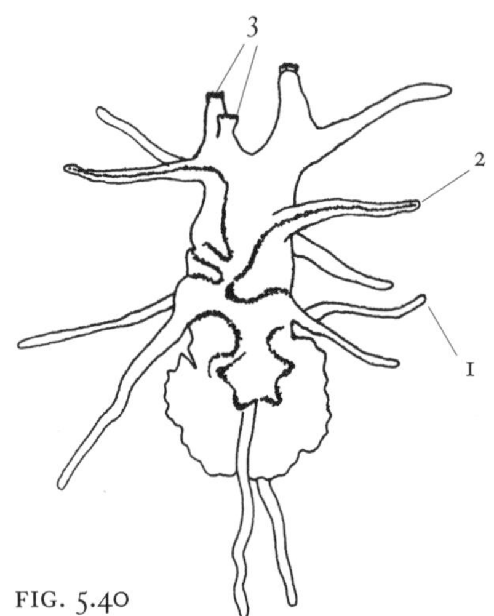

FIG. 5.40

1. Flexibility of obvious appendages (FIG. 5.40)
   - A. Appendages *bend* individually — be*
   - B. Appendages do *not* bend; supported by skeletal rods — x
2. Appearance of appendages (FIG. 5.40)
   - A. Major appendages extend from *basal* points of pyramidlike larvae — ba
   - B. All appendages *clustered* at one end — cl
   - C. All appendages form nearly *circular* ring towards the middle of the body — ci
   - D. Appendages arise along *sides* of body — si*
3. Presence of three stubby, adhesive arms, much shorter and wider than other arms (FIG. 5.40)
   - A. *Adhesive arms* present — aa*
   - B. Such adhesive arms *absent* — x

5.18

Flex	Location	Adhesive Arms	Type
x	ba	x	Pluteus larva of echinoderms; pyramid shape; swims with arms directed forward (often difficult to distinguish): 4–6 pairs of arms; pre-oral arms present (may be tiny): Echinopluteus larva of sea urchins OR 4 pairs of arms; pre-oral arms absent: Ophiopluteus larva of brittle stars
be	si	aa	Brachiolaria larva of sea stars
be	cl	x,aa	Pentacularia larva of sea cucumbers
be	ci	x	8 or fewer ciliated lobes: Müller's larva of some polyclad flatworms OR 8–24 elongate appendages: Actinotrocha larva of phoronid worms

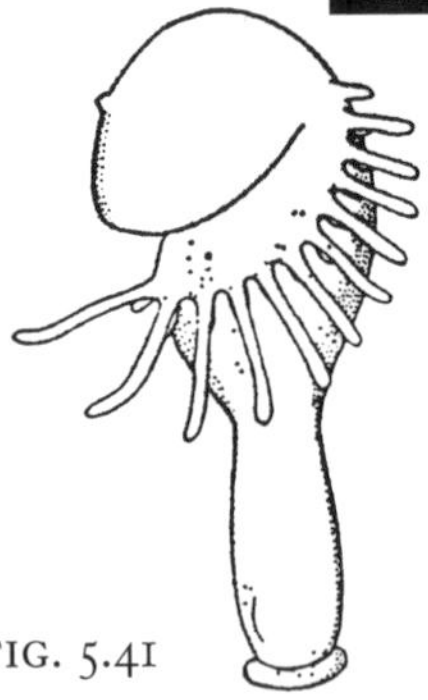

FIG. 5.41

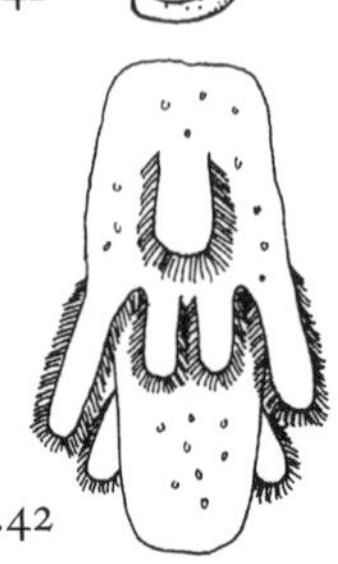

Actinotrocha larva of phoronid worms (Ph: Phoronida). (FIG. 5.41)

Brachiolaria larva of sea stars (Ph: Echinodermata, Cl: Asteroidea). Advanced stage near metamorphosis and about to settle out of the plankton. (FIG. 5.40)

Müller's larva of polyclad flatworms (Ph: Platyhelminthes, Cl: Turbellaria, Or: Polycladida). (FIG. 5.42)

FIG. 5.42

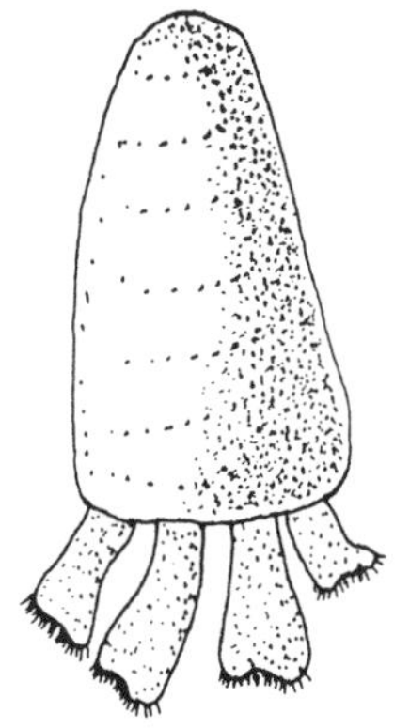

FIG. 5.43

Pentacularia larva of sea cucumbers (Ph: Echinodermata, Cl: Holothuroidea). (FIG. 5.43)

Pluteus larva

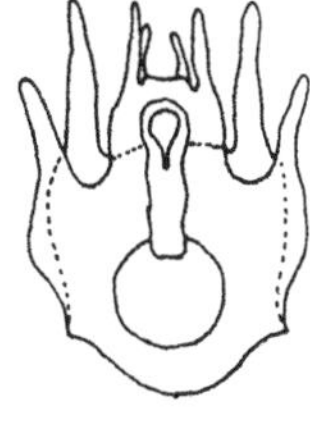

FIG. 5.44

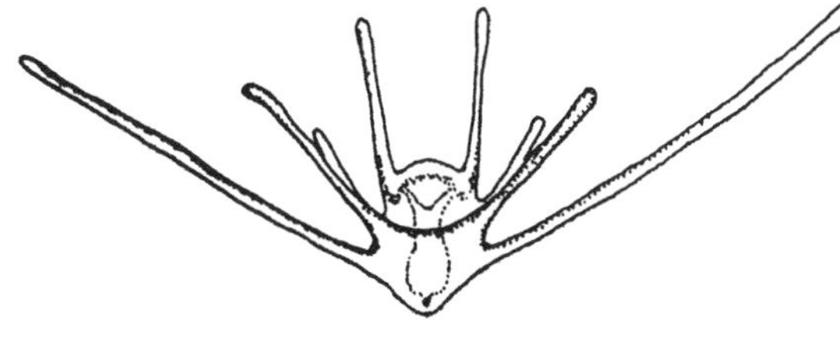

FIG. 5.45

Echinopluteus larva of sea urchins (Ph: Echinodermata, Cl: Echinoidea). (FIG. 5.44)

Ophiopluteus larva of brittle stars (Ph: Echinodermata, Cl: Ophiuroidea). (FIG. 5.45)

5.19 Larvae with Ciliated Lobes (from 5.2)

1. Shell or other hard parts
   - A. Tiny, *spiral,* single-piece shell present, resembles tiny snail or gastropod — sp
   - B. Tiny, clam or *bivalved* shell present — bi
   - C. Shell present, not as above, but with distinct lateral and/or posterior *points* — po
   - D. Paired clusters of bristles or *setae* along body — se
   - E. Shell or hard parts *absent* — x
2. Deployment of primary locomotory cilia
   - A. *Simple,* complete rings, horizontally oriented, encircling animal (minimally, one around the anterior end) — si
   - B. *Complexly* contorted, ciliated bands, do not encircle animals, but follow margins of lobes — co
3. Number of lobes bearing cilia: Provide *number* of ciliated lobes — __
4. Retractility of lobed structures:
   - A. Lobes *retractable* structure (i.e., only sometimes deployed) — re
   - B. Lobes *non*retractable structures (i.e., permanently deployed) — xr

5.20

Hard Parts	Cilia Bands	No. Lobes	Retractable	Type
se	si	>1	x	One equatorial ciliated ring: Trochophore larva OR More than one ciliated ring: Polytrochophore larva
sp	co	2,4,6	re	Veliger larva of snail
sp,po	co	2	re	Shelled pteropods
bi	co	2,4	re	Veliger larva of clam or mussel
x	co	2	x	Pilidium or helmet larva of ribbonworm
x	co	>2	x	Certain shell-less, multilobed larvae, 5.21

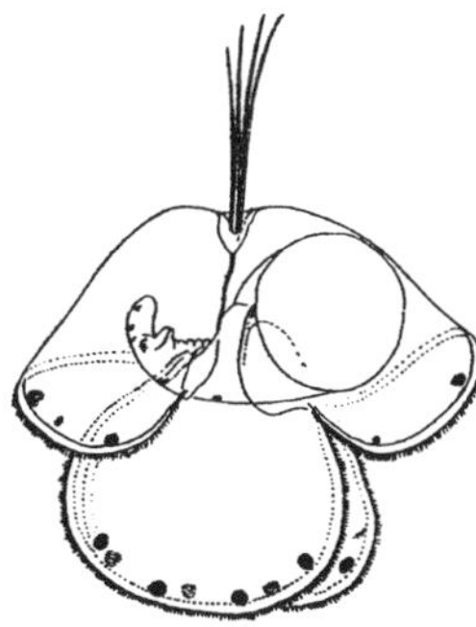

FIG. 5.46

Pilidium or helmet larva of ribbonworm (Ph: Nemertea). Lobes like earflaps on a football helmet; known to occur in *Cerebratulus,* some *Micrura,* and some *Lineus.* (FIG. 5.46)

Shelled pteropods or sea butterflies (Ph: Mollusca, Cl: Gastropoda, Or: Thecosomata). Locomote by flapping paired foot lobes. 13.71. (FIG. 5.47)

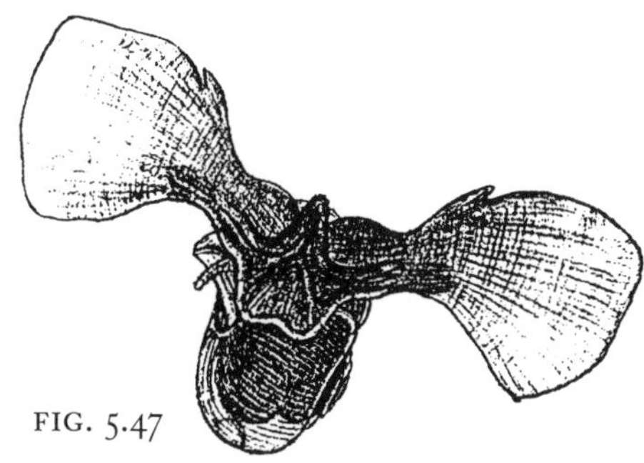

FIG. 5.47

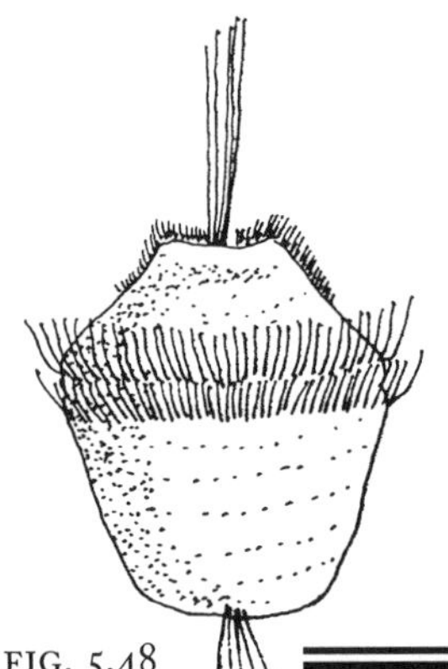

FIG. 5.48

Trochophore larva. Several phyla have similar trochophore larvae. This and other similarities in early development unify these phyla as "protostomes." See descriptive comments, 5.24 (FIG. 5.48)

Veliger larva of snail or bivalve (Ph: Mollusca, Cl: Gastropoda and Bivalvia). Typically with four-lobed, retractable velum for locomotion. In some, called the "D-shaped" larva, valves form straight margin on side opposite locomotory velum. (FIG. 5.49)

FIG. 5.49

5.21 Certain Shell-less, Multilobed Larvae (from 5.20)

1. General orientation of lobes
   - A. *Horizontally* oriented, as ring around the larva — ho
   - B. *Vertically* oriented along sides of larva — ve
2. Single or multitrack ciliary bands
   - A. *One* continuous band, regardless of how complexly contorted it may be — 1
   - B. At least *two* ciliary bands — 2

5.22

Orientation	Bands	Type
ho	2	Müller's larva of some polyclad flatworms
ve	2	Bipinnaria larva of sea stars
ve	1	Auricularia larva of sea cucumbers

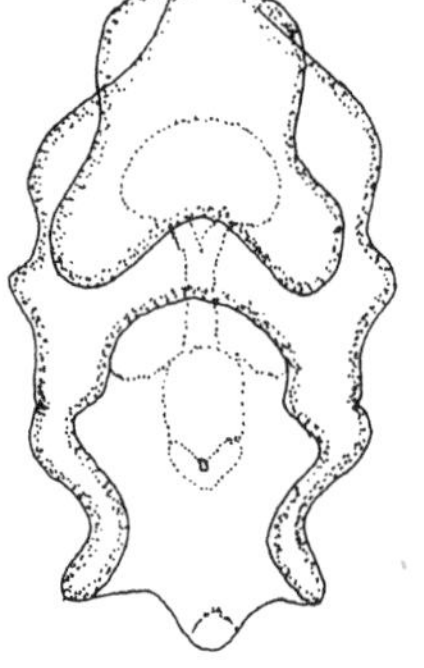

FIG. 5.50

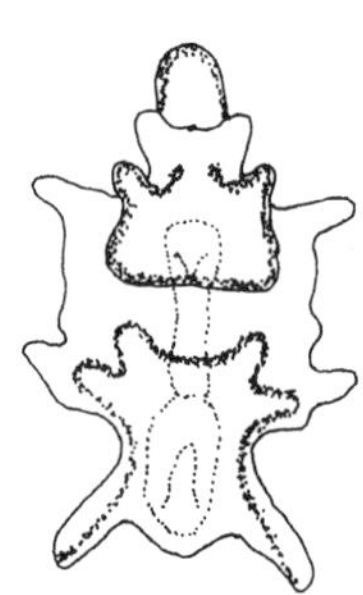

FIG. 5.51

Auricularia larva of sea cucumber (Ph: Echinodermata, Cl: Holothuroidea). Eventually develops into a barrel-shaped doliolaria larva. (FIG. 5.50)

Bipinnaria larva of sea star (Ph: Echinodermata, Cl: Asteroidea). During development, paired lobes elongate. (FIG. 5.51)

Müller's larva of some polyclad flatworms (Ph: Platyhelminthes, Cl: Turbellaria, Or: Polycladida). Apical tuft of sensory cilia; ventral surface flattened. (FIG. 5.52)

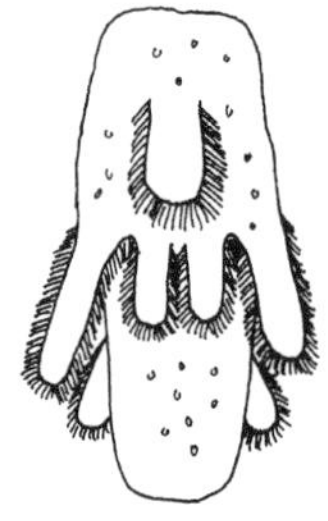

FIG. 5.52

5.23 Nonlobed, Ciliated Larvae (from 5.2)

1. Overall shape
   - A. Cylindrical or *egg*-shaped — eg
   - B. Basically *triangular* in lateral view — tr
   - C. *Spherical* — sp
2. General orientation of ciliary tracts
   - A. *Horizontally* oriented, as ring around the larva — ho
   - B. *Vertically* oriented along sides of larva — ve
3. Description and number of separate ciliary tracts
   - A. Ciliary tracts *simple* bands around the body; provide *number* of bands — s__
   - B. Ciliary tracts *complexly* contorted, no simple bands — co
   - C. Ciliary tracts a combination of one *simple* band plus additional *complexly* contorted bands — sc
4. Body with externally visible segments
   - A. Body *segmented* — se
   - B. Segmentation *absent* — x

5.24

Shape	Orientation	No. Bands	Segments	Type
tr	ho	s1	x	Cyphonautes larva of Ectoprocta
eg,sp	ho	s1,2	se,x	Trochophore larva
eg,sp	ho	s5	x	Doliolaria or barrel larva of sea cucumbers
eg,sp	ho,ve	sc	x	Tornaria larva of acorn worms
eg	ho	s>2	se	Polytrochophore larva of polychaete worms
eg,sp	ve	s8	x	Immature comb jelly (Ph: Ctenophora), 9.1
eg	ve	co	x	Early bipinnaria larva of sea star

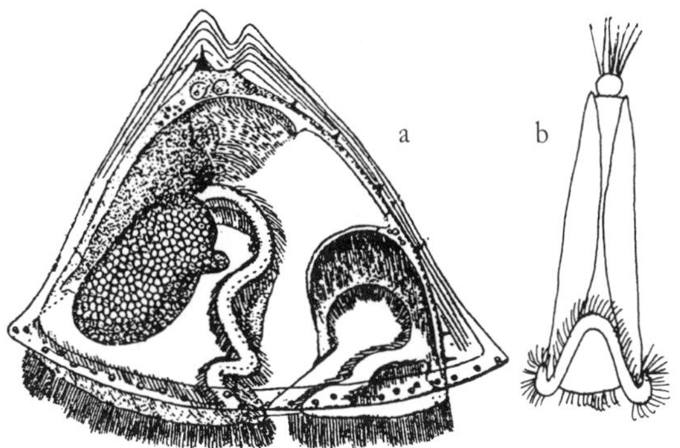

FIG. 5.53

FIG. 5.54

FIG. 5.55

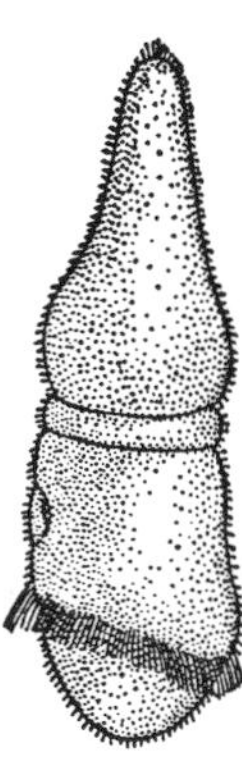

FIG. 5.56

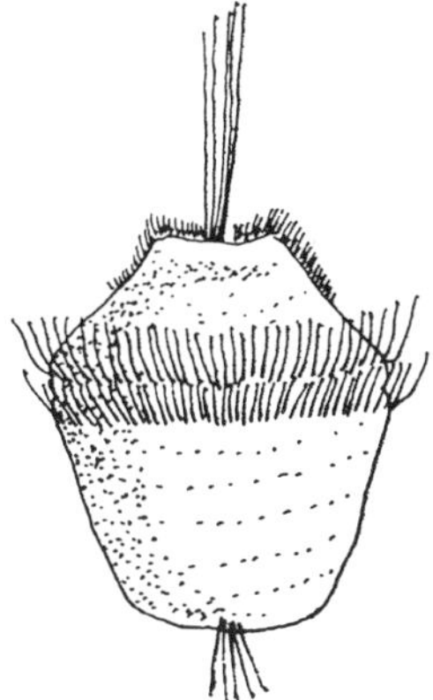

FIG. 5.57

Early bipinnaria larva of sea star (Ph: Echinodermata, Cl: Asteroidea).

Cyphonautes larva of ectoproct (Ph: Ectoprocta). (FIG. 5.53. a, Lateral view; b, end view.)

Doliolaria or barrel larva of sea cucumbers (Ph: Echinodermata, Cl: Holothuroidea). (FIG. 5.54)

Polytrochophore larva of polychaete worms (Ph: Annelida, Cl: Polychaeta). (FIG. 5.55)

Tornaria larva of acorn worms (Ph: Hemichordata). (FIG. 5.56)

Trochophore larva, earliest larval stage of several invertebrate phyla, known collectively as *Protostomia.* Polychaete annelids or mollusks are the most common. In development, polychaete trochophores elongate and add segments with ciliary bands (polytrochophore larva, above), eventually to be replaced by setae (nectochaeta larva, 5.2), and then parapodia (5.16). In mollusks, the prototroch or ciliary band of the trochophore expands into a lobed, locomotory *velum* in the veliger larva (5.20). (FIG. 5.57)

Chapter 6 ■ Eggs and Egg Masses

6.1 Egg masses encountered in the field serve as evidence for the presence of their producers. In this key, some distinctive types of eggs are identified. Information on this topic is sparse, and this listing is far from complete. Annotations include a hyphen (-) in the category for trophic type.

I. Appearance of egg mass
   - A. Eggs within tough outer *capsule* — ca
   - B. Eggs imbedded in *gelatinous* material or transparent covering — ge
   - C. Upright, collar-like egg mass encrusted with *sand* — sa

6.2

Covering	Group
ca	Encapsulated egg masses, 6.3
ge	Gelatinous egg masses, 6.7
sa	"Sand Collar" of naticid snails (moon snails)

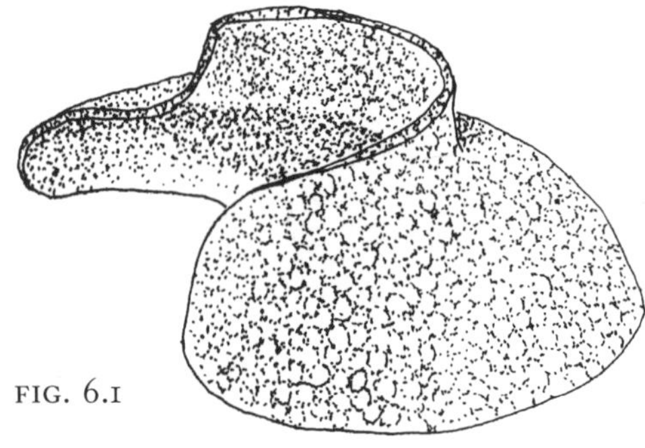

FIG. 6.1

"Sand Collar" of naticid snails (moon snails); 40 mm (1.6 in) tall by 90 mm (3.5 in) diameter; ☺. Tiny eggs embedded within a coiled, gelatinous sheet, encrusted with sand; becomes fragile and crumbles when dried. Bo,Sa–Mu,Li–Su,-,4.1 cm (1.6 in) height × 8.9 cm (3.5 in) diameter, ☺. Adult snails in family Naticidae, 13.53. (FIG. 6.1)

6.3 Encapsulated Egg Masses (from 6.2)

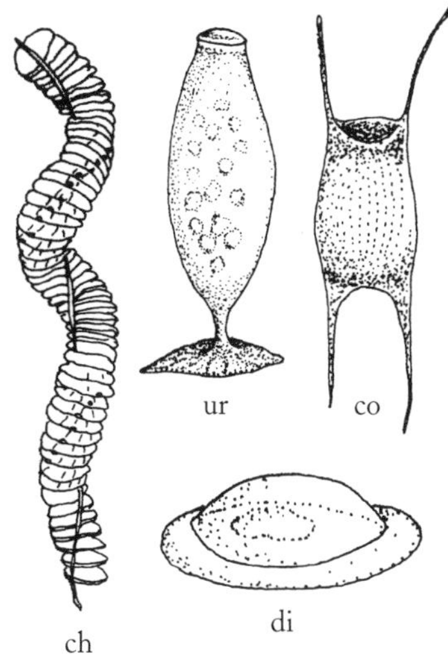

FIG. 6.2

1. Overall shape of egg mass (FIG. 6.2)
   - A. *Chain* of disklike capsules oriented on edge and joined to one another by slender stalk — ch
   - B. Upright *urn*-shaped capsules — ur
   - C. Low *discoid* capsules — di
   - D. Flattened case with elongated *corners* — co
2. Texture of capsule surface
   - A. Firm and *smooth* — sm
   - C. Firm with sharp *ridges,* often in irregular patterns; opaque — ri
   - D. *Leathery,* fibrous texture but not with sharp ridges; opaque — le
3. Attachment to one another
   - A. Laid *individually* — in
   - B. Laid singly but often in *groups,* although not attached to one another — gr
   - C. Laid as a *mass,* several connected directly to one another — ma
4. Attachment to substrate
   - A. *Loose* or one end buried — lo
   - B. Attached by *stalk* at one end — st
   - C. Attached by *surface* of capsule itself — su
   - D. Free-*floating* in the water column — fl
5. Habitat where egg mass was found
   - A. Attached to *hard, intertidal* substrates — hi
   - B. Attached to *hard, subtidal* substrates — hs
   - C. Attached to shallow water *vegetation* (seagrass, seaweeds) — ve
   - D. Deposited freely on *soft substrates* — ss
   - E. In the *water column* — wc

6.4

Shape	Surface Texture	Indiv./ Group/ Mass	Attachment to Substrate	Habitat	Species or Type
ch	ri	ma	lo	ss	Melongenid whelk egg cases; to 36 in long, ☺: Flat edges to capsule disks: *Busycon carica,* knobbed whelk OR Sharp edge to capsule disks: *Busycotypus canaliculatus,* channeled whelk
ur	sm,ri	gr,in	st	hi,hs,ve	Urn-shaped egg capsules; <10 mm (0.4 in) tall, ☺, 6.5
di	sm	in	fl	wc	*Littorina littorea,* periwinkle
di	sm	ma	su	hs	*Colus* whelk egg cases
di	le	ma	su	ss,hs	*Buccinum undatum,* waved whelk
co	sm	in	su	ss,ve	Skate's egg case, mermaid's purse

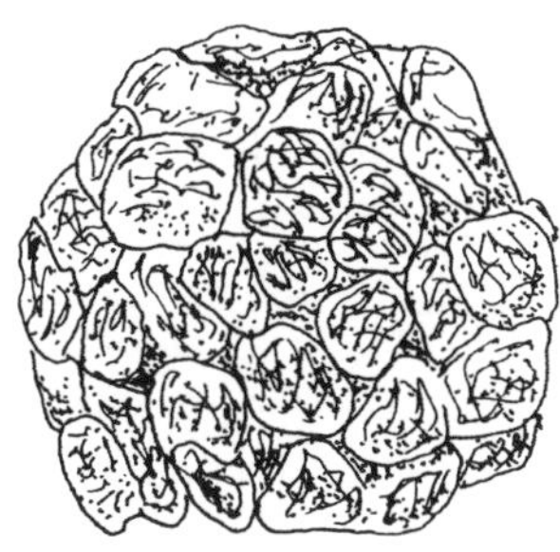
FIG. 6.3

Buccinum undatum, waved whelk† (Buccinidae), sailor's wash ball. Yellowish, ball-shaped, tough-walled capsules in irregularly rounded masses. May be from several females. Several 100 capsules per ball; several 100 eggs per capsule. First hatching young cannibalize weaker ones. Ac–,Ha,Su,-,12 mm (0.4 in) capsule diameter, 10.2 cm (4 in) capsule mass diameter. Adults, 13.23. (FIG. 6.3)

Colus spp., colus whelks. Flattened translucent capsules. Bo,Ha,Su,-,<10 mm (0.4 in) capsule diameter, often in multilayered sheets (<75 mm [3.5 in] sheet diameter). Adults, 13.23. (FIG. 6.4)

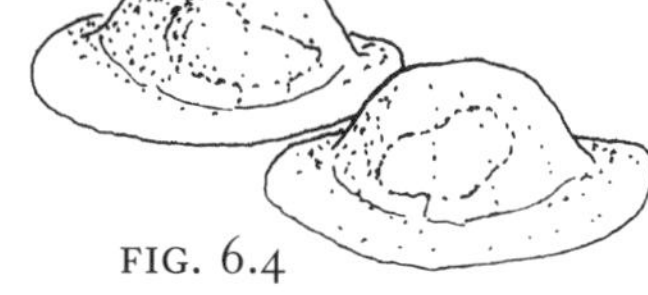
FIG. 6.4

FIG. 6.5

Littorina littorea, common periwinkle† (Littorinidae). Hat-shaped egg capsule with 1–5 pinkish eggs released into water column. Ac–,Pe,Ps,-,12 mm (0.5 in) capsule diameter, ☺. Adults, 13.42. (FIG. 6.5)

Melongenid whelk egg cases, called Venus necklaces; to 36 in long; ☺. Open a capsule to see tiny developing whelks inside.

Busycon carica, knobbed whelk†. String of capsules, each double-keeled laterally. String initially attached to stone or shell, capsules small and widely spaced; later, capsules larger and crowded. Up to 100 cases per string. More common from New Jersey southward. Vi,Sa,Li–Su,-,91.4 cm (3 ft) chain length. Adults, 15.48. (FIG. 6.6)

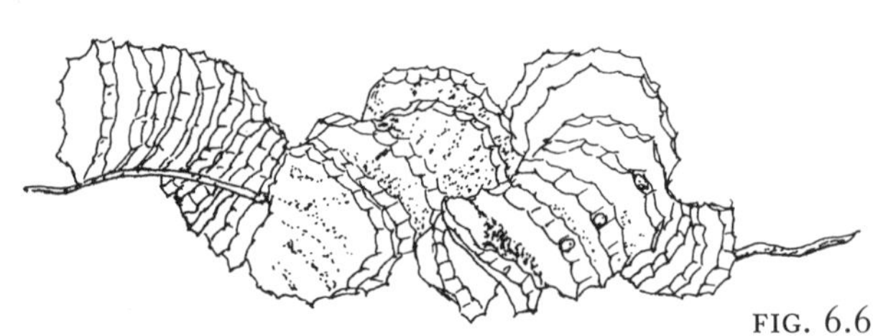

FIG. 6.6

Busycotypus canaliculatus, channeled whelk. String of capsules, each single-keeled laterally. More common from Cape Cod to New Jersey. Vi,Sa,Li–Su,-,91.4 cm (3 ft). Adults 13.48 (FIG. 6.7)

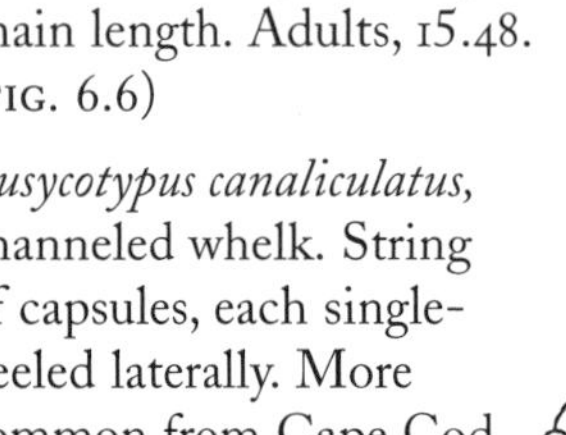

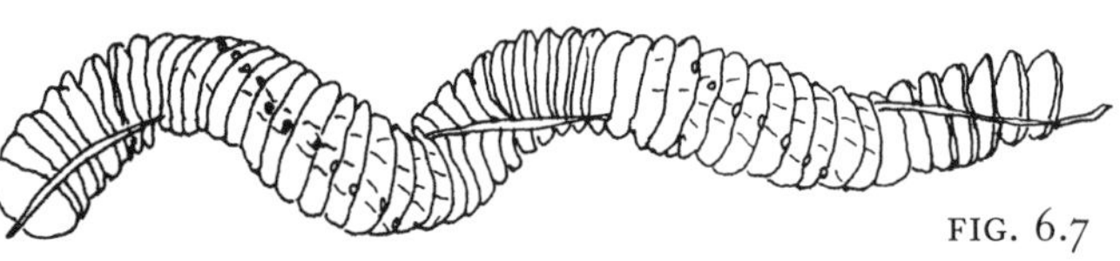

FIG. 6.7

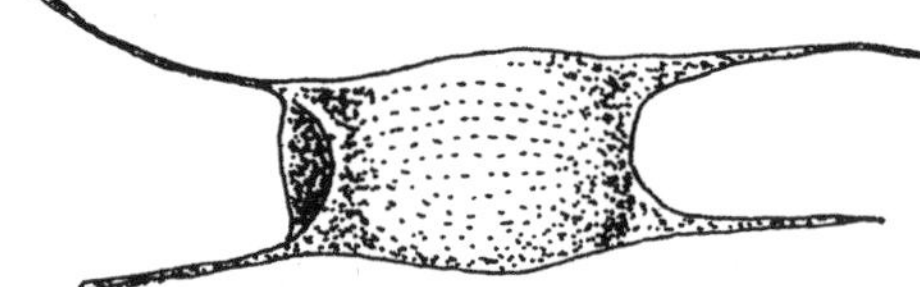

Skate's egg case, mermaid's purse; ☺ (Rajidae). One egg per translucent brownish capsule. Capsule darkens with age, especially as sun dries it on shore. Corner tendrils intended to entangle egg case in vegetation. Adults, 17.21. (FIG. 6.8)

FIG. 6.8

6.5 Urn-Shaped Egg Capsules (from 6.2)

All belong to gastropod snails, unless otherwise noted.

1. Cross-sectional shape (FIG. 6.9)
   - A. Nearly *circular* in cross-section (may be somewhat flattened on one side) ci
   - B. Distinctly *flattened* laterally fl*
2. Location of opercular opening through which snails emerge (FIG. 6.9)
   - A. Opening forms the *apex* of capsule ap
   - B. Opening asymmetrical, occupies a portion of the top surface of the capsule but off to one *side* si
   - C. Opening centered along top edge of capsule but offset to one *end* en*
3. Surface sculpture of capsule (FIG. 6.9)
   - A. *Smooth* sm*
   - B. *Ridged* ri
4. Substrate
   - A. *Hard* surfaces (e.g., rock) ha
   - B. *Vegetation* (e.g., seagrass blades, algae) ve
   - C. *Gills* of horseshoe crab gi

FIG. 6.9

6.6

Cross Section	Opening	Surface	Substrate	Species
ci	ap	sm	ha	*Nucella lapillus,* northern dog whelk, ☺
ci	ap	sm	gi	*Bdelloura candida,* limulus leech
fl	ap	sm	ha	*Urosalpinx cinerea,* oyster drill, ☺
fl	ap	sm	ve	*Nassarius vibex,* eastern dog whelk
fl	si	ri	ve	Side view, mid-lateral ridges form roughly straight vertical line: *Ilyanassa* (=*Nassarius*) *trivittata,* three-lined whelk OR Side view, mid-lateral ridges form irregular pattern—no straight vertical line: *Ilyanassa obsoleta,* mud dog whelk, ☺
fl	en	sm	ha,ve	*Eupleura caudata,* thick-lip drill

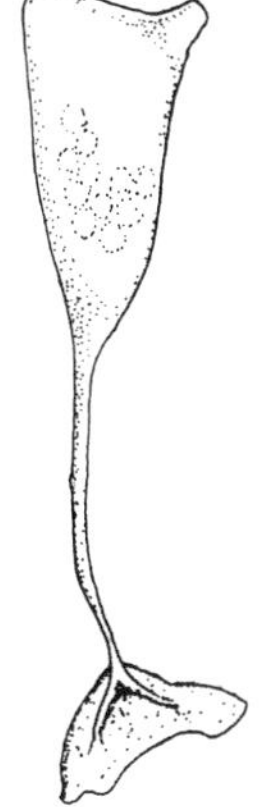
FIG. 6.10

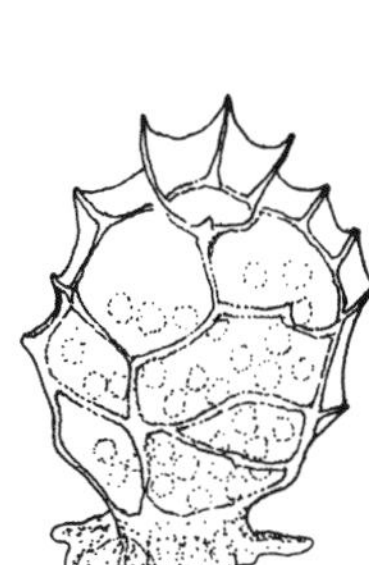
FIG. 6.11

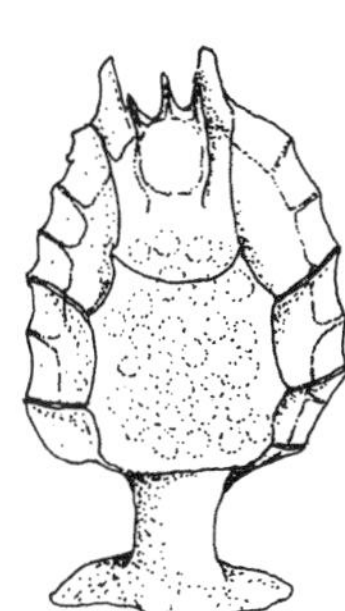
FIG. 6.12

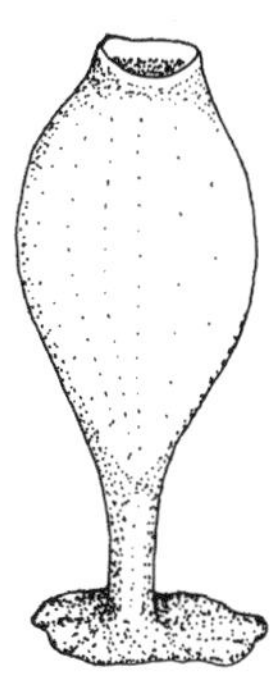
FIG. 6.13

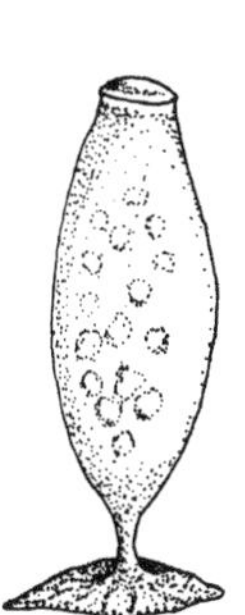
FIG. 6.14

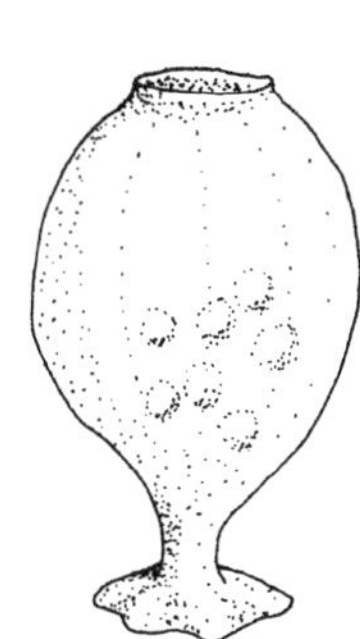
FIG. 6.15

Bdelloura candida, limulus leech (Platyhelminthes, Turbellaria). Small dark, stalked capsules attached to gill surface of horseshoe crab. Bo,Co,Li–Su,<5 mm (0.24 in). Adults, 6.6.

Eupleura caudata, thick-lip drill (Muricidae). Flattened, leathery capsules with opercular opening off to one side. Vi,Ha,Li–Su,-,6 mm (0.25 in). Adults, 13.50. (FIG. 6.10)

Ilyanassa obsoleta, eastern mudsnail† (Nassariidae). Capsules with sharp-ridged sculpturing; often in dense groups on eel grass, *Zostera.* Bo,SG,Li–Su,-,3 mm (0.12 in), ☺. Adults, 13.52. (FIG. 6.11)

Ilyanassa (=*Nassarius*) *trivittata,* three-lined basketsnail† (Nassariidae). Sharp-ridged, aperture to one side. 50–200 embryos/capsule. Bo,SG–Ha,Li–Su,-,2.2 mm (0.09 in) capsule height. Adults, 13.52. (FIG. 6.12)

Nassarius vibex, bruised nassa† (Nassariidae). Capsule broadly based and lanceolate in cross-section. Vi,Ha–SG,Li,-,1.5 mm (0.06 in). Adults, 13.52. (FIG. 6.13)

Nucella lapillus, Atlantic dogwinkle† (Muricidae). Often in clumps on undersurfaces of stones. More than 1 hour to produce each capsule, 10 in 24 hours. 6-31 capsules/individual. Ac+,Ha,Li–,-,8.4 mm (0.33 in) capsule height, ☺. Adults, 13.50. (FIG. 6.14)

Urosalpinx cinerea, Atlantic oyster drill† (Muricidae). Flattened urn-shaped capsules; attached in clumps on undersurfaces of stones or on shells. To 35 eggs/capsule; 2 months to crawling embryos. Vi+,Ha,Li–,-,6.4 mm (0.25 in) capsule height, ☺. Adults, 13.50. (FIG. 6.15)

6.7 Gelatinous Egg Masses (from 6.2)

1. Overall shape
   - A. Finger or *spindle*-shaped — sp
   - B. Slender, *tangled strands* — ts
   - C. Slender, regularly *coiled strands* — cs
   - D. *Ribbon* or bandlike sheet — ri
   - E. Flattened, *discoid* mass — di
   - F. *Doughnut*-shaped mass — do
   - G. *Balloon*-like, bulbous top attached by slender stalk — ba
   - H. *Amorphous* gelatinous mass — am
2. Attachment
   - A. Attached at *end(s)* (i.e., most of its mass unattached) — en
   - B. Attached by flattened *surface* (i.e., most of its mass attached to substrate) — su
   - C. Ribbon attached along one *edge* — ed
   - D. Free, attachment *absent* — x
3. Attachment substrate
   - A. *Hard* objects (e.g., rocks) — ha
   - B. *Vegetation* (e.g., algae, seagrass blades) — ve
   - C. *Soft sediment* — ss
   - D. Free, attachment *absent* — x
4. Dominant color
   - A. *Greenish* — gr
   - B. *Pinkish* — pi
   - C. *Yellowish* — ye
   - D. *Whitish* — wh
   - E. *Purplish* — pu
   - F. *Transparent* — tr

6.8

Shape	Attachment	Substrate	Color	Species or Type
sp	en	ss	tr,ye	*Loligo pealei*, squid
do	su	ve,ha	gr,wh	Snails (Gastropoda): *Lacuna vincta*
cs	su	ha,ve	pi	Nudibranchs & sea slugs (Gastropoda): *Aeolidia papillosa*
cs	su	ve	ye	Nudibranchs & sea slugs (Gastropoda): *Scyllaea pelagica*
cs,ts	su,x	ha,ve	wh,tr	Nudibranchs & sea slugs (Gastropoda): unidentified aeolid nudibranch
ts	x	ss	gr	*Limulus polyphemus*, horseshoe crab
ri	ed	ha,ve	wh,pi	Nudibranchs & sea slugs (Gastropoda): Dorid
ri	x	x	pu	*Lophius americanus*, goosefish
di	su	ha	wh,ye	Snails (Gastropoda): *Colus* spp.
ba	en	ha	tr	Snails (Gastropoda): *Crepidula fornicata*
di	su	ve	tr,wh	Snails (Gastropoda): *Littorina obtusata*
am	en,x	ss	pi,pu	*Arenicola*, lug worm

Arenicola, lug worm (Ph: Annelida, Cl: Polychaeta, Arenicolidae). Burrowing worm in shallow, muddy sand; eggs in purple-brown, mucous mass, extending into water column from burrow opening. Bo,Sa–Mu,Li–Su,-,60 cm (2.4 ft) length, ☺. Adults, 14.22.

Limulus polyphemus, horseshoe crab† (Ph: Arthropoda, SP: Chelicerata, Cl: Merostomata). In late spring, mass matings occur during nights of full moon in sandy shallows of protected beaches. Thousands of greenish eggs are deposited in scooped out depressions. Bo,Sa,Li,-,3 mm (0.12 in). Adults, 15.105.

FIG. 6.16

Loligo pealei, long-finned squid† (Ph: Mollusca, Cl: Cephalopoda, Loliginidae). Individual capsules to 10.2 cm (4 in), masses to 30.5 cm (1 ft) across. Washed onto beaches as large gelatinous masses with entangled algae and sand. Vi+,Sa,Su,-,10.2 cm (4 in) capsule length. Adults, 13.162. (FIG. 6.16)

Lophius americanus, goosefish† (Ph: Chordata, Cl: Osteichthyes, Lophiidae). Eggs embedded in large, violet to gray, free-floating sheets or veils of mucus; can be 10.9 m (36 ft) long × 0.91 m (3 ft) wide. Bo,Pe,Ps,-,10.9 m (36 ft). Adults, 17.68.

Nudibranchs and sea slugs (Ph: Mollusca, Cl: Gastropoda)

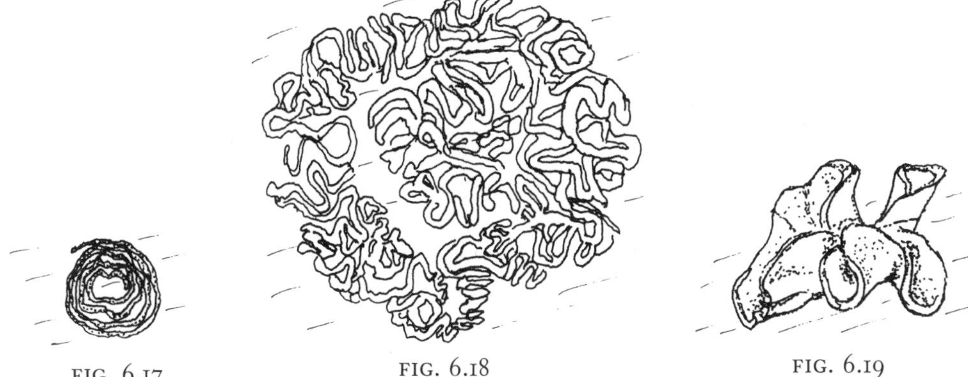

FIG. 6.17 FIG. 6.18 FIG. 6.19

Unidentified aeolid nudibranchs. Thin coiled strands either attached to hard substrates or entwined in hydroid colonies or on algae. Several species possible. Bo,Ha–Al,Li–Su,-,to 15 mm (0.6 in) coil diameter. Adults, 13.78. (FIG. 6.17)

Aeolidia papillosa, shag-rug aeolis†. Especially in vicinity of small anemones; April–May. Ac,Ha,Li–Su,-,15 mm (0.6 in) coil diameter, ☺. Adults, 13.81. (FIG. 6.18)

Unidentified dorid nudibranch. Upright, coiled ribbons with dozens of eggs. Attached to hard substrates, April–May. Bo,Ha–Al,Li,Su,-,15 mm (0.6 in) egg-mass diameter. Adults, 13.90. (FIG. 6.19)

Scyllaea pelagica, sargassum nudibranch†. On *Sargassum.* Vi,Pe–Al,Ps,-,?. Adults, 13.87.

Snails (Ph: Mollusca, Cl: Gastropoda)

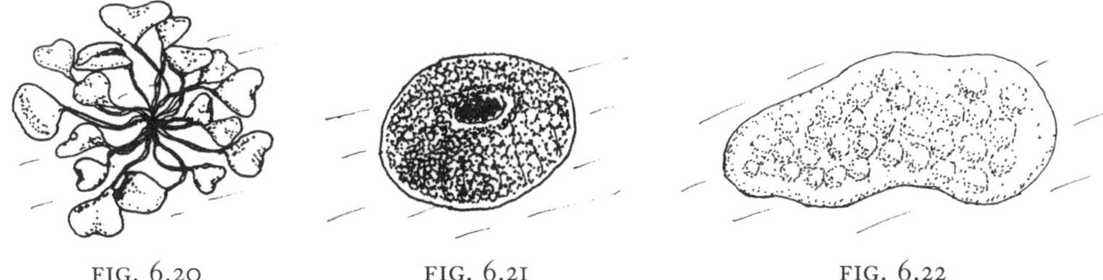

FIG. 6.20 FIG. 6.21 FIG. 6.22

Crepidula fornicata, common Atlantic slippersnail† (Calyptraeidae). Capsules are heart-shaped balloons, attached by stalk to hard substrate, often in vicinity of adults; often in clusters of 70 or more. Young hatch as veligers. Bo,Sa,Li–,-,10 mm (0.4 in) cluster diameter. Adults, 13.27. (FIG. 6.20)

Lacuna vincta, northern lacuna† (Lacunidae). Doughnut egg masses with 1000–1200 eggs; hatch as veligers in 2-3 weeks; from January through early summer. Often on *Laminaria* or various red algae. Ac,Ha–Al,Li–Su,-,6 mm (0.3 in) egg-mass diameter, ☺. Adults, 13.39. (FIG. 6.21)

Littorina obtusata, yellow periwinkle† (Littorinidae). Flattened, oval, circular, or kidney-shaped gelatinous mass on fucoid algae; 90–150 eggs; 2–3 weeks to hatch. Ac–,Al,Li,-,3 mm × 7 mm (0.12 in × 0.28 in), ☺. Adults, 13.42. (FIG. 6.22)

Chapter 7 ■ Phylum Porifera, Sponges

7.1 Sponges are a common, often colorful, but taxonomically difficult group of marine invertebrates. Many of their biological processes are dominated by water flow through their tissues. Seawater enters by way of many tiny openings or *ostia,* passes through internal channels and spaces, and collects at fewer but larger excurrent openings or *oscules* before exiting to the outside. This flow is driven by the beating of flagellated *choanocytes* or collar cells, often concentrated in clusters referred to as *flagellated chambers.* Most sponges possess skeletal support from needlelike *spicules.*

Variability in body shape and coloration make straightforward identification by the nonspecialist impossible in some cases. Be flexible in answering questions offered here, and check tentative identifications against more complete descriptions referred to in species annotations. For most species, even tentative identification requires examination of skeletal spicules. To observe spicule shapes, add household bleach (i.e., sodium hypochlorite) to a small piece of sponge tissue in a shallow dish in order to dissolve the organic material and leave behind only the needlelike spicules. Wash the spicules in a small quantity of tap water and then carefully transfer a drop or two of spicules to a microscope slide. Add a coverslip and observe them using a compound microscope. Note that spicules fall into two size categories, the large *megascleres* and much smaller *microscleres.*

After describing the growth form and morphology of the skeletal spicules, find the color section that most closely corresponds to the specimen. When the spicules have not been examined, shape and color characters may help to narrow the range of possibilities. If an appropriate match is not in this section, look under groups with which sponges may be confused, such as colonial tunicates (17.10) or eggs and egg masses (6.1).

Additional reference: (H) Hartman, W. D. 1958.

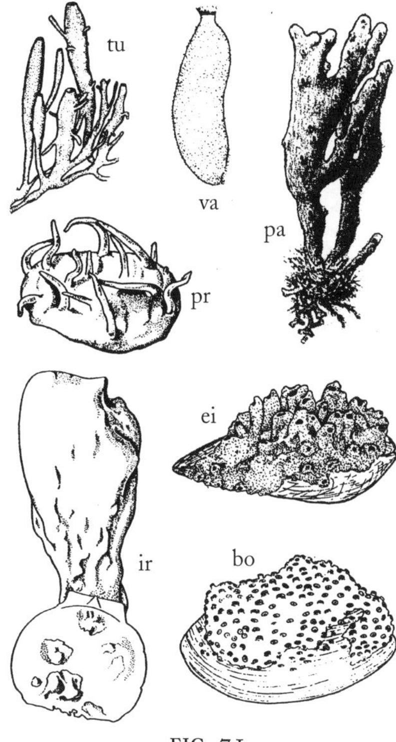

FIG. 7.1

1. Choose the growth form that best describes the sponge (FIG. 7.1)
   - A. *Tubular;* roughly parallel sides, with or without branches, single opening at tips — tu
   - B. *Vase*-shaped; sides bowed, single opening at tip — va
   - C. *Palmate,* upright, branched — pa
   - D. Basal pad with *projecting* fingers or papillae; not open at tips — pr
   - E. *Irregularly* lobed masses — ir
   - F. Smoothly rounded or *egg*-shaped — eg
   - G. *Encrusting* with *volcano*like mounds surmounted by openings — ev
   - H. *Encrusting* with *irregular* relief, and major openings not necessarily at highest prominence — ei
   - I. Encrusting as flattened *sheet* on substrate — es
   - J. *Bores* into calcareous material such as bivalve shells; bumps of sponge tissue project from holes in shell material — bo

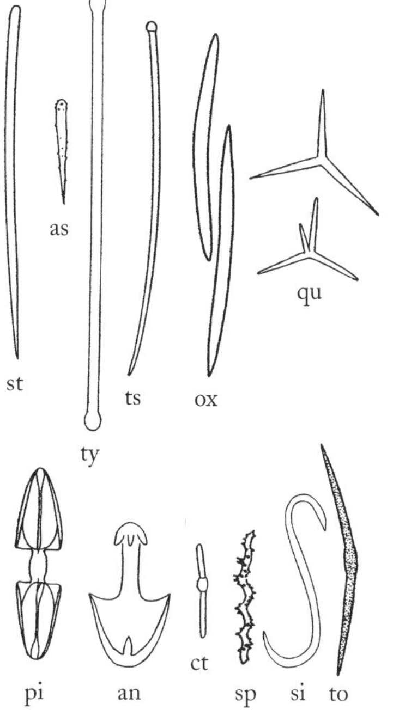

FIG. 7.2

2. Shape of megascleres or large spicules; list as many types as apply (see technique note in the introduction above) (compound microscope) (FIG. 7.2)
   - A. Smooth *styles* (one end round, one end pointed) — st
   - B. Spiny styles or *acanthostyles* (style with spines) — as
   - C. *Tylotes* (both ends bulbed) — ty
   - D. *Tylostyles* (one end bulbed, one end pointed) — ts
   - E. *Oxeas* (both ends pointed) — ox
   - F. Triradiate or *quadriradiate* (three or four pointed) — qu
   - G. Spicules *absent,* sponge feels slippery — x
3. Shape of microscleres or small spicules; list as many types as apply (compound microscope) (FIG. 7.2)
   - A. *Palmate isochelas* (curled ends approximately equal in size) — pi
   - B. *Anisochelae* (curled ends unequal in size) — an
   - C. *Centrotylotes* (one end bulbed, one end pointed, with mid-central bump) — ct
   - D. *Spirasters* (stout, bent, spiny shafts) — sp
   - E. *Sigmaspirae* (slender, bent, spiny shafts) — ss
   - F. *Sigmas* (slender with hooked ends) — si
   - G. *Toxas* (both ends pointed, bent in the middle) — to
   - H. Microscleres *absent* — x

7.2 Yellow Sponges

Growth Form	Megascleres	Microscleres	Organism
pr	ts	x	Fingerlike projections with pointed tips: *Polymastia robusta,* nipple sponge OR Short, knoblike projections: *Trichostemma hemisphericum*
pa	st+ty	pi+si	*Lissodendoryx isodictyalis,* garlic or stinking sponge
pa	ox+st	pi	*Isodictya* Mostly styles, few oxeas, few centrotylote oxeas with pointed ends and a central sphere: *I. deichmannae* OR Mostly oxeas, few styles: *I. palmata,* common palmate sponge
pa	st	x	*Haliclona oculata,* finger or eyed sponge
ev	ox+st	x	Mean length of megascleres, >200 microns: *Halichondria panicea,* crumb-of-bread sponge OR Mean length of megascleres, <150 microns: *Haliclona loosanoffi,* eroded sponge
ei	ox	x	*Halichondria bowerbanki,* crumb-of-bread or yellow sun sponge
ei	ts	an+si	*Mycale fibrexilis,* flabby sponge
bo	ts	x	*Cliona celata,* sulfur or boring sponge
bo	ts	sp	*Cliona* spp., sulfur or boring sponges
ir	as	pi+si	*Myxilla incrustans*
ir	ts+st+ox	x,ct	*Suberites ficus,* fig sponge

7.3 Red/Orange Sponges

Growth Form	Megascleres	Microscleres	Organism
pr	ts	x	*Polymastia robusta,* nipple sponge
pa	ox+st	pi	*Isodictya* Mostly styles, few oxeas, few centrotylote oxeas with pointed ends and a central sphere: *I. deichmannae* OR Mostly oxeas, few styles: *I. palmata,* common palmate sponge
pa	ox	x	*Haliclona (=Chalina) oculata.* Bright orange variant is found at Woods Hole on Cape Cod, Mass.
ev	ox	x	*H. loosanoffi*
pa,ir	st+as	pi+to	*Microciona prolifera,* red beard sponge
es	x	x	Uniform orange-brown: *Halisarca* sp. OR Bright orange, slippery coating on seaweeds, especially *Fucus*: may not be a sponge; try the colonial sea squirt, *Botrylloides diegensis,* 17.11 OR Translucent gelatinous masses with pink/orange bodies embedded within: also try colonial Ascidiacea, sea squirts, 17.4

7.4 Brown/Tan/Whitish Sponges

Growth Form	Megascleres	Microscleres	Organism
tu	qu	x	*Leucosolenia botryoides,* organ-pipe sponge
va	qu+ox	x	*Scypha ciliata,* little vase sponge
eg	ox+qu	ss	*Craniella gravida,* potato or egg sponge
lo	ts+st+ox	ct	*Suberites ficus*
pa	ox+st	pi	*Isodictya* Mostly styles, few oxeas, few centrotylote oxeas with pointed ends and a central sphere: *I. deichmannae* OR Mostly oxeas, few styles: *I. palmata,* common palmate sponge
pa	ox	x	*Haliclona* (=*Chalina*) *oculata*
pa,ir	st+as	pi+to	*Microciona prolifera*
ev	ox	x	*Halichondria bowerbanki*
es	x	x	*Halisarca* sp.
ei	ts	an	*Mycale fibrexilis*
ir	ts+st+ox	ct	*Suberites ficus*

7.5 Purplish Sponges

Growth Form	Megascleres	Microscleres	Organism
pa	ox	x	*Haliclona* (=*Chalina*) *oculata*
ev	ox	x	Smooth surface (other than large volcano-like mounds): *H. permollis* OR Rough surface: *H. loosanoffi*

7.6 Greenish/Greenish-Brown Sponges

Growth Form	Megascleres	Microscleres	Organism
pa	ox	x	*Halichondria panicea*
pa	st+ty	pi+si	*Lissodendoryx isodictyalis,* garlic or stinking sponge
pa	ox	x	*Haliclona* (=*Chalina*) *oculata*

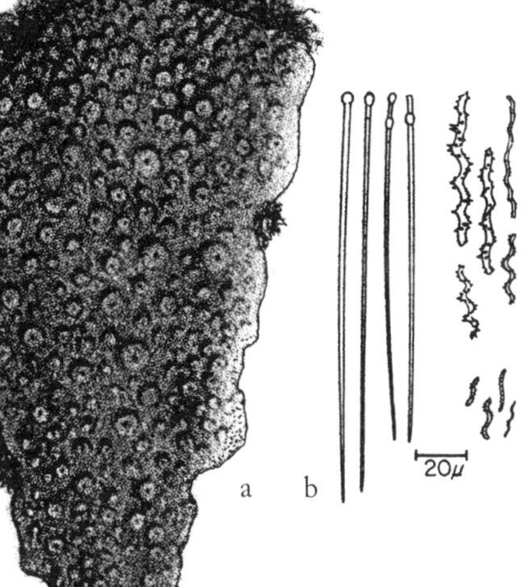

FIG. 7.3

Cliona celata, sulfur or boring sponge (Clionidae). This species matches the generic description that follows except that microscleres are absent, and it can live as a bright yellow mass after bivalve shell is entirely overgrown or eroded away. Considered a pest by oyster farmers. Bo,shells–Ha,Es–Li–Su,ff,15.2 cm (6 in), (G69,H16,NM330), ☺. (FIG. 7.3. a, colony; b, spicules.)

Cliona spp., sulfur or boring sponges (Clionidae). Bright–pale yellow. Papillalike oscules protrude above mass. Invades calcareous materials, especially bivalve shells; weakened shells vulnerable to predators. Yellow oscules protrude from numerous small holes in shell. Several difficult species limited to within-shell growth form only. Bo,shells–Ha,Es–Li–Su,ff,15.2 cm (6 in), (G69,H16,NM330).

Craniella gravida, potato or egg sponge (Craniellidae). Smooth with few obvious openings visible. Vi,Sa,Li–Su,ff,10.1 cm (4 in). (FIG. 7.4)

FIG. 7.4

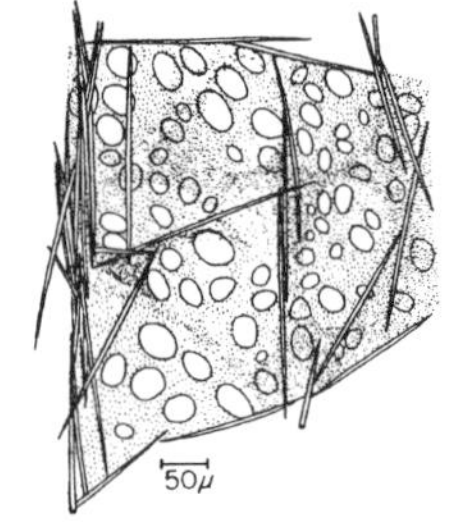

FIG. 7.5

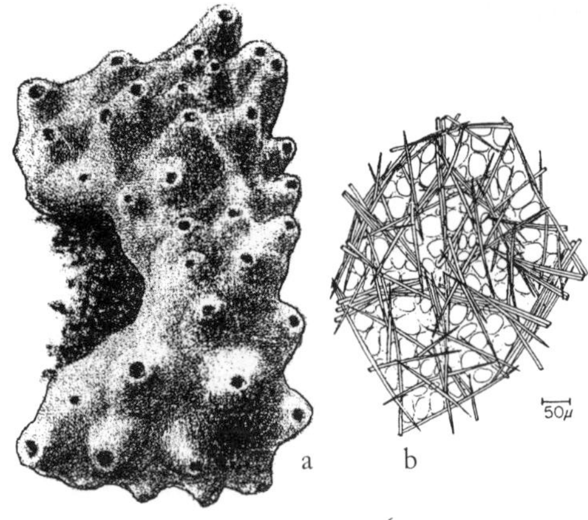

FIG. 7.6

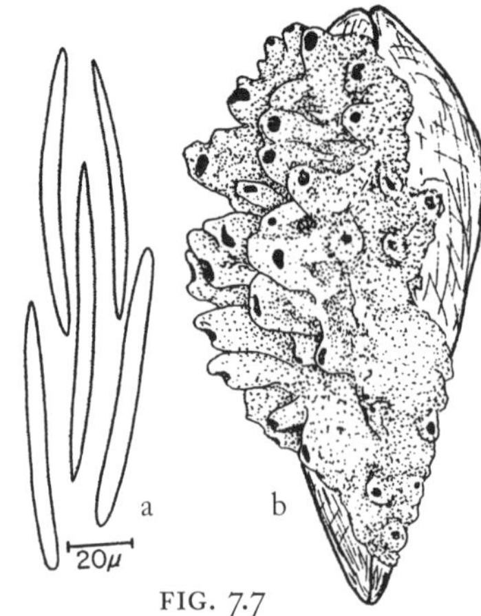

FIG. 7.7

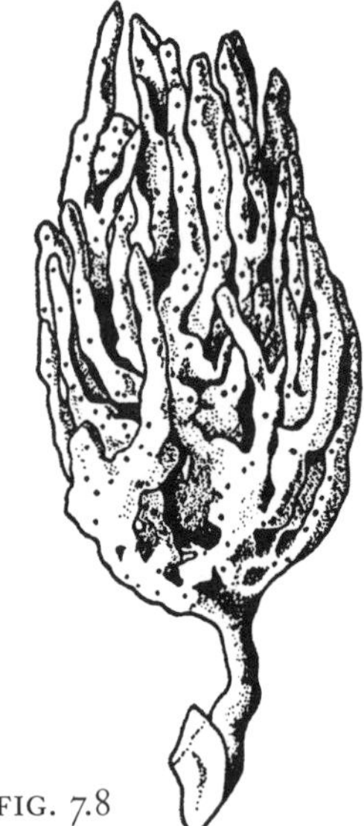

FIG. 7.8

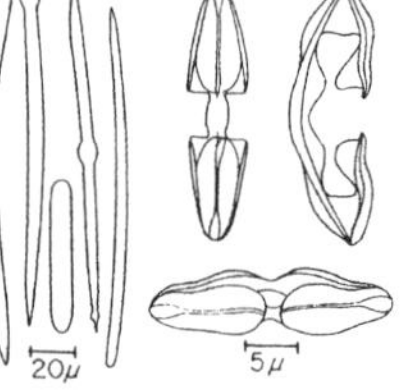

FIG. 7.9

Halichondria bowerbanki, crumb-of-bread or yellow sun sponge (Halichondriidae). Very similar to *H. panicea,* but with less prominent and regularly spaced oscules; more irregular and branching. Not as green; soft texture. Sulfur odor faint. Protruding parts resist breaking when twisted. Bo,Ha,Es–Li–Su,ff,7.6 cm (3 in) tall by 30.5 cm (12 in) wide, (G67,H21), ☺. (FIG. 7.5. Magnified view of surface with dispersed pores and spicules.)

Halichondria panicea, crumb-of-bread sponge (Halichondriidae). Oscules regularly spaced, at tips of "volcano" mounds; open coast: greenish from zoochlorellae, also with more vertical relief than *H. bowerbanki;* protected areas: more yellowish, flatter. Texture rather tough; strong sulfur odor. Protruding parts break easily when twisted. Ac,Ha,Li–Su,ff,61 cm (24 in), (G67), ☺. (FIG. 7.6. a, colony; b, magnified surface showing dense pores and spicules.)

Haliclona loosanoffi, eroded sponge (Haliclonidae). Dark tan, golden to pinkish-purple. Rough surface; "volcano" mounds to 1 in. Vi,Ha,Es–Li,ff,7.6 cm (3 in), (G66,H62). (FIG. 7.7. a, spicules; b, colony.)

Haliclona (=Chalina) oculata, finger or eyed sponge (Haliclonidae). Oscules obvious, without raised rims; may form lines. Branches slender and rounded south of Cape Cod, flatter to the north. Bo,Ha,Es–Li–Su,ff,22.9 cm (9 in), (G65,H52), ☺. (FIG. 7.8)

Haliclona permollis, purple or tufted sponge (Haliclonidae). Encrusting as field of volcano-like or tubular structures. Protected areas. Bo,Ha,Li–,ff,91.4 cm (36 in), (G66,NM326).

Halisarca sp. (Halisarcidae). Uniform orange-brown, slippery coating on hard surfaces. No spicules. Nudibranch, *Cadlina laevis,* is a major predator. Ac,Ha–Al,Li–,ff,15.2 cm (6 in), ☺.

Isodictya deichmannae (Desmacidonidae). Yellow-brown, upright, flattened branches; oscules with raised rims; surface lumpy. Ac–,Ha,Su,ff,7.6 cm (3 in) × 37 mm (1.5 in), (G65,H45). (FIG. 7.9)

Isodictya palmata, palmate sponge (Desmacidonidae). Branches flattened and fused; oscules with raised rims and paler toward branch edges. Bo,Ha,Su,ff,30.5 cm (12 in), (G65,H46). (FIG. 7.10)

FIG. 7.10

Leucosolenia botryoides, organ-pipe sponge (Leucosoleniidae). Several difficult species form small, gray-brown tangled mats of branching tubules with terminal orifices. Spicules are calcareous (i.e., they dissolve in acid). Texture, delicate. Ac,Ha–Al,Li–Su,19 mm (0.75 in), (G65,NM323). (FIG. 7.11)

FIG. 7.11

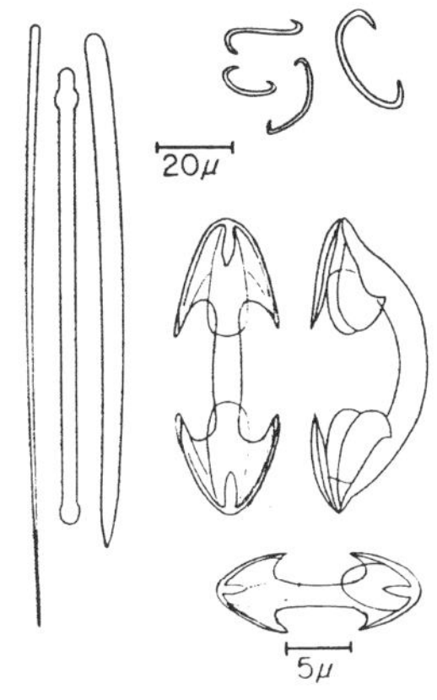

FIG. 7.12

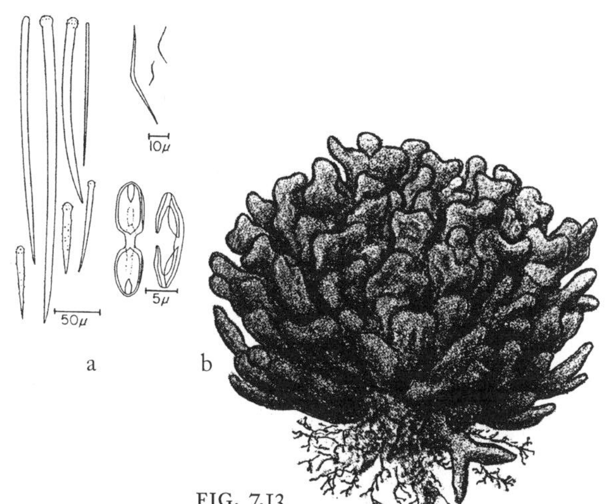

FIG. 7.13

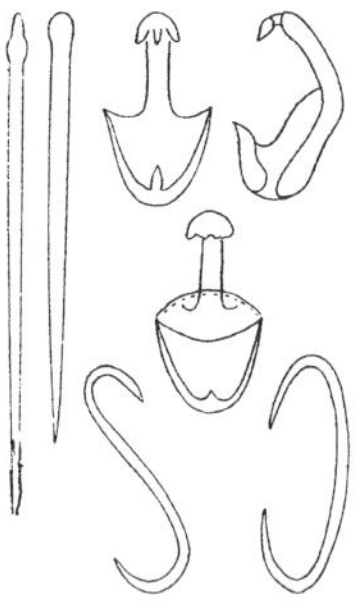

FIG. 7.14

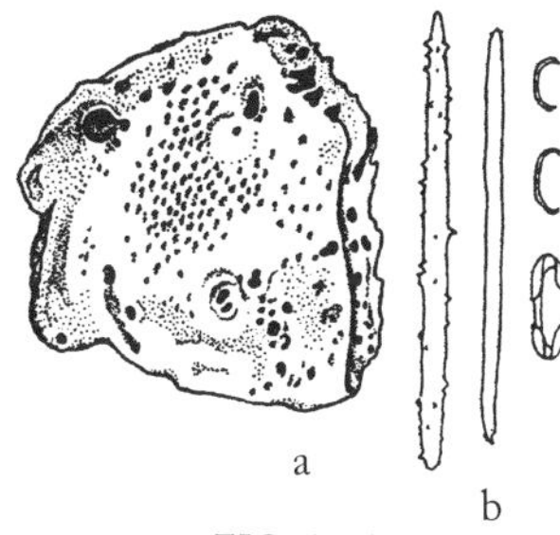

FIG. 7.15

Lissodendoryx isodictyalis, garlic or stinking sponge (Tedaniidae). Yellowish-gray. Distinctive, garliclike odor. Vi,Ha,Es–Li,ff,10.1 cm (4 in), (G68,H41). (FIG. 7.12. spicules.)

Microciona prolifera, red beard sponge (Clathriidae). Oscules or exit pores scattered and small. Tough texture; redder in summer; browner in winter. Encrusting in fast water and when young; erect and branching in slower, especially subtidal waters and with growth. Avoid contact; irritating surface chemicals. On pilings, oyster beds. Bo,Ha–SG,Es–Li–Su,ff,20.3 cm (8 in), (G68,H36,NM327), ☺. (FIG. 7.13. a, spicules; b, colony.)

Mycale fibrexilis, flabby sponge (Mycalidae). Large tylostyles slightly curved; usually thin encrustations; flimsy construction. On pilings, bivalve shells. Vi,Ha,Su,ff,5.6 cm (2.2 in), (G67). (FIG. 7.14)

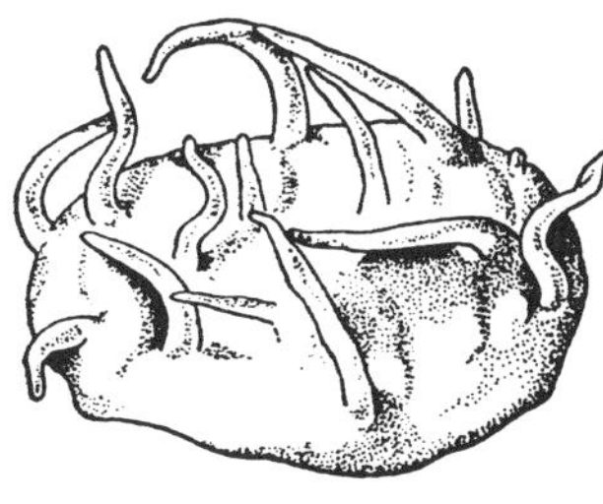

FIG. 7.16

Myxilla incrustans (Myxillidae). Cushion-shaped, smooth. Often on bivalve shells; cold water. Ac,Ha,Su–De,ff,19.7 cm (4.5 in), (G67). (FIG. 7.15. a, colony; b, spicules.)

Polymastia robusta, nipple sponge (Polymastiidae). Ac,Ha,Su,ff,10.1 cm (4 in) wide, (NM329). (FIG. 7.16)

Scypha ciliata, little vase sponge (Grantiidae). Large spicules surround osculum or major opening. Often attached to algae, hydroids, ectoprocts. Spicules are calcareous (i.e., they dissolve in acid). Texture, firm. Bo,Ha–Al,Li-Su,ff,25.4 mm (1 in), (G65,NM323). (FIG. 7.17)

Suberites ficus, fig sponge (Suberitidae). Varied growth forms: lobed, spheroid, encrusting. Surface smooth and firm, like chicken liver. Grayish outside, yellowish within. Bo,Ha,Li–Su,ff,35.6 mm (1.4 in) tall × 35.6 cm (14 in) wide, (G68). (FIG. 7.18. a, spicules; b, colony.)

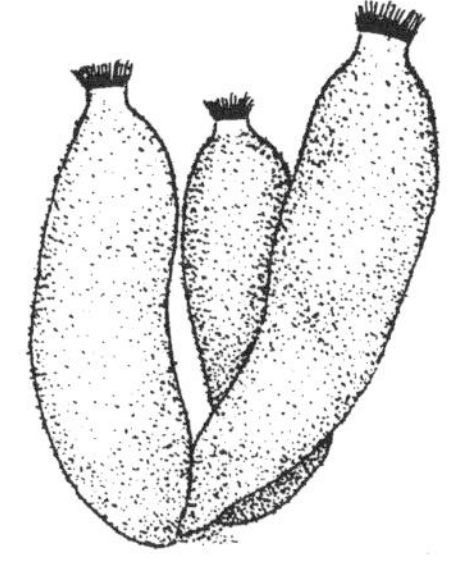

FIG. 7.17

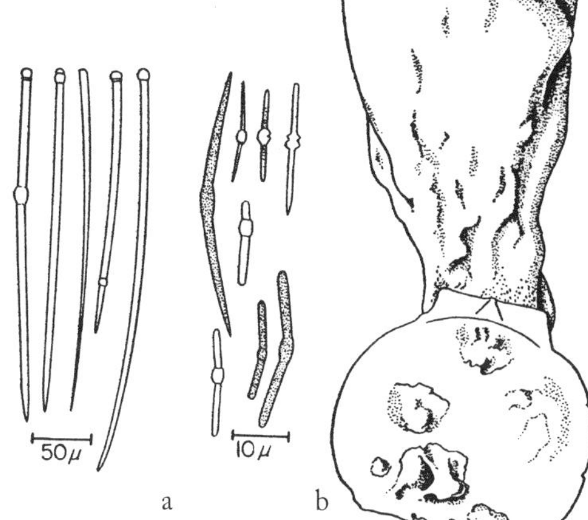

FIG. 7.18

Chapter 8 ■ Phylum Cnidaria

8.1 Cnidaria are relatively simple in construction, formed of discrete tissue layers, but lacking an interior body cavity; that is, they are *acoelomate* and function without reliance on true, complex organs. Generally, the body plan consists of two primary tissue layers, the epidermis and the gastrodermis, along with the *mesoglea,* a mostly noncellular, gelatinous layer of varying thickness. Food capture, and in some cases defense, is effected by tentacles bearing cells that produce unique stinging organelles called *cnidae* or *nematocysts.* In many cases, cnidae produce toxins capable of immobilizing prey. In a few cases (noted below), stinging cells are potent enough to be painful to humans. Tentacles surround a single opening that serves as mouth and anus for the digestive chamber, the gastrovascular cavity or radial canal network.

Cnidaria display *radial symmetry* (more than one imaginary plane can pass through the animal dividing it into mirror images) around an *oral–aboral* axis (defined by the surface that includes the mouth vs. the opposite surface). Many Cnidaria undergo a *polymorphic* (many shapes) life history, alternating between a sessile (attached) *polyp* stage and a free-swimming *medusa* (jellyfish) stage. Medusae are the sexually mature stage in the life cycle and have cup- or dish-shaped bodies with tentacles held downward. Polyps have columnar bodies, attached to the substrate basally, with the tentacled end uppermost. Polyps reproduce asexually to form additional polyps or, upon appropriate stimulation, to bud off young medusae. Especially in the class Hydrozoa (the hydroids), asexually produced polyps tend to remain attached, forming colonies (8.3). Generally, their tiny medusae (8.30) form seasonally and become members of the plankton community. The polyp stage of the life history of members of the class Scyphozoa (the jellyfish) is usually much reduced. Here, the medusa stage is dominant (8.38). Some Hydrozoa, in the order Siphonophora, form free-swimming colonies (8.38). Some medusalike Scyphozoa, in the order Stauromedusae, attach to seaweeds by a stalk (8.40). There is no medusa stage in the Class Anthozoa (anemone) (8.42). The more complexly constructed polyps in this class are generally capable of both asexual and sexual reproduction.

For reasons described in the next section, members of the phyla Entoprocta and Ectoprocta (Bryozoa) may also key out in this section.

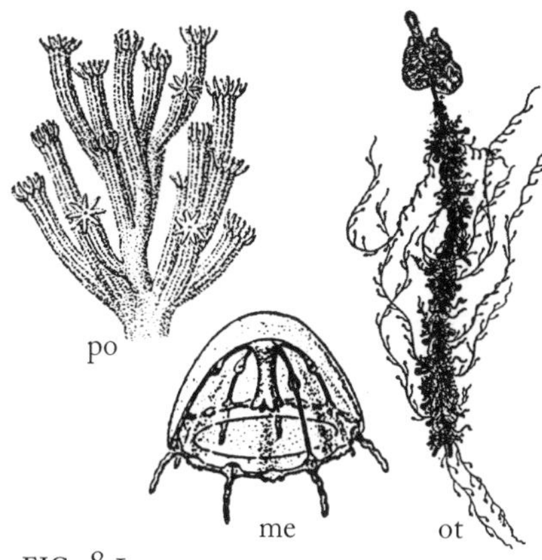

FIG. 8.1

1. Overall body form (FIG. 8.1)
   A. *Polyp* (cylindrical body column with tentacles surrounding mouth opening) (may be colony including many individual polyps) po
   B. *Medusa* (an individual cup-shaped or dish-shaped, pulsing bell, usually with tentacles associated with margin) me
   C. Shaped *otherwise;* either a delicate chain of paired units or a clustered mass of tentacles and other forms suspended beneath a gas-filled floatation sack ot
2. Coloniality
   A. Organism an *individual* polyp or medusa in
   B. Organism part of a *colony* formed by many individuals structurally connected to one another co
3. Lifestyle
   A. Found *attached* securely, though not necessarily permanently, to its substratum at
   B. Found freely *swimming* or floating in the medium sw
   C. *Burrowing* in particulate sediments (or, in rare cases, in tissues of other animals) bu
4. Height of individuals (this applies to single members of colonial forms)
   A. Body length *small,* less than 6 mm (¼ in) sm
   B. Body length *larger,* more than 6 mm (¼ in) la
5. Widest diameter of individual stalk or body column
   A. Diameter *slender,* less than 6 mm (¼ in) sl
   B. Diameter *stout,* more than 6 mm (¼ in) st

8.2

Form	Colony	Style	Height	Diameter	Group
po	co,in	at,bu	sm	sl	Cl: Hydrozoa, Or: Hydroida, 8.3‡
po	co,in	at,bu	la	sl	Cl: Hydrozoa, Or: Hydroida, 8.3‡
po	co,in	at,bu	la	st	Cl: Anthozoa, 8.42‡
me	in	sw	sm,la	st	Jellyfish, 8.30
me	in	at	la	st	Cl: Scyphozoa, Or: Stauromedusae, 8.40
ot	co	sw	sm,la	st	Cl: Hydrozoa, Or: Siphonophora, 8.38

‡ These criteria imperfectly separate Hydrozoa from Anthozoa. Although they cover the majority of cases, there are a few large Hydrozoa and a few small Anthozoa. If you fail to find a suitable match, try the other group.

8.3 Class Hydrozoa: Order Hydroida, Hydroids

Hydroid colonies are not easy to identify. Try to collect fresh, clean, living material for study. Most hydroids form colonial groups of interdependent members or zooids. Many hydroids produce transparent, rigid, chitinous exoskeletal material, called the *perisarc,* to support their stalks. In members of the suborder Leptothecata, perisarc is universally present and also forms cup-like or tubular *hydrothecae* into which feeding individuals or *gastrozooids* (also called *hydranths*) can withdraw for protection. Although the perisarc may support colonies to varying degrees, protective hydrothecae are never present in the suborder Anthoathecata. Using dissecting microscope magnification, touch tentacled colony members, the hydranths, with a pin to note the degree to which stalks are supported by perisarc and whether hydrothecae are present. A temporary wet-mount (instructions in 0.7) and a compound microscope may be required to observe fine details.

The Leptothecata produce separate, but interconnected, feeding and reproductive individuals in the colony. Reproductive members, or *gonozooids,* are usually larger, more ovaloid in shape, and often more opaque than the more numerous feeding individuals, the gastrozooids. There is no such separation of colony members in most Anthoathecata. When conditions dictate, medusa buds form at species-specific locations on the feeding hydranth itself.

Hydroids are most often confused with colonial members of the Bryozoa, originally designated as a phylum, but whose members are now included in the two phyla, Entoprocta (treated below, 8.26) and the Ectoprocta (12.1). Entoprocts are tiny, stalked organisms with a whorl of ciliated feeding tentacles. Although they resemble hydroids superficially and thus, key out here, they are unrelated, possessing a more complex interior anatomy. Entoprocts are more closely allied to the Ectoprocta, which are still more advanced. Ectoprocts generally appear as branched, erect, hydroidlike colonies or form lacelike, encrusting colonies. Although internal features clearly distinguish hydroids and the erect Ectoprocta, the easiest way to use external characters for distinguishing these two groups is by looking for evidence of ciliary currents around the feeding tentacles. Ectoproct zooids use strong ciliary currents to sweep water and particles of food from the outside toward the inside of their tentacular cluster. In contrast, hydroid hydranths lack ciliary currents. They rely on their stinging cells (cnidae) to immobilize prey that bump into their outstretched tentacles.

Scientific names follow suggestions of Calder (personal communication) and Cairns et al. (1991). Common names listed with a dagger (†) have been designated as the preferred, standardized name by the Committee on Scientific and Vernacular Names of Cnidaria (Cairns et al. 1991). In some cases, additional, well-recognized common names are listed. These are included during this transitional period to clarify linkages to the proper scientific name and to the single common name designated for the particular species.

Lengths refer to colony length in colonial forms. A dissecting microscope is required for many observations in this key.

Additional references: (C) Calder, D. R. 1975. (F) Fraser, C. M. 1944.

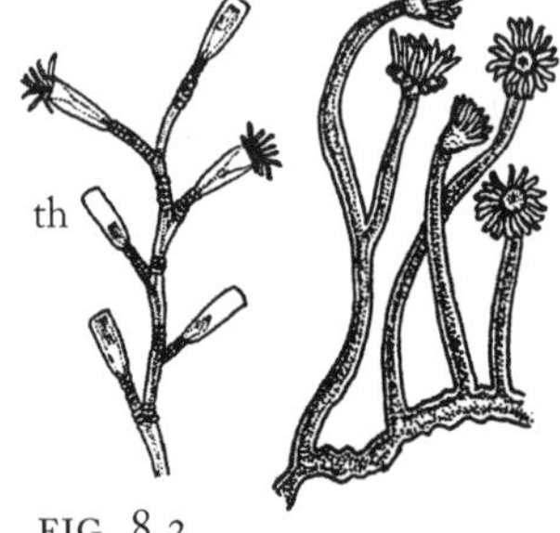

FIG. 8.2

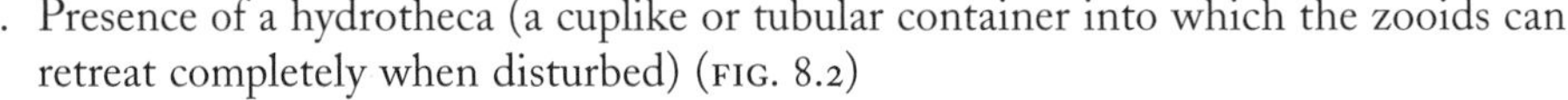

1. Presence of a hydrotheca (a cuplike or tubular container into which the zooids can retreat completely when disturbed) (FIG. 8.2)
   - A. *Hydrotheca* present, although it may be reduced and saucer shaped — th
   - B. Such a theca is *absent,* although other forms of perisarc support may be present — x

8.4

Theca	Group
th	Leptothecate Hydroids, 8.5
x	Anthoathecate Hydroids and Entoprocta/Ectoprocta, 8.7

8.5 **Leptothecate Hydroids** (from 8.4)

Leptothecate hydroids include both feeding gastrozooids and seasonally limited reproductive gonozooids as distinctive members on the same colony. They generally form small, bushy colonies attached to hard substrates or algae.

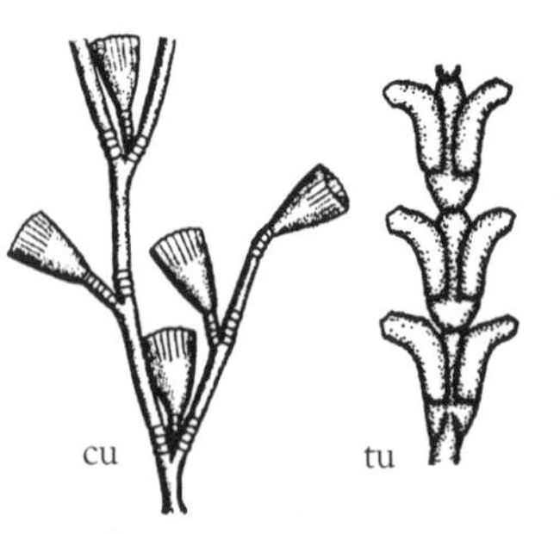

FIG. 8.3

1. Location of hydrothecae
   - A. Hydrothecae are *free* at tips of branches — fr
   - B. Hydrothecae are wholly or partly fused *directly* to the branch or stem — di
2. Shape of hydrothecae (FIG. 8.3)
   - A. Hydrotheca is *cup*like — cu
   - B. Hydrotheca is *tubular* — tu
   - C. Hydrotheca is *saucer*like, unable to entirely accomodate a retreating hydranth — sa

8.6

Location	Shape	Family
fr	cu	Campanulariidae (in part), 8.12
fr	tu	Campanulariidae (in part), 8.12
fr	sa	*Halecium halicinum*
di	cu	*Schizotricha tenella,* plumed hydroid†
di	tu	*Sertulariidae,* 8.22

FIG. 8.4

Halecium halicinum (=*gracile*) (Haleciidae). Stems without annulations; some rebranching to form pinnate (featherlike) colonies as opposed to bushy ones. Bo,Ha,Su,5.1 cm (2 in), (C297,G84,NM351). (FIG. 8.4. a, entire colony; b, section of colony; c, gonozooid.)

Schizotricha tenella, plumed hydroid[†] (Plumulariidae). Hydranths restricted to one side of branch; small, secondary tubular members interspersed with large hydranths. Colony white, feathery. On pilings. Bo,Ha–Al,Su,ff,10.1 cm (4 in), (G85). (FIG. 8.5. a, entire colony; b, portion of colony with gonozooid.)

FIG. 8.5

8.7 **Anthoathecate Hydroids and the Phyla Entoprocta/Ectoprocta** (from 8.4)

Anthoathecate hydroids range from tiny to conspicuous colonies several inches in height. They lack thecate cups. Medusa buds tend to form directly on feeding hydranths. Members of two other phyla, the *entoprocts* and the *ectoprocts,* also form very small colonies of tentacled individuals, but they lack stinging cells, characteristic of all Cnidaria, and use ciliary currents to sweep food particles within range of feeding tentacles.

1. Solitary or colonial
   - A. Hydroid occurs as *solitary* polyp — so
   - B. Hydroid forms a *colony* of individuals — co
2. Tentacle placement (tentacles may include larger basal and smaller distal ones) (FIG. 8.6)
   - A. Tentacles scattered *irregularly* on hydranth: provide *number* of tentacles — ir:__
   - B. Tentacles arise as distinct rings or *whorls*: provide *number* of whorls (not total number of tentacles) — wh:__
   - C. Both *whorls*—provide *number* of whorls; *irregularly* scattered—provide *number* of tentacles — wh__:ir__*
3. Shape of tentacles (FIG. 8.6)
   - A. Tentacles are *capitate* (possess swollen or bulbous tips) (do not confuse these with clusters of spherical medusa buds) — ca
   - B. Tentacles are *filiform* (uniformly slender or tapering) — fi
   - C. Both *filiform* and *capitate* tentacles present — fi + ca*
4. Presence of perisarc (nonliving, rigid covering) (use a pin to distinguish between soft, pliable body parts and stiffer, outer perisarc coverings) (FIG. 8.6)
   - A. *Perisarc* present — pe*
   - B. Perisarc *absent* — x

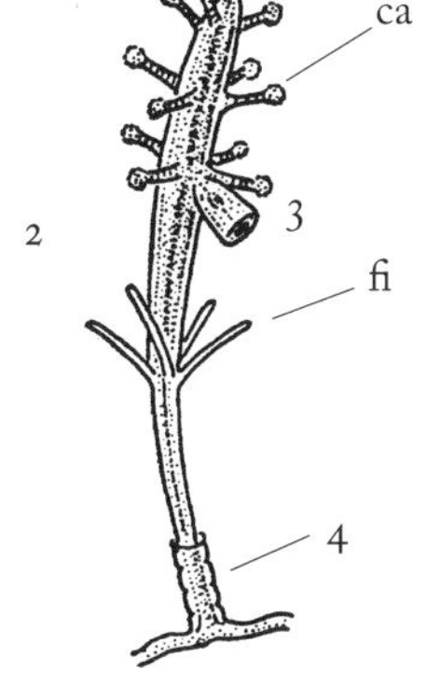

FIG. 8.6

5. Shape of hypostome (mouth cone)
   - A. Hypostome *trumpet*-shaped with flared "lip" — tr
   - B. Hypostome embedded within the distal cluster of *tentacles* so that its shape can not be seen — te
   - C. Hyopostome shaped *otherwise* (usually domed or rounded) — ot*
6. Ciliated tentacles (under magnification, look for cilia-driven water movements in the vicinity of the tentacles; this feature distinguishes the superficially similar but unrelated Entoprocta and Ectoprocta from the hydroids)
   - A. *Ciliated* tentacles present — ci
   - B. Ciliated tentacles *absent* — x*
7. Behavior of individuals (this feature is of greatest diagnostic value when it occurs; failure to observe it does not necessarily mean it can not occur)
   - A. Stalks *bend* voluntarily when provoked — be
   - B. Stalks do *not* bend voluntarily — x*

8.8

Form	Place	Shape	Perisarc	Hypostome	Cilia	Bend	Family
so,co	ir: ca. 30	ca	pe	ot	x	x	*Sphaerocoryne* (=*Linvillea*) *aqassizii*
so,co	ir: ca. 20	fi	pe	ot	x	x	*Aselomaris michaeli*
so,co	ir: ca. 10–20	ca	x	ot	x	x	*Sarsia* (=*Coryne*) *tubulosa,* clapper hydroid†
so	wh1:ir: 50+	fi + ca	pe	ot	x	x	*Acaulis primarius*
so	wh:3+	fi	pe	te,ot	x	x	*Corymorpha pendula,* nodding bouquet hydroid†
so,co	wh:2	fi	pe	te	x	x	Tubulariidae, 8.24
so,co	wh:1	fi	pe,x	ot	ci	be,x	Entoprocta and Ectoprocta, 8.26
co	wh:1	fi	pe	ot	x	x	Bougainvilliidae, 8.10
co	wh:1	fi	pe	tr	x	x	Eudendriidae, 8.16
co	wh:1,2	fi	x	ot	x	x	Hydractiniidae, 8.19
co	wh:3–6	fi + ca	pe	ot	x	x	*Pennaria disticha* (=*tiarella*), feather hydroid†
co	wh:3	fi	x	ot	x	x	*Clava multicornis* (=*leptostyla*), club hydroid†
co	ir: ca. 10–15	fi	x	ot	x	x	Branching evident: *Cordylophora caspia,* freshwater hydroid† OR No branching: *Rhizogeton fusiformis*
co	ir: ca. 20–30	fi	x	ot	x	x	*Clava multicornis* (=*leptostyla*), club hydroid†

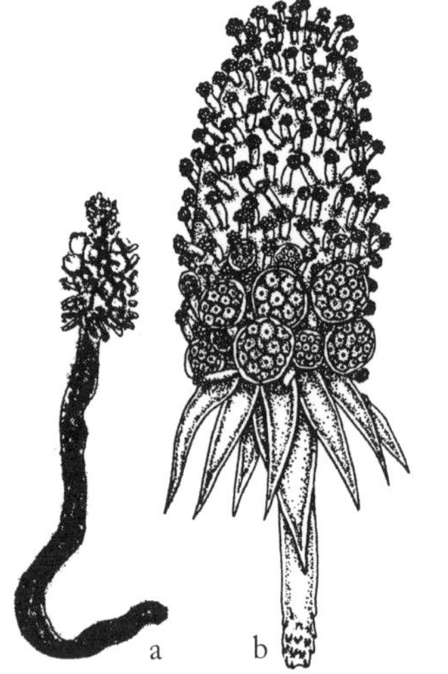

FIG. 8.7

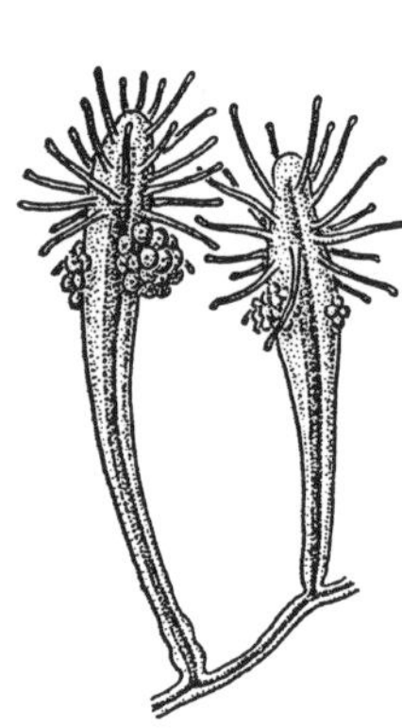
FIG. 8.8

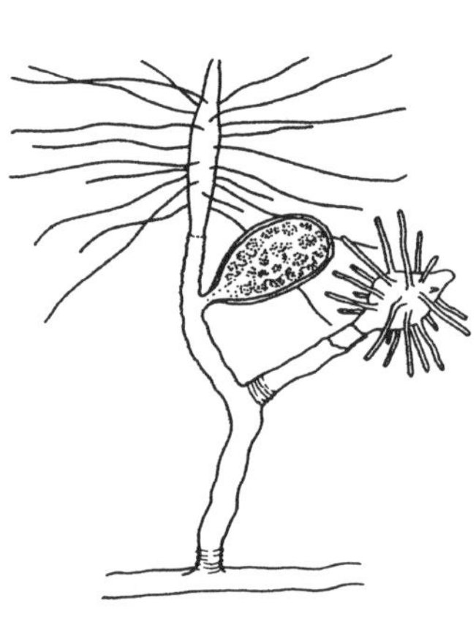
FIG. 8.9

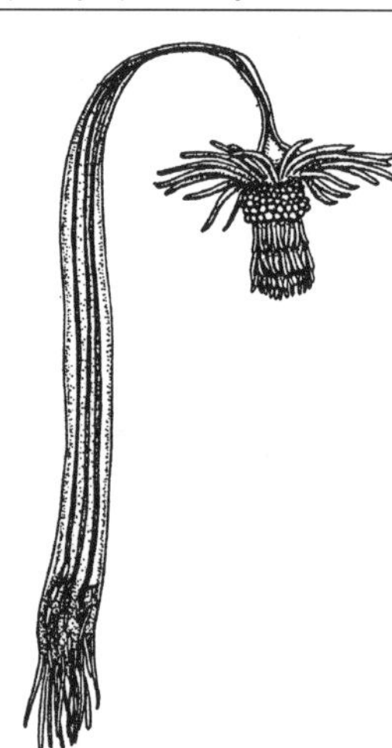
FIG. 8.10

Acaulis primarius (Acaulidae). Solitary, vermiform hydranth. Thin-skinned, gelatinous perisarc with embedded sand. 5–8 basal filiform tentacles. Ac,Ha,Su,ff,3 cm (1.2 in), (C294,F87). (FIG. 8.7. a, individual; b, enlarged distal end.)

Aselomaris michaeli (Bougainvilliidae). Separated individuals arise from stolon base. Gonophores from perisarc covered stalk. Ac–,Ha–SG,Su,ff,15.2 mm (0.6 in).

Clava multicornis (=*leptostyla*), club hydroid† (Clavidae). Small, pink, unbranched clusters under rocks or often on algae, especially *Fucus, Ascophyllum, Laminaria,* or *Chondrus.* Ac–,Ha–Al,Li–Su,ff,20.1 mm (0.8 in), (F33,G76,NM343). (FIG. 8.8)

Cordylophora caspia (=*lacustris*), freshwater hydroid† (Clavidae). Bushy, white with annulated stalks; spindle-shaped polyps. Very dilute brackish to freshwater only. Bo,Ha,Es–Su,ff,5.1 cm (2 in), (G77). (FIG. 8.9)

Corymorpha pendula, nodding bouquet hydroid† (Tubulariidae). Solitary hydranth in sand or mud. Long stalk; rhizoids near base. One whorl of ca. 30 long filiform tentacles, then clusters of short filiform tentacles. Ac,Sa–Mu–Al,Su,ff,10.1 cm (4 in), (C292,F89,G75,NM340). (FIG. 8.10)

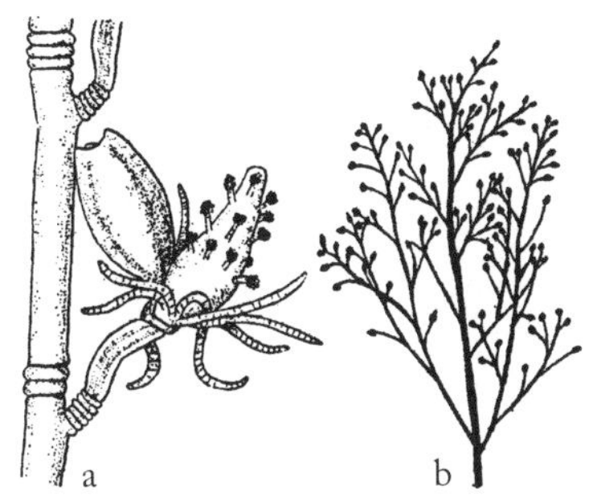

FIG. 8.11

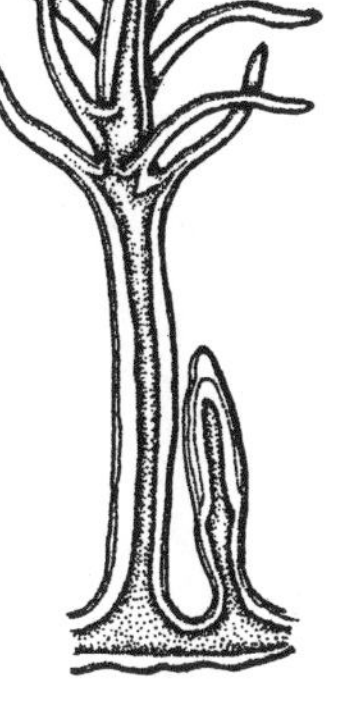

FIG. 8.12

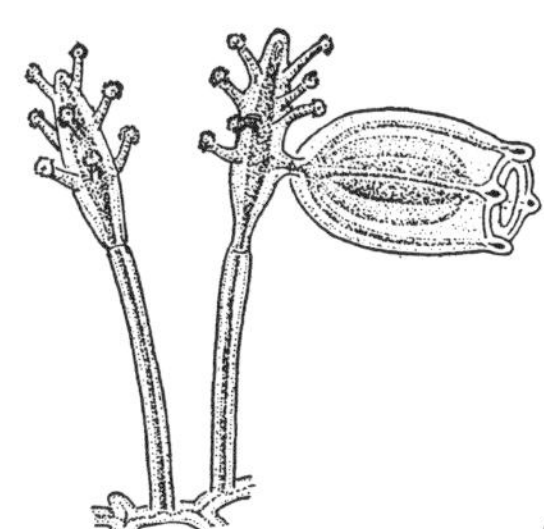

FIG. 8.13

Pennaria disticha (=*tiarella*), feather hydroid† (Halocordylidae). Brown, feathery colonies with pink hydranths; in moderate water currents. Vi?,SG–Ha,Su,ff,200 mm (8 in), (F84,G76), ☺. (FIG. 8.11. a, trophozooid budding medusa; b, colony form.)

Rhizogeton fusiformis (Clavidae). Soft stalk; especially on mussels, lower littoral. Bo,Ha,Li-Su,ff,8 mm (0.3 in), (F35,G77). (FIG. 8.12)

Sarsia (=*Coryne*) *tubulosa,* clapper hydroid† (Corynidae). Colorless to pink; from stolonate base. Individuals may be widely separated and appear to be solitary. Bo,Ha,Su,ff,20.1 mm (0.8 in), (C294,G76,NM342). (FIG. 8.13. Pair }of polyps and advanced medusa bud.)

Sphaerocoryne (=*Linvillea*) *agassizii* (Corynidae). Elongate hypostome or mouth cone. Grows among sponges or on shells. Vi,Ha,Su,ff,17.8 mm (0.7 in), (F39,G76). (FIG. 8.14)

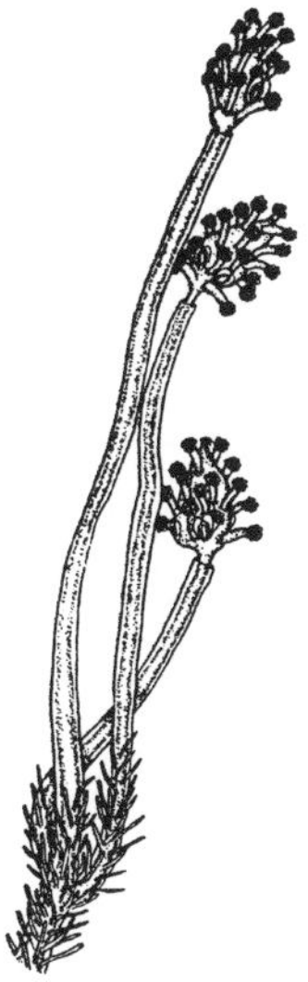

FIG. 8.14

Families of Hydroids (arranged alphabetically)

8.9 **Acaulidae** (see 8.8)

8.10 **Bougainvilliidae** (from 8.8)

Hydroids form branched, bushy, upright colonies; perisarc extends part way up hydranth; one whorl of filiform tentacles. (NM344)

1. Total number of tentacles per hydranth: provide total *number* of tentacles (FIG. 8.15) —
2. Adjacent stalks fascicled (partially fused to one another) (FIG. 8.15)
   - A. At least some stalks *fascicled* (with more than 1 tube) fa
   - B. Fascicled stalks *absent* (with 1 tube) x*
3. Annulations (rings) around stalks (FIG. 8.15)
   - A. At least some *annulation* present on stalks an
   - B. Annulation *absent* from stalks x*

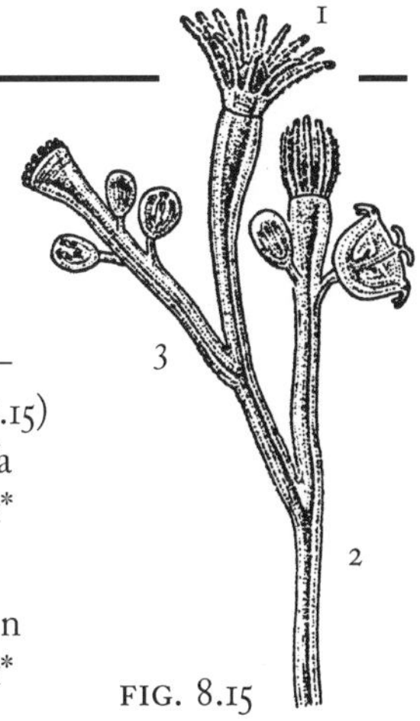

FIG. 8.15

8.11

Tentacles	Fascicled	Annulations	Species
8–10	x	an	*Garveia franciscana,* rope grass
8–10	fa	x	*Bougainvillia rugosa*
10–12	fa	an	*B. carolinensis*
15–20	x	an	*B. superciliaris*

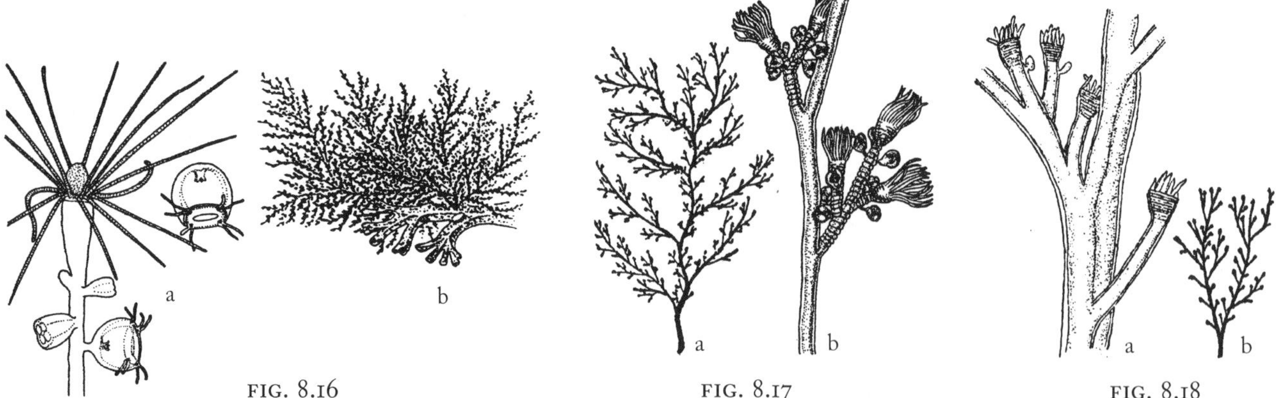

FIG. 8.16 FIG. 8.17 FIG. 8.18

Bougainvillia carolinensis. Stems intertwined and partially fused (fascicled); slightly greenish perisarc, reddish hydranths. Bo,Ha,Li–Su,ff,30.5 cm (12 in) [7.6 cm (3 in)], (F50,G78), ☺. (FIG. 8.16. a, hydranth and medusa; b, colony form.)

Bougainvillia rugosa. Stalks may be wrinkled or smooth but not annulated. Vi,Ha,Li–Su,ff,25.4cm (10 in), (G78). (FIG. 8.17 a, colony form; b, portion of colony.)

Bougainvillia superciliaris. Bo,Ha,Li–Su,ff,5.1 cm (2 in), (F53,G78). (FIG. 8.18. a, portion of colony; b, colony form.)

Garveia franciscana, rope grass. Annulated at base of each hydranth and gonopore stalk. Forms brownish mats on solid objects; stalks intertwine but do not fuse; appear "ropelike." Ca+,Ha,Su,ff,7.6 cm (3 in), (G78).

8.12 Campanulariidae, Wine-Glass Hydroids and Lafoeidae (1 species included here) (from 8.6)

Hydroids with bell-shaped or somewhat cylindrical, pedestalate hydrothecae; no opercular flap to cover hydranth opening. Shape of the *gonotheca* (perisarc covering of scattered, seasonal reproductive members) is used in species distinctions.

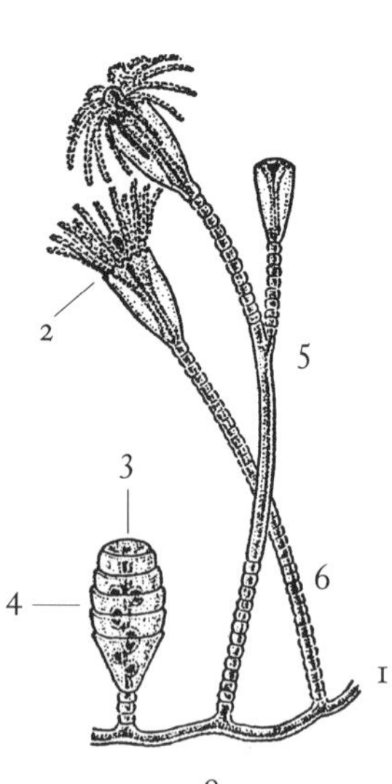

FIG. 8.19

1. Colony form (FIG. 8.19)
   - A. Colony primarily *vertical,* as upright stalks, branched or unbranched, but with several to many hydranths ve*
   - B. Colony primarily *horizontal,* with stolen-like runners producing short upright stalks, each with one or a few hydranths ho
2. Presence of "teeth" around the margin of the hydrotheca (FIG. 8.19)
   - A. *Teeth* present te*
   - B. *Cusped teeth* present (each primary tooth is secondarily notched) ct
   - C. Teeth *absent* x
3. Size of gonothecal opening (FIG. 8.19)
   - A. *Large* opening (approximately equal to diameter of gonotheca itself) la*
   - B. Unique *curled* tip covers opening cu
   - C. *Small* opening (obviously smaller than diameter of gonotheca) sm
4. Shape of gonotheca (FIG. 8.19)
   - A. *Smooth* outline sm
   - B. *Ribbed* outline ri*
5. Aspects of growth form (FIG. 8.19)
   - A. *Fascicled* (several upright stems twist and/or fused together) fa
   - B. *Geniculate* (straight sections end in hydranths and alternate with one another in zigzag fashion) ge
   - C. These features *absent,* upright stems individual and growth pattern does not zigzag x*

6. Frequency of branching stems (branches are "stems" from which hydranths bud; do not confuse simple, individual hydranth stalks with true branches) (FIG. 8.19)
   - A. Branching *extensive,* more than three branches per main stem — ex
   - B. Branching limited or *absent,* fewer than three branches per upright stem — x*
7. Colony size (this character is most useful for hydroids exceeding minimal sizes listed; young colonies of larger forms can be misleading)
   - A. Colony *small,* less than 1.25 cm (½ in) tall — sm
   - B. Colony *large,* more than 7.25 cm (3 in) tall — la
   - C. Colony *intermediate,* 1.25–7.5 cm (½–3 in) tall — in

8.13

Form	Teeth	Gono Size	Gono Shape	Growth	Branch	Size	Species
ho	te	sm	ri	x	ex,x	sm,in	*Clytia hemisphaerica* (=*edwardsi, johnstoni*)
ve	te	sm	sm	x	x	sm	*Campanularia groenlandica*
ve	ct	sm	sm	fa	ex	la	Hydrotheca cuplike, expanding sides: *Hartlaubella* (=*Campanularia*) *gelatinosa* OR Hydrotheca cylindrical: *Obelia* (=*Clytia*) *longicyatha*
ve	ct	la	ri	fa,x	ex	in	*O. bidentata* (=*bicuspidata*)
ve	x	sm	sm	x	ex	la	*O. longissima* (=*commissuralis*), bushy wine-glass hydroid*
ve	x	sm	sm	x	ex	sm	*Laomedea* (=*Campanularia*) *amphora*
ve	x	sm	sm	fa	ex	in	*Lafoea fruticosa*
ve	x	sm	sm	ge	x	in	*Obelia geniculata,* knotted threadhydroid†, zig-zag wine-glass hydroid
ve	x	sm	sm	x	x	in	*Obelia dichotoma,* sea thread-hydroid†
ve	x	sm	sm	x	x	sm	*Campanularia volubilis*
ve	x	cu	sm	ge	ex	in	*Laomedea* (=*Campanularia*) *calceolifera*
ve	x	la	sm	x	ex	in	*Laomedea* (=*Campanularia*) *flexuosa*

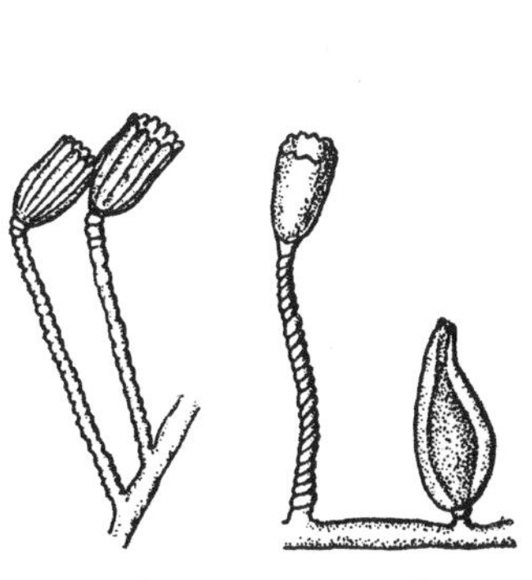

FIG. 8.20 FIG. 8.21 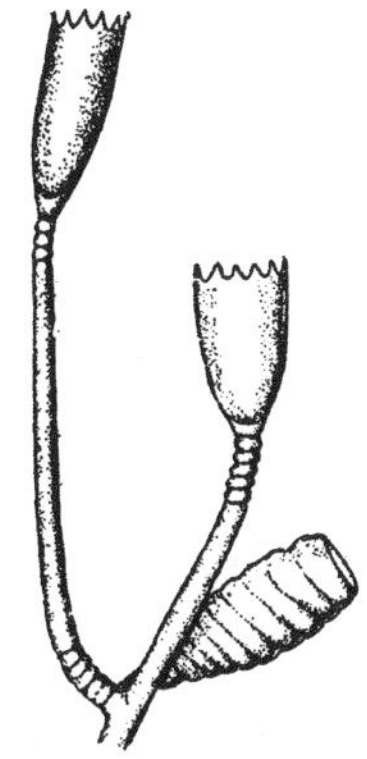 FIG. 8.22

Campanularia groenlandica. Thecal cup with longitudinal ribs. Stalk spirally twisted. Ac–,Ha,Su,ff,7.6 mm (0.3 in). (FIG. 8.20)

Campanularia volubilis. Spiral twist to stalks; cylindrical hydrothecae. Ac–,Ha–Al,Li–De,12.7 mm (0.5 in), (F131). (FIG. 8.21)

Clytia hemisphaerica (=*edwardsi, johnstoni*). Small, white colonies with occasional branching, 3 or fewer polyps/stalk; 12–14 hydrothecal teeth. Bo,Ha,Li–De,ff,25.4 mm (1 in), (C300,G80). (FIG. 8.22)

Hartlaubella (=*Campanularia*) *gelatinosa.* Each tooth is bicuspid, i.e., comprises two smaller teeth. Branching from all sides of annulate stem. Bo,Ha,Li–De,ff,25.4 cm (10 in), (F118,G80). (FIG. 8.23. a, colony form; b, portion of colony.)

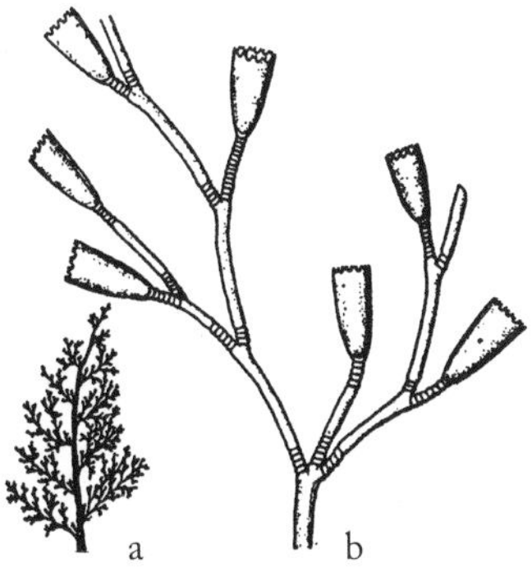

FIG. 8.23

Lafoea fruticosa (Lafoeidae). Large branches, sometimes all from one side. Twisted pedicels support hydranths. Ac–,Ha,Su–De,ff,5.1 cm (2 in), (F223). (FIG. 8.24. a, portion of colony with fascicled stems; b, colony form.)

Laomedea (=*Campanularia*) *amphora.* Ac–,Ha,Su,ff,17.8 mm (0.7 in), (F113). (FIG. 8.25. a, portion of colony with gonozooid; b, colony form.)

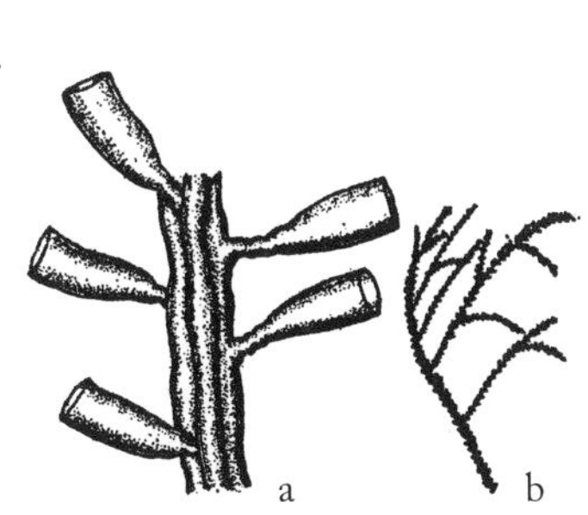

FIG. 8.24

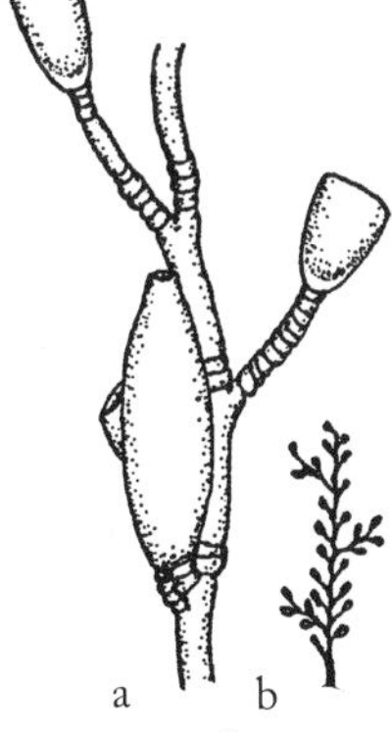

FIG. 8.25

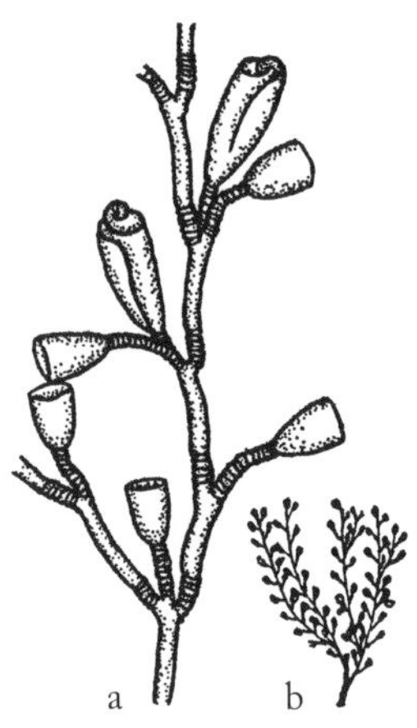

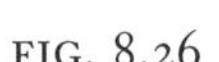
FIG. 8.26

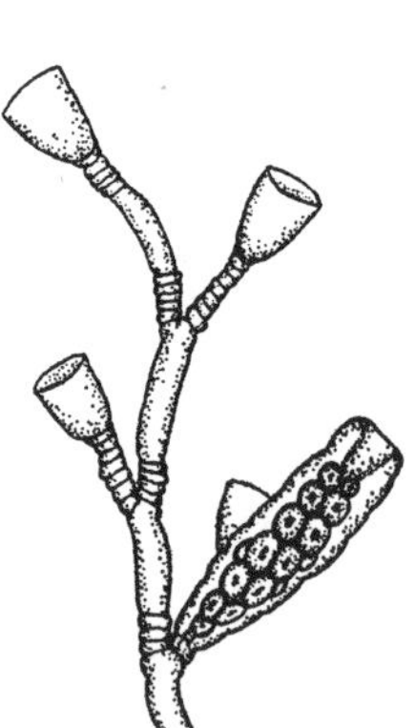
FIG. 8.27

FIG. 8.28

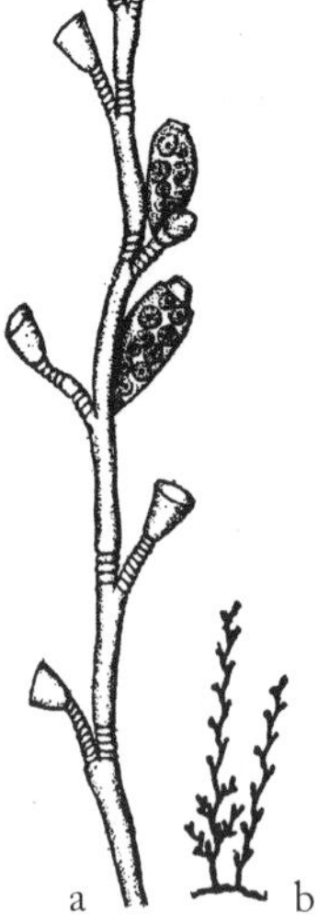

FIG. 8.29

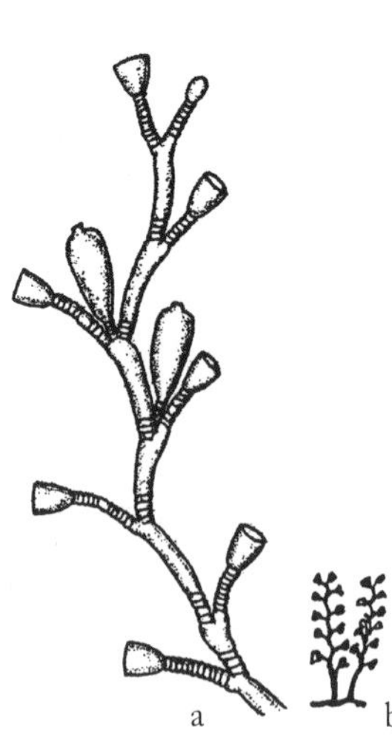

FIG. 8.30

FIG. 8.31

Laomedea (=*Campanularia*) *calceolifera.* On algae, mussels, pilings, etc. Bo,Ha,Es–Li–Su,ff,38.1 mm (1.5 in), (F115,G80). (FIG. 8.26. a, portion of colony with gonozooid; b, colony form.)

Laomedea (=*Campanularia*) *flexuosa.* Brown stems; never releases medusae. Hydrothecal length and width about equal. Lots of annulation. Mostly on fucoid algae. Ac,Ha,Li–Su,ff,33 mm (1.3 in), (F116,G80), ☺. (FIG. 8.27. Colony with gonozooid.)

Obelia bidentata (=*bicuspidata*). 14–20 hydrothecal teeth. Bo,Ha,Su,ff,12.7 mm (0.5 in), (F153,G80). (FIG. 8.28. a, colony form; b, portion of colony.)

Obelia dichotoma, sea threadhydroid†. Stalked hydranths form straight (not zigzag), periodically annulated stem. Some branching may occur. Tentacles 1.2 mm long. Bo,Ha–Al,Li–Su,ff,25.4 mm (1 in), (C303,F155,NM348). (FIG. 8.29. a, portion of colony with gonozooid; b, colony form.)

Obelia geniculata, knotted threadhydroid†, zig-zag wine-glass hydroid. Unbranched colony grows in alternating, zigzag pattern, hydranths lean first to the left, then to the right. Tentacles 0.7 mm long. Bo,Ha–Al,Li–Su,ff,25.4 mm (1 in), (C303,F158,NM348), ☺. (FIG. 8.30. a, portion of colony with gonozooid; b, colony form.)

Obelia (=*Clytia*) *longicyatha.* Vi,Ha,Es–Su,ff,25.4 cm (10 in) [7.6 cm (3 in)], (F142). (FIG. 8.31 a. portion of colony with gonozooids; b. colony form)

Obelia longissima (=*commissuralis*), bushy wine-glass hydroid†. Whitish; abundant branching. Can form large, bushy colonies. Bo,Ha,Li–Su,ff,20.3 cm (8 in), (C304,F154,G80), ☺. (FIG. 8.19)

8.14 **Clavidae** (see 8.8)

8.15 **Corynidae** (see 8.8)

8.16 **Eudendriidae, Stickhydroids** (from 8.8)

Hydroids form bushy colonies or erect, branching stems; no thecal cup for hydranths to withdraw into, but perisarc covers stalk to base of hydranth; with 20 to 30 tentacles per hydranth; do not release medusae.

1. Stems fasciculated (intertwined and partially fused together)
   - A. Stems *fasciculated* fa
   - B. Fasciculation *absent* x

2. Color of hydranths
   - A. *Red* re
   - B. *Pink* pi
   - C. *White* wh
3. Size of colony
   - A. Colony *small*, less than 1.25 cm (½ in) tall sm
   - B. Colony *large*, less than 2.5 cm (1 in) tall lg

8.17

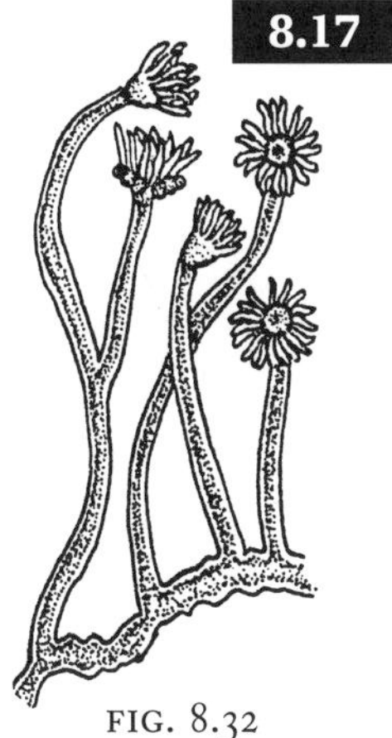
FIG. 8.32

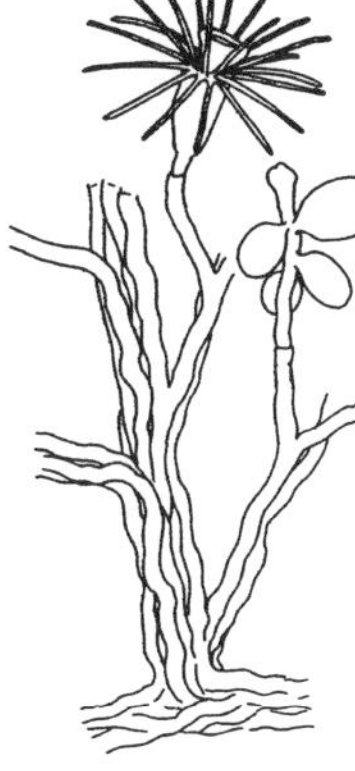
FIG. 8.33

Stems	Color	Size	Species
fa	pi	la	*Eudendrium ramosum*, stickhydroid†
fa	re	la	*E. carneum*, red stickhydroid†
x	wh	sm	*E. album*, white stickhydroid†

Eudendrium album, white stickhydroid†. Seldom more than two hydranths per stalk. 26–32 tentacles. Vi+,Ha,Su,ff,12.7 mm (0.5 in), (F61,NM345). (FIG. 8.32)

Eudendrium carneum, red stickhydroid†. Hydranths red. Branched; 24 tentacles. Bo,Ha,Su,ff,12.7 cm (5 in), (F64,G79,NM345). (FIG. 8.33. Colony with fascicled stems.)

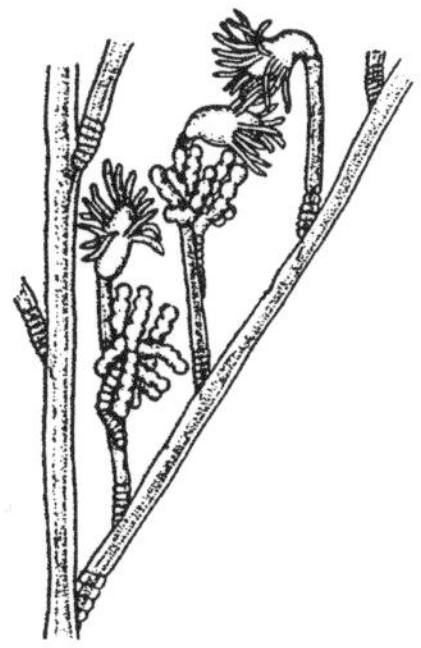

Eudendrium ramosum, stickhydroid†. White or pink with greenish tones. Branching mostly in one plane. About 20 tentacles per polyp. Bo,Ha,Su,ff,15.2 cm (6 in), (F72,G79,NM345). (FIG. 8.34)

FIG. 8.34

8.18 **Haleciidae** (see 8.6)

8.19 **Hydractiniidae, Snail Fur** (from 8.8)

Small, encrusting hydroids, often with polymorphic members; no perisarc but often with hard pointed basal spines. Typically on gastropod shells.

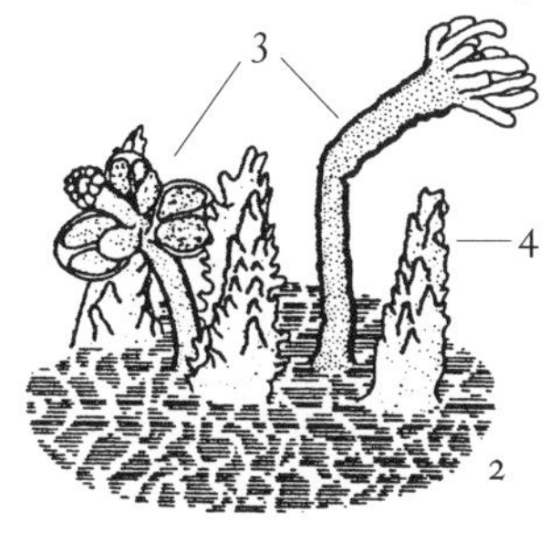

FIG. 8.35

1. Number of tentacle whorls: provide *number* (1 or 2) of tentacle whorls (FIG. 8.35) __
2. Solid, matlike base (FIG. 8.35)
   - A. *Mat*like base present ma*
   - B. Matlike base *absent* x
3. Strongly polymorphic hydranths (2 morphologically distinct types of individuals, not simply differing in size) (FIG. 8.35)
   - A. Colony strongly *polymorphic* po*
   - B. Polymorphic colony members *not* conspicuously different; all individuals similar but differ in size x

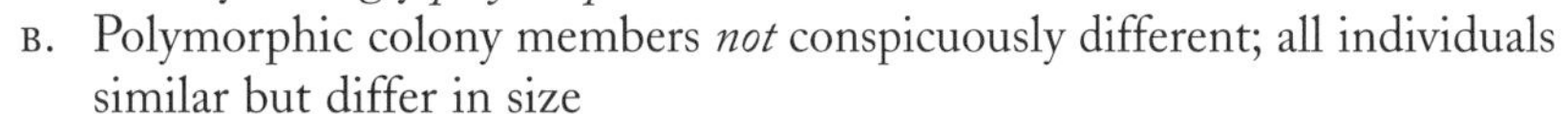

4. Texture of spines projecting from basal layer
   - A. Basal spines *rough* textured ro*
   - B. Basal spines *smooth* textured sm
5. Habitat
   - A. Found on gastropod shells inhabited by *hermit crabs* hc
   - B. Found in shell of live *Nassarius*, mud snail na
   - C. Found attached to *other* types of hard objects ot

8.20

Whorls	Mat	Polymorph	Spines	Habitat	Species
1	ma	po	ro	hc,ot	*Hydractinia echinata*, snail fur†
1	ma	x	sm	hc,na,ot	*Podocoryna* (=*Podocoryne*) *carnea*
2	x	x	sm	na,ot	*Stylactaria* (=*Stylactis*) *arge*

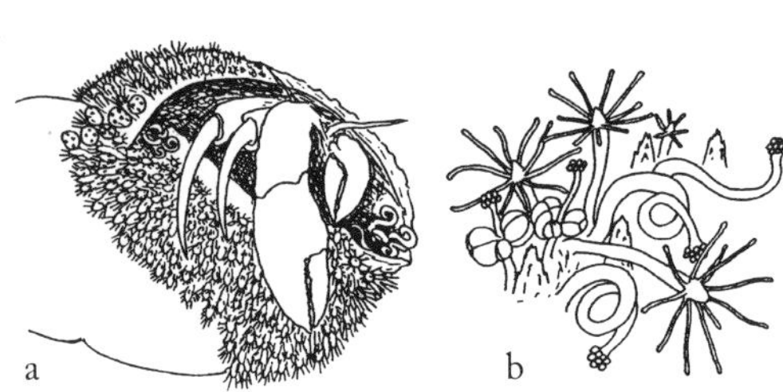
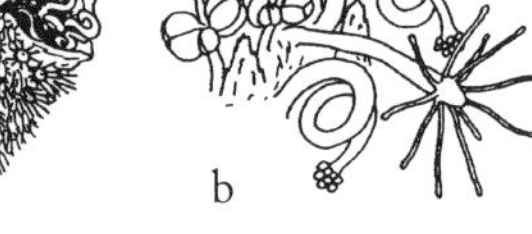

FIG. 8.36

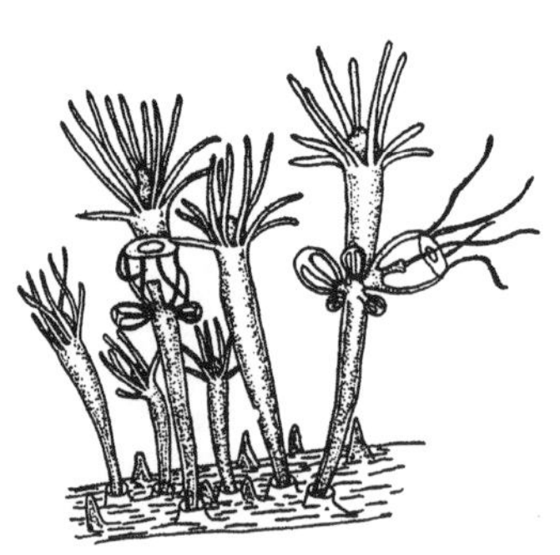

FIG. 37

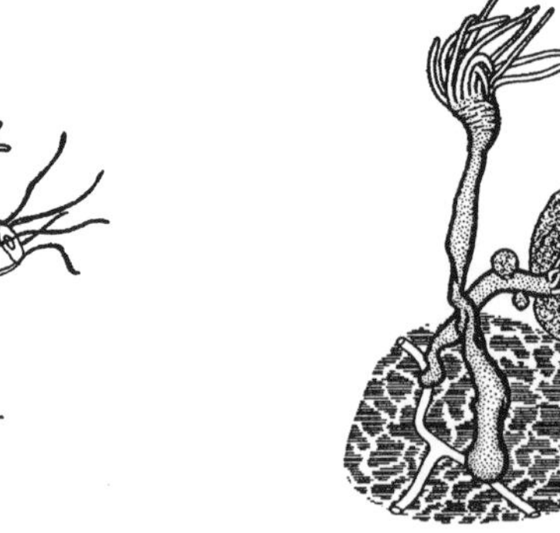

FIG. 38

Hydractinia echinata, snail fur†. White to salmon color, hard spiny base. Includes 15–20 tentacled feeding gastrozooids, reproductive gonozooids with medusa buds, and long, coiled, offensive and defensive spiral zooids. Bo,Ot,Li–Su,ff,7.6 mm (0.3 in), (C295,F78,G77,NM343). (FIG. 8.36. a, growth on shell occupied by hermit crab; b, colony enlarged.)

Podocoryna (=*Podocoryne*) *carnea.* Much less common than *Hydractinia.* Often on shells of *Ilyanassa trivittata.* Ac–,Ha,Li–Su,ff,5.1 mm (0.2 in), (F82,G77). (FIG. 8.37. Colony with medusa buds.)

Stylactaria (=*Stylactis*) *arge.* Especially on shells of living mud snails, *Ilyanassa obsoleta.* Vi,Ha,Su,ff,20.3 mm (0.8 in), (C295,F83). (FIG. 8.38)

8.21 **Plumulariidae** (see 8.6)

8.22 **Sertulariidae, Fern Garland Hydroids** (from 8.6)

Sturdy, upright colonies with paired or alternating tubular hydranths directly attached to main stalks; *hydrothecae* (perisarc covering on feeding members) often with pointed margin and one or several opercular flaps. Shape of *gonotheca* (perisarc covering on scattered, seasonal reproductive members) is also used. Sertularian hydroids are relatively common, often attached to intertidal rockweeds or subtidal kelps.

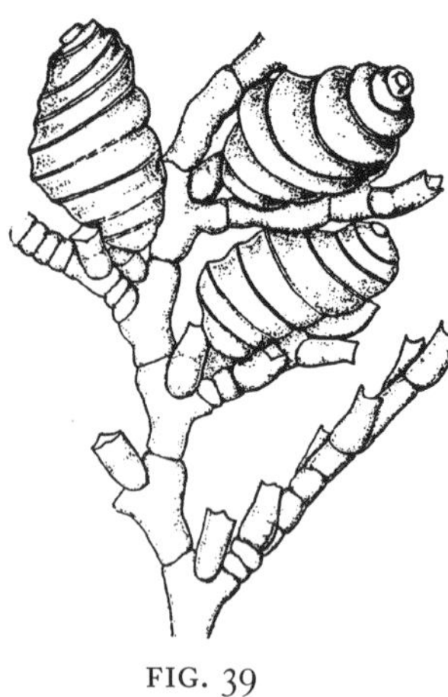

FIG. 39

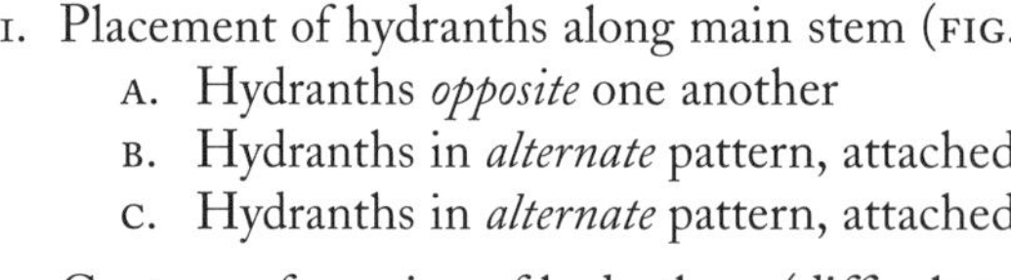

1. Placement of hydranths along main stem (FIG. 8.39)
   - A. Hydranths *opposite* one another — op
   - B. Hydranths in *alternate* pattern, attached along *both* sides of stalk — ab*
   - C. Hydranths in *alternate* pattern, attached along a *single* face of the stalk — as
2. Contour of opening of hydrotheca (difficult to see; look at fresh, clean specimens; beware of old eroded hydrothecae) (FIG. 8.39)
   - A. Opening *smooth* — sm
   - B. Opening rim with *two* raised *teeth* — 2te
   - C. Opening rim with *three* raised *teeth* — 3te*
   - D. Opening rim with *four* raised *teeth* — 4te
3. Shape of gonotheca (FIG. 8.39)
   - A. Gonotheca basically *smooth* outline — sm
   - B. Gonotheca distinctly *ribbed* outline — ri*
4. Colony size (this character is most useful for hydroids exceeding minimal sizes listed; young colonies of larger forms can be misleading)
   - A. Colony *small,* less than 3.7 cm (1 ½ in) tall — sm
   - B. Colony *large,* more than 7.5 cm (3 in) tall — la
   - C. Colony *intermediate,* 3.7 cm (1 ½–3 in) tall — in

8.23

Place	Opening	Shape	Size	Species
op	2te,sm	sm	in	*Dynamena (=Sertularia) pumila,* sea oak†
op	2te	ri	sm	*D. (=Sertularia) cornicina,* golden garland hydroid
ab	sm	sm	la	*Abietinaria abietina,* sea fur†
ab	2te	sm	la	*Sertularia (=Thuiaria) argentea,* silver garland hydroid or white hair
ab	3te	ri	la	*Symplectoscyphus (=Sertularella) tricuspidatus*
ab	4te	ri	sm	*Sertularella rugosa,* snail trefoil hydroid†
ab	4te	ri	la	*Sertularella polyzonias,* great tooth hydroid†
as	sm	sm	la	*Hydrallmania falcata,* sickle hydroid†

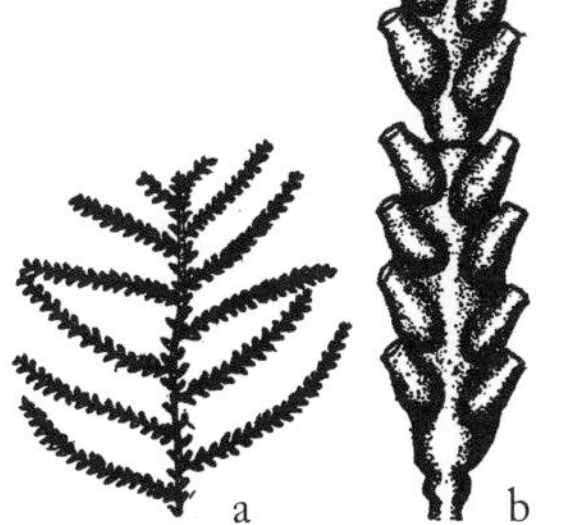

FIG. 8.40

Abietinaria abietina, sea fur†. Pinnate colonies; all branches in a single plane and somewhat zigzag. Ac,Ha,Su,ff,30.5 cm (12 in), (F238,NM352). (FIG. 8.40. a, colony form; b, portion of colony.)

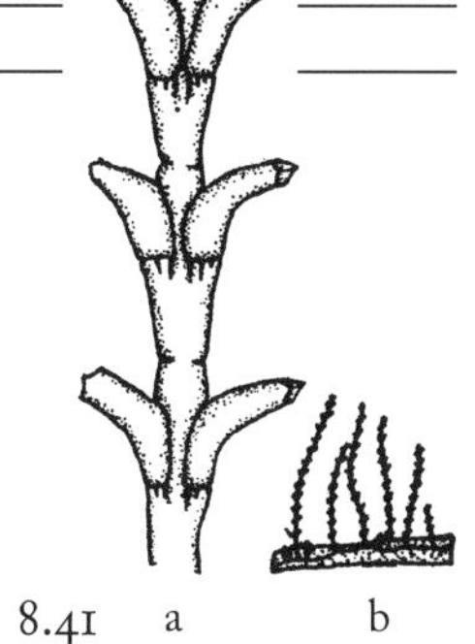

FIG. 8.41

Dynamena (=Sertularia) cornicina, golden garland hydroid. Stolonlike base along seagrass blades, gives rise to upright branches with hydranths. Ca+,SG,Es–Su,ff,12.7 mm (0.5 in), (F279,G85). (FIG. 8.41. a, portion of colony; b, colony form.)

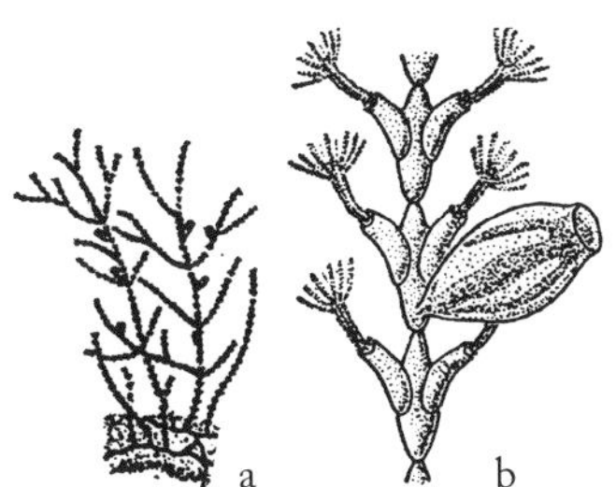

FIG. 8.42

Dynamena (=Sertularia) pumila, sea oak†. Stiff tannish colonies; branching in one plane. Common on brown algae. Bo,Ha–Al,Es–Li–Su,ff,5.1 cm (2 in), (C305,F286,G84,NM352), ☺. (FIG. 8.42. a, colony form; b, portion of colony with gonozooid.)

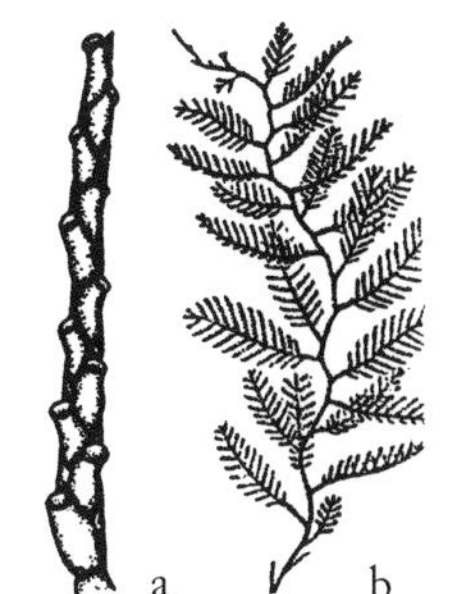

FIG. 8.43

Hydrallmania falcata, sickle hydroid†. Gonotheca with four denticles visible on inside wall near its opening. Spirally twisted stem; base of hydrothecae slightly swollen. Ac–,Ha–Al,Su–De,ff,30.5 cm (12 in), (C305,F250). (FIG. 8.43. a, portion of colony; b, colony form.)

FIG. 8.44

Sertularella polyzonias, great tooth hydroid†. Hydrothecae wider at base than tip; regularly branched. Ac–,Ha,Li–De,ff,15.2 cm (6 in), (C397,F268). (FIG. 8.44. a, colony form; b, portion of colony; c, gonozooid.)

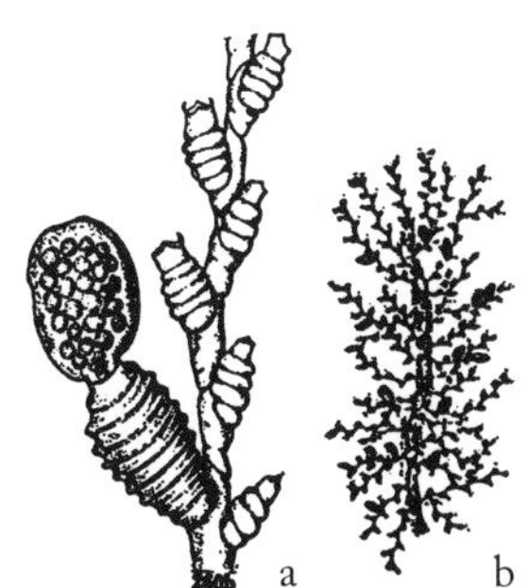

FIG. 8.45

Sertularella rugosa, snail trefoil hydroid†. Thick hydrothecae; sparsely branched. Ac–,Ha–Al,Su,ff,25.4 mm (1 in), (F271), ☺. (FIG. 8.45. a, portion of colony with medusa emerging from gonozooid; b, colony form.)

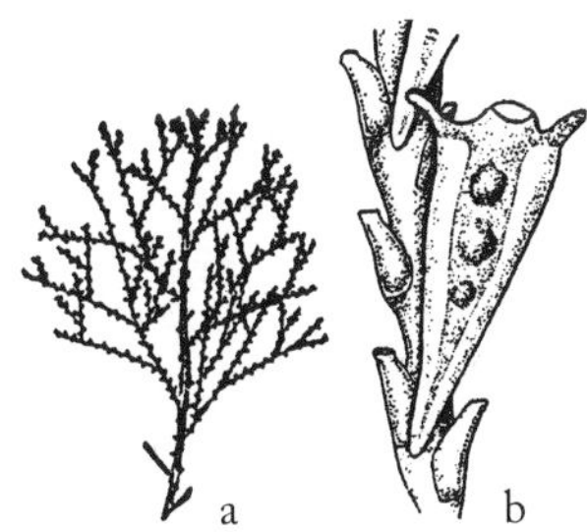

FIG. 8.46

Sertularia (=Thuiaria) cupressina (=argentea), silver garland hydroid or white hair. All branching in a single plane; feathery. Dense, whitish to silvery growths on solid objects. Abundant in Chesapeake Bay during winter. Sometimes dried and died green for sale as "sea fern." Bo,Ha,Li–De,ff,30.5 cm (12 in), (F293). (FIG. 8.46. a, colony form; b, portion of colony with gonozooid.)

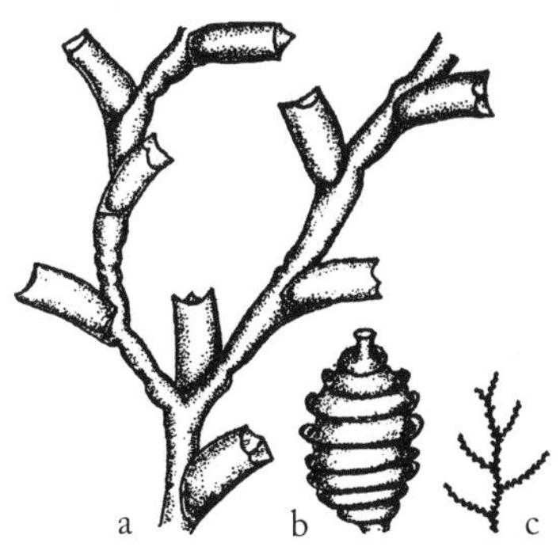

FIG. 8.47

Symplectoscyphus (=Sertularella) tricuspidatus. Hydrothecae smooth; branching in one plane. Ac–,Ha,Li–De,ff,15.2 cm (6 in), (C307,F274). (FIG. 8.47. a, portion of colony; b, gonozooid; c, colony form.)

8.24 **Tubulariidae** (from 8.8)

Hydroids often form large, colorful colonies on pilings or under intertidal and subtidal rocks. Their tangled stalks provide a convenient refuge for a mini-community of invertebrates including a variety of nudibranch gastropods, pycnogonid sea spiders and tiny water mites, polychaete worms, and several types of juvenile gastropods and bivalves. Positive identification requires attention to the shape of seasonally limited gonophores (character 4). If the number of tentacles on largest members is consistently smaller than the smallest numbers offered here, the colony may be immature and a positive identification cannot be made.

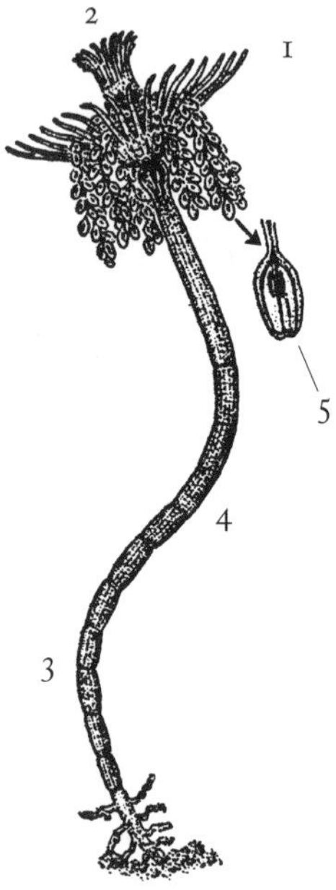

FIG. 8.48

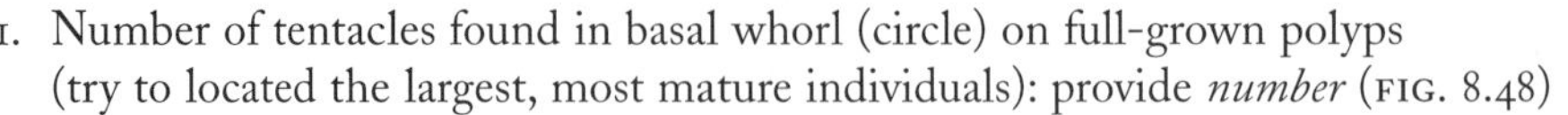

1. Number of tentacles found in basal whorl (circle) on full-grown polyps (try to located the largest, most mature individuals): provide *number* (FIG. 8.48) __

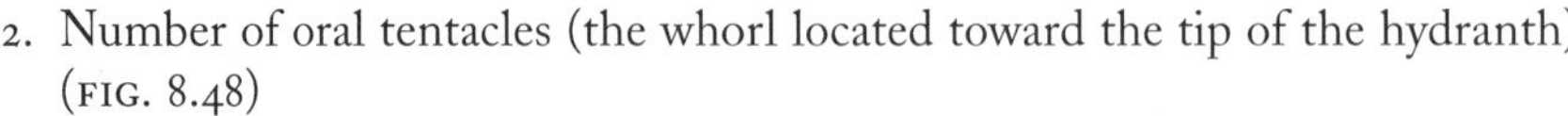

2. Number of oral tentacles (the whorl located toward the tip of the hydranth) (FIG. 8.48)

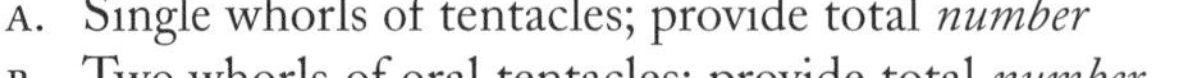

   A. Single whorls of tentacles; provide total *number* __
   B. Two whorls of oral tentacles; provide total *number* 2× __
3. Extent of branching among the stalks, exclusive of their bases (FIG. 8.48)

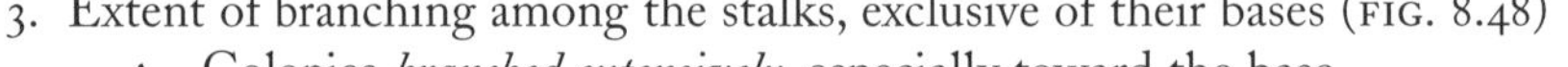

   A. Colonies *branched extensively,* especially toward the base — be
   B. Branching sparse or *absent* — x*
4. Texture of stalk (FIG. 8.48)
   A. Stalk with occasional but distinct *wrinkles* and/or annulations — wr
   B. Stalk basically *smooth,* or only a few wrinkles — sm
5. Shape of tip of gonophore (this character is not always available; clusters of gonophores appear seasonally on the hydranth between the basal and oral tentacle whorls) (FIG. 8.48)
   A. Apical tip of gonophore *conical* — co
   B. Apical tip of gonophore *flattened* — fl
   C. Apical tip on gonophore *absent* — x

8.25

Basal Tentacles	Oral Tentacles	Branches	Stalk	Gonophore Tip	Species
25–40	20–30	x	sm	x	*Tubularia indivisa* (=*couthouyi*), tall tubularian†
20–25	35	be	wr	co	*Ectopleura* (=*Tubularia*) *larynx* (=*tenella*, =*spectabilis*), ringed tubularian†
20–25	20–25	x	sm	fl	*E.* (=*Tubularia*) *crocea,* pink-hearted hydroid
30	2 × 24–25	x	wr	x	*E. dumortieri,* tube hydroid
16–30	2 × 16	x	wr	x	*Hybocodon prolifer*

FIG. 8.49

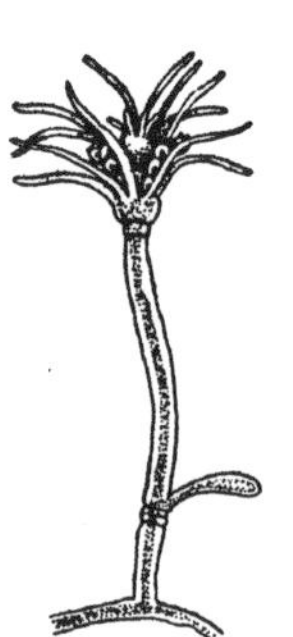

FIG. 8.50

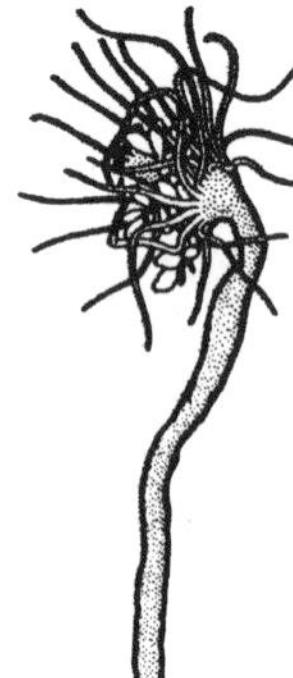

FIG. 8.51

Ectopleura (=*Tubularia*) *crocea,* pink-hearted hydroid. Rose-colored hydranths, tangled base. More common south of Cape Cod. Bo,Ha,Li–Si,ff,12.7 cm (5 in), (F97,G75,NM341). (FIG. 8.49)

Ectopleura dumortieri, tube hydroid. Often solitary. Grapelike gonophores absent; produces free-swimming medusae. Vi,Ha,Es–Su,ff,5.1 cm (2 in), (F92). (FIG. 8.50)

Ectopleura (=*Tubularia*) *larynx* (=*tenella*, =*spectabilis*), ringed tubularian†. Stalk is colorless; hydranths pink. Ac–,Ha,Su,ff,5.1 cm (2 in), (C293,F99,NM341), ☺. (FIG. 8.51)

FIG. 8.52

Hybocodon prolifer. Hydranths, orange; stalk, red-orange. Small colony, sometimes solitary. In tidal pools. Ac–,Ha,Li,ff,5.1 cm (2 in), (F106,G75). (FIG. 8.52)

Tubularia indivisa (=*couthouyi*), tall tubularian†. Stalks become entangled near base. Hydranths pink-orange; gonophores red; stalk yellow. May have longitudinal striations on stalk. Ac–,Ha,Li–De,ff,30.5 cm (12 in) [15.2 cm (6 in)], (F98,NM341). (FIG. 8.53)

FIG. 8.53

8.26 Unrelated Groups Superficially Similar to Hydroids
(from 8.8)

The phyla, *Entoprocta* and *Ectoprocta*, are similar in general appearance to polyp stages of Cnidaria in that they possess a tentacle crown atop a thin body attached to the substrate. In fact, members of both these phyla are considerably more complexly constructed and use water currents created by tentacular cilia to filter feed. Regional entoprocts produce creeping stolons that give rise periodically to single, upright, narrowly stalked individuals ("zooids"). The fact that their mouth and anus both occur within the ring of tentacles (the origin of the phylum name), along with other anatomical distinctions, separate entoprocts from ectoprocts, which have the anus outside the tentacle ring. Small size requires observation using a dissecting microscope.

1. Growth form
   - A. *Stolon* (runner) gives rise to spatially separated, isolated, individual members — st
   - B. Growth form *otherwise*, including branching colonies or close-packed attached individuals — ot
2. Individual zooid shape
   - A. Bulbous, tentacle-topped portion connects to basal stolon by narrow *stalk* — st
   - B. Shaped *otherwise* — ot
3. Pattern of ciliary current flow (compound microscope)
   - A. Currents pass from *outside* tentacle crown toward its center (i.e., particles would be trapped on outer surface of tentacles) — ou
   - B. Currents pass from *inside* tentacle crown toward the outside (i.e., particles would be trapped on inside surface of tentacles) — in

8.27

Form	Shape	Currents	Group
st	st	ou	Entoprocta, 8.28
st	ot	in	Ectoprocta, 12.1
ot	ot	in	Ectoprocta, 12.1

8.28 Entoprocta (from 8.27)

Tentacles tend to curl and uncurl; animals can bend their stalks voluntarily when disturbed.

1. Shape of upright stalk (FIG. 8.54)
   - A. Stalk with distinctly *swollen* base — sw*
   - B. Stalk with projecting *spicules* — sp
2. Rigidity of stalk (disturb an individual, perhaps with a very fine pin) (FIG. 8.54)
   - A. Stalk rather *stiff* — st
   - B. Stalk clearly *flexible* — fl

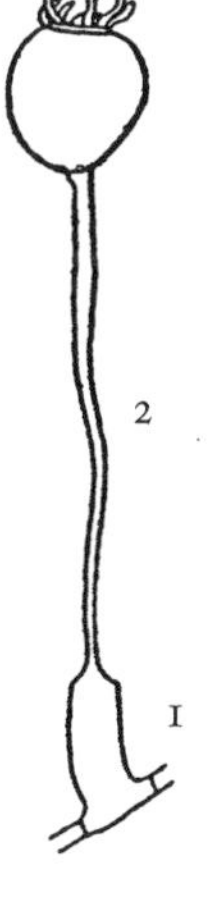

FIG. 8.54

8.29

Base	Rigidity	Species
sw	st	*Barentsia major,* thick-based entoproct
sw	fl	*B. laxa,* thick-based entoproct
x	fl	*Pedicellina cernua,* bowing entoproct

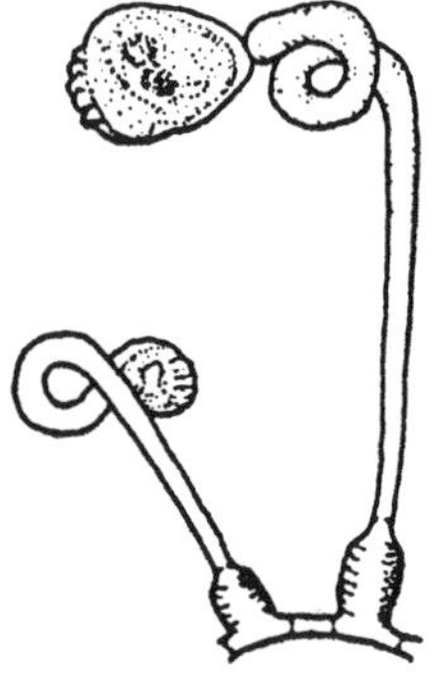

FIG. 8.55

FIG. 8.56

Barentsia laxa, thick-based entoproct (Barentsiidae). Forms patches on hard substrates, including bivalve shells. Vi,Ha,Su,ff,7.6 mm (0.3 in), (G110,NM718). (FIG. 8.55)

Barentsia major, thick-based entoproct (Barentsiidae). On hard substrates; reported from legs of spider crabs and horseshoe crabs. Ac,Ha,Su,ff,10.1 mm (0.4 in), (G110,NM718).

Pedicellina cernua, bowing entoproct (Pedicellinidae). On hard substrates, often attached to ectoprocts. Bo,Ha,Es–Li–,ff,7.6 mm (0.3 in), (G110,NM718). (FIG. 8.56)

8.30 Class Hydrozoa and Class Scyphozoa, Jellyfish

Jellyfish are fragile and are easily damaged in captivity or preservation. It is best to make careful observations while the organism is alive, in the field or in a dish of fresh, cold seawater. Look closely for a thin, shelflike *velum* membrane that may surround the underside of the *bell* in *hydromedusae* (jellyfish of Hydrozoa), but which is absent from *scyphomedusae* (scyphozoan jellyfish). Be careful handling any medusa larger than 4 in diameter. Often there is a *manubrium* (handlelike projection of tissue, sometimes taking the form of four or more *oral lobes*) hanging beneath the bell and including the animal's mouth. Their stinging cells (*cnidae* or *nematocysts*) may be capable or inflicting a painful, though not especially dangerous, sting. Measurements given refer to diameter of bell.

Additional reference: (s) Shih, C. T. 1977.

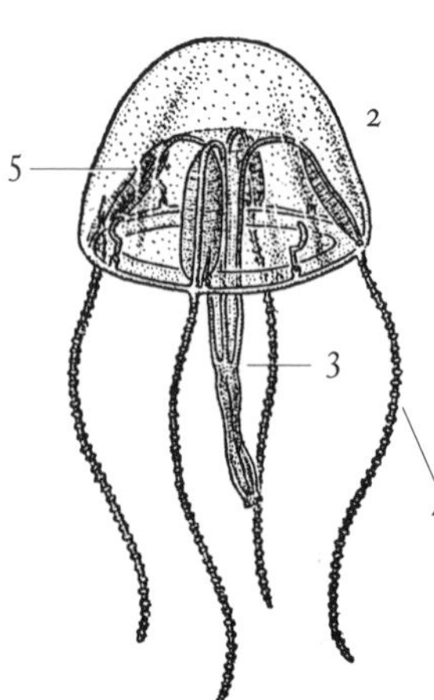

FIG. 8.57

1. Presence of velum (shelf-like membrane) extending toward the interior from the margin of the bell (FIG. 8.57)
   A. *Velum* present — ve*
   B. Velum *absent* — x

8.31

Velum	Group
ve	Cl: Hydrozoa (Hydromedusae), 8.32
x	Cl: Scyphozoa (Scyphomedusae), 8.36

8.32 Hydrozoan Jellyfishes (from 8.31)

The free-swimming, sexually mature, medusa stage in hydrozoan life histories includes *anthomedusae* (cup-shaped jellyfish typical of the hydroids in the suborder Anthoathecata), and *leptomedusae* (dish-shaped jellyfish produced by hydroids of the suborder Leptothecata). Usually the gastrovascular cavity involves a network of tubes, often including four (sometimes more) *radial canals* that extend from the aboral apex of the bell to a *ring canal* that follows the bell's perimeter.

FIG. 8.58

1. Approximate diameter of bell (this character is most useful if the specimen exceeds sizes given; obviously juvenile specimens can be much smaller than maximal sizes listed here): provide *diameter* in cm (inches) — __
2. General shape of jellyfish (FIG. 8.58)
   A. Bell *cup*-shaped (width less than 1.5 times its height) — cu*
   B. Bell *dish*-shaped (width more than 1.5 times its height) — di
3. Presence of lobes or other nontentacle structures below margin of the bell (FIG. 8.58)
   A. Such *lobes* present — lo*
   B. Such lobes as may be present do *not* extend below the bell margin — x

4. Approximate number of tentacles around the margin of the bell (tentacles are easily broken off in damaged specimens) (FIG. 8.58)
   A. *More than 100* tentacles >100
   B. If fewer than 100, provide *number* —
5. Number of radial canals extending from the apex of the bell to its periphery: provide *number* of radial canals (FIG. 8.58) —

8.33

Size (cm)*	Shape	Lobe	No. of Tentacles	Radial Canals	Organism
30.5 (12 in)	di	x	>100	4	*Staurophora mertensi,* whitecross jellyfish†
13.4 (6 in)	cu,di	x	50–300	>20	*Aequorea aequorea,* many-ribbed jellyfish
10.2 (4 in)	cu	lo	30–35	4	*Halitiara* (=*Tima*) formosa, elegant hydromedusa
3.5 (1.3 in)	cu	x	50–300	4	*Tiaropsis multicirrata* (=*diademata*), blackeye hydromedusa
3 (1.2 in)	cu	lo	4	4	*Liriope tetraphylla*
3 (1.2 in)	cu	x	50–100	8	*Aglantha digitalis*
2.5 (1 in)	cu	x	12–48	4	*Catablema vesicarium,* constricted jellyfish†
2.6 (1 in)	cu	x	60–80	4	*Gonionemus vertens,* clinging jellyfish†, angled hydromedusa
<2.6 (<1 in)	cu,di	x,lo	0–100	4	Small hydromedusae, 8.34

* Number in parentheses is the size in inches.

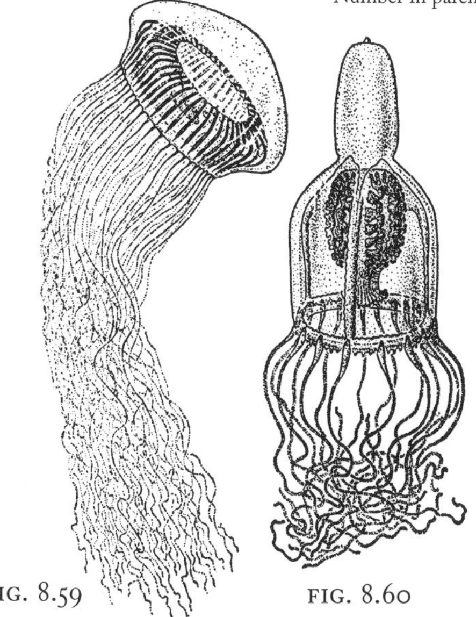

FIG. 8.59

FIG. 8.60

Aequorea aequorea, many-ribbed jellyfish (Aequoreidae). Eighty or more white, radial canals visible through its thick jelly. Tentacles easily broken on damaged specimens, which are found washed ashore as gelatinous disks. Luminescent. Bo,Pe,Ps,pr,17.8 cm (7 in), (G83,NM346). Other species of *Aequorea* may occur in northern waters (see S37). (FIG. 8.59)

Aglantha digitale (see 8.35)

Catemblema vesicarium, constricted jellyfish† (Pandeidae). Distinctive bulbous aboral projection; tentacle bases golden with red ocelli (eyespots). Ac,Pe,Ps,pr,25.4 mm (1 in), (G86,NM345). (FIG. 8.60)

Gonionemus vertens, clinging jellyfish†, angled hydromedusa (Olindiidae). With suckers part way along tentacle; clings to surfaces, leaving tentacle tips bent. Colored parts orange to reddish. Ac,Pe–SG,Ps,pr,25.4 mm (1 in), (G74,NM355). (FIG. 8.61)

FIG. 8.61

Halitiara (=Tima) formosa, elegant hydromedusa (Pandeidae). Long, frilled manubrium (oral lobes) in adults. Bo,Pe,Ps,pr,10.1 cm (4 in), (G84,NM350,S39). (FIG. 8.62)

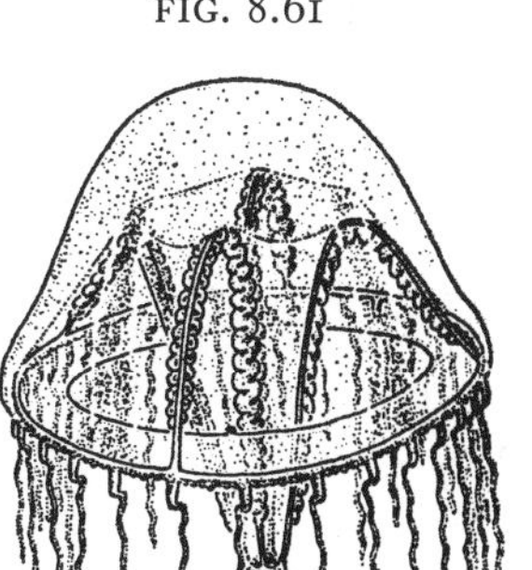

FIG. 8.62

Liriope tetraphylla (see 8.35)

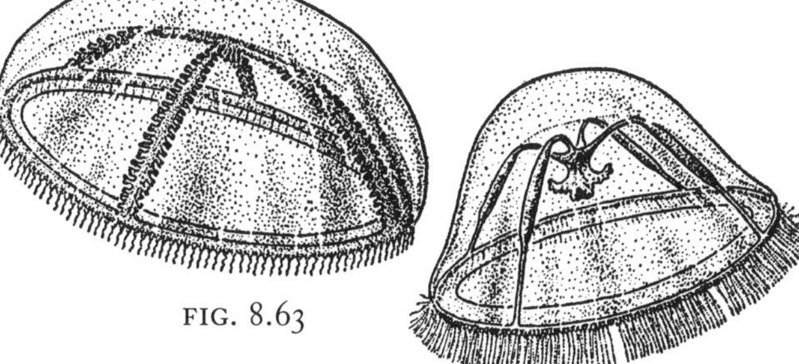

FIG. 8.63

FIG. 8.64

Staurophora mertensi, whitecross jellyfish† (Laodiceidae). Radial canals of gastrovascular cavity form a conspicuous X, viewed from above. Several thousand tentacles on large specimens. Nearer surface at night, deeper by day. Eats crustacea and medusae. Ac,Pe,Ps,pr,30.5 cm (12 in) [22.9 cm (9 in)], (G83,NM349,S37). (FIG. 8.63)

Tiaropsis multicirrata (=diademata), blackeye hydromedusa (Tiaropsidae). Black ocelli (eye spots) alternate with tentacles around margin. Ac,Pe,Ps,pr,30.5 mm (1.2 in), (G82,S40). (FIG. 8.64)

8.34 **Small Hydromedusae** (from 8.33)

Many velum-bearing hydromedusae approach or exceed the minimal size limit set for this guide, that is, visible to the unaided eye. Since hydromedusae often are conspicuous and important members of plankton communities, however, a key to the most common types is provided for those equipped to observe these small jellyfish. A dissecting or compound microscope may be necessary.

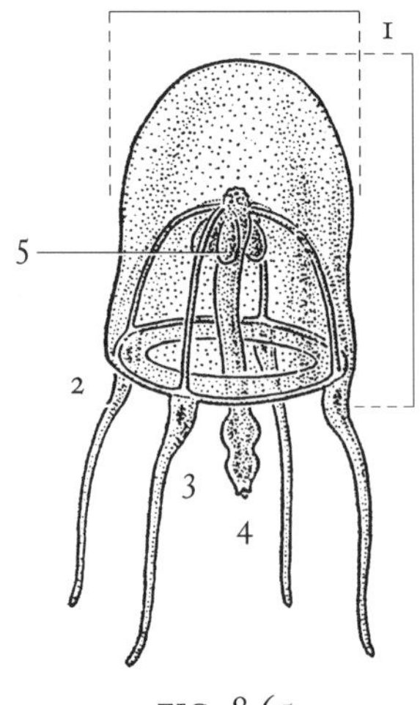

FIG. 8.65

1. Relative proportions of medusa (FIG. 8.65)
   - A. Medusa *tall,* height clearly much greater than width — ta*
   - B. Medusa *wide,* width clearly much greater than height — wi
   - C. Medusa more or less *equal* in dimensions — eq
2. Placement of tentacles around the bell margin (FIG. 8.65)
   - A. Tentacles deployed individually around the margin; provide their *number* — __*
   - B. Tentacles in clusters around the margin; provide the *number* of *clusters*: and the *number* of tentacles per cluster — __c:__
3. Relation of manubrium (oral lobes) to lower margin of the bell (FIG. 8.65)
   - A. *Manubrium* extends well beyond the lower margin of the bell — ma*
   - B. Manubrium does *not* extend beyond the lower margin of the bell — x
4. Oral tentacles, associated with the mouth at the end of the manubrium (FIG. 8.65)
   - A. Manubrium with distinct *tentacles* present — te
   - B. Manubrium simple, with distinct tentacles *absent* — x*
5. Location of gonads (usually pigmented: brown, orange, or greenish) (gonads may be seasonal; if no gonads are observed, ignore this question)
   - A. Gonads associated with the *radial canals* (found in four groups following the axis of the inside of the bell) — rc
   - B. Gonads associated with the *manubrium* (attached to the oral lobes hanging from the center of the bell) — ma

8.35

Shape	Tentacles	Manubrium Bell	Oral Tentacles	Gonads	Species
wi	4	ma	x	rc	*Liriope tetraphylla*
wi	16–34	ma,x	x	rc	Gonads in round clusters along radial canals: *Obelia* spp. OR Gonads arranged in elongate fashion along radial canals: *Clytia* (=*Phialidium*) spp.
ta	4	ma	x	ma	*Sarsia tubulosa,* clapper hydromedusa†
ta	to 100	x	x	rc	*Aglantha digitalis*
ta,eq	1c:3	x	x	rc	*Hybocodon prolifer*
ta,eq	4c:26	x	te	rc	*Nemopsis bachei*
ta,eq	8c:2–5	x	te	rc	*Rathkea octopunctata*
eq	4c:3	x	te	ma	*Bougainvillia rugosa*
eq	4c:6–9	x	te	ma	*Bougainvillia carolinensis*
eq	2–4	x	x	rc	*Sphaerocoryne* (=*Linvillea*) *agassizi*
eq	4	ma	x	rc	Tentacles longer than manubrium: *Liriope tetraphylla* OR Tentacles shorter than manubrium: *Sarsia tubulosa*
eq	4	x	te	ma	*Podocoryna* spp.
eq	4	x	x	rc	*Ectopleura dumortieri*
eq	16–20	x	x	rc	Eucheilota ventricularis
eq	80–90	x	x	ma	*Turritopsis nutricula*
eq	x	x	x	ma	*Pennaria* (=*Halocordyle*) *disticha*

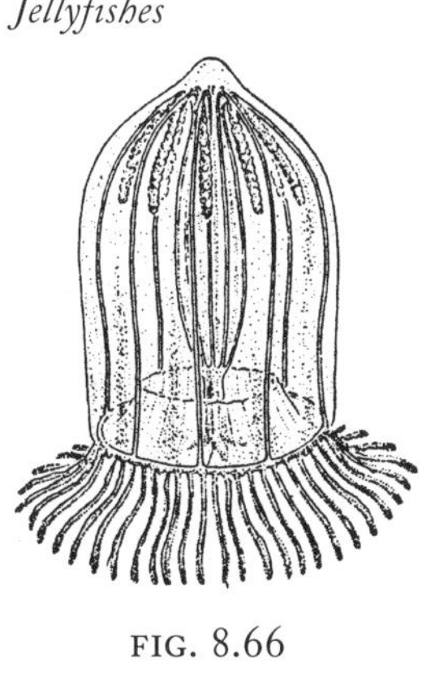
FIG. 8.66

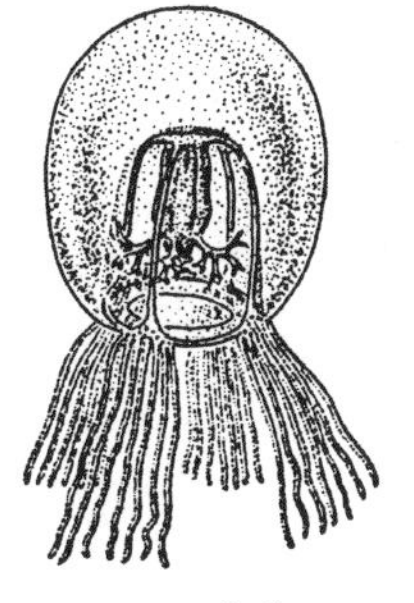
FIG. 8.67

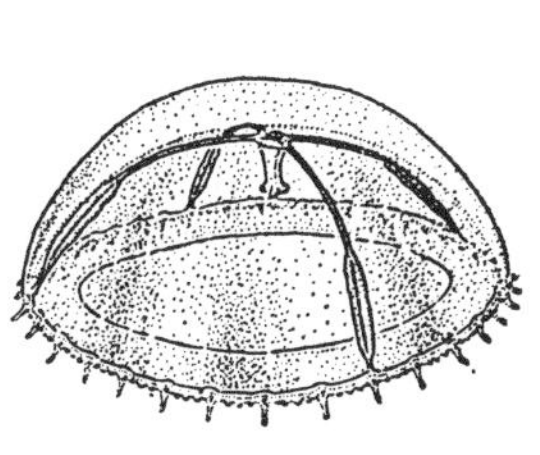
FIG. 8.68

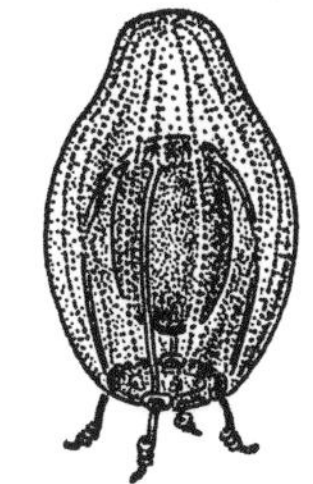
FIG. 8.69

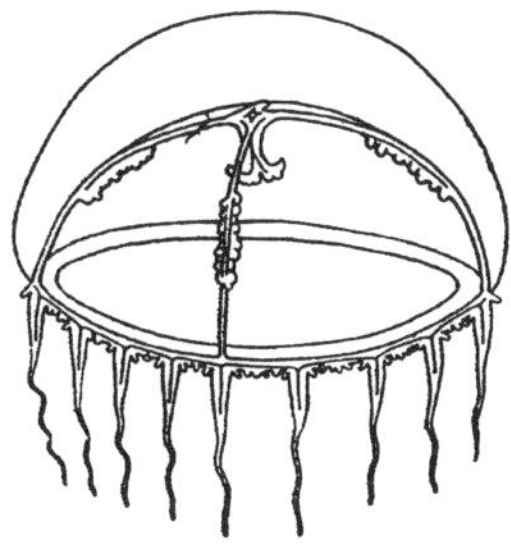
FIG. 8.70

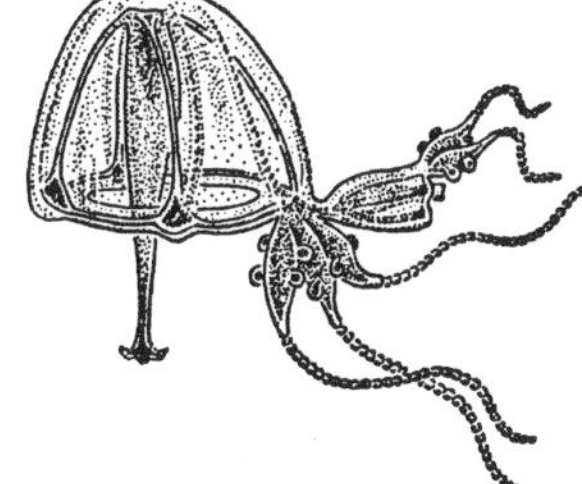
FIG. 8.71

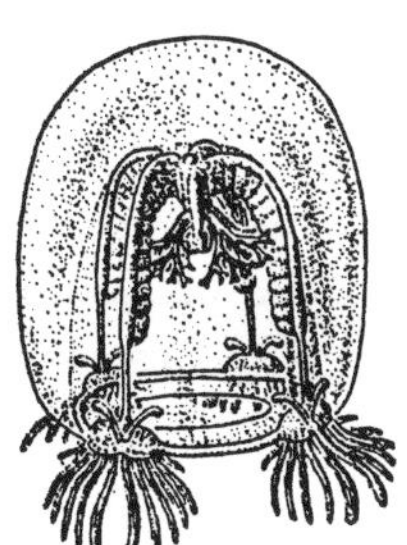
FIG. 8.72

FIG. 8.73

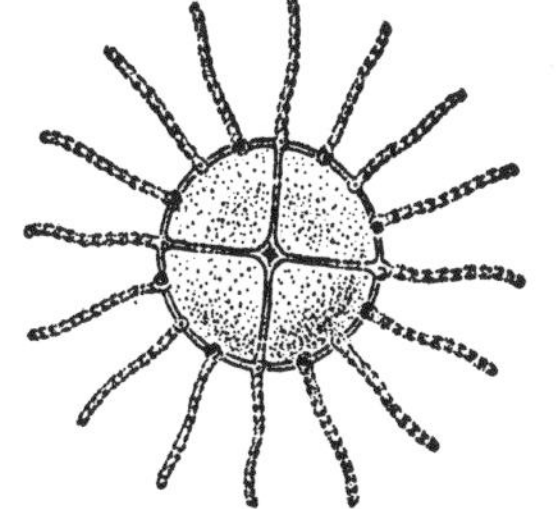
FIG. 8.74

FIG. 8.75

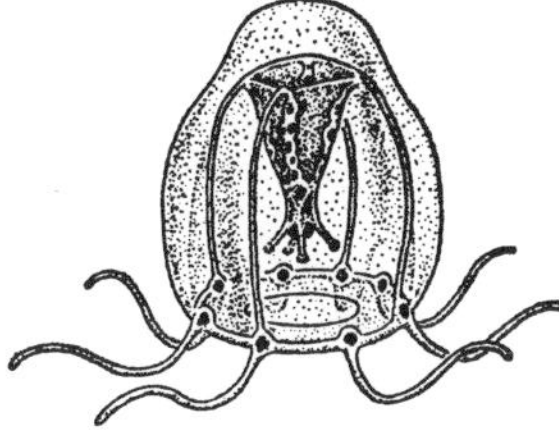
FIG. 8.76

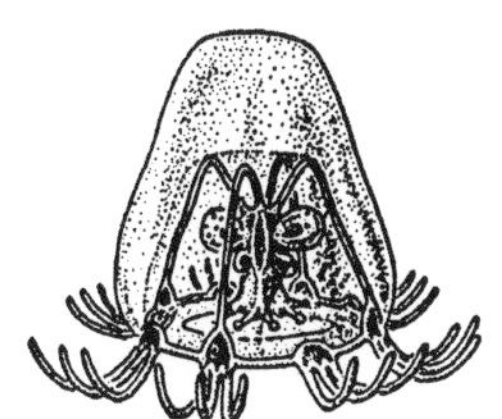
FIG. 8.77

Aglantha digitalis (Rhopalonematidae). Small apical dome; 8 radial canals. Ac–,Pe,Ps,pr,3 cm (1.25 in), often smaller, (c76,s41). (FIG. 8.66)

Bougainvillia carolinensis (Bougainvilliidae). Each oral tentacle divides up to four times. Vi,Pe,Es–Ps,pr,5.1 mm (0.2 in), (c35,s32). (FIG. 8.67)

Bougainvillia rugosa (Bougainvilliidae). Thick jelly layer. Oral tentacles present but not branched. Vi,Pe,Es–Ps,pr,<2 mm (0.1 in), (c35,s32).

Clytia (=*Phialidium*) spp. (Campanulariidae). Bo,Pe,Ps,pr,<2 mm (0.1 in), (s38). (FIG. 8.68)

Ectopleura dumortieri (Tubulariidae). Spherical. Vi,Pe,Es–Ps,pr,<2 mm (0.1 in), (c23). (FIG. 8.69)

Eucheilota ventricularis (Lovenellidae). Bo,Pe,Es–Ps,pr,12.7 mm (0.5 in), (c61). (FIG. 8.70)

Hybocodon prolifer (Tubulariidae). Only one of four bulbs around margin develops tentacles. Ac–,Pe,Ps,pr,<2 mm (0.1 in), (c24,s30). (FIG. 8.71)

Liriope tetraphylla (Geryoniidae). Generally offshore. Vi+,Pe,Es–Ps,pr,3 cm (1.25 in), often smaller, ☺. (FIG. 8.72)

Nemopsis bachei (Bougainvilliidae). Especially common toward southern portion of range. Bo,Pe,Es–Ps,pr,<2 mm (0.1 in), (c42,s32), ☺. (FIG. 8.73)

Obelia spp. (Campanulariidae). Several species; cannot be identified to species using gross characters. Gonads form round clusters along canals. Bo,Pe,Es–Ps,pr,<2 mm (0.1 in), (s38). (FIG. 8.74)

Pennaria (=*Halocordyle*) *disticha* (Pennariidae). Oval shape. Vi,Pe,Es–Ps,pr,2.5 mm (0.1 in), (c25). (FIG. 8.75)

Podocoryna spp. (Hydractiniidae). Ca+,Pe,Es–Ps,pr,<2 mm (0.1 in), (c34,s31). (FIG. 8.76)

Rathkea octopunctata (Rathkeidae). Each oral tentacle splits at tip. Bo,Pe,Es–Ps,pr,<2 mm (0.1 in), (c34,s32). (FIG. 8.77)

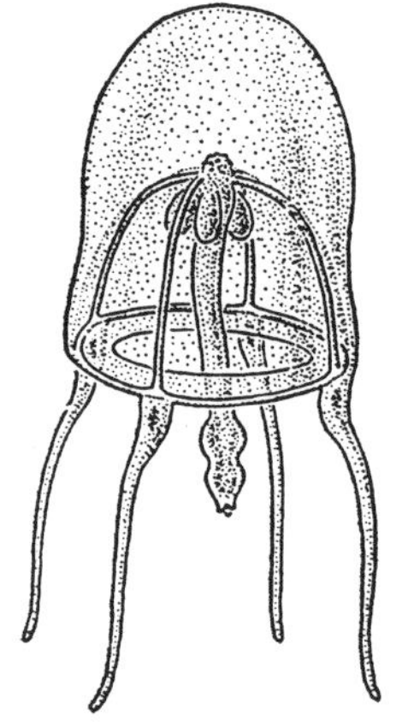

Sarsia tubulosa, clapper hydromedusa[†] (Corynidae). Bo,Pe,Es–Ps,pr,<2 mm (0.1 in), (C27,S30). (FIG. 8.78)

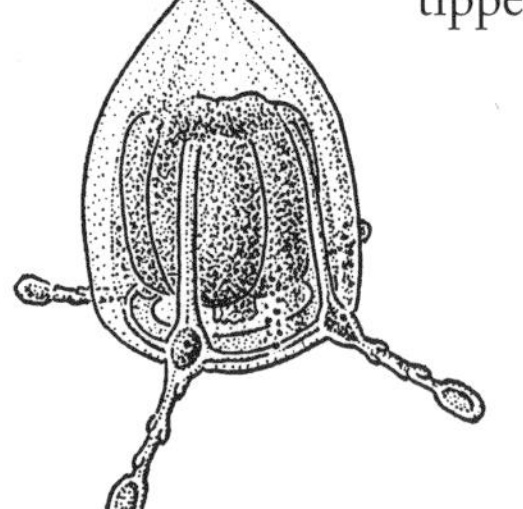

Sphaerocoryne (=*Linvillea*) *agassizi* (Corynidae). Has capitate (bulb tipped) tentacles. Vi,Pe,Es–Ps,pr,<2 mm (0.1 in), (C28). (FIG. 8.79)

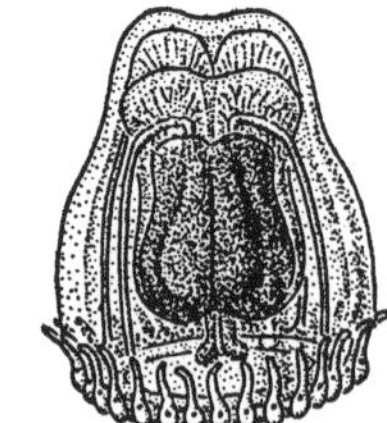

Turritopsis nutricula (Clavidae). Vi,Pe,Es–Ps,pr,5.1 mm (0.2 in), (C30). (FIG. 8.80)

FIG. 8.78

FIG. 8.79

FIG. 8.80

8.36 **Scyphozoan Jellyfishes** (from 8.31)

True jellyfishes lack a membranous velum. They tend to be larger than hydromedusae and are typically the most conspicuous stage in their life histories (i.e., their polyp stages are tiny in comparison). As important pelagic predators, most are well-armed with stinging cnidae or nematocysts, some of which are capable of inflicting a memorable sting on humans.

Additional reference: (L) Larson, R. J. 1976.

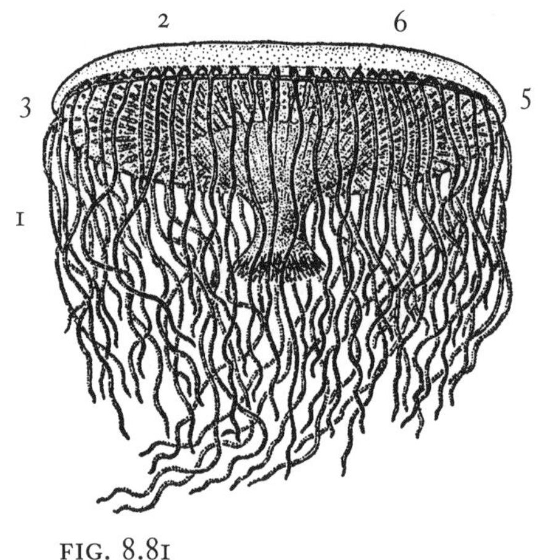

FIG. 8.81

1. Length of tentacles around the bell margin compared to height of bell (FIG. 8.81)
   - A. Tentacles *three or more* times the height of the bell — >3*
   - B. Tentacles *less than twice* the height of the bell — <2
   - C. Tentacles *absent* — x
2. General shape of jellyfish (FIG. 8.81)
   - A. Bell *cup*-shaped (width less than 1.5 times its height — cu
   - B. Bell *dish*-shaped (width more than 1.5 times its height — di*
3. Point at which tentacles attach to bell (FIG. 8.81)
   - A. Tentacles arise just *above* margin of the bell — ab*
   - B. Tentacles arise on *undersurface* of bell, away from margin — un
   - C. Tentacles arise *at* margin of bell, between scalloped lobes — at
   - D. Tentacles arise in *four* distinct clusters — 4
   - E. Tentacles *absent* — x
4. Approximate total number of tentacles (tentacles are easily broken off in damaged specimens) (FIG. 8.81)
   - A. *More* than 100 tentacles — >100*
   - B. If fewer than 100, provide *number* — ___
   - C. Tentacles *absent* — x
5. Ring canal around periphery of bell (FIG. 8.81)
   - A. *Ring canal* present — rc*
   - B. Ring canal *absent* — x
6. Approximate diameter of bell (this character is most useful if specimen exceeds sizes given; obviously juvenile specimens can be much smaller than maximal sizes listed here): provide *diameter* in cm (inches) — ___

8.37

Tentacles	Shape	Attachment Site	No. of Tentacles	Canal	Size (cm)*	Organism
>3	cu	at	8	x	5.1 (2 in)	*Pelagia noctiluca,* purple jellyfish
>3	cu	at	24-40	x	20.3 (8 in)	*Chrysaora quinquecirrha,* sea nettle†, golden-fringed jellyfish
>3	di	un	>100	x	14.2 (36 in)	*Cyanea capillata,* lion mane†, sun jellyfish
<2	di	ab	>100	rc	30.5 (12 in)	*Aurelia aurita,* moon jelly†, white jellyfish
<2	cu	4	50	x	10.2 (4 in)	*Chiropsalmus quadrumanus,* sea wasp
x	cu	x	x	x	30.5 (12 in)	*Rhopilema verrilli,* mushroom-cap jellyfish

* Number in parentheses is the size in inches.
† Official common name.

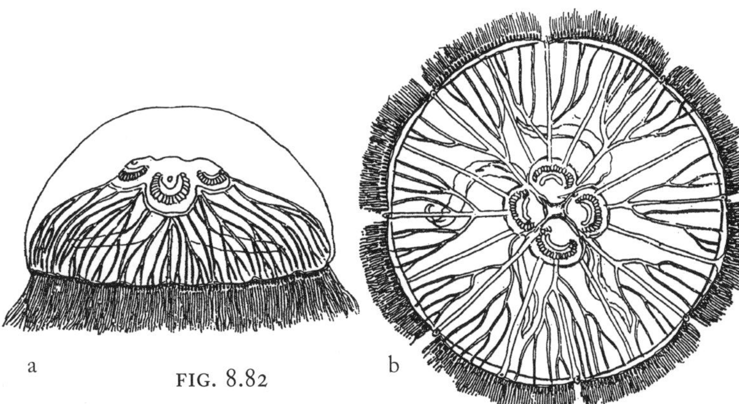
a FIG. 8.82 b

Aurelia aurita, moon jelly†, white jellyfish (Ulmaridae). Bluish white, pink, or brownish; small tentacles. Gonads form four white or tan horseshoe-shaped rings. Gravid medusae in June. Important planktivore. Look for parasitic amphipod, *Hyperia galba,* in its tissues. Mild sting. Bo,Pe,Es–Ps,pr,40.6 cm (16 in) [20.3 cm (8 in)], (G90,L15,NM363,S80). (FIG. 8.82. a, lateral; b, aboral view.)

Chiropsalmus quadrumanus, sea wasp (Chirodropidae). Capable of delivering a nasty but not dangerous sting. Ca+,Pe,Ps–Pd,pr,10.2 cm (4 in). (FIG. 8.83)

FIG. 8.83

Chrysaora quinquecirrha, sea nettle†, golden-fringed jellyfish (Pelagiidae). Long yellow tentacles from notches along bell margin. Often with red-brown radiating stripes with four long oral lobes. Painful stinger, especially problematic in Chesapeake Bay. Vi,Pe,Es–Ps,pr,25.4 .cm (10 in) [10.1 cm (4 in)], (G90,L15,NM361), ☺. (FIG. 8.84)

Cyanea capillata, lion mane†, sun jellyfish (Cyaneidae). Purplish, reddish, orange to brown; small ones lighter. Tentacles in 8 clusters of 70–150 each around margin. Painful sting. Look for parasitic amphipod, *Hyperia galba,* in its tissues. Bo,Pe,Ps,pr,243.8 cm (96 in) [30.5 cm (12 in)], (G89,L15,NM362,S80), ☺. (FIG. 8.85)

Pelagia noctiluca, purple jellyfish (Pelagiidae). Pink to purple. Long, reddish tentacles alternate with knobby sense organs. Bell surface with bumps. Luminescent. Vi,Pe,Ps,pr,7.6 cm (3 in), (G90,L15,NM361,S79). (FIG. 8.86)

Rhopilema verrilli, mushroom-cap jellyfish (Rhizostomatidae). White with oral arms mottled in brown or yellow. No tentacles; captures food using oral lobes. Ca+,Pe,Ps,pr,30.5 cm (12 in), (G89,L16). (FIG. 8.87)

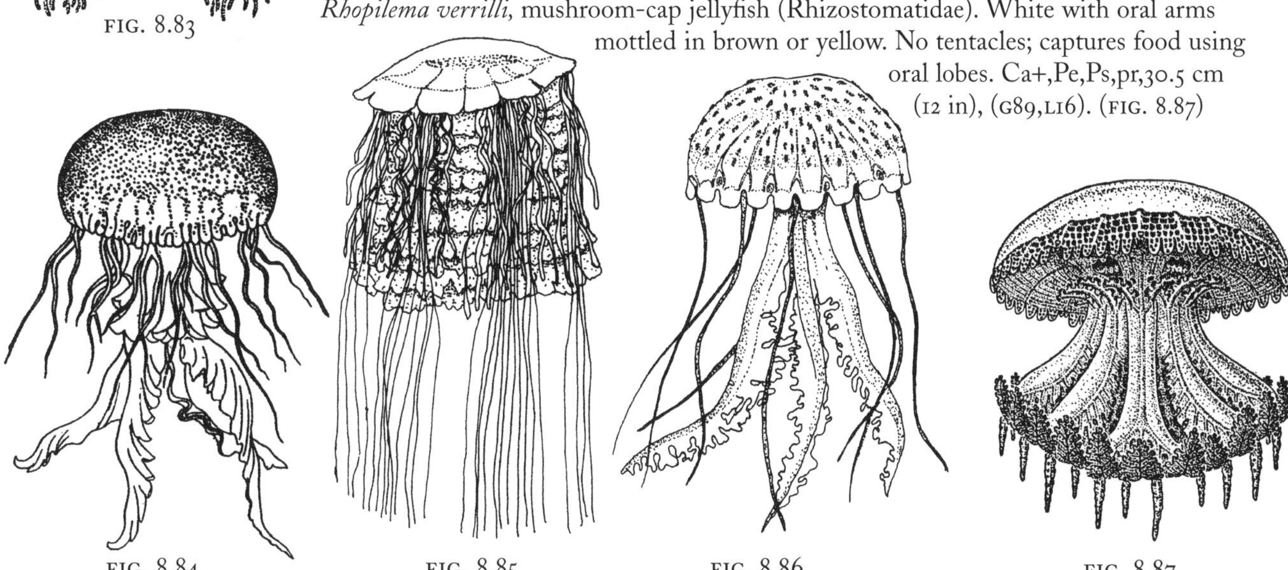
FIG. 8.84 FIG. 8.85 FIG. 8.86 FIG. 8.87

8.38 Class Hydrozoa: Order Siphonophora

Siphonophores are floating or swimming colonies comprising several types of specialized individuals functioning as a unit. Some colony members provide locomotion, whereas others may engage in food capture, digestion, or reproduction. Because they are organically interconnected to one another, the task(s) performed by each individual benefits all members.

These delicate members of the plankton community must be observed primarily in the field because they do not last well in aquaria or in preservation. Caution: at least one species, *Physalia*, the Portuguese man-of-war, has potent nematocysts capable of inflicting a painful and potentially serious (though not often fatal) sting. Only the two most common types are listed here. Many others are possible, although they are rarely encountered in routine collections.

1. Typical location
   - A. Animal at *surface* of the water; usually with gasbag extending above the surface — su
   - B. Animal within *water column* — wc
2. Primary locomotory device
   - A. Pinkish-blue inflated gasbag (the *pneumatophore*) provides floatation — pn
   - B. Several small, pulsing units (*nectophores*) form the clear, cylindrical anterior end of the colony and draw it through the water — ne

8.39

Location	Locomotion	Organism
su	pn	*Physalia physalis* (=*physalia*), Portuguese man-of-war[†]
wc	ne	*Nanomia* (=*Stephanomia*) *cara*, chain siphonophore[†]

Nanomia (=*Stephanomia*) *cara*, chain siphonophore[†] (Agalmatidae). Delicate colonies occasionally common in summer. Bo,Pe,Ps,pr,7.6 cm (3 in), (G87,NM356,S44). (FIG. 8.88)

FIG. 8.88

Physalia physalis (=*physalia*), Portuguese man-of-war[†] (Physaliidae). Blue, pink, and purple tones to inflatable float. Long, trailing tentacles. Severe stings, even from detached tentacles; potentially dangerous. Southern form sometimes drifts ashore from Gulf Stream. Vi+,Pe,Ps,pr,30.5 cm (12 in) (float length), (G87,NM356,S44). (FIG. 8.89)

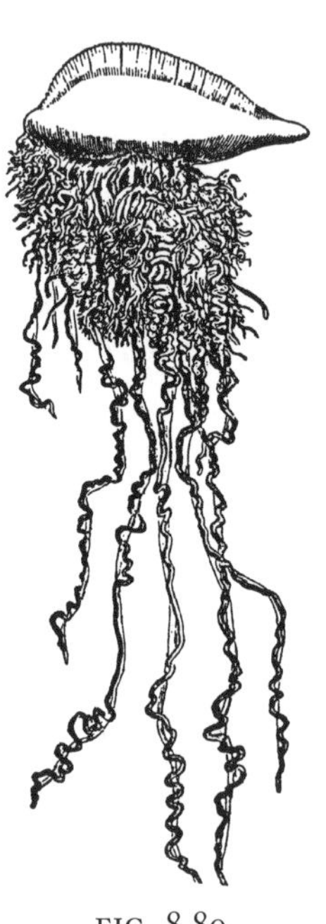

FIG. 8.89

8.40 Class Scyphozoa: Order Stauromedusae

The *Stauromedusae* are delicate, stalked jellyfish typically attached by an aboral stalk to algae swaying in low intertidal surf wash. Their coloring varies but is often similar to the alga to which they are attached.

Additional reference: (L) Larson, R. J. 1976.

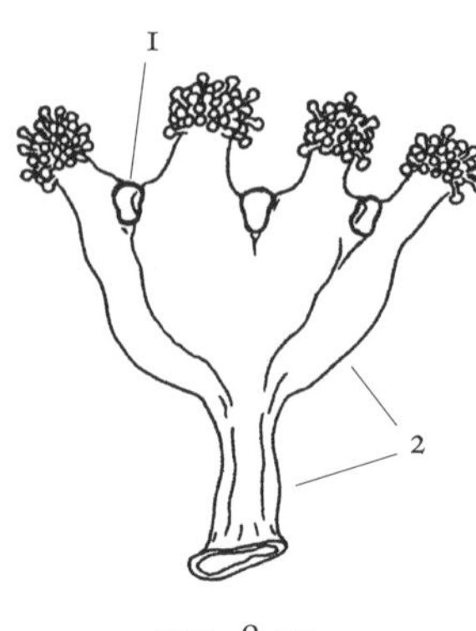

FIG. 8.90

1. Presence of anchors (swellings) between each tentacle cluster (FIG. 8.90)
   - A. *Cushion*-shaped *anchors* present — ca*
   - B. *Trumpet*-shaped *anchors* present — ta
   - C. Anchors with central *knobs* present — kn
   - D. Anchors *absent* — x
2. Length of stalk (s) compared to height of bell (b), viewed from the side (FIG. 8.90)
   - A. *Stalk longer than bell* height — s>b
   - B. *Stalk about equal to bell* height — s=b*
   - C. *Stalk much shorter than bell* height — s<b
3. Most typical algal substrate (this character is less reliable)
   - A. Associated with bladelike brown algae, *Laminaria* or *kelp* — ke
   - B. Associated with *rockweeds* — rw
   - C. Associated with *red* algae, especially epiphytes like *Ceramium* — re

8.41

Anchor	Stalk:Bell	Algae	Organism
kn	s=b	ke	*Manania (=Thaumatoscyphus) atlanticus*
ca	s=b	re,rw,ke	*Haliclystus auricula,* eared stalked-jellyfish
ta	s=b	re,ke	*H. salpinx,* trumpet stalked-jellyfish
x	s>b	ke	*Lucernaria quadricornis,* horned stalked-jellyfish
x	s<b	rw	*Craterolophus convolvulus,* goblet stalked-jellyfish

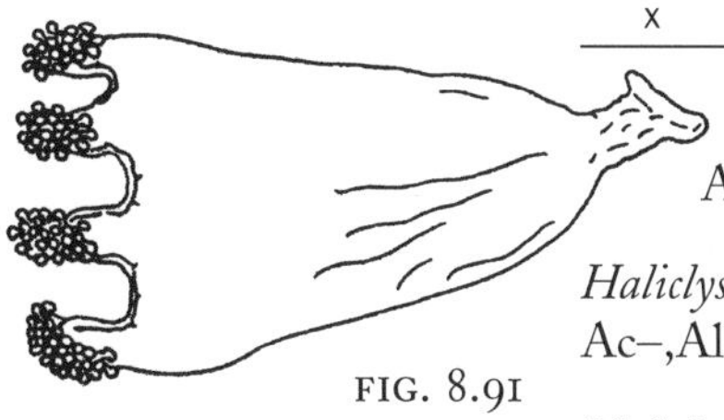

FIG. 8.91

Craterolophus convolvulus, goblet stalked-jellyfish (Depastridae). Ac,Al,Li–Su,pr,3.8 cm (1.5 in), (G91,L15,NM359). (FIG. 8.91)

Haliclystus auricula, eared stalked-jellyfish (Lucernariidae). Ac–,Al,Li–Su,pr,3.3 cm (1.3 in), (G91,L14,NM360). (FIG. 8.92)

Haliclystus salpinx, trumpet stalked-jellyfish (Lucernariidae). On *Ceramium* when young; on *Laminaria* or seagrass as adults. Ac–,Al–SG,Li–Su,pr,25 mm (1 in), (G92,L14,NM359). (FIG. 8.93)

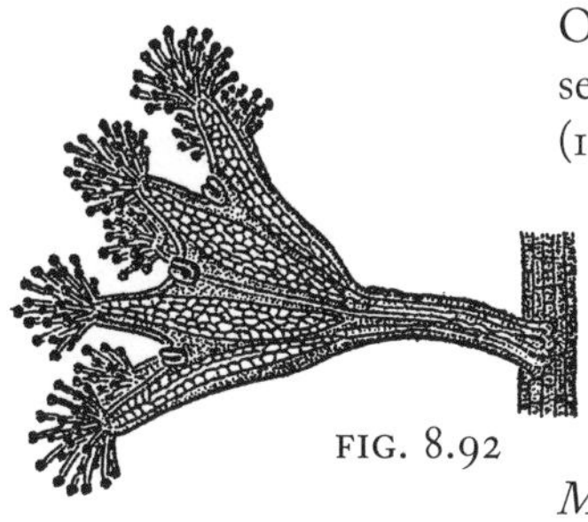

FIG. 8.92

Lucernaria quadricornis, horned stalked-jellyfish (Lucernariidae). Olive-brown color. Ac,Al,Li–Su,pr,7.6 cm (3 in), (G91,L14,NM359). (FIG. 8.94)

Manania (=Thaumatoscyphus) atlanticus (Depastridae). Cold-water form. Ac,Al,Su,pr,25 mm (1 in), (G92,L15).

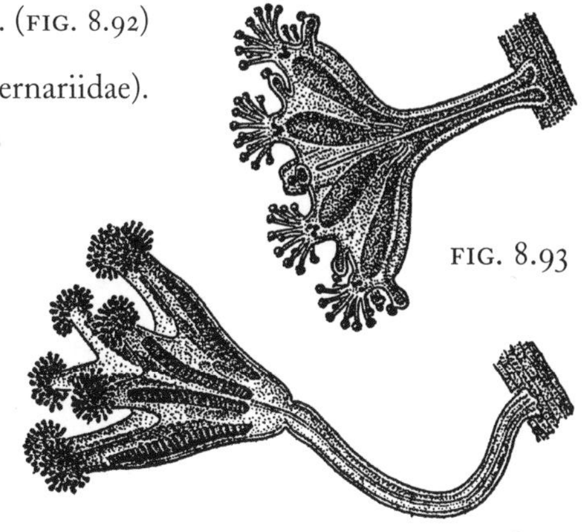

FIG. 8.93

FIG. 8.94

8.42

Class Anthozoa, Colonial and Solitary Anemones

Because anemones tend to be large and attached cnidarians, they are frequently encountered by intertidal observers. A flexible, columnar body clings to hard substrates or burrows into particulate sediments basally and gives rise to cnida-bearing feeding tentacles that surround the mouth at the oral surface. Many are solitary, but others form groups or even organically interconnected colonies. The medusa stage is absent in this class. Internally, a spacious gastrovascular cavity is partitioned by radially arranged mesentarial septae.

During low-tidal exposure or when disturbed, anemones can withdraw tentacles and contract their bodies to form nearly unrecognizable, soft masses that are slimy to the touch. It may require patience to wait for them to relax enough to reveal tentacles and color patterns. Anesthetization (see Appendix) may be helpful.

To observe fired and unfired cnidae, remove a small piece of tentacle to a drop of seawater on a glass slide. Add a coverslip with tiny clay pedestals (just touch the corners of the coverslip to a ball of modeling clay). Place the slide on the stage of a compound microscope and add a drop or two of 10% acetic acid at the interface between the coverslip and slide. Carefully bring the corner of a piece of paper towel into contact with the trapped fluid along the opposite side of the coverslip. Watch through the microscope as this procedure draws the acetic acid into irritating contact with the tentacle piece. You should be able to see threadlike cnidae discharge. Apply gentle pressure to the coverslip with a pin if necessary.

1. Growth form
   - A. *Colonial,* organically interconnected polyps — co
   - B. *Solitary,* although can be in close but not interconnected clusters — so
2. Habitat
   - A. Attached to *hard* substrates, usually with flattened basal disc — ha
   - B. *Burrowing* in sand or mud — bu
   - C. *Parasitic,* often in gelatinous tissue of Cnidaria or Ctenophora — pa

8.43

Form	Habitat	Group
co	ha,bu	Colonial anemones, 8.44
so	ha	Solitary attached anemones, 8.46
so	bu	Solitary burrowing anemones, 8.48
so	pa	12 tentacles; usually in tissues of the jellyfish, *Cyanea*: *Peachia parasitica* OR 16 tentacles; usually in tissues of the ctenophore, *Mnemiopsis*: *Edwardsia leidyi*

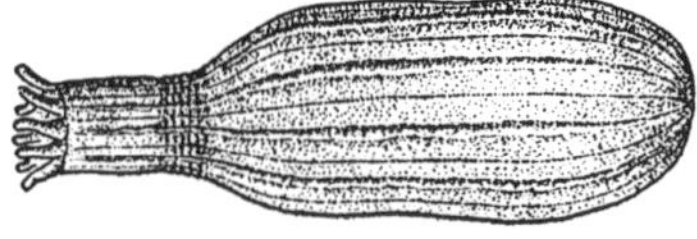

FIG. 8.95

Edwardsia leidyi (Edwardsiidae). Larval stage is parasitic; pink color. Adults are slender, brownish, and burrowing; 16 tentacles. Ac,Ot,Ps,pa,3 mm (1.2 in). (FIG. 8.95)

Peachia parasitica (Haloclavidae). Young forms only are parasitic; brownish body color. Adults unknown. Ac,Ot,Ps,pa,?. (FIG. 8.96)

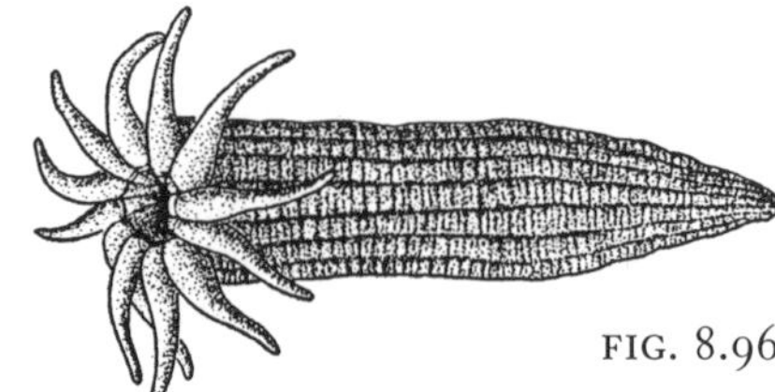

FIG. 8.96

8.44

Colonial Anemones (from 8.43)

Colonial anemones are not commonly encountered in the Northeast because most are limited to deeper waters. This mixed assemblage of species is included within several anthozoan orders, as indicated. The northern stony coral, *Astrangia astreiformis*, and the soft coral, *Alcyonium digitatum*, are the more common shallow species.

1. Presence of hard, or at least firm, skeletal material
   - A. *Hard* or firm skeletal material present — ha
   - B. Hard parts absent, entire colony is *soft* and flexible — so

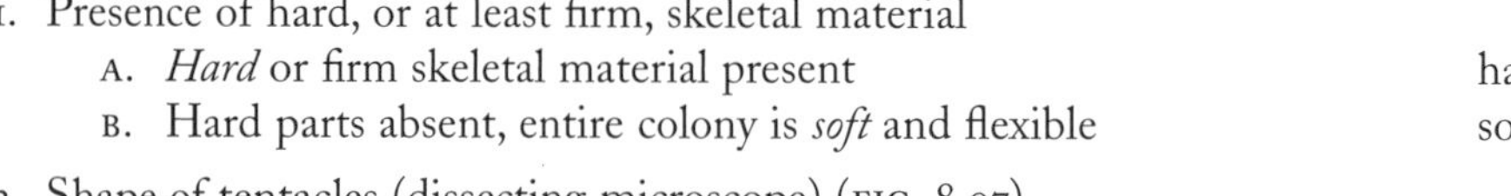

2. Shape of tentacles (dissecting microscope) (FIG. 8.97)
   - A. Tentacles *pinnate* (feather-branched) — pi*
   - B. Tentacles shaped variously, but branching *absent* — x
3. Shape of colony (FIG. 8.97)
   - A. Entire colony shaped like a *feather* — fe
   - B. Colony forms thin, stiffly flexible *stalks* like bare tree twig or branch — st
   - C. Colony forms a cushion or *mound* — mo
   - D. Colony forms amorphous branching *lobes* — lo*
   - E. Colony arises from *stolon* (runner) base — st
   - F. Colony forms encrusting *sheet* of interlinked, tiny anemonelike individuals — sh

FIG. 8.97

8.45

Hard/Soft	Tentacles	Colony	Organism
so	pi	mo,lo	Yellow-orange fleshy masses: *Alcyonium digitatum* (=*carnium*), dead man's fingers OR Main stem with red spicules: *Gersemia rubiformis*, red soft coral
so	x	st	*Clavularia modesta*
so	x	sh	*Epizoanthus incrustatus*
ha	x	sh,mo	*Astrangia astreiformis* (=*danae*), northern stony coral, star coral
ha	pi	st	*Leptogorgia* Many branches: *L. virgulata*, violet sea whip OR Unbranched: *L. setacea*, straight sea whip
ha	pi	fe	*Pennatula aculeata*, sea pen or sea feather

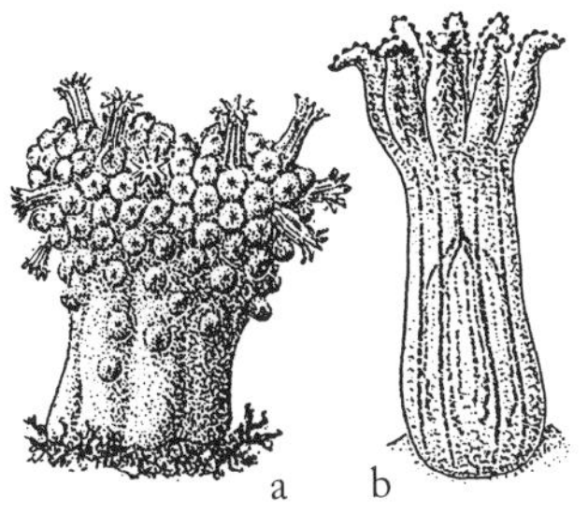

FIG. 8.98

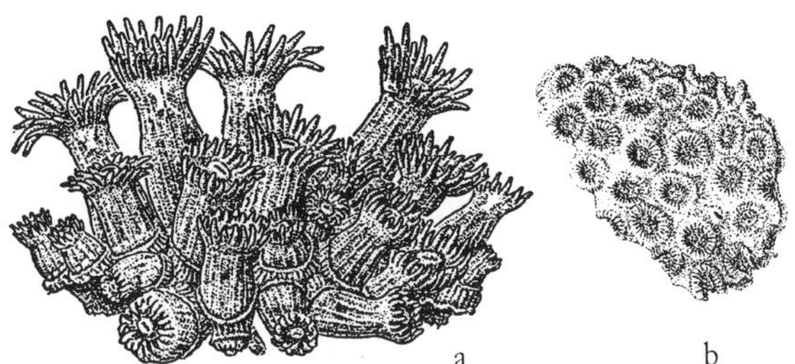

FIG. 8.99

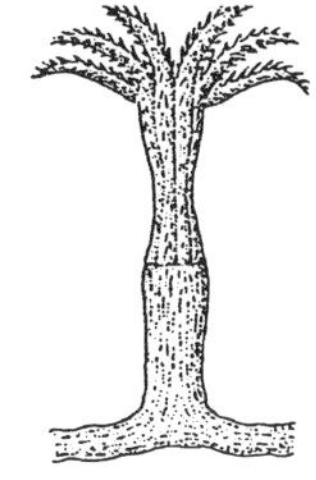

FIG. 8.100

Alcyonium digitatum (=*carnium*), dead man's fingers (Or: Alcyonacea, Alcyoniidae). Predators include nudibranchs, *Flabellina verrucosa* and *Tritonia.* Ac–,Ha,Su,ff,20.3 cm (8 in), (G92,M54). (FIG. 8.98. a, colony; b, polyp.)

Astrangia astreiformis (=*danae*), northern stony coral, star coral (Or: Scleractinia, Rhizangiidae [=Astrangiidae]). White or pink, crowded polyps; typically golf-ball-sized patches on edges of subtidal rocks. See Cairns (1981) on the taxonomy of this species. Vi,Ha,Su,ff,12.7 cm (5 in) (colony diameter), (G93,M58,NM391). (FIG. 8.99. a, polyps; b, skeleton.)

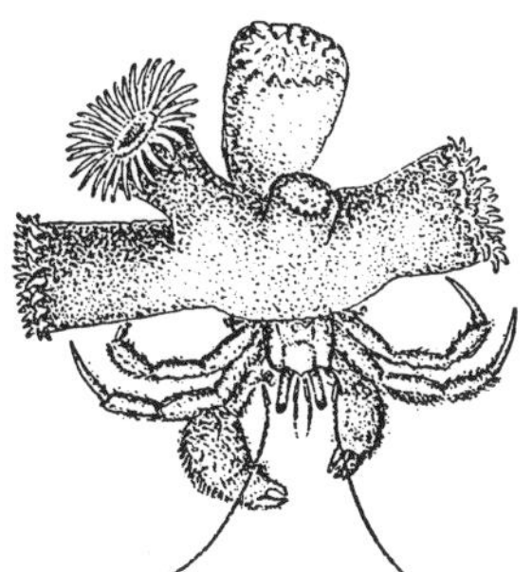

FIG. 8.101

Clavularia modesta (Or: Alcyonaria, Clavulariidae). Ac,Ha,Su–De,ff,17.8 mm (0.7 in). (FIG. 8.100)

Epizoanthus incrustatus (Or: Zoantharia, Epizoanthidae). Usually on hermit crab shells in deep water. Bo,Ha,De,ff–sc,12.7 mm (0.5 in). (FIG. 8.101. colony growing on hermit crab shell.)

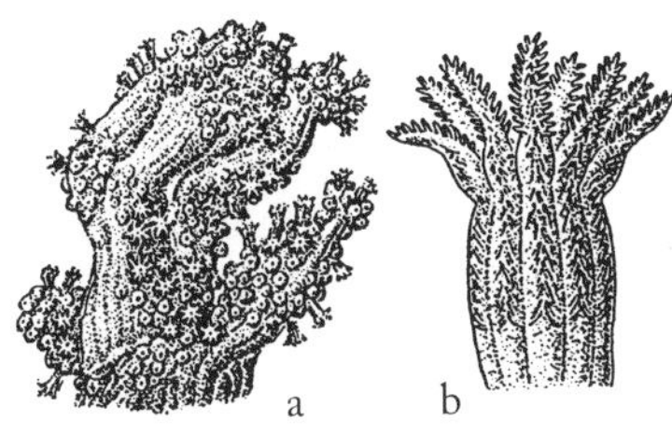

FIG. 8.102

Gersemia rubiformis, red soft coral (Or: Alcyonacea, Nephtheidae). Usually in deeper water than *A. digitatum.* Ac,Ha,Su,ff, 15.2 cm (6 in), (M56,NM364). (FIG. 8.102. a, colony; b, polyp.)

Leptogorgia setacea, straight sea whip (Or: Gorgonacea, Gorgoniidae). Ca+,Ha,Su,ff,182.9 cm (72 in), (G93).

Leptogorgia virgulata, violet sea whip (Or: Gorgonacea, Gorgoniidae). Polyps evenly spread; eight transparent white tentacles. Purple, red, orange, yellow or white. Vi,Ha,Su,ff,91.4 cm (36 in), (G93,NM365).

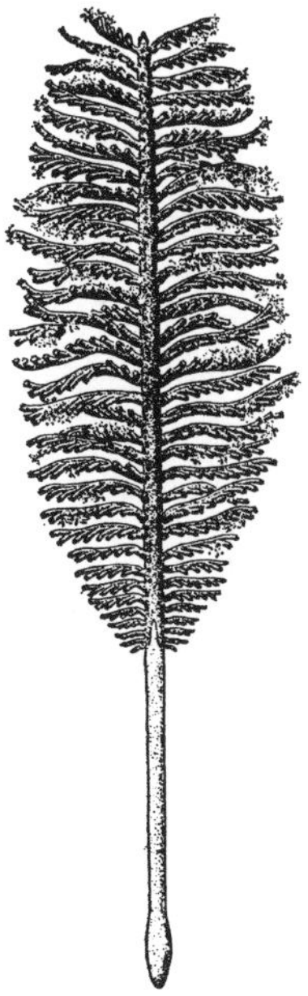

Pennatula aculeata, sea pen or sea feather (Or: Pennatulacea, Pennatulidae). Stalk orange; branches, reddish. Deep water. Bo,Sa–Mu,De,ff,10.1 cm (4 in), (G93). (FIG. 8.103)

FIG. 8.103

8.46 Solitary Attached Anemones (Or: Actiniaria) (from 8.43)

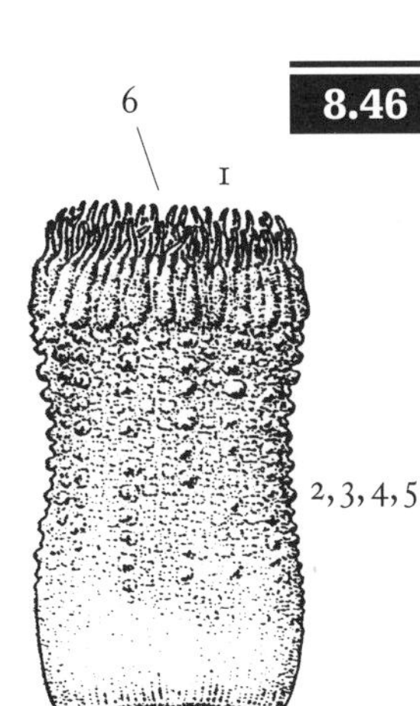

FIG. 8.104

Metridium senile is by far the most common anemone in our region. In fact, one must first prove that an anemone is not *M. senile* before proceeding with the identification process. Small specimens of this species lack the thousands of tiny tentacles on lobose bases that characterize larger individuals. Any anemone needs plenty of time in fresh, cold seawater to extend its tentacular crown following disturbance or handling.

1. Approximate number of tentacles (dissecting microscope) (FIG. 8.104)
   - A. *Many* tentacles, >1000 — ma
   - B. Comparatively *few* tentacles, <60 — fe*
   - C. *Intermediate* number of tentacles, 60–150 — in
2. Dominant color of body column, regardless of streaks or blotches (FIG. 8.104)
   - A. Some shade of *green* or olive green — gr
   - B. Has some real *red* coloring, not including pink — re
   - C. Green or distinct red coloring *absent* (often pink-cream-orange-brown) — x

3. Presence of distinct lines along body column (FIG. 8.104)
   A. *Lines* present li
   B. Lines *absent* x
4. Presence of blotches or streaks on body column (FIG. 8.104)
   A. *Blotches* or streaks distinct on body column bl
   B. Blotches or streaks *absent,* body column uniform in color x
5. Presence of bumps or papillae on body column (FIG. 8.104)
   A. Bumps or *papillae* present on body column pa*
   B. Body column basically smooth, bumps *absent* x
6. Presence of distinct colored lines radiating across oral disk from mouth (FIG. 8.104)
   A. Distinct *lines* present on oral disk li
   B. Distinct lines *absent* on oral disk; disk more or less uniform in color x
7. Possesses acontial threads (white threads often extruded from the mouth or pores in the body column when animal is disturbed; normally acontia are found within the gastrovascular cavity at the bottom of the pharyngeal tube). (Insistent poking to disturb the animal into deploying acontia; dissection may be necessary to be certain they are absent.) (This character is valuable but may not be necessary for adequate identification.)
   A. *Acontia* present ac
   B. Acontia *absent* x

8.47

Tentacles	Color	Body Lines	Body Blotch	Body Papilla	Disk Lines	Acontia	Organism
fe	gr	li	x	x	x	ac	*Diadumene lineata* (=*Haliplanella* (=*Sargartia*) *luciae*), green-striped anemone†
fe	x	x	x	pa,x	x	ac	*Diadumene leucolena,* pallid or ghost anemone
fe	x	x	x	x	x	ac	*Fagesia lineata,* lined anemone†
fe,in	x	x	bl,x	x	x	ac	*Metridium senile,* frilled anemone†, small, probably young, specimen; tentacles oftenwhite tipped; see notes below; ☺
in	or,re,x	x	bl	x	x	x	*Stomphia coccinea,* red stomphia†
in	gr,x	x	x	pa	li	x	*Bunodactis stella,* gem or silver-spot-ted anemone
in	re,gr	x	bl,x	pa,x	li	x	*Urticina* (=*Telia*) *felina* (=crassicornis), northern red or dahlia anemone
in	x	x	x	pa	x	ac	*Hormathia nodosa* (=*Actinauge rugosa*), knobby anemone
in,ma	x	x	x	x	x	x	*Bolocera tuediae*
ma	x	x	bl,x	x	x	ac	*Metridium senile,* frilled anemone†, large, older specimen

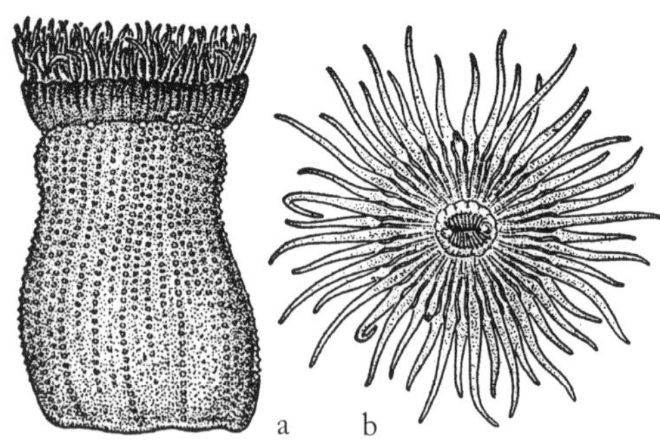

FIG. 8.105

Bolocera tuediae (Boloceroididae). Long tentacles, can detach at bases. Pink, orange, brown colors. Bo,Ha,Su,ff–pr,20.3 cm (8 in). (FIG. 8.105. a, lateral; b, oral view.)

Bunodactis stella, gem or silver-spotted anemone (Actiniidae). White disk lines or spots at tentacle bases. Partly buried in sediment in tidal pools; sand adheres to column. Ac,Ha–Sa,Li–,ff,38.1 mm (1.5 in), (G97,M58,NM375). (FIG. 8.106)

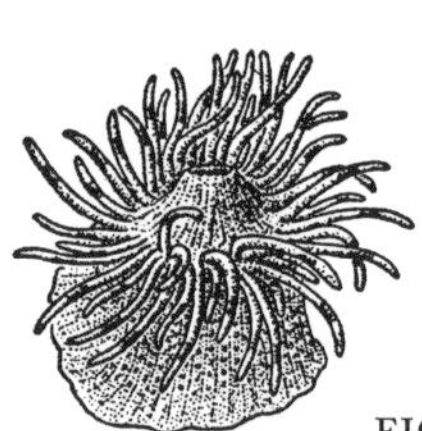
FIG. 8.106

Diadumene leucolena, pallid or ghost anemone (Diadumenidae). Almost transparent or pinkish; can see longitudinal septae through skin. Up to 60 tentacles. Found into low salinity waters; most common anemone in Chesapeake. Vi+,Ha,Es–Li–,ff,39.1 mm (1.5 in), (G98,M66,NM383), ☺. (FIG. 8.107)

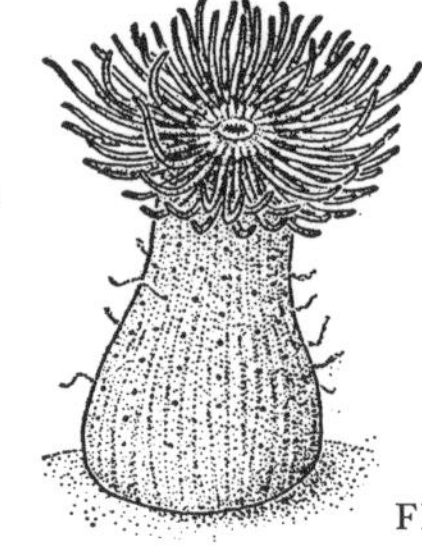
FIG. 8.107

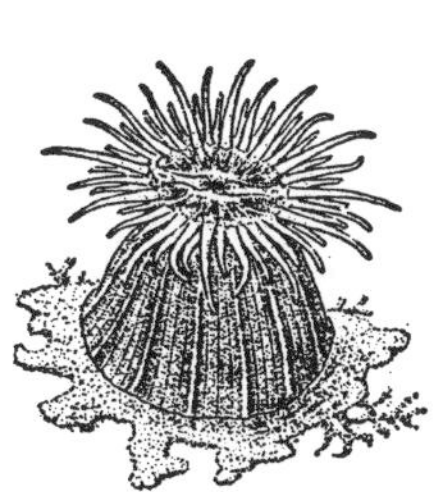

FIG. 8.108

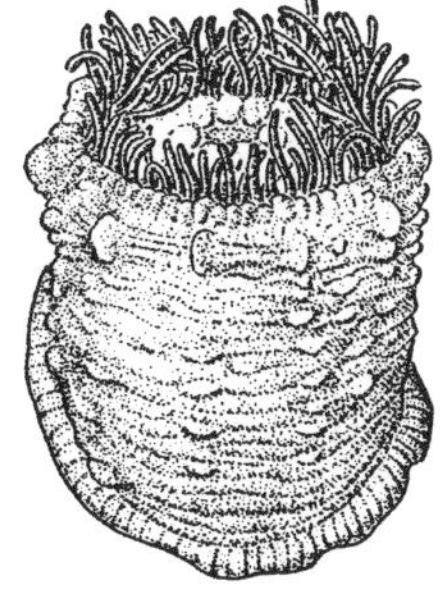

FIG. 8.109

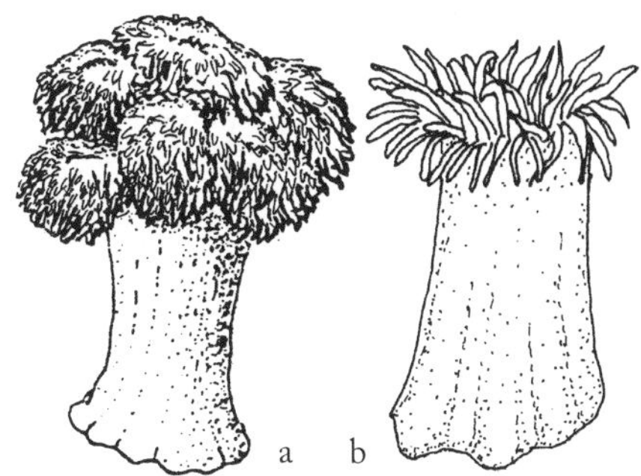

FIG. 8.110

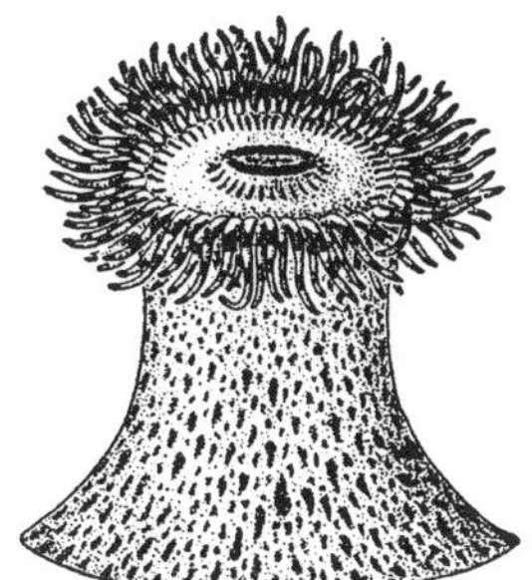

FIG. 8.111

Diadumene lineata (=*Haliplanella* [=*Sargartia*] *luciae*), green-striped anemone† (Aiptasiomorphidae). Orange or white lines up column; small size. Sometimes in mud or on vegetation in marshes; protected locations. Bo,Mu–Sa–SM–Ha,Es–Su,ff,20.3 mm (0.8 in), (G97,NM382), ☺. (FIG. 8.108)

Fagesia (=*Edwardsia*) *lineata*, lined anemone† (Edwardsiidae). White, delicate; 40 tentacles. Build mucus tube. Vi,Ha,Su,ff,38.1 mm (1.5 in), (M56,NM371).

Hormathia nodosa (=*Actinauge rugosa*), knobby anemone† (Hormathiidae). Column papillae are large lumps. Ac-,Ha,Su–De,ff,12.7 cm (5 in), (M68). (FIG. 8.109)

Metridium senile, frilled anemone† (Metridiidae). Many, tiny tentacles, on lobed bases in large individuals; up to several inches in height. White, cream, pink, orange, tan, brown varieties. Preference for fast currents. Bo,Ha,Li–De,ff,45.7cm (18 in), (G97,M64,NM382), ☺. (FIG. 8.110. a, adult with tentacle lobes; b, immature without lobed tentacle bases.)

Stomphia coccinea, red stomphia† (Actinostolidae). Small; reddish, banded tentacles. Ac,Ha,Su–De,ff,5.1 cm (2 in), (M62,NM378). (FIG. 8.111)

Urticina (=*Telia*) *felina* (=*crassicornis*), northern red or dahlia anemone (Actiniidae). Red disk lines; tentacles banded red and white. Mid or lower shore in gravel or shell debris. Ac,Ha,Su,ff,12.7 cm (5 in) [7.6 cm (3 in)], (G96,M60,NM373). (FIG. 8.112. a, animal; b, oral view.)

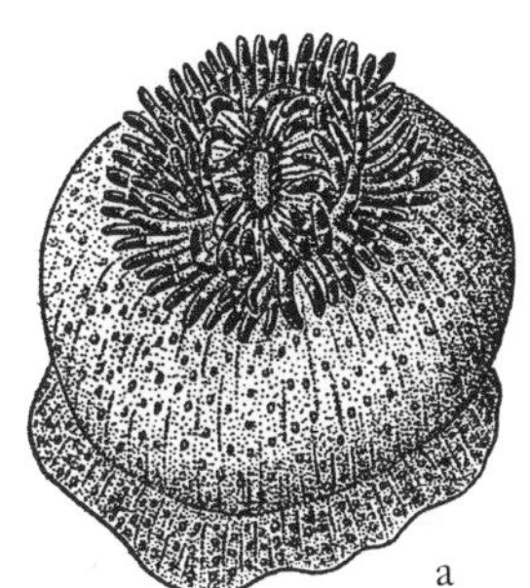

FIG. 8.112

8.48 **Solitary Burrowing Anemones** (from 8.43)

All except those indicated otherwise are in the Order Actiniaria.

1. Number of tentacles: provide *number* (FIG. 8.113) —
2. Surface of body column (FIG. 8.113)
   - A. Body column with *longitudinal ridges* (follows long axis of body) lr*
   - B. Body column perhaps wrinkled by contraction, but at least distinct ridges *absent* x
3. Presence of bumps or "warts" along body column (FIG. 8.113)
   - A. *Bumps* present bu
   - B. Bumps *absent*, column smooth x*
4. Presence of mucous tube
   - A. Anemone lives in distinct *mucous tube* mt
   - B. Such a mucous tube *absent*, anemone burrows freely in sediment x

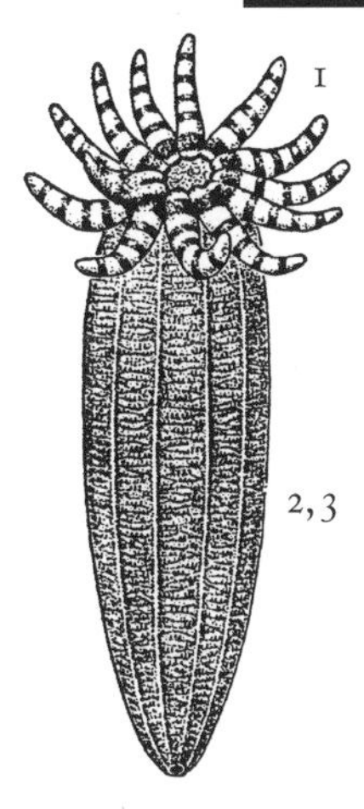

FIG. 8.113

8.49

Tentacles	Surface	Bumps	Tube	Organism
12	lr	x	x	*Halcampa duodecimcirrata*
12–16	x	x	x	*Nematostella vectensis*
15–16	lr	bu	x	*Edwardsia elegans*
20	lr	bu	x	*Haloclava producta*
40	x	x	x	*Fagesia lineata*
50–100	x	x	x	*Actinothoe modesta*
100+	x	bu	x	*Actinostola callosa*
100+	lr	x	x	*Paranthus rapiformis*, sea onion
100+	x	x	mt	Cerianthid or worm anemones. Tentacles in two whorls; up to 18" long; body gray to brown. Body tapers to a point; no basal disk. Thick mucous tubes. Subtidal. Tentacles differ in size, inner whorl shorter: *Cerianthus borealis*, northern cerianthid OR Tentacles all equal in length: *Ceriantheopsis americanus*, southern cerianthid

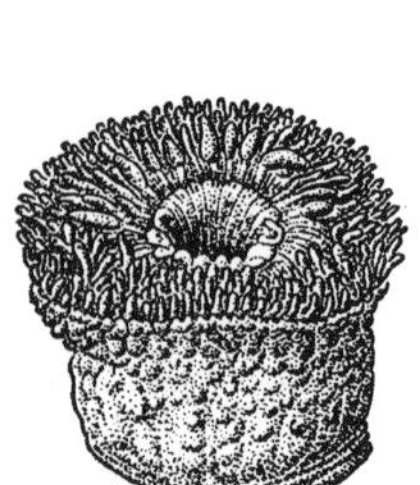
FIG. 8.114

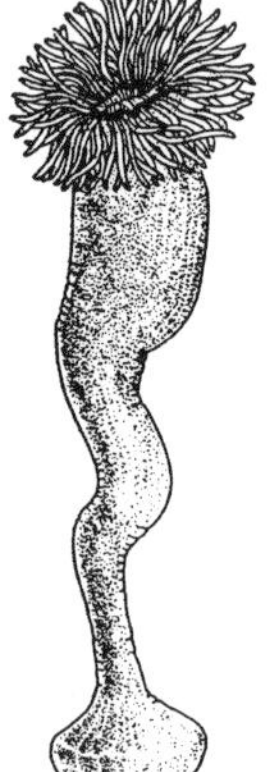
FIG. 8.115

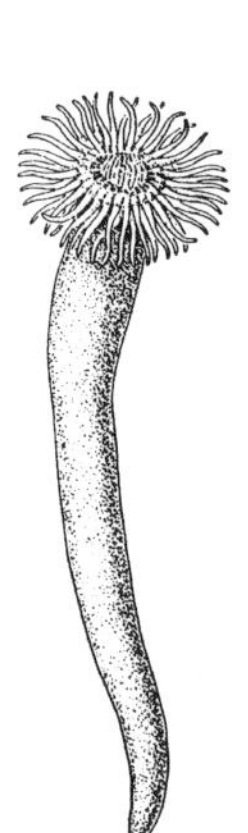
FIG. 8.116

FIG. 8.117

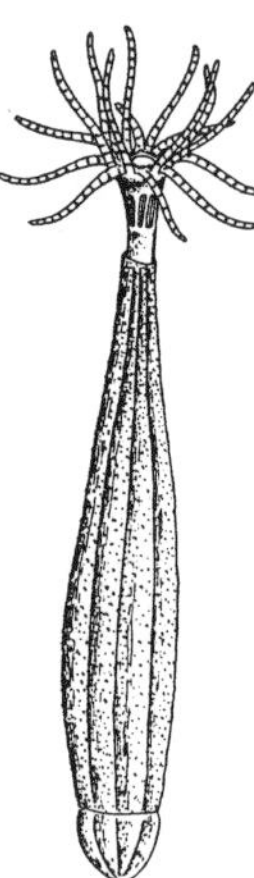
FIG. 8.118

Actinostola callosa (Actinostolidae). Salmon or orangish color. Large to 25 cm (10 in) across. Bo,Sa–Mu,Su–De,?,18.8 cm (7.5 in). (FIG. 8.114)

Actinothoe modesta (Sagartidae). Flattened base attaches to buried stones; white to pinkish-tan color. In mud-sand flats; Cape Cod south. Vi,Gr–Sa–Mu,Su?,df?,6.4 cm (2.5 in), (G96). (FIG. 8.115)

Ceriantheopsis americana, southern cerianthid (Or: Ceriantharia, Cerianthidae). Inner whorl of tentacles orangish; outer whorl, bluish pink. Vi,Mu,Li–Si,ff,20.3 cm (8 in), (G98), ☺. (FIG. 8.116)

Cerianthus borealis, northern cerianthid (Or: Ceriantharia, Cerianthidae). Ac,Mu,Su–De,ff,45.7 cm (18 in), (G98,M68,NM383). (FIG. 8.117)

Edwardsia elegans (Edwardsiidae). Yellowish pink; elaborately colored tentacles and upper body; oral disk with radiating lines. Sand adheres to column. Intertidal under stones or buried in gravel. Bo,Sa–Mu,Su–De,df?,3.3 cm (1.3 in), (G95,NM372). (FIG. 8.118)

Fagesia (=Edwardsia) lineata (Edwardsiidae). White, small, delicate; in thin mucous tube. Vi,Ha,Su,ff–df?,3.3 cm (1.3 in), (NM372).

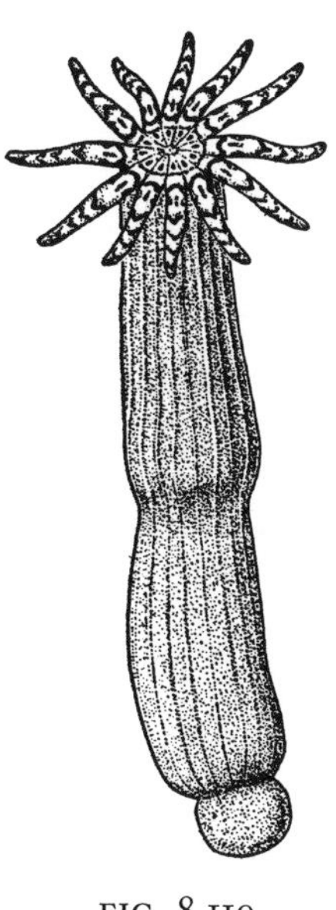

FIG. 8.119

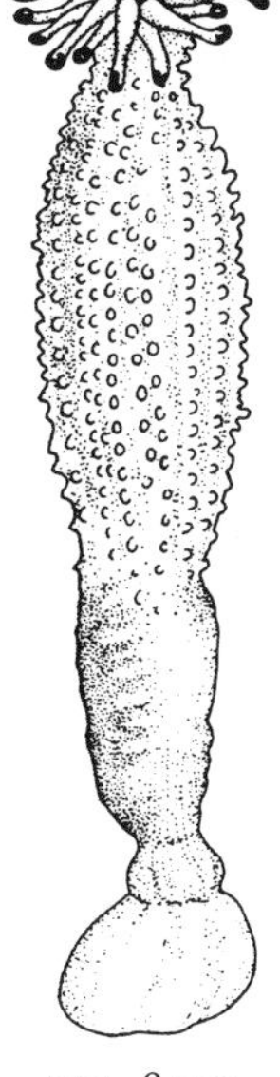

FIG. 8.120

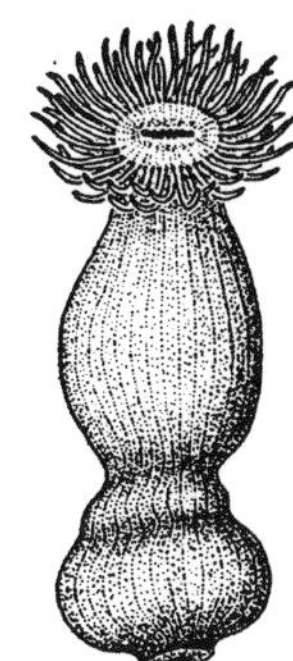

FIG. 8.121

Halcampa duodecimcirrata (Halcampidae). Purple spots surround mouth and tentacles. Ac,Sa,Su–De,pr?,25 mm (1 in). (FIG. 8.119)

Haloclava producta (Haloclavidae). Knobby tentacles; 20 rows of column bumps. Vi,Sa,Li,df?,15.2 cm (6 in), (NM373). (FIG. 8.120)

Nematostella vectensis (Edwardsiidae). Brownish; eight longitudinal body grooves. Vi+,Sa,Es,df–ff?,20 mm (0.8 in), (G95).

Paranthus rapiformis, sea onion (Actinostolidae). Yellow to pinkish. Contracts to squat body form. Vi,Sa,Li–Su,df–ff,7.6 cm (3 in), (G96). (FIG. 8.121)

Chapter 9 ■ Phylum Ctenophora, Comb Jellies

9.1

Comb jellies, like the medusae they resemble, are delicate planktonic organisms best observed in the field or in a dish of fresh, cold seawater. Locomotion in these predatory animals involves synchronous beating of rows of cilia (ctenes) arranged along eight meridional tracts (top to bottom) over the body. A dull, greenish bioluminescence is typical of several ctenophores. In the Northeast, *Mnemiopsis,* in particular, bioluminesces late in summer. The common *Pleurobrachia* does not bioluminesce.

1. Presence of a pair of retractile tentacles (FIG. 9.1)
   - A. Retractile *tentacles* present (when fully deployed they reveal themselves to be long and branched) — te*
   - B. Retractile tentacles *absent* — x
2. Body coloring
   - A. While basically transparent, body has distinct *pinkish* color, especially to internal canals — pi
   - B. Body basically transparent; pinkish color *absent* — x
3. Shape of top and bottom of animal's body (FIG. 9.1)
   - A. *Rounded* top: *rounded* bottom (i.e., spherical) — r:r
   - B. *Rounded* top: *straight* bottom (i.e., flat ended) — r:s
   - C. *Rounded* top: *lobed* bottom — r:l
   - D. *Pointed* top: *lobed* bottom — p:l
4. Compare body length to body width, including lobes but excluding tentacles, if present (FIG. 9.1)
   - A. Body length *more than twice* body width — >2*
   - B. Body length *less than twice* body width — <2

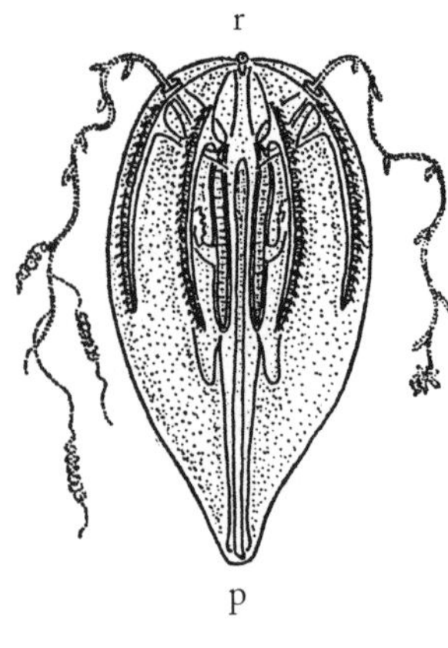

FIG. 9.1

9.2

Tentacles	Color	Top:Bottom	Width vs. Length	Organism
te	x	r:r	<2	*Pleurobrachia pileus,* sea grape or walnut or gooseberry
x	pi	r:s	<2	*Beroe cucumis,* pink slipper comb jelly
x	x	r:l	<2	*Mnemiopsis leidyi,* sea walnut
x	x	p:l	>2	*Bolinopsis infundibulum,* northern comb jelly

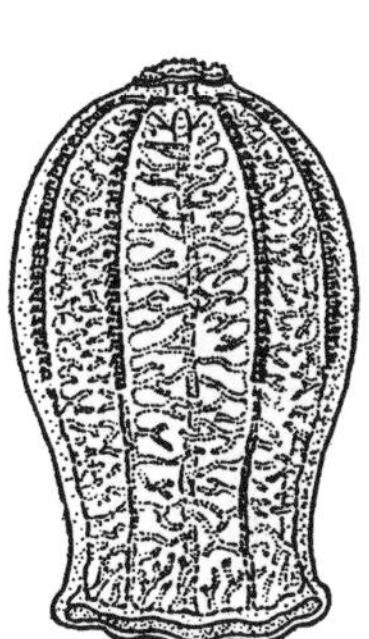

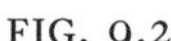

FIG. 9.2

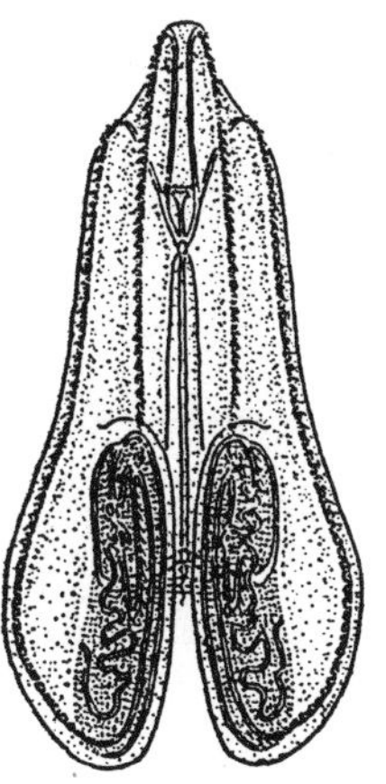

FIG. 9.3

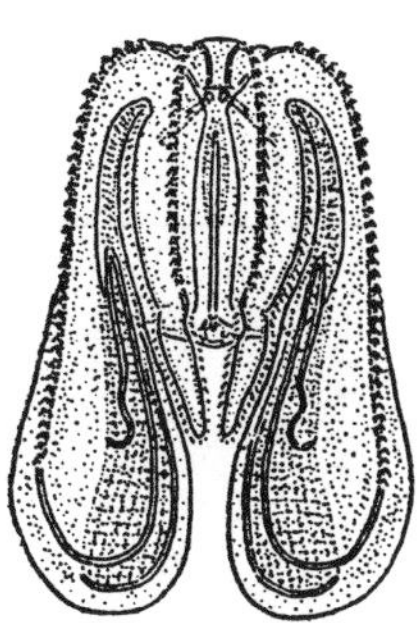

FIG. 9.4

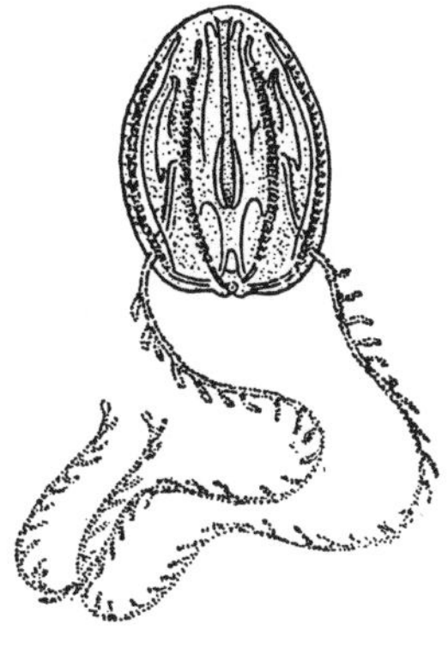

FIG. 9.5

Beroe cucumis, pink slipper comb jelly (Beroidae). Flattened sack with wide mouth. Often occurs in groups. Eats medusae and other comb jellies. Bo,Pe,Ps,pr,11.4 cm (4.5 in), (G100,M74,NM398). (FIG. 9.2)

Bolinopsis infundibulum, northern comb jelly (Bolinopsidae). Elongate; common, north of Cape Cod in summer. Ac,Pe,Ps,pr,15.2 cm (6 in), (G100,M72,NM397), ☺. (FIG. 9.3)

Mnemiopsis leidyi, sea walnut (Mnemiidae). Translucent; deep lateral furrows (view from above). Common south of Cape Cod in summer. Look for thin, pink, parasitic anemones, *Edwardsia leidyi,* in its tissues. Vi,Pe,Ps,pr,10.1 cm (4 in), (G100,M72,NM397), ☺. (FIG. 9.4)

Pleurobrachia pileus, sea grape or walnut or gooseberry (Pleurobrachiidae). Tentacles can extend 15–20 times body length. Bo,Pe,Es–Ps,pr,30.5 mm (1.2 in), (G99,M72,NM396), ☺. (FIG. 9.5)

Chapter 10 ■ Phylum Platyhelminthes: Class Turbellaria, Flatworms

10.1 Because of their small size and slow movements, free-living *flatworms* (Class Turbellaria) are frequently overlooked in field observations. More often they are encountered "incidentally" as they cruise into binocular microscopic view while other, larger specimens are being observed. Their flattened body places all interior cells within diffusion range to effect respiratory exchange. The digestive system is incomplete in that a single opening serves as both mouth and anus. The pattern of blind-ended branches of the gut have been used to define taxonomic orders within the class, differentiating the Order Polycladida (with several such branches), the Order Tricladida (with three such branches), and the Order Acoela (whose members lack a clearly defined gut) from several small orders characterized by a single, sacklike gut. In most cases, listings include ordinal as well as familial designations.

10.2 Flatworms shorter than 6 mm (<¼ in) in body length (small flatworms) are treated in section 10.3. Flatworms longer than 6 mm (>¼ in) (larger flatworms) are treated in section 10.5.

10.3 Small Flatworms (from 10.2)

Less than 6 mm (<¼ in) in body length. Identifications made using this key should be considered especially tentative until confirmed using more detailed references. Compound microscope required. Directions for making temporary wet mounts can be found in the Introduction.

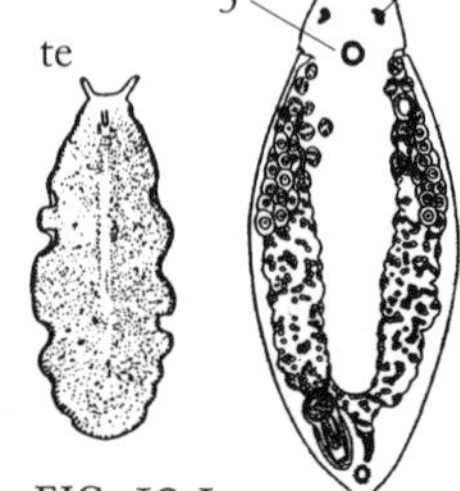

FIG. 10.1

1. Arrangement of eyespots (FIG. 10.1)
   - A. Tiny eyes in two distinct *clusters* — cl
   - B. Provide *number* of distinct eyespots — __*
   - C. Eyespots or clusters *absent* — x
2. Tentacles toward anterior end of body (tentacles may be retractile; they are best viewed with lighting from the side) (FIG. 10.1)
   - A. *Tentacle* pair present — te
   - B. Tentacles *absent* — x

3. Statocyst (for balance or orientation sensation) toward anterior end of body (appears as tiny, middorsal, dark—sometimes silvery—sphere) (FIG. 10.1)
   A. *Statocyst* present st*
   B. Statocyst *absent* x
4. Shape of posterior margin of body (especially in comparison with anterior margin)
   A. *Rounded* ro
   B. *Pointed* po*
   C. *Triangular* tr
   D. *Squared* sq
   E. *Indented,* with several contractile, caudal cirri (posterior tentacles) in
5. Maximal length (this character is useful for specimens that exceed maximal sizes for some species; small individuals of normally larger species can be problematic here): estimate body *length* (in mm) —

10.4

Eyes	Tentacles	Statocyst	Posterior Margin	Length	Species
cl	te	x	po	10	*Gnesioceros sargassicola,* gulfweed flatworm
4	x	x	po	1.5	*Monoophorum* sp.
2	te	x	ro	4.5	*Procerodes littoralis (=wheatlandi)*
2	x	x	ro	7	*Uteriporus vulgaris*
2	x	x	po	2.5	*Plagiostomum* sp.
2	x	x	sq	1.5	*Macrostomium* sp.
1–2	x	st	ro,tr	3	*Monocelis* sp.
x	x	st	ro	4.5	*Anaperis gardineri*
x	x	st	po	1	*Childia groenlandica*
x	x	st	in	4	*Polychoerus caudtus*

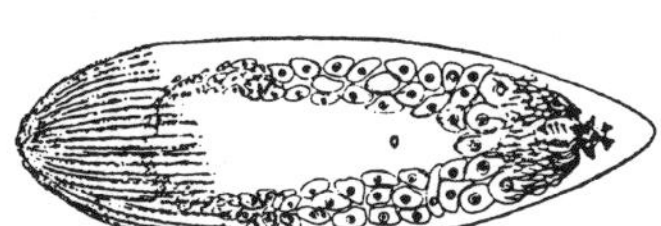
FIG. 10.2

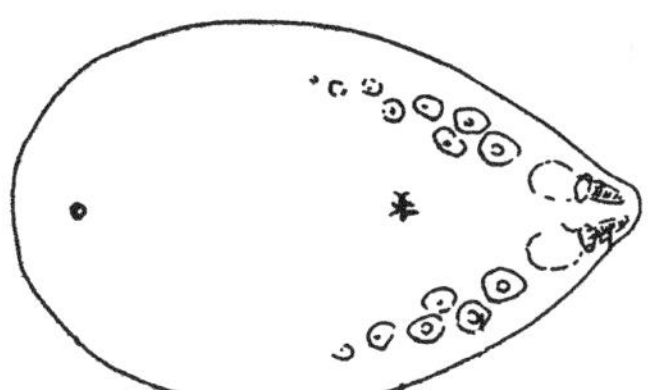
FIG. 10.3

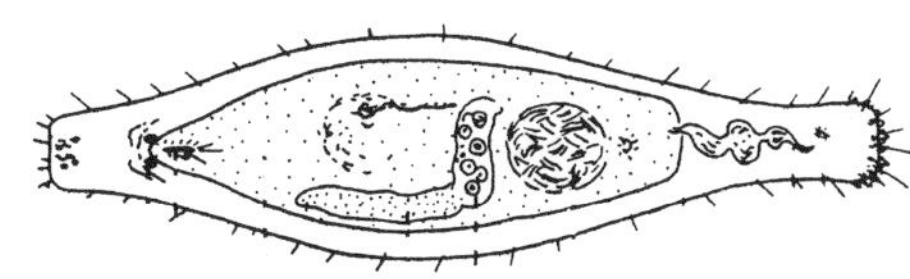
FIG. 10.4

Anaperis gardineri (Or: Acoela, Anaperidae). Orange-reddish brown. Vi,Sa–Mu,Su,?,5.1 mm (0.2 in). (FIG. 10.2)

Childia groenlandica (Or: Acoela, Childiidae). Vi,Mu–Al,Li–Su–Ps,?,1 mm (0.1 in). (FIG. 10.3)

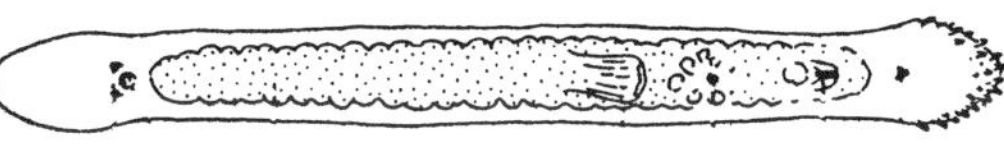
FIG. 10.5

Gnesioceros sargassicola, gulfweed flatworm (Or: Polycladida, Planoceridae). Eyespots on tentacles and in two clusters anteriorly. Drifts ashore on *Sargassum.* Vi,Al,Ps,pr?,10 mm (0.4 in).

Macrostomium sp. (Or: Macrostomida, Macrostomidae). Head end also squared. Setae scattered on body. On *Fucus.* Vi?,Al,Li,pr?,1 mm (0.1 in). (FIG. 10.4)

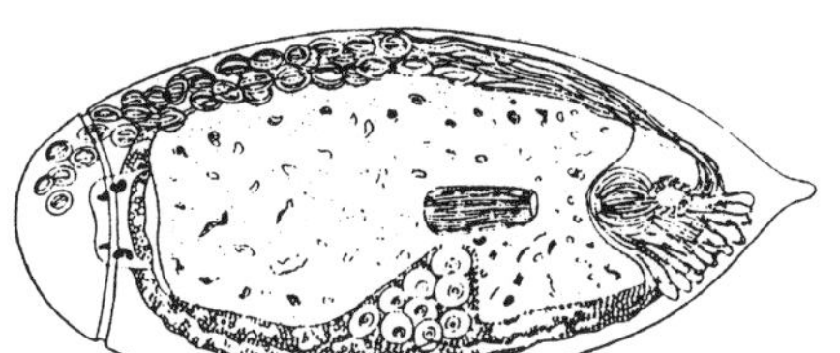
FIG. 10.6

Monocelis sp. (Or: Alloeocoela). Slender, brownish; may have anterior sensory hairs. Under intertidal rocks, on algae. Vi?,Ha–Al,Li,pr,5.1 mm (0.2 in). (FIG. 10.5)

Monoophorum sp. (Or: Alloeocoela). Vi?,Al,Li–Su,pr?,1.5 mm (0.1 in). (FIG. 10.6)

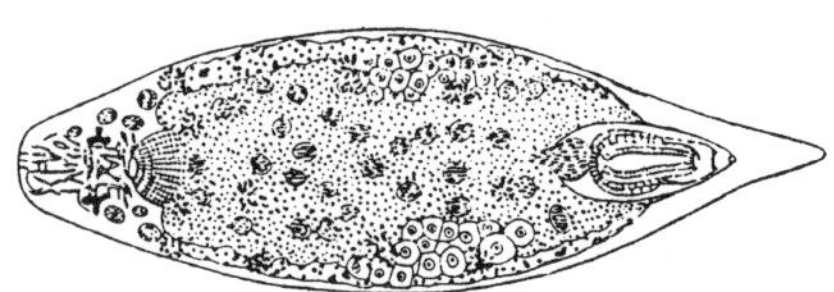
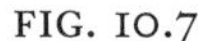

FIG. 10.7

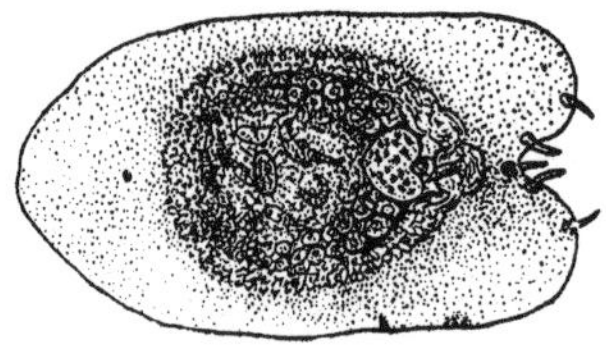

FIG. 10.8

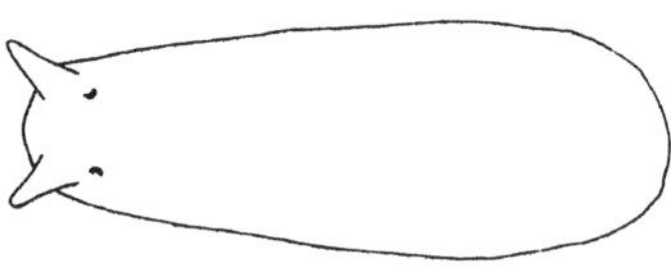

FIG. 10.9

Plagiostomum sp. (Or: Prolecithophora, Plagiostomidae). *P. album.* Whitish to yellow; lanceolate shape. Under stones, on sea lettuce, *Ulva.* Ac,Al–Ha,Li,pr,5.1 mm (0.2 in). (FIG. 10.7)

Polychoerus caudatus (Or: Acoela,?). Orange reddish. On eelgrass, *Zostera,* and sea lettuce, *Ulva.* Bo,SG–Al,Li–Su,pr,5.1 mm (0.2 in). (FIG. 10.8)

Procerodes littoralis (=*wheatlandi*) (Or: Tricladida, Uteriporidae). Brown with eyes in white streaks. Ac,Ha–Gr–Al,Es–Li–Su,pr?,8 mm (0.3 in). (FIG. 10.9)

Uteriporus vulgaris (Or: Tricladida, Uteriporidae). Pale white, yellow, orange; strap-shaped; under intertidal rocks. Ac,Ha,Li,pr?,7 mm (0.3 in).

10.5 **Larger Flatworms** (from 10.2)

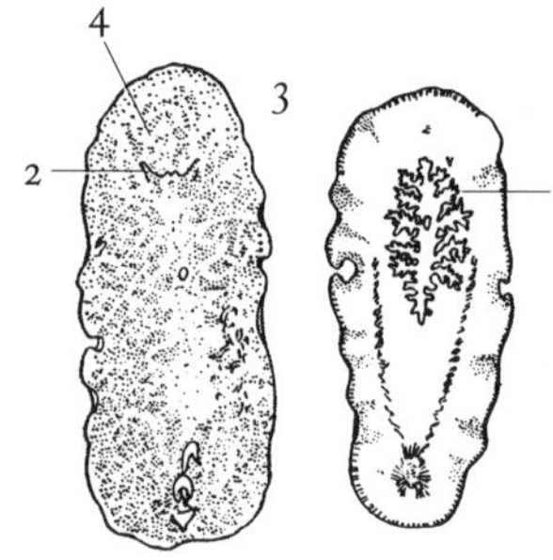

FIG. 10.10

Worms 6 mm (>¼ in) or larger. Dissecting microscope may be necessary.

1. Morphology of digestive tract (use strong transmitted light; i.e., from below) (FIG. 10.10)
   - A. *Three* primary branches (each with side branches), one extends anteriorly and two extend posteriorly from body midpoint — 3
   - B. *Many* branches extend from an slender central cavity — ma *
2. Pair of middorsal tentacles toward anterior portion of body (some tentacles are retractile when animal is disturbed) (FIG. 10.10)
   - A. *Tentacle* pair present — te*
   - B. Tentacles *absent* — x
3. Distinct marginal eyespots around anterior periphery of body (look for clusters of tiny, dark dots) (FIG. 10.10)
   - A. *Marginal eye*spots present — me*
   - B. Marginal eyespots *absent* — x
4. Arrangement of four, middorsal clusters of eyespots (FIG. 10.10)
   - A. Posterior (or lateral) clusters clearly *smaller* in size than anterior or medial clusters — sm
   - B. Posterior (or lateral) clusters about *equal* in size to anterior clusters — eq
   - C. Posterior cluster is on the *tentacle* pair — te
   - D. Clusters of eyespots *absent* (although individual eyespots may be present) — x
5. Number of individual eyespots
   - A. Only *two* discrete eyespots present — 2
   - B. *Fewer than 10* (but more than 2) eyespots present — <10
   - C. Middorsal eyespots equal to or *more than 10* — >10

10.6

Gut	Tentacles	Marginal Eyes	Eyespots	Number of Eyes	Species
ma	te	me	eq	<10	*Stylochus* Marginal eyes around anterior half of body: *S. ellipticus,* oyster flatworm OR Marginal eyes around entire periphery of body: *S. zebra*
ma	te	x	te	>10	*Gnesioceros* spp.
ma	x	me	eq	>10	*Coronadena mutabilis,* mutable flatworm
ma	x	me	sm	>10	*Discocelides ellipsoides*
ma	x	x	sm	>10	*Notoplana atomata,* speckled flatworm
ma	x	x	sm	<10	*Euplana gracilis,* oyster flatworm
3	te	x	x	2	*Procerodes littoralis*
3	x	x	x	2	Found among gill flaps of *Limulus,* the horseshoe crab: Posterior branches of gut do not join with one another: *Bdelloura candida,* Limulus leech OR Posterior branches of gut fuse with one another *Syncoelidium pellucidum* OR Free-living: *Foviella affinis*

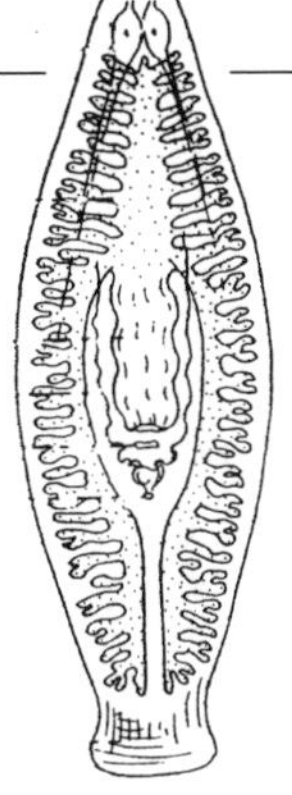

FIG. 10.11

Bdelloura candida, limulus leech (Or: Tricladida, Bdellouridae). White to yellowish; squared posterior with attachment sucker. Bo,Co,Li–Su,pr,14 mm (0.55 in), (GI02). (FIG. 10.11)

Coronadena mutabilis, mutable flatworm (Or: Polycladida, Discocelidae). Ovaloid; grayish to yellowish brown. Estuarine, especially in Chesapeake Bay. Vi,Ha–Al,Es–Li–Su,pr,20 mm (0.8 in), (GI02). (FIG. 10.12)

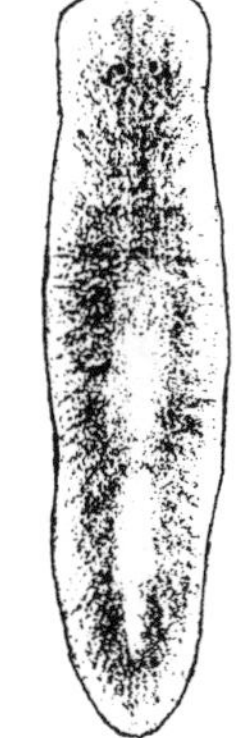

FIG. 10.14

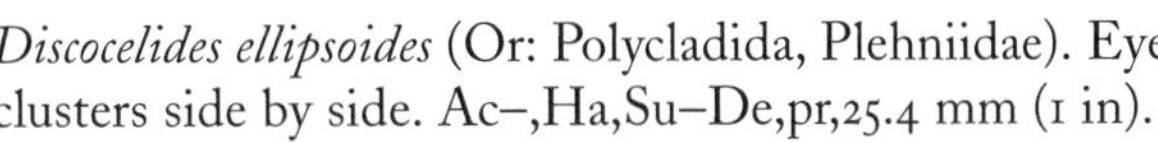

Discocelides ellipsoides (Or: Polycladida, Plehniidae). Eye clusters side by side. Ac–,Ha,Su–De,pr,25.4 mm (1 in).

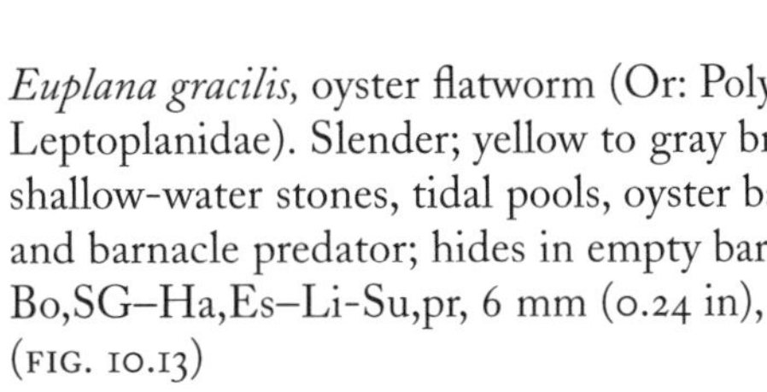

Euplana gracilis, oyster flatworm (Or: Polycladida, Leptoplanidae). Slender; yellow to gray brown. Under shallow-water stones, tidal pools, oyster bars. Serious oyster and barnacle predator; hides in empty barnacle plates. Bo,SG–Ha,Es–Li-Su,pr, 6 mm (0.24 in), (GI02,NM401). (FIG. 10.13)

FIG. 10.12

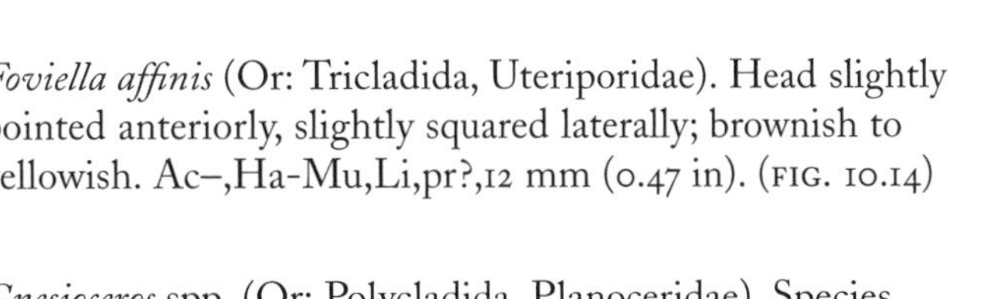

Foviella affinis (Or: Tricladida, Uteriporidae). Head slightly pointed anteriorly, slightly squared laterally; brownish to yellowish. Ac–,Ha-Mu,Li,pr?,12 mm (0.47 in). (FIG. 10.14)

Gnesioceros spp. (Or: Polycladida, Planoceridae). Species differ in details of reproductive system. *G. verrilli* is shown. Brownish, speckled. Vi,Al,Ps,pr,10 mm (0.4 in)?, (GI03). (FIG. 10.15)

FIG. 10.13

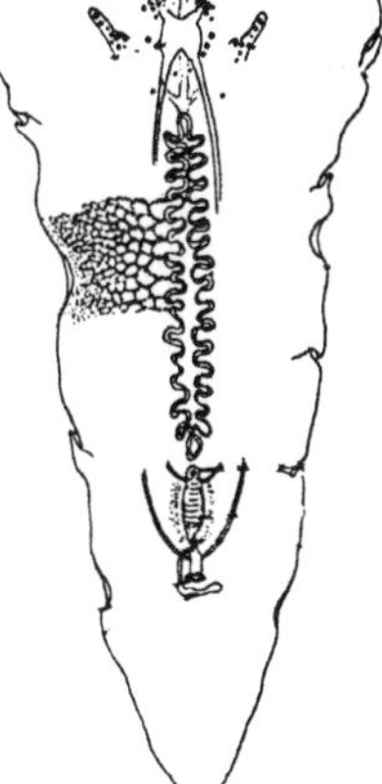

FIG. 10.15

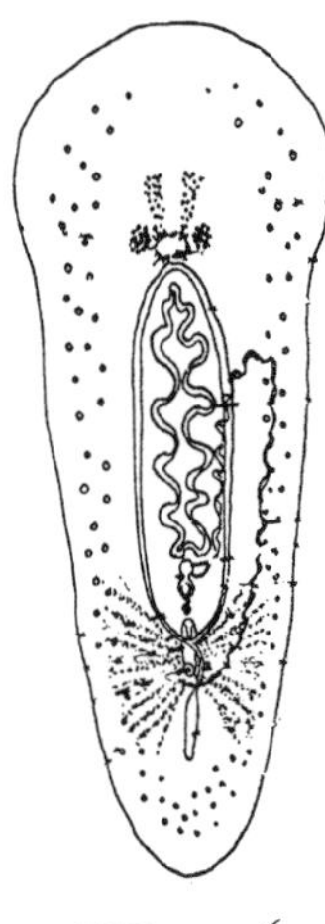

FIG. 10.16

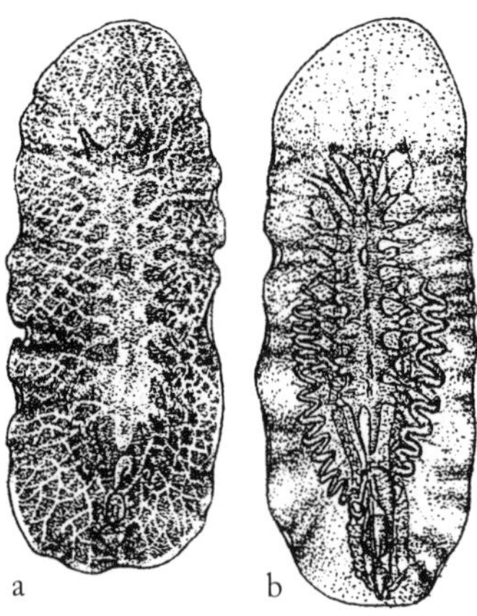

FIG. 10.17

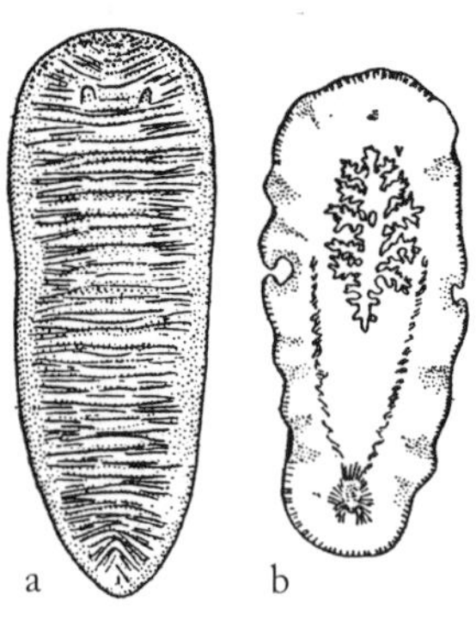

FIG. 10.18

Notoplana atomata, speckled flatworm (Or: Polycladida, Leptoplanidae). Chocolate brown with darker blotches, some with lighter center line longitudinally. Under intertidal and subtidal rocks, on algae, in tidal pools. Ac–,Ha–Al,Li–Su,pr,20 mm (0.8 in), (GI02,NM402), ☺. (FIG. 10.16)

Stylochus ellipticus, oyster flatworm (Or: Polycladida, Stylochidae). Cream, yellow, brownish. Under stones, shallow-water tidal pools. Eats oysters and barnacles; often hides in empty barnacle plates. Bo,Ha–Al,Li–Su,pr,25.4 mm (1 in), (GI02,NM401), ☺. (FIG. 10.17. a, dorsal; b, ventral.)

Stylochus zebra (Or: Polycladida, Stylochidae). White with narrow, brown stripes. Often on *Busycon* shells occupied by hermit crab *Pagurus pollicaris* (feeds on *Crepidula plana*); also free-living on stones and pilings; sluggish. Vi,Co–Ha–Al,Li–Su,pr,18 mm (0.7 in), (GI02,NM402). (FIG. 10.18. a, dorsal; b, ventral.)

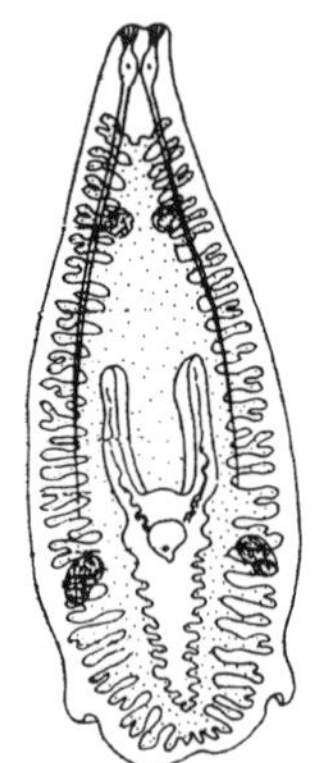

FIG. 10.19

Syncoelidium pellucidum (Or: Tricladida, Bdellouridae). Among gill flaps of *Limulus,* the horseshoe crab; similar to more common, *Bdelloura candida,* but smaller in size. Bo,Co,Li–Su,pr,5 mm (0.2 in). (FIG. 10.19)

Chapter 11 ■ Phylum Nemertea (Rhynchocoela), Ribbon Worms

11.1

Nemerteans, or ribbon worms, are predators living in soft sediments or bivalve beds. They are noted for their elongate, flexible, flattened, nonsegmented, and often colorful bodies. They are sometimes called the phylum Rhynchocoela, which emphasizes the unusual chamber surrounding the proboscis used in food capture. Identification requires observation of head details such as presence and location of tiny dark eyespots and of ciliated, sensory grooves. It can be difficult to collect intact nemerteans because they tend to fragment when handled. A dissecting microscope may be necessary to view details. Sometimes gentle coverslip pressure on a wet-mount preparation will enable you to see tiny eyespots that tend to be masked by body pigments.

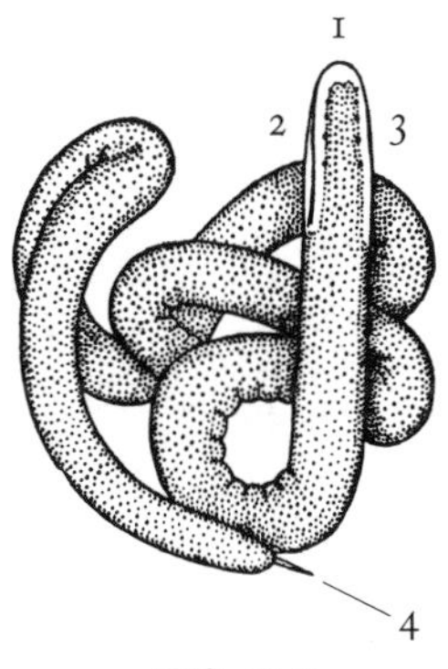

FIG. 11.1

1. Head shape (FIG. 11.1)
   - A. *Pointed* — po
   - B. *Rounded* — ro*
   - C. *Notched* — no
2. Grooves on head (FIG. 11.1)
   - A. Grooves are *horizontal* along sides of head — ho*
   - B. Single pair of grooves are *angled* and can be seen from above (look carefully at corners of head) — an
   - C. *Two* pairs of grooves are *angled* towards the back of the head — 2a
   - D. Grooves *absent* (look carefully from above and the side) — x
3. Presence of eyespots (tiny black spots; sometimes masked by body pigment) (FIG. 11.1)
   - A. *Many eyespots* (more than 4) on dorsal surface of head — me*
   - B. *Four eyespots* present on dorsal surface of head — 4e
   - C. *Two eyespots* present on dorsal surface of head — 2e
   - D. Eyespots *absent* — x

4. Presence of a tail (these worms are easily broken; be sure the tail end is intact to evaluate this feature) (FIG. 11.1)
   - A. Short but distinct *tail* present — ta*
   - B. Terminal *sucker* present — su
   - B. Tail and sucker both *absent* — x
5. Predominant body color
   - A. *Purple* — pu
   - B. *Brown* — br
   - C. *Green* — gr
   - D. *Red* — re
   - E. *Pink* — pi
   - F. *Yellow* — ye
   - G. *White* — wh

11.2

Head	Grooves	Eyes	Tail	Color	Organism
po	x	x	x	wh	*Procephalothrix spiralis,* thread ribbon worm
po	x	x	ta	pi,wh	*Zygeupolia rubens,* sharp-headed ribbon worm
ro,po	ho	x	ta	re	*Micrura leidyi,* bright red ribbon worm
ro	ho	x	ta	br,pi,wh	*Cerebratulus lacteus,* milky ribbon worm
ro	ho	x	x	br,pi,wh	*C. lacteus,* milky ribbon worm
ro	ho	me	x	re	*Lineus* (Lineidae) Contracts when stimulated: *L. ruber* OR Coils when stimulated: *L. sanguineus* (=*socialis*)
ro	ho	me	x	gr	Color solid: *Lineus ruber* OR Green-gray with light middorsal stripe: *Tenuilineus bicolor*
ro	ho	4-6e	ta	re	*Micrura affinis,* bright red ribbon worm
ro	ho	4e	x	ye,pi	*Lineus arenicola,* sandy lineus
ro	x	4e	x	any	*Oerstedia dorsalis*
ro	x	2e	x	re,ye	*Carcinonemertes carcinophila,* crab nemertean
ro	x	x	x	ye,wh	*Tubulanus pellucidus,* tube nemertean
ro,no	x	x	su	any	*Malacobdella grossa,* leech ribbon worm
ro,no	an	me	x	pu,br	*Amphiporus angulatus,* chevron amphiporus
ro,no	2a	me	x	gr,ye,wh	*Zygonemertes virescens,* green ribbon worm
ro,no	an	4e	x	ye,gr	*Tetrastemma elegans,* four-eyed ribbon worm
ro,no	an	4e	x	gr	Green with brown stripes: *Cyanophthalma cordiceps* (=*Tetrastemma vittatum*), striped, four-eyed ribbon worm OR Uniform greenish, reddish, yellowish: *Tetrastemma candidum,* green four-eyed ribbon worm

FIG. 11.2

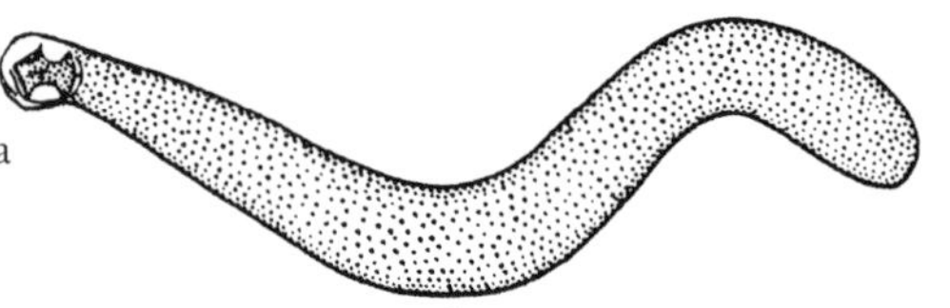

Amphiporus angulatus, chevron amphiporus (Amphiporidae). Light-colored V behind head. Dark dorsally, lighter ventrally. Ac–,Sa–Mu,Su,pr,150 mm (6 in). Several additional species of *Amphiporus* have been reported from the region. (FIG. 11.2. a, animal; b, head of *A. cruentatus.*)

Carcinonemertes carcinophila, crab nemertean (Carcinonemertidae). Slender, flattened; secretes sticky mucus on handling. On gills and eggs of swimming crabs, especially *Callinectes.* Pale, whitish or pinkish. Bo,Co,Sa,pr,40 mm (1.5 in), (G107)

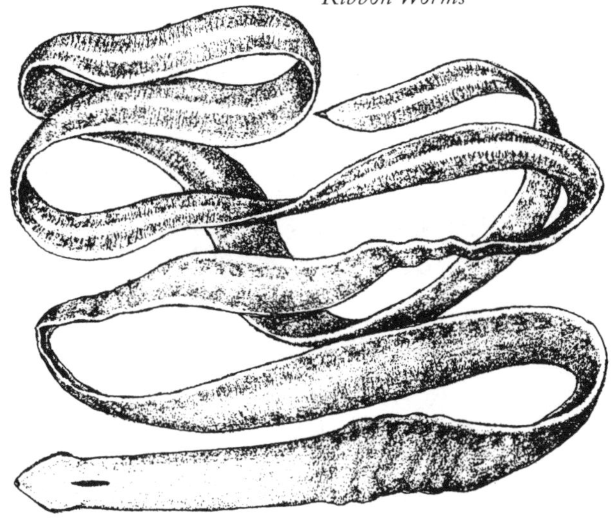

FIG. 11.3

Cerebratulus lacteus, milky ribbon worm (Lineidae). Lateral margins of trunk expanded for swimming. Tail is often missing or present as a regenerating bump. Long, sticky protrusible proboscis arises apart from mouth that forms a ventral slit. Posterior of body flattened; anterior round. Can be up to several feet long and ⅝ in. wide. Eats bivalves and polychaetes; capable of undulating swimming. Bo,Sa–Mu,Es–Li–Su,pr,60 cm (48 in), (GI05,NM408), ☺. (FIG. 11.3)

FIG. 11.4

Cyanophthalma cordiceps (*Tetrastemma vittatum*), striped four-eyed ribbonworm (Tetrastemmatidae). Ac–,Al–Mu,Li–Su,pr,30 mm (1.2 in), (GI06,NM410). (FIG. 11.4)

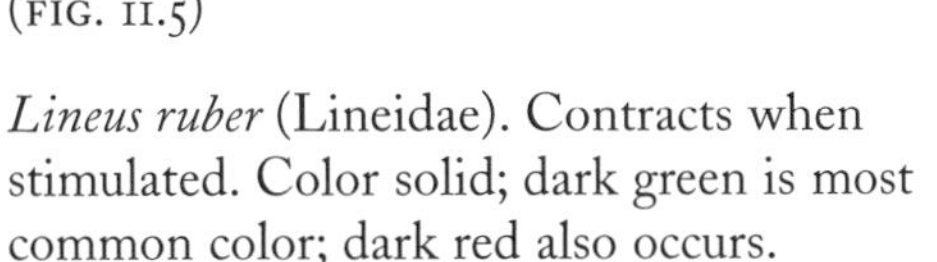

FIG. 11.5

Lineus arenicola, sandy lineus (Lineidae). Bo,Ha–Sa–Mu,Su,pr,102 mm (4 in), (GI04). (FIG. 11.5)

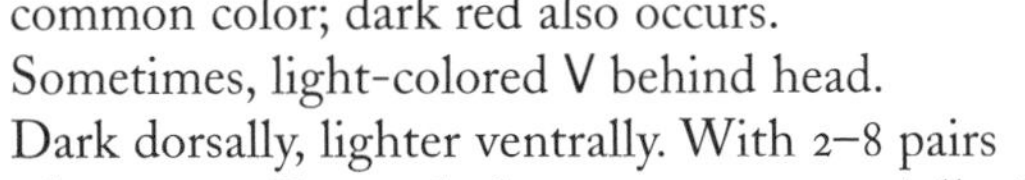

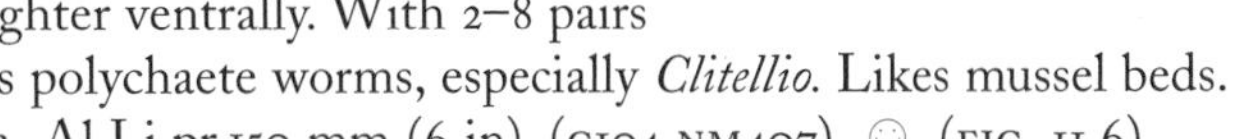

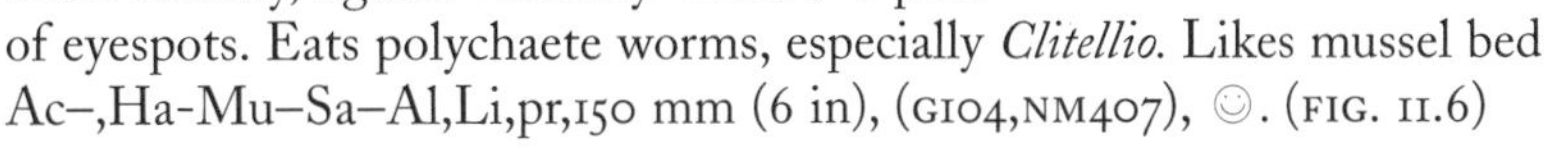

FIG. 11.6

Lineus ruber (Lineidae). Contracts when stimulated. Color solid; dark green is most common color; dark red also occurs. Sometimes, light-colored V behind head. Dark dorsally, lighter ventrally. With 2–8 pairs of eyespots. Eats polychaete worms, especially *Clitellio.* Likes mussel beds. Ac–,Ha-Mu–Sa–Al,Li,pr,150 mm (6 in), (GI04,NM407), ☺. (FIG. 11.6)

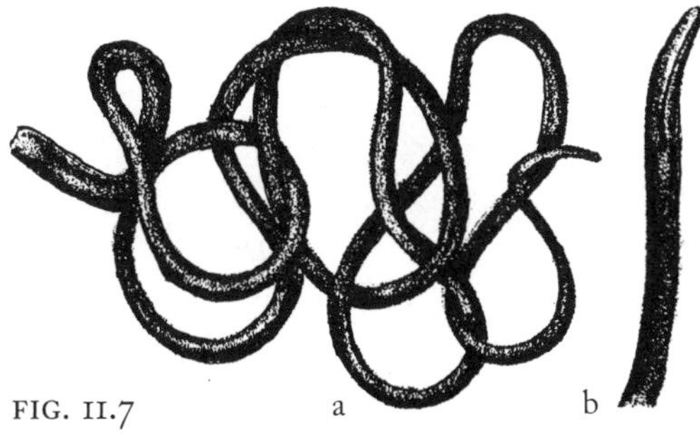

FIG. 11.7

Lineus sanguineus (=*socialis*) (Lineidae). Red to brown color, darker posteriorly, is most common. Trunk may be banded. Coils when stimulated. Sometimes in groups. Bo,Ha–Mu–Sa–Al,Li,pr,150 mm (6 in), (GI04,NM407). (FIG. 11.7. a, animal; b, head, lateral view.)

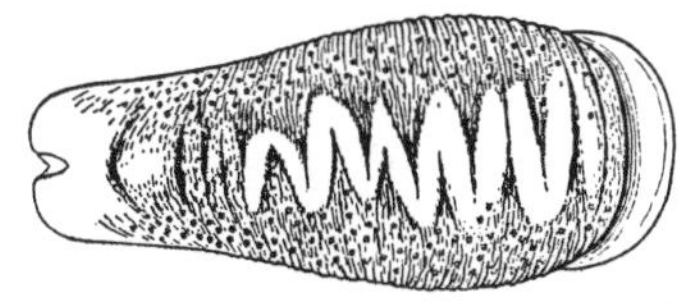

FIG. 11.8

Malacobdella grossa, leech ribbon worm (Macrobdellidae). Short, wide, pale-colored body. Commensal in mantle cavities of bivalves. Bo,Co,Li–Su,sc,75 mm (3 in), (GI07). (FIG. 11.8)

Micrura affinis, bright red ribbon worm (Lineidae). Bright red, cream-colored margins. Ac,Sa–Gr,Su,pr,15 cm (6 in).

FIG. 11.9

Micrura leidyi, bright red ribbon worm (Lineidae). Distinct neck; mouth a ventral, rounded hole. Lighter coloring toward head and body edges. Secretes mucus when handled. Darker red middorsal line. Vi,Sa–Mu,Es–Li,pr,30 cm (12 in), (GI05,NM408), ☺. (FIG. 11.9)

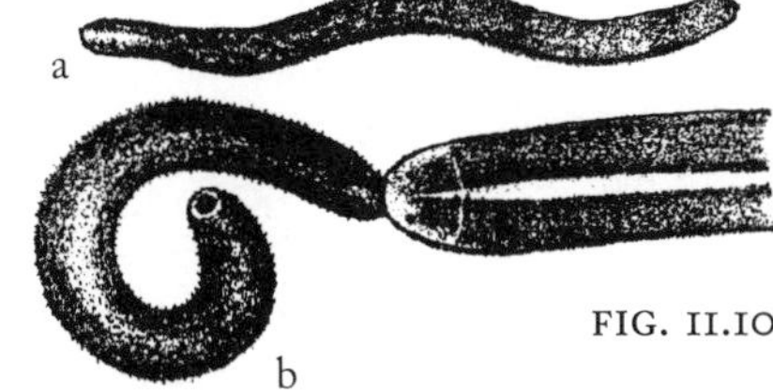

FIG. 11.10

Oerstedia dorsalis (Prosorchochmidae). Cylindrical; tapers anteriorly and posteriorly. Variable coloring, often lighter ventrally. Sluggish. Bo,SG–Al–Gr–Sa,Li–Su,pr,12 mm (0.5 in), (GI06). (FIG. 11.10. a, animal; b, head with proboscis extended.)

Procephalothrix spiralis, thread ribbon worm (Cephalothricidae). Very slender; coils; mouth ventral and well back from head tip. Ac–,Ha–Sa,Li,pr,102 mm (4 in), (GI07).

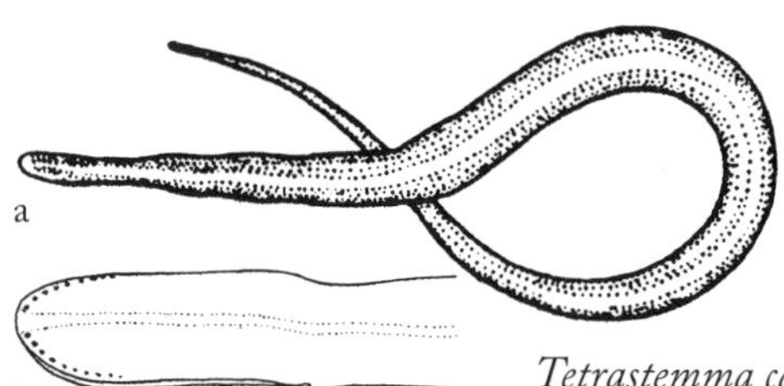

FIG. 11.11

Tenuilineus (=*Lineus*) *bicolor* (Lineidae). Green gray with light middorsal stripe, becoming tan toward posterior. Often with hydroids. Vi+,Ha–Sa,Es–Su,pr,50 mm (2 in), (G104,NM407). (FIG. 11.11. a, animal; b, head.)

Tetrastemma candidum, green four-eyed ribbon worm (Tetrastemmatidae). Distinct neck. Greenish brown. Bo,Gr–Ha–Al–SG,Es–Li–Su,pr,33 mm (1.3 in), (G106,NM410), ☺. (FIG. 11.12)

FIG. 11.12

FIG. 11.13

Tetrastemma elegans, four-eyed ribbon worm (Tetrastemmatidae). Distinct neck. Usually yellow (sometimes green especially when on eelgrass) with pair of full-length brown stripes. Vi,SG–Al–Ha,li,pr,12 mm (0.5 in), (G106,NM410). (FIG. 11.13)

Tubulanus pellucidus, tube nemertean (Tubulanidae). Some with orange stripes. In thin, mucous tubes in fouling community. Vi,Ha–Mu,Es–Li,pr?,12 mm (1 in), (G107,NM407).

Zygeupolia rubens, sharp-headed ribbon worm (Lineidae). Color ranges from white (head) to pinkish and then tan. Burrows in surfy sand beaches. Vi,Sa,Es–Li–Su,pr,77 mm (3 in), (G105), ☺.

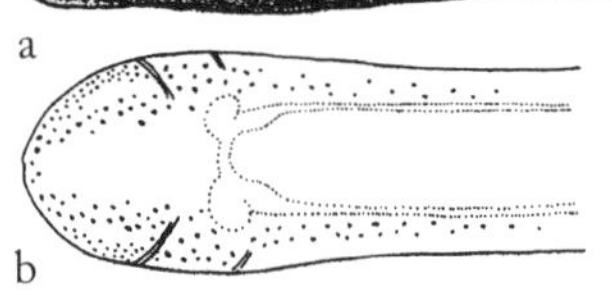

FIG. 11.14

Zygonemertes virescens, green ribbon worm (Amphiporidae). Eyes may extend well back onto body, past double pair of angular grooves. Active; eats amphipods. Bo,Ha–SG–Al,Li–Su,pr,50 mm (2 in), (G106). (FIG. 11.14. a, animal; b, head.)

Chapter 12 ■ Phylum Ectoprocta or Bryozoa

12.1 Formerly ectoprocts were grouped with the Entoprocta (treated here as superficially similar to cnidarian hydroids, 8.26) in the phylum Bryozoa. Substantial differences in their construction and development, however, place these animals as separate phyla. Although some continue to use Bryozoa, the older name, in a restricted sense to refer only to ectoprocts, I prefer the phylum name Ectoprocta to avoid this source of confusion.

Ectoprocts form colonies of tiny individuals called *zooids,* each encased in its own secreted covering, the *zooecium.* The *polypide* or living part of the zooid extends through an *orifice* (opening) to deploy its feeding tentacles. Some appear as flattened encrustations on firm substrates, whitish and calcareous, firm and leathery, or gelatinous in texture. Some ectoprocts grow as erect, bushy, or feathery colonies. The latter are easily confused with hydroid colonies. Fresh specimens, placed in seawater and observed using magnification, will extend their tentacular clusters and create water currents through the use of cilia. Healthy hydroids do not produce ciliary currents; rather, they rely on their stinging cnidae to immobilize prey.

There are many species of ectoprocts, and most are rather difficult to identify. Carefully examine several individuals from undamaged and uncluttered portions of a colony to look for such characters as spines and pores in their zooecial cases. Examine several individuals to evaluate each characteristic because individual zooids show considerable variation in morphological detail within the same species and even among members of the same colony. The orientational terms, *proximal* (closer to the colony base) and *distal* (farther from the colony base) are useful in describing characters in these animals.

In some Ectoprocta, certain zooids are modified and perform specialized functions. Small, pointy-tipped colony members with tiny biting jaws, *avicularia* are shaped like the head of a bird. These individuals protect the colony from small invaders and rely on the feeding members of the colony for sustenance. Look carefully all over the colony for these scattered individuals. They tend to be located near the zooecial orifice. Because they may appear seasonally or irregularly, their presence is useful information, but their absence is of more limited value. Another specialized zooid, the *ooecium,* functions as a brood chamber for a developing embryo.

These globular structures are found just above the major orifice only in some ectoprocts. As with avicularia, look widely over the colony for ooecia. Although their presence is a useful character to note, their absence may be only an artifact of their seasonality.

Regional marine ectoprocts are placed in three orders within two classes. The class Stenolaemata includes the order Cyclostomata whose members extend from circular, terminal orifices of their tubular, calcareous zooecia perforated with tiny pores. Most ectoproct species belong in two orders within the class Gymnolaemata. The order Cheilostomata includes zooids that construct boxlike, often calcareous zooecia, each with a frontal orifice provided with a tiny *operculum,* a trap door to seal in the withdrawn individual. Ectoprocts of the order Ctenostomata are tubular with cuticular, leathery, or gelatinous (but not calcareous) zooecial walls. Individual zooids are often interconnected along the substrate by a creeping *stolon* (runner), although in some species, zooids are tightly packed and form a crust.

In this treatment, assume that species belong to the Cheilostomata unless they are designated otherwise. Dimensions are for typical colony height for erect ectoprocts or colony width for those that encrust. Individuals are small enough to require dissecting microscope observations throughout and a compound microscope for details. Make temporary wet mounts following instructions in the Introduction.

Additional references: (RC) Rogick, M. D., and H. Croasdale 1949. (RH) Ryland, J. S., and P. J. Hayward 1991.

1. Overall growth form
   - A. Colonies *erect;* bushy or fernlike — er
   - B. Colonies *encrusting* (flattened along the substratum) — en
   - C. Individuals arise from *stolon*like runner along the substrate — st

12.2

Form	Group
en	Encrusting ectoprocta, 12.3
er	Erect ectoprocta, 12.11
st	Stolonate ectoprocta, 12.13

12.3 Encrusting Ectoprocta (from 12.2)

1. Compare area of the main opening or orifice in an individual zooecium to its surface area
   - A. Orifice occupies *less than half* the surface area — <½
   - B. Orifice occupies *most* of surface, well beyond half its area — mo
   - C. Orifice at *end* of tubular zooecium — en

12.4

Orifice	Group
<½	Encrusting ectoproct with small orifice, 12.5
mo	Encrusting ectoproct with large orifice, 12.7
en	Encrusting ectoproct with terminal orifice, 12.9

12.5 Encrusting Ectoprocts with Small Orifice (from 12.4)

1. Distribution of small pores (does not include the orifice itself) on zooecial surface (FIG. 12.1)
   - A. Pores *scattered* irregularly — sc*
   - B. Pores arranged in single row around the *border* of each zooecium — bo
   - C. Pores *absent;* zooecial surface solid — x

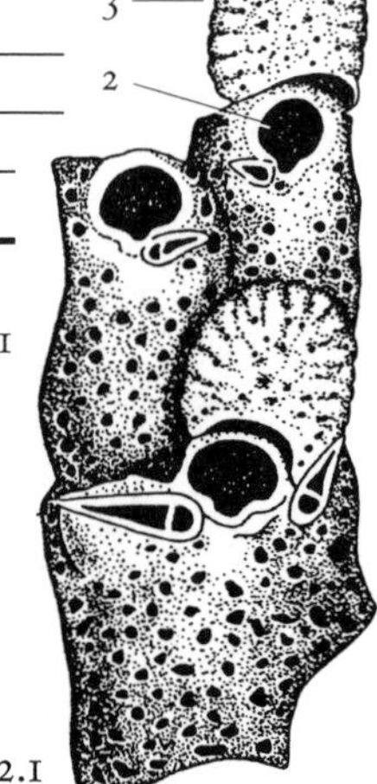

FIG. 12.1

2. Shape of orifice (examine after zooid has withdrawn; ignore "teeth" and spines for the moment) (FIG. 12.1)
   - A. Orifice rounded *oval,* circular, or square in shape — ov
   - B. Orifice *semicircular;* rounded at the top, straight along the bottom — sc
   - C. Orifice *key-hole* shaped: opening is framed by raised collar that does not meet (i.e., is notched) along its bottom rim — kh*
   - D. Orifice closes to form a pair of *flaps* — fl
   - E. Orifice closes to form *puckering* — pu
3. Surface features of the ooecium (globular brood pouch), if present (FIG. 12.1)
   - A. Ooecial surface with *pores* or openings — po
   - B. Ooecial surface *granular* and rough textured — gr
   - C. Ooecial surface ribbed, creating *fluted* margin to ooecium — fl*
   - D. Ooecia *absent* from colony — x
4. Presence of spines or "teeth" on the exposed surface ("spines" project upright; "teeth" project laterally into the orifice)
   - A. *One tooth* (the "lyrula") projects into orifice from below — 1t
   - B. *Two teeth* project into orifice from the sides — 2t
   - C. Series of several (provide *number*) of slender *spines* surround orifice — __sp
   - D. A broadly pointed *elevation* is formed toward the center of the zooecial surface — el
   - E. Teeth, spines and elevations *absent* from margin of orifice — x*

12.6

Pores	Orifice Shape	Ooecium	Teeth/ Spines	Species
sc	sc	gr	x	*Microporella ciliata,* micropore crust ectoproct
sc	kh	fl	el,x	*Schizoporella unicornis,* red crust ectoproct
sc	ov	x	x	*Cryptosula pallasiana,* orange crust ectoproct
sc	ov,kh	po	1t	*Cribrilina punctata*
bo	ov	?	2t	*Hippoporina contracta*
x	ov,kh	po	2t,x	*Hippothoa hyalina,* glassy ectoproct
x	ro,pu	x	x	*Alcyonidium,* rubbery ectoproct
x	fl	x	10sp	*Flustrellidra hispida,* bristly ectoproct

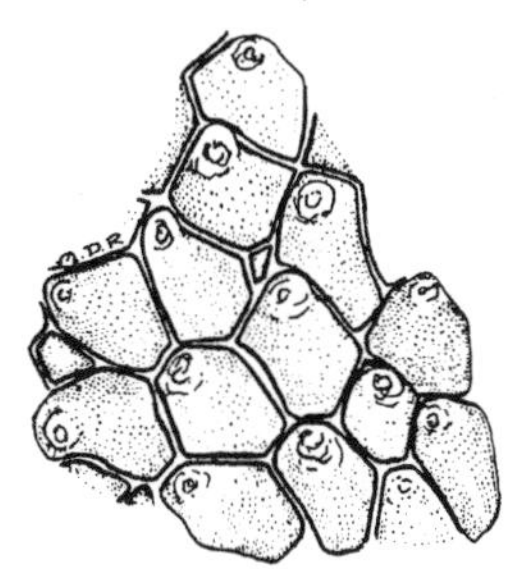

FIG. 12.2

Alcyonidium, rubbery ectoproct (Or: Ctenostomata, Alcyonidiidae). Yellow to brown; encrusting initially, more erect with added growth. Smooth and rubbery. Difficult species distinctions; comments here for "best guesses only":

Alcyonidium gelatinosum. 15–17 tentacles/zooid; soft, gelatinous. Ac,Ha,Es–Li–Su,ff,?, (G112).

Alcyonidium hirsutum. Zooid surface hairy, often muddy. Vi,Ha,Su,ff,?.

Alcyonidium parasiticum. Only found on animals. Bo,Ha,Es–Su,ff,?, (G112).

Alcyonidium polyoum. Flat growth; grayish, yellowish or reddish. On shells, crabs, algae. Bo,Ha,Es–Li–Su,ff,?, (G112,RC45). (FIG. 12.2)

FIG. 12.3

Alcyonidium verrilli, dead man's fingers. Palmate; yellowish pink; 16 tentacles/zooid. Dredged, Chesapeake Bay, abundant in winter. Vi,Ha,Es–Su,ff,40 cm (15.8 in), (G112,RH23). (FIG. 12.3)

Cribrilina punctata (Cribrilinidae). Front surface with plate of 11–13 fused ribs separated by lines of tiny pores. Ac,Ha–Al,Li–De,ff,17.8 mm (0.7 in), (G119,RC54). (FIG. 12.4)

FIG. 12.4

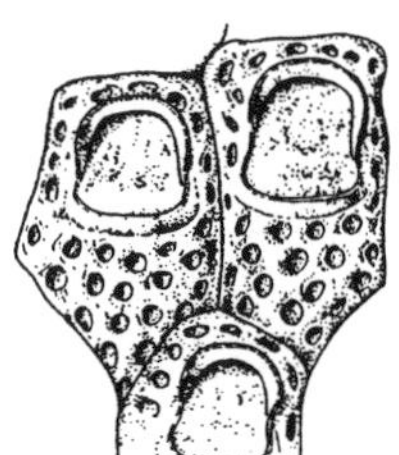

FIG. 12.5

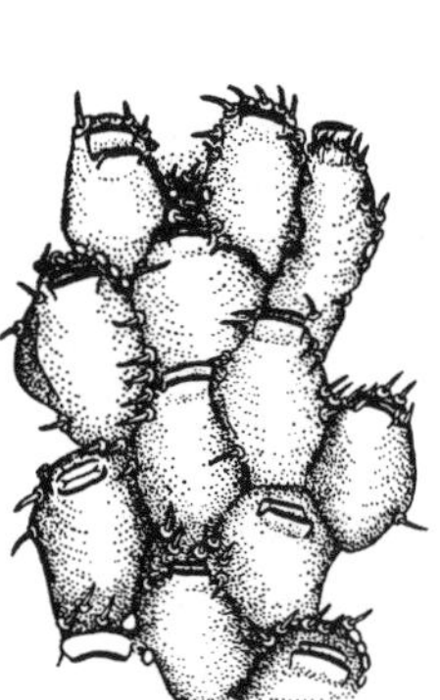

FIG. 12.6

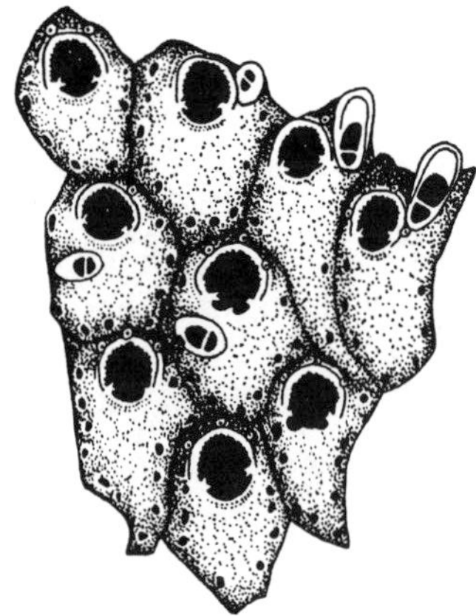

FIG. 12.7

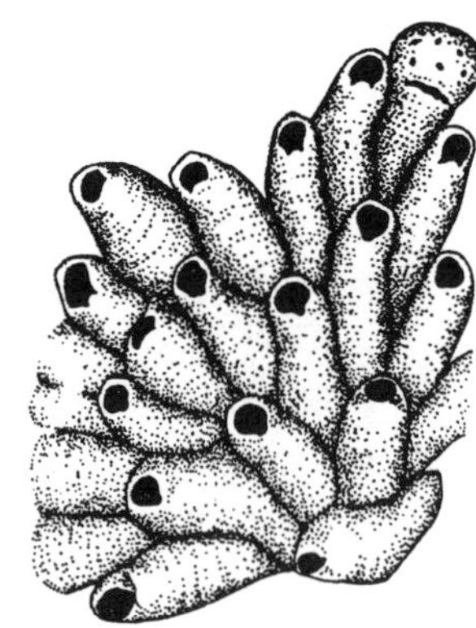

FIG. 12.8

Cryptosula pallasiana, orange crust ectoproct (Hippoporinidae). Orange, brownish, to pinkish. Circular patches. Bo,Ha–Al,Es–Li–,ff,30.5 mm (1.2 in), (GI18,RC55), ☺. (FIG. 12.5)

Flustrellidra hispida, bristly ectoproct (Order Ctenostomata, Flustrellidridae). Purple, red, or brown; horny or proteinaceous spines. Rubbery crust on algae, especially *Fucus* and *Gigartina.* Ac–,Al,Li–Su,ff,?, (GI12,RC56). (FIG. 12.6)

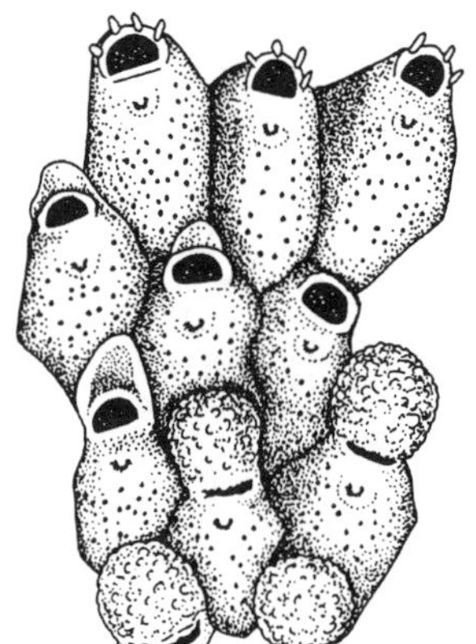

FIG. 12.9

Hippoporina contracta (Hippoporinidae). Small, crowded colonies; zooecia with peripheral border of pores. Top margin of orifice slightly beaded. Several species possible. Bo,Ha–Al,Li–De,ff,?, (RC57). (FIG. 12.7)

Hippothoa hyalina, glassy ectoproct (Hippothoidae). Frontal wall is glassy-iridescent to whitish, smooth. Small, circular colonies on stones, *Laminaria,* and other algae. Bo,Ha–Al,Li–De,ff,5 mm (0.2 in), (GI18,RC57). (FIG. 12.8)

Microporella ciliata, micropore crust ectoproct (Microporellidae). Small, white silvery patches. Bo,Ha–Al,Es–Li–De,ff,?, (GI19,RC63). (FIG. 12.9)

Schizoporella unicornis, red crust ectoproct (Schizoporellidae). Pink, reddish, or whitish coloring. Zooids rectangular with bump just below orifice. Ooecium unusually shaped. Vi+,Ha–Al,Es–Li–Su,ff,?, (GI18,RC66), ☺. (FIG. 12.10)

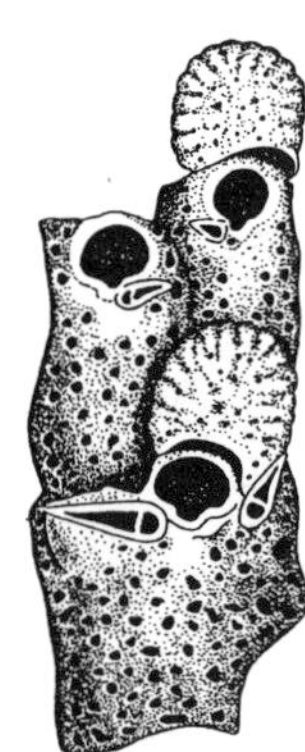

FIG. 12.10

12.7 **Encrusting Ectoproct with Large Orifice** (from 12.4)

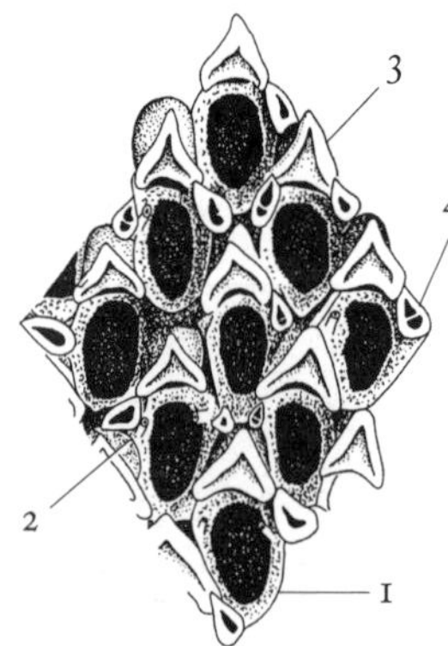

FIG. 12.11

1. Distribution of small pores (does not include the orifice itself) on zooecial surface (FIG. 12.11)
   - A. Pores *scattered* irregularly — sc
   - B. Pores *absent;* zooecial surface solid — x*
2. Presence of spines on the exposed surface (FIG. 12.11)
   - A. One or more (provide *number*) distinct spines on zoecial surface — __*
   - B. A *blunt knob* located in at least two corners of zooecium — bk
   - C. Spines and knobs *absent* from zoecial surface — x
3. Surface features of the ooecium (globular brood pouch) (FIG. 12.11)
   - A. Ooecial surface bears distinctive *chevron*-shaped ridge — ch*
   - B. Ooecial surface with *ridge,* although not chevron-shaped — ri
   - C. Ooecia *absent* from colony — x
4. Presence of avicularia (tiny, pointy-tipped colony members with tiny biting jaw) (FIG. 12.11)
   - A. *Avicularia* present — av*
   - B. Avicularia *absent* — x

12.8

Pores	Spines	Ooecium	Avicularia	Species
sc	3–13	x	x	*Electra pilosa*
x	12–15	ri	av	*Callopora craticula*
x	10–18	x	x	*Electra monostachys* (=*hastingsi*)
x	7–8	x	x	*Conopeum tenuissimum*, lacy crust ectoproct
x	2–4	ch	av	*Callopora aurita*
x	1–8	x	x	*Electra crustulenta*
x	bk,x	x	x	*Membranipora* spp., coffin box ectoproct

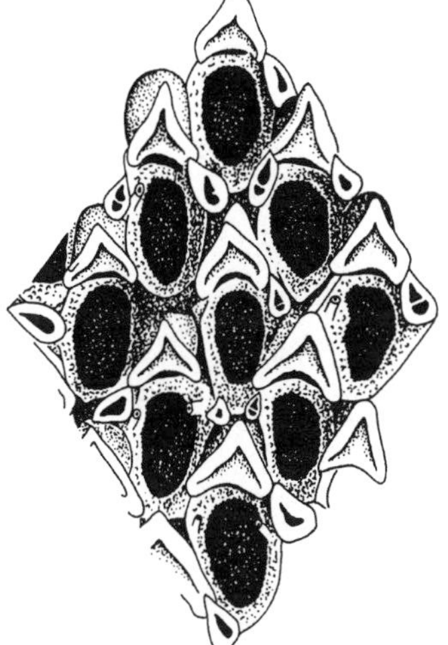
FIG. 12.12

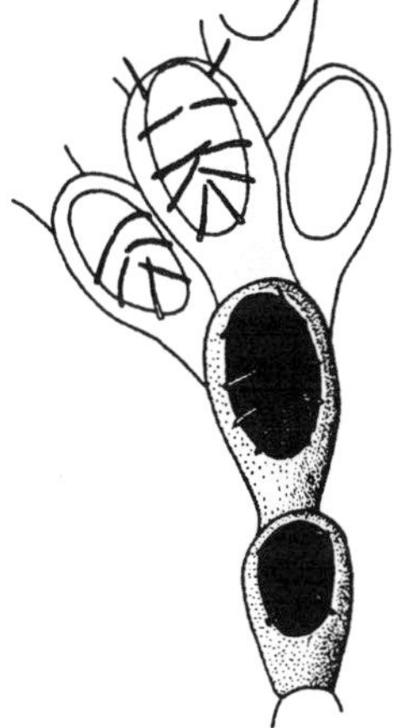
FIG. 12.13

FIG. 12.14

FIG. 12.15

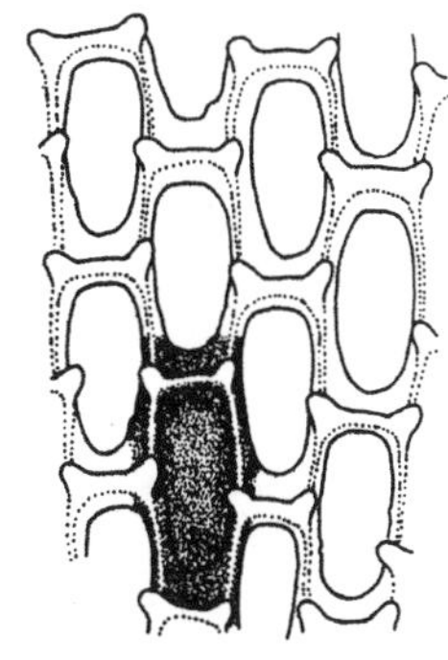
FIG. 12.16

Callopora aurita (Calloporidae). Four spines near opening; one enlarged. Small, white colonies. Ac–,Ha–Al,Li–Su,ff,?, (RC53). (FIG. 12.12)

Callopora craticula (Calloporidae). Strong spines slant toward top of orifice; two pairs of spines stand upright. Small, white colonial patches. Ac–,Ha–Al,Su–De,ff,?, (GI16).

Conopeum tenuissimum, lacy crust ectoproct (Membraniporidae). Vi,Ha–Al–SG,Es–Su,ff,?.

Electra crustulenta (Membraniporidae). Orifice covers nearly entire surface; no shelf. Not clear in the literature: described either with several sharp spines rimming the orifice (but fewer than *E. monostachys*), or with only one bottom, medial tooth. Can form large multilayered clumps. Chesapeake: most abundant shallow-water ectoproct. Important competitor with oysters for settling space. Bo,Al,Es–Li–Su,ff,?, (GI16), ☺. (FIG. 12.13)

Electra monostachys (=*hastingsi*) (Membraniporidae). Orifice with several, delicate marginal spines. Small, white, dendritic colonies on rocks, shell, or algae. Bo,Ha–Al,Li–Su,ff,20 mm (0.8 in), (RC54). (FIG. 12.14)

Electra pilosa (Membraniporidae). Orifice occupies two-thirds of surface; porous shelf covers bottom third. Median spine often elongated; marginal spines are shorter and weaker. Colonies grayish and often stellate (starlike). Especially on *Laminaria* and other algae. Bo,Al,Es–Li–Su,ff,30.5 cm (12 in) [7.6 cm (3 in)], (GI16,RC56), ☺. (FIG. 12.15)

Membranipora spp., coffin box ectoproct (Membraniporidae). Zooids rectangular, often with bumps or tubercles. *Membranipora membranacea* with stout tubercles in each corner. Forms extensive lacy colonies with smooth borders on seaweeds, especially on kelp. Recent invader from Europe (see Berman et al. 1992). Vi,Al,Li–,ff,7.6 cm (3 in), (GI15,RC59–63), ☺. (FIG. 12.16)

12.9 **Encrusting Ectoproct with Terminal Orifice** (from 12.4)

1. Shape of orifice (examine after zooid has withdrawn; ignore "teeth" and spines for the moment)
   - A. Orifice rounded *oval* or circular in shape — ov
   - B. Orifice distinctly *squared* — sq
   - C. Orifice *keyhole* shaped (framed by raised collar which does not meet, i.e., is notched, along its bottom rim) — kh
2. Surface features of the *ooecium,* a globular brood pouch for developing embryos found just above the major opening only in some ectoprocts
   - A. Ooecial surface with *pores* or openings — po
   - B. Ooecia *absent* from colony — x
3. Orientation of colony members
   - A. Nearly all zooecia *horizontal,* lying flat — ho
   - B. Nearly all zooecia *erect,* upright — er
   - C. Zooecia *mixed;* horizontal toward periphery, more vertical toward center of colony — mi

12.10

Orifice Shape	Ooecium	Orientation	Species
ov,kh	po	ho	*Hippothoa hyalina*
ov,sq	x	mi	*Tubulipora liliacea,* panpipe ectoprocts
sq	x	er	*Bowerbankia* spp. (descriptions under 12.13)

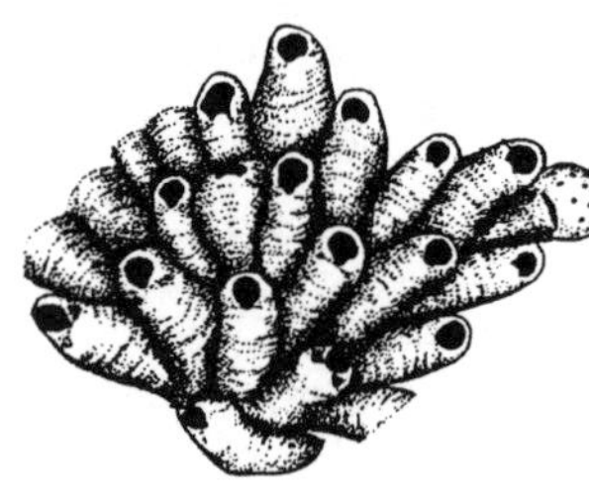

FIG. 12.17

Hippothoa hyalina, glassy ectoproct (Hippothoidae). Frontal wall is glassy-iridescent to whitish, smooth. Small, circular colonies on stones, *Laminaria* and other algae. Bo,Ha–Al,Li–De,ff,5 mm (0.2 in), (G118). (FIG. 12.17)

Tubulipora liliacea, panpipe ectoprocts (Cl: Stenolaemata, Or: Cyclostomata, Tubuliporidae). Tubular zooecia with terminal openings, fused along part or entire length. Radiate from center to form fan-shaped colonies with central zooids largest. Especially common on hydroids. Ac,Ha,Su,ff,12 mm (0.5 in), (G114). (FIG. 12.18)

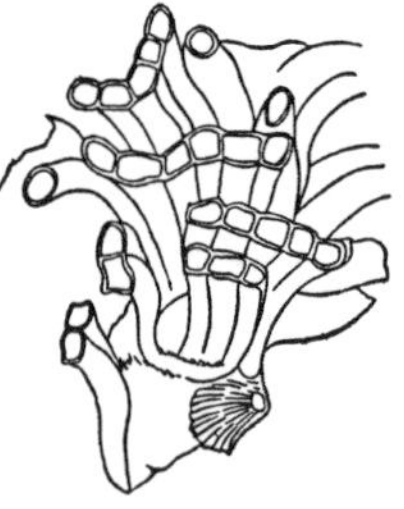

FIG. 12.18

12.11 **Erect Ectoprocta** (from 12.2)

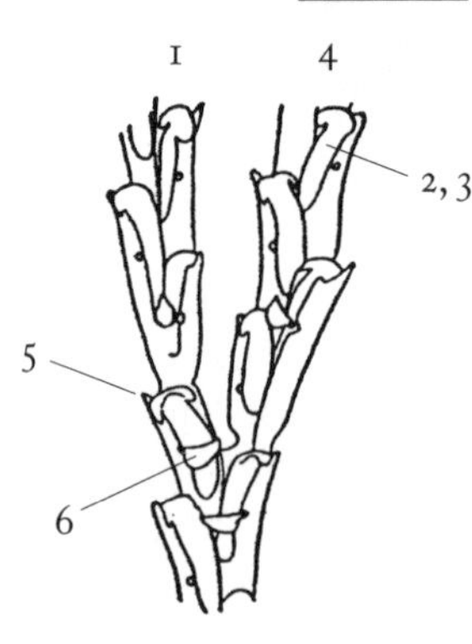

FIG. 12.19

1. Placement of zooids in colony (FIG. 12.19)
   - A. Zooids attached *end* to *end,* forming erect chains of individuals — ee
   - B. Zooids *paired* and *fused* together to form upright branches, two zooids in width — pf*
   - C. Zooids *grouped* and *fused* together to form branches of more than two zooids in width — gf
   - D. Soft mass of mud-covered branches with zooids limited to branch *tips* — ti
2. Size of major opening or orifice (FIG. 12.19)
   - A. Orifice *small,* about equal to diameter of zooecium — sm
   - B. Orifice *large,* occupies half or more of lateral surface — la*
3. Shape of orifice (FIG. 12.19)
   - A. Orifice *rounded* — ro
   - B. Orifice *elongate* in shape — el*
4. Number of side-by-side zooids which form width of typical branch: provide *number* of adjacent zooids (FIG. 12.19) — __

5. Structures around margin of orifice (does not include spines elsewhere on zooecium, e.g., at corners) (FIG. 12.19)
   - A. *Spines* present around orifice, provide *number* — __sp
   - B. A single, very long, spinelike member, the *vibraculum,* on each zooid — vi
   - C. Triangular or fan-shaped flap, the *scutum,* partially covers the orifice — sc
   - D. Both terminal *spines* (provide *number*) and *scutum* present — sp__:sc
   - E. Spines *absent* from margin of orifice — x*
6. Tiny, bird-head-shaped avicularium present
   - A. *Avicularium* present — av*
   - B. Avicularium *absent* — x

12.12

Placement	Opening Size	Opening Shape	No. Zooids	Spines	Avicularia	Species
gf	sm	ro	3–7	x	x	*Crisia eburnea,* jointed-tube ectoprocts
gf	la	el	4–12	4–5sp	av	*Dendrobeania murrayana*
gf	la	el	2–4	vi	av	*Caberea ellis*
gf	la	el	3–6	x	av	*Bugula simplex,* fan bugula
gf	la	el	2	5–6sp	av	*B. fulva*
pf	la	el	2	2–3sp	av	*B. turrita,* bushy bugula
pf	la	el	2	sp2–3:sc	av	*Scrupocellaria scabra*
pf	la	el	2	x	x	*Eucratea loricata,* shelled ectoproct
pf	sm	ro	2	sp2–3:sc	av	*Tricellaria ternata*
pf	la,sm	ro	2	4–9sp	av	*Bicellariella ciliata*
ee	sm	ro,el	1	x	x	Stolon connects upright chains of individuals: *Scruparia chelata* OR Horizontal chains of zooids act as stolon to connect upright chains: *S. ambigua*
ti	sm	ro	1	x	x	*Anguinella palmata,* ambiguous ectoproct or "hair"

FIG. 12.20

Anguinella palmata, ambiguous ectoproct or "hair" (Or: Ctenostomata, Nolellidae). Common but unusual ectoproct. Outer surface with fuzzy, muddy appearance, obscures zooids within. Vi,Ha,Es–Li–Su,ff,7.6 cm (3 in), (G112,RH24), ☺. (FIG. 12.20)

Bicellariella ciliata (Bicellariellidae). Feathery, white; long, inwardly curving spines. Bo,Ha,Li–De,ff,3 cm (1.2 in), (RH32). (FIG. 12.21)

FIG. 12.21

Bugula fulva (Bugulidae). Yellow-brown fan; three spines on outside corners; two to three on inside corners. Bo,Ha,Li–Su,ff,3 cm (1.2 in). (RH33)

Bugula simplex (=*flabellata*), fan bugula (Bugulidae). Orifice occupies all of frontal margin; zooecia with one rounded spike at distal corners. Thick, tufted, fanlike, yellow-orange colonies; protected localities. Vi+,Ha,Li–Su,ff,25.4 mm (1 in), (G117), ☺. (FIG. 12.22. Typical colony, to six zooids wide. FIG. 12.23. Detail with avicularia.)

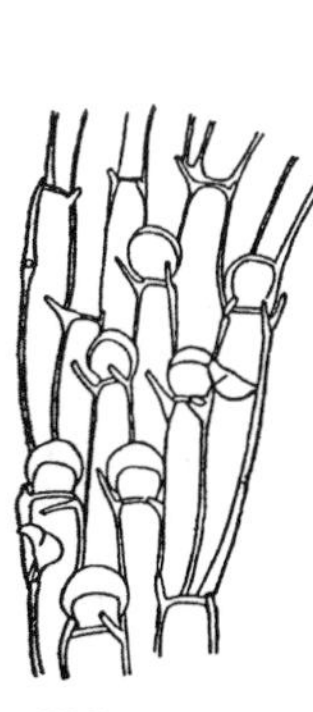

FIG. 12.22

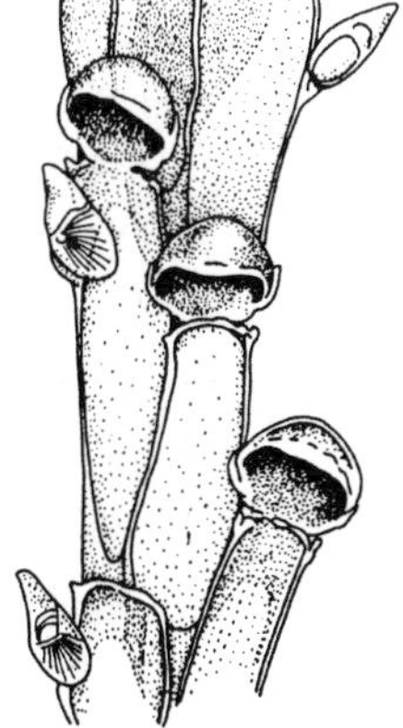

FIG. 12.23

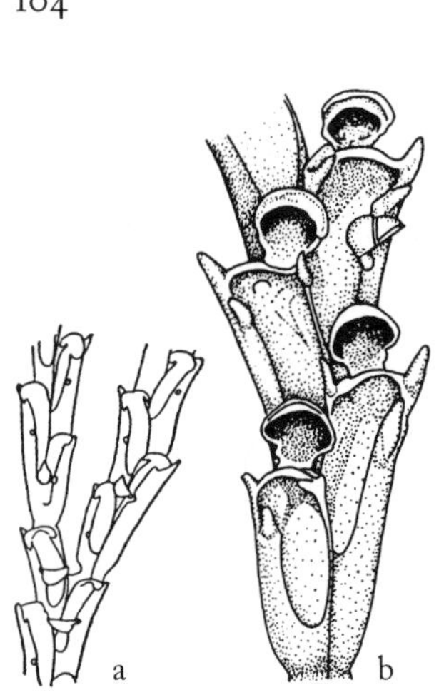

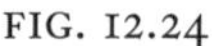

FIG. 12.24

FIG. 12.25

FIG. 12.26

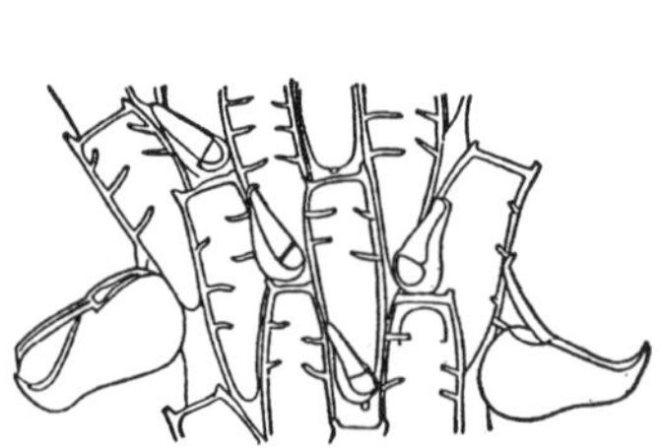

FIG. 12.27

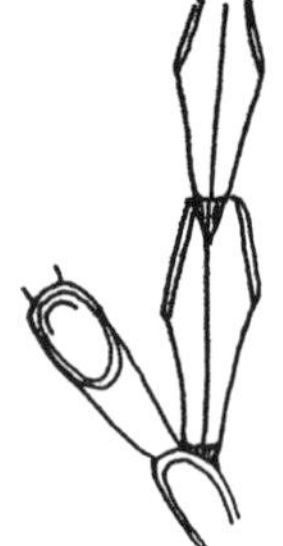

FIG. 12.28

FIG. 12.29

FIG. 12.30

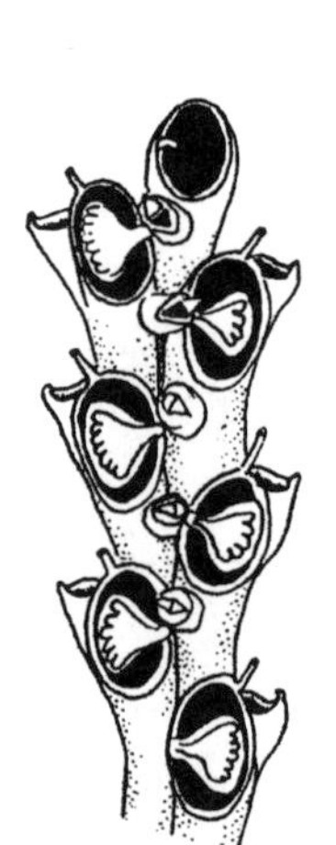

FIG. 12.31

Bugula turrita, bushy bugula (Bugulidae). All zooids extend from same surface; branches mostly two zooids wide, sometimes more. Orange-yellow color. Large, spirally branched colonies. Bo,Ha–SG,Es–Li–Su,ff,7.6 cm (3 in), (GI16,RC51). (FIG. 12.24. a, colony; b, portion of colony.)

Caberea ellisi (Scrupocellariidae). Stiff fan; ridgelike rhizoids along center line of colony. Large vibracula/zooid. Ac,Ha–Gr,Su,ff,3 cm (1.2 in), (RH29). (FIG. 12.25)

Crisia "eburnea," jointed-tube ectoprocts (Cl: Stenolaemata, Or: Cyclostomata, Crisiidae). White, twiglike, calcareous tubes, dotted with tiny pores. Incurved branches form from alternating sides. Shading from brownish-yellow at base to clear toward tips. Occasionally with bulb-shaped gonozooids, inflated distally, with fine pores. Dense tufts upright on algae, especially *Chondrus crispus.* Species in quotation marks because specimens from this region may have been misidentified as this European species (see RH19). Bo,Al,Es–Li–De,ff,20.3 mm (0.8 in), (GI14,RC54,RH19). (FIG. 12.26)

Dendrobeania murrayana (Bugulidae). Orifice covers most of frontal surface; two or more rounded spikes along distal margin. Branches variable in width, but can be broad. Ac,Ha,Su–De,ff,38 mm (1.5 in), (GI17). (FIG. 12.27)

Eucratea loricata, shelled ectoproct (Scrupariidae). Zooecia become wider distally and fused back to back. Dense, slender, bushy clumps; whitish to yellowish to brown. Ac–,Ha?,Li–De,ff,25.4 cm (10 in), (GI15,RH26). (FIG. 12.28)

Scruparia ambigua (Scrupariidae). Zooecia long, slender. Form end-to-end chains on algae, hydroids, ectoprocts (especially on *Bugula*), and shells. Vi,Ha–Al,Li–Su,ff,<3 mm (0.12 in), (RC66,RH25). (FIG. 12.29)

Scruparia chelata (Scrupariidae). Zooecia long, somewhat wider distally. Vi,Ha–Al,Li–Su,ff,<3mm (0.12 in), (RH26). (FIG. 12.30)

Scrupocellaria scabra (Scrupocellariidae). Ac,Ha,Su–De,ff,20 mm (0.8 in), (RH30). (FIG. 12.31)

Tricellaria ternata (Scrupocellariidae). Delicate, white tuft; tapered zooids; small triangular scutum. Ac,Su,Ha–Al,ff,3 cm (1.2 in), (RH32). (FIG. 12.32)

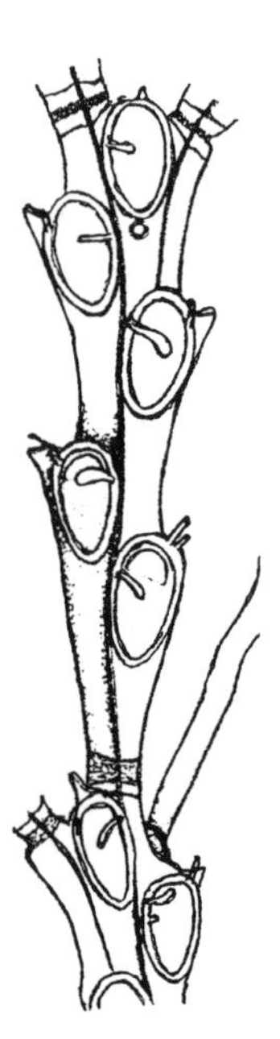

FIG. 12.32

12.13 **Stolonate Ectoprocta** (from 12.2)

All are members of the Order Ctenostomata.

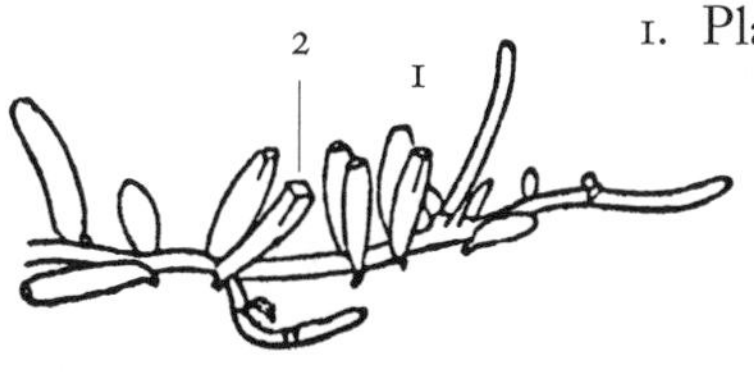

FIG. 12.33

1. Placement of zooids in colony (FIG. 12.33)
   - A. Zooids occur regularly spaced, *single* file along stolon — si
   - B. Zooids in scattered *clumps* along stolon — cl
   - C. Zooids *paired* along stolon; arise from short peduncles — pa*
   - D. *Clumps* of distinctly paired zooids occur along the stolon — pc
   - E. Zooids form dense, consolidated *mats;* stolons form basal network — ma
2. Shape of major opening or orifice (examine individuals with retracted tentacles) (FIG. 12.33)
   - A. Orifice *rounded* — ro
   - B. Orifice elongate *oval* — ov
   - C. Orifice *squared,* or at least distinctly angled at corners — sq*
3. Number of tentacles per zooid: provide *number* — —
4. Spines or other special features (list all that apply)
   - A. Small *spines* surround orifice — sp
   - B. Tiny *hook* present at base of peduncular attachment to stolon — ho
   - C. Zooids *bud* additional zooids, appear branched — bu
   - D. Colony grows in *spirals* around slender branches and stems of substrate — sr
   - E. These special features *absent* — x

12.14

Placement	Opening	Tentacles	Features	Species
pa	ro	8	ho,sp	*Aeverrillia setigera,* pinnate ectoproct
pa	ro	8	sp	*A. armata,* pinnate ectoproct
pa,cl	sq	8	x	*Bowerbankia gracilis,* creeping ectoproct
cl	sq	10	x	*B. imbricata,* creeping ectoproct
pc	sq	8?	sr	*Amathia vidovici,* spiral ectoprocts
ma	sq	8	bu	*Victorella pavida,* cushion moss ectoproct
si	ov	9–11	x	*Aetea recta*

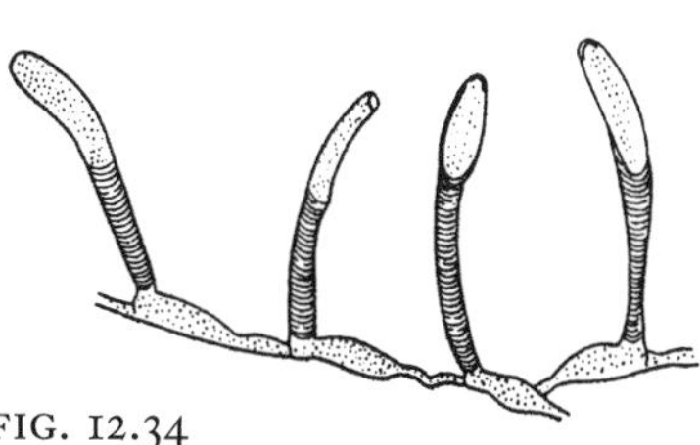

FIG. 12.34

Aetea recta (Aeteidae). Small, white, upright stalks from swellings along stolon. Stalks faintly annulated. With 9–11 tentacles. Vi,Ha–Al,Su,ff,0.6 mm (<0.1 in), (RC43). (FIG. 12.34)

Aeverrillia armata, pinnate ectoproct (Walkeriidae). Four spines around orifice of each zooecium. Vi+,Ha–Al,Es–Su,ff,7.6cm (3 in), (GI14,RC45). (FIG. 12.35)

Aeverrillia setigera, pinnate ectoproct (Walkeriidae). Paired, yellowish zooecia. Vi,Ha–Al,Su,ff,7.6 cm (3 in), (GI14,RC45).

Amathia vidovici, spiral ectoprocts (Vesiculariidae). Yellow to brown, with clusters of paired zooecia appearing darker. Vi,Ha–Al,Es–Su,ff,5.1 cm (2 in), (GI13,RH24). (FIG. 12.36)

FIG. 12.35

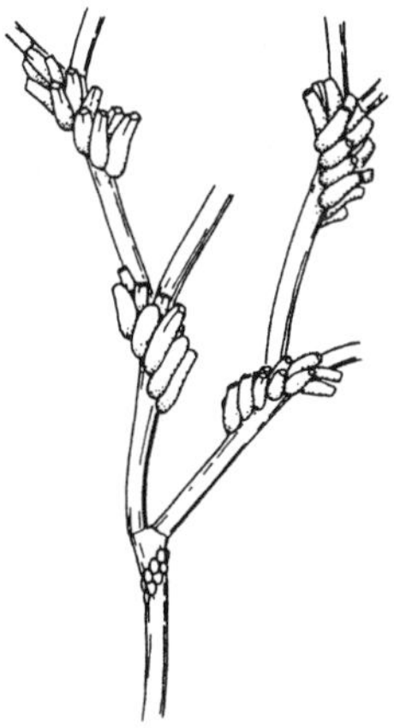

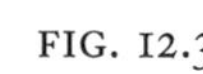

FIG. 12.36

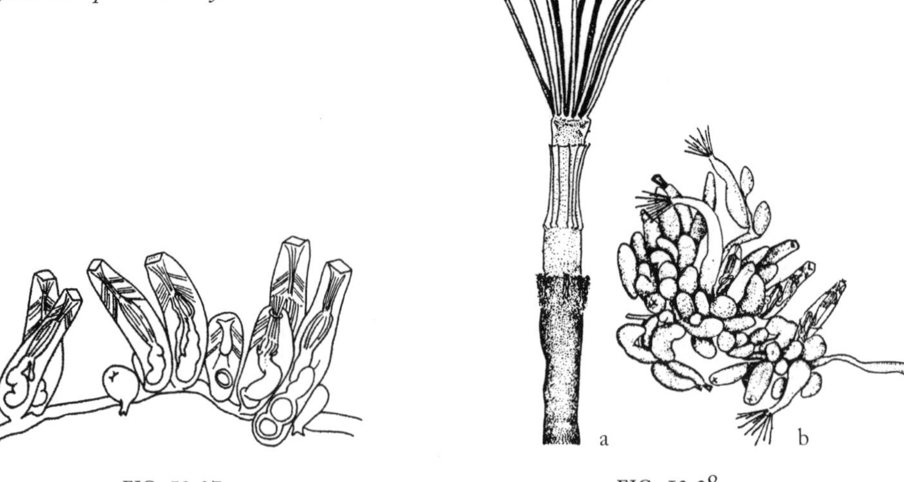

FIG. 12.37

FIG. 12.38

Bowerbankia gracilis, creeping ectoproct (Vesiculariidae). Transparent zooids, soft, flexible. Bo,Ha–Al-SG,Es–Li–Su,ff,2 mm (0.1 in), (GI13,RC47,RH25). (FIG. 12.37)

Bowerbankia imbricata, creeping ectoproct (Vesicularidae). Zooids soft, flexible. Vi,Ha–Al–SG,Li,ff,5 cm (2 in)?, (GI13,RC47,RH25). (FIG. 12.38. a, individual zooid; b, colony.)

Victorella pavida, cushion moss ectoproct (Victorellidae). Dense clusters can form on anything hard in brackish water. Vi,Ha–Sa–SG,Es-Su,ff,12 mm (0.41 in).

Chapter 13 ■ Phylum Mollusca

13.1 The classes of living mollusks from the Northeast may be distinguished most easily on the basis of overall construction of their shell. The class Bivalvia—including clams, mussels, and their relatives—are covered by a shell divided into two valves, hinged dorsally. Gastropoda (13.5), the snails and whelks, have a dorsally placed, single-piece shell that is often, but not always, spiraled. Chitons in the class Polyplacophora (13.3) are distinguished by a dorsal shell divided into eight distinct valves. The region's squid and octopods, the class Cephalopoda (13.161), are distinctive in lacking an external shell, but possessing a foot elaborated to form sucker-bearing tentacles. Finally, the nudibranchs or sea slugs (13.76) are shell-less Gastropoda, with a snaillike appearance but with various bumps, projections, and/or feathery gills adorning their dorsal surfaces.

In most cases, features of shells are obvious and dominate in these keys to molluscan groups. I have included, however, descriptions of soft parts where appropriate or useful. The order of queries places external features first and internal ones last. In many cases, it will be possible to identify intact, living specimens without turning to internal characters. Important external features include shell shape, the appearance of the outermost shell layer (the *periostracum*, which is soft and thin enough to be scratched with a fingernail but is not present in all specimens), and the texture of shell sculpturing. More class-specific notes follow. Most species designations follow Abbott (1993).

Standardized common names (indicated by a dagger, †) follow recommendations included in Turgeon et al. (1988). Additional unofficial but widely used, common names also are listed to facilitate linkage to approved names. Only the standardized names should be used subsequently.

Additional references include: (A) Abbott, R. T., 1993; (AM) Abbott, R. T., and P. A. Morris, 1995. (R) Rehder, H. A., 1981.

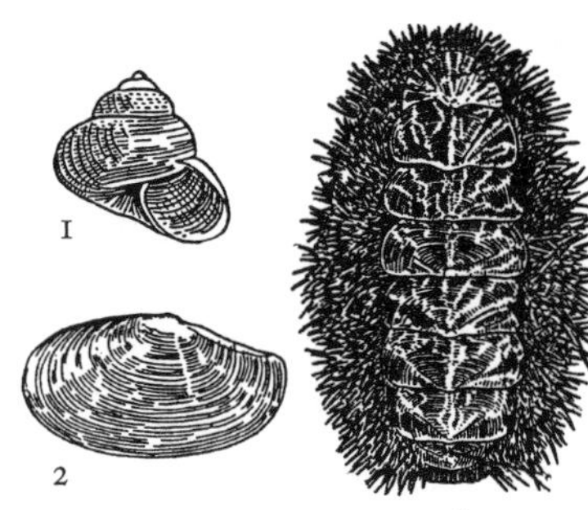

FIG. 13.1

1. Shell construction (FIG. 13.1)
   - A. Shell of living animal formed as *one* piece that may or may not be spiraled (be careful of "clam shell", i.e., a single, isolated valve from a dead, normally bivalved mollusk; see Bivalvia, 13.92) — 1
   - B. Shell formed of *two* pieces or valves (i.e., bivalved or clamlike) — 2
   - C. Shell formed of *eight* separate, curved, rectangular valves, normally covering dorsal surface of animal — 8
   - D. Shell *absent* — x
2. Foot construction
   - A. Animal glides or attaches using a ventrally *flattened* muscular foot — fl
   - B. Muscular foot forms *wedge* or blade shape that animal uses to burrow through or into the substratum — we
   - C. Foot formed into sucker-bearing *tentacles* — te
   - D. Foot forms flaplike *lobes* used for swimming — lo
   - E. Foot *not* visible; animal completely withdrawn into shell — x

13.2

Shell	Foot	Group
1	fl,x	Cl: Gastropoda (snails, whelks), 13.5
1	lo	Cl: Gastropoda, Or: Thecosomata (shelled pteropods), 13.71
2	we,x	Cl: Bivalvia (clams, mussels, etc.), 13.92
8	fl	Cl: Polyplacophora (chitons), 13.3
x	fl	Cl: Gastropoda, Or: Nudibranchia (nudibranchs), Saccoglossa (sea slugs), 13.76
x	te	Cl: Cephalopoda (octopus, squid), 13.161
x	lo	Cl: Gastropoda, Or: Gymnostomata (sea butterflies), 13.70

13.3 Class Polyplacophora, Chitons (from 13.2)

Chitons possess a shell comprising eight valves embedded in a muscular mantle or "girdle." The surface texture of the girdle is an important diagnostic characteristic. In some cases, for positive identification, it may be necessary to remove the hindmost valve and examine its underside for coloration and for the number of slits occurring along its posterior margin. Behavioral characteristics of chitons include their ability to cling tenaciously to the substrate (a drop of two of some irritating fluid helps to dislodge them) and their tendency to roll up armadillo-style when disturbed. As omnivorous grazers, chitons will ingest any encrusting biota, plant, or animal.

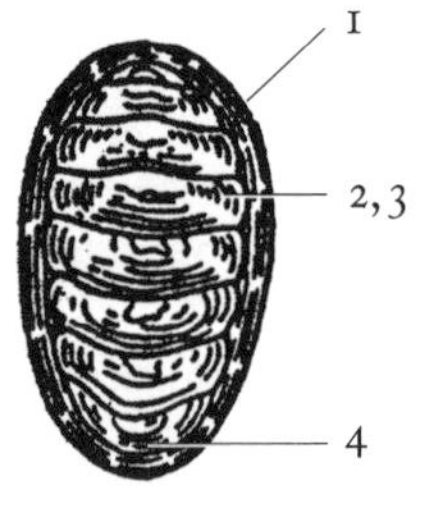

FIG. 13.2

1. Appearance of girdle surface (FIG. 13.2)
   - A. Smooth or *leathery* — le
   - B. Granular or *scaled* — sc
   - C. *Hairy,* with tiny, scattered translucent hairs — ha
2. Color of valves (FIG. 13.2)
   - A. Valves with *red* blotches or lines — re
   - B. Valves *bluish black* — bbl
   - C. *Grayish* or chestnut — gy
3. Texture of valve surfaces (FIG. 13.2)
   - A. Valves with lines of tiny, raised *beads* — be
   - B. Valves *smooth* or with growth lines only — sm
4. Number of slits in posterior valve (remove posteriormost valve and examine the under surface for these slits): provide *number* of slits (FIG. 13.2) — __
5. Color of inside of valves or plates
   - A. *Pink* — pi
   - B. *Gray-white* — gw

13.4

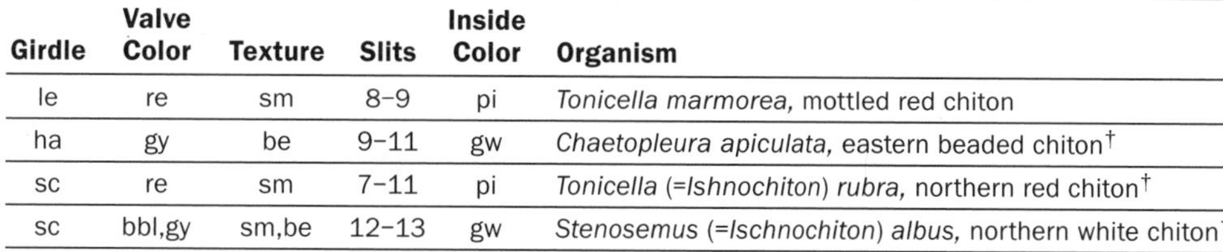

Girdle	Valve Color	Texture	Slits	Inside Color	Organism
le	re	sm	8–9	pi	*Tonicella marmorea,* mottled red chiton
ha	gy	be	9–11	gw	*Chaetopleura apiculata,* eastern beaded chiton†
sc	re	sm	7–11	pi	*Tonicella* (=*Ishnochiton*) *rubra,* northern red chiton†
sc	bbl,gy	sm,be	12–13	gw	*Stenosemus* (=*Ischnochiton*) *albus,* northern white chiton†

FIG. 13.3

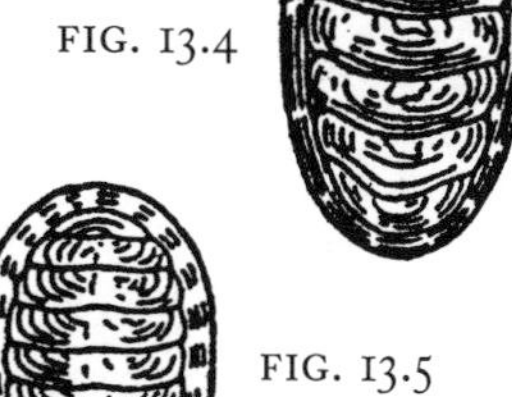

FIG. 13.4

FIG. 13.5

Chaetopleura apiculata, eastern beaded chiton† (Chaetopleuridae). Lateral regions beaded. Valve interior, white to gray. Shallows. Often with *Crepidula.* Vi,Ha,Su,om,20 mm (0.8 in), (AM294,G123,NM467,R335), ☺. (FIG. 13.3)

Stenosemus (=*Ischnochiton*) *albus,* northern white chiton† (Ischnochitonidae). Dark coloring rubs off preserved specimens leaving animal white. Ac,Ha,Li–Su,om,16 mm (0.6 in), (AM294,G123,NM463,R326).

Tonicella marmorea, mottled red chiton (Ischnochitonidae). Valves rosy pink inside. Plates with small posterior projection; 19–26 pairs of gills. Eats attached algae, sponges, hydroids, ectoprocts, etc.. Ac,Ha,Su,om,38 mm (1.5 in), (A110,AM293,NM461). (FIG. 13.4)

Tonicella (=*Ishnochiton*) *rubra,* northern red chiton† (Ischnochitonidae). Plates strongly keeled. Valve interior, bright pink. Ac,Ha,Li–De,om,25.4 mm (1 in), (A110,AM293,G123,NM461,R333), ☺. (FIG. 13.5)

13.5 Classes Gastropoda (Snails and Whelks) and Scaphopoda (Tooth Shells)

Shelled gastropods—snails, whelks, limpets, etc.—are commonly encountered. Try to observe specimens alive in seawater. Note especially the color of their soft tissues, the presence or absence of tentacles both on the head and elsewhere (such as around the shell margin), and the presence or absence of an anterior *siphon* (tubelike extension of the mantle; if present, visible over the animal's left shoulder) above the head. Observe or collect specimens covering the range of colors, shapes, and sizes characteristic of the species. Be cautious of identifications on damaged, worn, or juvenile specimens.

The majority of gastropods treated in these pages have a well-developed head with eyes and tentacles, a flattened ventral foot, and a shell secreted dorsally by the mantle. They also possess one or two gills attached to the roof of a mantle chamber, filled with fluid from the surrounding medium. (This area is sometimes partially visible externally by peering over the the left shoulder of an animal held upside down in a dish of seawater.) In some cases, the shell is markedly reduced, as in some sea butterflies and bubble shells. (Shell-less gastropods such as nudibranchs or sea slugs and some sacoglossa are covered in a subsequent section.) The pulmonate snails comprise terrestrial forms, with two pairs of tentacles and a mantle cavity modified to serve as a gill-less lung for breathing air.

Directional orientation in gastropods is determined by holding the shell (see FIG. 13.6a) with the *aperture* (ap) (opening for the soft parts) toward the viewer, with the *spire* (sp) (pointed tip) directed upward. In most cases, the aperture is located on the right side of this view (which also corresponds to the snail's right side, because the aperture is, in fact, ventral and anterior). In some gastropods, the aperture includes a notch or groove to accommodate the siphon mentioned above (e.g., FIG. 13.6b). In others, the aperture contour is not interrupted in this way (e.g., FIG. 13.6c). The basal and usually largest spiral beginning at the outer lip of the aperture is referred to as the *body whorl,* (bw in FIG. 13.6d), whereas the smaller spirals toward the apex may be called *spire whorls* (sw). The contour line that describes the profile shape of whorls is referred to as the *shoulder* and it may be straight, smoothly rounded, or sharply angled. The central axis around which most gastropod shells spiral is called the *columella.* In some cases, this central axis forms a hollow tube running the length of the snail, visible externally as an *umbilicus* (small hole located to the side of the aperture). Note the presence and composition of the *operculum* (the trap-door), which in many species closes the aperture as the soft parts

FIG. 13.6

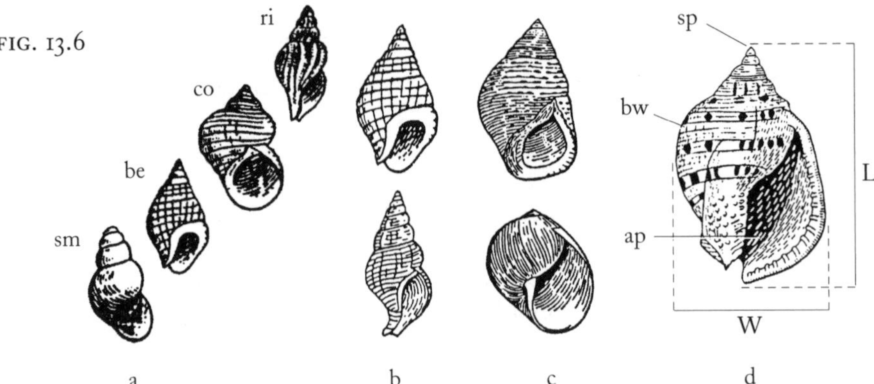

are withdrawn. It will either be absent, or if present, will be calcareous (in a very few cases) or horny (usually a thin, crisp, brown material) in most others.

Snail shells grow incrementally with new material laid down as *growth lines* along the outer apertural lip. This is evidenced by fine striations following the main or body whorl. Such growth lines are not considered sculpture, and shells with *only* these fine lines are considered to be *smooth* (FIG. 13.6a, sm). *Cords* (FIG. 13.6a, co) are raised areas following the spiraling contour of the shell. Heavy growth lines, narrow ribs, or broader *ridges* (FIG. 13.6a, ri) are raised areas that run axially, following the long axis of the shell. In some gastropods, cords and ridges occur with about equal prominence, giving the shell a *beaded* appearance (FIG. 13.6a, be).

The *shape index* may be necessary for identifying the shells. To make this measurement, hold the shell with its apertural opening facing you and its spire (if present) pointing upward (FIG. 13.6d). Measure the distance from the tip of the spire to the greatest extension of the shell along the opposite edge as its length (L), and divide that measurement by the greatest width (W) taken at approximately right angles to the length line. Possibly the easiest way to do this is to place the shell aperture downward on a piece of paper. Imagine a projection of its greatest dimensions onto the paper and mark the end points of its length and width lines with four dots. Measure these distances using a ruler in millimeters for greater accuracy.

Despite listings for many species as subtidal in habitat, empty shells are frequently encountered nearshore or on beaches, especially following storms at sea. Also included here are a few members of the related class, Scaphopoda (tooth shells) and a single, shell-bearing species of the class Cephalopoda.

FIG. 13.7

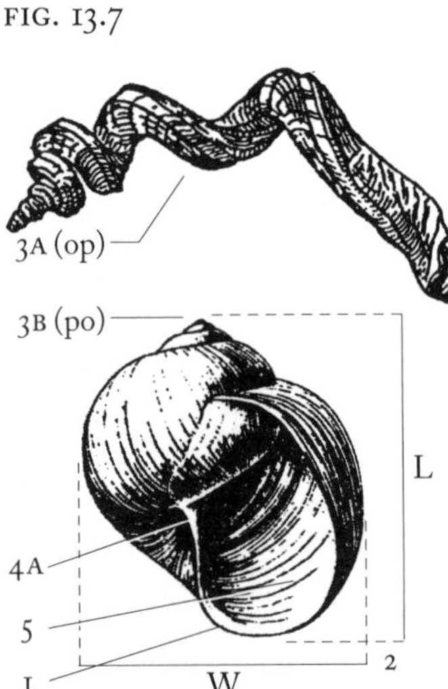

1. Shape described by line traced around the aperture
   - A. Basically a *smooth* circle or oval line (FIG. 13.6c) — sm*
   - B. Distinct short or long *break,* formed at end opposite spire by a canal or notch to accommodate the siphon (FIG. 13.6b) — br
2. Shell index (see introduction) (FIG. 13.7)
   - A. Shell *width greater than length* — l<w
   - B. Shell *length from one to two times its width* — l=w*
   - C. Shell *length more than twice its width* — l>w
3. Shell spiraling (FIG. 13.7)
   - A. Shell spiraling apparent but *open* (i.e., whorls are not fully fused to one another) — op*
   - B. Shell spiraling apparent and closed; spire distinctly *pointed* (i.e., typical snail shell) — po*
   - C. Shell spiraled and closed, but spire is *flattened* and, at most, only slightly higher than body whorl — fl
   - D. Shell spiraling *absent* — x
4. Presence of "umbilicus" (small opening located along the shell center line, just beside the aperture) (FIG. 13.7)
   - A. *Umbilicus* present as distinct rounded opening — um*
   - B. Umbilicus present but reduced to small but distinct *groove* — gr
   - C. Umbilicus *absent,* no hole present next to aperture — x

5. Shell opening to the right (dextral) or to the left (sinistral) (hold apex upward and aperture toward you) (FIG. 13.7)
   A. Shell opening *dextral* — de*
   B. Shell opening *sinistral* — si
   C. Opening *not* distinctly dextral or sinistral — x

13.6

Aperture	Shape	Spiral	Umbilicus	Opening	Group or Organism
sm	l=w	x,fl	x	de	Gastropod Group 1, 13.7
sm	l=w	x	x	x	Gastropod Group 1, 13.7
sm	l=w	po	x	de	Gastropod Group 2, 13.9
sm	l=w	po	x	si,x	Order Thecosomata (shelled pteropods), 13.71
sm	l=w	po,fl	um	de	Naticidae (moon snails), 13.53
sm	l=w	po	gr	de	Smoothly rounded shoulders: *Lacuna vincta,* northern lacuna†, banded chink snail; see description under the family Lacunidae, 13.39 OR Sharply angled shoulders, deep groove follows whorls: *Trichotropis borealis,* boreal hairysnail; see description under the family Trichotrophidae, 13.60
sm	l=w	op	x	de	*Spirula spirula,* ram's horn squid; see description under the Cl: Cephalopoda, family Spirulidae, 13.73
sm	l>w	po	x	de	Gastropod Group 3, 13.11
sm	l>w	op	um	de	*Vermicularia spirata,* West Indian wormsnail; see description under the family Turritellidae, 13.65
sm	l>w	x	x	de	Gastropod Group 4, 13.13
sm	l<w	po	x	de	Shell thin; purple color: *Janthina janthina,* janthina†, purple seasnail; see description under the family Janthinidae, 13.38 OR Shell yellow-orange-brown color: *Littorina obtusata,* yellow periwinkle†; see description under the family Littorinidae, 13.41
sm	l<w	po	um	de	Trochidae (sun snails), 13.62
sm	l<w	fl	um	de	*Skeneopsis planorbis,* flat skenia†, orbsnail; see description under the family Skeneopsidae, 13.58
br	l=w	x	x	si,x	Or: Thecosomata (shelled pteropods), 13.71
br	l=w	po	x	de	Gastropod Group 5, 13.15
br	l=w	po	x	si	*Busycon perversum,* perverse whelk; see description under the family Melongenidae, 13.47
br	l>w	po	x	de	Gastropod Group 6, 13.17

13.7 **Gastropod Group 1 (Limpets, Slippersnails, Bubbles)** (from 13.6)

Gastropods with nonspiraled shells, without an umbilicus, but with a circular aperture and with shell length one to two times the width.

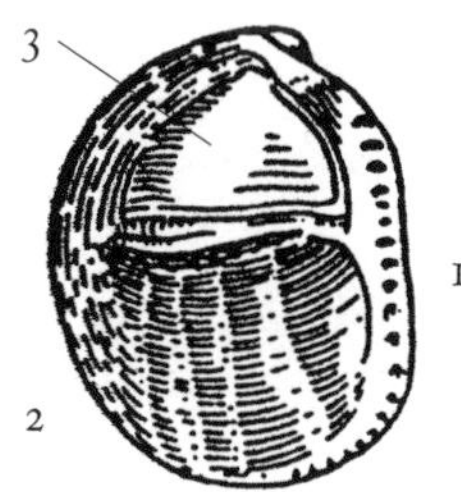

FIG. 13.8

1. Second opening (in addition to main aperture) at or near top of shell (FIG. 13.8)
   A. *Opening* present at or near apex of shell — op
   B. Second opening *absent* — x*
2. Appearance of ventral surface of shell (hold aperture toward you to describe the ventral surface view) (FIG. 13.8)
   A. Aperture occupies entire *ventral* surface — ve*
   B. View of ventral surface also includes *body whorl* — bw
3. Presence and shape of projections from internal surface of shell (on live specimens, use a pin or probe to feel for such a shelf) (FIG. 13.8)
   A. Distinct shelf or flaplike *projection* extends into interior of shell — pr*
   B. Shelf or flap *absent* — x

13.8

Opening	Aperture	Interior	Organism
op	ve	x	Fissurellidae (keyhole limpets), 13.35
x	ve	x	*Tectura* (=*Acmaea*, =*Notoacmaea*) *testudinalis*, tortoiseshell or Atlantic limpet; see description under family Lottiidae, 13.43
x	ve	pr	Calyptraeidae (boatsnails or slippersnails), 13.26
x	bw	x	Acteonidae, Atyidae, and Scaphandridae (bubbles), 13.19

13.9

Gastropod Group 2 (from 13.6)

Gastropods with spiraled shells, without an umbilicus, but with a circular aperture and with shell length one to two times the width.

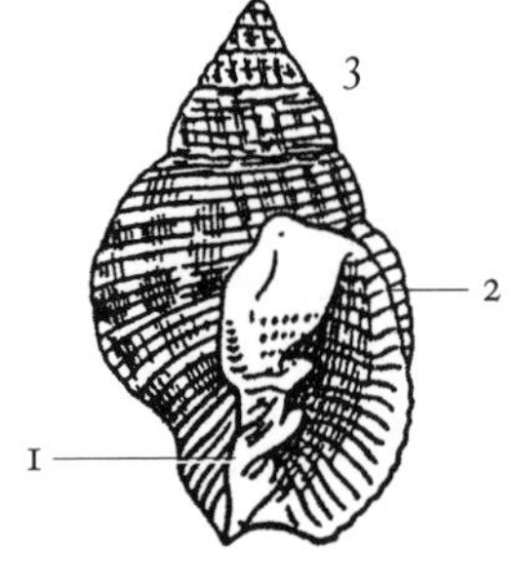

FIG. 13.9

1. Folds or ridges along medial (toward the center line) side of aperture margin (FIG. 13.9)

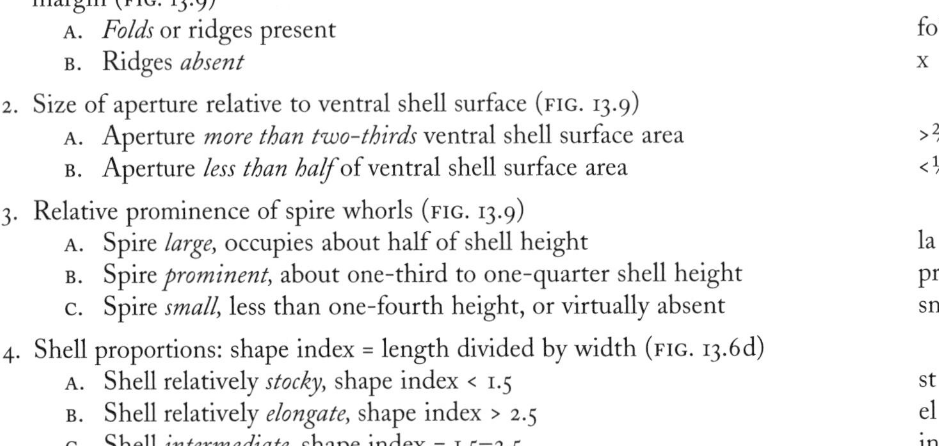
   - A. *Folds* or ridges present — fo*
   - B. Ridges *absent* — x

2. Size of aperture relative to ventral shell surface (FIG. 13.9)
   - A. Aperture *more than two-thirds* ventral shell surface area — >⅔
   - B. Aperture *less than half* of ventral shell surface area — <½*

3. Relative prominence of spire whorls (FIG. 13.9)
   - A. Spire *large*, occupies about half of shell height — la
   - B. Spire *prominent*, about one-third to one-quarter shell height — pr*
   - C. Spire *small*, less than one-fourth height, or virtually absent — sm

4. Shell proportions: shape index = length divided by width (FIG. 13.6d)
   - A. Shell relatively *stocky*, shape index < 1.5 — st
   - B. Shell relatively *elongate*, shape index > 2.5 — el
   - C. Shell *intermediate*, shape index = 1.5–2.5 — in

13.10

Ridges	Aperture	Spire	Shape	Group or Organism
fo	<½	pr	in	Olive or brown shell: Melampodidae, 13.45 OR Yellow to white shell with rows of tiny punctations (depressions): *Rictaxis* (=*Acteon*) *punctostriatus*, pitted baby-bubble; see description under family Acteonidae, 13.19
x	<½	pr	st,in	Littorinidae (periwinkles), 13.41
x	<½	sm	in	*Prunum* (=*Marginella*) *roscidum*, seaboard marginella†, boreal or dewy marginella; see description under family Marginellidae, 13.44
x	<½	sm	st	With distinct callus pad (flattened shell growth) covering umbilicus area (adjacent to the aperture): Naticidae (moon snails), 13.53 OR Without a callus in umbilical area: *Littorina obtusata*, yellow periwinkle; see description under family Littorinidae, 13.41
x	<½	la	st	Trochidae (topsnails), 13.62
x	<½	la	el	*Bittiolum* (=*Bittium*, *Diastoma*); see descriptions under family Cerithiidae and Cerithiopsidae, 13.29
x	>⅔	pr	st,in	Velutinidae (velvetsnails), 13.67
x	>⅔	sm	st,in	With internal shelf in shell: *Crepidula fornicata*, common Atlantic slippersnail; see description under family Calpytraeidae, 13.26 OR Without internal shelf in shell: *Sinum perspectivum*, white baby-ear†; Very flat, white shell; see description under family Naticidae, 13.53 OR *Lamellaria* (=*Marsenina*) *perspicua*, transparent lamellaria. Transparent shell; no periostracum; see description under family Lamellariidae, 13.40

13.11 **Gastropod Group 3** (from 13.6)

Gastropods with spiraled shell, no umbilicus, with circular aperture and are slender, with length more than twice the width.

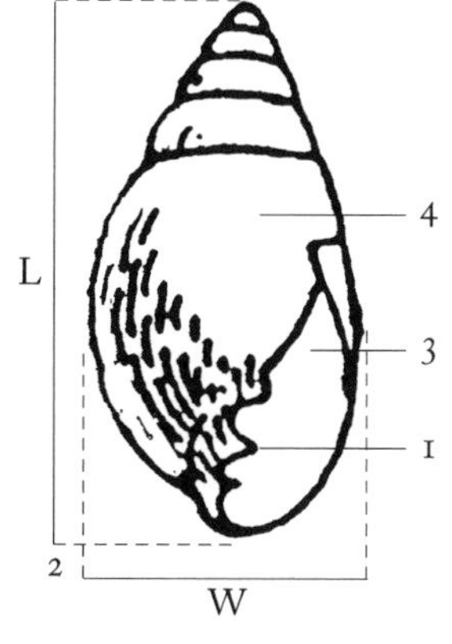

FIG. 13.10

1. Presence of folds or ridges along medial side of aperture margin (FIG. 13.10)
   - A. *Folds* or ridges present — fo*
   - B. Such ridges *absent* — x
2. Shell proportions: shape index = length divided by width (FIG. 13.10)
   - A. Relatively *stocky;* shape index ≤ 2.3 — st*
   - B. More *elongate;* shape index > 2.3 — el
3. Extent of apertural opening along the side of the shell (FIG. 13.10)
   - A. Aperture *long,* extends half way or more up the right side of the shell — lo*
   - B. Aperture *short,* extends much less than half way up the shell — sh
4. Maximal shell length (tip of spire to opposite apertural lip)
   - A. *Tiny,* ≤ 7 mm (⅓ in) — ti
   - B. *Larger,* > 7 mm (⅓ in) — la
5. Dominant sculpture on body whorl (see FIG. 13.6a)
   - A. Strong longitudinal or axial *ribs* or ridges — ri
   - B. Tiny scratchlike *striations* follow long axis of shell — st
   - C. Distinct spiraling *cords* — co
   - D. Shell *smooth* or with growth lines only — sm*

13.12

Folds	Index	Aperture	Size	Sculpture	Organism
x	st	sh	ti	sm	Hydrobiidae, 13.37
x	st	lo	ti	sm	Acteonidae, Atyidae, and Scaphandridae (bubbles), 13.19
fo	st	lo	ti	st	*Ovatella myosotis,* mouse melampus†; see description under family Melampodidae, 13.45
fo	el	sh	ti	sm	Pyramidellidae (mostly ectoparasites or predators on other mollusks), 13.55
fo	el	sh	ti	ri	*Turbonilla* spp.; see description under family Pyramidellidae, 13.55
x	el	sh	ti	sm,st	Rissoidae, 13.56
x	el	sh	la	co	*Tachyrhynchus erosus,* eroded turretsnail†; see description under family Turitellidae, 13.65
x	el	sh	la	st	Epitoniidae (wentletraps), 13.33

13.13 **Gastropod Group 4** (from 13.6)

Gastropods with shells, longer than wide, with circular aperture, no spiraling, and no umbilicus.

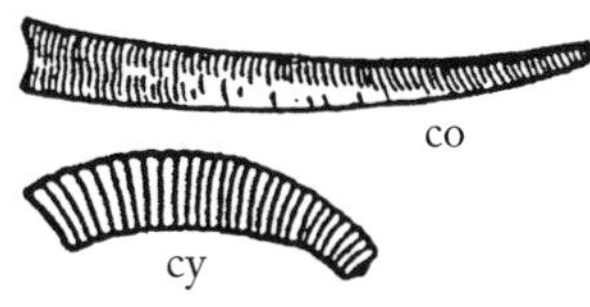

FIG. 13.11

1. Shape of shell (FIG. 13.11)
   - A. Tapering *cone* — co
   - B. Round-ended *bulb* — bu
   - C. *Cylindrical,* open at both ends — cy
2. Size of shell
   - A. *Tiny,* ≤ 12 mm (0.5 in) — ti
   - B. *Larger,* > 12 mm (0.5 in) — la

13.14

Shape	Size	Group
co	la	Class Scaphopoda, tusk or toothsnails, 13.74
bu	ti	*Cuvierina columnella,* cigar pteropod; see description under Or: Thecosomata (shelled pteropods), 13.71
cy	ti	Caecidae, 13.24

13.15 Gastropod Group 5 (from 13.6)

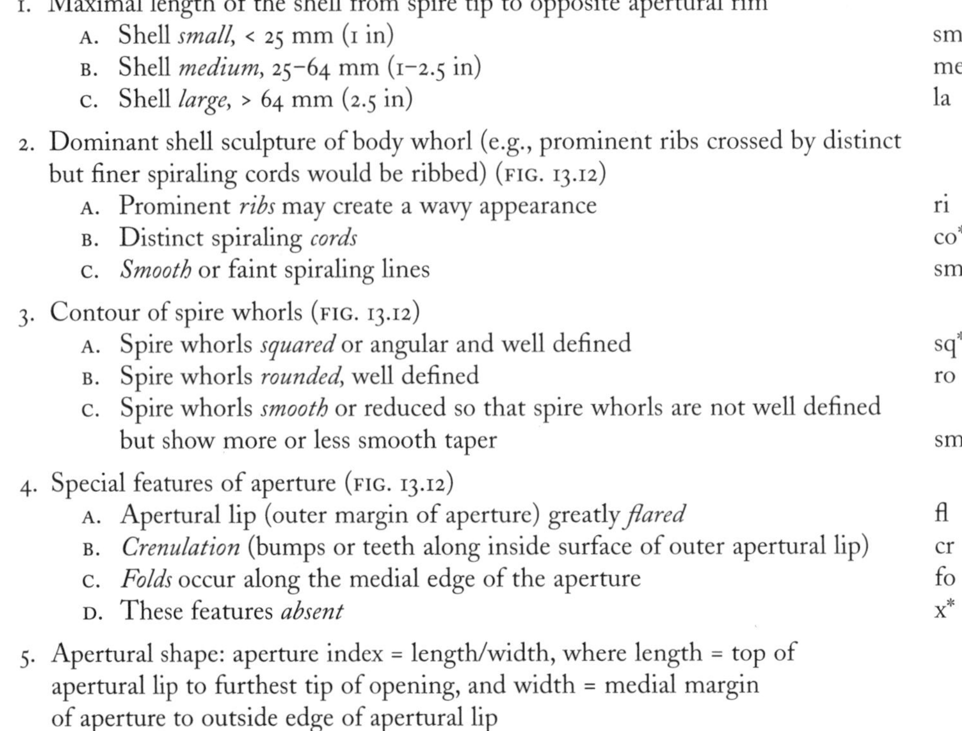

Gastropods with spiral shells and an apertural line interrupted by a siphonal notch or canal. Umbilicus is lacking and length is one to two times the width.

1. Maximal length of the shell from spire tip to opposite apertural rim
   - A. Shell *small*, < 25 mm (1 in) — sm
   - B. Shell *medium*, 25–64 mm (1–2.5 in) — me
   - C. Shell *large*, > 64 mm (2.5 in) — la
2. Dominant shell sculpture of body whorl (e.g., prominent ribs crossed by distinct but finer spiraling cords would be ribbed) (FIG. 13.12)
   - A. Prominent *ribs* may create a wavy appearance — ri
   - B. Distinct spiraling *cords* — co*
   - C. *Smooth* or faint spiraling lines — sm
3. Contour of spire whorls (FIG. 13.12)
   - A. Spire whorls *squared* or angular and well defined — sq*
   - B. Spire whorls *rounded*, well defined — ro
   - C. Spire whorls *smooth* or reduced so that spire whorls are not well defined but show more or less smooth taper — sm
4. Special features of aperture (FIG. 13.12)
   - A. Apertural lip (outer margin of aperture) greatly *flared* — fl
   - B. *Crenulation* (bumps or teeth along inside surface of outer apertural lip) — cr
   - C. *Folds* occur along the medial edge of the aperture — fo
   - D. These features *absent* — x*
5. Apertural shape: aperture index = length/width, where length = top of apertural lip to furthest tip of opening, and width = medial margin of aperture to outside edge of apertural lip
   - A. Aperture *elongate*, index ≥ 2.4 — el
   - B. Aperture more *symmetrical*, index < 2.4 — sy

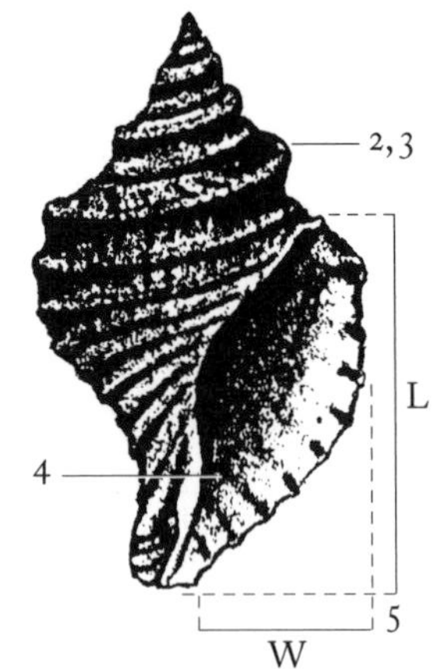

FIG. 13.12

13.16

Size	Sculpture	Shoulder	Aperture	Aperture Index	Organism
sm	sm	sm	cr	el	*Astyris lunata*, lunar dovesnail†; see description under family Columbellidae, 13.31
sm	co	ro	x	el	*A. rosacea*, rosy northern dovesnail†; see description under family Columbellidae, 13.31
sm	co	ro	fo	el	*Admete couthouyi*, nutmeg snail; see description under family Cancellariidae, 13.28
sm	ri	ro	cr,x	el	Muricidae, 13.49
sm	ri	sm	cr	sy	*Nassarius vibex*, bruised nassa†, mottled dog whelk; see description under family Nassariidae, 13.51
me	sm,co	sm	cr,x	el	*Nucella lapillus*, Atlantic dogwinkle†; see description under family Muricidae, 13.49
me	co	ro	x	sy	*Colus ventricosus*, ventricose whelk†, fat colus; see description under family Buccinidae, 13.22
me	ri	ro	cr	el	Muricidae, 13.49
me	ri	ro	x	sy	Buccinidae, 13.22
me,l,a	ri	ro	fl	sy	*Aporrhais occidentalis*, American pelicanfoot; see description under family Aporrhaidae, 13.21
la	co	ro,sq	x	sy	*Beringius (=Neoberingius) turtoni;* see description under family Buccinidae, 13.22
la	ri	ro	x	sy	Buccinidae, 13.22

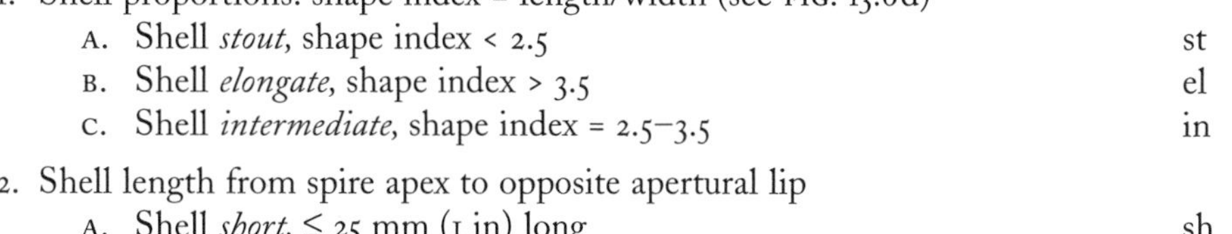

13.17 Gastropod Group 6 (from 13.6)

Gastropods with spiral shells, no umbilicus, a siphonal notch or canal, and slender, with length more than twice the width.

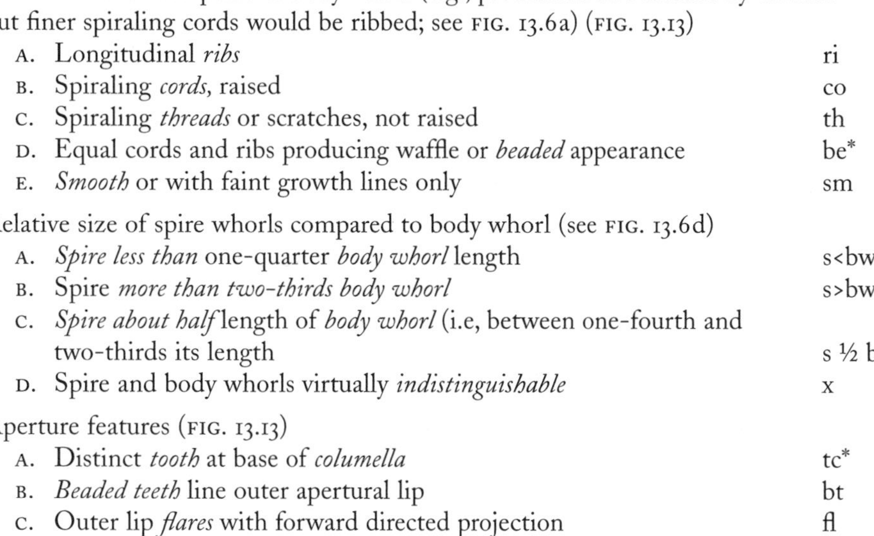

1. Shell proportions: shape index = length/width (see FIG. 13.6d)
   - A. Shell *stout,* shape index < 2.5 — st
   - B. Shell *elongate,* shape index > 3.5 — el
   - C. Shell *intermediate,* shape index = 2.5–3.5 — in
2. Shell length from spire apex to opposite apertural lip
   - A. Shell *short,* ≤ 25 mm (1 in) long — sh
   - B. Shell *long,* > 64 mm (2.5 in) long — lo
   - C. Shell *medium* length, 25–64 mm (1–2.5 in) long — me
3. Dominant shell sculpture of body whorl (e.g., prominent ribs crossed by distinct but finer spiraling cords would be ribbed; see FIG. 13.6a) (FIG. 13.13)
   - A. Longitudinal *ribs* — ri
   - B. Spiraling *cords,* raised — co
   - C. Spiraling *threads* or scratches, not raised — th
   - D. Equal cords and ribs producing waffle or *beaded* appearance — be*
   - E. *Smooth* or with faint growth lines only — sm
4. Relative size of spire whorls compared to body whorl (see FIG. 13.6d)
   - A. *Spire less than* one-quarter *body whorl* length — s<bw
   - B. Spire *more than two-thirds body whorl* — s>bw
   - C. *Spire about half* length of *body whorl* (i.e, between one-fourth and two-thirds its length — s ½ bw*
   - D. Spire and body whorls virtually *indistinguishable* — x
5. Aperture features (FIG. 13.13)
   - A. Distinct *tooth* at base of *columella* — tc*
   - B. *Beaded teeth* line outer apertural lip — bt
   - C. Outer lip *flares* with forward directed projection — fl
   - D. Three to four *folds* appear along the *columella* — fc
   - E. *None* of these features present — x

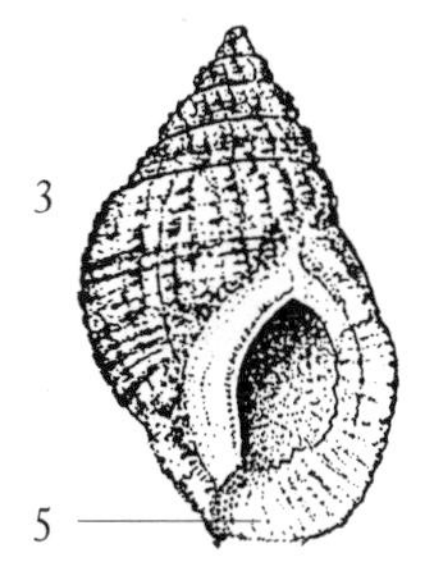

FIG. 13.13

13.18

Index	Size	Sculpture	Spire	Aperture	Organism
st	lo	sm	s½bw	x	Buccinidae (whelks), 13.22
st	lo	sm	s<bw	x	*Volutopsius norvegicus,* Norway whelk†; see description under family Buccinidae, 13.22
st	lo	co	s½bw	x	*Neptunea despecta,* disreputable whelk†; see description under family Buccinidae, 13.22
st	me,lo	th	s½bw	x	Buccinidae, 13.22
st	lo	ri,th	s<bw	x	Melongenidae, 13.47
st	me,lo	ri	s½bw	fl	*Aporrhais occidentalis,* American pelicanfoot; see description under family Aporrhaiidae, 13.21
st	me	ri	s½bw	x	Muricidae, 13.49
st	sh	ri,co	s<bw	bt	*Eupleura caudata,* thick-lip drill†; see description under family Muricidae, 13.49
st	sh	ri	s½bw	x	Line from outer apertural lip where it meets the body whorl projects toward spire: Columbellidae, 13.31 OR Line from outer apertural lip projects into body whorl: Turridae, 13.64
st	sh	th	s½bw	bt	*Astyris* (=*Mitrella*) *lunata,* lunar dovesnail†, crescent dovesnail; see description under family Columbellidae, 13.31

(continued)

13.18 *(continued)*

Index	Size	Sculpture	Spire	Aperture	Organism
st	sh	th	s½bw	x	1.9 cm (¾ in) shell: *Colus pygmaeus,* pygmy whelk†; see description under family Buccinidae, 13.22 OR <1.25 cm (½ in) shell: *Astyris* (=*Mitrella*) *rosacea,* rosy northern dovesnail†; see description under family Columbellidae, 13.31
st	sh	be	s½bw	tc	*Ilyanassa* (=*Nassarius*) *trivittata,* three-lined mudsnail†, New England dog whelk or basket snail; see description under family Nassariidae, 13.51
st	sh	sm	s½bw	tc	*Ilyanassa* (=*Nassarius*) *obsoleta,* eastern mudsnail†, mud dog whelk, eroded basket shell; see description under family Nassariidae, 13.51
st,in	sh	sm	s<bw	fc	*Volutomitra groenlandica,* Greenland miter; see description under family Volutomitridae, 13.69
in	sh	be	s>bw	x	*Triphora nigrocincta,* black-lined triphora†; see description under family Triphoridae, 13.61
in	sh	ri	s>bw	x	*Cryoturris* (=*Mangelia*) *cerinella* (=*cerina*); see description under family Turridae, 13.64
el	sh	co	x	x	*Seila adamsi,* wood-screw snail; see description under family Cerithiidae & Cerithiopsidae, 13.29
el	sh	be	s<bw	x	*Cerithiopsis greeni,* green cerith; see description under family Cerithiidae & Cerithiopsidae, 13.29
el	sh	ri	s>bw	x	*Terebra dislocata,* eastern auger†; see description under family Terebridae, 13.59

Families of Gastropoda (arranged alphabetically)

13.19 **Acteonidae, Atyidae, and Scaphandridae, Bubbles** (from 13.8, 13.10, 13.12)

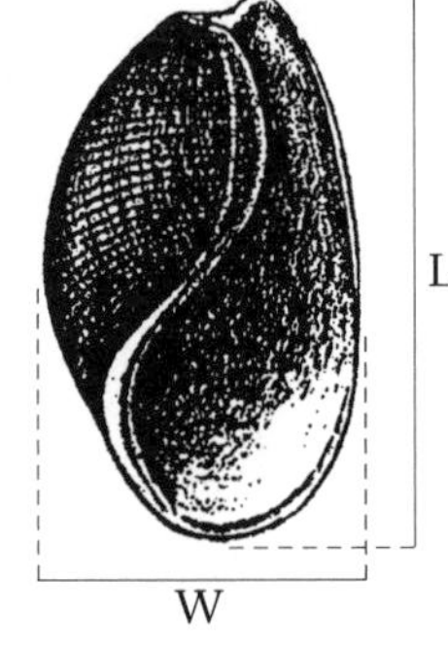

FIG. 13.14

Thin shell engulfed by mantle of active, living animal.

1. Visibility of spire (FIG. 13.14)
   - A. *Spire* clearly visible — sp
   - B. Spire *not* visible, covered by body whorl — x*
2. Shell proportions: shape index = length/width (FIG. 13.14)
   - A. Shell *slender,* shape index > 2.0 — sl
   - B. Shell more *stout,* shape index ≤ 2.0 — st*
3. Appearance (FIG. 13.14)
   - A. Shell *uniform* in color — un*
   - B. Lower half of body whorl with rows of dark *dots* (dirt-filled punctations) — do

13.20

Spire	Index	Color	Species
sp	st	un	*Rictaxis* (=*Acteon*) *punctostriatus,* pitted baby-bubble†
sp	st,sl	do	*Acteocina* (=*Retusa*) *canaliculata,* channeled barrel-bubble†
x	st	un	*Haminoea solitaria,* solitary glassy-bubble
x	sl	un	*Scaphander punctostriatus,* giant canoe bubble†

Acteocina (=*Retusa*) *canaliculata,* channeled barrel-bubble† (Scaphandridae). Aperture extends nearly to shell apex. Glossy white, often stained with orange; cylindrical shell; living tissue is pink. Bo,Mu,Li–Su,pr,6 mm (0.24 in), (AM269,R635).

FIG. 13.15

Haminoea solitaria, solitary glassy-bubble† (Atyidae). Spiral scratches, shiny white, yellow or brown; inside, dull white to amber. Vi+,SG–Mu–Sa,Su,pr,12 mm (0.5 in), (A100,AM268,G134,NM518), ☺. (FIG. 13.15)

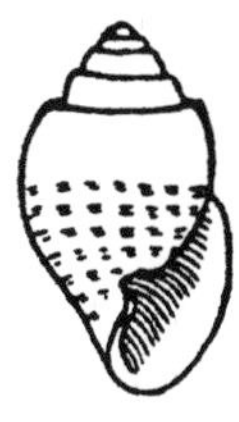

FIG. 13.16

Rictaxis (=*Acteon*) *punctostriatus*, pitted baby-bubble† (Acteonidae). Yellow to white; lower part of body whorl with rows of tiny dots. Medial aperture rim with twisted fold. Vi,Sa,Li–Su,pr,6 mm (0.24 in), (AM265,R633). (FIG. 13.16)

Scaphander punctostriatus, giant canoe bubble† (Scaphandridae). Smooth, yellow to brown outside; white inside. Bo,Sa,Su–De,pr,38 mm (1.5 in), (A99,AM271). (FIG. 13.17)

FIG. 13.17

13.21 **Aporrhaidae (Duckfoot Snails)** (from 13.16, 13.18)

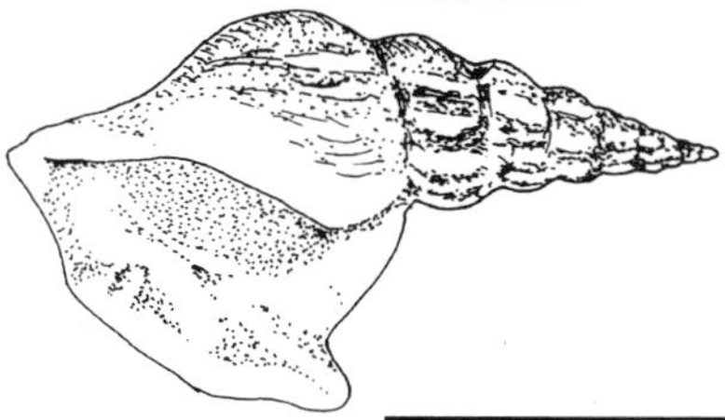

Thick, heavy shell with greatly flared lip (outer margin of aperture) in adults; horny (proteinaceous; fingernail-like material) operculum.

Aporrhais occidentalis, American pelicanfoot†. Striking apertural wing less developed in young. Shiny white inside aperture. Bo,Mu,Su–De,df,64 mm (2.5 in), (A60,AM183,R469). (FIG. 13.18)

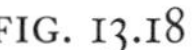

FIG. 13.18

13.22 **Buccinidae, Whelks** (from 13.16, 13.18)

Large, substantial shells with large openings; sculpture often includes wavy, axial ridges; horny operculum.

1. Dominant shell sculpture of body whorl (e.g., prominent ribs crossed by distinct but finer spiraling cords would be ribbed; see FIG. 13.6a)
   - A. Longitudinal *ribs* — ri
   - B. Spiraling *cords*, raised — co
   - C. Spiraling *threads* or scratches, not raised or barely raised — th
   - D. *Smooth* or with faint growth lines only — sm
2. Shell proportions: shape index = length/width (see FIG. 13.6d)
   - A. Shell *stout*, shape index ≤ 2.0 — st
   - B. Shell *elongate*, shape index > 2.0 — el
3. Coloration
   - A. *Dark* periostracum, dark brown — da
   - B. *Light* periostracum, olive to gray, yellow to brown — li
4. Shell length from spire apex to opposite apertural lip (this character is most useful for specimens that exceed minimal lengths; smaller individuals of larger species may cause confusion)
   - A. Shell *short*, < 56 mm (2¼ in) long — sh
   - B. Shell *long*, > 75 mm (3 in) long — lo
   - C. Shell *intermediate* length, 56–75 mm (2¼–3 in) long — in

13.23

Sculpture	Index	Color	Length	Species
ri	st	li	lo	*Buccinum undatum*, waved whelk†, buckie
ri	st	li	sh	*B. totteni*, thin whelk†
ri	st	da	sh	*B. scalariforme*, ladder whelk†, silky waved whelk
th	el	li,da	lo	*Colus stimpsoni*, Stimpson whelk†
th	el	da	in	*C. pubescens*, hairy whelk†
th	el	li	sh	*C. pygmaeus*, pygmy whelk†
th	st	li	sh	*C. ventricosus*, ventricose whelk†, fat colus
th	st	li	in,lo	*Neptunea despecta*, disreputable whelk†
co	st	li	lo	*N. lyrata decemcostata*, wrinkle whelk†, New England ten-ridge whelk
co	el	li	lo	*Beringius turtoni*, turton neptune
sm	el	li	lo	*Volutopsius norvegicus*, Norway whelk†

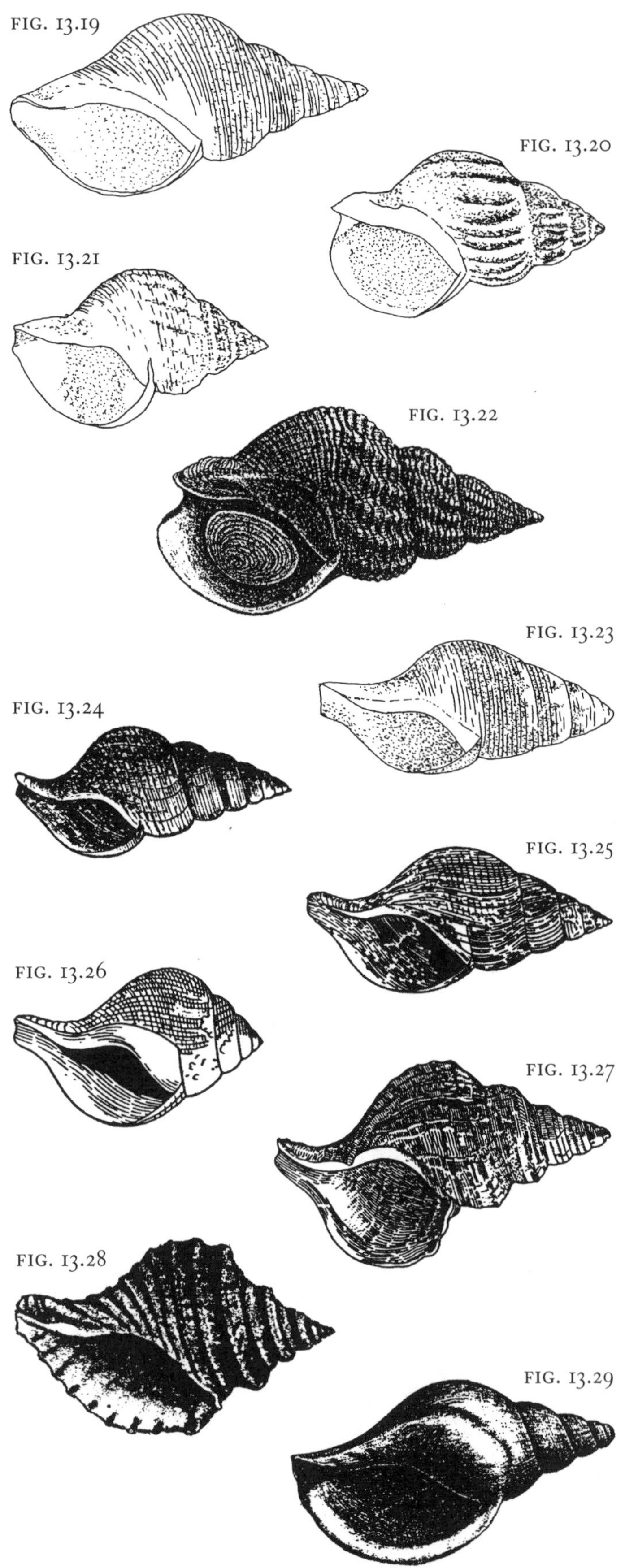

Beringius turtoni, turton neptune. Siphonal canal less well developed. Polar species, barely included in our range. Po–,?,Su,om,127 mm (5 in), (A79). (FIG. 13.19)

Buccinum scalariforme, ladder whelk†, silky waved whelk. Rugged shell. Ac,?,Su–De,pr?,51 mm (2 in), (A77). (Fig.13.20)

Buccinum totteni, thin whelk†. Outer apertural lip thin, fragile. Ac,?,Su,sc?,51 mm (2 in), (A76). (FIG. 13.21)

Buccinum undatum, waved whelk†, buckie. Outer apertural lip thick; gray yellow with thick gray greenish periostracum; somewhat flared outer apertural lip, used as a wedge between the valves of bivalve victims. Animal, white with black spots. Eggs in ball of overlapping capsules (see 6.4). Ac–,Ha–Sa,Su–De,pr–sc,140 mm (5.5 in), (A76,AM222,GI32,R549), ☺. (FIG. 13.22)

Colus pubescens, hairy whelk†. Spiral threads more distinct than *C. stimpsoni.* Bo,?,Su–De,pr–sc?,63.5 mm (2.5 in), (A79,GI32,R552). (FIG. 13.23)

Colus pygmaeus, pygmy whelk†. Spiral threads more distinct than *C. stimpsoni.* Bo,?,Su–De,pr–sc?,22 mm (0.87 in), (A79,AM222,GI32,R551). (FIG. 13.24)

Colus stimpsoni, Stimpson whelk. Large, chestnut to dark periostracum; spiral threads faint. Soft parts white with black spots. Egg capsules, small discs with transparent covers (see 6.4). Bo,Ha–Sa,Su–De,pr–sc,101 mm (4 in), (A81,AM223,GI32,R551), ☺. (FIG. 13.25)

Colus ventricosus, ventricose whelk†, fat colus. Aperture extends two-thirds of length of shell. Ac,?,De,om,51 mm (2 in), (A81). (FIG. 13.26)

Neptunea despecta, disreputable whelk. Rounded aperture. Ac,?,Su–De,pr,76 mm (3 in), (A82). (FIG. 13.27)

Neptunea lyrata decemcostata, wrinkle whelk†, New England ten-ridge whelk. Thick shell; grayish with brown cords. Body white with large black blotches. Bo,Ha,Su–De,pr,127 mm (5 in), (A82,AM223,GI33,R551). (FIG. 13.28)

Volutopsius norvegicus, Norway whelk†. Body, yellow with dark speckling. Polar species, barely within our range. Ac,Mu,?,pr,127 mm (5 in), (A78). (FIG. 13.29)

13.24 Caecidae (from 13.14)

Tiny, cylindrical, nonspiraled shells in sandy sediment; horny operculum.

1. Shell sculpture (see FIG. 13.6a)
   - A. Faint *ribs* follow long axis of shell — rb
   - B. Strong *rings* — rn
   - C. *Smooth* — sm

13.25

Sculpture	Species
rn	*Caecum pulchellum,* beautiful caecum
rb	*C. cooperi,* Cooper caecum
sm	*C. johnstoni*

Caecum cooperi, Cooper caecum. Vi,Sa,Su,pr,8 mm (0.3 in), (AM153).

Caecum johnstoni. Vi,Sa,Su,pr,8 mm (0.3 in), (GI35).

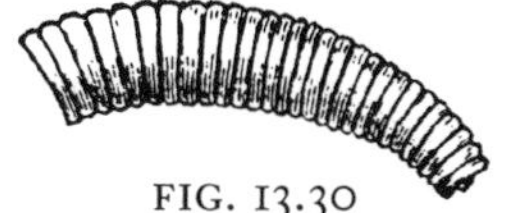
FIG. 13.30

Caecum pulchellum, beautiful caecum. Feeds on protists in sediments. Vi,Sa–SG,Su,pr,8 mm (0.3 in), (AM153,GI35,R420). (FIG. 13.30)

13.26 Calyptraeidae, Slippersnails or Boatsnails (from 13.8, 13.10)

Large aperture; attach firmly to hard substrate; internal shelf within shell; no operculum. Unusual as filter-feeding gastropod.

1. Arch of shell in profile
   - A. Shell comes to a dorsal apex or *point* — po
   - B. Shell is *arched* but lacks a dorsal point — ar
   - C. Shell very *flattened,* no arch or point — fl
2. Shape of internal shelf
   - A. *Flap*like shelf extends into interior of shell from one side — fl
   - B. Distinct *shelf* completely covers one-fourth to one-half of interior of shell — sh
3. Coloring
   - A. *White* — wh
   - B. *Tan* or gray, sometimes with brown mottling — ta
   - C. Rather *dark,* red-brown to purple-brown — da

13.27

Profile	Shelf	Color	Species
po	fl	ta	*Crucibulum striatum,* striate cup-and-saucer†
ar	sh	ta	*Crepidula fornicata,* common Atlantic slippersnail†, quarter-deck
ar	sh	da	*C. convexa,* convex slippersnail†
fl	sh	wh	*C. plana,* eastern white slippersnail†

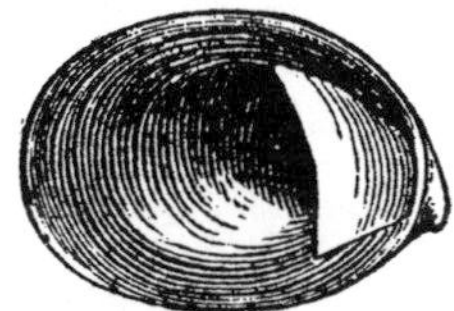
FIG. 13.31

Crepidula convexa, convex slippersnail†. Brown inside shell. Distinct apex toward anterior. Some on eelgrass; most attach to anything hard. Bo,SG–Sa–Ot,Li–Su,ff,12.7 mm (0.5 in), (A58,AM180,GI28,R465). (FIG. 13.31)

Crepidula fornicata, common Atlantic slippersnail†, quarter-deck. Outside, light with red to brown blotches and streaks. Forms stacks; large female on bottom, small males on top; hermaphrodites in middle; sometimes attached to other shells. Large filter-feeding gill. Bo,Sa–Ot,Li+,ff,50 mm (2 in), (A58,AM181,GI28,NM487,R464), ☺. (FIG. 13.32)

FIG. 13.32

Crepidula plana, eastern white slippersnail†. Flattened shells, often on whelk or moon snail shells or on *Limulus.* Does not stack; females large, surrounding males small. Bo,Sa–Ot,Li+,ff,30.5 mm (1.2 in), (A28,AM181,GI29,NM487,R466), ☺. (FIG. 13.33)

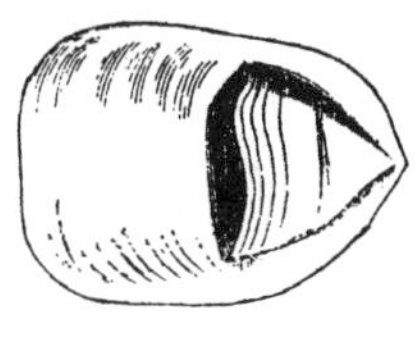

FIG. 13.33

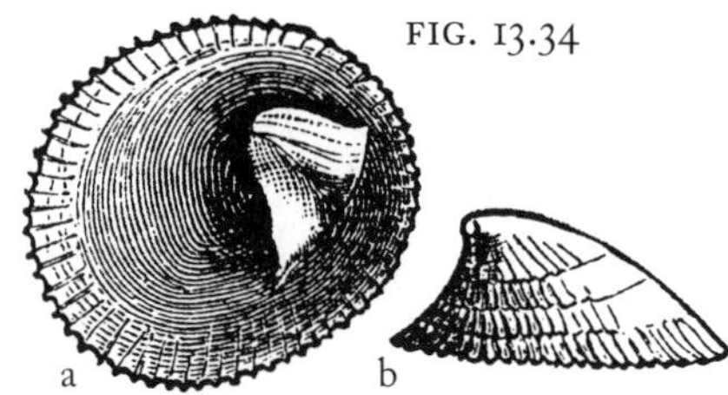

FIG. 13.34

Crucibulum striatum, striate cup-and-saucer†. Apex slightly tipped. Small males ride on larger females. Bo,Ha,Su–De,ff,35 mm (1.3 in), (AM180,G129,463). (FIG. 13.34. a, ventral view with flap; b, lateral view.)

13.28 **Cancellariidae, Nutmeg Snails** (from 13.16)

Strong, beaded shells with marginal folds but no operculum.

Admete couthouyi, northern admete†, nutmeg snail. Yellow to brown. Ac,Al,Su,gr,19 mm (0.75 in), (A93,AM243). (FIG. 13.35)

FIG. 13.35

13.29 **Cerithiidae and Cerithiopsidae, Ceriths or Hornsnails** (from 13.10, 13.18)

Small, elongate, knobby shells; horny operculum.

1, 2

3

FIG. 13.36

1. Dominant shell sculpture of body whorl (FIG. 13.36)
   - A. Raised, spiraling, *smooth* cords — sm
   - B. Raised, spiraling, *beaded* cords — be*
2. Rib running axially along the dorsal surface of the body whorl (FIG. 13.36)
   - A. A single *rib* present — ri
   - B. Such a rib *absent* — x*
3. Siphonal notch at anterior of apertural opening (FIG. 13.36)
   - A. Distinct siphonal *notch* present — no*
   - B. Siphonal notch barely visible or *absent* — x

13.30

Sculpture	Rib	Notch	Species
sm	x	no	*Seila adamsi,* wood screw snail
be	ri	x	*Bittiolum (=Bittium, Diastoma) varium,* grass cerith†, variable bittium
be	x	x	*Bittium (=Diastoma) alternatum,* alternate bittium
be	x	no	*Cerithiopsis greeni,* green cerith

FIG. 13.37

Bittium (=Diastoma) alternatum, alternate bittium (Cerithidae). Reddish brown color; tiny but common. Some individuals have slightly notched apertural lip; most do not. Bo,SG–Al–Sa–Mi,Li–Su,gr,8 mm (0.32 in), (G127,AM166,NM485,R440), ☺. (FIG. 13.37)

FIG. 13.38

Bittiolum (=Bittium, Diastoma) varium, grass cerith†, variable bittium (Cerithidae). Grayish color; smaller, more slender than preceding species. Gregarious, forms large groups. Sometimes with snail fur or the hydroid *Hydractinea.* Vi,SG–Al,Su,gr,8 mm (0.32 in), (G127,AM166,NM485,R440).

Cerithiopsis greeni, green cerith (Cerithiopsidae). Dull brown. Vi,Sa–Ha,Li–,gr,6 mm (0.24 in), (AM168,G136). (FIG. 13.38)

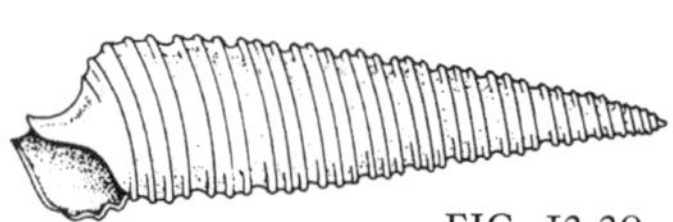
FIG. 13.39

Seila adamsi, wood-screw snail (Cerithiopsidae). Ten to twelve whorls; brown. Vi,SG–Al–Ha,Li–Su,gr,12.7 mm (0.5 in), (AM168,G136,R445). (FIG. 13.39)

13.31 Columbellidae, Dovesnails (from 13.16, 13.18)

Modest sizes, somewhat elongate; horny operculum.

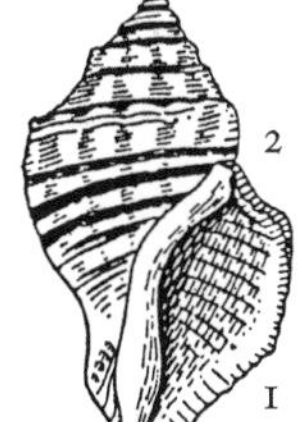

FIG. 13.40

1. Crenulation (i.e., bumps or teeth along inside surface of outer apertural lip) (FIG. 13.40)
   - A. *Crenulation* present — cr*
   - B. Crenulation *absent* — x
2. Dominant shell sculpture of body whorl (e.g., prominent ribs crossed by distinct but finer spiraling cords would be ribbed; see FIG. 13.6a) (FIG. 13.40)
   - A. Longitudinal *ridges* — ri
   - B. Ribs crossed by cords to produce *beaded* effect — be
   - C. Spiraling *threads* or scratches, not raised or barely raised — th*
   - D. *Smooth* or with faint growth lines only — sm

13.32

Crenulation	Sculpture	Species
cr	ri	*Costoanachis avara*, greedy dovesnail†
cr	be	*C.* (=*Anachis*) *lafresnayi* (=*translirata*), well-ribbed dovesnail†
cr	sm	*Astyris* (=*Mitrella*) *lunata*, lunar dovesnail†, crescent dovesnail
x	th	*A.* (=*Mitrella*) *rosacea*, rosy northern dovesnail†

FIG. 13.41

Astyris (=*Mitrella*) *lunata*, lunar dovesnail†. Narrow aperture with four small teeth inside outer lip. Brownish with lighter spots and/or dark zigzag lines. Eats invertebrates, especially erect ectoprocts. Bo,Al–Ha–SG,Li–,pr,8 mm (0.32 in), (A68,AM219,G136,R545). (FIG. 13.41)

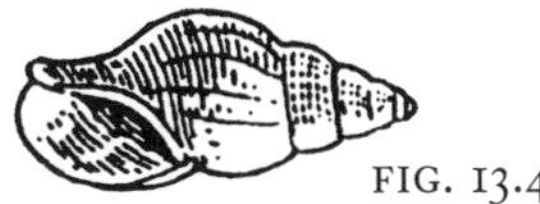

FIG. 13.42

Astyris (=*Mitrella*) *rosacea*, rosy northern dovesnail†. No teeth inside apertural lip. Spire whorls are slightly rounded and better defined than *A. lunata*. Ac–,Sa–Gr,Su–De,pr,10 mm (0.4 in), (A68,R545). (FIG. 13.42)

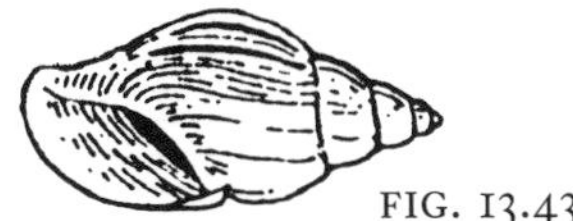

FIG. 13.43

Costoanachis (=*Anachis*) *avara*, greedy dovesnail†. Shell stouter, length/width ratio = 2.6. Ridges only on apical half of body whorl. Crenulated; inside surface of outer edge of apertural lip with tiny teeth. Brown with darker bands, often spotted with white. Vi,Ha,Su,pr?,12.7 mm (0.5 in), (AM218,R537). (FIG. 13.43)

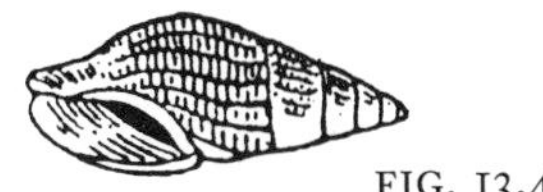

FIG. 13.44

Costoanachis (=*Anachis*) *lafresnayi* (=*translirata*), well-ribbed dovesnail†. Compared to preceding species, shell more slender, length/width ratio = 2.8. Ridges more numerous and may extend length of body whorl. Yellowish brown. Vi,Ha,Su,pr?,12.7 mm (0.5 in), (AM218,R536). (FIG. 13.44)

13.33 Epitoniidae, Wentletraps (from 13.12)

Slender shells with rounded whorls; axial ribs; reported to feed on anemones; no operculum.

1. Number of whorls: provide *number* — __
2. Number of axial ribs on body whorl: provide *number* — __
3. Shell length (in inches): provide *length* — __

13.34

No. Whorls	No. Ribs	Length	Species
8	0	1	*Acirsa borealis*, chalky wentletrap†, northern white wentletrap
10	0	1.2	*A. costulata*, northern costate wentletrap
6–8	9–10	1	*Epitonium angulatum*, angulate wentletrap†
8	16–19	1	*E. multistriatum*, many-rib wentletrap†
9	8–9	0.8	*E. humphreysii*, Humphrey wentletrap
10	12–18	0.5	*E. rupicola*, brown-band wentletrap†
12	9–12	1.5	*Boreoscala* (=*Epitonium*) *greenlandica*, Greenland wentletrap†

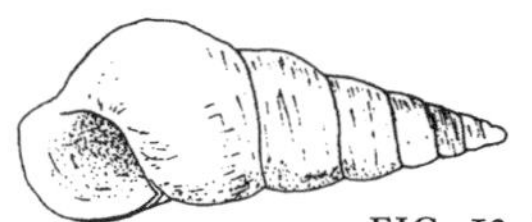
FIG. 13.45

Acirsa borealis, chalky wentletrap†, northern wentletrap. A bit more stout than next species. Ac,?,Su,pr,25.4 mm (1 in), (A51,AM171). (FIG. 13.45)

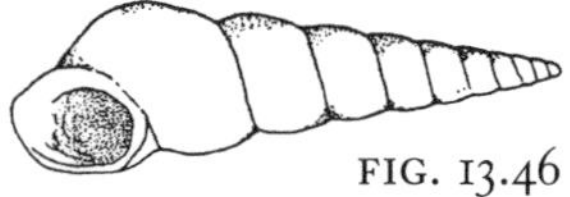
FIG. 13.46

Acirsa costulata, northern costate wentletrap. Thin apertural lip. Ac,?,Su,pr,30 mm (1.2 in), (A51). (FIG. 13.46)

FIG. 13.47

Boreoscala (=*Epitonium*) *greenlandica,* Greenland wentletrap†. Ac–,Gr,Su–De,pr,38 mm (1.5 in), (A52,AM173,GI28,NM487). (FIG. 13.47)

Epitonium angulatum, angulate wentletrap†. Vi,Sa,Su,pr,25.4 mm (1 in), (A52,AM172,GI28,NM486,R453).

FIG. 13.48

Epitonium humphreysii, Humphrey wentletrap. Vi,Sa,Li–Su,pr,22 mm (0.87 in), (A52,AM173,GI28,R452). (FIG. 13.48)

Epitonium multistriatum, many-rib wentletrap†. Vi,Sa?,Li–Su,pr,25.4 mm (1 in), (A52,AM174,GI28). (FIG. 13.49)

Epitonium rupicola, brown-band wentletrap†. Vi,Sa?,Li–Su,pr,12.7 mm (0.5 in), (AM174,GI28,R455).

FIG. 13.49

13.35 Fissurellidae, Keyhole Limpets (from 13.8)

Conical shells with prominent ribs; slit or hole at or near apex. No operculum.

1. Location of second opening (in addition to the large aperture)
   - A. Opening at *apex* of shell — ap
   - B. Opening *near* but not at apex — ne
2. Appearance of ribbed sculpturing on shell
   - A. Ribs approximately *equal* in size — eq
   - B. Every *fourth* rib considerably enlarged compared to the rest — fo

13.36

Opening	Ribs	Species
ne	eq	*Puncturella noachina,* diluvian puncturella†, keyhole limpet
ap	eq	*Diodora tanneri,* tanner keyhole limpet
ap	fo	*D. cayenensis,* cayenne keyhole limpet†

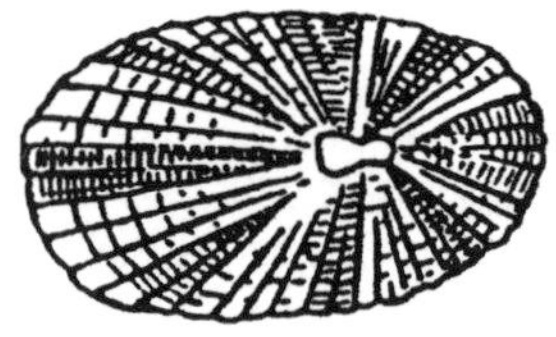
FIG. 13.50

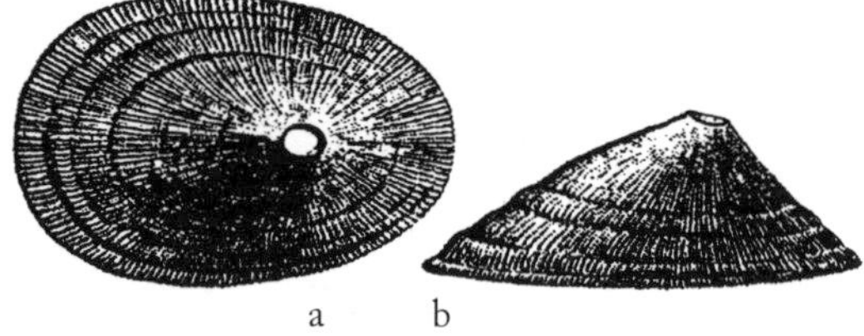

FIG. 13.51

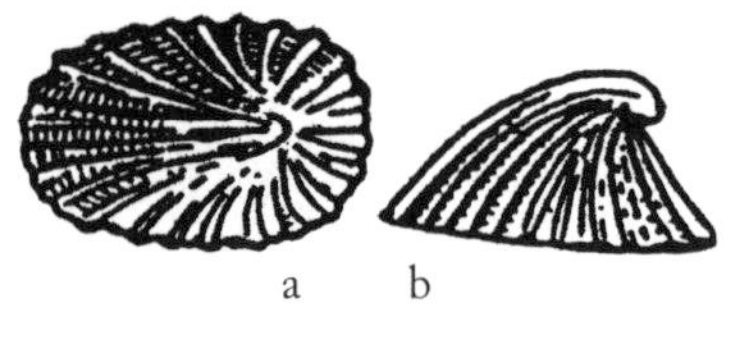

FIG. 13.52

Diodora cayenensis, cayenne keyhole limpet†. Many radiating ribs crossed by growth lines giving corrugated appearance. Often covered with attached algae. Ca+,Ha,Li–Su,gr,51 mm (2 in), (AM126,NM469,R349). (FIG. 13.50)

Diodora tanneri, tanner keyhole limpet. Many radiating ribs. Large, gray. Ca+,Sa–Mu,Su–De,gr,51 mm (2 in), (A30). (FIG. 13.51. a, dorsal; b, lateral.)

Puncturella noachina, diluvian puncturella†, keyhole limpet. Shell with curved apex, laterally compressed; glossy white inside; 20+ ribs. Raised collar around inside surface of slitlike apical opening. Ac,Ha,Li–De,gr,12.7 mm (0.5 in), (A28,AM129,GI25,R348). (FIG. 13.52. a, dorsal; b, lateral.)

13.37 **Hydrobiidae, Swamp Snails** (from 13.12)

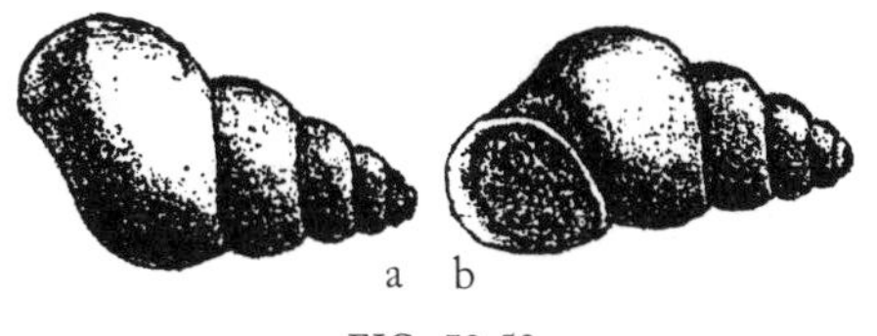

FIG. 13.53

Tiny, thin, rounded shells; horny operculum.

Hydrobia spp. Several species differ in shell proportions. Commonly reported is *Hydrobia totteni,* minute hydrobia†, seaweed snail. Aperture occupies lower third of ventral shell surface. Yellow to brown; deep sutures; apex often eroded away. Ac–SM–Mu–Al,Li,gr,6 mm (0.24 in), (AM150,GI36,R416), ☺. (FIG. 13.53. a, dorsal; b, ventral.)

13.38 **Janthinidae, Seasnails** (from 13.6)

Thin shell, wide aperture, no operculum; pelagic predators.

Janthina janthina, janthina†, purple sea snail. Thin lavender shell for pelagic species that floats by secreted bubbles; eats siphonophoran jellyfish. Vi,Po,Ps,pr,25.4 mm (1 in), (A48,AM176,GI28,NM486,R448). (FIG. 13.54)

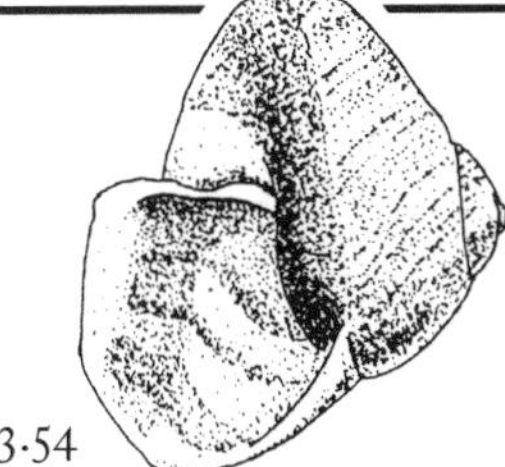
FIG. 13.54

13.39 **Lacunidae, Lacunas** (from 13.6)

FIG. 13.55

Small, thin shell with umbilicus reduced to a groove near medial apertural margin; horny operculum.

Lacuna vincta, northern lacuna†, banded lacunal. Thin shell with large aperture; dark color or, rarely, with white bands. Usually on algae: fucoids, kelps, *Ceramium, Polysiphonia.* Egg mass, tiny white doughnut on seaweed (see 6.8). Ac–,Ha–Al–SG,Li–Su,gr,12.7 mm (0.5 in), (A47,AM146,GI26,NM477,R410), ☺. (FIG. 13.55)

13.40 **Lamellariidae, Widemouth Snails** (from 13.10)

Thin and fragile shell associated with larger, sluglike animal; no operculum.

Lamellaria (=Marsenina) perspicua, transparent lamellaria†. Transparent shell; no periostracum. Bo,Ha,Li–Su,pr,12.7 mm (0.5 in), (A59,AM186). (FIG. 13.56)

FIG. 13.56

13.41 **Littorinidae, Periwinkles** (from 13.6, 13.10)

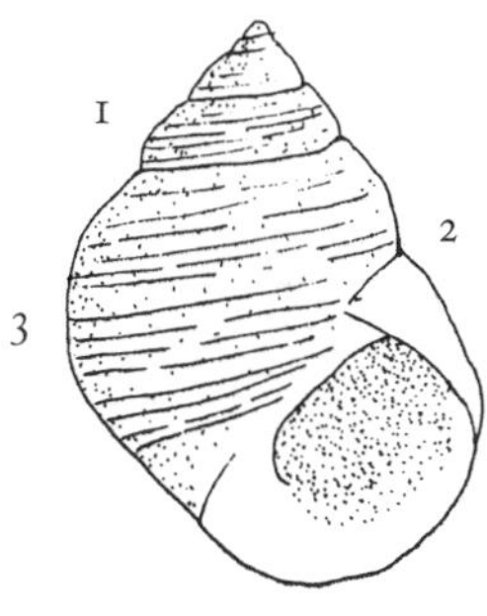

FIG. 13.57

Strong, compact shell; horny operculum.

1. Contour of spire whorls (FIG. 13.57)
   - A. Whorls *round* shouldered — ro*
   - B. Whorls smoothly sloped (i.e., *no* shoulders) — x
2. Shape of outer apertural lip where it meets the body whorl (FIG. 13.57)
   - A. Line from lip projects toward *spire* — sp
   - B. Line from lip projects into *body whorl* — bw*
3. Body whorl sculpture (FIG. 13.57)
   - A. Distinct, regular spiral *grooves* flecked with red-brown spots — gs
   - B. Distinct, regular spiral *grooves* with colored flecks *absent* — gx*
   - C. Irregular, faint spiral lines, but *not* forming regular grooves — x
4. Habitat
   - A. *Salt marsh* — sm
   - B. On rocks in *upper* half of *intertidal* — ui
   - C. On *rock-weeds* in mid to low intertidal — rw
   - D. On rocks in *lower* half of *intertidal* — li
   - E. On sand or mud *flat* — fl

13.42

Spire	Lip	Sculpture	Habitat	Species
x	sp	x	sm,ui,li,fl	*Littorina littorea,* common periwinkle†
x	sp	gs	sm	*L. irrorata,* marsh periwinkle†
ro	bw	gx	ui	*L. saxatilis,* rough periwinkle†
ro,x	bw	x	rw	*L. obtusata,* yellow periwinkle†, smooth periwinkle

FIG. 13.58

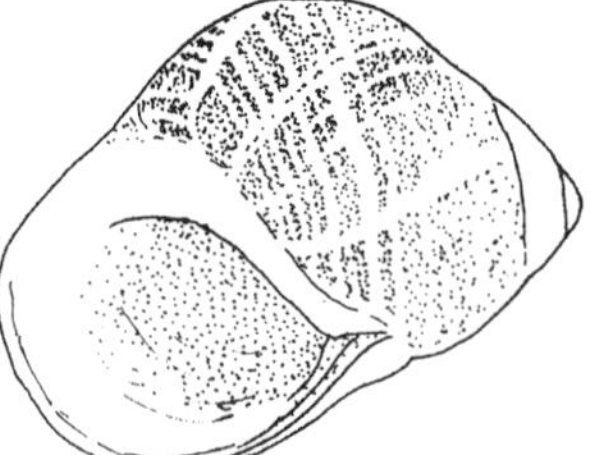
FIG. 13.59

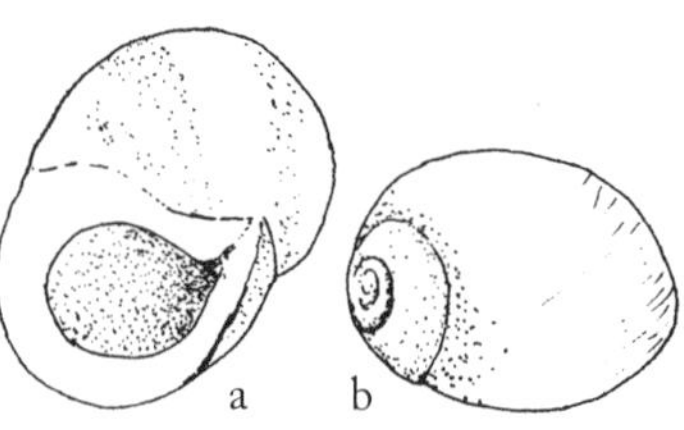

FIG. 13.60

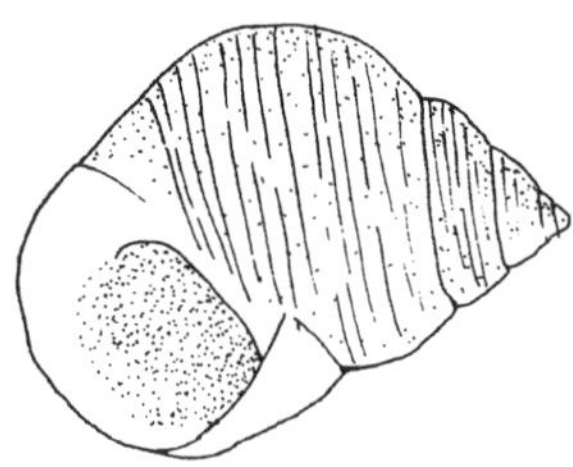
FIG. 13.61

Littorina irrorata, marsh periwinkle†. Medial edge of aperture orange. Especially common among marsh sedges. Climbs marsh grass blades to avoid immersion in water and/or predation by crabs. Vi,SM–Al–Ha,Es–Li,gr,25.4 mm (1 in), (A45,AM146,G126,NM479,R406), ☺. (FIG. 13.58)

Littorina littorea, common periwinkle†. Whorl sutures not well developed; transverse black stripes on tentacles and head. Inside of outer apertural lip is white; outside is often dark. Introduced from Europe certainly by mid-1800s but perhaps well before. Ubiquitous, omnivorous grazers prefer tender algae such as *Ulva, Enteromorpha,* or other ephemeral greens; may play keystone role in some settings. Moves deeper in cold months. Bo,Ha–Sa–Mu–SM–SG–Al,Es–Li,gr,43 mm (1.7 in), (A45,AM147,G126,NM478,R401), ☺. (FIG. 13.59)

Littorina obtusata, yellow periwinkle†, smooth periwinkle†. Smooth, shiny brown, olive, yellow, or banded. More yellowish in full, clean marine setings; browner in stagnant and brackish waters. Often on rockweeds, especially *Ascophyllum.* Eats decaying plants. Ac–,Ha–Al,Li,gr,12.7 mm (0.5 in), (A45,AM148,G126,NM479,R402), ☺. (FIG. 13.60. a, ventral; b, dorsal.)

Littorina saxatilis, rough periwinkle†. Deep sutures between whorls; gray, yellowish, reddish, to dark-colored. Longitudinal stripes on tentacles. High intertidal. Ac–,Ha,Li,gr,12.7 mm (0.5 in), (A46,AM148,G126,NM480,R402). (FIG. 13.61)

13.43 **Lottiidae, Limpets** (from 13.8)

Conical shell with large apertural opening; no operculum.

Tectura (=Acmaea, =Notoacmaea) testudinalis, plant limpet†, tortoiseshell or Atlantic plate limpet. Interior is bluish with brown apex. Eats coralline algae. Rocky shore and tidal pools. Bo,Ha–Al,Li–,gr,51 mm (2 in). (A35,AM131,G125,NM471,R363), ☺. (FIG. 13.62. a, lateral; b, ventral.)

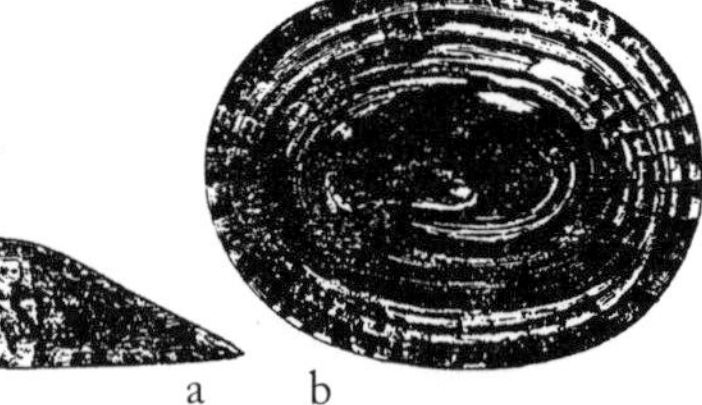

FIG. 13.62

13.44 **Marginellidae** (from 13.10)

Solid, glossy shell with little or no spire but long, thin aperture.

Prunum (=Marginella) roscidum, seaboard marginella†, boreal or dewy marginella. Outer apertural lip with dark spots. Shiny brown with white spots. Vi,Sa,Su–De,pr,16 mm (0.63 in), (A94). (FIG. 13.63)

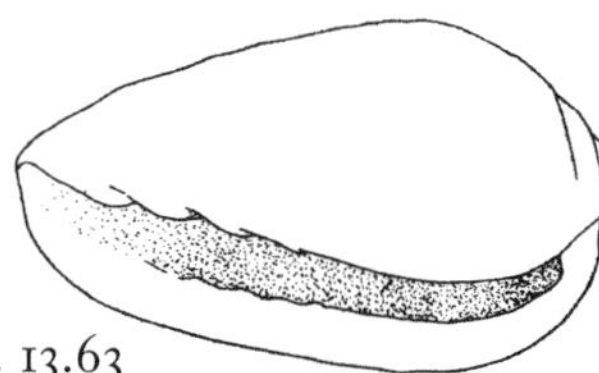
FIG. 13.63

13.45 Melampodidae, Marsh Pulmonates (from 13.10, 13.12)

Most are stubby shells with small pointed spire and long, thin aperture with folds along medial apertural margin; no operculum.

1. Teeth or folds along inside surface of outer apertural rim
   - A. *Teeth* present — te
   - B. Teeth *absent* from outer apertural rim — x
2. Spire whorls compared to total length of shell
   - A. Spire whorls nearly *one-third* shell length — ⅓
   - B. Spire whorls *less than one-quarter* shell length — <¼

13.46

Teeth	Spire	Species
te	⅓	*Phytia (=Ovatella) myosotis,* mouse melampus
x	<¼	*Leucophytia (=Melampus) bidentata,* eastern melampus†, salt marsh pulmonate

FIG. 13.64

FIG. 13.65

Leucophytia (=*Melampus*) *bidentata,* eastern melampus†, salt marsh pulmonate. Bean shaped; olive green to brown outside: gray inside. Air-breathing. Bo,SM,Li,ot,15 mm (0.6 in), (A103,AM274,G135,NM532,R645), ☺. (FIG. 13.64)

Phytia (=*Ovatella*) *myosotis,* mouse melampus. Dark brown. Two to four folds or ridges along columella. Pulmonate air-breather. Introduced from Europe. Bo,SM–Al,Li,gr,9 mm (0.35 in), (A103,G135,R650). (FIG. 13.65)

13.47 Melongenidae, Crown Conchs and Whelks (from 13.6, 13.18)

Large, solid shells with well developed siphonal canal and expanded body whorl; horny operculum.

1. Features related to whorls
   - A. *Channel*like furrow marks spire whorls — ch
   - B. Shoulder of body whorl marked by *sharp knobs* — sk
2. Color of shell interior
   - A. *Red to orange* — ro
   - B. *Yellowish* — ye
3. Color of soft tissues
   - A. *Dark* — da
   - B. *Light* — li

13.48

Whorls	Inside	Tissue	Species
ch	ye	li	*Busycotypus (=Busycon) canaliculatus,* channeled whelk†
sk	ro	da	*Busycon carica,* knobbed whelk†

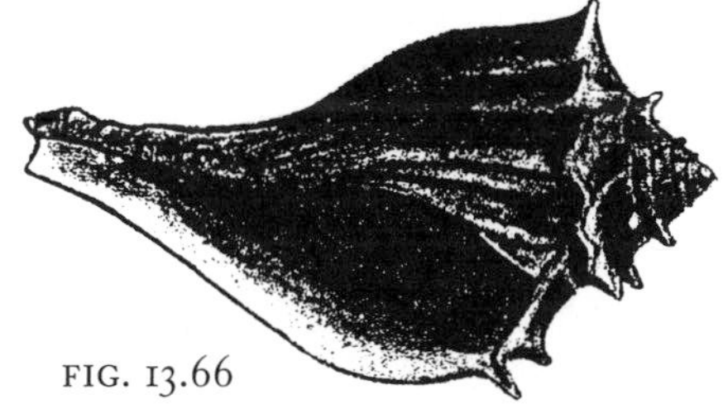

FIG. 13.66

Busycon carica, knobbed whelk†. Shoulders form knobs. Young with faint purple stripes. Often red-orange tones inside. Soft parts dark. Eats clams. Sold in markets as scungili. For distinctive egg cases, see 6.4. Vi,Sa,Li–Su,pr,22.9 mm (9 in), (A86,AM227,G133,R562), ☺. (FIG. 13.66)

Busycotypus (=*Busycon*) *canaliculatus,* channeled whelk. Deep channel follows sutures between whorls. Periostracum hairy; yellowish inside aperture; soft parts light. Sold in markets as scungili. For distinctive egg cases, see 6.4. Vi,Sa,Li–Su,pr,203 mm (8 in), (A86,AM227,G133,R563), ☺. (FIG. 13.67)

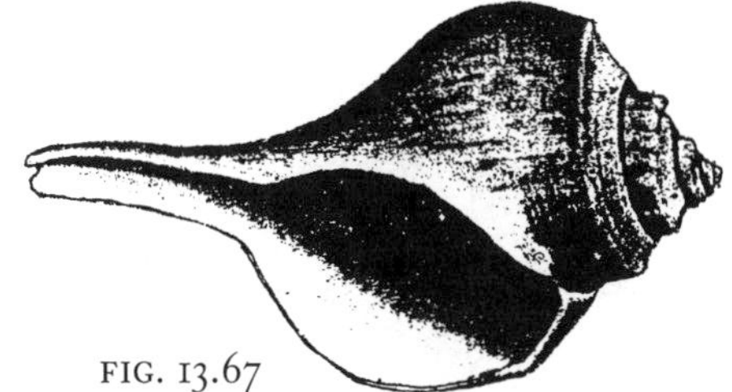

FIG. 13.67

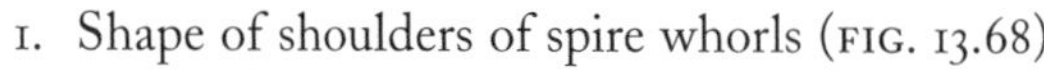

13.49 **Muricidae, Rock Snails** (from 13.16, 13.18)

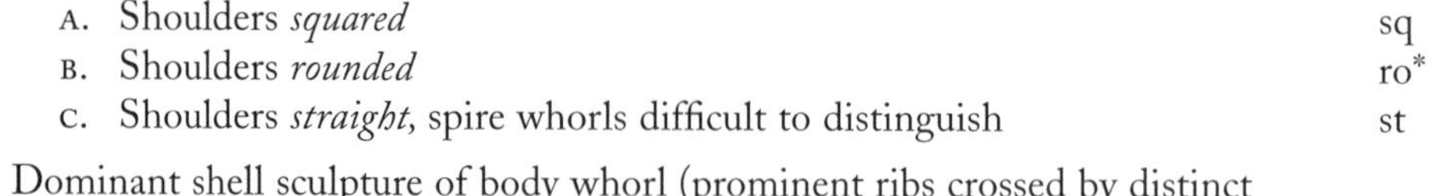

Solid shells with axial ridges in most species; siphonal canal moderately to strongly developed; horny operculum.

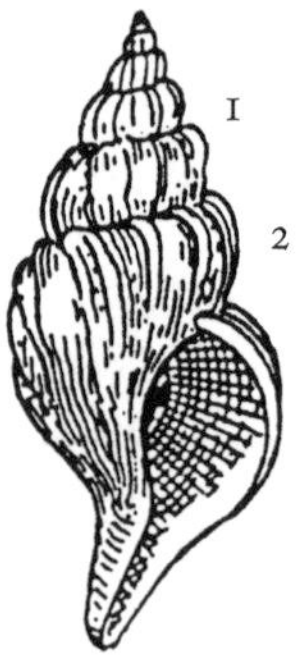

1. Shape of shoulders of spire whorls (FIG. 13.68)
   - A. Shoulders *squared* — sq
   - B. Shoulders *rounded* — ro*
   - C. Shoulders *straight,* spire whorls difficult to distinguish — st
2. Dominant shell sculpture of body whorl (prominent ribs crossed by distinct but finer spiraling cords would be ribbed; see FIG. 13.6a) (FIG. 13.68)
   - A. Wide axial *ridges* — ri
   - B. Narrow axial *ribs,* raised as *blades* — rb*
   - C. Narrow axial *ribs, not* raised as blades — rx
   - D. Spiraling *cords,* either as scratches or somewhat raised — co
   - E. *Smooth* or with faint growth lines only — sm
3. Color of median apertural lip
   - A. *Dark,* brown to purplish — da
   - B. *Light,* cream, white, gray, yellow — li

FIG. 13.68 (labels: 1, 2)

13.50

Shoulders	Sculpture	Aperture	Species
ro	rb	li	Shell also shows spiraling cords: *Boreotrophon craticulatum,* latticed trophon† OR Shell lacks spiraling cords: *Boreotrophon clathratus,* clathrate trophon†
ro	rx	li	*B. truncatus,* bobtail trophon†
ro	ri	da	*Urosalpinx cinerea,* Atlantic oyster drill†
ro,st	co,sm,ri	li	*Nucella lapillus,* Atlantic dogwinkle†, dog whelk
sq	rx	li	*Eupleura caudata,* thick-lip drill

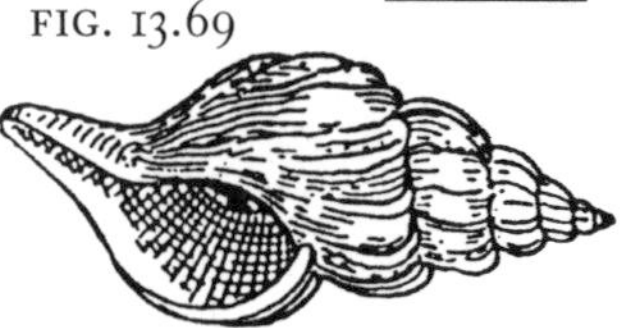

FIG. 13.69

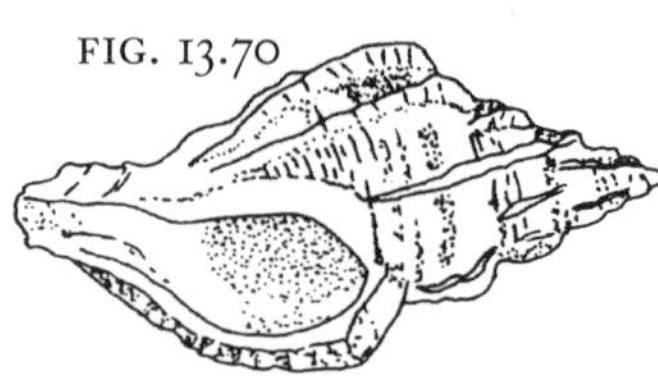

FIG. 13.70

Boreotrophon clathratus, clathrate trophon†. Ac,Mu,Su,pr,40 mm (1.6 in), (A74,AM213,R523). (FIG. 13.69)

Boreotrophon craticulatum, latticed trophon†. Polar species, barely within our range. Ac,?,Su,pr,25.4 mm (1 in), (A75). (FIG. 13.70)

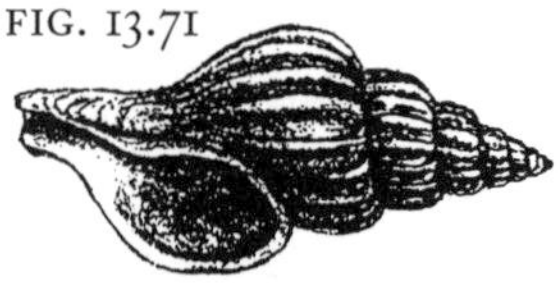

Boreotrophon truncatus, bobtail trophon. Ac,Sa?,Su,pr,15 mm (0.6 in), (A75). (FIG. 13.71)

FIG. 13.71

Eupleura caudata, thick-lip drill†. Eats oysters. Eggs laid in flattened, shield-shaped capsules attached to rock or shell by stalks (see 6.6). Vi,Ha–Sa–SG,Li–Su,pr,25.4 mm (1 in), (A72,AM212,G130,R514). (FIG. 13.72)

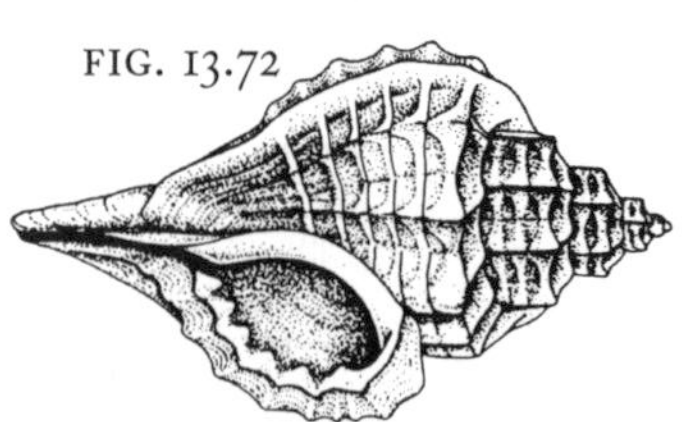

FIG. 13.72

Nucella lapillus, Atlantic dogwinkle†, dog whelk. Variable color: dark to light yellow, orange, brown, or purple bands. Variable thickness: thinner at exposed sites; thicker in protected areas. Variable in sculpture: smooth, ringed, wavy ridges, combinations. Implicated as keystone species by harvest of young mussels. Also eats barnacles. Eggs in bowling-pin-shaped capsules on rock (see 6.6). Ac+,Ha,Li–,pr,44 mm (1.7 in), (A73,AM216,G131,R526), ☺. (FIG. 13.73)

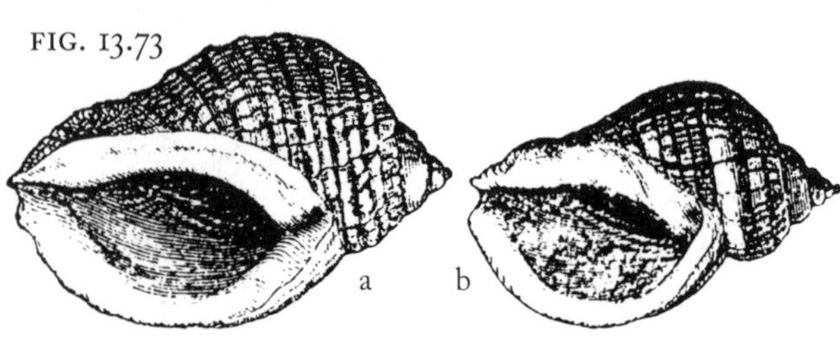

FIG. 13.73 (a, b)

Urosalpinx cinerea, Atlantic oyster drill†. Primarily south of Cape Cod. Yellow-brown-gray-white outside; inside often with hint of purple. Vase-shaped egg cases on rock (see 6.6). Preys heavily on juvenile oysters, other mollusks, barnacles. Vi+,Ha,Li–,pr,20.3 mm (0.8 in), (A71,AM211,G180,R519), ☺. (FIG. 13.74)

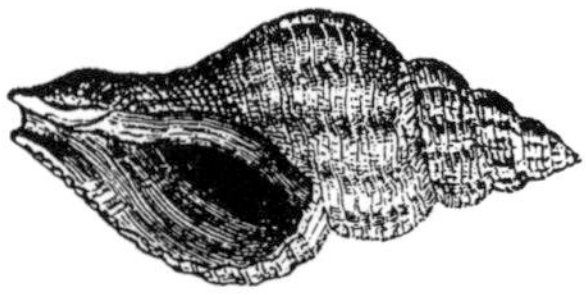

FIG. 13.74

13.51 Nassariidae, Dog Whelks (from 13.16, 13.18)

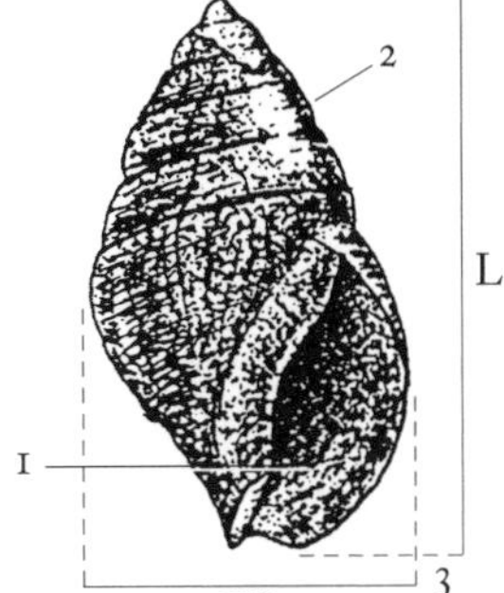

FIG. 13.75

Solid shell with pointed spire and smooth or beaded sculpture; predator/scavengers; horny operculum.

1. Color of aperture (FIG. 13.75)
   - A. *Brown* — br*
   - B. *White* — wh
2. Contour of spire whorls (FIG. 13.75)
   - A. Spire whorls *rounded,* well differentiated — ro
   - B. Spire whorls *not* rounded, and not well differentiated — x*
3. Shell proportions: shape index = length/width (FIG. 13.75)
   - A. Shell *chunky,* shape index < 1.8 — ch
   - B. Shell more *slender,* shape index > 2.0 — sl*

13.52

Aperture	Spire	Index	Species
br	x	sl	*Ilyanassa* (=*Nassarius*) *obsoleta,* eastern mudsnail†, mud-dog whelk, eroded basketsnail
wh	x	ch	*Nassarius vibex,* bruised nassa†, eastern or mottled dog whelk
wh	ro	sl	*Ilyanassa* (=*Nassarius*) *trivittata,* three-lined basketsnail†, New England dog whelk

FIG. 13.76

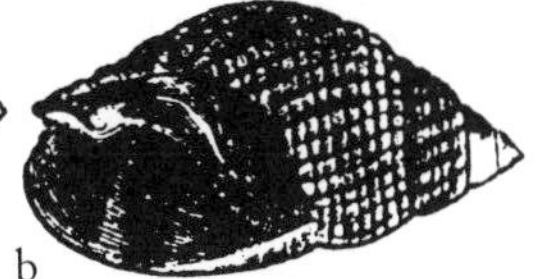

Ilyanassa (=*Nassarius*) *obsoleta,* eastern mudsnail†, mud dog whelk, eroded basketsnail. Apex often eroded; dark color. Gregarious. Intermediate host for larvae of some trematode (fluke) parasites. Eggs laid in capsules (see 6.6) on grass blades. Bo,Mu–SG,Li–Su,ot,25 mm (1 in), (A88,AM220,G131,R571), ☺. (FIG. 13.76. a, smooth, eroded shell form; b, beaded shell form.)

FIG. 13.77

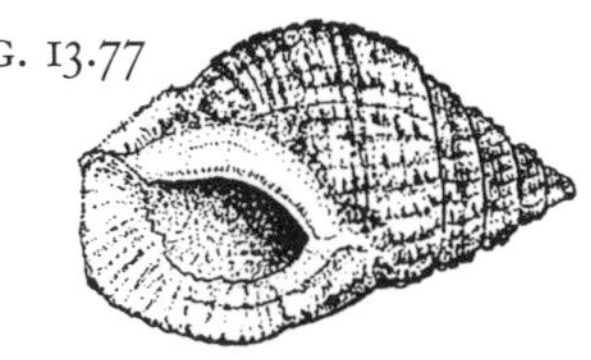

Ilyanassa (=*Nassarius*) *trivittatus,* three-lined basketsnail†, New England dog whelk. Deep sutures between whorls; sharp shoulders. Color light; medial aperture lip, white. Two pointed processes at posterior of foot. Detects dead organisms. Eggs laid in capsules (see 6.6). Bo,Sa–SG,Li–Su,sc,25 m (1 in), (A89,AM221,G132,R567). (FIG. 13.77)

FIG. 13.78

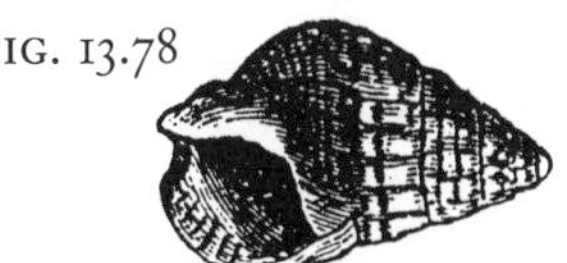

Nassarius vibex, bruised nassa†, eastern or mottled dog whelk. Two pointed processes at posterior of foot. White with gray or brown blotches outside. Eats polychaete eggs and carrion. Eggs laid in capsules (see 6.6). Vi,Sa–Mu–SG,Li,sc–ot,18 mm (0.7 in), (A89,AM221,G132,R566). (FIG. 13.78)

13.53 Naticidae, Moon Snails (from 13.6, 13.10)

Predatory moon snails have a globose shell and a very large foot. They drill symmetrical access holes through shells of other mollusks and use long tubular probosces to feed. Operculum calcareous in a few; horny in most.

FIG. 13.79

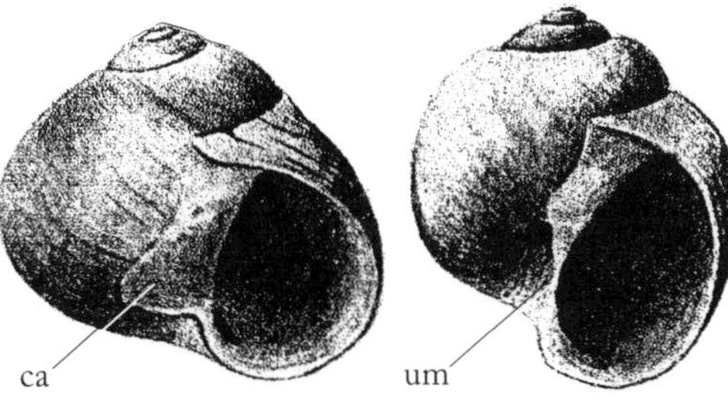

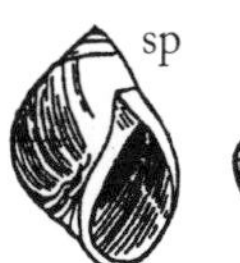

1. Callus (flattened shell patch medial to aperture) bordering, but not covering over, the umbilicus (FIG. 13.79)
   - A. *Callus* present — ca
   - B. Callus *absent* — x
2. Appearance of umbilicus (opening representing the axis around which the shell spirals; if present, located along midline, adjacent to the aperture) (FIG. 13.79)
   - A. Distinct *umbilicus* visible (even if partially blocked by callus) — um
   - B. Umbilicus *absent* — x
3. Shape of outer apertural lip where it meets the body whorl (FIG. 13.79)
   - A. Line from lip projects toward *spire* — sp
   - B. Line from lip projects into *body whorl* — bw

4. Shell length (spire to anterior lip of aperture) (this character is useful if specimen exceeds a smaller limit; young individuals of larger species could be misleading)
   A. Shell *large,* > 40 mm (1.5 in) long — la
   B. Shell *small,* < 12 mm (0.5 in) long — sm
   C. Shell *intermediate,* 12–40 mm (0.5 –1.5 in) long — in

13.54

Callus	Umbilicus	Lip	Size	Species
x	um	bw	la	*Euspira (=Lunatia) heros,* northern moon snail†
x	um	sp	sm	*Lunatia triseriata,* spotted moon snail†
x,ca	um	sp	in	*Neverita (=Polinices) duplicata,* shark eye†, lobed moon snail
ca	um	sp	sm	Shell with rows of dark streaks: *Tectonatica (=Natica) pusilla* OR Shell uniformly white: *Euspira (=Polinices) immaculata,* immaculate moon snail†
ca	x	sp	in	*Bulbus smithi,* Smith moon snail
ca	x	bw	la	*Natica clausa,* Arctic moon snail†
x	x	bw	in	*Amauropsis islandica,* Iceland moon snail†
x	x	sp	la	*Sinum perspectivum,* white baby-ear†

FIG. 13.80

FIG. 13.81

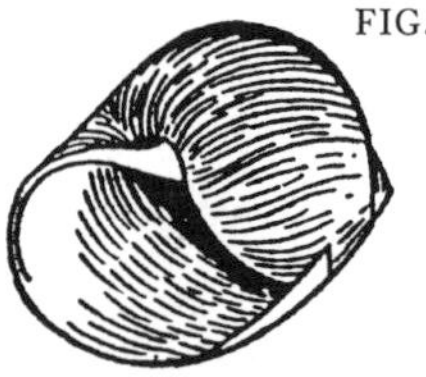

FIG. 13.82

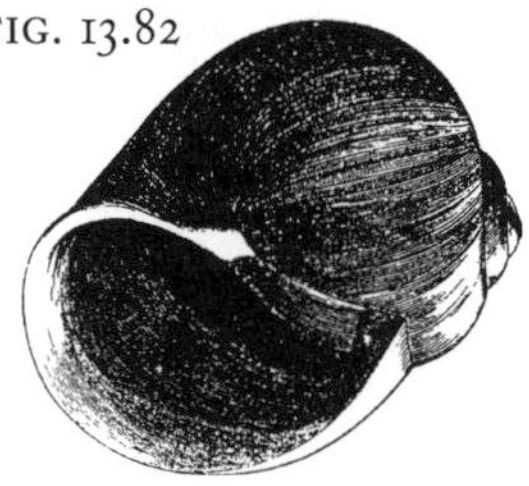

FIG. 13.83

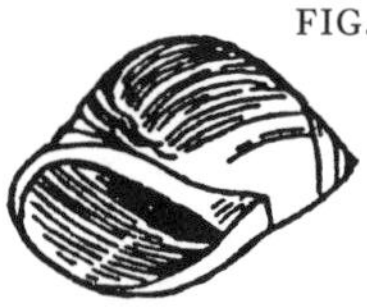

FIG. 13.84

FIG. 13.85

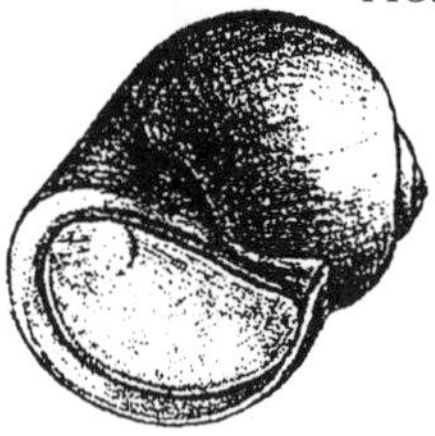

FIG. 13.86

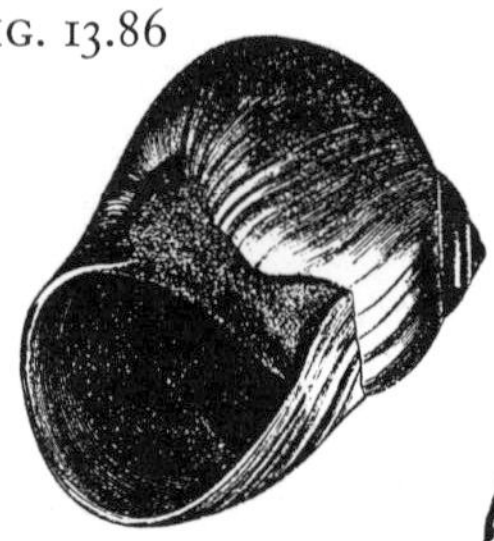

FIG. 13.87

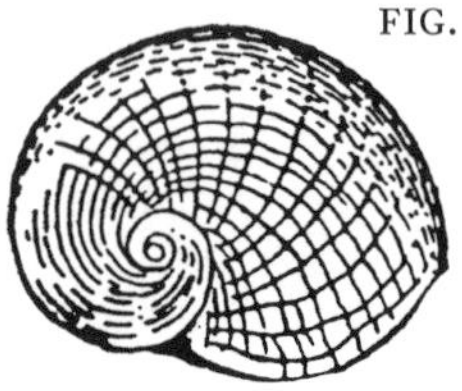

Amauropsis islandica, Iceland moon snail†. Bo,Sa?,Su–De,pr,33 mm (1.3 in), (A65,AM190). (FIG. 13.80)

Bulbus smithi, Smith moon snail. Ac,Sa,Su,pr?,25 mm (1 in), (A65). (FIG. 13.81)

Euspira (=Lunatia) heros, northern moon snail†. Dull white to dirty gray. Huge gray foot. Drills neat hole to eat bivalves. Produces "sand collar" egg mass (see 6.2). Bo,Sa,Li–De,pr,114 mm (4.5 in), (A64,AM189,G129,NM491,R487), ☺. (FIG. 13.82)

Euspira (=Polinices) immaculata, immaculate moon snail†. Small, milky white. Bo,Sa?,Su,pr,11 mm (0.4 in), (AM189,G130). (FIG. 13.83)

Euspira (=Lunatia) triseriata, spotted moon snail†. Body whorl with three rows of brownish square spots. White foot, black tentacles. Produces "sand collar" egg mass (see 6.2). Bo,Sa,Li–De,pr,31 mm (1.2 in), (A63,AM190,G130,NM491,R487). (FIG. 13.84)

Natica clausa, Arctic moon snail†. Operculum calcareous; pale brown. Important fish food. Bo,Sa,Li-De,pr,51 mm (2 in), (A66,AM191,R491). (FIG. 13.85)

Neverita (=Polinices) duplicata, shark eye†, lobed moon snail. Medium brown, hint of blue. Large, obvious brown callus. Vi,Sa,Li–,pr,7.6 cm (3 in), (A63,AM188,G129,NM490,R485), ☺. (FIG. 13.86)

Sinum perspectivum, white baby-ear†. Very flat, white shell. Body of living animal nearly engulfs shell. No operculum. Eats bivalves. Ca+,Sa,Li–Su,pr,51 mm (2 in), (A66,AM193,NM492,R487). (FIG. 13.87)

Tectonatica (=Natica) pusilla. Vi,Sa,Su,pr,6 mm (0.3 in).

13.55 **Pyramidellidae, Pyramid Snails** (from 13.12)

Small, white, slender shells with folds along medial apertural margin; many are parasitic; horny operculum.

FIG. 13.88

Odostomia spp. One ridge along columella. Many difficult species; see references. Bo,Sa–SG–Mi,Li–Su,pa–pr,7 mm (0.3 in), (AM263,GI36,R627–629).

Pyramidella spp. Two to three ridges along columella. Many difficult species; see references. Bo,Sa-Mu,Li–Su,pa–pr,7 mm (0.3 in), (AM261,GI36,R625–626). (FIG. 13.88)

FIG. 13.89

Turbonilla spp. Many difficult species; see references. Bo,Ot,Li–Su,pa–pr,7 mm (0.3 in), (AM261–263,GI37,R592). (FIG. 13.89)

13.56 **Rissoidae, Rissos** (from 13.12)

Many difficult species of tiny snails, associated with algae; horny operculum; see references. (MI35,R411).

13.57 **Scaphandridae** (See Acteonidae, Atyidae, and Scaphandridae, 13.19)

13.58 **Skeneopsidae, Orbsnail** (from 13.6)

Tiny, flat coils with deep umbilicus.

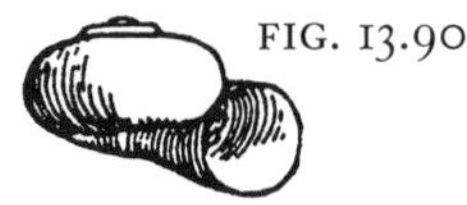

FIG. 13.90

Skeneopsis planorbis, flat skenea†, orbsnail. Smooth, dull brown; tiny, found among stones and algae. Bo,Ha–Al,Li–Su,gr,6 mm (0.24 in), (AM154). (FIG. 13.90)

13.59 **Terebridae, Augersnails** (from 13.18)

Very long and pointed shells; horny operculum.

FIG. 13.91

Terebra dislocata, eastern auger†. Long, thin, pointed. Gray to pale orange. Ca+,Sa,Su,pr,51 mm (2 in), (AM252,GI24,R611). (FIG. 13.91)

13.60 **Trichotropidae, Hairysnails** (from 13.6)

Thin shells with grooved umbilicus.

FIG. 13.92

Trichotropis borealis, boreal hairysnail†. Aperture pointed anteriorly; shell with cords. Dull yellow brown periostracum. Ac,?,Su,ff,20 mm (0.8 in), (AM178,GI31,R460). (FIG. 13.92)

13.61 **Triphoridae, Left-Handed Snails** (from 13.18)

Small, slender shells with aperture on left side (sinistral).

FIG. 13.93

Triphora nigrocincta, black-line triphora†. Dark band between whorls; smooth, beaded surface. Vi,Ha–Al,Li–,pr,8 mm (0.3 in), (AM169,GI36,R446). (FIG. 13.93)

13.62 **Trochidae, Topsnails or Sunsnails** (from 13.6, 13.10)

With spiraled shells and circular apertures, prominent umbilicus and width distinctly greater than length; horny operculum.

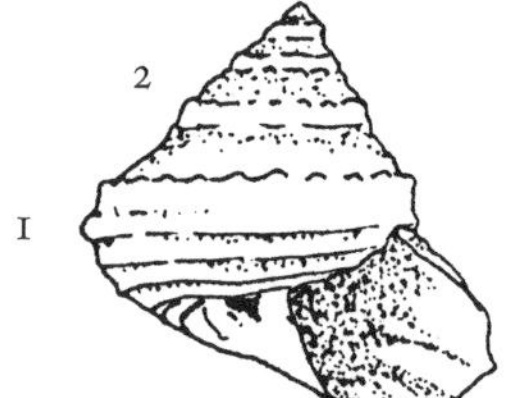

FIG. 13.94

1. Sculpture of body whorl (FIG. 13.94)
   - A. *Smooth* or faint growth lines only — sm
   - B. Raised *smooth,* spiral *cords;* provide *number* — sc:__
   - C. Raised *beaded* spiral *cords;* provide *number* — bc:__*
2. Contour of shoulders on spire whorls (FIG. 13.94)
   - A. *Angled* shoulders, formed by spiral cords — an*
   - B. Shoulder *smoothly* rounded — sm
   - C. Shoulders *straight;* spire whorls not well defined — st
3. Appearance of body whorl
   - A. *Pearly* — pe
   - B. *Dull white* to rose colored — dw
   - C. *Reddish brown* — rb
   - D. *Light* brown with *reddish* spots — lr

13.63

Sculpture	Shoulder	Body	Species
sm	sm	pe	*Margarites helicinus,* spiral margarete†, smooth topsnail
sc:4–5	sm	rb	*M. groenlandicus,* Greenland margarete†
sm,sc:3–4	an	dw,pe	*Solariella obscura,* obscure solarelle†
sc:8–10	an	dw	*Margarites costalis,* boreal rosy margarite†
bc:4–5	an	pe	*Lischkeia ottoi,* otto spiny topsnail
bc:5	an	pe	*L. regularis,* spiny topsnail
bc:7–8	st	lr	*Calliostoma bairdi,* Baird topsnail
sc:3–6	st	pe	*C. occidentale,* boreal topsnail†

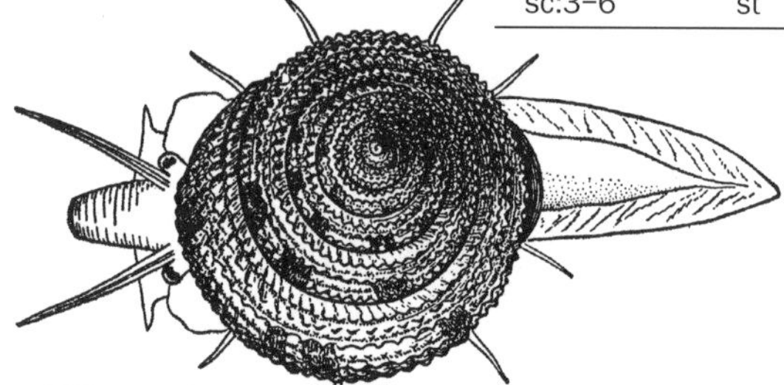
FIG. 13.95

FIG. 13.96

FIG. 13.97

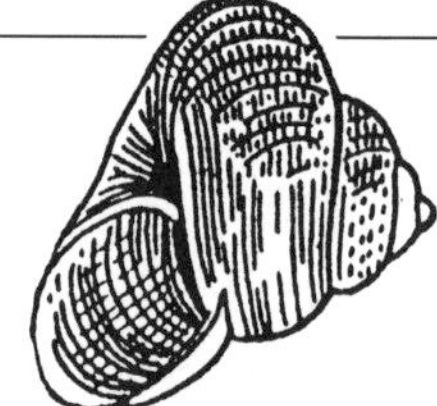
FIG. 13.98

Calliostoma bairdi, Baird topsnail. Sutures between whorls indistinct. Strong beaded cords; beads largest toward top of whorls. Bo,Ha–Sa,Su,om,25.4 mm (1 in), (A38). (FIG. 13.95. Dorsal view; note head, foot, and marginal tentacles.)

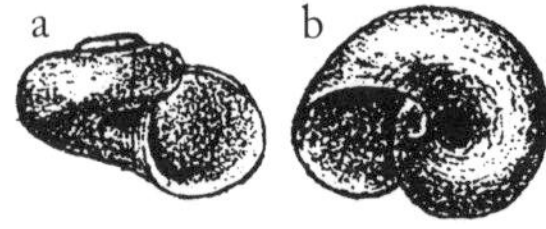

FIG. 13.99

Calliostoma occidentale, boreal topsnail†. Sutures between whorls apparent. Three to six strong spiraling cords; may be slightly beaded toward apex; pearly color; iridescent aperture. Eats hydroids and alcyonarians, but grazes algae too. Ac–,Ha–Sa,Su–De,om,18 mm (0.71 in), (A38,AM136,NM475,R377). (FIG. 13.96)

Lischkeia ottoi, otto spiny topsnail. Bo,?,Su–De,gr,18 mm (0.71 in), (A37,AM135). (FIG. 13.97)

Lischkeia regularis, spiny topsnail. Ac,?,Su,gr,18 mm (0.71 in), (A37).

Margarites costalis, boreal rosy margarite†. Outer aperture lip crenulated; rosy color. Ac,Ha,Su–De,gr,25.4 mm (1 in), (A36,R369).

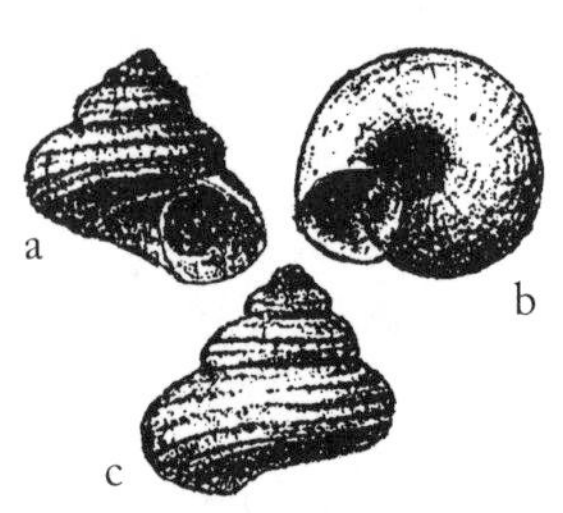

FIG. 13.100

Margarites groenlandicus, Greenland margarite†. Rosy tan color. Many spiral lines. Ac,Ha–Sa–Al,Su–De,gr,13 mm (0.5 in), (A36,AM132,G136,R370). (FIG. 13.98)

Margarites helicinus, spiral margarite†, smooth topsnail. Shiny yellowish pink to brown. Ac,SG–Al,Su,gr,13 mm (0.5 in), (A36,AM133,G136,R368). (FIG. 13.99. a, lateral; b, ventral.)

Solariella obscura, obscure solarelle†. Bo,Al,Su–De,gr,10 mm (0.39 in), (A37,AM134,R373). (FIG. 13.100. a, lateral; b, ventral; c, dorsal.)

13.64 Turridae, Turretsnails (from 13.18)

Many small genera (e.g., *Oenopta* or *Lora*) and species; see references. (M248–9,R615).

Cryoturris (=*Mangelia*) *cerinella* (=*cerina*), waxy mangelia†. Pale yellowish white; no operculum. Vi,Sa–Gr,Su–De,sc?,13 mm (0.5 in), (M249,R622).

Pyrogocythara (=*Mangelia*) *plicosa*, plicate mangelia†, spindle or ribbed turretsnail. Bo,SG–Sa–Mu,Su,sc?,8 mm (0.3 in), (AM258,R623).

13.65 Turritellidae, Turretsnails (from 13.6, 13.12)

Long, thin shells; horny operculum.

1. Overall shape of shell
   - A. Shell with tightly coiled apex but remainder appears *unwound* — un
   - B. Shell *compactly* spiraled; typically snaillike — co
2. Number of raised cords following spiral line around the shell: provide *number* of spiraling cords — __

13.66

Shape	Cords	Species
un	1–2	*Vermicularia spirata*, West Indian wormsnail†
co	5–6	*Tachyrhynchus erosus*, eroded turretsnail†
co	3	*Turitellopsis acicula*, needle turretsnail†

FIG. 13.101

FIG. 13.102

FIG. 13.103

Tachyrhynchus erosus, eroded turretsnail†. White with brown periostracum; five to six spiral cords. Ac,Mu,Su,ds–ff,25.4 mm (1 in), (A50,AM156,R425). (FIG. 13.101)

Turitellopsis acicula, needle turretsnail†. White to brown shell; three spiral cords. Ac,?,Su,?,13 mm (0.5 in), (FIG. 13.102)

Vermicularia spirata, West Indian wormsnail†. Grows attached to rocks, or embedded in sponges or algae; horny operculum. Vi,Sa–Mu,Su,?,152 mm (6 in), (AM157,GI27,NM482,R425). (FIG. 13.103)

13.67 Velutinidae, Velvetsnails (from 13.10)

Thin, fragile shell, like related Lamellariidae, with soft parts as sluglike animal.

1. Shell sculpture
   - A. Spiraling *grooves* present — gr
   - B. Shell *smooth* — sm
2. Coloration
   - A. *Yellow* periostracum — ye
   - B. Light *brown* periostracum — br

13.68

Sculpture	Color	Species
gr	br	*Velutina velutina*, striped lamellaria†, striped velvetsnail
sm	ye	*V. undata*, smooth lamellaria†, smooth velvetsnail

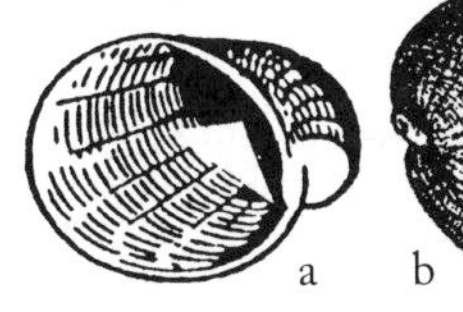

FIG. 13.104

Velutina undata, wavy lamellaria†. Transparent shell; thin periostracum. Living animal is sluglike; shell is internal: Ac,Ha,Li,pr,15 mm (0.6 in), (FIG. 13.104. a, ventral; b, dorsal.)

Velutina velutina, smooth lamellaria†, striped velvetsnail. Translucent shell; thick periostracum. Eats solitary ascidians. Ac,Ha,Li,pr,17 mm (0.67 in), (A59,AM186). (FIG. 13.105)

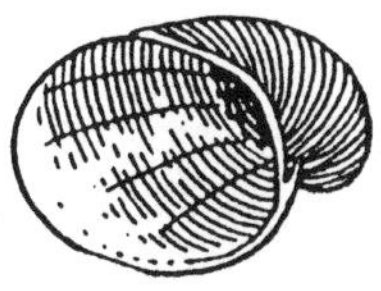
FIG. 13.105

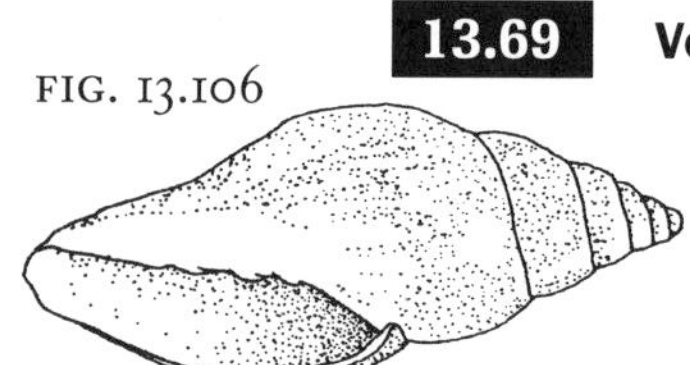
FIG. 13.106

13.69 **Volutomitridae, Mitersnails** (from 13.18)

Slender with large aperture bearing folds along medial margin.

Volutomitra groenlandica, Greenland miter†. Ac,?,?,?,25 mm (0.98 in), (A92). (FIG. 13.106)

Other Molluscs with One-Piece Shells, Class Gastropoda (continued)

13.70 **Order Gymnostomata, Sea Butterflies** (from 13.16)

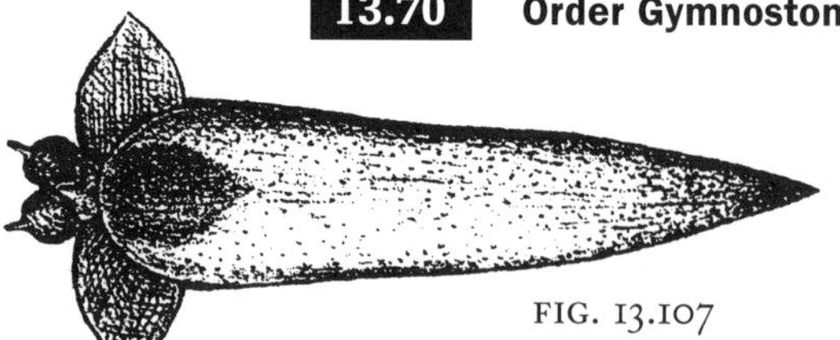
FIG. 13.107

Pelagic, shell-less, sluglike animals which swim with winglike parapodial flaps.

Clione limacina, naked sea butterfly (Clionidae). Semitranslucent, gray, pink, or yellow pteropod. One pair simple tentacles, on pair eye tubercles. Bo,Pe,Ps,pr,25.4 mm (1 in), (AM281,G137). (FIG. 13.107)

13.71 **Order Thecosomata, Shelled Pteropods** (from 5.16, 13.6, 13.14)

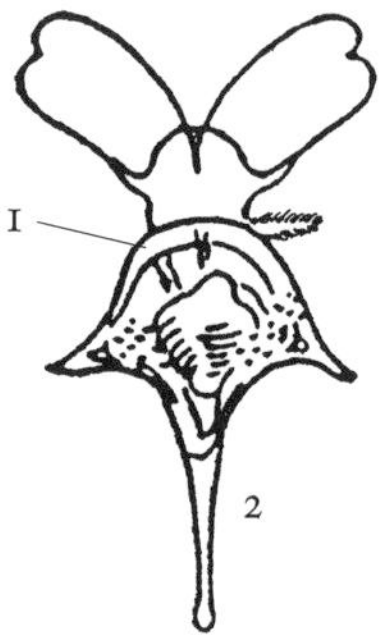

Pelagic, offshore animals with reduced shells. They swim with flaplike mantle extensions. All species listed are in the family Cavoliniidae.

1. Shape of shell bordering aperture (lateral view) (FIG. 13.108)
   - A. *Rounded* — ro*
   - B. *Pointed* — po
   - C. *Straight,* forms smooth, circular opening — st

2. Size of points along side opposite apertural opening (FIG. 13.108)
   - A. *Central* point much *longer than lateral* points — c>l*
   - B. *Lateral* points *longer than central* point — l>c
   - C. *Central* point about *equal to lateral* points — c=l
   - D. Points *absent* — x

FIG. 13.108

13.72

Shell	Points	Species
ro	c=l	*Cavolinia uncinata,* uncinate cavoline†
ro	c>l	*Diacria trispinosa,* three-spined cavoline†
po	l>c	*Clio pyramidata,* pyramid clio†
st	x	*Cuvierina columnella,* cigar pteropod

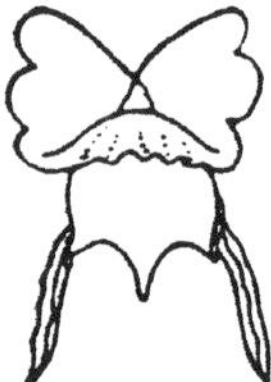
FIG. 13.109

Cavolinia uncinata, uncinate cavoline†. Bo,Pe,Ps,pl,13 mm (0.5 in), (A102,AM278). (FIG. 13.109)

Clio pyramidata, pyramid clio†. Bo,Pe,Ps,pl,18 mm (0.7 in), (A101,AM280).

Cuvierina columnella, cigar pteropod†. Bo,Pe,Ps,ff,10 mm (0.4 in), (A102,AM282). (FIG. 13.110)

Diacria trispinosa, threespine cavoline†. Bo,Pe,Ps,pl,13 mm (0.5 in), (A101,AM279). (FIG. 13.108)

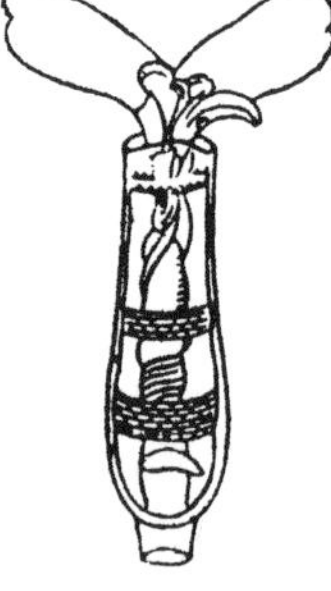
FIG. 13.110

13.73 **Class Cephalopoda (in part)** (from 13.6)

Head with tentacles; shell typically reduced, internal, or lacking; well-developed organ systems, especially the nervous, cirulatory, and locomotory systems.

Spirula spirula, ram's horn squid† (Or: Sepioidea, Spirulidae). Chambered shell from small, deepwater squid. Open spiral (not fused), flat, white shell found on southern beaches. In life, shell embedded toward posterior of squidlike body. Ca+,Pe,Pd,pr,25.4 mm (1 in) (shell diam.), (AM300,G163). (FIG. 13.111)

FIG. 13.111

13.74 Class Scaphopoda (Or: Dentaliida, Dentaliidae), Tooth Snails (from 13.14)

The tooth or tusk snails are not Gastropoda but are members of the class Scaphopoda. They all feed primarily on benthic foraminiferan protists living in sandy sediments, using a unique set of feeding tentacles, the captaculae.

1. Shape of imaginary cross section through shell
   - A. Shell cross section is *circular* — ci
   - B. Shell cross section is *ovoid* — ov
2. Dominant shell sculpture
   - A. Distinct or faint *ribs* follow the long axis — rb
   - B. Shell basically *smooth*, or irregularly wrinkled — sm

13.75

Cross Section	Sculpture	Species
ci	rb	*Antalis entale occidentale,* occidental tuskshell†, ribbed tuskshell
ci	sm	*A. (=Dentalium) entalis stimpsoni,* Stimpson tuskshell†
ov	rb	*Fissidentalium meridionale,* meridian tuskshell†

FIG. 13.112

Antalis (=Dentalium) entale stimpsoni, Stimpson tuskshell. Bo,Sa,Su–De,df,51 mm (2 in), (A107,G122,R658). (FIG. 13.112)

FIG. 13.113

Antalis entale occidentale, occidental tuskshell†, ribbed tuskshell. Bo,Sa,Su-De,df,30 mm (1.2 in), (A106,AM298,G122). (FIG. 13.113)

Fissidentalium meridionale, meridian tuskshell†. Vi,Sa,Su–De,df,8.9 cm (3.5 in), (A107,G122).

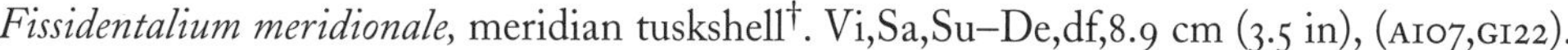

13.76 Class Gastropoda: Subclass Opisthobranchia (in part); Orders Sacoglossa and Nudibranchia, Sea Slugs

Regional sea slugs represent two orders within the subclass Opisthobranchia. Some, members of the order Sacoglossa, are small herbivores with a single pair of *rhinophores* (sensory tentacles) on the head. Most, however, belong to the carnivorous Nudibranchia, which includes a more diverse array of species grouped into three suborders. Nudibranchs in the suborder Aeolidacea are generally slender with sensory rhinophores on the head and, in many cases, a second pair of more anterior *oral tentacles.* The dorsal surface of Aeolidacea tends to be covered with elongate *cerata* (fingerlike structures that include diverticulae of the digestive gland) and often are colored brightly by pigments from their typically cnidarian prey. The related suborder Dendronotacea includes nudibranchs with branched cerata and tentacles. Finally, the absence of cerata but presence of a circle of feathery gills surrounding a posterior-dorsal anus characterizes members of the suborder Doridacea. Dorids are usually domed and oval-shaped with a single pair of rhinophores anteriorly. (Shelled opisthobranchs, including the Pyramidellidae and the shelled Thecosomata or pteropods, are treated within the Gastropoda, which they resemble; 13.5).

Most nudibranchs are delicate. Their specialized diets are difficult to accommodate in captivity. They also do not preserve well. It is best to observe them alive in the field or in a dish of fresh, cold seawater, taking special note of features apt to deteriorate with time, such as coloration, appearance of their cerata (if present), and their gills (if present). Their small size and activity level may require the use of a dissecting microscope and perhaps of anesthetization (see Appendix).

Species listed are in the order Nudibranchia unless noted otherwise.

Additional reference: (M) Martinez, A. J. 1994.

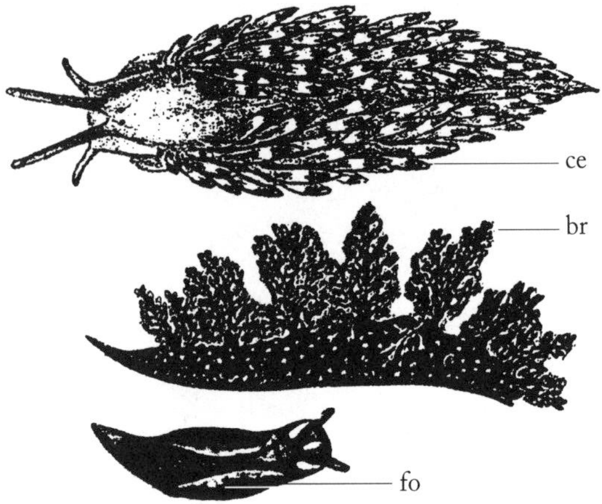

FIG. 13.114

1. Presence of distinctive features on the dorsal surface (FIG. 13.114)
   - A. Long fingerlike, flaplike, or bulbous *cerata* in rows or clusters along dorsal surface (does not include small bumps or one to two pairs of anterior tentacles) ce
   - B. Arborescent (*branched*) cerata all along dorsal surface (does not include a single clump of branched structures located midway or at posterior end) br
   - C. Lateral *folds,* wings or parapodia bend dorsally to nearly envelope the dorsal surface fo
   - D. Cerata *absent* x
2. Circle of feathery (i.e., branched) gills visible middorsally or toward the posterior end (permit specimen to relax because gills can be retracted) (FIG. 13.115)
   - A. *Gills* present gi*
   - B. Gills *absent* x
3. Number of pairs of anterior tentacles (as distinct from cerata): provide *number* of pairs (1 or 2) (FIG. 13.115) __

FIG. 13.115

13.77

Cerata	Gills	Tentacles	Group
br	x	2	*Dendrontus* *D. frondosus,* frond-aeolis†, bushy-backed sea slug OR *D. robustus,* robust frond-aeolis†, bushy-backed sea slug
ce	x	2	Opisthobranch Group 1 (Aeolidacea [in part]), 13.78
ce	x	1	Opisthobranch Group 2, 13.86
ce	gi	1	Opisthobranch Group 3 (Nudibranchia, Doridacea [in part]), 13.88
fo	x	1	*Elysia* Lateral folds extend ⅔ body length: *E. catula,* eelgrass or cat sea slug OR Lateral folds extend to posterior tip of foot: *E. chlorotica,* emerald elysia†
x	x	1	*Doridella obscura,* obscure corambe†, limpet nudibranch
x	gi	1	Opisthobranch Group 4 (Nudibranchia, Doridacea [in part]), 13.90

FIG. 13.116

Dendronotus frondosus, frond-aeolis†, bushy-backed sea slug (Dendronotidae). Whitish with reddish mottling toward north of range; brownish mottling toward south; occasionally white. Second pair of tentacles, the rhinophores, sheathed. Associated with hydroids *Ectopleura crocea, Dynamena pumila,* and *Obelia* spp., also on ectoprocts and tunicates. March–June. Ac–,Ha–Al,Li–Su,pr,11.4 cm (4.5 in), (AM291,GI40,MI06,NM549), ☺. (FIG. 13.116)

Dendronotus robustus, robust frond-aeolis†, bushy-backed sea slug (Dendronotidae). Large; red with white spots. May–August. Ac,Ha–Al,Su,pr,10.2 cm (4 in).

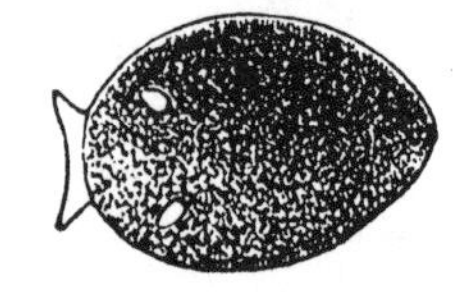

FIG. 13.117

Doridella obscura, obscure corambe†, limpet nudibranch (Or: Sacoglossa, Corambidae). Dorid shape but gills are underneath mantle edge and not visible dorsally. Light to dark color, often mottled with pigmented spots. Eats encrusting ectoprocts, especially *Alcyonidium.* Vi,SG–Ha,Li–,pr–om,13 m (0.5 in), (GI39). (FIG. 13.117. Dorsal view.)

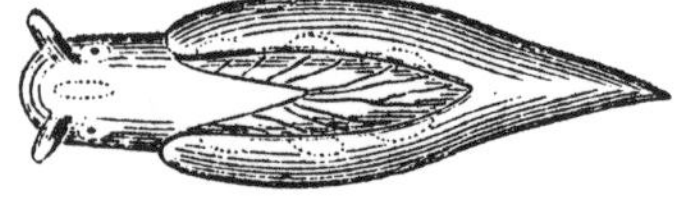

FIG. 13.118

Elysia catulus, cat elysia or eelgrass sea slug (Or: Sacoglossa, Elysiidae). Lateral folds extend two-thirds body length. Brownish with green tones and light patches on head; small triangular "ears." Eats *Zostera,* eelgrass. Found year-round. Vi+,SG,Su,gr,12.5 mm (0.5 in), (GI39). (FIG. 13.118)

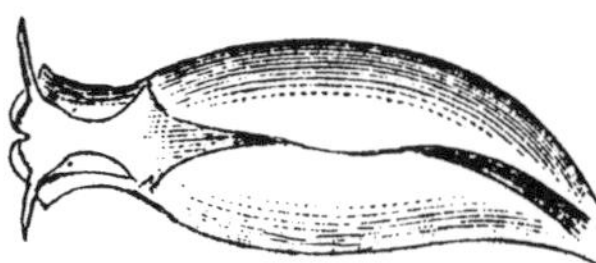
FIG. 13.119

Elysia chlorotica, emerald elysia† (Or: Sacoglossa, Elysiidae). Bright green or brown with light and sometimes red dots. Tissues incorporate functional chloroplasts extracted from food. Found in marshes on green algae, *Cladophora* or *Vaucheria;* August–October. Bo,SM–Al,Es–Li,gr,31 mm (1.2 in), (GI39). (FIG. 13.119)

13.78 Opisthobranch Group 1 (Aeolidacea) (from 13.77)

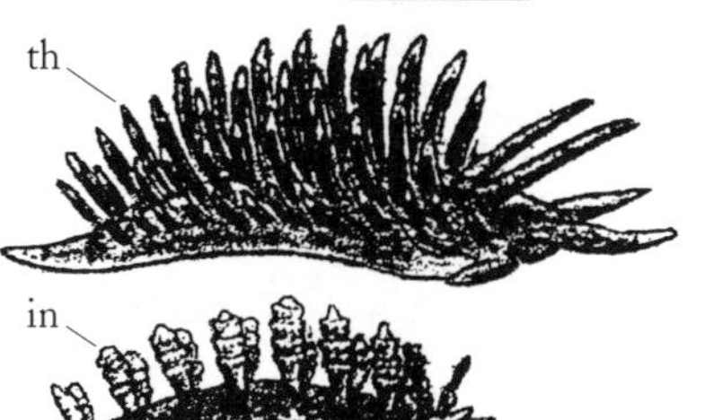

FIG. 13.120

Nudibranchs possessing elongate cerata and lacking clustered, branched gills. The digestive glands of *aeolid nudibranchs* form diverticulate branches that typically project along their dorsal surface in fingerlike cerata. They are often associated with cnidarians, especially hydroids, which they eat.

1. Shape of cerata or other dorsal projections (FIG. 13.120)
   - A. *Inflated* cerata—distinctly "swollen" appearance — in*
   - B. *Thin* cerata—uniform thickness along its length, tapering gradually toward tip (FIG. 13.120) — th*

13.79

Cerata	Group or Organism
th	Aeolids with thin cerata, 13.80
in	Aeolids with inflated cerata, 13.84

13.80 Aeolids with Thin Cerata (from 13.79)

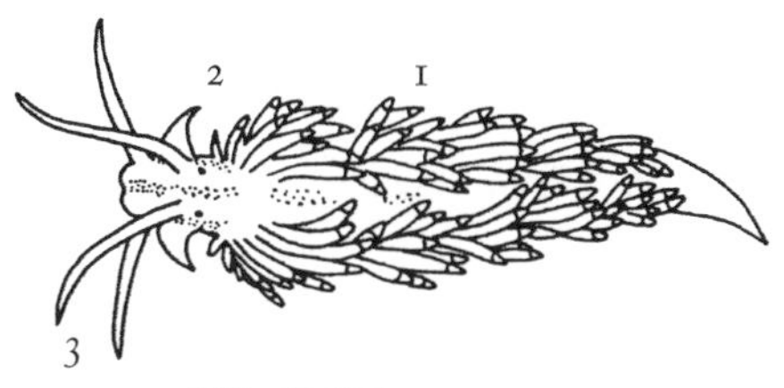

FIG. 13.121

1. Approximate number of cerata on one side of body: provide *number* — __
2. Shape of leading corners of foot (encourage the animal to walk across a dish and look for this feature from beneath) (FIG. 13.121)
   - A. Foot sharply *angled* — an*
   - B. Foot *rounded* — ro
3. Shape of pairs of tentacles on head—first pair (oral): second pair (rhinophore) (FIG. 13.121)
   - A. First pair *smooth:* second pair *smooth* — s:s*
   - B. First pair *smooth:* second pair *ringed* or irregular along posterior margin — s:r

13.81

No. Cerata	Foot	Tentacle 1:Tentacle 2	Organism or Group
20–25	an	s:r	*Facelina bostonensis,* Boston facelina†
25–30	an,ro	s:s	*Cuthona concinna*
30–50	ro	s:s	*Catriona gymnota (=aurantia)*
30–100	an	s:s,s:r	Grayish green body with whitish margins and 3 longitudinal reddish stripes on head: *Cratena pilata,* striped nudibranch OR Lacks coloring above: Flabellinid (=Coryphellid) eolids, ☺, 13.82
100+	an	s:s	*Aeolidia papillosa,* shag-rug aeolis†, maned nudibranch

FIG. 13.122

Aeolidia papillosa, shag-rug aeolis†, maned nudibranch (Aeolidiidae). Cerata, grayish; body, pale orange. With small anemones, *Metridium.* Egg mass, a coil of pink eggs in mucus (see 6.8). April–August. Ac–,Ha,Li–Su,pr,10 cm (4 in), (AM290,GI42,MI14), ☺. (FIG. 13.122)

Cratena pilata, striped nudibranch (Facelinidae). On hydroids, *Ectopleura crocea, E. larynx, Obelia geniculata, Pennaria disticha.* Also eats jellyfish polyps. Late summer–fall. Ac–,Ha–SG,Li,pr,33 mm (1.3 in), (GI41).

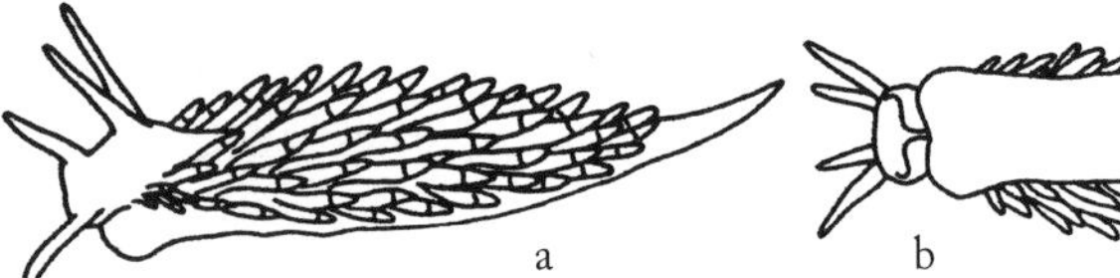

FIG. 13.123

Catriona gymnota (=*aurantia*) (Tergipedidae). With hydroids, *Ectopleura larynx, E. crocea.* May–August. Ac–,Ha,Li–Su,pr,18 mm (0.7 in), (GI41). (FIG. 13.123. a, dorsal; b, anterior end, ventral.)

Cuthona concinna (Tergipedidae). Body tends toward bluish purple color. On hydroids, *Sertularia argentea* or *Dynamena pumila.* Ac–,Ha,Li–Su,pr,18 mm (0.7 in), (GI41). (FIG. 13.124. a, dorsal; b, anterior end, ventral.)

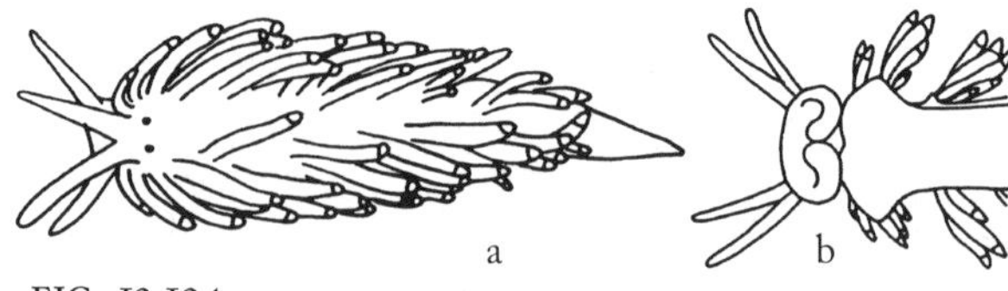

FIG. 13.124

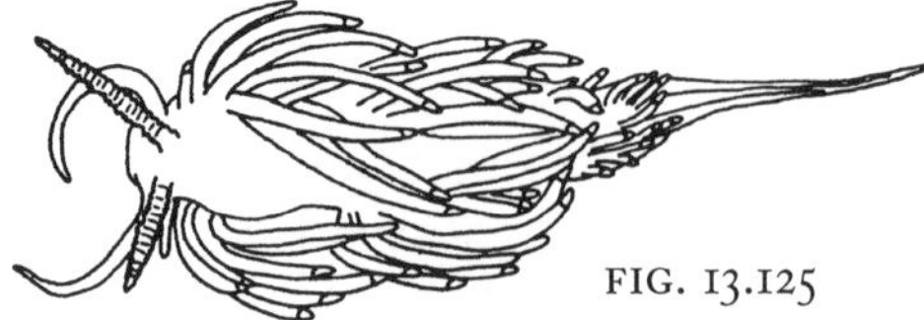

FIG. 13.125

Facelina bostonensis, Boston facelina[†] (Facelinidae). Often with blue stripe around foot. With hydroids, especially *Ectopleura larynx* or on stauromedusae. October–December. Ac–,Ha,Li–Su,pr,25.4 mm (1 in), (AM331,GI41). (FIG. 13.125)

13.82 Flabellinid (=Coryphellid) Aeolid or Red-Gilled Nudibranchs (Coryphellidae) (from 13.81)

The former *Flabellina* (=*Coryphella*) *rufibranchialis* of older literature has been divided into the first three species listed below.

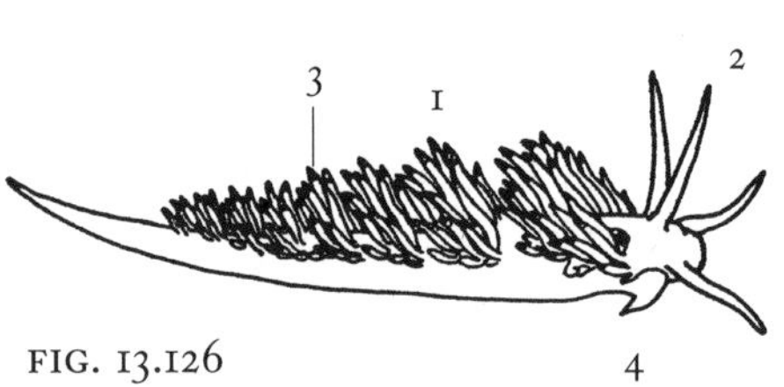

FIG. 13.126

1. Distribution (clumping) of fingerlike cerata groups (FIG. 13.126)
   - A. Cerata appear to be *continuous* along back — co
   - B. Cerata in clusters: provide *number* of clusters — __*
2. Length of second pair of tentacles (rhinophores) compared to first pair (oral tentacles) (FIG. 13.126)
   - A. *Rhinophores equal* in length to *oral* tentacles — r=o*
   - B. *Rhinophores larger than oral* tentacles — r>o
   - C. *Rhinophores shorter than oral* tentacles — r<o
3. Color of ceratal tip (FIG. 13.126)
   - A. Solid *white* tip — wh
   - B. Tip *translucent* with white ring — tr
4. Length of foot corners (auricles) (FIG. 13.126)
   - A. Auricles *large,* greater than foot width — la
   - B. Auricles *small,* less than half of foot width — sm*
   - C. Auricles *intermediate,* about equal to foot width — in
5. Body length: provide *length* in mm — __

13.83

Clumps	Tentacle 1:Tentacle 2	Tip	Foot	Length	Organism
4-6	r=o,r>o	wh	la	3–12	*Flabellina gracilis*
6-7	r=o,r<o	tr	sm	15	*F. verrucosa rufibranchialis,* red-finger aeolis[†]
8-10	r>>o	wh	in,la	20–30	*F. pellucida,* pellucid aeolis[†]
co	r<o	tr	sm	25–40	*F. salmonacea,* salmon aeolis[†]

Flabellina gracilis. Cerata, orange/crimson. White pigment spots on tentacles and tail. Anus near first row of cerata. On *Eudendrium.* Egg mass an undulating coil. Ac–,Ha–Al,Li–Su,pr,33 mm (1.3 in), (GI41,NM529).

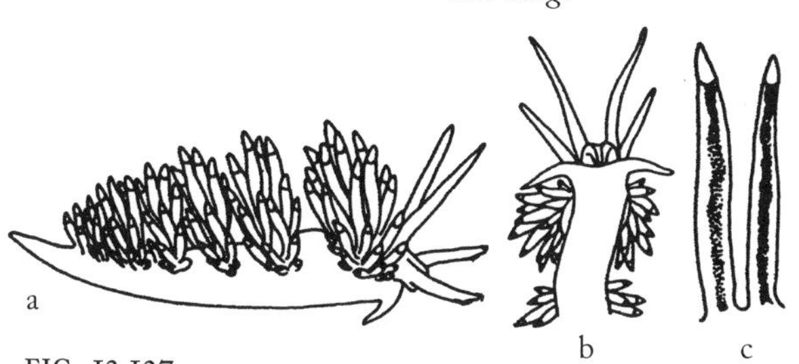

FIG. 13.127

Flabellina pellucida, pellucid aeolis†. Crimson cerata. Thin, undulating egg mass. Rhinophores twice the length of oral tentacles. Winter–spring. Ac–,Ha–Al,Li–Su,pr,33 mm (1.3 in), (MI10,NM529). (FIG. 13.127. a, lateral; b, anterior end, ventral; c, two cerata.)

Flabellina salmonacea, salmon aeolis†. Cerata orange to reddish brown. On hydroid, *Dynamena* or tunicate, *Aplidium.* Ac–,Ha?,Li–Su,pr,33 mm (1.3 in), (MI13,NM529). (FIG. 13.128)

FIG. 13.128

FIG. 13.129

Flabellina verrucosa rufibranchialis, red-finger aeolis†. On variety of hydroids and tunicates. Anus near third row of cerata. Egg mass a smooth coil. Ac–,Ha–Al,Su,pr,33 mm (1.3 in), (GI41,MI08–113,NM529), ☺. (FIG. 13.129. a, lateral; b, anterior end, ventral; c, two cerata.)

13.84 Aeolids with Inflated Cerata (from 13.79)

1. Approximate number of cerata on one side of body: provide *number* —
2. Dominant body color
   - A. Body *white* or translucent — wh
   - B. Body light with dark *spots* or blotches — sp

13.85

Cerata	Color	Organism
4–7	wh	*Tergipes tergipes* (=*despectus*)
5–10	sp	*Eubranchus* (=*Capellinia*) *exiguus,* dwarf balloon aeolis†, club-gilled nudibranch
30–50	sp	*E. pallidus,* club-gilled nudibranch

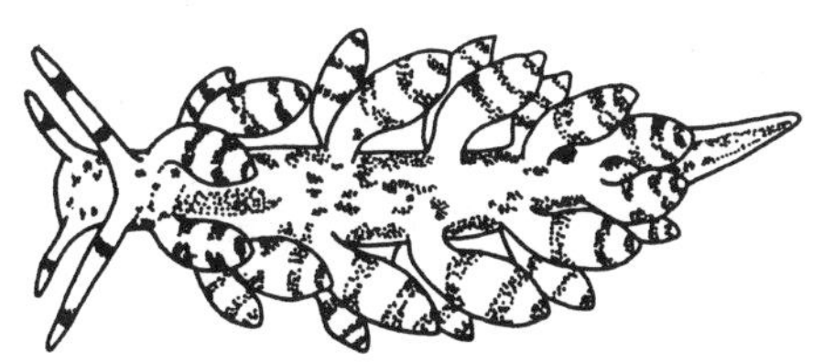

FIG. 13.130

FIG. 13.131

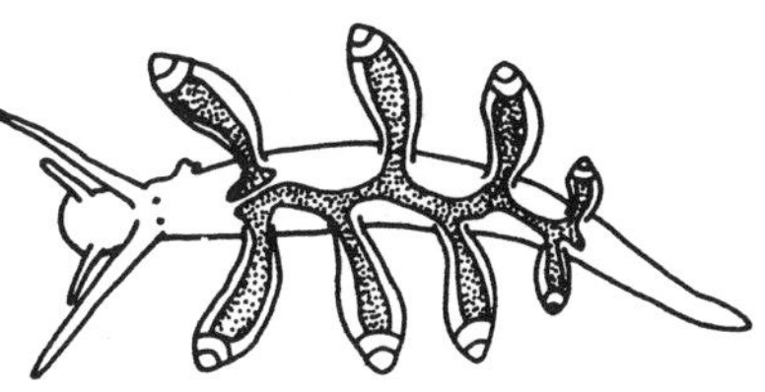

FIG. 13.132

Eubranchus (=*Capellinia*) *exiguus,* dwarf balloon aeolid†, club-gilled nudibranch (Eubranchidae). Spots are brown or green; tentacles banded with brown or green. Especially with *Obelia longissima* growing on eelgrass, *Zostera.* March–June. Ac–,SG,Su,pr,8 mm (0.3 in), (GI41,MI08). (FIG. 13.130)

Eubranchus pallidus, club-gilled nudibranch (Eubranchidae). With hydroids, especially *Ectopleura.* December–May. Ac–,Ha–Gr,Li–Su,pr,13 mm (0.5 in), (GI41). (FIG. 13.131)

Tergipes tergipes (=*despectus*) (Tergipedidae). Body clear, sometimes with reddish anterior streaks. On campanularian hydroids, e.g., *Obelia geniculata* growing on eelgrass, *Zostera,* or kelp, *Laminaria.* Year-round. Ac–,SG,Su,pr,8 mm (0.3 in), (GI41). (FIG. 13.132)

13.86 **Opisthobranch Group 2 (Nudibranchs)** (from 13.77)

Nudibranchs with cerata, no gills, and with only one pair of anterior tentacles, the rhinophores.

1. Shape of cerata
   - A. *Inflated* cerata: distinctly swollen appearance — in
   - B. Flattened *blade*like structures; dorsal body surface otherwise with many small branched filaments; on *Sargassum* weed — bl
   - C. *Thin,* fingerlike cerata, tapering toward tip, perhaps slightly constricted toward base — th
2. Number of cerata along one side of body: provide *number* — —
3. Shape of anterior tentacles or rhinophores
   - A. Tentacles arise from distinct *collar*like base — co
   - B. Tentacles are flattened or *blade*like, similar to cerata — bl
   - C. Tentacles are bifid or *branched* at tip — br
   - D. Tentacles *lack* these distinctive shapes — x

13.87

Cerata Shape	No. Cerata	Tentacle Shape	Species
in	5–8	co	*Doto (=Idulia) coronata,* crown doto†
in	ca. 10	x	*Tenellia fuscata*
bl	2	bl	*Scyllaea pelagica,* sargassum nudibranch†
th	10–15	x	*Ercolania (=Stiliger) nigra (=fuscatus),* dusky stiliger†
th	19–24	br	*Hermaea cruciata,* cross-bearer sea slug

FIG. 13.133

Doto (Idulia) coronata, crown doto† (Dotonidae). Yellow, rose, or cream body with brown dots; cerata with prominent tubercles. On hydroids, especially *Sertularia argentae* or *Dynamena pumila.* April–June. Ac–,Ha,Li–Su,pr,13 mm (0.5 in), (G142). (FIG. 13.133)

Ercolania (=Stiliger) nigra (=fuscatus), dusky stiliger† (Sacoglossa, Stiligeridae). Small; black/charcoal body with dark cerata with light tips. Leading corners of foot are angled. On green algae, *Bryposis, Chaetomorpha, Cladophora.* June–November. Bo,SM–Al,Li,gr,8 mm (0.3 in).

Hermaea cruciata, cross-bearer sea slug (Sacoglossa, Hermaeidae). Digestive gland extensions into translucent cerata form dark cross pattern at ceratal tips. Coloring includes tiny white spots. Leading corners of foot are rounded. Often with seaweed *Codium* and seagrasses. Vi,SG–Al,Su,pr,10 mm (0.4 in), (FIG. 13.134. Ventral view.)

Scyllaea pelagica, sargassum nudibranch† (Scyllaeidae). Brown to bright yellow or olive with brown and white flecks; sides streaked with white. Preys on hydroids on floating (but not attached) *Sargassum* drifting toward shore from Gulf Stream. North to Vineyard Sound and Cape Cod. Vi,Pe–Al,Ps,pr,33 mm (1.3 in), (G140). (FIG. 13.135. Lateral view, anterior to right.)

Tenellia fuscata (Tergipedidae). Tiny. On algae, e.g., *Cladophora,* or hydroids, *Bougainvillia, Obelia, Ectopleura;* can tolerate low salinity (to 16 °/oo). June–September. Ac,Ha–Al,Es–Li,pr,8 mm (0.3 in), (G142).

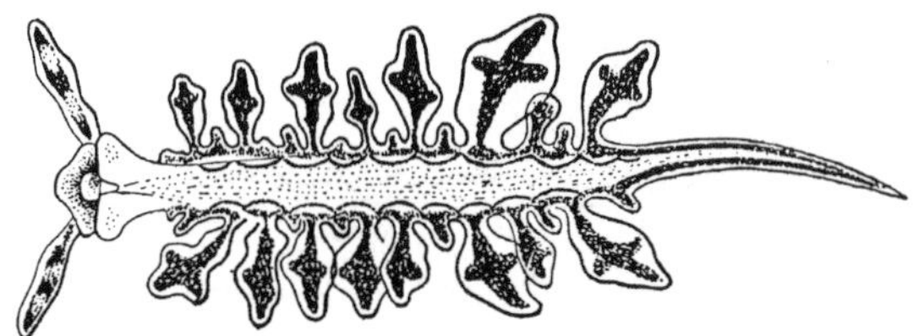

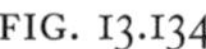

FIG. 13.134

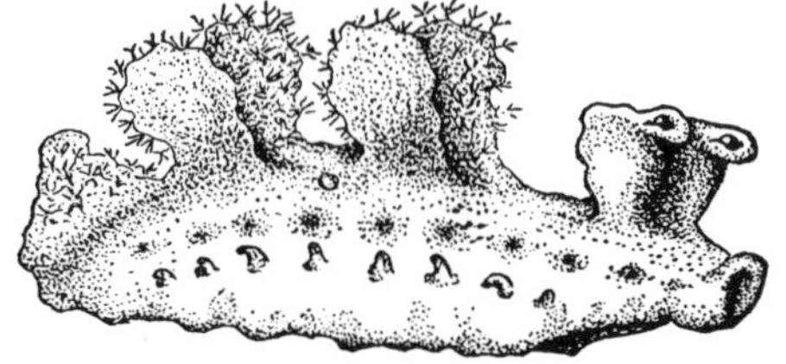

FIG. 13.135

13.88 **Opisthobranch Group 3 (Ridge-Backed Nudibranchs, Doridacea, [in part])** (from 13.77)

Nudibranchs with branched gills and fingerlike projections.

1. Distribution of nonbranched, fingerlike projections
   - A. Projections develop from lateral *ridges* along body — ri
   - B. Projections form a middorsal *circle* around branched gills — ci
2. Number of branching gills: provide *number* — __

13.89

Projections	No. Gills	Organism
ci	3	*Ancula gibbosa,* Atlantic ancula†
ri	3	*Polycerella conyma,* ridge-back nudibranch
ri	6–8	*Polycera dubia* (=*Palio lessoni*), ridged sea slug

Ancula gibbosa, Atlantic ancula† (Goniodorididae). White with yellow tips of tentacles and bumps. First tentacles short; second tentacles with double points plus central knob. Eats *Botryllus* and other tunicates or ectoprocts. Ac,Al,Li–Su,pr,13 mm (0.5 in), (GI40,NM522).

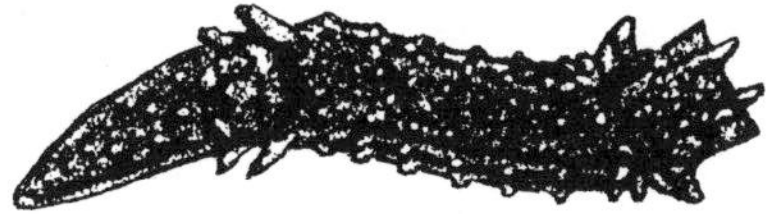

FIG. 13.136

Polycera dubia (=*Palio lessoni*), ridged sea slug (Polyceridae). Yellowish green with scattered yellow tubercles. On ectoprocts, *Bowerbankia, Bugula,* or *Cryptosula.* Egg mass, light pink. May-July. Ac–,Ha–Gr,Li–Su,pr,20.3 mm (0.8 in), (GI40,MI04). (FIG. 13.136)

Polycerella conyma, ridge-back nudibranch (Polyceridae). Translucent, yellowish green highlights, a few tiny dark spots. Eats the ectoproct, *Bowerbankia.* Ca+,Ha–SG,Su,pr,5 mm (0.2 in).

13.90 **Opisthobranch Group 4 (Doridacea, [in part])** (from 13.77)

Dorid nudibranchs lack cerata but show a circle of feather-shaped gills surrounding a posterio-dorsal anus. Surface may be covered by low bumps.

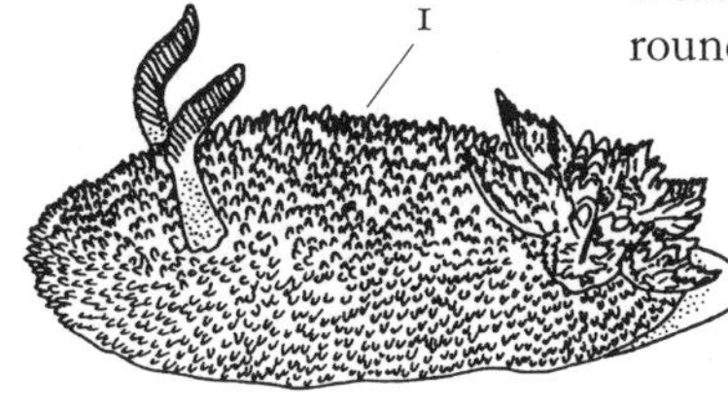

FIG. 13.137

1. Shape of bumps or papillae on dorsal surface (FIG. 13.137)
   - A. Bumps *rounded* — ro
   - B. Bumps *pointed* — po*
2. Number and branching pattern of gills (FIG. 13.138)
   - A. Provide *number:* gills *singly* pinnate (with stalk plus single branches) — __:si
   - B. Provide *number:* gills *complexly* pinnate (branches are branched) — __:co*
3. Dominant body color
   - A. Body color *light* and uniform — li
   - B. Body color *dark* and uniform — da
   - C. Body color basically light but with dark *blotches* or patches — bl

FIG. 13.138

13.91

Bumps	Gills	Color	Organism
po	5–9:co	li,da	Gills contract individually if touched: *Acanthodoris pilosa,* hairy spiny doris† ☺, OR Gills contract as group if individual gill is touched: *Cadlina laevis,* white Atlantic cadlina†
ro	6:si	li	*Doris verrucosa,* rough-backed sponge nudibranch
ro	11:si	li	White; gills and somewhat flattened rhinophores, transparent: *Onchidoris muricata* (=*aspera*), fuzzy onchidoris† OR Yellowish; gills and finger-like rhinophores, slightly darker than back: *Adalaria proxima,* yellow false doris†
ro	16–32:si	bl	*Onchidoris bilamellata* (=*fusca*), barnacle-eating onchidoris†, rough-mantled nudibranch

FIG. 13.139

Acanthodoris pilosa, hairy spiny doris† (Onchidorididae). White to yellow or brown to black. Often with ectoproct, *Flustrellidra* on *Fucus* or *Cryptosula pallasiana* on rocks; midtidal to subtidal. Egg mass, white and ribbonlike. Ac–,Ha–Al,Li–Su,pr,3.3 cm (1.3 in), (G139,M102,NM521), ☺. (FIG. 13.139)

Adalaria proxima, yellow false doris† (Onchidorididae). Yellowish. Up to 12 gills surround anus. Eats encrusting ectoprocts, e.g., *Electra pilosa.* Ac,Ha,Li–Su,pr,17 mm (0.6 in).

Cadlina laevis, white Atlantic cadlina† (Cadlinidae). White with tiny bumps and yellow speckles. On sponge, *Halisarca* and others. Ac,Ha,Li–Su,pr,32 mm (1.3 in), (NM524).

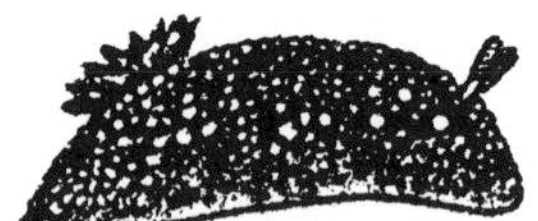

FIG. 13.140

Doris verrucosa, rough-backed sponge nudibranch (Dorididae). Coloring varies with sponges eaten. Gray, yellow, or dull orange. Uncommon on mud bottoms; with sponge, *Halichondria bowerbanki.* August–October. Vi,Mu,Su,pr,4.6 cm (1.8 in), (G139).

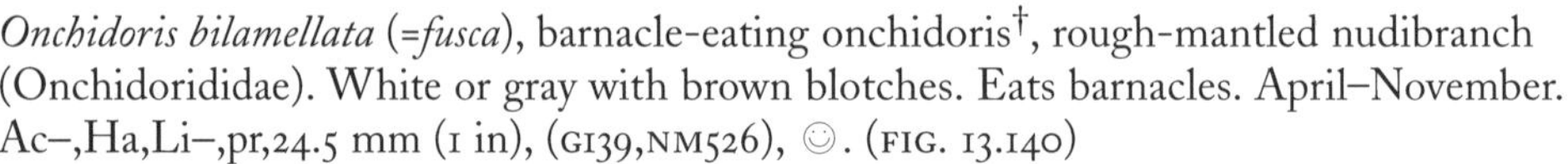

Onchidoris bilamellata (=fusca), barnacle-eating onchidoris†, rough-mantled nudibranch (Onchidorididae). White or gray with brown blotches. Eats barnacles. April–November. Ac–,Ha,Li–,pr,24.5 mm (1 in), (G139,NM526), ☺. (FIG. 13.140)

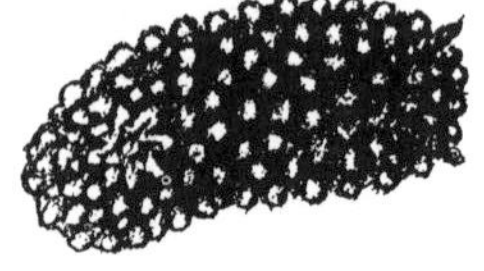

FIG. 13.141

Onchidoris muricata (=aspera), fuzzy onchidoris†, rough-mantled nudibranch (Onchidorididae). White. Often with ectoprocts, *Electra, Cryptosula,* or *Parasmittina;* also on sponges. April–November. Ac–,Ha,Li–Su,pr,13 mm (0.5 in), (G139,M102), ☺. (FIG. 13.141)

13.92 Class Bivalvia, Clams and Mussels

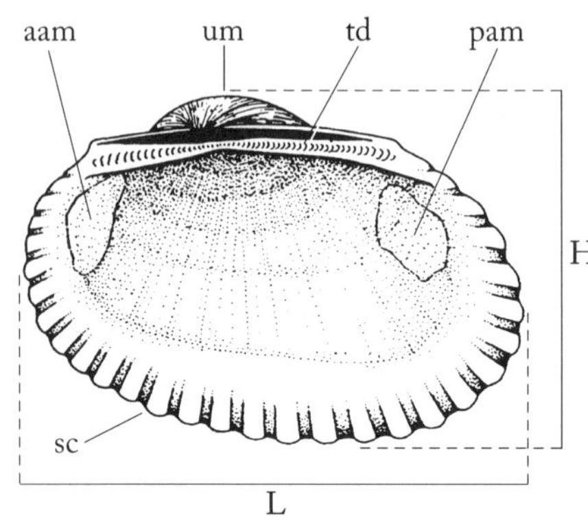

FIG. 13.142

Bivalves—clams, mussels, scallops, oysters, etc.—are among the most frequently encountered marine organisms. Because some variability occurs in shell features, be lenient in using this key. Allow animals to relax in a dish of fresh, cold seawater to see features such as the siphon, mantle fringe, and extended foot. Along the dorsal edge, the two valves that make up the shell are hinged and in most cases there are distinct dorsal umbos or beaks, (FIG. 13.142, um) (points around which growth lines are centered). The umbo often tilts toward the anterior end. Siphons extend posteriorly and the foot, ventrally. In some cases, internal characteristics are useful or necessary for definitive identification. These characters are listed last in each question series, anticipating that most identifications will be attempted on living, intact organisms using features that appear first in the sequence. If viewing characteristics of the internal surface of valves is necessary, a sharp knife can be inserted between the valves and used to cut through the anterior and posterior adductor muscles, freeing the valves to open. Frequently, empty shells of dead bivalves can be found at your field site. Features of the inside surface of the shell can be especially useful here, although heavy abrasion may modify or obscure them.

Useful characteristics of the interior surface of valves include the dentition or interlocking structures associated with the hinge. Although various tooth patterns occur along the inside of bivalve hinges, two basic types are encountered most frequently. *Taxodont dentition* (FIG. 13.142, td) consists of a row of comblike teeth located anterior to and another row posterior to the umbo. In other bivalves, *heterodont dentition* includes one to several strong, vertical or angled ridges comprising the *hinge teeth.* Strong anterior and posterior adductor muscles (FIG. 13.142, aam and pam, respectively) pull valves together tightly, with hinge teeth serving to keep the valves aligned properly. A tough, slightly elastic, proteinaceous ligament forces the valves apart when the adductors relax. (When bivalves die and the adductor muscles are inoperative, the valves gape open.) External ligaments stretch across the outside dorsal rims of valves in some species, whereas in other species, pads of ligament material, known as the *resilium,* may be internal within the hinge mechanism (i.e., not visible from the outside)

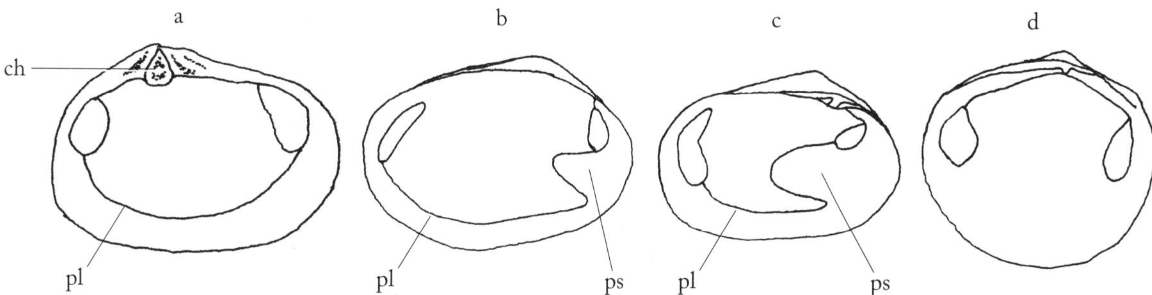

FIG. 13.143

to wedge valves open. In some bivalves (especially the family Mactridae), one or both valves bears a small spoon-shaped or teardrop-shaped shelf, the *chondrophore* (FIG. 13.143, ch), which accommodates the resilium.

Another useful feature found on the inside of valves is the shape of the *pallial line,* the scar left from the attachment of the mantle to the inner surface of the valve (FIG. 13.143, pl). When present, this scar extends as a semicircular loop between the scars that mark the sites of attachment of the anterior and posterior adductor muscles. Just before reaching the posterior adductor muscle, the pallial line may include a triangular or rectangular indentation, the *pallial sinus* (FIG. 13.143, ps), which accommodates the posteriorly placed siphons when they are withdrawn into the shell. At several points, keys will refer to Figure 13.143, which illustrates four general types of pallial scars found in bivalves: a, pallial line, no pallial sinus; b, small pallial sinus; c, large pallial sinus; and d, no pallial line or sinus. Bivalves with large siphons usually have a large pallial sinus and tend to burrow deeply in sediment. Those with shorter siphons need more modest accommodation. Bivalves living on or near the surface of the substrate tend to lack siphonal extension of the mantle and thus tend to lack a pallial sinus; the pallial line is uninterrupted. A pallial sinus is particular diagnostic of the families Veneridae and Tellinidae.

Shell sculpture is a useful feature as well (see FIG. 13.144). Narrow *ribs* or broader *ridges* extend axially, that is from the umbo toward the shell periphery. *Cords* or *raised rings* are concentric with the umbo. Less distinct *growth lines* also follow this concentric pattern. The texture and color of the outermost shell layer, the *periostracum,* prominent in many bivalves, can be useful. Typically, this proteinaceous material is thin enough to be scratchable by a finger nail. It can be extensively (or even completely) abraded away in old specimens. Finally, in some species, useful sculpture is associated with internal, ventral edges of the valve. *Crenulation* is formed by the presence of tiny teeth along this surface, whereas *scalloping* (FIG. 13.142, sc) produces more obvious undulation to the ventral shell edge of thinner shells, usually reflecting the shell's overall axial ribs or ridges.

The *shape index* is an estimate of shell proportions. Calculation of this index involves dividing the shell length (or width) by its height (FIG. 13.142, L and H). Hold the hinge upright and use a ruler (with a millimeter scale for greater accuracy) or calipers to note the longest anterior to posterior dimension (length) and the maximal dorsal to ventral dimension (height) taken at approximately right angles to the length line. Now divide the length by this height to determine the shape index. It may be easiest to determine these dimensions by placing the shell on its side on a sheet of paper (outside surface facing you if you are working with a single valve). Looking directly down on the shell use a pencil to place dots on the paper representing the end points of the visual projection of the shell's maximal height and length onto the paper. Note that the length often runs from shell margin to shell margin, but maximal height will often run from the projection of the umbo to the opposite shell margin.

Although many species live subtidally, their empty valves are frequently encountered nearshore or on beaches, especially following storms at sea. Habitat listings are based on preferences of live individuals.

FIG. 13.144

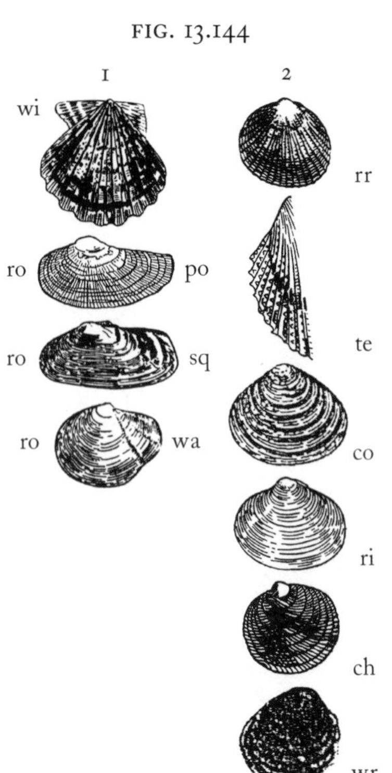

1. General shape of valve viewed from the side, holding hinge upward—one end: other end (FIG. 13.144)
   - A. Rounded edges but with *wings* on one or both sides of the umbo — wi*
   - B. *Rounded* or squared along one edge: *pointed* at the other edge — ro:po*
   - C. *Squared* along one edge: *squared* along the other — sq:sq
   - D. *Rounded* along one edge: *squared* along the other — ro:sq*
   - E. *Rounded* along one edge: *rounded* along the other — ro:ro
   - F. *Rounded* along one edge: posterior margin *wavy* — ro:wa*
   - G. Shape *irregular* or not matching those described above — ir
2. The most prominent shell sculpture (FIG. 13.144)
   - A. *Raised ribs* or ridges radiate axially or from umbo toward edges of shell — rr*
   - B. Some or all raised ribs form rows of distinct *teeth* — te*
   - C. Shell with strong raised concentric *cords* (< 30) — co*
   - D. Obvious to faint growth *rings*, concentric with the umbo, or smooth — ri*
   - E. Shell with distinct *chevron*like scratches — ch*
   - F. Shell surface irregularly *wrinkled* — wr*
3. Shell proportions: shape index is the length (or width) divided by height (hold hinge surface up; use ruler with millimeter scale; measure by imagining the projection of the shell's dimensions onto a flat surface)
   - A. Roughly *equal* proportions, index = 0.5–1.5 — eq
   - B. Shell *oval* or triangular, index = 1.5–2.5 — ov
   - C. Shell *elongate*, index > 2.5 — el

13.93

Shape	Sculpture	Proportions	Group, Family, or Species
wi	rr	eq	Pectinidae (scallops), 13.144
ir	wr,rr	eq	Anomiidae (jingles), 13.114
ir	wr	ov,el	Bivalve Group 1, 13.94
ro:ro	ri	eq	Bivalve Group 2, 13.96
ro:ro	ri	ov,el	Bivalve Group 3, 13.98
ro:ro	rr,te	eq	Bivalve Group 4, 13.100
ro:ro	rr,te	ov,el	Bivalve Group 5, 13.102
ro:ro	wr,rr	eq	Anomiidae (jingles), 13.114
ro:ro	wr	ov,el	Bivalve Group 1, 13.94
ro:ro	co	eq	Astartidae, 13.118
ro:ro	ch	eq	*Divaricella quadrisulcata,* cross-hatched lucine†; see description under family Lucinidae, 13.130
ro:sq	ri	eq,ov	Bivalve Group 6, 13.104
ro:sq	rr	ov	Bivalve Group 7, 13.106
ro:po	ri	eq	Bivalve Group 8, 13.108
ro:po	ri,sm	ov	Bivalve Group 9, 13.110
ro:po	rr	ov	Bivalve Group 10, 13.112
ro:po	te	ov,el	Petricolidae and Pholadidae (piddocks), 13.147
ro:wa	ri	eq	*Thyasira trisinuata,* Atlantic cleftclam†; see description under family Thyasiridae, 13.158
sq:sq	ri	ov,el	Bivalve Group 3, 13.98

13.94 **Bivalve Group 1** (from 13.93)

Bivalves with rounded or irregularly shaped ends or with irregularly wrinkled surfaces because they adopt the contour of the hard surface to which they are attached.

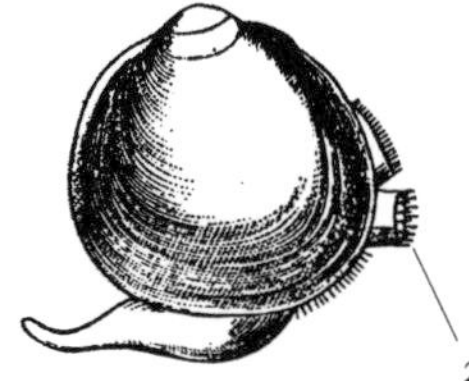

FIG. 13.145

1. Texture of irregular shell
   - A. Thick shell formed of many, distinct, flattened *layers* — la
   - B. Thin shell with wrinkles and/or rings but flattened layers *absent* — x
2. Siphons protrude from posterior end of valves (allow animal to relax in seawater to observe) (FIG. 13.145)
   - A. *Siphons* present — si*
   - B. Siphons *absent* — x

13.95

Shell	Siphons	Species
la	x	*Crassostrea virginica,* eastern oyster†, Virginia oyster; see description under family Ostreidae, 13.142
x	si	*Hiatella arctica,* arctic rock borer†, red-nosed clam; see description under family Hiatellidae, 13.127

13.96 **Bivalve Group 2** (from 13.93)

Bivalves of basically circular or rounded-triangular shape, with rings or no sculpturing.

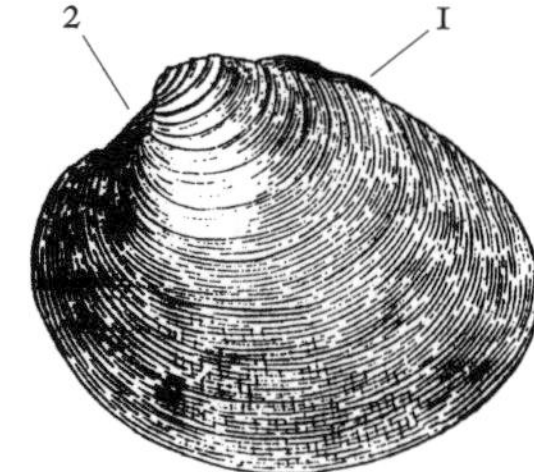

FIG. 13.146

1. Position of the ligament (dark, nonshelly, proteinaceous material); may be missing from abraded specimens, especially from isolated valves (FIG. 13.146)
   - A. Ligament material clearly visible *externally* along the middorsal line — ex*
   - B. Ligament material *not* visible externally — x
2. Lunule or heart-shaped area incised into shell just beneath (anterior to) umbo along midline (half a heart if there is only one valve) (FIG. 13.146)
   - A. *Lunule* present — lu*
   - B. Lunule *absent* — x
3. Presence of gaps or gape between valves on intact animals (i.e., valves do not press together around all margins)
   - A. Valves gape at *both ends* of shell — 2e
   - B. Valves gape at only *one end* of shell — 1e
   - C. Valves fit tightly together, gape *absent* — x
4. Dentition or ridgelike teeth associated with the hinge along inside of dorsal edge of shell
   - A. *Taxodont* dentition (rows of several small teeth both anterior and posterior to umbo) present (FIG. 13.142, td) — ta
   - B. *Strong teeth* present in well-developed hinge area near dorsal inside edge of shell near umbo — st
   - C. Hinge area not well developed; teeth weak or absent — x
5. Presence of scoop-shaped pocket or shelf, the chondrophore (FIG. 13.143a, ch), at inside surface of dorsal edge of shell, at umbo location (serves to support the resilium, a pad of ligament material used to wedge the valve apart)
   - A. Scooplike *chondrophore* present — ch
   - B. Chondrophore *absent* — x
6. Shape of pallial line or scar left by mantle attachment along inside of shell (between adductors), specifically emphasizing the pallial sinus: provide *letter* from Figure 13.143 that most closely corresponds — __
7. Length of shell (size is more definitive for small-sized shells; during growth, large shells can be smaller than sizes listed here but small shells cannot be larger)
   - A. *Small,* < 12 mm (0.5 in) — sm
   - B. *Medium,* 12–51 mm (0.5–2 in) — me
   - C. *Large,* > 51 mm (2 in) — la

13.97

Ligament	Lunule	Gape	Dentition	Chordrophore	Pallial Line	Size	Family
ex	x	2e,x	st	ch	B	la	Mactridae (surfclams), 13.132
ex	x	1e	st	x	A	me	Cardiidae (cockles), 13.120
ex	x	x	st	x	A	la	Arcticidae, 13.117
ex	lu	x	st	x	B	la	Veneridae (venus clams), 13.159
ex	lu	x	st	x	A	me	Astartidae, 13.118
ex	lu	x	st	x	A	la	Lucinidae, 13.130
x	x	2e,x	st	ch	B	la	Mactridae (surfclams), 13.132
x	x	1e	x	ch	C	me,la	Myidae, 13.135
x	lu	x	st	x	A	la	Lucinidae, 13.130
x	lu,x	x	x	x	B	sm	*Gemma gemma,* amethyst gem-clam†; see description under family Veneridae, 13.159
x	x	x	x	x	D	me	Anomiidae (jingles), 13.114
x	x	x	ta	x	A	sm	Limopsidae, 13.129
x	x	x	ta	ch	D	sm	Nuculidae (nutclams), 13.140

13.98 **Bivalve Group 3** (from 13.93)

Bivalves elongated with rounded or squared ends and smooth or ringed sculpturing.

1. Shell proportions: shape index = length (or width)/height (hold hinge surface up; use ruler with millimeter scale; measure by imagining the projection of the shell's dimensions onto a flat surface)
   - A. Shell *long,* index = 1.5–2.0 — lo
   - B. Shell *longer,* index = 2.0–3.0 — lr
   - C. Shell *longest,* index > 3.0 — lt

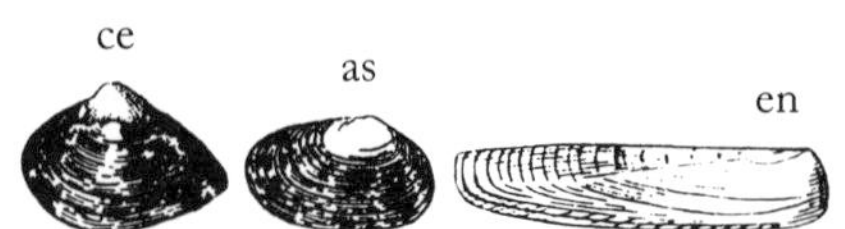

FIG. 13.147

2. Relative position of the umbo (FIG. 13.147)
   - A. Umbo approximately in *center* of dorsal surface — ce
   - B. Umbo distinctly *asymmetrical,* off to one side of center — as
   - C. Umbo all the way at one *end* of dorsal surface — en
3. Presence of gaps or gape between valves on intact animals (i.e., valves do not press together around all margins)
   - A. Valves gape at *both ends* of shell — 2e
   - B. Valves gape at *only one* end of shell — 1e
   - C. Valves fit tightly together, gape *absent* — x

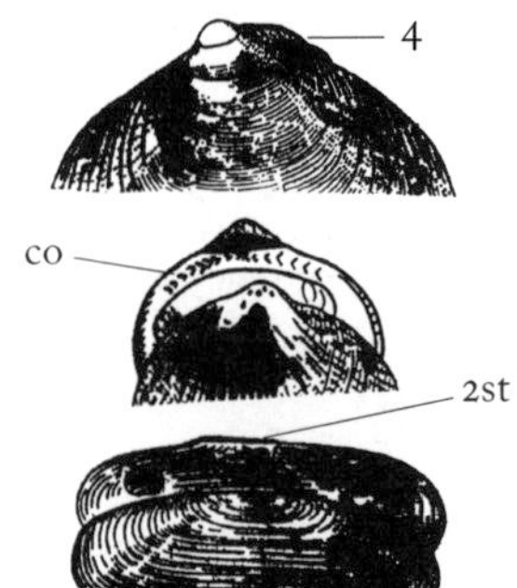

FIG. 13.148

4. Position of the ligament (dark, proteinaceous, nonshelly material); may be missing on abraded specimens, especially on isolated valves (FIG. 13.148)
   - A. Ligament material clearly visible *externally* along the middorsal line — ex*
   - B. Ligament material *not* visible externally — x
5. Dentition or ridgelike teeth associated with the hinge along inside of dorsal edge of shell (FIG. 13.148)
   - A. Lateral teeth *comb*like — co
   - B. *Two small teeth* project into shell's interior from just below umbo on inside edge of shell — 2st
   - C. Hinge area not well developed; teeth weak or *absent* — x
6. Presence of scoop-shaped pocket or shelf, the chondrophore, at inside surface of dorsal edge of shell, at umbo location (serves to support the resilium, a pad of ligament material used to wedge the valve apart; FIG. 13.143a, ch)
   - A. Scooplike *chondrophore* present — ch
   - B. Chondrophore *absent* — x

7. Reinforcing rib extends from umbo toward margin on inside surface of shell
   A. A *strong rib* is present, extending at least half the distance across the shell — sr
   B. A *weak rib* is present; extends less than half the distance across the shell — wr
   C. Reinforcing rib is *absent* — x
8. Shape of pallial line or scar left by mantle attachment along inside of shell (between adductors), emphasizing shape of pallial sinus: provide *letter* from Figure 13.143 that most closely corresponds — __

13.99

Index	Umbo	Gape	Ligament	Dentition	Chondrophore	Rib	Pallial Line	Species
lo	ce	x	x	x	ch	wr	B	Periplomatidae (spoon clams), 13.146
lo	ce	1e	x	x	ch	x	C	Myidae (soft-shelled clams), 13.135
lo	as	x	x	x	x	x	A	*Musculus discors;* see description under family Mytilidae, 13.136
lo	as	x	x	co	ch	x	B	*Mesodesma arctatum,* wedgeclam. See description under family Mesodesmatidae, 13.134
lo	ce	x	ex	x	x	x	C	*Macoma tenta;* see description under family Tellinidae, 13.155
lr	as	2e	ex	x	x	sr	C	Solenidae (razors), 13.153
lt	en	2e	ex	x	x	x	C	Solenidae (razors), 13.153
lt	ce	2e	ex	2st	x	wr	C	Solecurtidae (little razors), 13.151

13.100 **Bivalve Group 4** (from 13.93)

Bivalves of equal width and height and with rounded ends but with ridges (broad) or ribs (narrow) radiating from the umbo toward the opposite side of the shell.

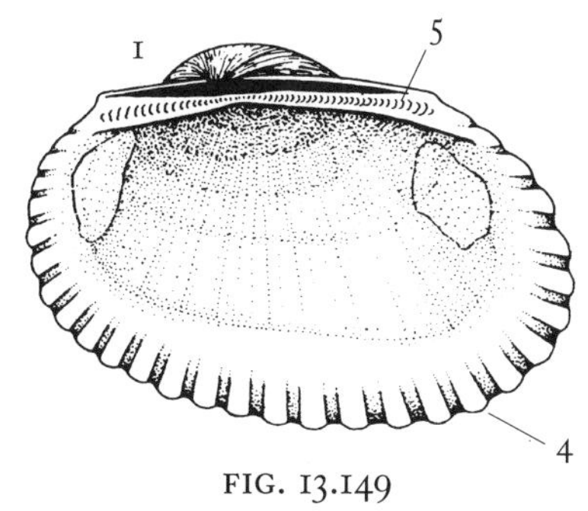

FIG. 13.149

1. Symmetry of the hinge line or dorsal edge of shell (FIG. 13.149)
   A. Hinge line is *straight* — st*
   B. Hinge line is *curved* — cu
2. Thickness of periostracum or outermost thin shell layer
   A. Periostracum *thick* and often abraded away in places — tk
   B. Periostracum *thin* or absent entirely — tn
3. Lunule or heart-shaped line incised into shell valves just below (or anterior to) umbo (look for half-a-heart if there is only one valve)
   A. *Lunule* present — lu
   B. Lunule *absent* — x
4. Features of inside ventral edge of shell (FIG. 13.149)
   A. *Crenulations* (small bumps or teeth on inside ventral edge of shell; see FIG. 13.164); unrelated to outer shell sculpturing — cr
   B. Inside edge *scalloped,* directly related to outer shell sculpturing — sc*
   C. Both these features *absent* — x
5. Dentition or teeth as part of hinge mechanism on inside dorsal margin of shell (FIG. 13.149)
   A. *Taxodont* dentition (row of many comblike teeth anterior and posterior to umbo) present — ta*
   B. One to three *strong* ridgelike *teeth* present at well-developed hinge area — st
   C. Distinct teeth *absent* — x
6. Symmetry between the two valves
   A. Valves *asymmetrical;* lower valve flat, with opening for byssus attachment to substrate — as
   B. Valves basically *symmetrical* and equivalent, or only weakly asymmetrical — sy

13.101

Hinge	Periostracum	Lunule	Margins	Dentition	Symmetry	Family or Species
st	tk	x	sc	ta	sy	Arcidae (arks), 13.115
cu	tk	x	sc	st	sy	Cardiidae (cockles), 13.120
cu	tk	lu	sc	st	sy	Carditidae (heartclams), 13.122
cu	tn	x	cr	x	sy	*Crenella glandula,* glandular bean-clam†; see description under family Mytilidae, 13.136
cu	tn	x	x	x	sy	Lyonsiidae (paperclams), 13.131
cu	tn	x	x	x	as	Anomiidae (jingles), 13.114

13.102 **Bivalve Group 5** (from 13.93)

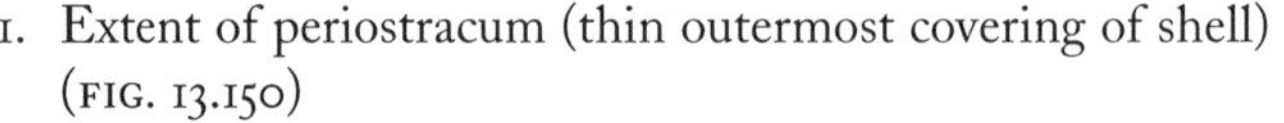

Bivalves with oval or elongate shells and with rounded ends but with ridges (broad) or ribs (narrow) radiating from the umbo toward the opposite side of the shell.

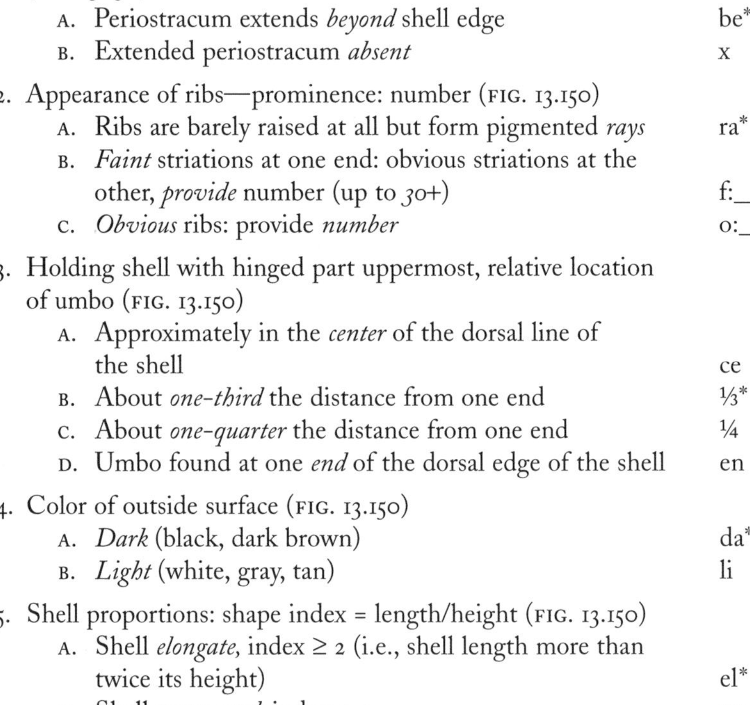

1. Extent of periostracum (thin outermost covering of shell) (FIG. 13.150)
   - A. Periostracum extends *beyond* shell edge — be*
   - B. Extended periostracum *absent* — x
2. Appearance of ribs—prominence: number (FIG. 13.150)
   - A. Ribs are barely raised at all but form pigmented *rays* — ra*
   - B. *Faint* striations at one end: obvious striations at the other, *provide* number (up to *30+*) — f:__
   - C. *Obvious* ribs: provide *number* — o:__
3. Holding shell with hinged part uppermost, relative location of umbo (FIG. 13.150)
   - A. Approximately in the *center* of the dorsal line of the shell — ce
   - B. About *one-third* the distance from one end — ⅓*
   - C. About *one-quarter* the distance from one end — ¼
   - D. Umbo found at one *end* of the dorsal edge of the shell — en
4. Color of outside surface (FIG. 13.150)
   - A. *Dark* (black, dark brown) — da*
   - B. *Light* (white, gray, tan) — li
5. Shell proportions: shape index = length/height (FIG. 13.150)
   - A. Shell *elongate,* index ≥ 2 (i.e., shell length more than twice its height) — el*
   - B. Shell more *oval;* index < 2 — ov

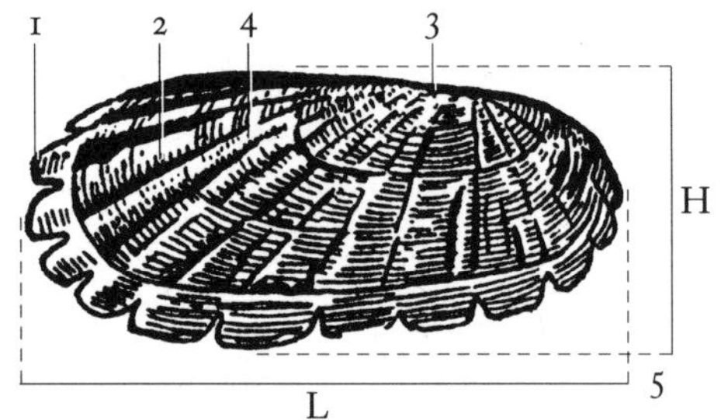

FIG. 13.150

13.103

Periostracum	Ribs	Umbo	Color	Index	Family or Species
be	ra	⅓	da	el	Solemyidae (awningclams), 13.152
x	f:30+	en	da	ov,el	Mytilidae (mussels), 13.136
x	f:30+	⅓	li	ov	*Lyonsia arenosa,* sand lyonsia†; see description under family Lyonsiidae, 13.131
x	f:8	⅓	li	el	Petricolidae (false angelwings), 13.147
x	o:25–30	ce,¼	li	el	Pholadidae (borers or piddocks), 13.147

13.104 **Bivalve Group 6** (from 13.93)

Bivalves with one rounded and one squared end, possessing smooth surfaces or rings as shell sculpture.

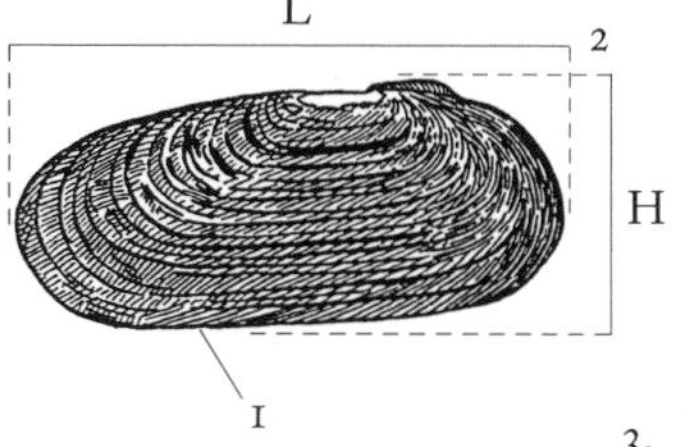

FIG. 13.151

1. Contour of ventral edge of shell (FIG. 13.151)
   - A. Considerable portion of ventral edge *straight* st*
   - B. Ventral edge distinctly *curved* cu
2. Shell proportions: shape index = length divided by height (FIG. 13.151)
   - A. Shell comparatively *elongated*, index > 2 el*
   - B. Shell more *compact*, index < 2 co
3. Presence of taxodont dentition (row of small teeth along inside of dorsal edge of shell) (FIG. 14.142, td)
   - A. *Taxodont* dentition present ta
   - B. Taxodont dentition *absent* (although other sorts of dentition may be found there) x
4. Presence of tear-shaped or scooplike pocket, the chondrophore (FIG. 14.143a, ch), at inside surface of dorsal edge of shell, at umbo location on left valve only
   - A. Scooplike *chondrophore* present ch
   - B. Chondrophore *absent* x
5. Shape of pallial line or scar left by mantle attachment along inside of shell (between adductors), emphasizing the pallial sinus: provide *letter* from Figure 13.143 that most closely corresponds __

13.105

Ventral Edge	Index	Taxodont	Chondrophore	Pallial Line	Family or Species
st	co,el	x	x	C,D	Hiatellidae, 13.127
st	co	x	ch	C	Myidae (soft-shelled clams), 13.135
cu	co	x	ch	C	Myidae (soft-shelled clams), 13.135
cu	co	ta	ch	D?	*Yoldia thraciaeformis,* ax yoldia†; see description under family Nuculanidae, 13.138
cu	co	x	x	B,C	Thraciidae, 13.157

13.106 **Bivalve Group 7** (from 13.93)

Oval bivalves with one rounded and one squared end, possessing ribs that radiate from the umbo to the opposite perimeter of the shell.

1. Shell construction
   - A. Shell *fragile* fr
   - B. Shell *solid* so
2. Appearance of shell exterior
   - A. Dark *periostracum* covers at least some of the shell pe
   - B. Periostracum absent, *semitranslucent* shell st
3. Shell length (this character is useful primarily if the specimen exceeds maximal lengths of species under consideration)
   - A. Shell ≤ 25 mm (1 in) long <1
   - B. Shell > 25 mm (1 in) long >1

13.107

Surface	Exterior	Length	Species
fr	st	<1	*Lyonsia hyalina,* glassy lyonsia†; see description under family Lyonsiidae, 13.131
so	pe	>1	*Anadara transversa,* transverse ark†; see description under family Arcidae, 13.115

13.108 **Bivalve Group 8** (from 13.93)

Bivalves of approximately equal width and length, and with one round and one pointed end, and either smooth surfaced or with rings as only sculpture.

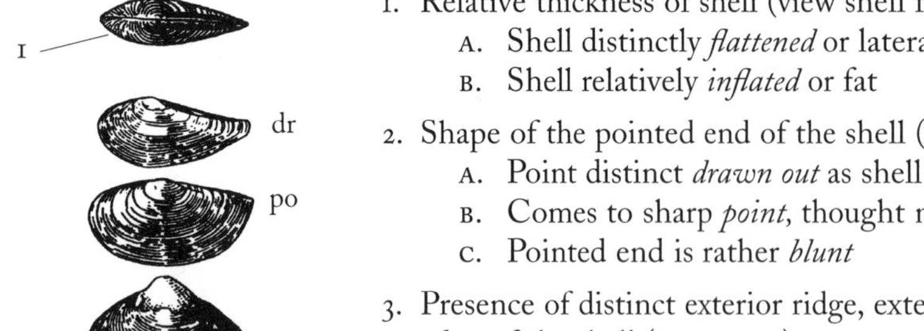

1. Relative thickness of shell (view shell from dorsal [hinged surface] view) (FIG. 13.152)
   - A. Shell distinctly *flattened* or laterally compressed — fl*
   - B. Shell relatively *inflated* or fat — in
2. Shape of the pointed end of the shell (FIG. 13.152)
   - A. Point distinct *drawn out* as shell extension — dr
   - B. Comes to sharp *point,* thought not drawn out into extension — po
   - C. Pointed end is rather *blunt* — bl
3. Presence of distinct exterior ridge, extending from umbo toward the posterior edge of the shell (FIG. 13.152)
   - A. Distinct posterior *ridge* present — ri*
   - B. Posterior ridge *absent* — x

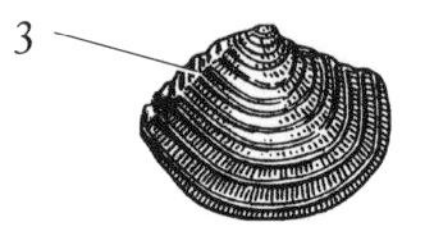

FIG. 13.152

4. Presence of tear-shaped or scooplike pocket, the chondrophore (FIG. 14.143a, ch), at inside surface of dorsal edge of shell, at umbo location on left valve only
   - A. Scooplike *chondrophore* present — ch
   - B. Chondrophore *absent* — x
5. Shape of pallial line or scar left by mantle attachment along inside of shell (between adductors), and especially, the pallial sinus: provide *letter* from Figure 13.143 that most closely corresponds — __

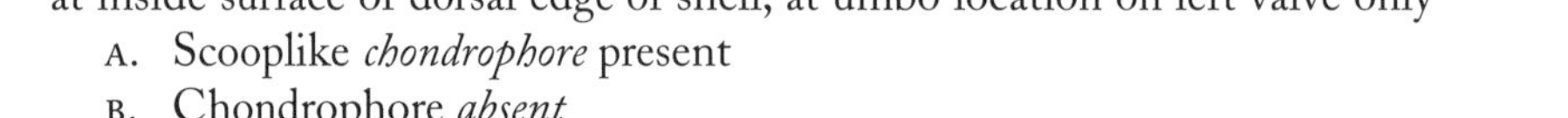

13.109

Width	Point	Ridge	Chondrophore	Pallial Line	Family or Species
in	dr	ri	ch	A	Cuspidariidae (dipperclams), 13.124
in	po	x	x	C	*Cumingia tellinoides,* common cumingia†; see description under family Semelidae, 13.150
fl	po	x	x	C	*Pandora gouldiana,* Gould pandora†; see description under family Pandoridae, 13.143
in	bl	ri	ch	B,C	*Mulinia lateralis,* little surfclam†; see description under family Mactridae, 13.132
fl	bl	ri	x	B,C	Tellinidae (tellins), 13.155
fl	bl	x	x	C	Tellinidae (tellins), 13.155

13.110 **Bivalve Group 9** (from 13.93)

Wide bivalves with one round and one pointed end, and either smooth surfaced or with rings as sculpture.

1. Holding shell with hinged part uppermost, relative location of umbo (FIG. 13.153)
   - A. Approximately in the *center* of the dorsal line of the shell — ce
   - B. About *one-third* the distance from one end — ⅓
   - C. Umbo found at one *end* of the shell — en
   - D. Umbo *near end;* occupies only top half of pointed end — ne
2. Distinct exterior ridge extending from umbo toward the posterior edge of the shell
   - A. *Ridge* present — ri
   - B. Ridge *absent* — x

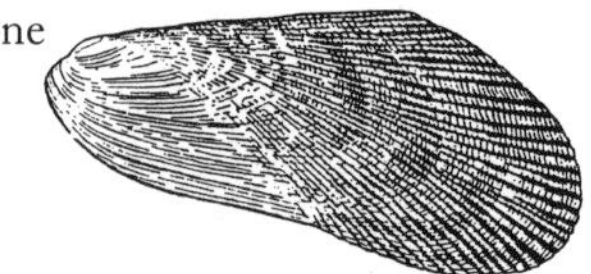

FIG. 13.153

3. Dentition or teeth as part of hinge mechanism on inside dorsal margin of shell
   - A. *Taxodont* dentition (row of many comblike teeth anterior and posterior to umbo) present (FIG. 13.142, td) — ta
   - B. One to three *strong* ridgelike *teeth* present at well developed hinge area — st
   - C. *Weakly* developed *teeth* and hinge area — wt
   - D. Distinct teeth *absent* — x

4. Features along inside edge of hinge area
   - A. Scooplike pocket or *chondrophore* present (FIG. 13.143a, ch) — ch
   - B. Small but distinct *shelf* covers inside of pointed umbo region — sh
   - C. These features *absent* — x
5. Shape of pallial line or scar left by mantle attachment along inside of shell (between adductors), and especially, the pallial sinus: provide *letter* from Figure 13.143 that most closely corresponds —

13.111

Umbo	Ridge	Dentition	Hinge	Pallial Line	Family or Species
⅓	ri	x	ch	A	Cuspidariidae (dipperclams), 13.124
⅓	ri	ta	x	D	*Nuculana tenuisulcata,* sulcate nutclam†; see description under family Nuculanidae, 13.138
⅓	x	st	x	C	*Pandora gouldiana,* Gould pandora†; see description under family Pandoridae, 13.143
ce	ri	wt	ch	A	Corbulidae (basketclams), 13.123
ce	ri,x	ta	x	D	Nuculanidae (nutclams), 13.138
en	x	x	sh	A,D	Dreissenidae (false mussels), 13.126
en,ne	x	wt,x	x	A,D	Mytilidae (mussels), 13.136

13.112 **Bivalve Group 10** (from 13.93)

Bivalves of oval shape with one rounded and one pointed end but also with smooth ridges (wider) or ribs (narrower) radiating from umbo toward opposite perimeter of shell.

1. Holding shell with hinged part uppermost, relative location of umbo (see FIG. 13.153)
   - A. About *one-third* the distance from one end — ⅓*
   - B. At or *near end;* umbo occupies only top half of pointed end — ne
   - C. Precisely at *end;* umbo occupies entire pointed end — en
2. Relative size of radiating ribs (FIG. 13.154)
   - A. *Ribs* are obvious features, rough to the touch — ri*
   - B. Ribs are subtle *striations,* fine lines — st
3. Features along bottom inside margin of shell (FIG. 13.154)
   - A. *Crenulations* (tiny teeth; see FIG. 13.164) present — cr
   - B. Inside edge *scalloped,* directly related to outer shell sculpturing — sc
   - C. Both these features *absent* — x
4. Color of interior of shell toward rounded end
   - A. Light *purplish* to rosy — pu
   - B. *Pearly* — pe
   - C. Dark *blue,* purple, or black — bl
   - D. Creamy *white* — wh

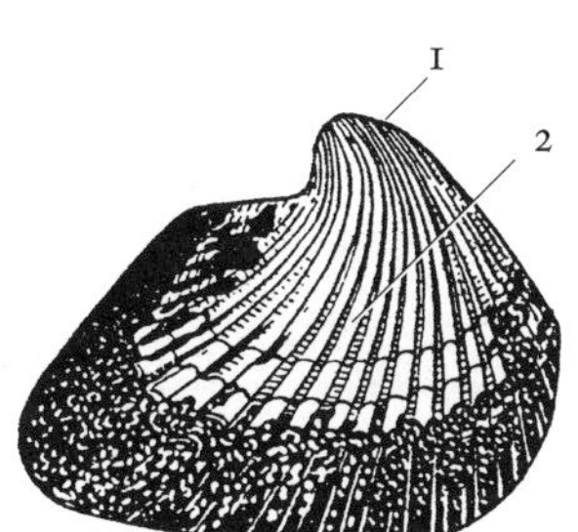

FIG. 13.154

13.113

Umbo	Ribs	Margin	Color	Family or Species
⅓	ri	sc	wh	*Noetia ponderosa,* ponderous ark†; see description under family Arcidae, 13.115
⅓	st	cr	pe,pu	Donacidae (donax clams), 13.125
⅓	st	x	pe	*Lyonsia hyalina,* glassy lyonsia†; see description under family Lyonsiidae, 13.131
en,ne	st,ri	x	bl,pu	Mytilidae (mussels), 13.136

Families of Bivalvia (arranged alphabetically)

13.114 Anomiidae, Jingles (from 13.93, 13.97, 13.101)

Semitranslucent, pearly, strong, thin shell; attaches by byssus threads (tough strands of secreted material) through opening in asymmetrically flattened ventral valve; pallial line indistinguishable.

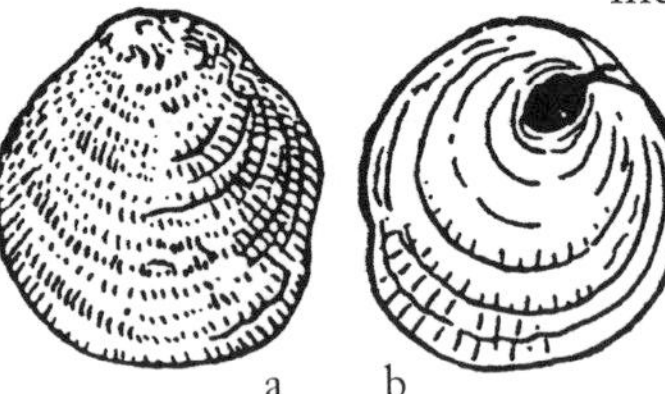

FIG. 13.155

Anomia squamula (=*aculeata*), prickly jingle†, rough jingle shell. Fragile. Shell light yellow or purplish color; upper valve convex. Shell surface with spines or hairs. Bo,Ha,Su–De,ff,17 mm (0.66 in), (A135,AM38,G149,NM547,R715), ☺. (FIG. 13.155. a, upper valve; b, lower valve.)

FIG. 13.156

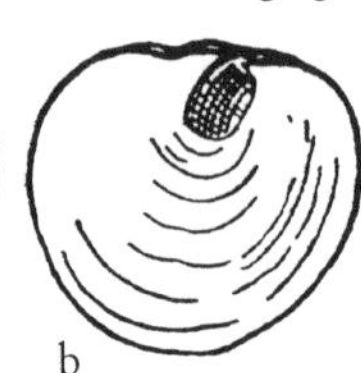

Anomia simplex, common jingle†, mermaid's toenail. Roughly circular; yellow, orange, or silvery; smooth or wrinkled surface. Spines and hairs absent from shell surface. Bo,Ha,Li–Su,ff,25 mm (1 in), (A135,AM38,G149,NM546,R714), ☺. (FIG. 13.156. a, upper valve; b, lower valve.)

13.115 Arcidae, Arks (from 13.101, 13.107, 13.113)

Sturdy, inflated shells with relatively straight hinged-edge and coarse, dark periostracum; taxodont dentition; no siphons; surface dwellers, attach to hard substrata or pebbles by byssus threads.

1. Ventral margin of shell
   - A. Ventral margin roughly *parallel* to dorsal or hinged margin — pa
   - B. Ventral margin *rounded,* not parallel to dorsal margin — ro
2. Umbo orientation
   - A. Umbos closer to and bent toward *anterior* end — an
   - B. Umbos about *centered* and bent slightly toward posterior end — ce
3. Grooves along ribs
   - A. Flattened ribs have central *grooves,* especially toward ventral margin — gr
   - B. Flattened ribs with central grooves *absent* — x

13.116

V-Margin	Umbo	Groove	Species
pa	ce	gr	*Noetia ponderosa,* ponderous ark†
pa	an	gr	*Anadara transversa,* transverse ark†
ro	an	x	*A. ovalis,* blood ark†

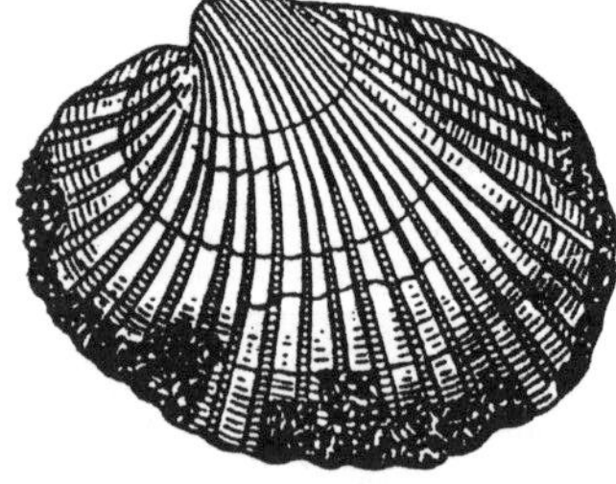
FIG. 13.157

Anadara ovalis, blood ark†. Shell thick; left valve slightly overlaps right valve. Gray/brown hairy periostracum outside; white inside. Has red blood and flesh; 26–35 ribs. Vi,Sa,Su,ff,6.3 cm (2.5 in), (A123,AM11,G145,NM535,R670). (FIG. 13.157. Left valve.)

Anadara transversa, transverse ark†. Dorsal margin of shell straight—forms sharp corners at margins. Solid, white shell with light brown, hairy periostracum, 30–35 thick, flat-topped ribs. Left valve slightly overlaps the right valve. White inside. Vi,Mu–SG,Su,ff,3.8 cm (1.5 in), (A123,AM11,G145,R670). (FIG. 13.158. Left valve.)

FIG. 13.158

Noetia ponderosa, ponderous ark†. Heavy shell with 30 large squared ribs; eroded furry brown periostracum. Heart-shaped, end view. Soft parts, orange-red color. Ca+,Sa,Su,ff,6.3 cm (2.5 in), (A124,AM11,G145,NM536,R672). (FIG. 13.159. Left valve.)

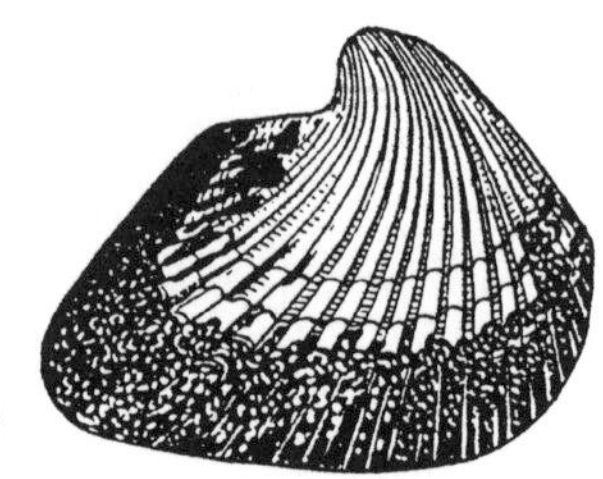
FIG. 13.159

13.117 **Arcticidae, Ocean Quahogs** (from 13.97)

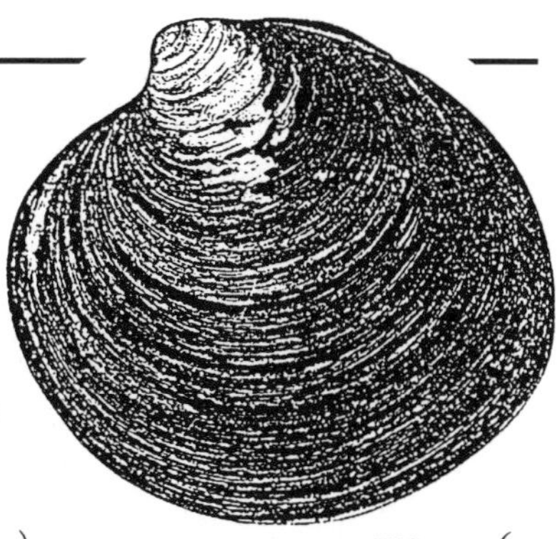

FIG. 13.160

Rounded, strong shells with thick dark periostracum, white interior; very short siphons; no pallial sinus.

Arctica islandica, queen quahog†, black or ocean clam. White with yellowish brown periostracum in young; brownish black in adults; interior is white. Dark periostracum and lack of pallial sinus helps differentiate this from *Mercenaria mercenaria.* Shallow burrower in silty sand. Edible. Bo,Mu–Sa,Su–De,ff,12.7 cm (5 in), (A163,AM39,G149,NM552,R792), ☺. (FIG. 13.160. Left valve.)

13.118 **Astartidae** (from 13.93, 13.97)

Solid, oval or triangular shells with ringed sculpture; tough darkish periostracum; prominent umbos; attach by byssus threads (tough, secreted fibers); no pallial sinus.

1. Number of cords or rings on shell surface
   - A. Sculpture consists of *more than 25* concentric rings of modest size — >25
   - B. Sculpture consists of fewer than 25 large raised cords; provide *number* — __
2. Crenulation (tiny teeth; see FIG. 13.164) along inside surface of ventral margin of shell
   - A. *Crenulation* present — cr
   - B. Crenulation *absent* — x
3. Periostracum color
   - A. *Dark* reddish brown — da
   - B. *Lighter,* yellowish brown — li
4. Ring or cord pattern
   - A. Rings definitely more pronounced toward *umbo* than toward ventral margin — um
   - B. Rings approximately *equally* developed from umbo to margin — eq

13.119

Cords	Crenulation	Color	Pattern	Species
10–15	cr	da	eq	*Astarte undata,* wavy astarte†
10–15	cr	li	eq	*A. subaequilatera,* lentil astarte†
>25	x	da	um	*Tridonta (=Astarte) borealis,* boreal tridonta†
>25	x	li,da	um	*T. (=Astarte) montagui,* Montagu tridonta†
>25	x	li	eq	*Astarte castanea,* smooth astarte†, chestnut astarte

FIG. 13.161

Astarte castanea, smooth astarte†, chestnut astarte. Glossy brown periostracum; white inside. Shell smooth with fine concentric growth lines. Bright orange foot. Ac–,Sa–Mu,Li–Su,ff,25 mm (1 in), (A143,AM41,G150,NM551,R741). (FIG. 13.161. Left valve.)

FIG. 13.162

Astarte subaequilatera, lentil astarte†. Thick yellowish brown periostracum; white interior. Bo,Ha–Gr–Sa–Mu,Su,ff,3.8 cm (1.5 in), (AM42).

Astarte undata, wavy astarte†. Chestnut-brown periostracum; white interior. Much like *A. subaequilatera* but fewer, heavier, concentric ridges. Bo,Mu–Sa–Gr,Su,ff,3.8 cm (1.5 in), (A144,AM42,G150,NM550,R741), ☺. (FIG. 13.162. Left valve.)

FIG. 13.163

Tridonta (=Astarte) borealis, boreal tridonta†. Tough, yellowish to dark-brown, fibrous, often frayed periostracum. Ac,Sa–Mu–Gr,Su–De,ff,5.1 cm (2 in), (AM41,NM550,R740). (FIG. 13.163. Left valve.)

Tridonta (=Astarte) montagui, Montagu tridonta†. Dull tannish brown periostracum. Fine growth lines apparent, especially near the umbo. Ac,Mu–Sa,Su–De,ff,25 mm (1 in), (A143,R741).

13.120 **Cardiidae, Cockles** (from 13.97, 13.101)

Inflated shells, margins scalloped, gapes at one end; large foot, short siphons; no pallial sinus.

1. Appearance of shell
   - A. Radiating *ribs:* provide *number* — r:__
   - B. Smooth or growth rings but with darker *zigzag* line — zz
   - C. Smooth or growth rings only, ribs and zigzag lines *absent* — x
2. Features of inside ventral edge of shell
   - A. *Crenulations* (small bumps or teeth; see FIG. 13.164) line inside ventral edge of shell; they appear to be unrelated to outer shell sculpturing — cr*
   - B. Inside edge *scalloped,* directly related to outer shell sculpturing (see FIG. 13.142, sc) — sc
   - C. Both these features *absent* — x
3. Maximal size of shell: provide *length* in millimeters (or inches) — __

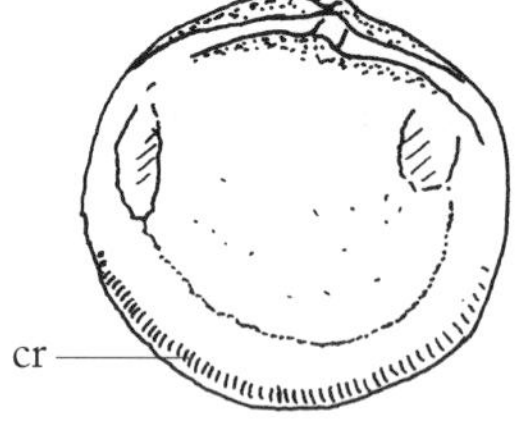

FIG. 13.164

13.121

Shell	Margin	Size	Species
r:20–28	sc	17 mm (0.7 in)	*Cerastoderma pinnulatum,* northern dwarf-cockle†, little cockle
r:32–38	sc	60 mm (3 in)	*Clinocardium ciliatum,* hairy cockle†, Iceland cockle
zz	cr	25 mm (1 in)	*Laevicardium mortoni,* Morton eggcockle†
x	x	101 mm (4 in)	*Serripes groenlandicus,* Greenland cockle†

FIG. 13.165

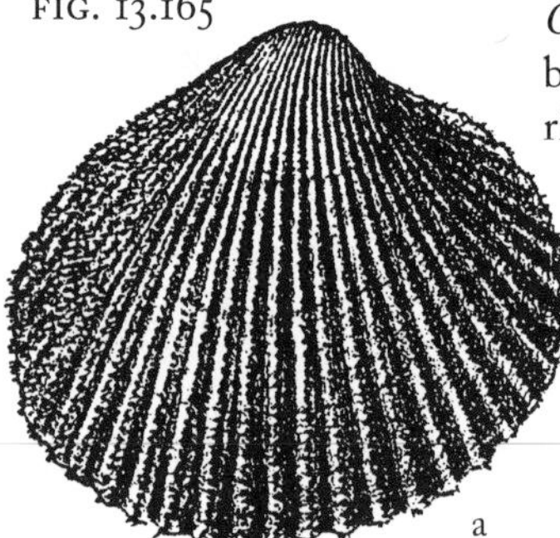

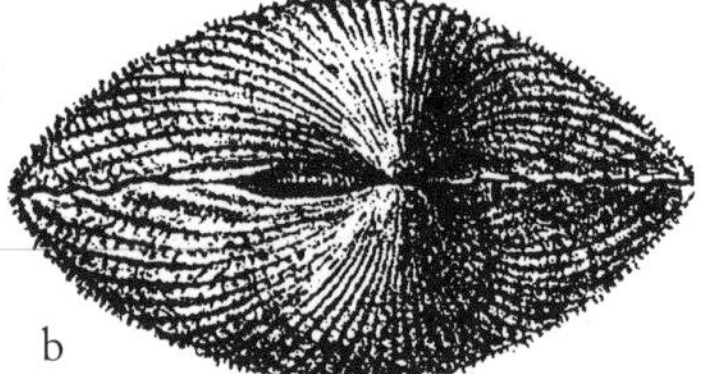

Cerastoderma pinnulatum, northern dwarf-cockle†, little cockle. Fat shell; whitish or light brown outside; glossy white inside. Some ribs may have teeth (pointed projections along ribs) especially toward sides. Bo,Sa–Gr,Li–Su,ff,17 mm (0.67 in), (A147,G150).

Clinocardium ciliatum, hairy cockle†, Iceland cockle. Gray to brown, fuzzy periostracum; light yellow outside; white inside. Margin of shell sometimes slightly scalloped. Ac,Sa–Gf,Su,ff,7.6 cm (3 in), (A147,AM58,G151,NM556,R749). (FIG. 13.165. a, right valve; b, dorsal.)

FIG. 13.166

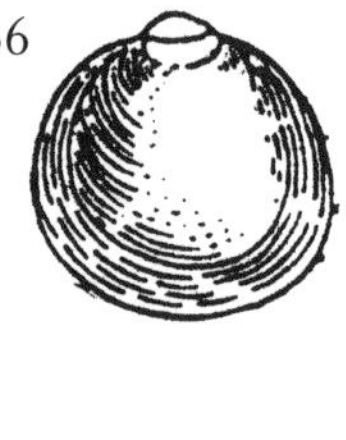

Laevicardium mortoni, Morton eggcockle†. Zigzag brown lines on outside of shell; inside yellowish, often with spot of brown or purple. Common duck food. Vi,Sa–Mu–Gr,Su,ff,25 mm (1 in), (A146,AM59,G152,NM557,R747). (FIG. 13.166. Left valve.)

Serripes groenlandicus, Greenland cockle†. Brown to yellow outside, white inside; foot is mottled red. Some with axial ribs at anterior and posterior ends. Ac,Gr–Sa–Mu,Sa–De,ff,10.1 cm (4 in), (A146,AM60,G152). (FIG. 13.167. Left valve.)

FIG. 13.167

13.122 **Carditidae, Heartclams** (from 13.101)

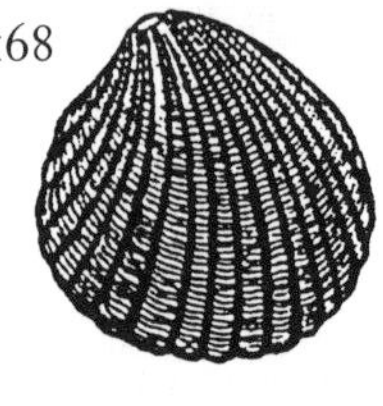

Stout shell with large, bent umbos; crenulate shell margin; attach by byssus threads (tough secreted fibers).

FIG. 13.168

Cyclocardia (=Venericardia) borealis, northern cyclocardia†, heartclam. Thick, solid shell; heavy, shaggy brown periostracum; white inside. About 20 radial ribs. Under stones. Ac–,Sa–Gr–SG,Li–De,ff,3.8 cm (1.5 in), (AM45,G150,R739). (FIG. 13.168. Left valve.)

13.123 Corbulidae, Marsh or Basketclams (from 13.111)

Small, inflated, tough shells with left valve flatter and smaller than right valve; rough periostracum; chondrophore present inside hinge; concentric sculpture; no pallial sinus.

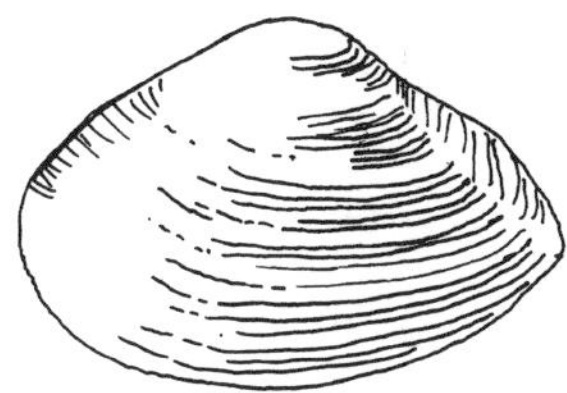

FIG. 13.169

Corbula swiftiana, swift corbula†, box or basketclam. Posterior drawn out to short projection. Single, posterior, radial ridge on left valve; double ridge on right valve. Vi+,Sa–Mu?,Su–De,ff,7 mm (0.3 in), (A173,AM105). (FIG. 13.169. Left valve.)

13.124 Cuspidariidae, Dipperclam (from 13.109, 13.111)

Small, light shells with inflated anterior and drawn out, pointed posterior (dipper handle); chrondrophore inside hinge; no pallial sinus.

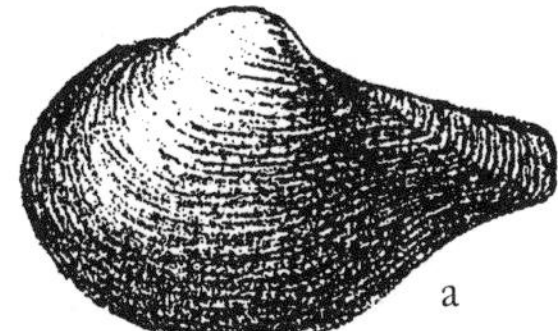

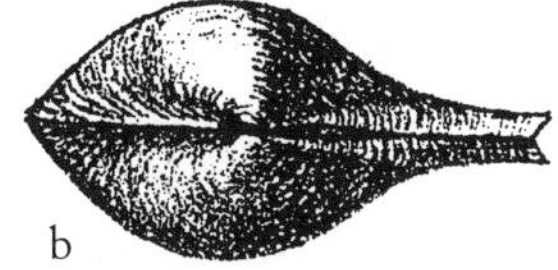

FIG. 13.170

Cuspidaria glacialis, glacial dipperclam†. Posterior end with much shorter projection than *C. rostrata.* Ac,Mu,Su–De,ff,3.8 cm (1.5 in), (A183,AM118,G160). (FIG. 13.170. a, left valve; b, dorsal.)

Cuspidaria rostrata, rostrate dipperclam†. Posterior end as long (0.5 length of shell) tubular projection. Bo,Mu,Su–De,ff,25 mm (1 in), (A183,AM119,G160). (FIG. 13.171. a, left valve; b, dorsal.)

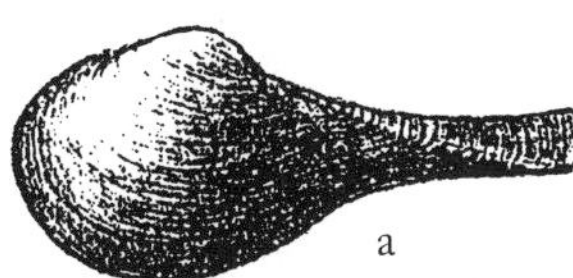

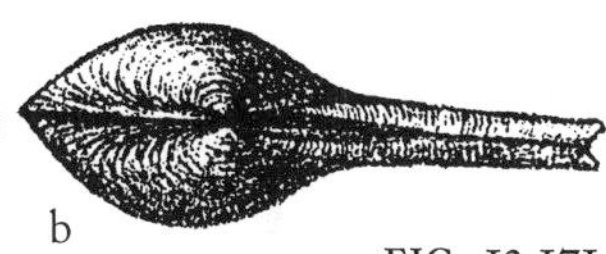

FIG. 13.171

13.125 Donacidae, Donax Clams (from 13.113)

Small, blade-shaped shell with elongated, rounded posterior end; anterior is abruptly sloped; tropical family only occasionally within our range.

FIG. 13.172

Donax variabilis, variable coquina†. Somewhat flattened shell; shiny, often with sunburst of darker, axial lines from umbo. Color variable, but often bright; interior white or dark purple. Can form dense aggregations at surfy beaches. Vi,Sa,Li,ff,20 mm (0.8 in), (A158,AM91,NM556,R779). (FIG. 13.172. Right valve.)

13.126 Dreissenidae, False Mussels (from 13.111)

Shell with unique internal shelf at umbo-hinge; long, thin posterior muscle scar inside shell; well developed siphons; pallial line is indistinguishable or no pallial sinus.

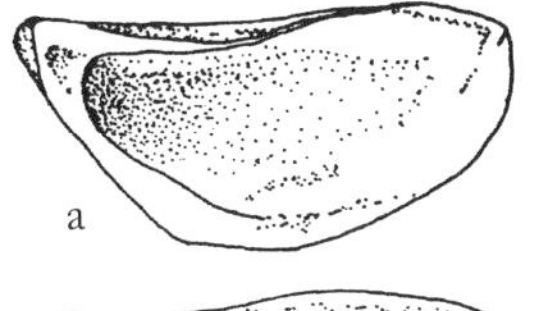

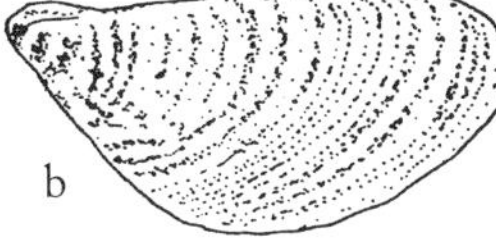

FIG. 13.173

Dreissena polymorpha, zebra mussel†. Outside tan with dark apical stripes; inside, white with brown lines. Freshwater species, included here as an alert because of fouling problems its recent introduction from Asia, via Europe, has caused in Great Lakes area. First reported from upper Hudson River in 1991. (A164,AM95,R790). (FIG. 13.173. a, left valve interior; b, right valve exterior.)

Mytilopsis (=*Congeria*) *leucophaeata,* dark falsemussel†, Conrad false or platform mussel. Outside brown with thin, glossy periostracum; inside, white with bluish tones. Small triangular shelf along inside of pointed end. Attaches with byssus threads in clumps in fresh and brackish water. Vi,Ha–Es,Li–Sa,ff,20 mm (0.8 in), (A164,AM95,R791). (FIG. 13.174. a, left valve interior; b, right valve exterior.)

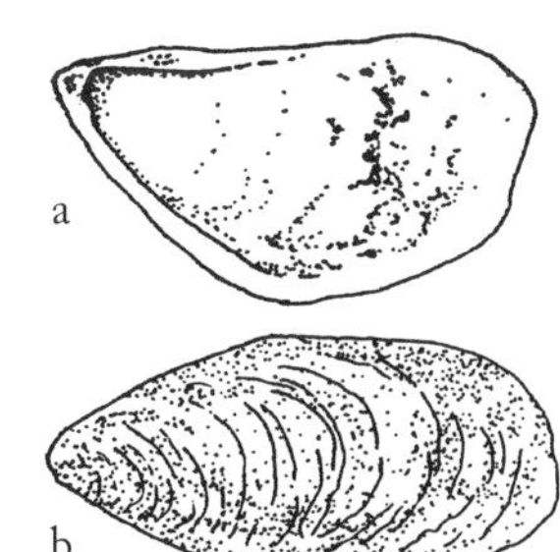

FIG. 13.174

13.127 **Hiatellidae, Rock Borer Clams** (from 13.95, 13.105)

Irregularly wrinkled as growth follows contours of habitat; dull colors; often pallial line is indistinguishable, but if visible, shows large pallial sinus to accommodate siphons.

1. Color of periostracum or thin outer covering of shell
   - A. Dark brown to *black* — bl
   - B. *Yellowish* brown — ye
2. Shell proportions: shape index = width/height
   - A. Shell *elongate,* shape index > 2 — el
   - B. Shell *short,* shape index < 2 — sh
3. Features of the interior
   - A. *Red* to orange siphon from posterior end (visible on intact animals if allowed to relax in cold seawater) — rs
   - B. Large, irregular, thickened, shelly *callus* toward center of interior shell surface — ca
   - C. These features *absent* — x

13.128

Color	Index	Inside	Species
bl	el	ca	*Cyrtodaria siliqua,* northern propellerclam†
ye	el	rs	*Hiatella arctica,* Arctic hiatella†, rock borer or red-nosed clam
ye	sh	x	*Panomya arctica,* Arctic roughmya†

FIG. 13.175

FIG. 13.176

FIG. 13.177

Cyrtodaria siliqua, northern propellerclam†. Slightly twisted valves gape (valves do not completely meet) at both ends; heavy, chalky, bluish white shell with dark, flaky periostracum. Ac–,?,Su,ff,6.3 cm (2.5 in), (A170,AM100). (FIG. 13.175. Right valve.)

Hiatella arctica, Arctic hiatella†, rock borer or red-nosed clam. White shell; gray, thin, flaky periostracum, red to orange siphons. External rib from umbo to posterior margin often visible. Rocky shore, tide pools, holdfasts. Shell irregular, takes on contour of habitat. Bo,Ha–Al,Li–Su,ff,3.8 cm (1.5 in), (A170,AM100,G158,NM571,R818), ☺. (FIG. 13.176. Left valve.)

Panomya arctica, Arctic roughmya†. Resembles distorted *Mya* but with light brown flaky periostracum. Bo,Mu,Su,ff,6.3 cm (2.5 in), (A171,AM101). (FIG. 13.177. Left valve.)

13.129 **Limopsidae** (from 13.97)

Small oval shell with taxodont dentition and furry periostracum.

Limopsis sulcata, sulcate limops†. Crenulate margin; ca. 18 hinge teeth. Dull white; thick periostracum extends beyond ventral margin. Vi,Sa,Su–De,ff,12 mm (0.5 in), (A124,AM12).

13.130 **Lucinidae, Lucine Clams** (from 13.93, 13.97)

Circular, flattened, small umbo; long, narrow anterior adductor muscle scar inside shell; short siphons; no pallial sinus.

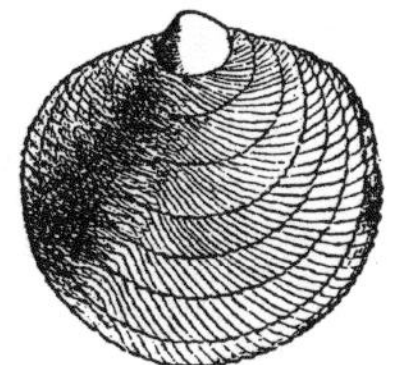

FIG. 13.178

Divaricella quadrisulcata, cross-hatched lucine†. Glossy white shell; finely sculptured criss-crossed lines. Vi,Sa,Su,ff,25 mm (1 in), (AM53,GI51,NM553,R724). (FIG. 13.178. Left valve.)

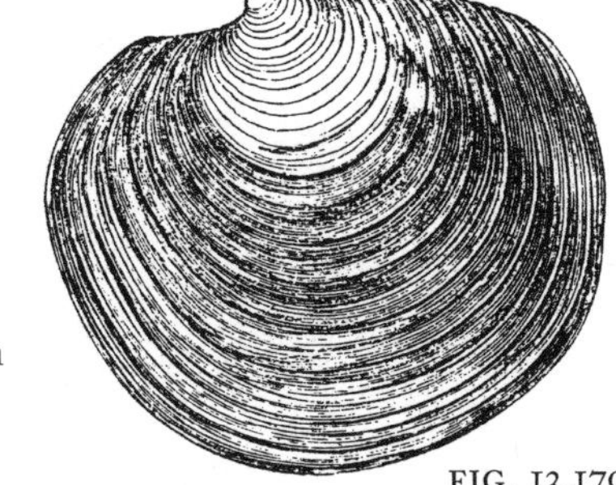

Lucinoma filosa, northeast lucine†. Exterior white with thin, yellow periostracum. Sharp, raised growth lines in adults. Bo,?,Su–De,ff,6.3 cm (2.5 in), (AI38,AM49). (FIG. 13.179. Left valve.)

FIG. 13.179

13.131 **Lyonsiidae, Paperclams** (from 13.101, 13.103, 13.107, 13.113)

Small, fragile, semi-translucent shells with lots of tiny ribs and often with sand grains adhering to shell.

FIG. 13.180

Lyonsia arenosa, sandy lyonsia†. Shell more oval or rectangular than next species. Valves fit together without gape in posterior. Ac,Sa,Li–Su,ff,12 mm (0.5 in), (AI77,AM114,R829). (FIG. 13.180. Left valve.)

Lyonsia hyalina, glassy lyonsia†. Shell tapers to narrowed, squared, or pointed posterior. Valves chalky white, fragile, and gape somewhat at posterior end. Except for umbo area, shell nearly covered by sand; shiny and smooth inside. Bo,Mu–Sa–SG,Li–Su,ff,20 mm (0.8 in), (AI77,AM114,GI59,R829). (FIG. 13.181. Left valve.)

FIG. 13.181

13.132 **Mactridae, Surfclams** (from 13.97, 13.109)

Substantial, triangular shell with chondrophore inside hinge surface; short siphons; modest pallial sinus provision for siphons.

1. Size and shape of umbo
   - A. Umbo relatively *small* and oriented *straight* up — ss
   - B. Umbo *large* and *bent* toward the anterior — lb
2. Presence of distinct exterior ridge extending from umbo to create flattened surface toward the posterior edge of the shell
   - A. Distinct *ridge* present — ri
   - B. Distinct ridge *absent* — x

13.133

Umbo	Ridge	Species
ss	ri	*Mulinia lateralis,* dwarf surfclam†, little coot or surfclam
ss	x	*Spisula solidissima,* Atlantic surfclam†, bar or hen clam
lb	x	*Rangia cuneata,* Atlantic rangia†, brackish-water wedgeclam

FIG. 13.182

Mulinia lateralis, dwarf surfclam†, little coot or surfclam. Anterior and posterior margins of shell are flattened. Thick shell with faint growth lines; thin white-yellow periostracum, white inside. Duck/goose food. Bo,Mu,Su,ff,20 mm (0.8 in), (AM77,GI54,R754), ☺. (FIG. 13.182. Left valve.)

Rangia cuneata, Atlantic rangia†, brackish-water wedgeclam. Umbos large, give clam heart-shaped cross-section. Outside white with yellowish brown periostracum; glossy white to bluish gray inside. Ca+,SM–Mu–Sa,Es–Li–Su,ff,6.3 cm (2.5 in), (AM78,R755). (FIG. 13.183. Left valve, outside and inside.)

FIG. 13.183

Spisula solidissima, Atlantic surfclam†, bar or hen clam. Triangular shape; yellowish-white with shiny yellowish brown periostracum. Edible. Exposed beaches to offshore. Tolerates lower salinity. Can leap if distressed. Moon snails are major predator. Common shell on beaches. Bo,Sa,Li–Su,ff,20 cm (8 in), (A149,AM76,G154,NM569,R753), ☺. (FIG. 13.184. Left valve.)

FIG. 13.184

13.134 Mesodesmatidae, Wedgeclams (from 13.99)

Tough, wedge- or blade-shaped shell with short posterior end; chondrophore present inside hinge.

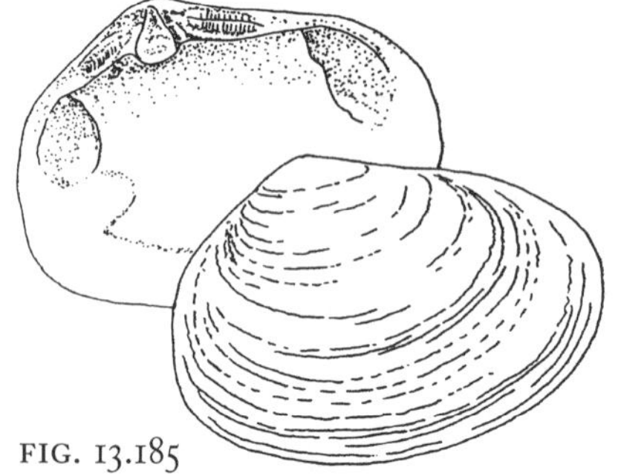

FIG. 13.185

Mesodesma arctatum, Arctic wedgeclam†. Umbos not especially enlarged. White with shiny yellowish periostracum; tan to white interior. Bo,Sa,Li–Su,ff,3.8 cm (1.5 in), (A151,AM79,G154,R758). (FIG. 13.185. Inside right valve; outside left valve.)

13.135 Myidae, Soft-Shelled Clams (from 13.97, 13.99, 13.105)

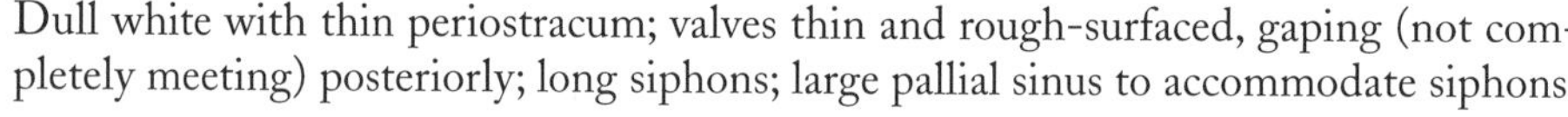

Dull white with thin periostracum; valves thin and rough-surfaced, gaping (not completely meeting) posteriorly; long siphons; large pallial sinus to accommodate siphons.

Mya arenaria, softshell†, steamer, long-necked clam, nanny nose. Thin shell; grayish to straw-colored periostracum. Important commercial species. Bo,Sa–Mu,Li–,ff,15.2 cm (6 in), (A172,AM102,G158,NM572,R811), ☺. (FIG. 13.186. a, Left valve; b, animal in natural orientation; siphons extend to sediment surface, digging foot beneath.)

FIG. 13.186

Mya truncata, truncate softshell†, blunt clam. Valves gape widely posteriorly; yellowish brown periostracum. Siphons extend well beyond shell. Edible. Ac,Sa–Mu,Su,ff,7.6 cm (3 in), (A172,AM102). (FIG. 13.187)

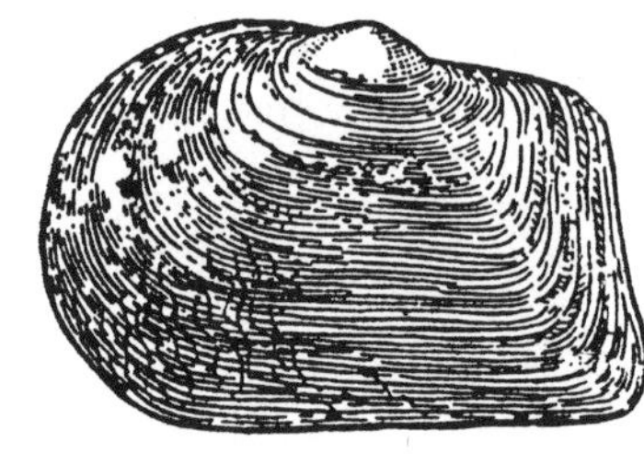

FIG. 13.187

13.136 Mytilidae, Mussels (from 13.99, 13.101, 13.103, 13.111, 13.113)

Shells typically pointed anteriorly; long hinge line; attach by byssus threads (tough, secreted fibers), often in large colonies; pallial line difficult to see or if visible, lacks pallial sinus.

1. Holding shell with hinged part uppermost, relative location of umbo (FIG. 13.188)
   A. About *one-third* the distance from one end along the dorsal surface — ⅓
   B. At or *near end;* umbo occupies only top half of pointed end — ne
   C. Precisely at *end;* umbo occupies entire pointed end — en
2. Dominant shell sculpture (FIG. 13.188)
   A. Obvious *ribs* radiate from umbo toward periphery of shell — rr
   B. Shell with faint radiating *striations* or scratches — st
   C. Shell may have growth rings but otherwise is *smooth* — sm

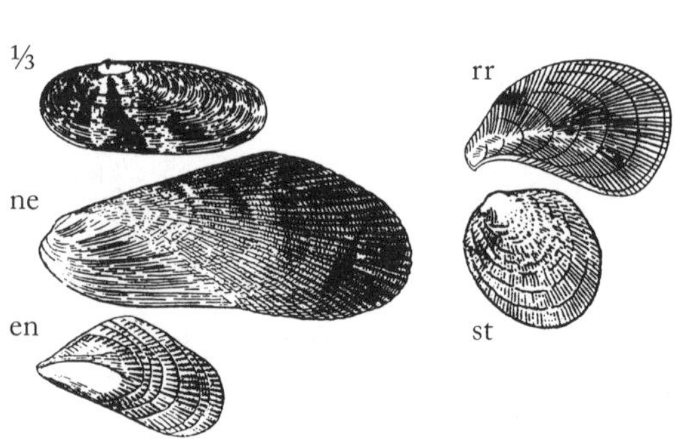

FIG. 13.188

3. Color of inside surface toward rounded end of shell
   - A. Light *purplish* to rosy — pu
   - B. *Pearly* white — pe
   - C. Dark *blue,* purple, or black — bl
   - D. *Whitish* — wh
4. Features along bottom inside margin of shell
   - A. *Crenulations* (tiny teeth; see FIG. 13.164) present — cr
   - B. Inside edge *scalloped,* directly related to outer shell sculpturing (see FIG. 13.142, sc) — sc
   - C. Both these features *absent* — x

13.137

Umbo	Sculpture	Color	Margin	Species
en	rr	pu	x	*Ischadium* (=*Brachidontes*) *recurvum,* hooked mussel†, bent mussel
en	sm,st	bl	x	*Mytilus edulis,* blue mussel†
ne	rr	pu	sc	*Geukensia* (=*Modiolus*) *demissa,* ribbed mussel†
ne	sm	wh	x	*Modiolus modiolus,* northern horsemussel†
ne	rr	pu	sc	*Musculus niger,* black mussel†
ne	st	pe	x	*M. discors,* discordant mussel†
⅓	st	pe,wh	cr	*Crenella glandula,* glandular crenella†, bean mussel

FIG. 13.189

FIG. 13.190

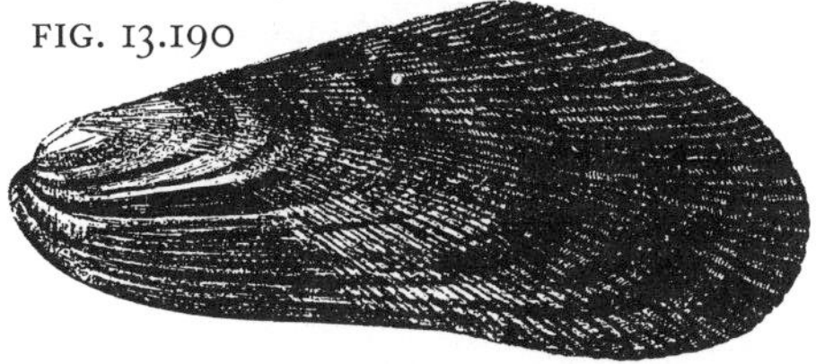

FIG. 13.191

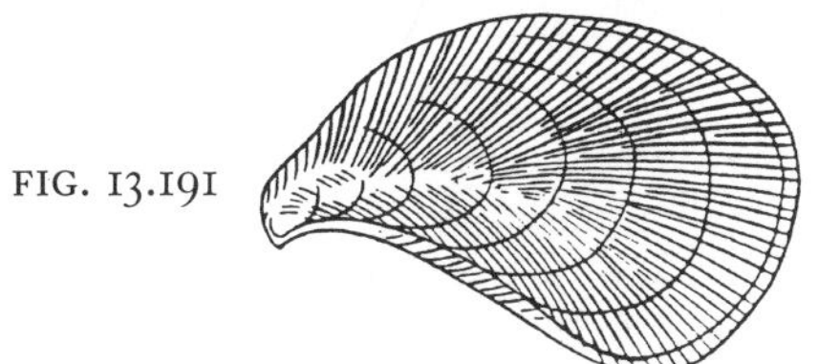

FIG. 13.192

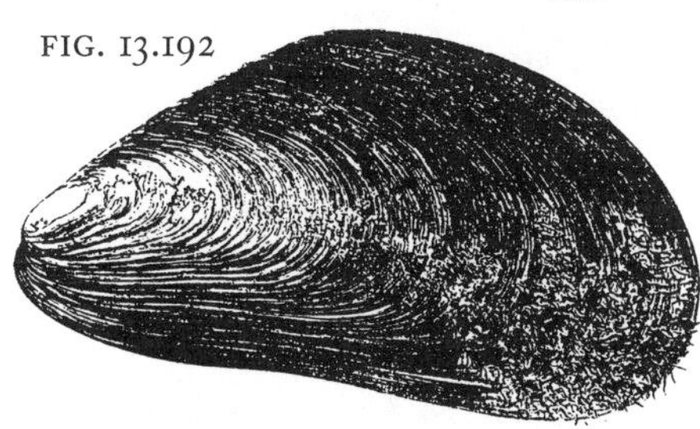

FIG. 13.193

FIG. 13.194

Crenella glandula, glandular crenella†, bean mussel. Brown or yellowish outside; pearly white inside. Builds nests in algal holdfasts. Bo,Ha–Al–Mu,Li–Su,ff,12 mm (0.5 in), (A129,AM19,G150). (FIG. 13.189. Left valve.)

Geukensia (=*Modiolus*) *demissa,* ribbed mussel†. Interior silvery white with purplish posterior edge; yellowish, brownish green, or blackish brown outside. Tissues yellow. Often in brackish and/or polluted waters. Bo,SM–Mu,Es–Li,ff,10.1 cm (4 in), (A128,AM16,G146,R687), ☺. (FIG. 13.190. Left valve.)

Ischadium (=*Brachidontes*) *recurvum,* hooked mussel†, bent mussel†. Grayish brown outside; purplish brown with bluish gray margin inside. Strong radial sculpture. Primarily Chesapeake southward. Vi,Ha–Es–Li,ff,5.1 cm (2 in), (A127,AM17,G146,NM538,R680), ☺. (FIG. 13.191. Left valve.)

Modiolus modiolus, northern horsemussel†. Thick reddish brown flaky periostracum, with long smooth hairs; chalky shell beneath; interior is pearly. Shell often hairy posteriorly. Attached by tough byssus threads; kelps often attached to large ones via their holdfasts. Bo,Ha–Sa–Gr,Li–Su,ff,16.5 cm (6.5 in), (A128,AM15,G147,NM537,R685), ☺. (FIG. 13.192. Left valve.)

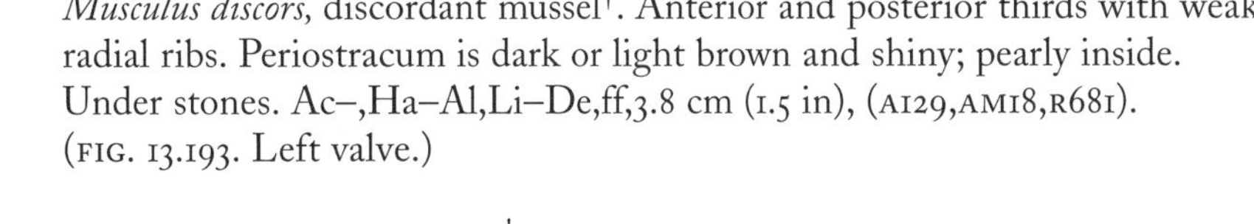

Musculus discors, discordant mussel†. Anterior and posterior thirds with weak radial ribs. Periostracum is dark or light brown and shiny; pearly inside. Under stones. Ac–,Ha–Al,Li–De,ff,3.8 cm (1.5 in), (A129,AM18,R681). (FIG. 13.193. Left valve.)

Musculus niger, black mussel†. Both ends of shell with axial ribs; central section lacks them. Interior pinkish. More active mover than most mussels, although can use byssus threads to attach. Bo,Ha,Li–Su,ff,5.1 cm (2 in), (A129,AM19,R681). (FIG. 13.194. Left valve.)

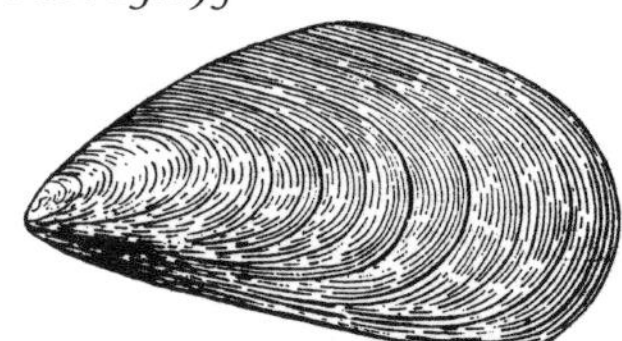
FIG. 13.195

Mytilus edulis, blue mussel†. Younger specimens are brighter-greenish, banded or rayed. Attach by strong byssus threads. Slender brown foot. Often competitive dominant in lower intertidal if unchecked by predators. Groups held together by byssus threads host a community of smaller invertebrates. Commercial species. Warning: becomes toxic when exposed to red tide organisms. Bo,Ha–Mu,Li,ff,10.1 cm (4 in), (AI26,AM14,GI46,NM538,R676), ☺. (FIG. 13.195. Left valve.)

13.138 Nuculanidae, Nutclams (from 13.105, 13.111)

Anterior end round, posterior drawn into point; taxodont dentition; chondrophore present; shiny white inside.

ro

sq

po

FIG. 13.196

1. Shape of the posterior end of the shell (FIG. 13.196)
   - A. Both ends are nearly equally *rounded* — ro
   - B. Posterior end is *squared* off — sq
   - C. Posterior end *pointed* — po
2. Holding shell with hinged part uppermost, relative location of umbo
   - A. About *one-third* the distance from one end — ⅓
   - B. Umbo approximately in the *center* of dorsal surface — ce
3. Features extending from umbo toward anterior or posterior (pointed) end of shell
   - A. Distinct *ridge* toward posterior end, creates flattened upper posterior surface — ri
   - B. *Groove* runs from umbo to *posterior* margin — pg
   - C. Distinct *groove* runs from umbo to *anterior* margin — ag
   - D. Ridge and groove both *absent* — x
4. Number of teeth in taxodont dentition (see FIG. 13.140, ta): provide *number* anterior to umbo: provide *number* posterior to umbo — __:__
5. Thickness of shell
   - A. Shell *inflated,* rather chubby — in
   - B. Shell *flattened,* thin — fl

13.139

Posterior	Umbo	Ridge Groove	Dentition	Width	Species
po	ce	ri	20:20	fl	*Yoldia limatula,* file yoldia†
po	ce	ri	16–19:16–19	in	*Nuculana acuta,* pointed nutclam†
po	ce	pg	50:50	fl	*Yoldia sapotilla,* short yoldia†
po	ce	ag	18:18	fl	*Y. amygdalea,* almond yoldia†
po,ro	ce	x	12:12	fl	*Y. myalis,* oval yoldia†
po	⅓	ri	15:36	fl	*Nuculana tenuisulcata,* thin nutclam†
sq	ce,⅓	ri	12:12	fl	*Yoldia thraciaeformis,* broad yoldia†, ax yoldia

Nuculana acuta, pointed nutclam†. Concentric rings; white with yellowish periostracum. Vi,Sa,Su–De,df,7 mm (0.3 in), (AM4,R663).

FIG. 13.197

Nuculana tenuisulcata, thin nutclam†. Thin shell; white with shiny yellowish brown periostracum. Ac-,Mu,Su–De,df,25 mm (1 in), (AM5,NM533,R663), ☺. (FIG. 13.197. Left valve.)

Yoldia amygdalea, almond yoldia†. Golden brown periostracum. Small concave depression on anterior ventral margin. Ac,Mu,Su,df,50 mm (2 in), (AI21).

Yoldia limatula, file yoldia†. Greenish tan to light chestnut outside, glossy white inside. Strong foot; can lurch forward. Bo,Mu,Su,df,5.1 cm (2 in), (AI22,AM6,GI44,NM534,R664), ☺. (FIG. 13.198. Left valve.)

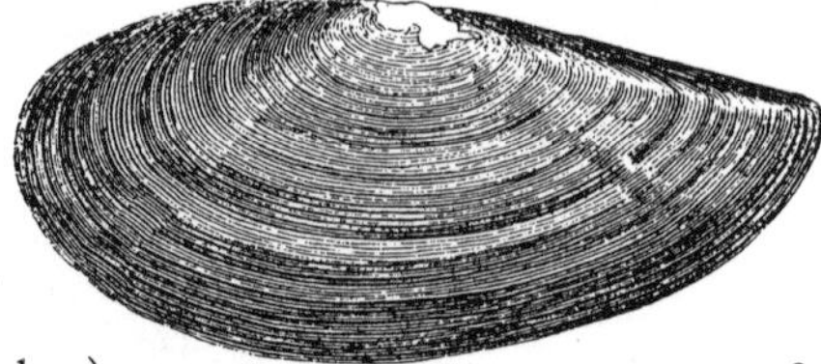
FIG. 13.198

FIG. 13.199

FIG. 13.200

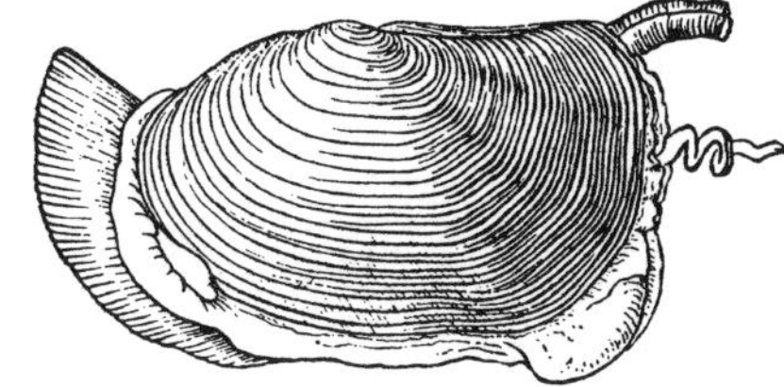
FIG. 13.201

Yoldia myalis, oval yoldia†. White with yellowish brown periostracum, alternating dark and light; white inside. Ac,Mu?,Su,df,25 mm (1 in), (AM6,GI45,R665). (FIG. 13.199)

Yoldia sapotilla, short yoldia†. Smooth, thin shell. White with yellowish green periostracum: white interior. Bo,Mu,Su,df,3.8 cm (1.5 in), (AM6,GI45,R664), ☺. (FIG. 13.200. Live animal, anterior foot to left, siphons and palps to right.)

Yoldia thraciaeformis, broad yoldia†, ax yoldia. Upturned posterior end; dark periostracum. Bo,Mu–Sa,Su–De,df,5.1 cm (2 in), (AI22,AM7,GI45). (FIG. 13.201. Live animal, anterior foot to left, siphons and palps to right.)

13.140 **Nuculidae (Nutclams)** (from 13.97)

Tiny, fragile shell with taxodont dentition; olive green outside, pearly interior, and pallial line not visible or no pallial sinus. Important as fish food.

1. Shell shape; hold umbo upward, describe perimeter of shell
   - A. Shell *triangular* in outline — tr
   - B. Shell more *oval* in outline — ov
2. Crenulation or minute teeth along inside ventral margin of shell
   - A. *Crenulation* present — cr
   - B. Crenulation *absent* — x
3. Shell sculpture
   - A. In addition to growth rings, minute *axial striations* run from umbo to the opposite perimeter (striations are faint; look carefully) — as
   - B. Growth rings only, axial striations totally *absent* — x

13.141

Shape	Crenulation	Sculpture	Species
tr	cr	as	*Nucula proxima,* Atlantic nutclam†
tr	x	x	*N. delphinodonta,* dolphintooth nutclam†
ov	x	x	*N. tenuis,* smooth nutclam†, thin nutclam

FIG. 13.202

Nucula delphinodonta, dolphintooth nutclam†. Olive brown, ovate, shell with coarse, concentric, growth lines. Bo,Sa–Mu,Su,ff,4 mm (0.2 in), (AM3,GI44,R661). (FIG. 13.202. Left valve.)

Nucula proxima, Atlantic nutclam†. Greenish gray with irregular brownish, concentric lines. Quite variable. Tolerates sludge and organic pollution. Bo,Sa–Mu,Su,df,10 mm (0.4 in), (AI18,AM3,GI44,NM533,R660). (FIG. 13.203. Left valve.)

FIG. 13.203

FIG. 13.204

Nucula tenuis, smooth nutclam†, thin nutclam. Shiny, smooth, olive green with darker irregular growth lines. Taxodont pattern: eight teeth along upper inside surface of shell anterior to umbo; four to five teeth posterior to umbo. Bo,Sa–Mu,Su,df,4 mm (0.2 in), (AI18,AM3,GI44), ☺. (FIG. 13.204. Left valve.)

13.142 **Ostreidae, Oysters** (from 13.95)

Large, heavy, irregular valves; lower valve cemented to substrate, upper valve smaller than lower.

Crassostrea virginica, eastern oyster†, Virginia oyster. Variable shape. Grayish outside; glossy white inside with single purple muscle scar. Valve edges can be very sharp. Commerical species; becoming scarce. Bo,Ha–SM,Es–Li–Su,ff,25.4 cm (10 in), (A137,AM37,G148,NM547,R699). (FIG. 13.205)

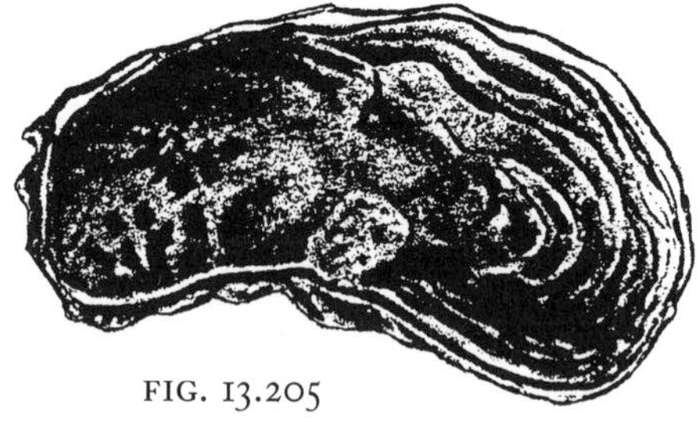

FIG. 13.205

13.143 **Pandoridae** (from 13.109, 13.111)

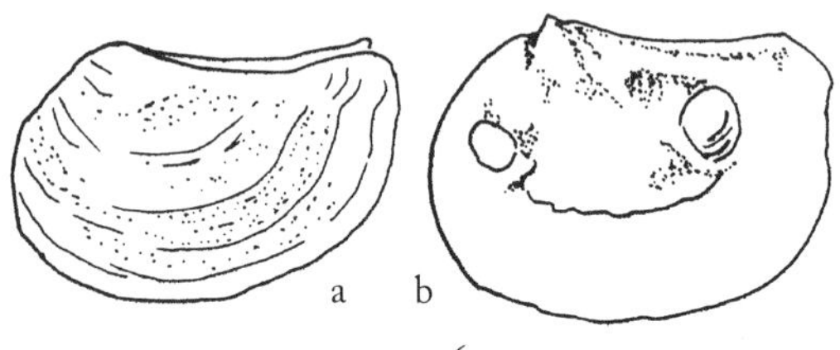

FIG. 13.206

Strongly compressed and fragile; right valve flatter than left valve; pointed anteriorly; large pallial sinus to accommodate siphons.

Pandora gouldiana, Gould pandora†. Extremely flattened shell; chalky white to rust colored. Upturned anteriorly; left convex valve overlaps right flattened valve at ventral margin. Inside, pearly; outside often eroded. Bo,Sa–Mu,Li–De,ff,3.8 cm (1.5 in), (A179,AM115,G160,R832). (FIG. 13.206. a, left valve, outside; b, left valve, inside.)

13.144 **Pectinidae, Scallops** (from 13.93)

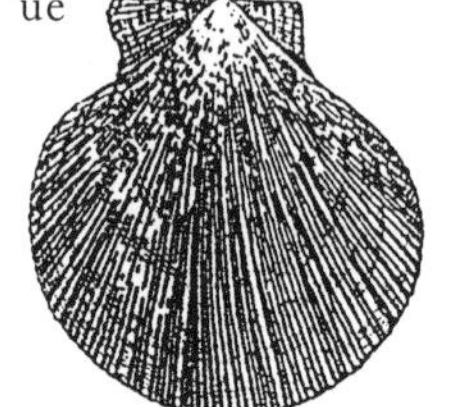

FIG. 13.207

With wings on either side of the umbo. Eyes located along mantle edge. Can open and close valves fast enough to swim short distances to elude predators such as sea stars.

1. Size of wings on either side of umbo (FIG. 13.207)
   - A. About *equal* in size — eq
   - B. *Unequal* in size, reduced or absent on one side — ue
2. Approximate number of radiating ribs
   - A. Provide *number* — __
   - B. Many fine radiating ribs (*80 or more*) — 80+
   - C. Radiating ribs *absent* — x

13.145

Wings	Ribs	Organism
eq	<20	*Argopecten (=Aequipecten) irradians,* bay scallop†
eq	80+,x	*Placopecten magellanicus,* ocean scallop†, deep sea or giant scallop
ue	50	*Chlamys islandica,* Iceland scallop†

FIG. 13.208

Argopecten (=Aequipecten) irradians, bay scallop†. Lower valve usually lighter color. Bright blue eyes. Source of commercial "bay scallops." Becoming scarce. Bo,SG–Sa,Su,ff,7.6 cm (3 in), (A134,AM30,G147,NM542,R707). (FIG. 13.208. Left valve.)

Chlamys islandica, Iceland scallop†. Gray to cream, tinged with purple, red, orange, both inside and out. Black eyes along mantle edge. Ac–,Sa–G,Su–De,ff,10.1 cm (4 in), (A132,AM27,G147,NM542,R705). (FIG. 13.209. Left valve.)

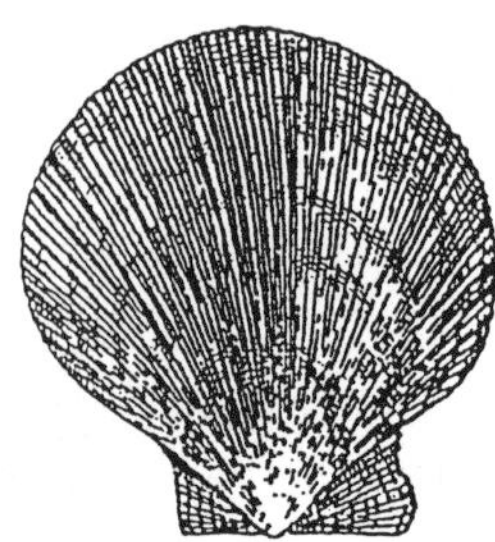

FIG. 13.209

Placopecten magellanicus, ocean scallop†, deep sea or giant scallop. Circular shell; upper valve convex, pinkish to reddish brown; lower valve is flat, pinkish white. Interior is glossy white; eyes gray. Sold as ocean scallops. Bo,Sa–Gr,Su,ff,22.9 cm (9 in), (A131,AM28,G148,NM544,R710), ☺. (FIG. 13.210. Left valve.)

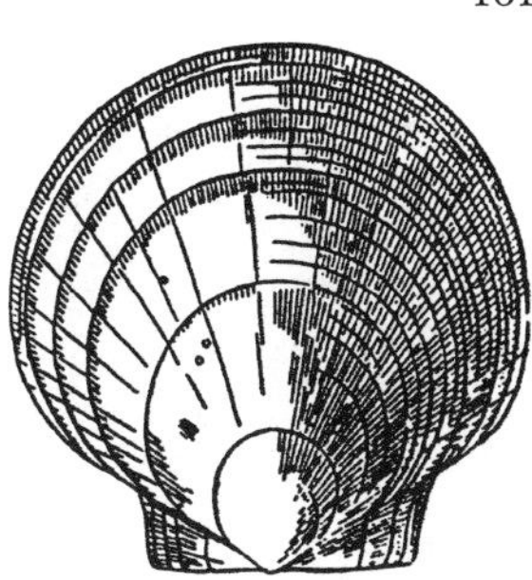

FIG. 13.210

13.146 **Periplomatidae, Spoonclams** (from 13.99)

FIG. 13.211

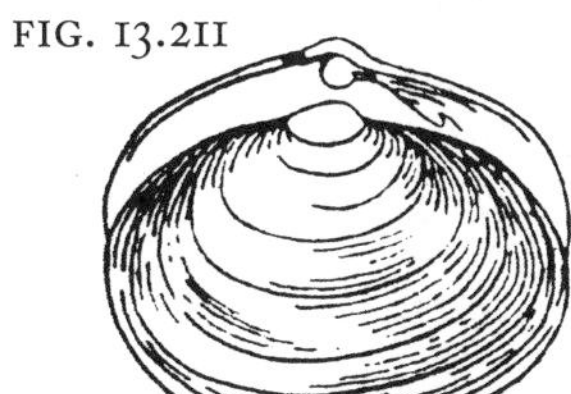

Small, oval shell; chondrophore and small reinforcing rib inside hinge; pallial line with modest pallial sinus for siphons.

Periploma leanum, Lea spoonclam†, lantern shell. Brittle; left valve nearly flat, right valve only slightly convex. Thin yellow periostracum, white inside. Inside rib is small but distinct. Bo,Sa–Mu,Su–De,ff,3.8 cm (1.5 in), (A182,AM117,G161,R837). (FIG. 13.211. Left valve.)

13.147 **Petricolidae and Pholadidae, Piddocks or Borers** (from 13.93, 13.103)

Thin, fragile shells tapering posteriorly but with abrasive rows of teeth that permit animal to bore into peat, clay, or even soft stone.

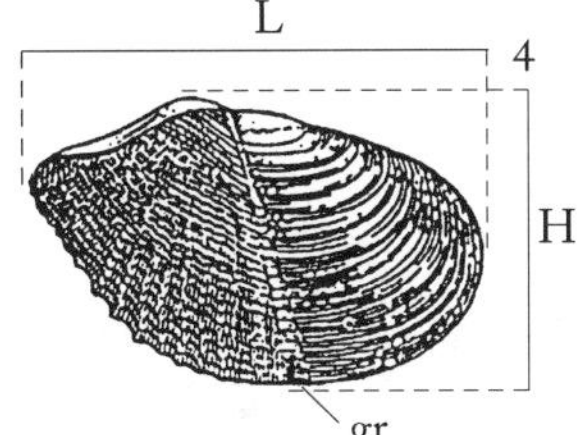

FIG. 13.212

1. Presence of distinct groove separating toothed sculpture from remainder of shell (FIG. 13.212)
   - A. Distinct *groove* present — gr*
   - B. Distinct groove *absent* — x
2. Presence and color of thin periostracum over posterior end of shell
   - A. Thin *rust*-colored *periostracum* present — rp
   - B. Periostracum thin and *grayish* — gr
   - C. Periostracum *absent* — x
3. Gaping (space left between valves when pressed together)
   - A. Widely gaping at *both ends* — be
   - B. Slightly gaping at *one end* only — 1e
4. Shell proportions: shape index = length divided by height
   - A. Shell *elongate,* index > 2.5 — el
   - B. Shell relatively *shorter,* index < 2.5 — sh

13.148

Groove	Periostracum	Gape	Index	Organism
gr	x	be	sh	*Zirfaea crispata,* great piddock†
x	gr	be	el	*Cyrtopleura costata,* angelwing†
x	rp	be	sh	*Barnea truncata,* Atlantic mud-piddock†, truncate borer or fallen angelwing
x	rp,x	be	el	*Pholas dactylus,* European piddock†
x	x	1e	el	*Petricola pholadiformis,* false angelwing†, American piddock

Barnea truncata, Atlantic mud-piddock†, truncate borer or fallen angelwing. Very thin and fragile. Reduced posterior sculpture. Peat, mud, or wood. Vi,SM–Mu,Li,ff,7.6 cm (3 in), (AM108,G160,R821). (FIG. 13.213. Left valve.)

FIG. 13.213

Cyrtopleura costata, angelwing. White shell with grayish periostracum. Thin and fragile. Spoon-shaped tooth from inside of hinge. Siphons appear huge for size of shell. Burrow 0.3–0.9 m (1–3 ft) deep in mud; siphons look like a pink "tulip bud" at surface. Southern form, rare north of Virginia. Vi,Sa–Mu,Su,ff,15.2 cm (6 in), (A174,AM107,G159,NM573,R822). (FIG. 13.214. Left valve.)

FIG. 13.214

Petricola pholadiformis, false angelwing†, American piddock. Tiny teeth along obvious ribs. Fragile shell, chalky white outside and inside. More common form in this region. Bores cylindrical burrow into mud or peat. Bo,Mu–SM,Es–Li–Su,df,5.1 cm (2 in), (AM74,G159,NM562,R810). (FIG. 13.215. Left valve.)

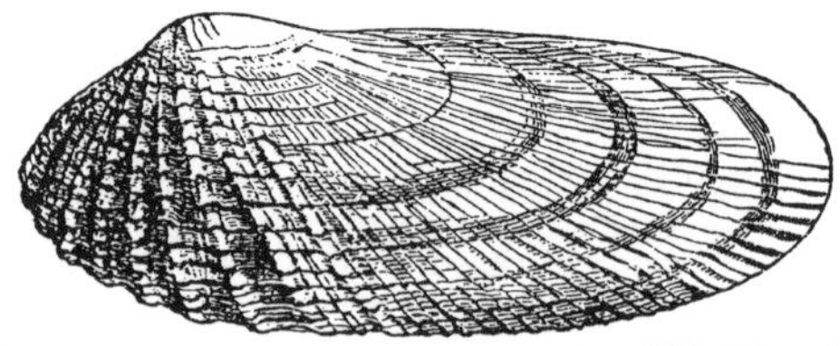

FIG. 13.215

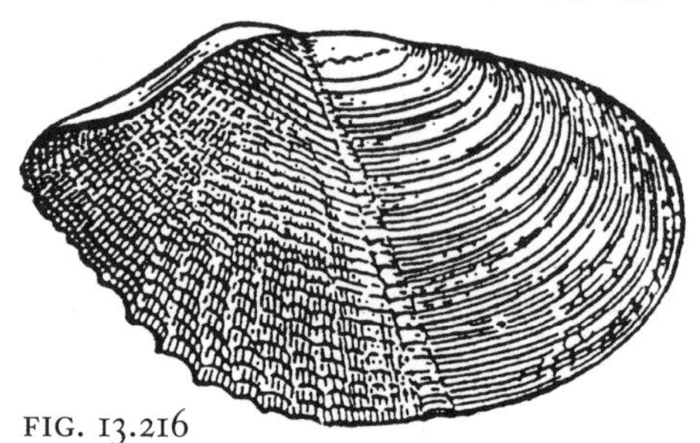

Pholas dactylus, European piddock†. Northern form, south to Delaware. Bores into sand, mud, peat, or soft stone. Bo,Mu–Sa,Su,ff,12.7 cm (5 in), (A175).

Zirfaea crispata, great piddock†. Sturdy shell; widely gaping. Reduced posterior sculpture. Grayish-white. Foot white. Burrows into soft substrates. Bo,Mu–Sa–Ha,Li–Su,ff,7.6 cm (3 in), (AM108,G160,NM574,R823). (FIG. 13.216. Left valve.)

FIG. 13.216

13.149 Pholadidae (see Petricolidae)

13.150 Semelidae (from 13.109)

Laterally compressed shell with pointed posterior and large siphons, large pallial sinus.

Cumingia tellinoides, tellin semele†, common cumingia. Thin shell, dull outside, glossy inside. Has spoonlike chondrophore inside umbo. Bo,Sa–SG,Su,df,20 mm (0.8 in), (AM97,G155,R786), ☺. (FIG. 13.217. Left valve.)

FIG. 13.217

13.151 Solecurtidae, Little Razors (from 13.99)

Rectangular shells that gape at both ends; two long, separate siphons, large pallial sinus; persistent periostracum.

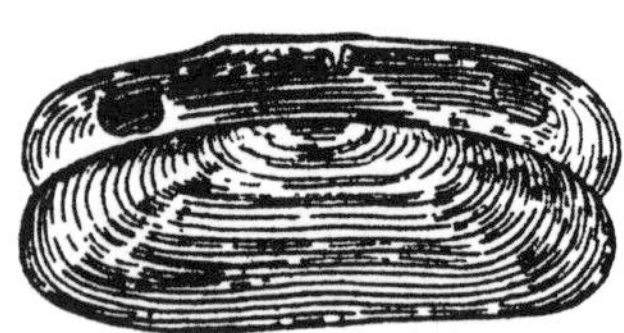

FIG. 13.218

Tagelus divisus, purplish tagelus†, purple razor. Shell purple inside with supporting rib perpendicular to long axis; thin and shiny, with chestnut brown periostracum outside. Vi,Mu–Sa,Li,ff,3.8 cm (1.5 in), (A160,AM93,G157,R788). (FIG. 13.218. Left valve.)

Tagelus plebeius, stout tagelus†, little razor, jackknife, or spit clam. Shell with easily eroded yellowish brown periostracum; white inside. Squirts water from burrow when disturbed. Anterior squared. Vi,Mu–Sa,Li+,ff,10.1 cm (4 in), (A160,AM93,G157,NM567,R789), ☺. (FIG. 13.219. a, right valve; b, normal orientation; siphons extend to sediment surface, digging foot beneath.)

FIG. 13.219 a b

13.152 **Solemyidae, Swimming Clams or Awningclams** (from 13.103)

Elongate shells with overly large, shiny, yellowish brown periostracum; rely primarily on sulfur-oxidizing chemosynthetic bacteria in gill filaments for sustenance; able to swim short distances.

Solemya borealis, boreal awningclam†. Grayish blue inside. Shell surface with radiating color pattern but actually smooth surface. More compressed than *S. velum.* More common. Ac–,Mu,Su,ot,6.3 cm (2.5 in), (A117,AM1,G144), ☺.

FIG. 13.220

Solemya velum, Atlantic awningclam†. Shell with slightly raised, broad ribs axially from umbo toward margins. Light brown, radial bands sometimes present, shiny outside: light purple or grayish inside. Bo,Sa–Mu,Li–,ot,3.8 cm (1.5 in), (A117,AM2,G144,NM534,R666). (FIG. 13.220. Right valve.)

13.153 **Solenidae, Razors** (from 13.99)

Shell very long and thin; smooth, glossy periostracum; large siphons, large pallial sinus.

1. Shell proportions: shape index = width/length
   - A. Shell highly *elongate,* shape index ≥ 5 — el
   - B. Shell relatively *intermediate* in length, shape index of 4–5 — in
   - C. Shell relatively *shorter,* shape index ≤ 4 — sh
2. Holding shell with hinged part uppermost, relative location of umbo
   - A. About *one-third* the distance from one end along the dorsal surface of shell — ⅓
   - B. Umbo close to the *end* of the shell — en
3. Reinforcing rib extends from umbo toward opposite side of interior surface of shell
   - A. Distinct *rib* present — ri
   - B. Distinct rib *absent* — x

13.154

Index	Umbo	Rib	Species
el	en	x	*Ensis directus,* Atlantic jackknife†, common straight-razor
in	en	x	*Solen viridis,* green jackknife†
sh	⅓	ri	*Siliqua costata,* Atlantic razor†, ribbed pod

FIG. 13.221

Ensis directus, Atlantic jackknife†, common straight-razor. Long, lateral, triangular area from umbo to posterior end. Umbo close to end of shell. Gaping valves (i.e., do not fit tightly together at ends). White with smooth shiny olive brown periostracum. Fast burrower; stands upright in sand. Can swim briefly using jet propulsion. *Cerebratulus* (Ph: Nemertea) and moon snails are major predators. Bo,Sa–Mu,Li–Su,ff,25.4 cm (10 in), (A152,AM98,G157,NM569,R762), ☺. (FIG. 13.221. a, left valve; b, normal orientation; siphons extend to sediment surface, digging foot beneath.)

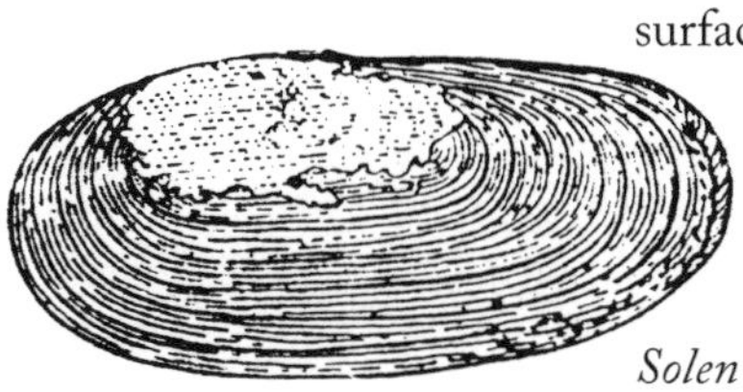

FIG. 13.222

Siliqua costata, Atlantic razor†, ribbed pod. Thin, oval shell; yellowish green periostracum; inside glossy and purplish white. Bo,Sa,Su,ff,5.1 cm (2 in), (A152,AM99,G157,NM568,R759). (FIG. 13.222. Left valve.)

Solen viridis, green jackknife†. Much smaller than *E. directus* and lacks triangular area. Umbo at end of shell. Shiny, greenish periostracum; inside glossy and purplish white. Long, fused, segmented siphons. Vi,Li,Su–Mu,ff,7.6 cm (3 in), (AM98,G157,R762). (FIG. 13.223. Right valve.)

FIG. 13.223

13.155 **Tellinidae, Tellins** (from 13.99, 13.109)

Smooth, laterally compressed, thin shells; long separate siphons for deposit feeding; somewhat pointed posteriorly; large pallial sinus for siphons.

1. Shape of posterior end of shell (viewed dorsally)
   - A. Posterior end slightly *twisted* toward the left side — tw
   - B. Posterior end *not* twisted (i.e., straight) — x
2. Shell proportion: shape index = width/length
   - A. Shell *elongate,* shape index > 1.5 — el
   - B. Shell more *rounded,* shape index < 1.5 — ro
3. Appearance of shell surface
   - A. Shell *shiny* white or translucent — sh
   - B. Shell *dull* white — du
4. Shape of indented portion of pallial line (the pallial sinus) on inside surface of shell
   - A. Top edge of pallial sinus roughly *parallel* to bottom edge of shell — pa
   - B. Top edge of pallial sinus *angled* with respect to bottom edge of shell — an

13.156

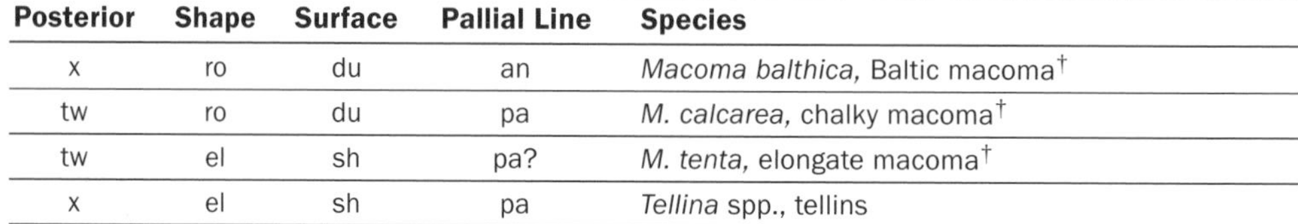

Posterior	Shape	Surface	Pallial Line	Species
x	ro	du	an	*Macoma balthica,* Baltic macoma†
tw	ro	du	pa	*M. calcarea,* chalky macoma†
tw	el	sh	pa?	*M. tenta,* elongate macoma†
x	el	sh	pa	*Tellina* spp., tellins

FIG. 13.224

Macoma balthica, Baltic macoma†. Chalky white to pinkish white with well-worn grayish periostracum; pinkish white interior. Duck/goose food. Bo,Sa–Mu,Es–Li–,df,3.8 cm (1.5 in), (A155,AM87,G156,NM564,R774), ☺. (FIG. 13.224)

Macoma calcarea, chalky macoma†. Shell chalky white, grayish periostracum on margins. Larger and more elongate than *M. balthica.* Ac–,Sa–Mu,Su,df,5.1 cm (2 in), (A155,AM87,G156,R775). (FIG. 13.225. a, right valve; b, dorsal view to show slight posterior twist to shell.)

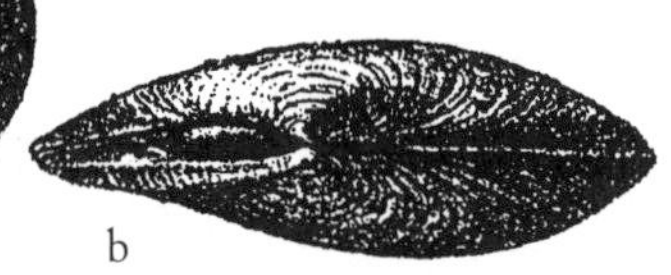

a b

FIG. 13.225

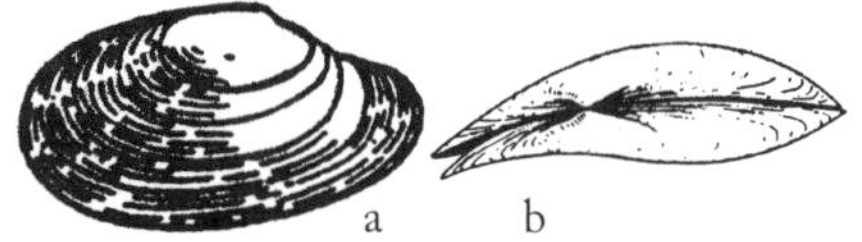

a b

FIG. 13.226

Macoma tenta, elongate macoma†. White with an iridescent sheen. Bo,Mu–Sa,Su,df,20 mm (0.8 in), (A156,AM88,G156,R775). (FIG. 13.226. a, left valve; b, dorsal view to illustrate slight twist to shell.)

FIG. 13.227

Tellina spp., tellin. Fragile pink to white shells. Burrow horizontally in silty sands. Prefers lower salinity. Siphons often bitten off by fish. For difficult species distinctions, see references. *Tellina agilis,* northern dwarf-tellin†, is the most commonly reported. Bo,Mu–Sa,Su,df,13 mm (0.6 in), (AM80,G153,NM537,R685), ☺. (FIG. 13.227. Left valve.)

13.157 **Thraciidae** (from 13.105)

Delicate, thin shell with irregular ventral contour due to axial ridge; large pallial sinus.

Thracia conradi, Conrad thracia†. Valves chalky white. Posterior end of shell with distinct but weak axial ridge. Ac–,Sa–Mu,Su,ff,10.1 cm (4 in), (A180,AM116,G161,R835). (FIG. 13.228. Left valve.)

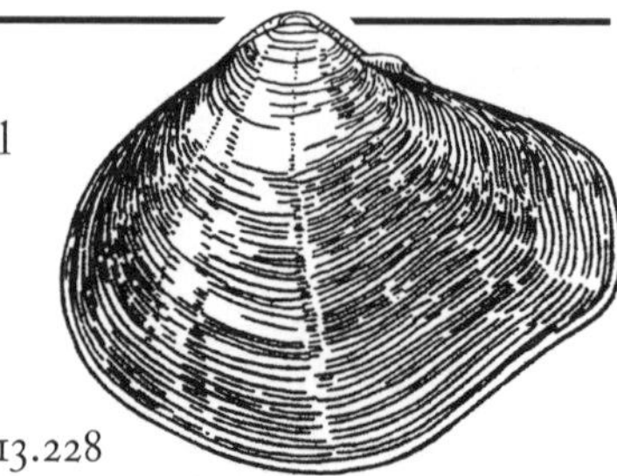

FIG. 13.228

13.158 **Thyasiridae, Cleftclams** (from 13.93)

Small, fragile shells with groove or cleft toward posterior edge.

Thyasira trisinuata, Atlantic cleftclam†. Deep groove from umbo to edge gives shell rippled posterior margin. Bo,Sa,Su–De,ff,12 mm (0.5). (AM47).

13.159 **Veneridae, Hard-Shelled Clams** (from 13.97)

Heavy, rounded shells with well-developed, and anteriorly bent umbos; heart-shaped lunule inscribed into shell anterior to umbo along midline; modest pallial sinus for siphons.

1. Crenulation (series of tiny teeth) along inside surface of ventral edge of shell (see FIG. 13.164)
   - A. *Crenulation* present — cr
   - B. Crenulation *absent* — x
2. Color of interior of shell
   - A. Interior *white* only — wh
   - B. Interior with at least some *purplish* tones, blotches, or borders — pu
3. Maximal shell length: provide *length* (in metric or inches) of largest specimen — __

13.160

Crenulation	Interior	Size	Species
cr	pu	4 mm (0.2 in)	*Gemma gemma,* amethyst gemclam†
cr	pu	11.4 cm (4.5 in)	*Mercenaria (=Venus) mercenaria,* northern quahog†
x	wh	5.1 cm (2 in)	*Pitar morrhuanus,* false quahog†

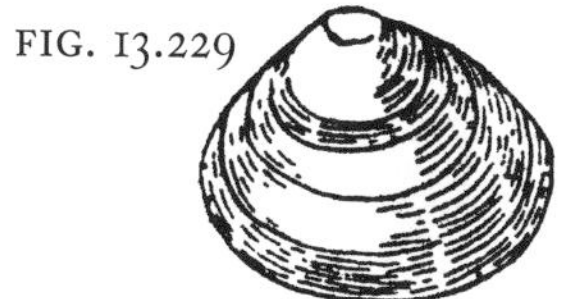
FIG. 13.229

Gemma gemma, amethyst gemclam†. Shiny valves are white to purplish gray; interior is white. Shallow bays to marshes and estuaries. Bo,Sa–SM,Li–Su,ff,4 mm (0.2 in), (A168,AM72,G153,R801), ☺. (FIG. 13.229. Left valve.)

Mercenaria (=*Venus*) *mercenaria,* northern quahog†, hardshell clam, little neck clam (up to 3.8 cm [1.5 in]), cherrystone clam (up to 5.1 cm [2 in]), chowder clam (up to 7.6 cm [3 in]). Thick solid shell; dull grayish white outside; interior, white with purplish border. Important commercially. Shell fragments served as money (wampum) for New England aboriginal people. Bo,Sa–Mu,Li–Su,ff,11.4 cm (4.5 in), (A165,AM62,G152,NM559,R806), ☺. (FIG. 13.230. Left valve.)

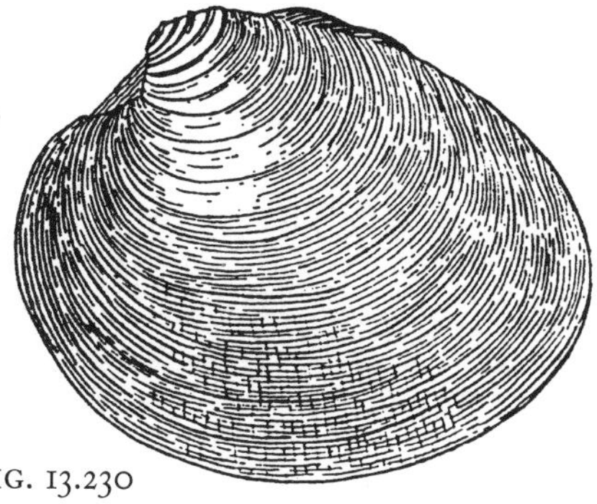

FIG. 13.230

FIG. 13.231

Pitar morrhuanus, false quahog†. Thinner than *Mercenaria.* Dull white with reddish brown patches; inside is white with no purple stains. Bo,Sa,Su,ff,5.1 cm (2 in), (A167,AM69,G152). (FIG. 13.231. Left valve.)

13.161 **Class Cephalopoda: Subclass Coleoidea, Squid and Octopus**

Squid and octopus are found in off-shore and subtidal locations, respectively. Squid are nektonic predators, sometimes visible in schools in the water column, deeper by day and more toward the surface at night. Octopi are benthic and solitary and are more apt to be encountered in dredged material, stomach contents of demersal fishes, or by divers. Regional squid are members of the Order Teuthoidea (shell as flattened, internal "pen"; eight arms and two longer tentacles) and Suborder Myopsida (a corneal membrane covers the eye). Octopods are in the Order Octopoda (soft body with eight arms) and the Suborder Incirrata (lack fins).

The deep-sea dwelling *Spirula* squid (Order Sepioidea) lives beyond the depth range of this manual. Its white, flat-coiled, chambered shell rinses ashore on southern beaches. This shell has been included among Gastropoda, with which it is most likely to be confused.

Additional references: (N) Nesis, K. N., 1987; (V) Vecchione, M., C.F.E. Roper, and M. J. Sweeney, 1989.

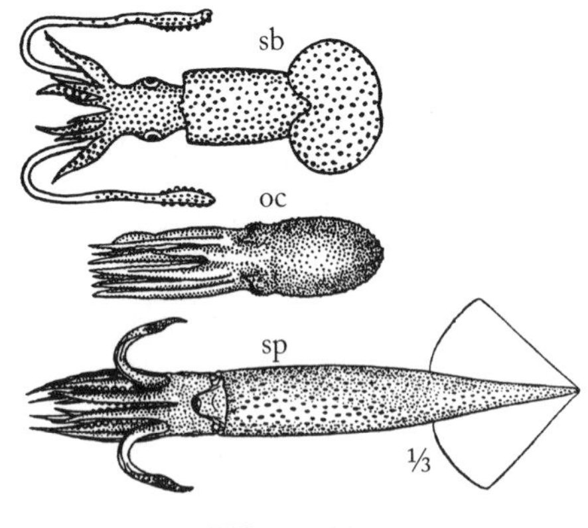

FIG. 13.232

1. Body shape (FIG. 13.232)
   - A. *Squid* shape (i.e., elongated with fins, with *pointed* posterior end) — sp
   - B. *Squid* shape with *blunt* posterior end — sb
   - B. *Octopus* shape (i.e., rounded body without fins) — oc
2. Size of fins (FIG. 13.232)
   - A. Fins about *a third* trunk length — ⅓
   - B. Fins nearly *half* trunk length — ½
   - C. Fins more than *three-fourths* trunk length — ¾
   - D. Fins *absent* — x
3. Presence of hornlike projection above eye, the supraocular cirrus
   - A. *Horn* present — ho
   - B. Horn *absent* — x
4. Presence of membranous covering over eye
   - A. *Membrane* present (myopsid eye) — me
   - B. Membrane covering eye *absent,* surrounding skin appears like an eye-lid (oegopsid eye) — x

13.162

Shape	Fins	Horn	Eye Cover	Species
oc	x	ho	me	*Bathypolypus arcticus,* offshore octopus†
oc	x	x	me	*Octopus vulgarus,* common Atlantic shore octopus†
sp	½	x	me	*Loligo pealei,* long-finned squid†, southern squid
sp	⅓	x	x	*Illex illecebrosus,* short-finned squid†, northern squid
sb	⅓	x	me	*Lolliguncula brevis,* brief squid†
sb	¾	x	x	*Semirossia tenera*

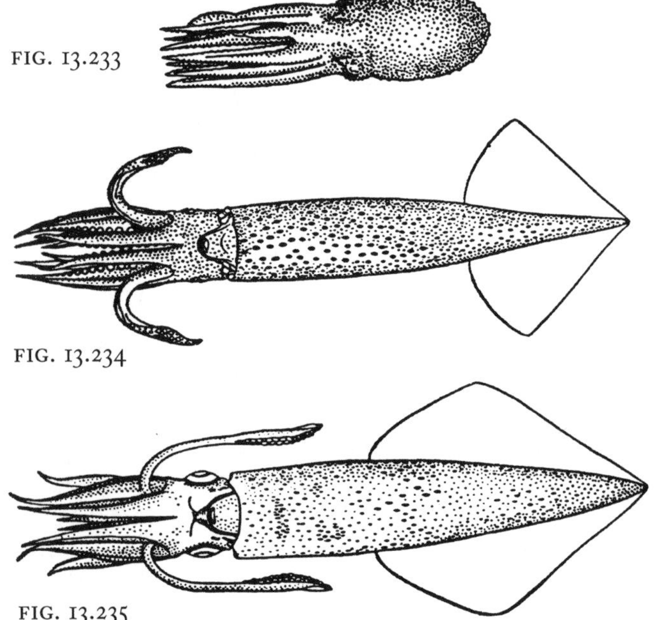
FIG. 13.233

FIG. 13.234

FIG. 13.235

Bathypolypus arcticus, offshore octopus† (Octopodidae). Body surface with wartlike bumps. Arms 1.5–3 times mantle length. Reproductive arm in males has a large paddle-shaped ending. Bo,Ha–Sa–Mu,Su–De,pr,61 cm (24 in), (G163,N315,V17). (FIG. 13.233. Anterior view.)

Illex illecebrosus, short-finned squid†, northern squid (Ommastrephidae). Darker color. Commercial species. Ac–,Pe,Ps–Pd,pr,38.1 cm (15 in), (AM303,G162,N234,NM578,V17), ☺. (FIG. 13.234. Posterior view.)

Loligo pealei, long-finned squid†, southern squid (Loliginidae). Commercially exploited for bait, food, and its giant axon for neurological research. Occasionally as far north as mid-Maine coast. Vi+,Pe,Ps,pr,61 cm (24 in), (AM302,G162,N147,NM577), ☺. (FIG. 13.235. Posterior view.)

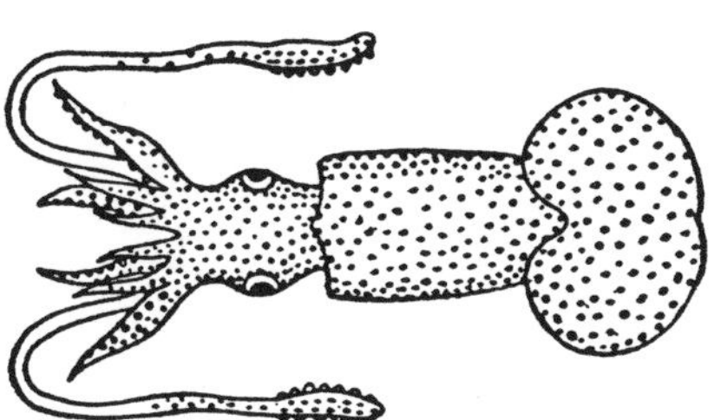

FIG. 13.236

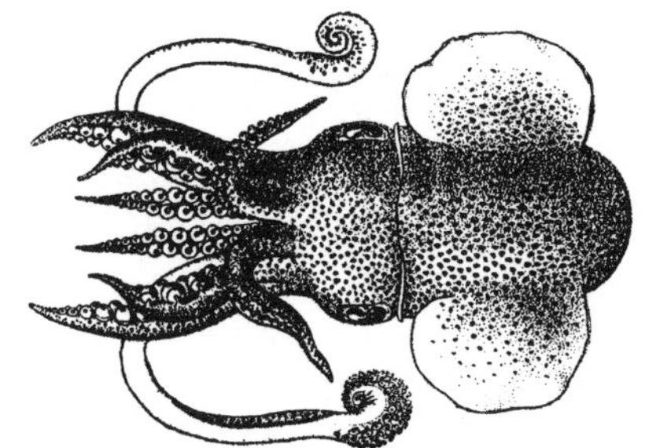

FIG. 13.237

Lolliguncula brevis, brief squid† (Loliginidae). Smaller, southern species with wide, short fins. Tolerates estuarine conditions. Ca+,Pe,Es–Ps,pr,22.9 cm (9 in), (AM303,GI62,NI46,NM577). (FIG. 13.236. Anterior view.)

Octopus vulgarus, common Atlantic shore octopus† (Octopodidae). Usually reddish brown; but color is variable. Body with small tubercles or warts. Arms 70–80% total length; webbed up to 20–25% of length. Barely extends into southern limit of range covered here. Vi,Ha–Sa–Mu,Su–De,pr,61 cm (24 in), (GI63,N304,NM579).

Semirossia tenera (Sepiolidae). Stubby trunk about equal in length to head and tentacles. Reported to be common in deeper waters from Maine to the Caribbean. Bo,Sa–Mu,De,pr,5 cm (1.9 in)-trunk, (VI6). (FIG. 13.237. Anterior view.)

Chapter 14 ■ Phylum Annelida

14.1 Segmented annelid worms are conspicuous members of marine, freshwater, and atmospheric communities. Members of the class Polychaeta (14.3) are mostly marine and are distinguished by possessing many pairs of lateral flaps or parapodia that usually include many embedded chaetae or setae. Often the head is adorned with sensory and food-gathering structures. They occur in a rich diversity of forms and in considerable numbers, reflecting their ecological importance. At the risk of defining the Oligochaeta (14.83) in negative terms, they lack parapodia, possess few setae per segment, and often are devoid of delicate, external cephalic structures. Terrestrial earthworms are well known to everyone, but comparatively small size and difficult distinguishing features leave aquatic oligochaetes less thoroughly studied. Nonetheless, they are abundant, important members of marine and freshwater communities. Finally, the class Hirudinea (14.88) includes leeches that may function as predators, ectoparasites, or commensals in marine and freshwater communities. Like the oligochaetes, these dorsoventrally flattened animals lack parapodia and sensory appendages on the head. Their segments are subdivided by annuli, and common regional species possess two suckers for attachment and locomotion.

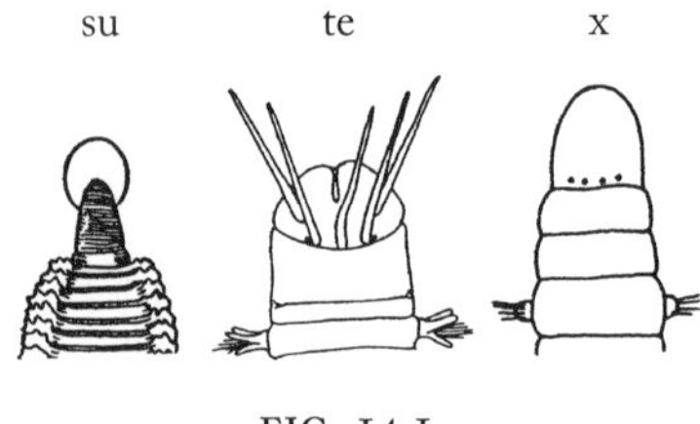

FIG. 14.1

1. Conspicuous features of body ends (FIG. 14.1)
   - A. A *sucker* occurs at each end of the worm; anterior sucker usually larger — su
   - B. *Tentacles,* lobes, long setae, or other structures (but not suckers) occur on anterior end — te
   - C. Anterior end variously rounded or pointed, but *without* obvious adornments — x
2. Segmentation
   - A. Well-developed *segmentation* — se
   - B. Body may be annulated but obvious segmentation *absent* — x
3. Presence of segmental parapodia or lateral flaps
   - A. *Parapodia* present — pa
   - B. Parapodia *absent* — x

14.2

Ends	Body	Parapodia	Group
su	x	x	Cl: Hirudinea (leeches), 14.88
te	se	pa,x	Cl: Polychaeta (seaworms, etc.), 14.3
x	se	pa	Cl: Polychaeta (seaworms, etc.), 14.3
x	se	x	Tips of some abdominal setae (last half of body) with curved and forked tips covered by tiny chitinous envelope (i.e., "hooded hooks"); some distinction between anterior thorax and posterior abdominal region: Cl: Polychaeta, Capitellidae, 14.24 OR No abdominal setae with hooded tips; body not regionalized: Cl: Oligochaeta, 14.83

14.3 Class Polychaeta

Polychaetes are abundant and diverse. The serious student of marine biology will spend a good deal of time working on the identification of polychaete worms. Many are fragile and should be isolated in protective containers in the field. Keep careful notes regarding habitat characteristics in which these worms are found. The activities of many polychaetes may interfere with the observation of characteristics necessary for identification. In such cases, it may be beneficial to immobilize them temporarily by treating them with an appropriate anesthetizing agent (refer to Appendix). Examine the anterior portion of the body most carefully. This may be difficult because headless pieces of these delicate worms are frequently encountered. (Quantitative studies should only count heads. Otherwise, a single, fragmented worm could be counted several times.) The length, number, and deployment of appendages found there is critical for identification. Use a scalpel or razor blade to trim off several of the parapodia or lateral flaps found on each segment to observe details of their morphology. Often study of features of polychaetes requires observation at the dissecting-microscope level.

A word on polychaete terminology is necessary. The anterior end of most polychaetes includes two important elements. Anteriormost is a partial segment that typically appears as an anterior lobe extending over the mouth area. This lobe, the *prostomium,* meaning "before the opening," houses the cephalic ganglia or brain of the worm and often is adorned with several types of sensory structures. Tapering antennae may occur, as may food-gathering protuberances called *palps.* One to several pairs of ocelli or eyespots are often found here, too. The next body segment that surrounds the mouth is termed the *peristomium,* literally "around the mouth." This segment, and sometimes others immediately following it, often lack the parapodia found on most subsequent body segments. When tapering, antennalike structures extend from the peristomium, they are called *tentacular cirri.* Many polychaetes can evert the anterior portion of the digestive tract, the *proboscis* or *pharynx,* through the mouth opening. Its shape and armaments can be useful indentifying characters. The second true segment or *buccal segment* follows. It may or may not be modified through loss of lateral parapodia that occur on most subsequent trunk segments. In some polychaetes, a thicker, anterior region, the *thorax,* is differentiated from a more slender, posterior portion, the *abdomen.* Fingerlike *anal cirri* may be associated with the terminal segment, the *pygidium.*

Construction of the *parapodium* or lateral flaplike appendages can also be useful. Parapodia are usually divided into dorsal, *notopodial lobes* and more ventral *neuropodial lobes.* A *biramous* parapodium is one on which both lobes are well developed, whereas one reduced to a single lobe is *uniramous.* Notopodial lobes often serve as a base for *dorsal cirri* of various shapes, and neuropodia often support *ventral cirri.* Gills or *branchiae* may occur attached to parapodial lobes. Heavy support rods, the *aciculae,* are often embedded within the parapodium but are visible from the outside.

Finally, there exists a bewildering array of descriptive terms for the many setae or bristles ("poly + chaeta" means "many bristles") extending from the parapodia. Although these terms are required for some species identifications, they are not for many of the common polychaetes. Refer to that vocabulary when consulting more advanced references for confirmation of identifications made here.

Additional references: (F) Fauchald, K., 1977; (P) Pettibone, M. H., 1963.

FIG. 14.2

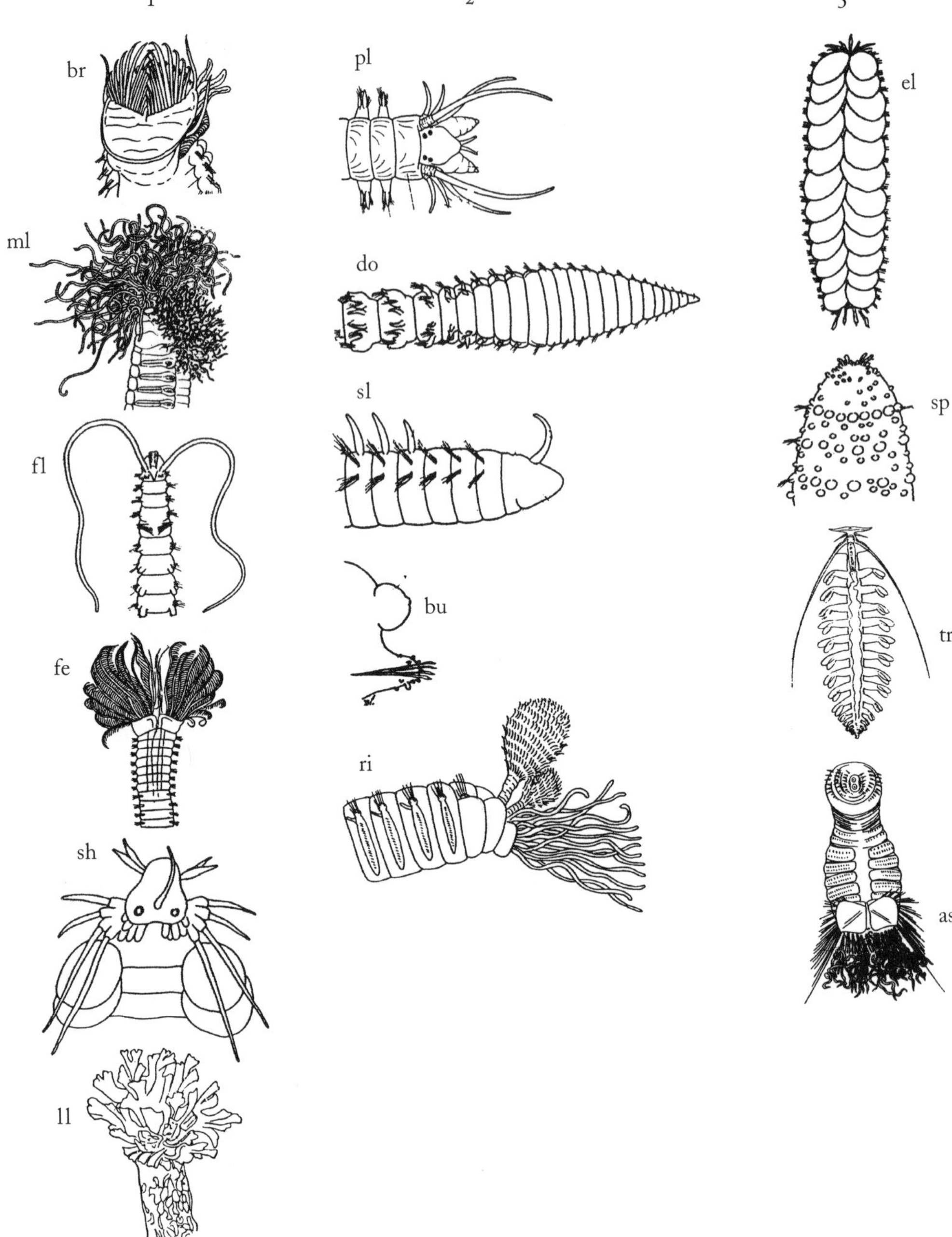

1. Description of dominant head appendages (FIG. 14.2)
   - A. Obvious stiff *bristles* or setae (the worm may or may not have cluster of tentacles as well) br
   - B. *Many* (four or more) *long* (at least three times the head length) threadlike tentacles on head and/or anterior end ml
   - C. *Fewer* than four *long*, often grooved tentacles on the head or anterior end fl
   - D. Appendages with branches, gives *feathery* appearance fe
   - E. Antennae or tentacles are *short* (less than 3 times the head length) but obvious sh
   - F. Head without tentacles or antennae but *body* bearing many, colored, threadlike *tentacles* bt
   - G. Head with flattened, *leafy lobes* surrounding the mouth ll
   - H. Head and body with bristles and tentacles *absent* x

2. Shape of parapodia or lateral flaps on anterior third of the body (does not include tufts of setae only) (FIG. 14.2)
   A. Parapodia distinct, flaplike, and *prominent* or at least obvious, arising *laterally* — pl
   B. Parapodia obvious, extend laterally on anterior half of body but *dorsally* on posterior half — do
   C. Parapodia reduced to setal tufts beneath one or two setaless *slender* fingerlike lobes and/or gills — sl
   D. Parapodia reduced to setal tufts beneath one or more *bulbous* growths — bu
   E. Parapodia reduced to distinct lateral *ridges* — ri
   F. Parapodia present but shape *otherwise* than described above — ot
   G. Parapodia *absent*, although tufts of setae alone may be present — x
3. Other conspicuous features of body construction (FIG. 14.2)
   A. Dorsal surface obscured by paired scales or *elytra* (in some, elytra may be partially hidden by a felty layer of bristles) — el
   B. Worm covered with rows and scattered *spherical papillae* — sp
   C. Planktonic worm, *transparent* with unusually large parapodia — tr
   D. Pair of large *anal shields* located near tentaclelike gills — as
   E. Worm encased in hard *calcareous* tube attached to substrate — ca
   F. Worm within tube made of *sand* grains — sa
   G. Worm with such distinctive features *absent* — x

14.4

Appendages	Parapodia	Features	Group
sh	pl	el	Polynoidae, Aphroditidae, Pholoidae, Sigalionidae (scaleworms), 14.58
sh	pl	x	Polychaete Group 1, 14.5
sh	ri	x	Ampharetidae, 14.17
fe	x,ri	ca	Serpulidae and Spirorbidae (featherduster worms), 14.64
fe	x,ri	x	Polychaete Group 2, 14.7
br	pl,ri	sa	Pectinariidae (ice-cream-cone worms), 14.54
br	pl, ri	x	Polychaete Group 3, 14.9
br	x	as	Sternaspidae, 14.73
bt	x	x	Cirratulidae, 14.27
ml	pl	x	Polychaete Group 1, 14.5
ml	ri,x	x	Polychaete Group 4, 14.11
fl	pl	x	Polychaete Group 1, 14.5
fl	ri,sl,x	x	Spionidae, 14.68
fl	ot	x	Chaetopteridae, 14.26
fl	pl	tr	Tomopteridae, 14.82
ll	ri	sa,x	Maldanidae and Oweniidae, 14.40
x	pl	x	Polychaete Group 5, 14.13
x	ri,x	x	Polychaete Group 6, 14.15
x	ri	sa	Maldanidae and Oweniidae, 14.40
x	bu	sp	Sphaerodoridae, 14.67
x	sl	x	Slender body, segments obvious: Paraonidae, 14.52 OR Chubby body, segments indistinct: Opheliidae, 14.48
x	do	x	Orbiniidae, 14.50

14.5 Polychaete Group 1

(Apistobranchidae, Eunicidae, Hesionidae, Nereididae, Onuphidae, Phyllodocidae, Syllidae) (from 14.4)

Polychaetes with prominent lateral parapodia and several, obvious anterior appendages.

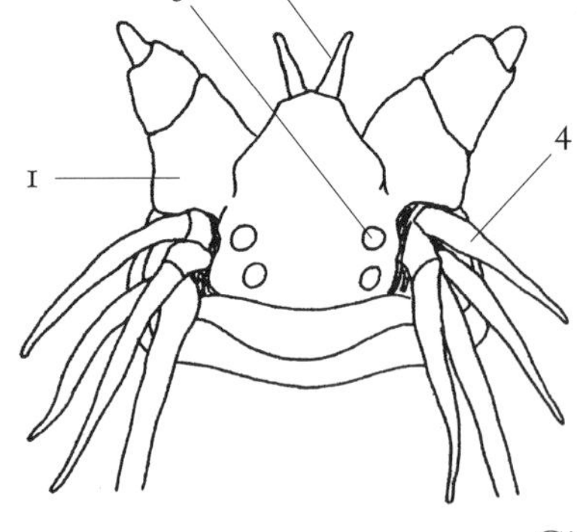

FIG. 14.3

1. Palps associated with the prostomium (palps are lobelike or thick, tentaclelike structures involved in food capture or manipulation) (FIG. 14.3)

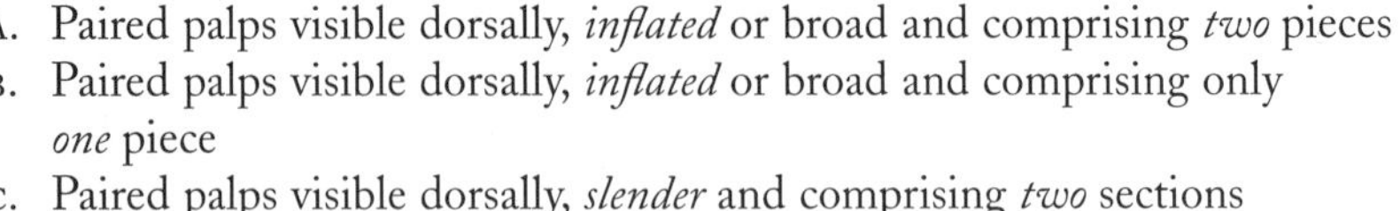

   A. Paired palps visible dorsally, *inflated* or broad and comprising *two* pieces — i2*
   B. Paired palps visible dorsally, *inflated* or broad and comprising only *one* piece — i1
   C. Paired palps visible dorsally, *slender* and comprising *two* sections — s2
   D. Paired palps visible dorsally, *slender* (although distinctly thicker than tentacles) and comprising a *single* section — s1
   E. Paired palps *ventral,* not visible dorsally, appear as two swollen *lips* — vl
   F. Palps *not* paired, but fused to form a single *lobe,* visible dorsally — xl
   G. Palps *absent* entirely — x

2. Gills covering a portion of dorsal surface

   A. Dorsal surface of first one third of body at least partially obscured by feathery *gills* — gi
   B. Dorsal surface mostly exposed; gill cover *absent* — x

3. Number of anterior segments without parapodia: provide *number* of parapodium-free segments — ___
4. Tentacular cirri on peristomium (the first full segment): provide the *number* of tentacular cirri (FIG. 14.3) — ___*
5. Number of ocelli (eye spots) on the prostomium: provide *number* or ocelli (FIG. 14.3) — ___*
6. Antennae on prostomium (anteriormost lobe): provide the *number* of antennae on prostomium (FIG. 14.3) — ___*

14.6

Palps	Gills	Without Parapodia	Cirri	Eyes	Antennae	Family
x	x	1–3	4–8	2	4–5	Phyllodocidae, 14.56
x	x	2	x	2	1,3,5	Eunicidae (in part), 14.30
x	x	0	x	0,4	2–4	Polychaete Group 5, 14.13
s2	x	3–6	4–6,9–16	2–4	1–3	Hesionidae, 14.36
s1,x	x	0	0,2	0	0	Apistobranchidae, 14.20
i2	x	0	6–8	4	2	Nereididae, 14.44
i1,xl	x	0,1	2–4	4	3	Syllidae, 14.74
i1	x	2	0,2	2	1,3,5	Eunicidae (in part), 14.30
i1	gi	2	0,2	2	1,3,5	Eunicidae (in part), 14.30
vl	gi,x	1	0,2	0,2	5	Onuphidae (in part), 14.46

14.7 Polychaete Group 2

(Sabellidae, Serpulidae, Spirorbidae, Featherduster Worms) (from 14.4)

Polychaetes with branched, pinnate, or feathery anterior appendages.

FIG. 14.4

1. Composition of tube (FIG. 14.4)
   A. *Coiled calcareous* tube; hard and dirty white — cc*
   B. *Irregularly* shaped *calcareous* tube; hard and dirty white — ic
   C. *Leathery* tube; flexible, sandy or muddy — le
   D. Tube *absent* — x
2. Presence of operculum (pluglike structure, located among the feeding tentacles; seals the tube as worm withdraws) (FIG. 14.4)
   A. *Operculum* present — op*
   B. Operculum *absent* — x

14.8

Tube	Operculum	Family
ic	op	Serpulidae, 14.64
cc	op	Spirorbidae, 14.72
le,x	x	Sabellidae, 14.61

14.9

Polychaete Group 3
(Flabelligeridae, Pectinariidae, and Sabellariidae) (from 14.4)

Polychaetes with obvious setae or bristles as dominant anterior appendages; with parapodia either well developed or reduced to ridges.

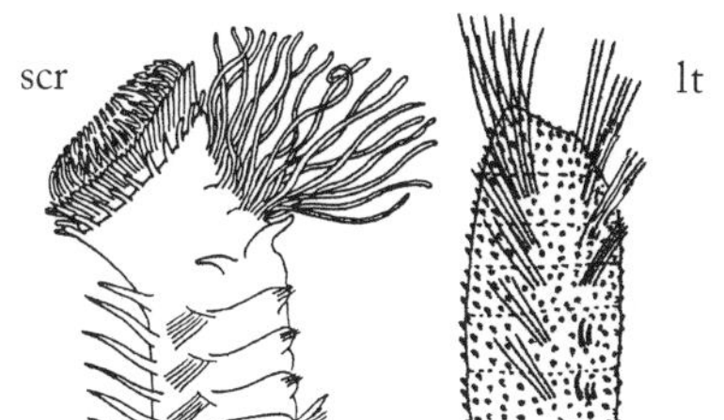

FIG. 14.5

1. Arrangement of bristles on head (FIG. 14.5)
   - A. *Stout* bristles arranged in two *concentric rings* around head — scr
   - B. *Stout* bristles in two *diagonal rows* across head — sdr
   - C. *Long* bristles in *tufts* (usually two) on head — lt
2. Tube construction
   - A. Symmetrical *conical* tube of sand grains; tube unattached to substrate or to other tubes — co
   - B. *Irregular* sand-grain tubes cemented to objects or to one another — ir
   - C. *No* tube, but body may be encased in jellylike coating — x

14.10

Bristles	Tube	Family
sdr	co	Pectinariidae (ice-cream-cone worms), 14.54
scr	ir	Sabellariidae, 14.60
lt	x	Flabelligeridae, 14.32

14.11

Polychaete Group 4
(Ampharetidae, Cirratulidae, Sternaspidae, Terebellidae, and Trichobranchidae) (from 14.4)

Polychaetes whose anterior end is dominated by many long tentacles; with parapodia reduced to ridges or lacking.

1. Location of tentacles
   - A. Tentacles arise from *body* segments — bo
   - B. Tentacles restricted to *head* — he
   - C. Tentacles are actually gills located *posteriorly* adjacent to a pair of anal shields or plates — po
2. Retractility of tentacles
   - A. Tentacles can be *retracted* into mouth when animal is disturbed — re
   - B. Tentacles are *not* retractile upon disturbance — x
3. Shape of gills as (red) colored structures on head, behind tentacle mass, or along body (FIG. 14.6)
   - A. Gills are *branched* — br
   - B. Gills are *simple,* not branched — si
   - C. Gills are *absent* or missing — x
4. Number of pairs of gills (colored structures just posterior to tentacles)
   - A. Provide *number* of pairs of gills — __
   - B. Gills *absent* — x

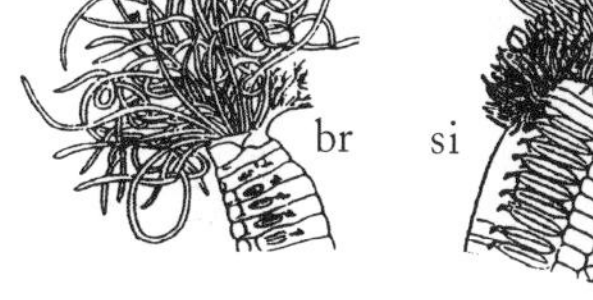

FIG. 14.6

14.12

Tentacles	Retractility	Gill Shape	Gill Pairs	Family
he	re	si	3–4	Ampharetidae, 14.17
he	x	br,x	1–3,x	Terebellidae and Trichobranchidae, 14.80
bo	x	si,x	x,>6	Cirratulidae, 14.27
po	x	si	>6	Sternaspidae, 14.73

14.13 Polychaete Group 5

(Glyceridae and Goniadidae, Lumbrineridae and Arabellidae, Nephtyidae) (from 14.04)

Polychaetes with anterior appendages as tiny antennae or lacking entirely; parapodia prominent.

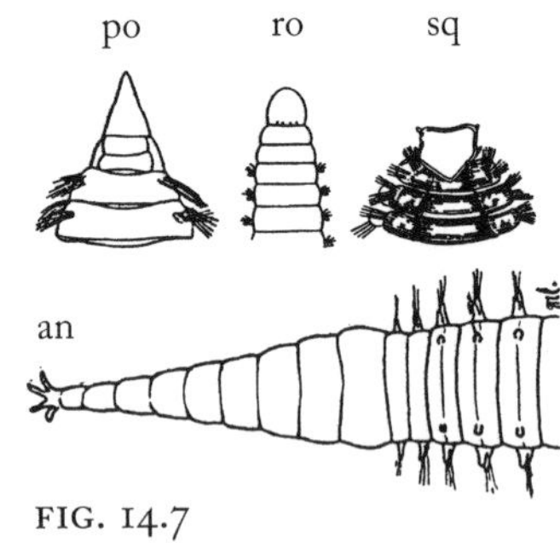

FIG. 14.7

1. Shape of anterior end of worm (FIG. 14.7)
   - A. Head is *pointed* — po
   - B. Head is *rounded* or conical — ro
   - C. Head is *squared* off or shield-shaped — sq
2. Presence of minute antennae on prostomium (look carefully) (FIG. 14.7)
   - A. Tiny *antennae* present at anterior tip or along anterior margin of head — an
   - B. Minute antennae (and all other head appendages) *absent* — x
3. Shape of parapodia
   - A. Parapodia *uniramous* (basically a single lobe) — un
   - B. Parapodia *biramous* (divided into distinct upper [notopodial] and lower [neuropodial] lobes) — bi

14.14

Head	Antennae	Parapodia	Family
po	an	bi	Glyceridae (bloodworms) and Goniadidae (chevronworms), 14.34
po,ro	x	un	Lumbrineridae and Arabellidae, 14.38
sq	an	bi	Nephtyidae, 14.42

14.15 Polychaete Group 6

(Arenicolidae, Capitellidae, Cossuridae, Maldanidae, and Scalibregmidae) (from 14.4)

Polychaetes lacking anterior appendages and with parapodia reduced to ridges or lacking entirely.

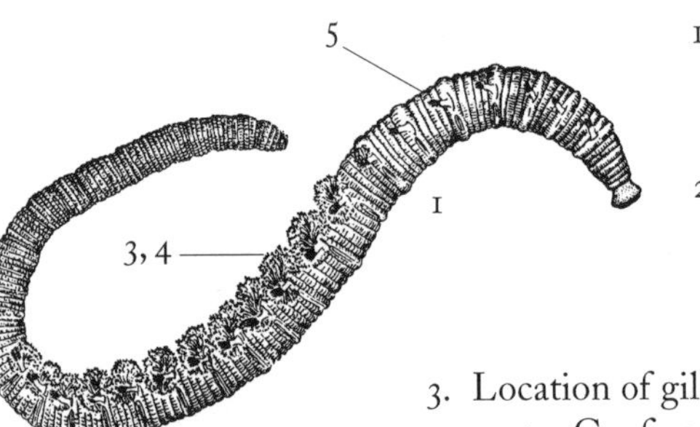

FIG. 14.8

1. Segment shape (FIG. 14.8)
   - A. Segments clearly *longer than wide* — l>w
   - B. Segments *not* longer than wide — x*
2. Shape of anterior end (FIG. 14.8)
   - A. Head slanted or *squared* off — sq
   - B. Head distinctly *pointed* — po
   - C. Head *rounded* — ro*
3. Location of gills or branchiae or gilllike structures (FIG. 14.8)
   - A. Confined to not more than *anterior* third of the body — an
   - B. Confined to *posterior* two-thirds of the body — po*
   - C. A *single* slender tentacle or palp arises middorsally, three to six segments back from the head — 1
   - D. Gilllike structures *absent* — x
4. Gill shape (FIG. 14.8)
   - A. Gills *branched* — br*
   - B. Gills *absent* — x
5. Segments subdivided by superficial annuli (FIG. 14.8)
   - A. Segments with distinct *annulations* or superficial rings on each distinct segment — an*
   - B. Segments distinct but annulations *absent* or infrequent — x

14.16

Segments	Head	Gill Location	Gill Shape	Annuli	Family
l>w	sq	x	x	x	Maldanidae and Oweniidae, (bamboo worms), 14.40
x	sq,ro	an	x	an	Scalibregmidae, 14.63
x	ro	po	br	an	Arenicolidae (lugworms), 14.22
x	po	1	x	x	Cossuridae, 14.29
x	po	x	x	x	Capitellidae, 14.24

Families of Polychaeta (arranged alphabetically)

14.17

Ampharetidae (Spaghetti-Mouth Worms) (from 14.4, 14.12)

Body in two parts: thorax slightly inflated, abdomen tapering; retractile tentacles that can be withdrawn into mouth; some with golden setae on prostomium. Build muddy tubes. (FI21)

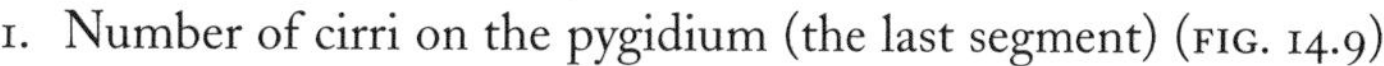

1. Number of cirri on the pygidium (the last segment) (FIG. 14.9)
   - A. Provide *number* of pygidial cirri — __*
   - B. Pygidial cirri *absent* — x
2. Shape of tentacles associated with the mouth (FIG. 14.9)
   - A. Tentacles *pinnate* with tiny side branches — pi
   - B. Tentacles *simple,* no branches — si*
3. Number of segments forming the more slender, abdominal region: provide *number* of abdominal segments (FIG. 14.9) — __
4. Presence of bundles of golden setae on prostomium
   - A. *Setae* present — se
   - B. Setae *absent* — x

FIG. 14.9

14.18

Cirri	Tentacles	Abdominal Segments	Setae	Species
10–14	pi	12	se	*Ampharete acutifrons*
2	pi	12–14	se	*A. arctica*
2	pi	?	se	*Asabellides oculata*
x	si	>50	x	*Melinna cristata*
x	pi	13	se	*Anobothrus gracilis*
x	si	20	se	*Hobsonia (=Hypaniola) florida (=grayi)*

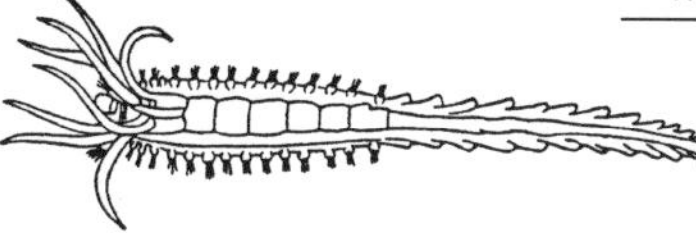

FIG. 14.10

Ampharete acutifrons. Thin muddy tube. Less common than *A. arctica.* Bo,Sa–Mu,Su,df,25 mm (1 in), (FI25,GI91). (FIG. 14.10)

Ampharete arctica. Stiff, muddy tube. Bo,Sa–Mu,Su,df,25 mm (1 in), (FI25,GI91).

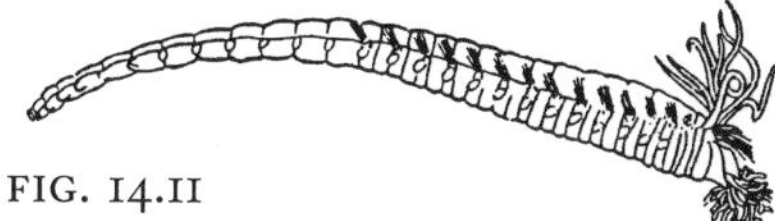

FIG. 14.11

Anobothrus gracilis. Ac–,Mu?,Su,df,3.8 cm (1.5 in), (FI25). (FIG. 14.11)

Asabellides oculata. Vi,Sa–Mu,Es–Su,df,20 mm (0.8 in), (FI25,GI91).

Hobsonia (=Hypaniola) florida (=grayi). Ac?,Mu?,?,df,17 mm (0.7 in), (FI25). (FIG. 14.12)

FIG. 14.12

Melinna cristata. Lower half of gills fused in groups by membrane. Yellowish-pink, gills red with green patches. Bo,Mu–SG,Es–Li–Su,df,5.1 cm (2 in), (FI26,GI91). (FIG. 14.13)

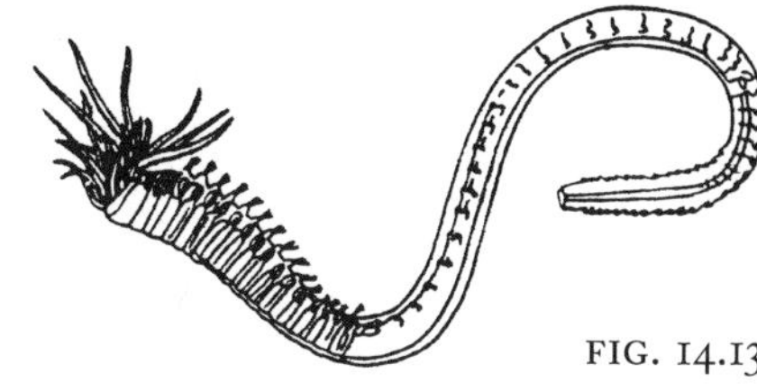

FIG. 14.13

14.19 **Aphroditidae** (see 14.58)

14.20 **Apistobranchidae** (from 14.6)

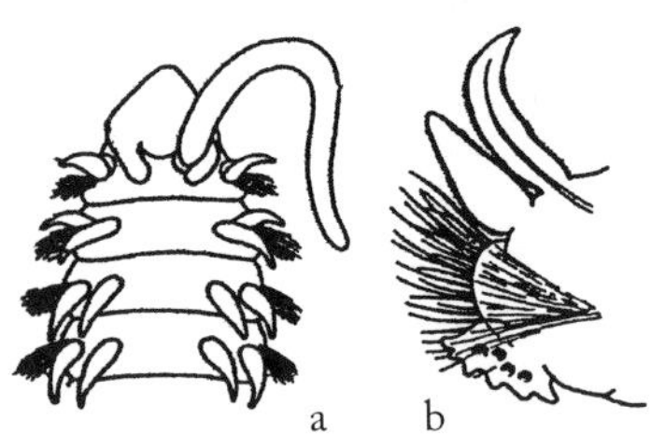

FIG. 14.14

Fragile body; prostomium with two long (but easily broken) palps. Body divided into two regions, a thicker thorax and a more slender abdomen. Neuropodia (lower portion of parapodia) on segments 4 and 7 modified as lobed flap with small papillae.

Apistobranchus tullbergi. Small, delicate, but can be abundant in deep water; tube dwellers. Ac,Mu,Su,df,13 mm (0.5 in), (F22,P295). (FIG. 14.14. a, anterior end; b, parapodium.)

14.21 **Arabellidae** (see 14.38)

14.22 **Arenicolidae, Lugworms** (from 14.16)

Head rounded, lacks appendages. Thick body in three regions: middle one with branched gills; proboscis rounded, unarmed; parapodia reduced, setal bundles only. Each segment further annulated. L-shaped burrow up to 30.5 cm (12 in) deep in sandy mud; tail end marked by large coils of fecal castings; slight depression marks head shaft. Nonselective deposit feeders. (F37)

1. Number of setigers (seta-bearing segments): provide *number* of setigers —
2. Number of pairs of gills (branchiae): provide *number* of gill pairs —
3. Basic body color
   A. *Greenish black* gb
   B. *Pinkish tan* pt

14.23

Setigers	Gills	Color	Species
17	11	pt	*Arenicola brasiliensis,* lugworm
16–18	11	gb	*A. cristata,* lugworm
19	12-13	gb	*A. marina,* northern lugworm

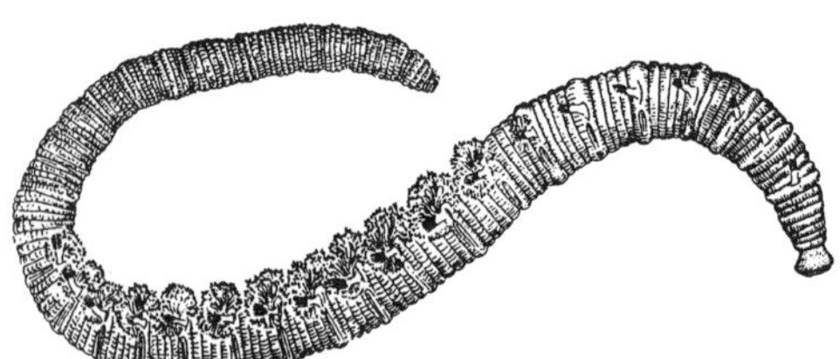
FIG. 14.15

Arenicola brasiliensis, lugworm. Less robust than other two species. May be smaller growth form of *A. cristata.* Vi,Sa–Mu,Li–,df–ff,15.2 cm (6 in), (F37,G185). (FIG. 13.8)

Arenicola cristata, lugworm. Vi,Sa–Mu–Gr,Li–,df,30.5 cm (12 in), (F37,G185,NM425), ☺. (FIG. 13.9)

Arenicola marina, northern lugworm. Ac–,Sa–Mu,Li–,df,20.3 cm (8 in), (F37,G185), ☺. (FIG. 14.15)

14.24 **Capitellidae, Threadworms** (from 3.2, 3.4, 14.16)

Earthwormlike, often red to purple; parapodia highly reduced but biramous; no head appendages; segments poorly defined; proboscis globular. Body divided between short anterior end with stubby segments and a longer posterior end with elongate segments. Networks of tubes in mud; tolerate organic-rich, low-oxygen, and polluted conditions. (F31)

1. Number of segments in mature adults with simple, single-pointed or capillary setae only (may have additional segments with more complex setae: provide *number* of segments —

2. Number of pairs of genital pores, appear as small dorsally placed openings
   - A. *Single* pair of pores between the eighth and ninth seta-bearing segments — 1
   - B. *Four* pairs of genital pores on the ninth through twelfth seta-bearing segments — 4
   - C. *Seven to 12* pairs of pores found on first several segments of posterior portion of body — 7–12
3. Presence of gills (branchiae)
   - A. *Gills* present — gi
   - B. Gills *absent* — x
4. Presence of thin, anal cirrus, lacking setae
   - A. *Anal cirrus* present — ac
   - B. Anal cirrus *absent* — x

14.25

Segments	Genital Pores	Gills	Anal Cirrus	Species
x	?	x	ac	*Amastogos caperatus*
4	?	x	ac	*Mediomastus* spp.
5	4	gi	x	*Heteromastus* spp.
5–7	1	x	x	*Capitella* spp.
11	7–12	gi,x	x	*Notomastus* spp.

Amastigos caperatus. Abdominal segments with unusual dorsal setae: wrinkled and hooded (tip covered by membrane) hooks (tips recurved). Bo?,Sa–Mu?,Su?,df,?.

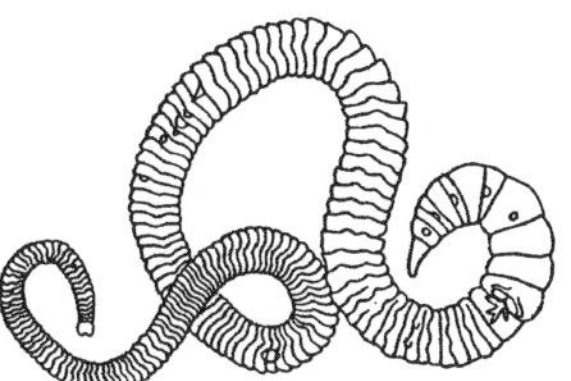

FIG. 14.16

Capitella spp. Marine–brackish; subsurface, any depth; any bottom type. Mucous galleries; an indicator species, tolerant to pollution and low oxygen. Several species may be intermixed in literature references to *C. capitata.* Bo,Mu–Sa,Es–Li–Su,df,10.1 cm (4 in), (F33,G184), ☺. (FIG. 14.16)

Heteromastus spp. *Heteromastus filiformis* is the species most frequently listed. Very slender. Subsurface burrower. Bo,Sa–Mu–SG,Es–Li–Su,df,5.1 cm (2 in), (F34,G184).

Mediomastus spp. *Mediomastus ambiseta* can be abundant especially in mussel beds. Widespread; reported from California and Florida as well as New England. Bo,Ha–Sa–Mu,Li–Su,df,?, (F34).

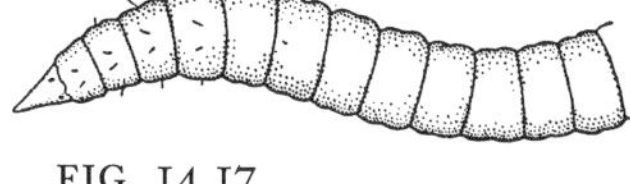

FIG. 14.17

Notomastus spp. *Notomastus latericeus* is most often listed. Abdominal area with a pair of parallel dorsal grooves. Present especially in shallow sand flats with lots of organic debris. Bo,Sa–SG,Es–Li–,df,30.5 cm (12 in), (F34,G184). (FIG. 14.17)

14.26 **Chaetopteridae, Parchmentworms** (from 14.4)

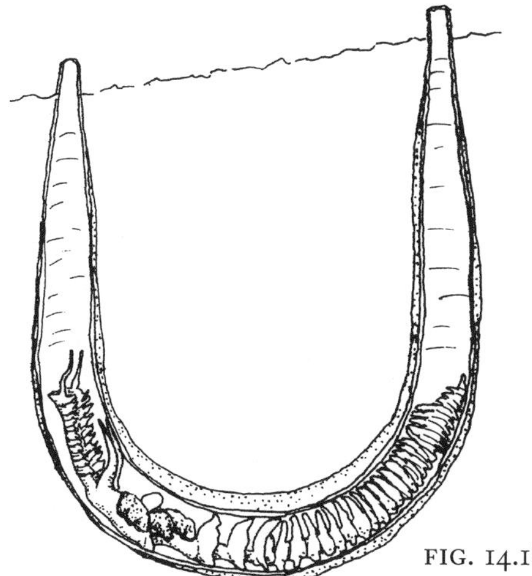

FIG. 14.18

Highly modified body divided into three distinct regions. Parapodia often highly modified. Produce parchmentlike tubes. (F28)

Chaetopterus variopedatus (parchmentworm). Body light in color. Head broad and flat; two tentacles followed by nine similar segments with ciliated gutter along dorsal surface. Next section, five segments: first segment with two large "wings" held dorsally, last three segments with flaplike rounded "fans." Final section, small, similar segments. Lives in narrow-mouthed, U-shaped, parchment tubes in mud. May release bioluminescent material from tube upon disturbance. Check for commensal crabs. Vi+,Sa–Mu–SG,Es–Su,ff,25.4 cm (10 in), (F28,G188). (FIG. 14.18. Worm in tube, lateral view.)

Spiochaetopterus oculatus is also found in our area but keys out with Spionidae (14.68).

14.27 **Cirratulidae, Fringeworms** (from 14.4, 14.12)

Cylindrical body; prostomium pointed. Parapodia biramous but reduced; many with one or more pairs of palps; all with tentacles (grooved) and tentaclelike gills (not grooved) on anterior segments. Some build muddy tubes attached to rocks or shell. (F29)

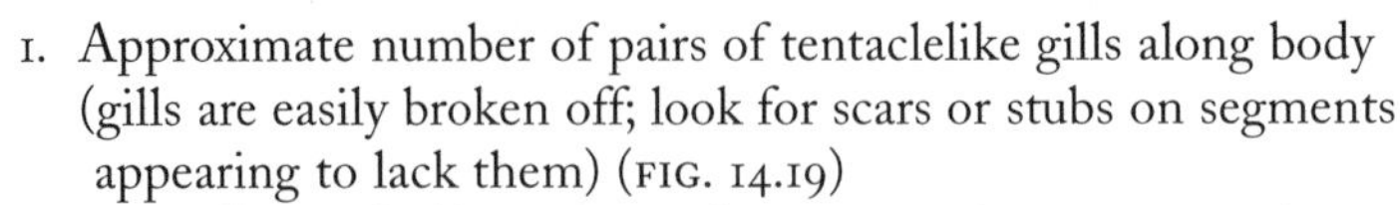

1. Approximate number of pairs of tentaclelike gills along body (gills are easily broken off; look for scars or stubs on segments appearing to lack them) (FIG. 14.19)
   - A. Pairs of gills on fewer than 10 *anterior segments* only — as*
   - B. Pairs of gills on virtually every segment up to the *entire body* — eb
2. Number of gill-less or tentacle-less segments directly behind the prostomium (FIG. 14.19)
   - A. Provide *number* of gill-less segments — —
   - B. Segments *unidentifiable* — x
3. Presence of two large, anterior, grooved palps, in addition to and distinctly thicker than, threadlike gills (FIG. 14.19)
   - A. Such *palps* present — pa
   - B. Such palps *absent* — x*
4. Presence of eyespots (look carefully; they may be located very close to the posterior margin of the prostomium) (FIG. 14.19)
   - A. *Eyes* present on prostomium as row or clumps of several tiny dark spots — ey
   - B. Eyes *absent* — x

FIG. 14.19

14.28

Gill Pairs	Gill-less Segments	Palps	Eyes	Species
as	0,x	pa	x,ey	*Dodecaceria:* Associated with the coral, *Astrangia* or shells; south of Cape Cod: *D. coralii* OR Associated with the coralline alga, *Lithothamnion* or shells; north of Cape Cod: *D. concharum*
eb	0	pa	x	*Chaetozone setosa*
eb	2	pa	x	*Tharyx:* Setae and threadlike gills begin on same segment: *T. setigera* OR First 3 gill-bearing segments lack setae: *T. acutus*
eb	3	x	x	*Cirriformia grandis*
eb	3	x	ey	*Cirratulus cirratus*

FIG. 14.20

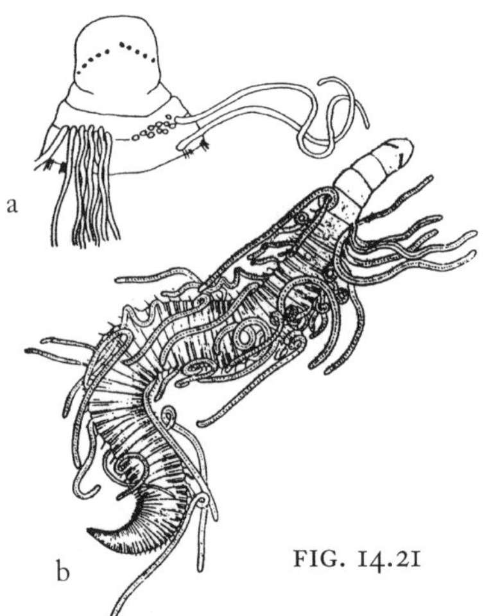

FIG. 14.21

Chaetozone setosa. Long body; palps on segment 1. Ac,?,Su,df,?, (F29). (FIG. 14.20. Anterior end.)

Cirratulus cirratus. Ac–,Ha,Li–Su,df,12.2 cm (4.8 in), (F29,G190). (FIG. 14.21. a, anterior end; b, animal.)

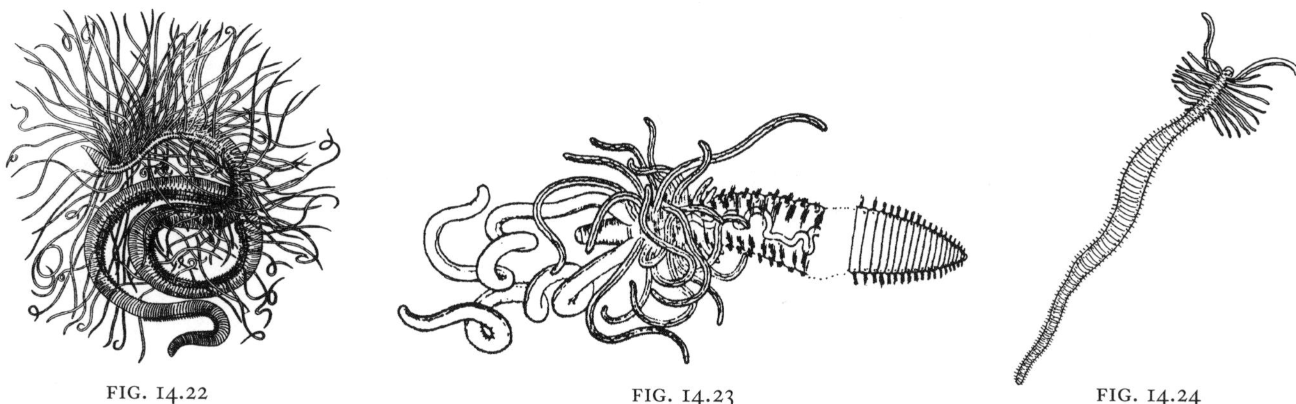

FIG. 14.22 FIG. 14.23 FIG. 14.24

Cirriformia grandis. Less common than *Cirratulus cirratus.*
Vi+,Mu–Sa,Li–Su,df,15.2 cm (6 in), (F30,G190). (FIG. 14.22)

Dodecaceria concharum. Found in galleries within sheets of the encrusting red alga, *Lithothamnion,* or on scallop or *Arctica* shells. Vi,Ha–Sa,Su,df,?.
(FIG. 14.23. Composite of animal.)

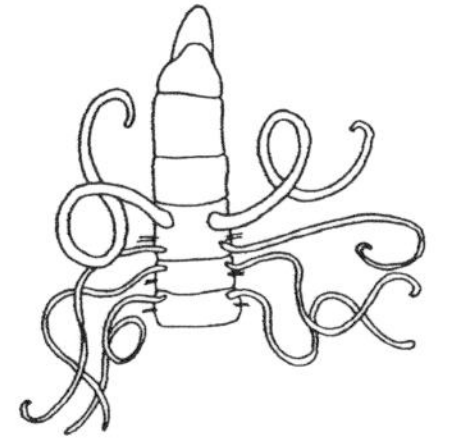

FIG. 14.25

Dodecaceria coralii. Small; dark color. Often with northern star coral, *Astrangia,* or on shells. Vi,Ha,Su,df,13 mm (0.5 in), (F30,G190). (FIG. 14.24)

Tharyx acutus. Pale coloring, small; palps on segment 3. Sublittoral.
Ac–,Sa–Mu,Li–Su,df,15 mm (0.6 in), (G190). (FIG. 14.25. Anterior end of *Tharyx* sp.)

Tharyx setigera. Ca+,Sa–Mu,Li–,df,15 mm (0.6 in), (G190).

14.29 **Cossuridae** (from 14.16)

Small burrowing worms with pointed, naked head, reduced parapodia, and a single tentacle-like palp arising middorsally from three to six segments back from head.

Cossura longocirrata. Threadlike, motile burrower. Single sensory tentacle from fourth seta-bearing segment. Ac,Sa–Mu,Su,df,12 mm (0.4 in).

14.30 **Eunicidae** (from 14.6)

Long, somewhat flattened worms; prostomium short and thick; one, three, or five antennae; stout, fused, cushionlike palps; two eyes. First two segments without parapodia or setae. Active predators; most live in tubes as adults. (F105)

1. Pair of small tentacular cirri on the peristomium or segment just behind the prostomium
   - A. *Tentacular cirri* present — tc
   - B. Nucchal tentacles *absent* — x
2. Number of the segment on which fingerlike gills begin: provide segment *number* — __
3. Gill construction
   - A. Gills a group of simple *filaments* — fi
   - B. Gills *branched* — br
4. Number of filaments or branches per gill: provide *number* of gill filaments or branches — __
5. Color of aciculae or supporting rods embedded inside parapodia but visible through the parapodial walls
   - A. Aciculae *black* — bl
   - B. Aciculae *yellow* — ye

14.31

Tentacular Cirri	No. of Gills	Filaments or Branches	No. of Gill Branches	Aciculae Color	Species
tc	3 (3–5)	br	16 (8–22)	ye	*Eunice pennata*
tc	8 (7–10)	br	6–8 (7–10)	bl	*E. norvegica*
x	12–15	br	11 (7–19)	bl	*Marphysa bellii*
x	20 (10–40)	fi	4 (2–8)	bl	*M. sanguinea*, red-gilled rock worm

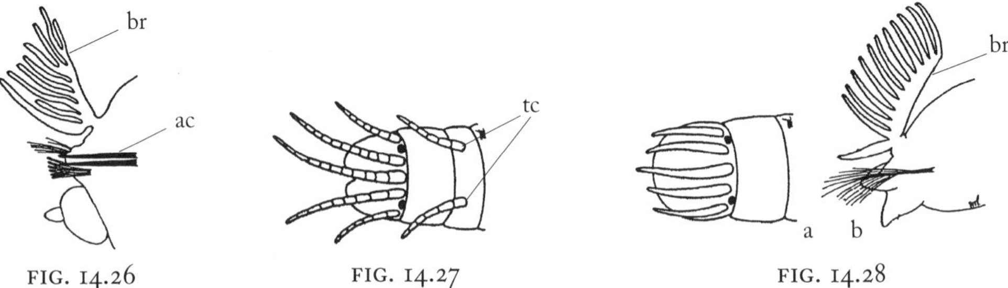

FIG. 14.26 FIG. 14.27 FIG. 14.28

Eunice norvegica. Gills continue to near the posterior end. Pink or brown, spotted with brown. Parchment tubes. Bo,Sa–Mu–Gr,Su,pr,20.3 cm (8 in), (F106,P240). (FIG. 14.26. Parapodium with gill [br] and embedded acicular rods [ac].)

Eunice pennata. Posterior region free of gills. Parchment tubes on shells. Bo,Sa–Mu–Ha,Su,pr,15.2 cm (6 in), (F106,G180,P242). (FIG. 14.27. Anterior end with tentacular cirri [tc].)

Marphysa bellii. Anterior margin of head appears smoothly rounded. Pink with red, comblike gills. Bo,Mu–Sa,Li–,pr,20.3 cm (8 in), (F106,G180,P238). (FIG. 14.28. a, anterior; b, parapodium with branched gill [br].)

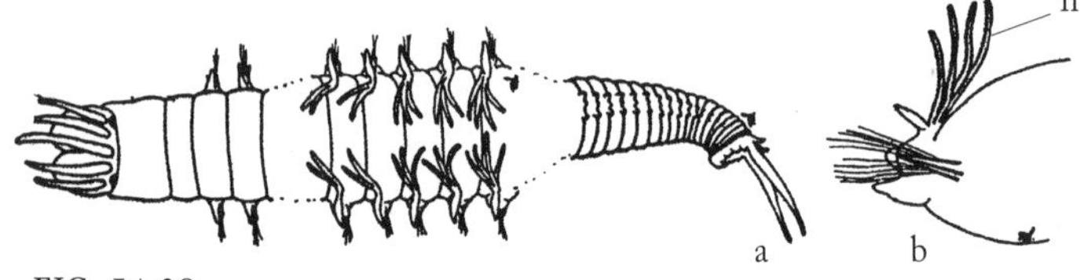

FIG. 14.29

Marphysa sanguinea, red-gilled or rock worm. Palps make anterior margin of head appear lobed. Iridescent yellow to orange with bright red gills; two caudal cirri. Worm breaks into pieces with rough handling. Prefers firm mud. Vi,SG–Mu–Sa–Ha,Li–Su,sc,30.5 cm (12 in), (F106,G180,P236). (FIG. 14.29. a, composite; b, parapodium with filamentous gill [fi].)

14.32

Flabelligeridae (from 14.10)

Body inflated anteriorly and covered with dense, adhesive papillae. Most with large setae forming a cephalic "cage"; two stout palps and short, retractile gills on head. Parapodia reduced. (F115)

1
3
2, 4

FIG. 14.30

1. Large anteriorly directed setae forming cephalic cage (FIG. 14.30)
   - A. *Cephalic cage* of setae present — cc*
   - B. Cephalic cage of setae *absent* — x
2. Body coating (FIG. 14.30)
   - A. Body encased in thick *mucus* — mu
   - B. Body with attached *sand* grains — sa
3. Number of pairs of cephalic gills: provide *number* of pairs of gills (FIG. 14.30) — __
4. Shape of body papillae (FIG. 14.30)
   - A. Papillae *stalked* — st
   - B. Papillae *elongated* but not stalked — el
   - C. Papillae *short* and not stalked — sh
5. Shape of setae in notopodial (upper lobe) versus neuropodial (lower lobe) of parapodium
   - A. Notopodial setae *long;* neuropodial setae short *hooks* — l:h
   - B. Notopodial setae *long;* neuropodial setae *long* — l:l

14.33

Cage	Coating	Gills	Papillae	Setae	Species
x	sa	>4	el	l:l	*Brada* spp.
cc	sa	4	el	l:h	*Pherusa plumosa*
cc	sa	4	sh	l:h	*P. affinis*
cc	sa	4	el	l:l	*Diplocirrus hirsutus*
cc	mu	20–30	st	l:h	*Flabelligera affinis*

FIG. 14.31

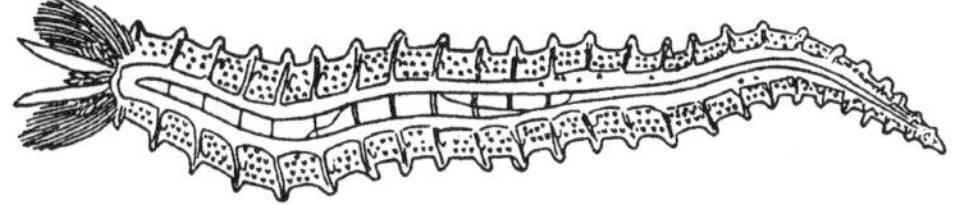

Brada spp. Short, stout, grublike body. Ac–,Mu,Su,df,3.8 cm (1.5 in), (F116,G194). (FIG. 14.31)

Diplocirrus hirsutus. Pale green. Ac–,Mu,Su,df,25 mm (1 in), (F116,G194).

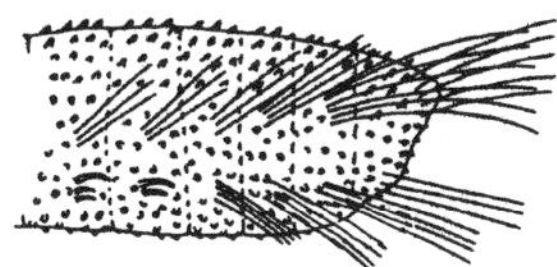

FIG. 14.32

Flabelligera affinis. Greenish with yellow-orangish palps. Young often commensal among spines of echinoids. Ac–,Mu–Ha–Al,Li–Su,df,6.4 cm (2.5 in), (F117,G194). (FIG. 14.32)

FIG. 14.33

Pherusa affinis. Large frilled palps. Large papillae secrete mucus to which sand sticks. Tolerates high organic detritus and sludge. Bo,Mu,Es–Li,df,5.8 cm (2.3 in), (F117,G193). (FIG. 14.33. Anterior with cephalic cage of setae.)

Pherusa plumosa. Dark greenish-brown; young are orangish. Mucus clings to setal cage. Sand or mud adheres to surface. Ac,Ha–Mu,Li–,df,6.4 cm (2.5 in), (F117,G194). (FIG. 14.34)

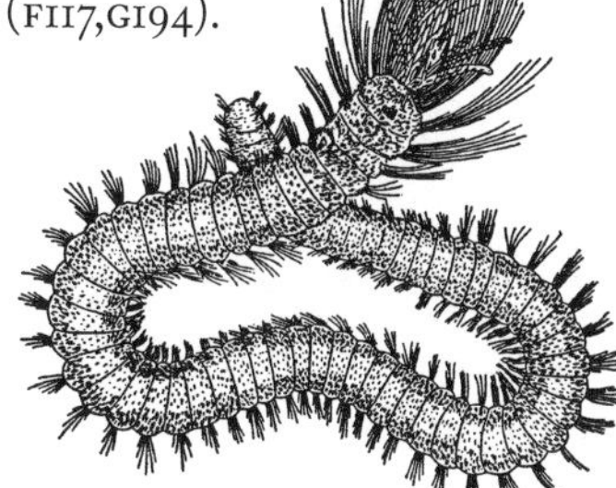

FIG. 14.34

14.34 **Glyceridae (Bloodworms) and Goniadidae (Chevronworms)** (from 14.14)

Elongate, cylindrical, pinkish body; tapered at both ends; annuli divide segments. Prostomium small, sharp-tipped with four minute antennae. Proboscis long, strong, four large jaws (their bite is painful, like a bee sting). Uses proboscis to burrow in sand or mud; carnivorous or detrital feeding. Sold for bait. Goniadidae have larger parapodia and more elaborate proboscis adornments. (F92)

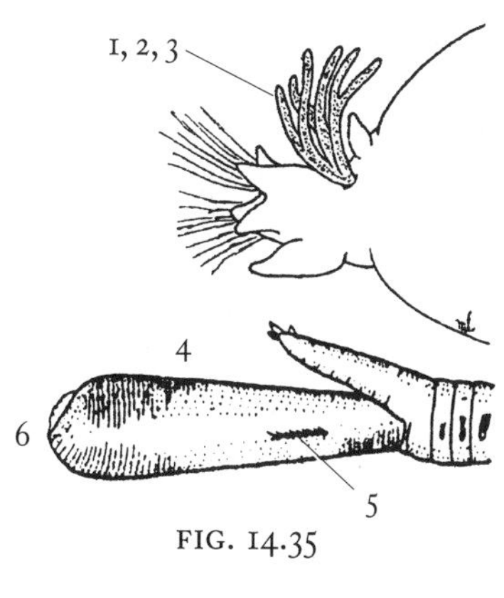

FIG. 14.35

1. Number and location of gills (branchiae) (FIG. 14.35)
   - A. *Two* gills per parapodium, located on top and bottom margins of parapodium — 2
   - B. *One* gill located on top margin of parapodium — 1*
   - C. Gills *absent* — x
2. Shape of gills (FIG. 14.35)
   - A. Gills *branched* — br*
   - B. Gills simple, *digitate* (fingerlike) — di
   - C. Gills lump-shaped (*blister*like) — bl
   - D. Gills *absent* — x
3. Retractility of gills (touch them with a pin) (FIG. 14.35)
   - A. Gills *retract* — re
   - B. Gills do *not* retract — x
4. Features of the proboscis (pressure on the anterior quarter of the body will often induce the worm to evert its proboscis) (FIG. 14.35)
   - A. Proboscis surface basically *smooth* — sm*
   - B. Proboscis with conspicuous patch of *horny spines* giving furry appearance — hs
5. Presence of small dark chevron teeth toward base of proboscis (FIG. 14.35)
   - A. Two short rows of *chevron* teeth present — ch*
   - B. Chevron teeth *absent* — x
6. Presence of ventral row of teeth along proboscis (FIG. 14.35)
   - A. Row of *ventral teeth* present along the proboscis — vt
   - B. Ventral teeth *absent* from proboscis — x*

14.35

No. of Gills	Gill Shape	Gill Retractility	Proboscis	Chevron Teeth	Ventral Teeth	Species
1	br	re	sm	x	x	*Glycera americana,* tufted bloodworm
1	bl	x	sm	x	x	*G. robusta*
2	di	x	sm	x	x	*G. dibranchiata,* two-gilled bloodworm or beak thrower
x	x	x	sm	x	x	*G. capitata,* bloodworm
x	x	x	sm	ch	x	*Goniada maculata,* chevronworm
x	x	x	hs	ch	x	*Goniadella gracilis*
x	x	x	hs	x	vt	*Glycinde solitaria,* chevronworm
x	x	x	hs	x	x	*Ophioglycera gigantea*

Glycera americana, tufted bloodworm (Glyceridae). Vi,Mu–Sa–SG,Li–Su,pr,35.6 cm (14 in), (F92,G172,P213), ☺. (FIG. 14.36. Parapodium with one, branched gill.)

FIG. 14.36

Glycera capitata, bloodworm (Glyceridae). Ac–,Mu–Sa–SG,Su,sc,10.1 cm (4 in), (F92,G172,P211). (FIG. 14.37. a, anterior end; b, parapodium lacking gill.)

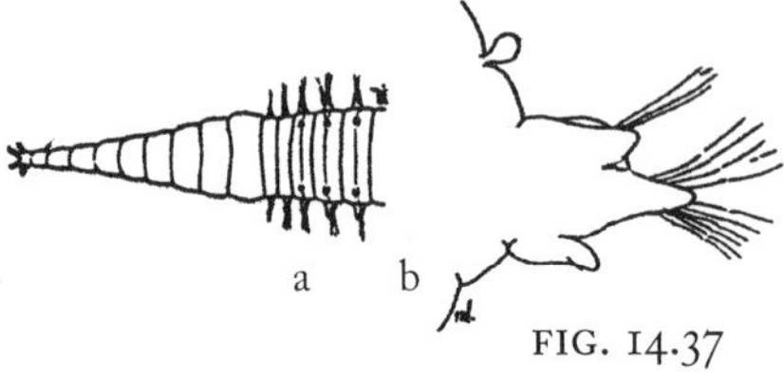

FIG. 14.37

Glycera dibranchiata, two-gilled bloodworm or beak thrower (Glyceridae). Bo,Mu–Sa–Gr–SG,Li–De,sc,38 cm (15 in), (F92,G171,P215). (FIG. 14.38. a, composite with proboscis extended; b, parapodium with gills [br].)

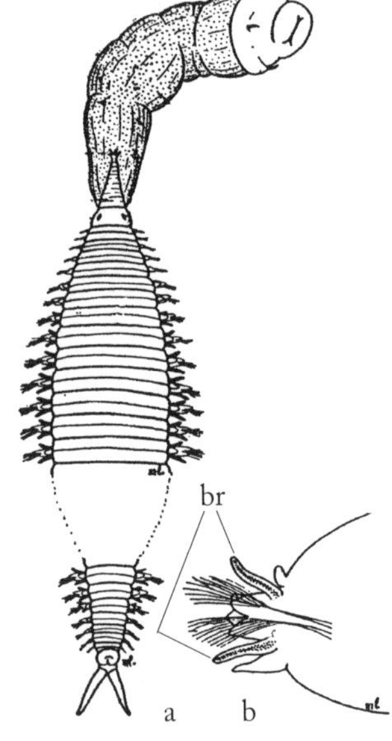

FIG. 14.38

Glycera robusta (Glyceridae). Burrows deeply into mud. Commercial bait worm. Bo,Mu–Sa,Li–Su,pr,50.8 cm (20 in), (G172,P218). (FIG. 14.39. Parapodium with blisterlike gills [br].)

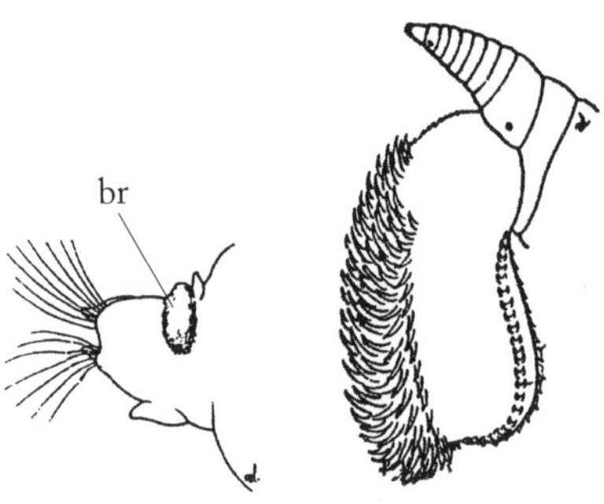

FIG. 14.39 FIG. 14.40

Glycinde solitaria, chevronworm (Goniadidae). Yellowish green color. Vi,Sa–Mu,Es–Li–Su,pr?,3.3 cm (1.3 in), (F93,G172,P222), ☺. (FIG. 14.40. Head with extended proboscis bearing horny spines.)

Goniada maculata, chevronworm (Goniadidae). Greenish anteriorly to orangish posteriorly. Fringed ring surrounds proboscis end. Ac–,Sa–Gr–Mu,Su,pr?,10.1 cm (4 in), (F93,G93,G172). (FIG. 14.41. a, anterior end; b, proboscis showing chevron teeth [ch]; c, fringed proboscis tip.)

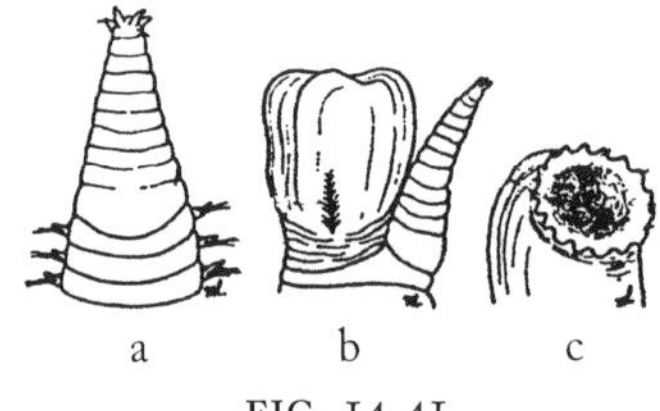

FIG. 14.41

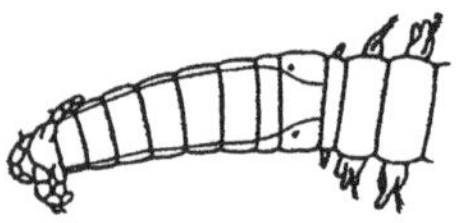

Goniadella gracilis (Goniadidae). Elongate, annulated prostomium; first pair of parapodia reduced. Tolerates low oxygen. Vi,Sa–Gr–Mu,Su–De,pr,5.1 cm (2 in), (F93,P220). (FIG. 14.42)

FIG. 14.42

Ophioglycera gigantea (Goniadidae). First pair of parapodia reduced; dorsal cirri flattened. Ac–,Mu–Gr,Li–,pr?,76.2 cm (30 in), (F94,P223). (FIG. 14.43. Parapodium lacking gills.)

FIG. 14.43

14.36 **Hesionidae** (from 14.6)

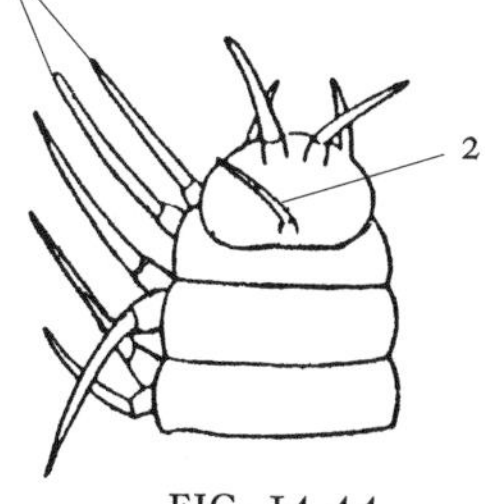

FIG. 14.44

Prostomium with two pairs of eyes, one pair of biarticulate (two-piece) palps; two or three pairs of antennae. Peristomium with six to eight pairs of tentacular cirri. Parapodia biramous, but with notopodium (upper lobe) reduced to dorsal cirri; dorsal and ventral cirri beaded (segmented). Proboscis a muscular tube. Body stout, about 16 segments, first one to four segments lack parapodia.

1. Number of prostomial eyes: provide total *number* of eyes —
2. Location of medial antenna on prostomium (FIG. 14.44)
   - A. Medial antenna along *anterior margin* of prostomium am
   - B. Medial antenna along *posterior margin* of prostomium pm*
3. Number of pairs of tentacular cirri: provide *number* of pairs (FIG. 14.44) —

14.37

No. of Eyes	Medial Antennae	Tentacular Cirri	Species
2	pm	6	*Microphthalmus aberrans*
4	am	6	*Podarke obscura,* swift-footed worm
4	am	8	*Podarkeopsis* sp. (=*Gyptis vittata*)

Microphthalmus aberrans. Found commensally with polychaete *Enoplobranchus sanguineus.* Ac,Sa–Gr,Li,pr,12 mm (0.4 in), (F76,P104).

FIG. 14.45

Podarke obscura, swift-footed worm. Dark brown with banding; three prostomial antennae; six pairs of tentacular cirri. Free-living or associated with sea urchins. Vi,SG–Mu–Sa,Su–De,pr,3.8 cm (1.5 in), (F77,G175,P104).
(FIG. 14.45. a, head; b, parapodium.)

Podarkeopsis sp. (=*Gyptis vittata*). Pale or lightly banded. Vi,Ha,Li–Su,pr,9 mm (0.3 in), (F76,P106).
(FIG. 14.46. a, head; b, parapodium.)

FIG. 14.46

14.38 **Lumbrineridae and Arabellidae, Threadworms** (from 14.14)

Long, slender worms in fine sand. Prostomium a smooth cone or lobe without appendages; first two segments lack parapodia; parapodia are uniramous. Bodies fragile; often fragment upon handling. Lumbrineridae (F107) differ from Arabellidae (F110) in construction of setae. Those requiring definitive identification of these difficult worms should refer to Frame (1992).

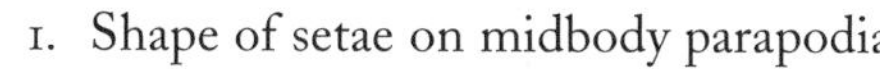

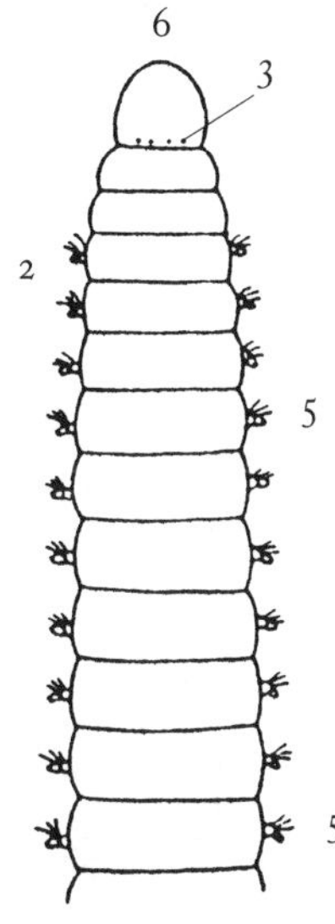

FIG. 14.47

1. Shape of setae on midbody parapodia
   - A. *One* type: all setae are sharp-tipped and similar (Arabellidae) 1
   - B. *Two* types: some setae are sharp-tipped and others are blunt or hooked (Lumbrineridae) 2
2. Presence of gills (branchiae) (FIG. 14.47)
   - A. Branched *gills* present toward anterior gi
   - B. Gills *absent* x*
3. Presence of eyes on prostomium (look carefully just where prostomium meets the peristomium) (FIG. 14.47)
   - A. *Eyes* present ey*
   - B. Eyes *absent* x
4. Color of aciculae (supporting bristles) embedded inside parapodia but visible through the parapodial wall
   - A. Aciculae are *yellow* ye
   - B. Aciculae are *black* bl

5. Relative sizes of parapodia along the anterior quarter of the worm compared to the middle of the worm
   - A. *Anterior* parapodia about *equal* in size to *middle* parapodia — a=m*
   - B. *Anterior* parapodia much *smaller than middle* parapodia — a<m

6. Prostomial proportions: length/width (FIG. 14.47)
   - A. Length *greater than twice* the width — >2
   - B. Length *less than twice* the width — <2*

14.39

Setae	Gills	Eyes	Aciculae Color	Parapodia	Prostomial Proportions	Species
2	gi	x	bl	a=m	<2	*Ninoe nigripes*
2	x	x	bl	a=m	>2	*Lumbrinerides* (=*Lumbrineris*) *acuta*
2	x	x	bl	a<m	<2	Outermost lobe of anterior parapodia conical: *Scoletoma acicularum* Outermost lobe of anterior parapodium ear- or flap-shaped: *S. fragilis*
2	gi	x	bl	a<m	<2	*Paraninoe* (=*Lumbrineris*) *brevipes*
2	x	x	ye	a=m	<2	*Scoletoma* (=*Lumbrineris*) *tenuis*
1	x	x	ye	a<m	<2	*Drilonereis longa*, threadworm
1	x	ey	ye	a=m	<2	*Arabella iricolor*, opal worm
1	x	x	ye	a=m	<2	*Drilonereis magna*

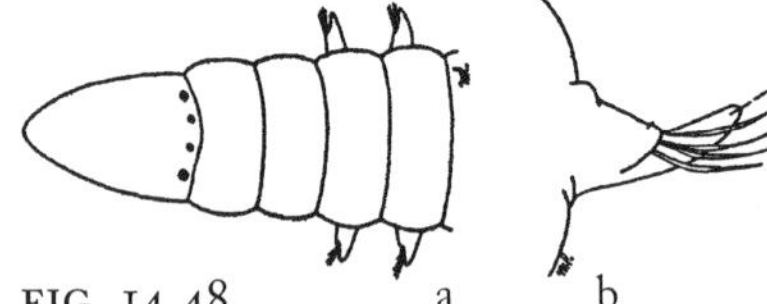

FIG. 14.48 a b

Arabella iricolor, opal worm (Arabellidae). Long, slender, shiny, greenish or reddish iridescent body. Four eyes in a row. Contracts in a ball; sand grains adhere to its body. Ac,Sa–Mu–SG,Es–Li,pr?,61 cm (24 in), (F110,G180,NM423,P269), ☺. (FIG. 14.48. a, head; b, parapodium lacking gills.)

Drilonereis longa, threadworm (Arabellidae). Red. Anterior parapodia reduced to bumps. Forms knotlike clusters when removed from sediment. Vi+,Sa–Mu–SG,Li–Su,df,10.1 cm (4 in), (F110,G181,P272). (FIG. 14.49. a, head; b, anterior parapodium; no gills.)

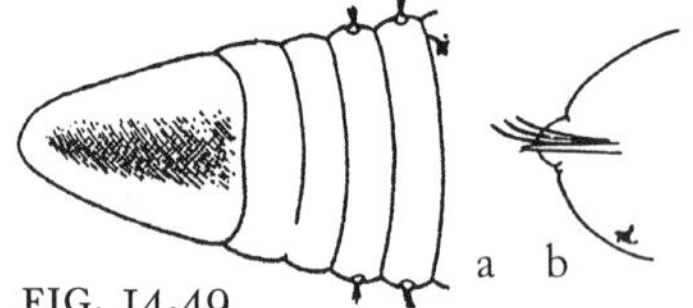

FIG. 14.49 a b

Drilonereis magna (Arabellidae). Bo,Gr–Sa–Mu,Li–Su,pr?,20.3 cm (8 in), (F110,P273).

FIG. 14.50 a b

Lumbrinerides (=*Lumbrineris*) *acuta* (Lumbrineridae). Prostomium two to three times longer than wide. Ac–,Mu–Sa,Li–Su,pr?,4.3 cm (1.7 in), (F109,P260). (FIG. 14.50. a, head; b, parapodium, lacking gills.)

Ninoe nigripes (Lumbrineridae). Mud and mucus tubes; motile burrower. Bo,Mu,Su,df,pr,10.1 cm (4 in), (F109,G182,P266). (FIG. 14.51. a, head; b, parapodium with gills.)

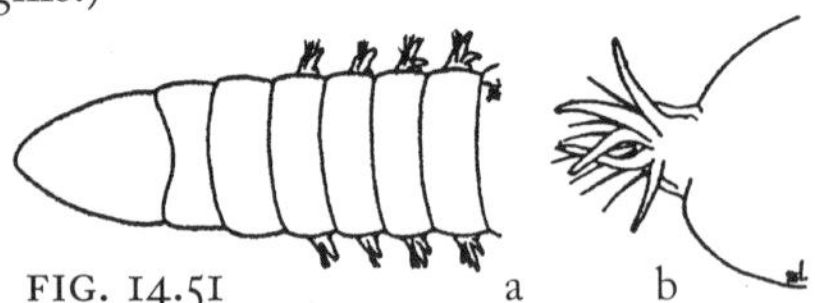

FIG. 14.51 a b

FIG. 14.52

Paraninoe (=*Lumbrineris*) *brevipes* (Lumbrineridae). Lower half of setae dark, outer half, yellow. Fragile; difficult to collect intact specimen. Motile burrower. Bo,Sa–Mu,Su,pr?,2.1 cm (0.8 in), (F109,P260). (FIG. 14.52. Parapodium, lacking gills.)

Scoletoma acicularum, fragile threadworm (Lumbrineridae). Iridescent reddish-yellow. Prostomium short. Burrower; eats benthic invertebrates. Bo,Mu–Sa–Gr,Li–De,pr,38.1 cm (15 in), (F109,G181,NM423,P262). (FIG. 14.53. a, head; b, parapodium, lacking gills.)

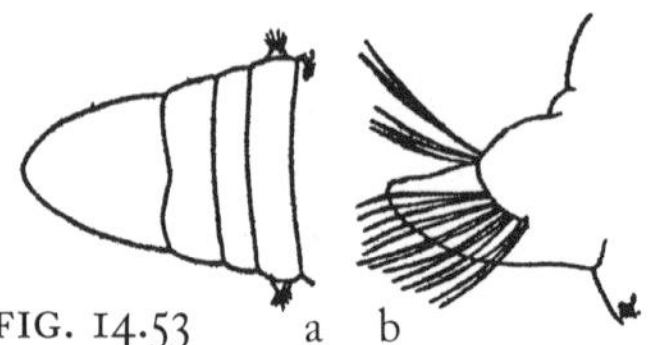

FIG. 14.53 a b

Scoletoma fragilis, fragile threadworm (Lumbrineridae). Iridescent pinkish tan. Midbody parapodia with long, thin setae. Ac,Mu–Sa,Su–De,pr,12 cm (4.7 in).

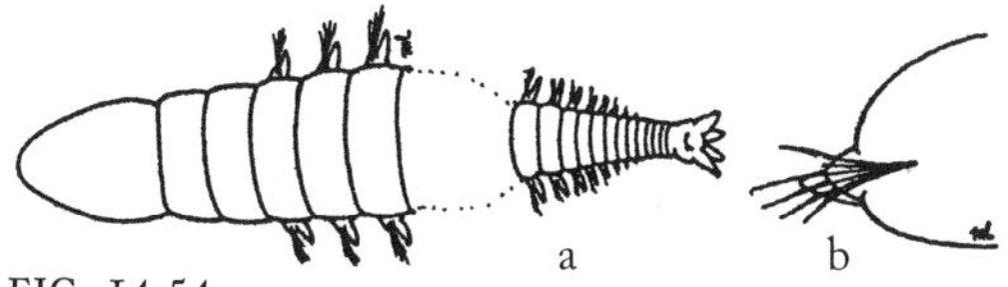

FIG. 14.54

Scoletoma (=*Lumbrineris*) *tenuis* (Lumbrineridae). Iridescent color. Often with tunicate, *Apylidium.* Bo,Gr–Sa–Mu,Li–Su,pr–df,15.2 cm (6 in), (F109,G182,P264). (FIG. 14.54. a, composite; b, parapodium, no gills.)

14.40 **Maldanidae and Oweniidae, Bambooworms** (from 14.4, 14.16)

Cylindrical worms with segments longer than wide; be certain that you are working with a complete specimen because these worms break easily. Head blunt; prostomium without appendages; proboscis bulbous; parapodia reduced, barely biramous in thoracic region; pygidium or last segment, distinctively shaped. Live head-down in upright sand tubes. Oweniids (F115) differ by possessing a rounded posterior end; their segments become shorter posteriorly. Worms below are Maldanidae (F37) unless indicated otherwise.

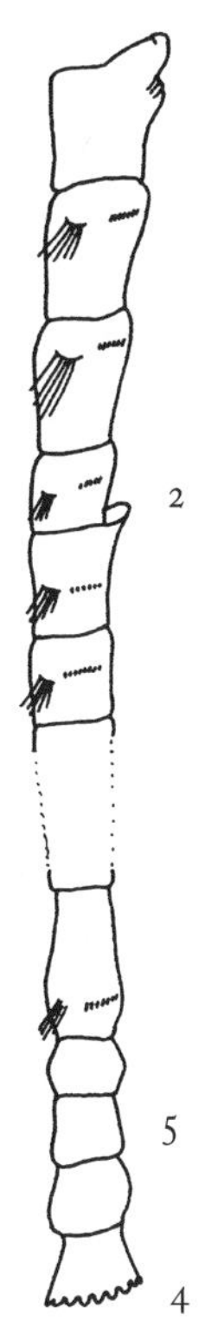

FIG. 14.55

1. Distinctive coloration
   - A. Body light with *darker* (reddish or greenish) *nodes* between segments — dn
   - B. Body intermediate in color (greenish yellow) with *lighter nodes* between segments — ln
   - C. Anterior end *speckled* with black dots — sp
   - D. Body relatively uniformly dark *red* — re
2. Distinctive body features (FIG. 14.55)
   - A. Flattened, *leafy lobes* surround the mouth — ll
   - B. Distinctive *collar* formed around anterior margin of segment four — co*
   - C. Both these features *absent* — x
3. Composition of tubes
   - A. Tubes of sand and shell fragments in overlapping, *shingle*like pattern — sh
   - B. Tubes comprise nonoverlapping *sand* grains and mucus — sa
   - C. Tubes comprise *mud* and mucus — mu
4. Shape of pygidium (terminal segment) (FIG. 14.55)
   - A. *Frilled* margin present on pygidial funnel — fr*
   - B. Pygidium *rounded* — ro
   - C. Pygidium *slanted* — sl
   - D. Pygidium flattened or *boot*like in shape — bo
5. Number of segments without setae just anterior to pygidial (terminal) segment: provide *number* of setaless segments — __

14.41

Color	4th Segment	Tube	Margin	Segments	Species
ln	ll	sh	ro	4–5	*Owenia fusiformis*
dn	co	sa	fr	3	*Clymenella torquata,* bambooworm
dn	x	sa	fr	3	*Euclymene* (=*Clymenella*) *zonalis,* bambooworm
re	x	mu	fr	4–5	*Praxillella praetermissa*
re	x	mu	bo	2	*Maldane sarsi,* bambooworm
sp	x	mu	sl	0	*Sabaco* (=*Asychis,* =*Maldanopsis*) *elongatus,* bambooworm

FIG. 14.56

Clymenella torquata, bambooworm. Cream-colored or greenish. Flat-topped head with opercular plate to seal sand tube. Grows in dense beds. Bo,Sa–Mu–SG,Es–Li–Su,df,15.2 cm (6 in), (F40,G186,NM426), ☺. (FIG. 14.55. Composite.)

Euclymene (=*Clymenella*) *zonalis,* bambooworm. Dark red bands near head. Bo,Sa–Mu,Es–Li–Su,df,20 mm (0.8 in), (F40,G186).

Maldane sarsi, bambooworm. Dark anteriorly; lighter posteriorly. Tube thick, encrusted with mud. Ac?,Mu–Sa–Gr,Li–Su,df,10.1 cm (4 in), (F40). (FIG. 14.56. a, head; b, posterior end.)

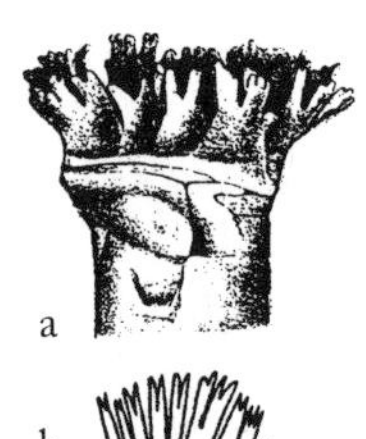

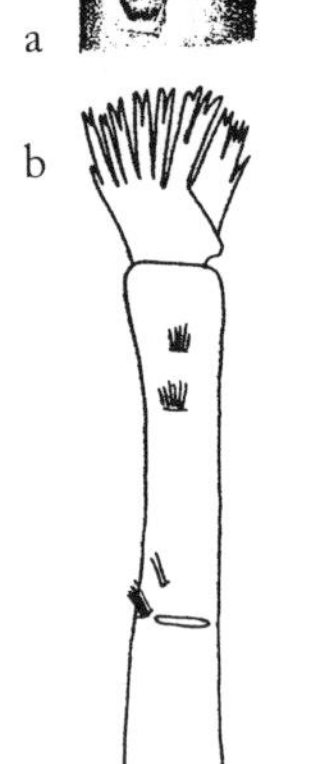

FIG. 14.57

Owenia fusiformis (Oweniidae). Twenty to 30 segments. Head lobes are distinctive. Ca+,Sa,Su,df–ff,10.1 cm (4 in), (G186). (FIG. 14.57. a, head; b, anterior end.)

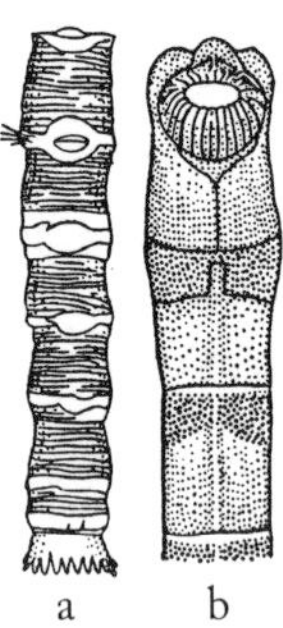

FIG. 14.58

Praxillella praetermissa. Brown with reddish nodes. Ac,Sa–Mu,Su–De,df,4.6 cm (1.8 in). (FIG. 14.58. a, posterior end; b, anterior end.)

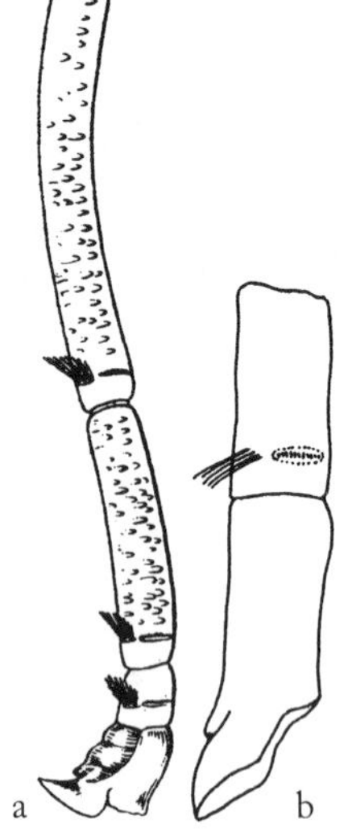

FIG. 14.59

Sabaco (=Asychis, =Maldanopsis) elongatus, bambooworm. Elongate body, rounded head. Ac–,Sa–Mu,Li,df,30.5 cm (12 in). (FIG. 14.59. a, worm; b, posterior end.)

14.42 **Nephtyidae: Red-Lined Worms, Catworms, or Shimmyworms** (from 14.14)

Elongate, stout, light-colored body, rectangular in cross-section, often with slight indentation along middorsal line. Prostomium flattened and pentagonal with four marginal antennae (or 1 pair antennae and 1 pair of tentaclelike palps); eyeless. Parapodia distinctly biramous with medially placed, curled gills. Shape of parapodial lobes and attachments used in species distinctions. Proboscis muscular; ends with fingerlike papillae when fully everted. Characteristic shimmy movement in water. (F96)

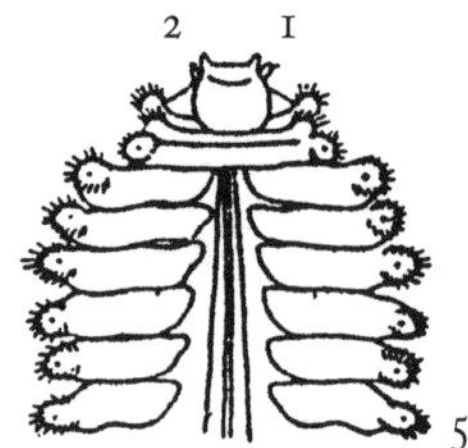

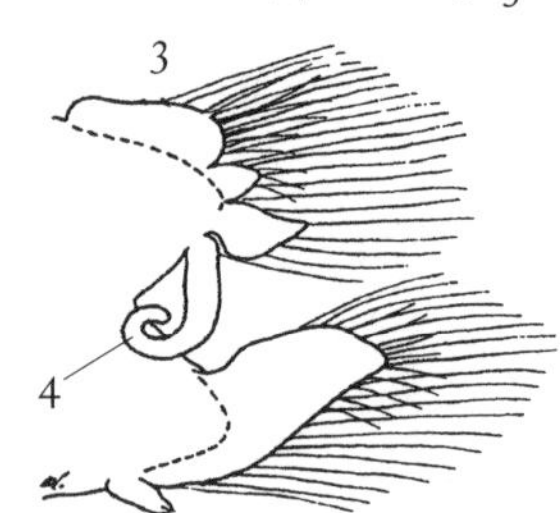

FIG. 14.60

1. Shape of head shield (between anteriormost antennae) (FIG. 14.60)
   - A. *Anterior* margin of shield distinctly *wider than posterior* margin — a>p
   - B. *Anterior* margin of shield approximately *equal* in width to *posterior* margin — a=p*
   - C. *Anterior* margin of shield distinctly *narrower than posterior* margin — a<p
2. Number of antennae at anterior corners of head shield: provide *number* of tiny antennae (2 or 4) (FIG. 14.60) — __
3. Shape of notopodial or uppermost lobe of parapodium (FIG. 14.60)
   - A. Notopodium distinctly *pointed* — po
   - B. Notopodium *rounded* — ro*
4. Location of gills: provide *number* of seta-bearing segment on which gills begin — __
5. Direction of curl for midparapodial gills (FIG. 14.60)
   - A. Gills curl *outward* — ou
   - B. Gills curl *inward* — in*

14.43

Shield	Antennae	Notopodium	Gills Start	Gill Curl	Species
a>p	2	ro	3	ou	*Nephtys bucera*
a=p	2	ro	4	ou	*N. picta*
a=p	2	po	4–6	ou	*N. caeca*, leafy shimmyworm
a=p	4	ro	6–8	ou	*N. incisa*
a>p,a=p	4	ro	5–7	in	*Aglaophamus verrilli*
a>p	4	po	2	in	*A. circinata*
a<p	4	ro	4–6	ou	*Nephtys ciliata*

FIG. 14.61 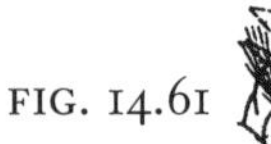

Aglaophamus circinata. Eyes absent. Ac–,Mu–Sa–Gr–Ha,Su,pr,5.1 cm (2 in), (F97,P192). (FIG. 14.60. Parapodium; FIG. 14.61. Head.)

FIG. 14.62

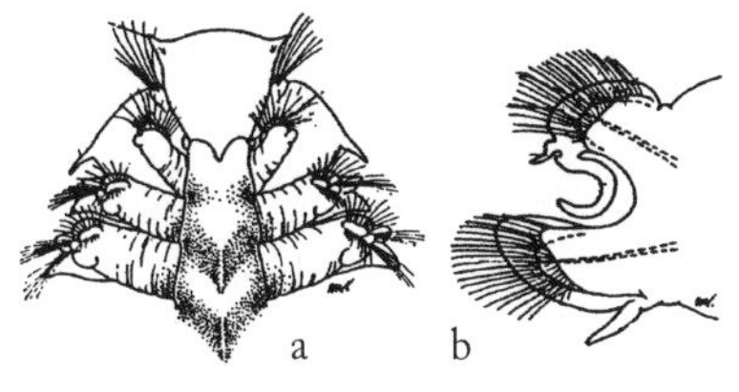

FIG. 14.63

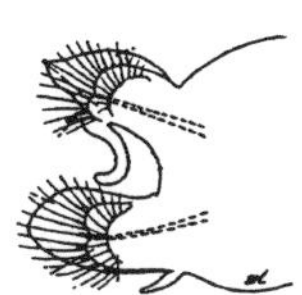

FIG. 14.64

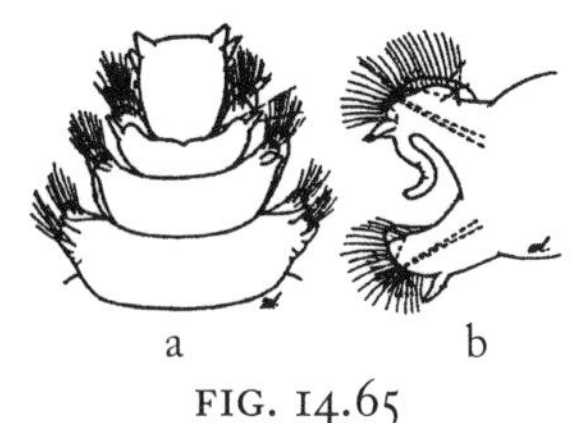

FIG. 14.65

FIG. 14.66

Aglaophamus verrilli. Two tiny eyes on prostomium. Ca+,Sa–Mu,Li–Su,pr,4.6 cm (1.8 in), (F97,P190). (FIG. 14.62. Parapodium.)

Nephtys bucera. Dorsal V-shaped bands on anterior segments. Setae light-colored. Eats small clams, crabs, worms in sandy sediment. Bo,Sa,Es–Li–Su,df–pr,30.5 cm (12 in), (F97,G173,NM419,P196), ☺. (FIG. 14.63. a, head; b, parapodium, midregion.)

Nephtys caeca, leafy shimmyworm. Setae light-colored. Ac–,Mu–Sa,Li–De,pr,20.3 cm (8 in), (F97,G173,NM419,P203). (FIG. 14.64. Parapodium, anterior region.)

Nephtys ciliata. Aciculae dark-tipped; setae light-colored. Ac–,Gr–Sa–Mu,Su,pr,30.5 cm (12 in), (F97,P202).

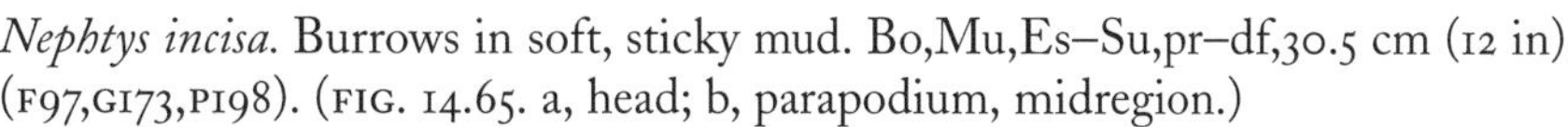

Nephtys incisa. Burrows in soft, sticky mud. Bo,Mu,Es–Su,pr–df,30.5 cm (12 in), (F97,G173,P198). (FIG. 14.65. a, head; b, parapodium, midregion.)

Nephtys picta. Setae darker basally. Vi,SG–Sa–Mu,Li–Su,pr,30.5 cm (12 in), (F97,G173,P195), ☺. (FIG. 14.66. a, head; b, gill, midregion.)

14.44 **Nereididae, Ragworms or Clamworms** (from 14.6)

Elongate body, cylindrical in cross-section. Prostomium with two pairs of eyes, one pair of stout, two-segmented palps, one pair of antennae; four pairs tentacular cirri from peristomium. Proboscis with one pair of jaws. Parapodia large, biramous; shape of lobes and cirri help differentiate species. (F85)

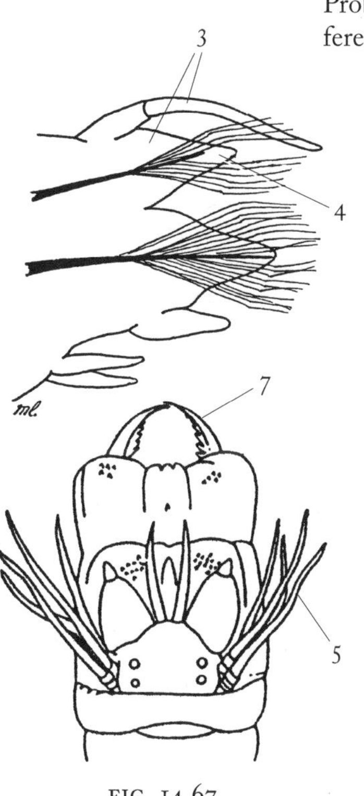

FIG. 14.67

1. Body shape
   - A. Body distinctly *thicker* anteriorly than posteriorly — th
   - B. Body *cylindrical,* approximately the same diameter throughout — cy
2. Parapodial symmetry
   - A. Parapodia of anterior third of body distinctly *different* in shape than parapodia from posterior third of body — di
   - B. Parapodia of anterior third of body about the *same* in shape as those from the posterior third of the body — sa
3. Length of fingerlike dorsal cirrus compared to the notopodium (upper parapodial lobe to which dorsal cirrus is attached) (FIG. 14.67)
   - A. Dorsal *cirrus* is equal to or *longer than notopodial* lobe — c>n*
   - B. Dorsal *cirrus* at least *one-half* the length of the *notopodium* — c½n
   - C. Dorsal *cirrus* is clearly *less than* one-half the length of the *notopodium* — c<n
4. Shape of the tip of the notopodium (dorsalmost parapodial lobe) (FIG. 14.67)
   - A. Notopodial tip is distinctly *pointed* — po*
   - B. Notopodial tip is *rounded* — ro
5. Length of longest tentacular cirri from peristomial segment (FIG. 14.67)
   - A. Tentacular *cirri more than five* times the length of the prostomium — c>5
   - B. Tentacular *cirri less than five* times the length of the prostomium — c<5*

6. Color of aciculae (support rods) embedded with the parapodium but visible through the parapodial wall (FIG. 14.67)
   - A. Aciculae are *dark* — da*
   - B. Aciculae are *light* — li
7. Color of teeth on proboscis (to evert its proboscis, irritate the worm or gently squeeze just behind the head; this more intrusive character may not be needed to distinguish species) (FIG. 14.67)
   - A. Teeth are *dark* in color — da
   - B. Teeth are *light* in color — li*

14.45

Shape	Parapodia	Dorsal Cirrus Lobe	Notopodial Tip	Cirri Length	Aciculae	Teeth	Species
th	sa	c<n	po,ro	c<5	da	li,da	*Nereis virens,* clamworm, ragworm
th	sa	c½n	po	c<5	da	da	*Hediste* (=*Nereis*) *diversicolor*
cy	di	c>n	po	c<5	da	li	*Neanthes* (=*Nereis*) *succinea,* common clamworm
cy	di	c>n	ro,po	c>5	da	li	*Platynereis dumerilii*
cy	sa	c>n	ro	c<5	da	da	Body banded: *Nereis zonata* OR Body not banded: *N. pelagica,* pelagic clamworm
cy	sa	c>n	po	c<5	da	da	*N. grayi*
cy	sa	c<n	po	c<5	li	da	*Neanthes* (=*Nereis*) *acuminata* (=*arenaceodonta*)

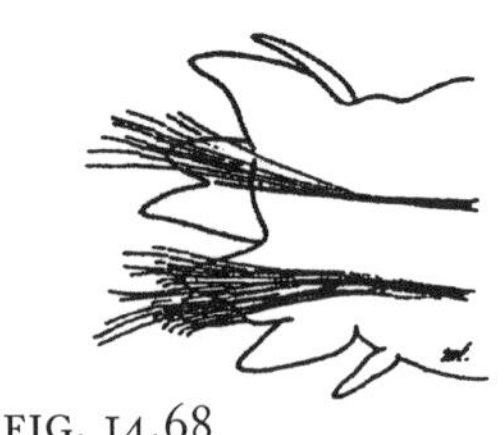
FIG. 14.68

Hediste (=*Nereis*) *diversicolor.* Body flattens posteriorly. Green, yellow, orange, reddish colors possible. Often two darker dorsal stripes along body. Ac,Mu–Sa,Es–Li–,pr–sc–ff,20.3 cm (8 in), (F90,G176,P174), ☺. (FIG. 14.68. Parapodium, anterior region.)

Neanthes (=*Nereis*) *acuminata* (=*arenaceodonta*). Muddy tubes; agile and active. Ac,Mu–Sa–Gr,Su,pr–gr,7.1 cm (2.8 in), (F90,P162). (FIG. 14.69. a, head; b, parapodium, midregion.)

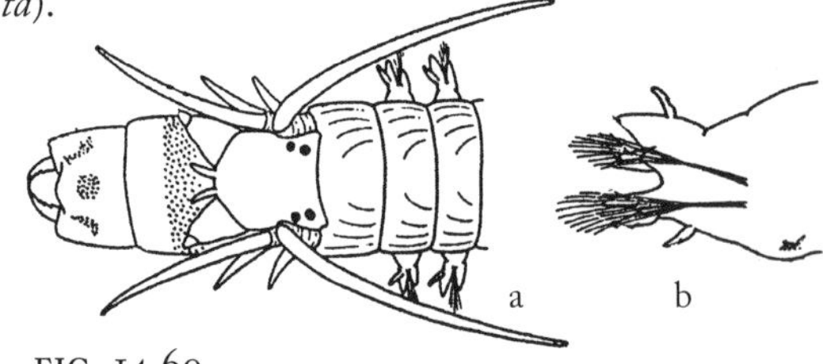

FIG. 14.69

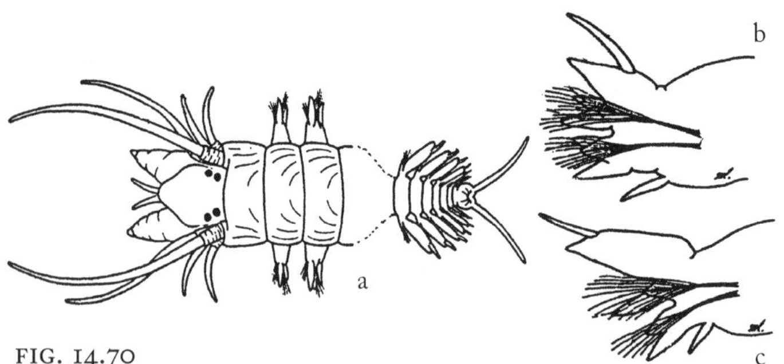

FIG. 14.70

Neanthes (=*Nereis*) *succinea,* common clamworm. Notopodium is straplike toward posterior. Greenish to brown anteriorly, yellowish to reddish posteriorly. U-shaped burrow in sandy mud. Heteronereid (seasonal, reproductive, swimming form) with enlarged parapodia; swarms at night in May; attracted to light held near water surface. Vi+,Ha–Sa–Mu–SG,Es–Li–,pr–sc–df,15.2 cm (6 in), (F90,G176,P165), ☺. (FIG. 14.70. a, head; b, parapodium, anterior region; c, parapodium, posterior region.)

Nereis grayi. Often with bambooworm, *Clymenella.* Vi,Sa–Mu–Co,Li–Su,pr–sc,5.8 cm (2.3 in), (F90,G176,P183). (FIG. 14.71. Parapodium, anterior region.)

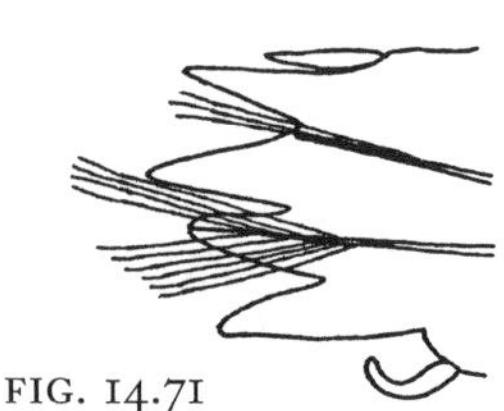
FIG. 14.71

Nereis pelagica, pelagic clamworm. Omnivore that eats invertebrates, algae, detritus. Especially on algae, pilings, in fouling community, and in shell beds. Ac–,Ha–Al,Es–Li–Su,pr–sc,15.8 cm (6.2 in), (F90,G176,NM421,P179), ☺. (FIG. 14.72. Parapodium, anterior region.)

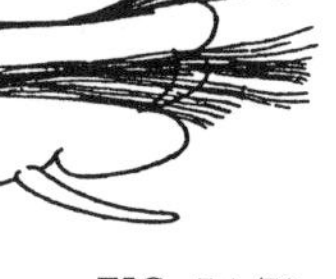
FIG. 14.72

FIG. 14.73

Nereis virens, clamworm, ragworm. Iridescent coloring. Makes irregular burrows; emerges at night to feed. An important fish food. Commercial market as bait. Reported to reach 91 cm (36 in). Bo,Sa–Mu–Ha,Es–Li–Su,gr–sc–om,30.5 cm (12 in), (F90,G176,NM420,P170), ☺. (FIG. 14.73. Parapodium, midregion.)

Nereis zonata. Body banded. Ac,Sa–Gr,Su,pr–sc,12.7 cm (5 in), (F90,G176,P181). (FIG. 14.74. Parapodium, anterior region.)

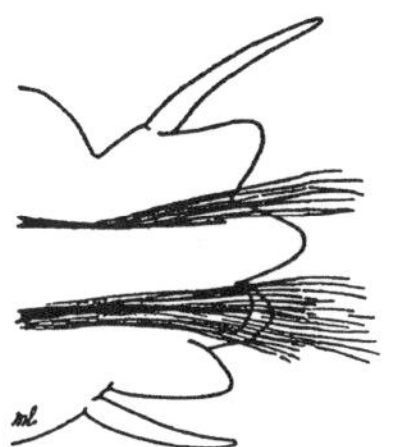

FIG. 14.74

Platynereis dumerilii. Iridescent. Transparent tubes among algae, especially gulfweed, *Sargassum.* Heteronereid (seasonal, reproductive, swimming form) has huge eyes. B,SG–Al–Pe,Li–Su,gr–pr,7.6 cm (3 in), (F90,G178,P154). (FIG. 14.75. a, anterior end; b, parapodium, midregion.)

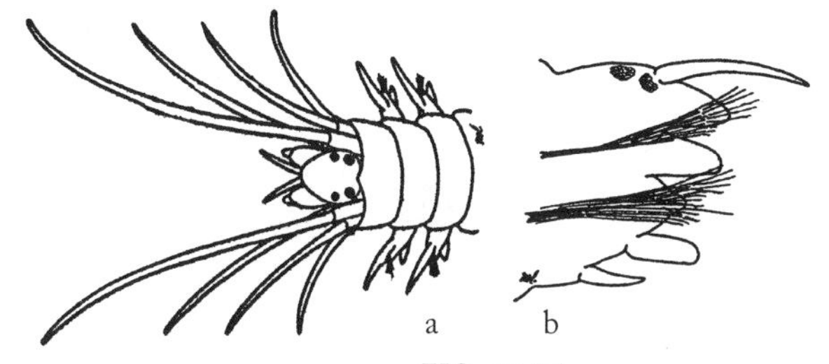

FIG. 14.75

14.46 **Onuphidae** (from 14.6)

Iridescent body with many segments. Prostomium with two liplike palps and five large antennae, each ringed toward the base. Peristomium with two tentacular cirri along anterior margin; without parapodia or setae. Parapodia uniramous; anterior pairs with fingerlike ventral cirri; posterior parapodia lack cirri. Carnivorous, parchment-tube-dwellers. (F104)

1. Shape of gills (branchiae) (FIG. 14.76)
   - A. Spiraling *whorls* of branches from axial stalk — wh*
   - B. Coupled with dorsal cirrus, gill forms *two finger*like projections — 2f
   - C. Coupled with dorsal cirrus, gills form *more than two finger*like projections — >2f
2. Segment on which gills begin: provide the *number* of the segment on which gills begin (FIG. 14.76) — __
3. Eyes or eyelike prostomial spots (FIG. 14.76)
   - A. Eye spots *large* — la*
   - B. Eye spots *small* — sm
   - C. Eye spots *absent* — x
4. Pygidial or anal cirri, located on terminal body segment (FIG. 14.76)
   - A. *Four* cirri, approximately *equal* in length — 4e*
   - B. *Four* cirri, *unequal* in length — 4u
   - C. *Two* cirri — 2

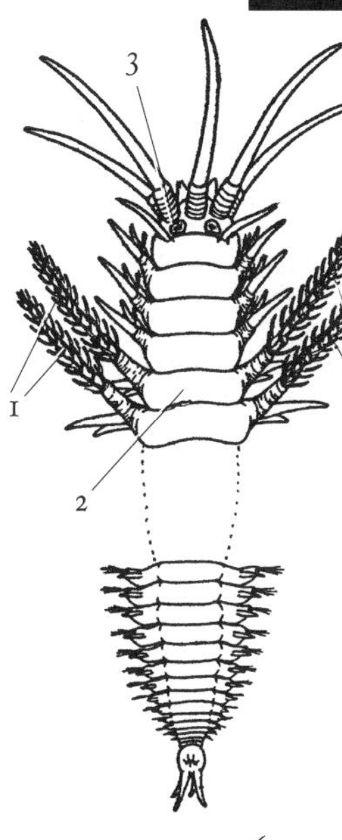

FIG. 14.76

14.47

Shape	Gills	Eyes	Anal Cirri	Species
wh	4–5	la	4e	*Diopatra cuprea,* plumed worm
>2f	5–6	sm,x	4u	*Paradiopatra (=Onuphis) quadricuspis*
2f	9–13	la	2	*Nothria (=Onuphis) conchylega,* mosaic worm
2f	1	x	4e	*Onuphis opalina*

Diopatra cuprea, plumed worm. Reddish brown iridescent; antennae with ringed bases. Chimneylike extension of leathery tube with bits of shell and other debris. Can leave tube to feed. Vi,Sa–Mu,Su,pr–sc,30.5 cm (12 in), (F105,G179,NM422,P250), ☺. (FIG. 14.76. Composite.)

FIG. 14.77

Nothria (=Onuphis) conchylega, mosaic worm. Drags membranous tube along with it. Can be very abundant in cold, deep waters. Bo,Mu–Sa–Gr,Su,pr–om?,15.2 cm (6 in), (F105,G179,P246). (FIG. 14.77. Parapodium, anterior region before gills begin.)

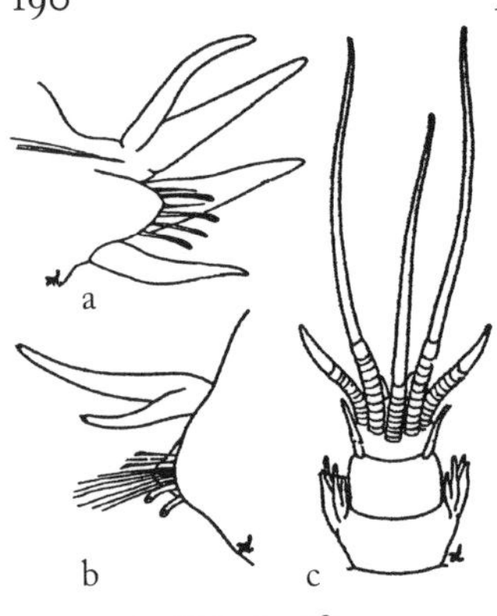

FIG. 14.78

Onuphis opalina. Permanent tube dweller. Bo,Mu–Sa–Gr,Su,pr,12.7 cm (5 in), (F105,P245). (FIG. 14.78. a, first parapodium; b, posterior parapodium; c, anterior end.)

Paradiopatra (=*Onuphis*) *quadricuspis.* Permanent tube dweller. Ac,SG–Mu–Sa–Gr,Su–De,pr,6.3 cm (2.5 in), (F105,P249). (FIG. 14.79. Parapodium, posterior region.)

FIG. 14.79

14.48 **Opheliidae, Sandbar Worms** (from 14.4)

Stout body, with glossy, thick skin and two lateral grooves; also with lengthwise ventral groove. Prostomium is conical without appendages. Parapodia are poorly developed, but bright red gills begin on about segment 10. Pygidium or posteriormost segment with anal lobes and papillae. Digests microflora from ingested sand. (F41)

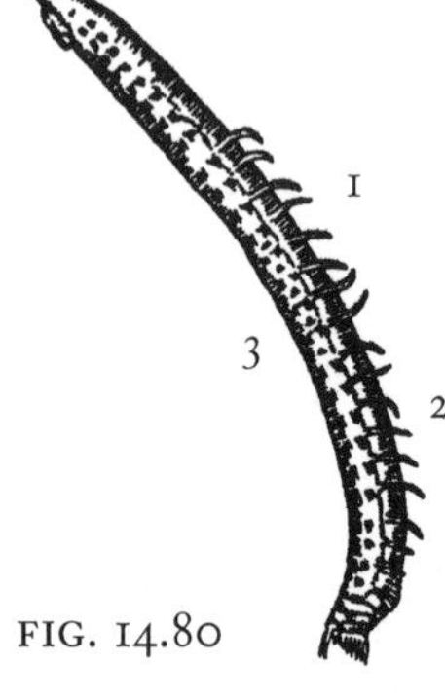

FIG. 14.80

1. Number of pairs of gills or branchiae: provide *number* of pairs of gills (FIG. 14.80) —
2. Appearance of gill surfaces (FIG. 14.80)
   - A. Gills *wrinkled* — wr
   - B. Gill surface *smooth* — sm
3. Length of groove along midventral surface (FIG. 14.80)
   - A. Groove extends for full *body length* — bl
   - B. Groove *absent* from *anterior* 10–12 segments but present thereafter — xa
   - C. Groove *absent* entirely — x

14.49

Gill Pairs	Gill Surface	Groove	Species
11–15	wr	xa	*Ophelia denticulata*
15	sm	x	*Travisia carnea*
18	sm	xa	*Ophelia bicornis*
40+	sm	bl	*Ophelina acuminata* (=*Ammotrypane aulogaster*)

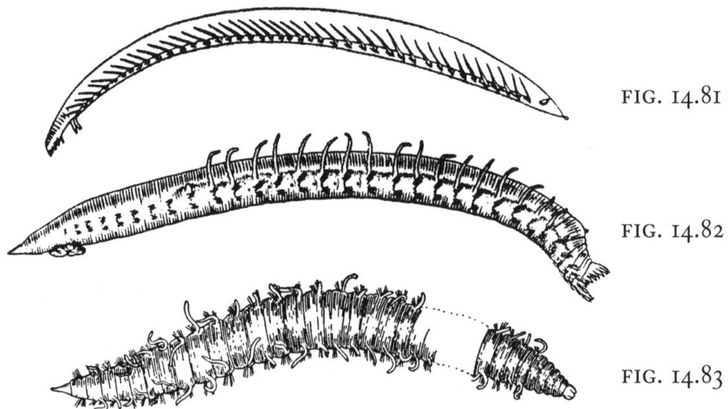
FIG. 14.81

FIG. 14.82

FIG. 14.83

Ophelina acuminata (=*Ammotrypane aulogaster*). Grayish color. Ac–,Sa–Mu,Li–Su,df,7.6 cm (3 in), (F43,G187). (FIG. 14.81. Anterior to right.)

Ophelia bicornis. Ac–,Sa,Es–Li–,df,6.4 cm (2.5 in), (F43,G187).

Ophelia denticulata. Bo,Sa,Es–Li–,df,7.6 cm (3 in), (F43,G187,NM427). (FIG. 14.82)

Travisia carnea. Maggotlike body; lacks anal cirri. Orients head down in sediment. Vi+,Sa,Su,df,7.6 cm (3 in), (F43,G187). (FIG. 14.83. Composite.)

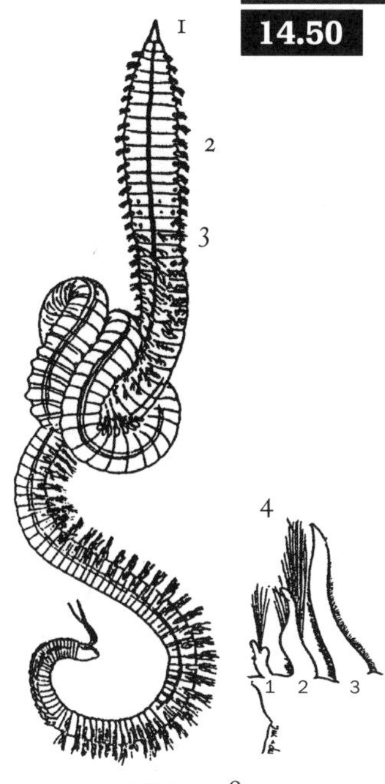

FIG. 14.84

14.50 **Orbiniidae** (from 14.4)

Long, slender body, slightly flattened anteriorly; fragile, break easily on handling. Thoracic area with lateral, biramous parapodia; abdominal area longer, with parapodia dorsally placed. Extended proboscis subglobular (flattened, spherical, or fistlike). Gills or branchiae simple, dorsal, and bright red. Burrowing deposit feeders. (FI4)

1. Shape of prostomium (FIG. 14.84)
   - A. Prostomium *pointed* po*
   - B. Prostomium *rounded* or hemispherical ro
2. Lower portion of parapodia (neuropodium) on thoracic region of body (FIG. 14.84)
   - A. Lower lobe of parapodia with distinct fringe of *papillae* pa
   - B. Papillar margin to lower lobe of parapodia *absent* x
3. Segment on which gills (branchiae) begin: provide *number* of segment on which gills begin (FIG. 14.84) —
4. Number of dorsally directed, fingerlike projections on parapodia from the abdominal region (detects presence of interramal cirri [non-seta-bearing projection between the fingerlike upper and lower parapodial lobes]) (FIG. 14.84)
   - A. *Four* projections: one gill, two seta-bearing projections with a non-seta-bearing interramal cirrus between them 4
   - B. *Three* projections: as above but non-seta-bearing interramal lobe absent 3*

14.51

Prostomium	Papillae	Gills	Projections	Species
ro	x	5 (4–5)	4	*Naineris quadricuspida*
po	pa	5 (4–6)	3	*Orbinia ornata,* ragged worm
po	x	10 (8–10)	4	*O. (=Haploscoloplos) riseri*
po	x	12 (9–17)	3	*Scoloplos* All setae of neuropodial or ventral lobe of anterior segments are the same, sharply pointed: *S. acutus* OR Some neuropodial setae are blunt-tipped in addition to many sharp-tipped ones: *S. armiger*
po	x	16 (11–23)	4	*Leitoscoloplos (=Haploscoloplos) fragilis*
po	x	24 (16–32)	4	*L. (=Haploscoloplos) robustus*

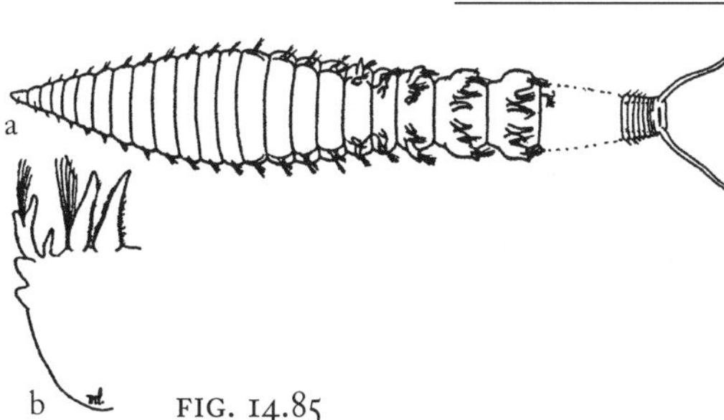

FIG. 14.85

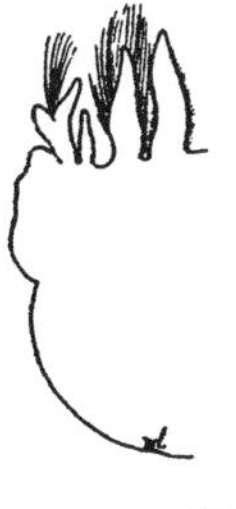
FIG. 14.86

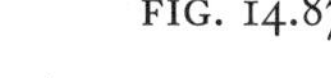
FIG. 14.87

Leitoscoloplos (=Haploscoloplos) fragilis. No spots along dorsal surface; forms aggregations. Bo,Sa–Mu,Es–Li–Su,df,15.2 cm (6 in), (FI6,GI83,P290). (FIG. 14.85. a, composite; b. parapodium; note four dorsal projections.)

Leitoscoloplos (=Haploscoloplos) robustus. Spots along dorsal surface; builds burrows. Bo,SG–Mu–Sa–Gr,Su,df,36.8 cm (14.5 in), (FI6,GI83,P288). (FIG. 14.86. Parapodium with four projections.)

Naineris quadricuspida. Brownish yellow. Rocky shore or offshore. Ac,Mu–Gr–Ha–Al,Li–Su,df,8.1 cm (3.2 in), (FI6,GI83,P279). (FIG. 14.87. Anterior end.)

Orbinia ornata, ragged worm. Dark red. Vi+,Mu–Sa–Gr,Su,df,25.4 cm (10 in), (FI6,GI83,P285). (FIG. 14.88. Thoracic parapodium.)

Orbinia (=Haploscoloplos) riseri. Pale green color. Vi,Sa,Li,df,6.3 cm (2.5 in), (FI6,P288). (FIG. 14.89. Parapodium.)

FIG. 14.88 FIG. 14.89

FIG. 14.90

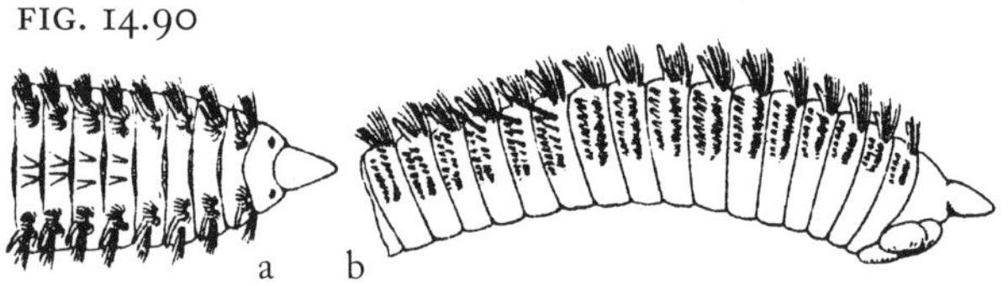

Scoloplos acutus. Bo,Mu–Sa–Gr,Su,df,3.8 cm (1.5 in), (F17,G183,P293).

Scoloplos armiger. Ac,SG–Gr–Sa–Mu,Li–,pr,12.2 cm (4.8 in), (F17,G292). (FIG. 14.90. a, anterior end, dorsal view; b, anterior end, lateral view.)

14.52 **Paraonidae** (from 14.4)

Threadlike with numerous segments. Extended proboscis subconical (roughly cone-shaped), unarmed (no teeth). Parapodia biramous with no ventral cirri; thoracic gills (branchiae) are straplike and begin on segment 4. Parapodial details are used in some species distinctions. Form spiral. Burrows in sand or mud; build tubes of mucus plus surrounding mud or sand. (F17)

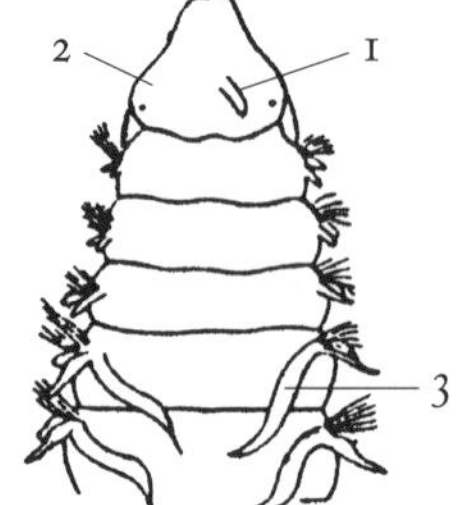

FIG. 14.91

1. Medial antenna on head (FIG. 14.91)
   - A. Medial antenna present on head — ma*
   - B. Medial antenna *absent* — x
2. Eyes (FIG. 14.91)
   - A. A pair of small *eyes* present — ey*
   - B. Eyes *absent* — x
3. Pairs of gills or branchiae: provide *number* of pairs of gills (FIG. 14.91) — ___*
4. Segment on which branchiae or gills begin: provide *number* of segment — ___

14.53

Medial Antenna	Eyes	Gills	Gill Segments	Species
ma	x	9–24	4	*Aricidea* spp.
x	ey	16–25	4	*Paraonis fulgens*
x	x	7–17	4–5	*Paradoneis (=Paraonis) lyra*
x	x	9–14	6–7	*Levinsenia (=Paraonis) gracilis*

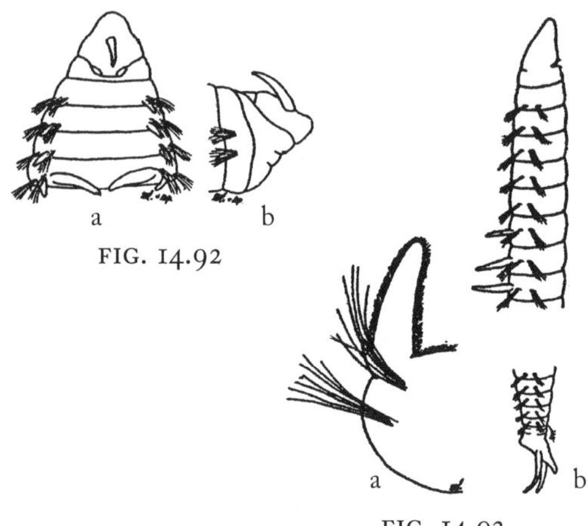

FIG. 14.92

FIG. 14.93

Aricidea spp. See Pettibone (1963, p. 303). Bo,Sa–Mu,Su,df,10.1 cm (4 in), (F18,G184,P303). (FIG. 14.92. a, anterior end, dorsal view; b, anterior end, lateral view.)

Levinsenia (=Paraonis) gracilis. Colorless. Two anal cirri. Motile burrower in soft mud or sand of high organic content. Ac,Gr–Sa–Mu,Su,df,25 mm (1 in), (P301). (FIG. 14.93. a, parapodium with gill; b, composite.)

Paradoneis (=Paraonis) lyra. Colorless. Three anal cirri. Ac,Sa–Mu,Su,df,20 mm (0.8 in), (P300).

Paraonis fulgens. Light with greenish tone. Thin sand tubes. Bo,Sa–Mu,Li–,df,3.3 cm (1.3 in). See Pettibone (1963, p. 303) for other species. (F18,G184,P302). (FIG. 14.94. Parapodium from gilled region.)

FIG. 14.94

14.54 **Pectinariidae** (from 14.4, 14.10)

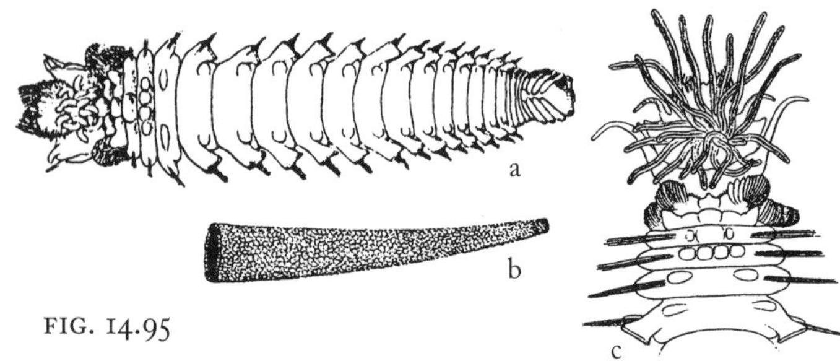

FIG. 14.95

Cone-shaped tube of sand grains. Large, golden setae form comblike arch across flattened head. Bright red gills and large, well-developed parapodia. Distinctions in the shape of body and parapodia divide into three regions. (F120).

Pectinaria (=Cistenides) gouldi, ice-cream-cone worm, trumpetworm. Lives head down in sediment, uses comblike setae and anterior tentacles for feeding. Fanlike gills. Bo,Sa–Mu–SG,Es–Su–De,df,5.1 cm (2 in), (F120,G191,NM433), ☺. (FIG. 14.95. a, worm; b, tube; c, anterior end, dorsal view.)

14.55 **Pholoidae** (see 14.58)

14.56 **Phyllodocidae, Paddleworms** (from 14.6)

Slender, often brightly colored. Anterior segments 1 through 3 with several pairs of tentacular cirri. Prostomium conical or ovaloid with two eyes and four to five antennae. Proboscis is tubular, muscular, and unarmed, although phyllodocids are known or assumed to be active carnivores. Parapodia uniramous with flattened, leaflike dorsal and ventral cirri. Shapes of cirri are useful species characters. A distasteful or toxic mucus is produced for defense. Aciculae (heavy supporting bristles) usually present within parapodia. (F45)

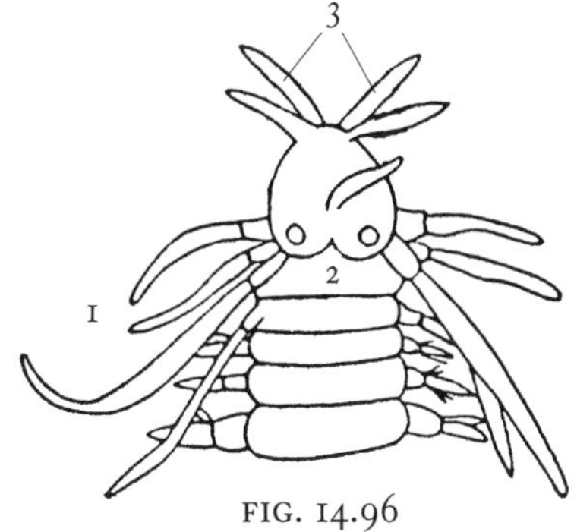

FIG. 14.96

1. Number of pairs of tentacular cirri on anterior segments: provide *number* of pairs (FIG. 14.96) __*
2. Number of segments bearing tentacular cirri that are distinctly visible dorsally: provide *number* of segments (FIG. 14.96) __
3. Number of prostomial antennae: provide *number* (FIG. 14.96) __*
4. Shape of prostomium (FIG. 14.96)
   - A. Distinctly *heart*-shaped, with notch or notches along posterior margin he*
   - B. Basically *triangular*, with pointed front, rounded corners, and no posterior notches tr
   - C. Basically a *straight*, squared, or slightly rounded anterior margin st
   - D. Prostomium *diamond*-shaped, tapers both anteriorly and posteriorly di
5. Overall distinctive color pattern, regardless of colors involved (because coloration tends to vary, compare the unknown to others close to matching this feature)
   - A. Worm is basically *uniform* in its coloration un
   - B. Worm's coloring shows distinct cross-*banding* patterns ba
   - C. Worm's coloring shows distinct *spots* or blotches sp
   - D. Worm's coloring includes distinct lengthwise *stripe* or stripes st

14.57

No. Tentacular Cirri	No. Segments	No. Antennae	Prostomium	Color Pattern	Species
2	1	4	tr	sp	*Eteone*, freckled paddleworms (F49) Second segment without setae: *E. lactea* OR Second segment with setae: *E. longa*
2	1	4	tr	ba	*Eteone trilineata*
2	1	4	tr	un	*E. heteropoda*
3	2	4	st	un	*Mystides borealis*
4	2	4	he	sp	*Phyllodoce maculata*
4	2	4	he	ba	*P. arenae*
4	2	4	he	st	*P. mucosa*
4	2	4	he	un	*P. groenlandica*
4	2	4	st	?	*Nereiphylla paucibranchiata*
4	2	4	di	sp	*Paranaitis speciosa*
4	2	5	he,tr	ba,sp	*Eumida sanguinea*
4	3	5	he,tr	ba	*Eulalia bilineata*
4	3	5	he,tr	sp	*E. viridis*

FIG. 14.97

Eteone heteropoda. Pale yellow or green. Eats nereids; also detritus. Bo,Sa–Mu,Li–Su,pr–df,9.4 cm (3.7 in), (F49,NM413,P72), ☺. (FIG. 14.97. Parapodium, midregion, leaflike dorsal cirrus.)

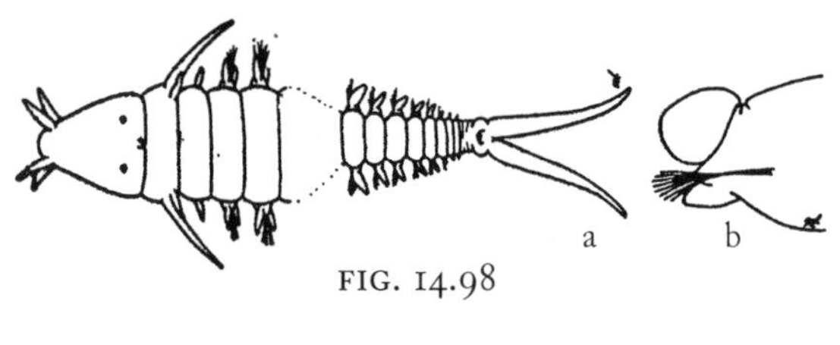

FIG. 14.98

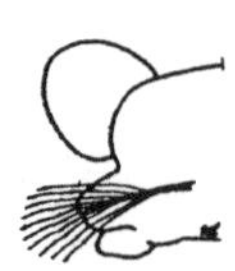
FIG. 14.99

FIG. 14.100

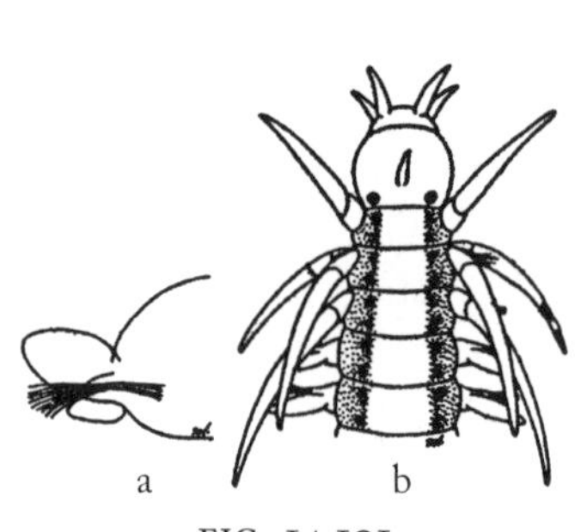

FIG. 14.101

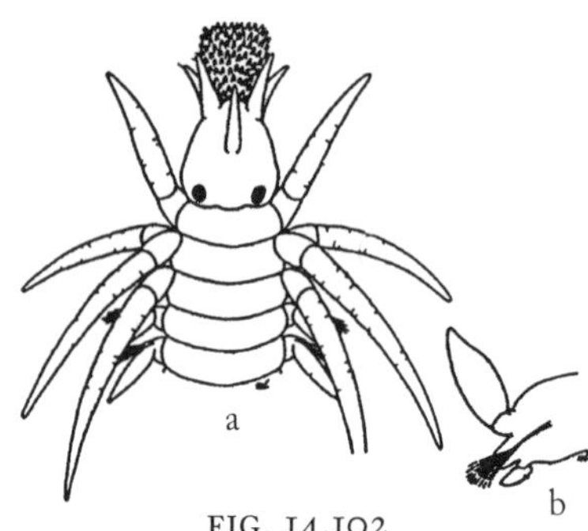

FIG. 14.102

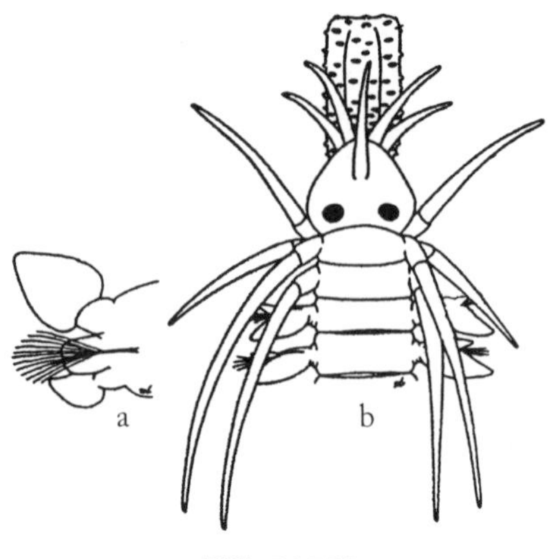

FIG. 14.103

FIG. 14.104

FIG. 14.105

Eteone lactea, freckled paddleworm. Pale or yellow with light brown spots; long, thin. A high intertidal mud burrower. Bo,Sa–Mu–SG,Li,pr,22.9 cm (9 in), (F49,G166,NM413,P70). (FIG. 14.98. a, composite; b, parapodium, midregion, with circular dorsal cirrus.)

Eteone longa, freckled paddleworm. Robust worm; dark green with lighter bands. Eats spionid worms. Bo,SG–Mu–Sa,Li–Su,pr,16.5 cm (6.5 in), (F49,G166,P73). (FIG. 14.99. Parapodium, midregion, with circular dorsal cirrus.)

Eteone trilineata. Yellow with three longitudinal brownish bands. Ac,Sa–Mu,Li–,pr,13 mm (0.5 in), (F49,G166,P71). (FIG. 14.100. Anterior end.)

Eulalia bilineata. Gray, yellowish green, or olivebrown with lateral brown bands. Bo,Ha–Al–Sa,Li–Su,pr,10.1 cm (4 in), (F49,G167,P86). (FIG. 14.101. a, parapodium with oval dorsal cirrus; b, anterior end.)

Eulalia viridis. Pale to dark green with darker spots. Ac–,Ha–Gr–Sa–Mu,Li–,pr–sc,15.2 cm (6 in), (F49,NM414,P85). (FIG. 14.102. a, anterior end with proboscis extended; b, parapodium with lanceolate dorsal cirrus.)

Eumida sanguinea. Color and pattern variable. Often with tunicate, *Apylidium.* Bo,Ha–Gr–Sa–Mu,Li–Su,pr,5.8 cm (2.3 in), (F49,P88). (FIG. 14.103. a, parapodium with heart-shaped dorsal cirrus; b, anterior end with proboscis extended.)

FIG. 14.106

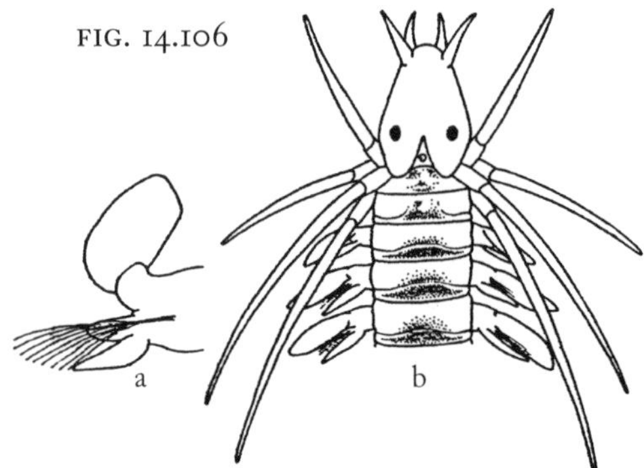

Mystides borealis. Yellowish brown or greenish with darker cirri. Ac,Ha–Sa,Su,pr,15 mm (0.6 in), (F49,P74). (FIG. 14.104. Anterior end.)

Nereiphylla paucibranchiata. Vi?,Mu–Sa,Su,?,?, (F49).

Paranaitis speciosa. Greenish yellow with reddish, dorsal spots. Mussel beds. Bo,Sa–Mu,Li–Su,pr,20 mm (0.8 in), (F50,G167,P75). (FIG. 14.105. Anterior end.)

Phyllodoce arenae. Light or greenish with darker bands and brown spots along midventral line. Ac–,Gr–Sa–Mu,Li–Su,pr,10.1 cm (4 in), (F50,G168,P82). (FIG. 14.106. a, parapodium with elongate oval dorsal cirrus, ventral cirrus slender and pointed; b, anterior end.)

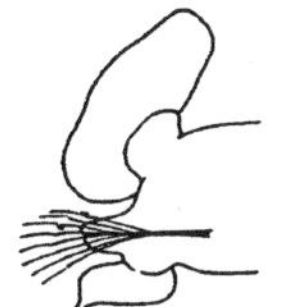
FIG. 14.107

Phyllodoce groenlandica. Uniformly green, grayish brown, or bluish. Bo,Sa–Al,Su,pr,45.7 cm (18 in), (F50,P80). (FIG. 14.107. Parapodium with elongate oval dorsal cirrus, pointed leaflike ventral cirrus.)

Phyllodoce maculata. Yellow or green with blotchy brown. Eats barnacles and polychaetes. Protected by foul-tasting mucus. Ac–,Al–Ha–Sa–Mu,Li–Su,pr,10.1 cm (4 in), (F50,G164,NM413,P78), ☺. (FIG. 14.108. a, parapodium with elongate to oval dorsal cirrus and oval to pointed ventral cirrus; b, anterior end.)

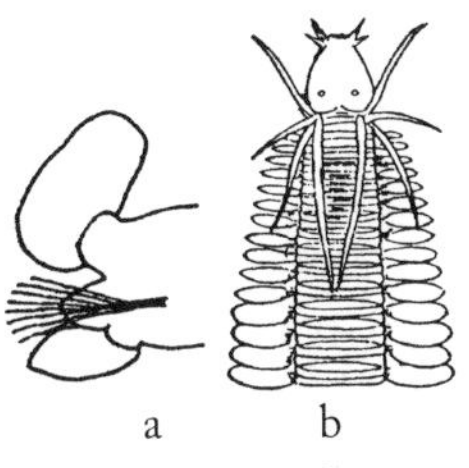

FIG. 14.108

Phyllodoce mucosa. Light with brown middorsal stripe. Female, bright green when gravid; male, light. Bright-green egg masses laid on vegetation. Buries in substrate; no tube. Ac–,Mu–Sa–Al,Li–Su,pr,15.2 cm (6 in), (F50,P81), ☺. (FIG. 14.109. a, anterior end with proboscis extended; b, parapodium with leaflike dorsal cirrus and slender, pointed ventral cirrus.)

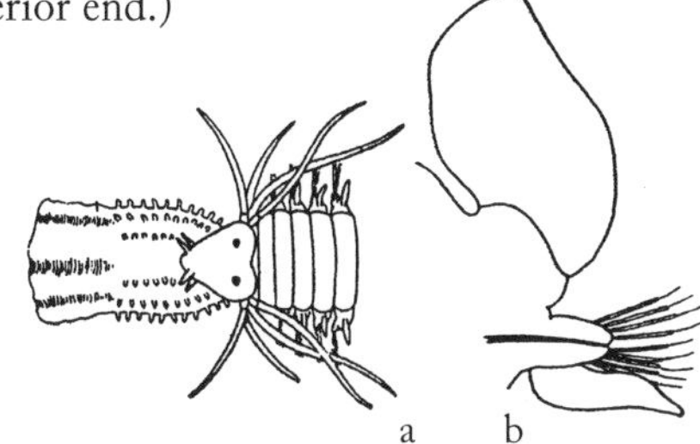

FIG. 14.109

14.58 Polynoidae, Pholoidae and Sigalionidae (Scaleworms), and Aphroditidae (Seamouse) (from 14.4)

FIG. 14.110

Scaleworms (Polynoididae [F55], Pholoidae, and Sigalionidae [F68]) have a flattened body, dorsal surface covered by exposed elytra or scales (actually, modified, flattened dorsal cirri that occur on every other or every few segments). Texture of elytral surface is useful in species indentification. Two-lobed prostomium, with two pairs of eyes, three antennae, and one pair of palps. Peristomium with two pairs of tentacular cirri. Muscular proboscis with two pairs of jaws and a circle of papillae. Parapodia biramous; used for crawling. Some live as commensals with other organisms.

Aphroditidae have short, broad bodies, pointed at both ends; dorsal surface convex, ventral surface flat; 30–40 iridescent segments. Prostomium: one to two pairs of eyes, two long palps, one median antenna, two pairs tentacular cirri. Parapodia biramous. Crawlers, subtidal in mud and silt. (F53)

Polynoidae and Aphroditidae are distinguished from Sigalionidae and Pholoidae by the presence of slender dorsal cirri on those parapodia that do not bear elytrae. All parapodial setae are simple in Polynoidae and Aphroditidae, but neurosetae (found low on parapodia) are compound (more complexly constructed) in Pholoidae and Sigalionidae.

1. Number of pairs of scales or elytra (beware of lost elytrae or incomplete specimens, though bumplike elytrophores on parapodia persist and can be counted even if some scales are lost): provide *number* of scale pairs (FIG. 14.110) —
2. Long, slender dorsal cirri extend from parapodia that do not bear elytra or scales (FIG. 14.110)
   - A. Slender *dorsal cirri* present and visible dorsally dc*
   - B. Slender dorsal cirri *absent* x
3. Presence/density of long, felty setae on dorsal surface (FIG. 14.110)
   - A. Long, felty setae *abundant;* obscures elytra or scales ab*
   - B. Long, felty setae *absent* x
   - C. Density of long, felty setae *intermediate* (i.e., present but do not obscure elytra or scales) in
4. Presence of branched or complex tubercles toward the center of elytra or scales (requires magnification) (FIG. 14.110)
   - A. *Large branched* tubercles in addition to small rounded ones lb*
   - B. *Small branched* (4-pointed) tubercles only sb
   - C. Branched tubercles *absent* x

14.59

Scale Pairs	Dorsal Cirri	Felty Setae	Tubercules	Species
12	dc	x	x	*Lepidonotus squamatus,* twelve-scaled worm
15	dc	ab	x	*Aphrodite hastata,* seamouse
14–18	dc	in	x	*Laetmonice filicornis*
15–16	dc	x	x	Fifteen-scaled worms (peer beneath or pry off anteriormost two scales to observe distribution of eyespots on prostomium) Two pairs of eyes visible on dorsal surface of prostomium: Larger elytral tubercles scattered over each elytrum or scale: *Harmothoe extenuata,* four-eyed, fifteen-scaled worm OR Larger elytral tubercles form single row toward the margin of each elytrum or scale: *Eunoe (=Harmothoe) nodosa* OR One pair of eyes is dorsal and one pair is ventral on prostomium (this ventral pair may be faintly visible through the translucent prostomial tissue): *Harmothoe imbricata*
15–16	dc	x	lb	*Eunoe (=Harmothoe) oerstedi*
40	x	x	x	*Pholoe minuta*
>50	x	x	x	*Sthenelais* Scales or elytrae with simple pointed projections along outer margin: *S. boa,* burrowing scaleworm OR Scales with branched points (resemble arrow tails) along outer margins: *S. limicola*
15	dc	x	sb	*Gattyana cirrosa*
12	dc	x	x	*Lepidonotus sublevis,* commensal twelve-scaled worm
40–50	dc	x	x	*Lepidametria commensalis,* commensal scaleworm

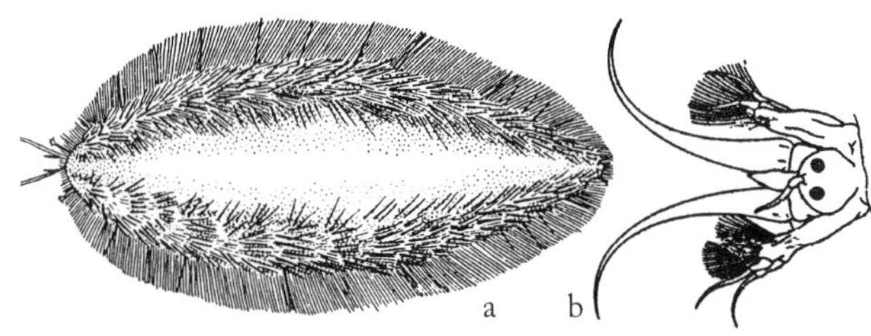

FIG. 14.111

Aphrodita hastata, sea mouse (Aphroditidae). Compare to *Laetomonice* below. Bo,Mu,Su–De,pr,15.2 cm (6 in), (F54,G169,NM414,P13). (FIG. 14.111. a, worm; b, anterior end.)

Eunoe (=Harmothoe) nodosa (Polynoidae). Ac–,Ha–Gr–Sa–Mu,Su,pr,8.9 cm (3.5 in), (F62,P44). (FIG. 14.112. Elytrum with marginal row of tubercles.)

Eunoe (=Harmothoe) oerstedi (Polynoidae). Mottled, dark. Ac,Ha–Gr–Sa–Mu–Al,Su,8.4 cm (3.3 in), (F62,P44). (FIG. 14.113. Elytrum with large, branched tubercles.)

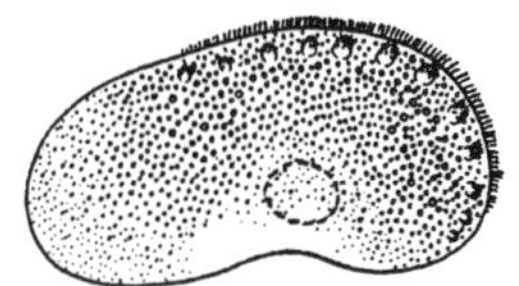

FIG. 14.112

Gattyana cirrosa (Polynoidae). Found among hard and muddy substrate mixtures; can be commensal with other animals. Bo,Ha–Gr–Mu–Al,Su,pr,5.2 cm (2 in), (F62,P28). (FIG. 14.114. Elytrum with small branched tubercles; see enlargement.)

Harmothoe extenuata, four-eyed, fifteen-scaled worm (Polynoidae). Tolerant of low salinity. Bo,Ha–Al–Gr–Sa–Mu,Li–Su,pr,7.6 cm (3 in), (F62,G170,NM418,P41). (FIG. 14.115. a, elytrum with scattered large tubercles; b, head; two pairs, dorsally placed eyes.)

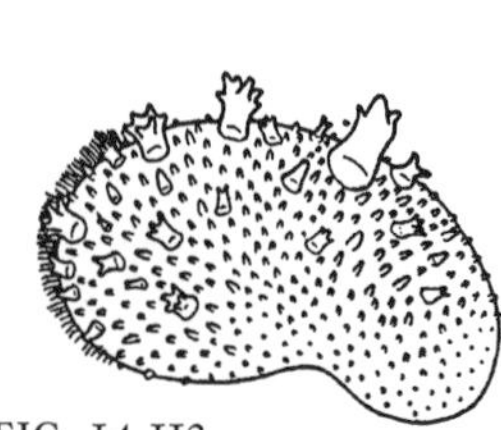

FIG. 14.113

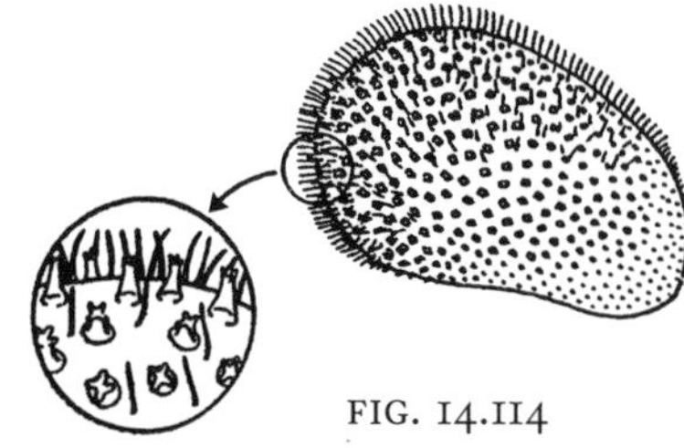

FIG. 14.114

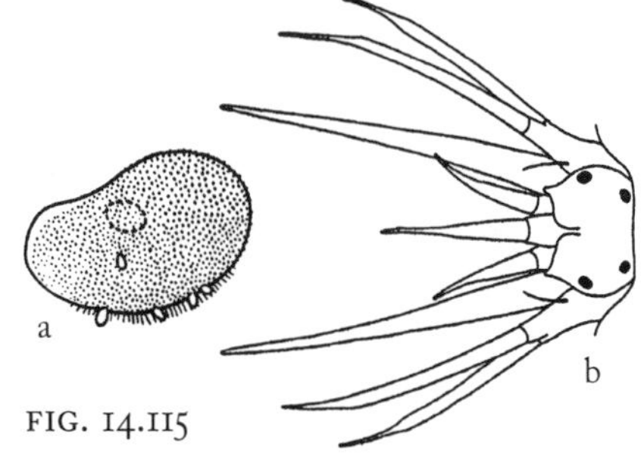

FIG. 14.115

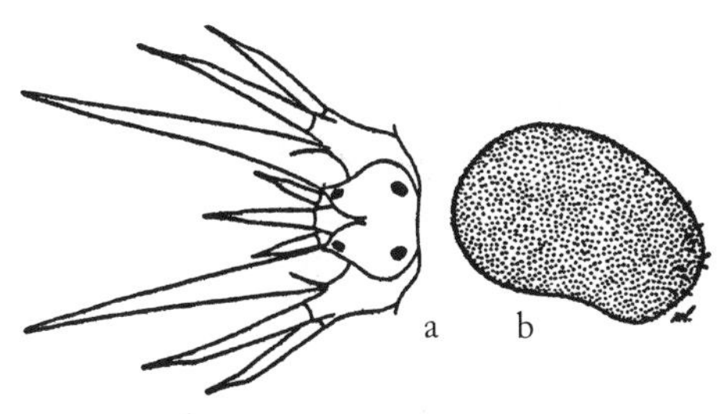

FIG. 14.116

Harmothoe imbricata (Polynoidae). Variable color; often brown, yellow, or reddish, or yellowish with brown median stripe. Eggs brooded under scales of female. Occasionally commensal with other polychaetes. Bo,Ha–Al–Gr–Sa–Mu–SG,Li–Su,pr–gr,6.6 cm (2.6 in), (F62,G170,NM417,P36), ☺. (FIG. 14.116. a, head with four eyes, anterior pair ventral, posterior pair, dorsal; b, elytrum.)

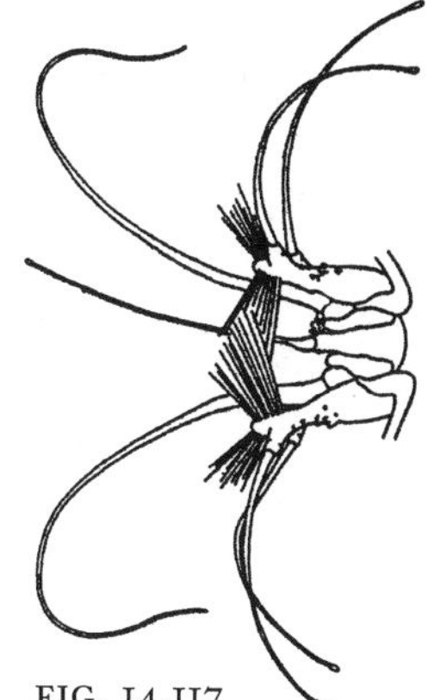
FIG. 14.117

Laetmonice filicornis (Aphroditidae). Felty setae barely cover or fail to cover dorsal surface. Deeper waters. Compare to *Aphrodite* above. Bo,Mu,Su–De,pr,8.9 cm (3.5 in), (F54,P11). (FIG. 14.117. Head.)

FIG. 14.118

Lepidametria commensalis, commensal scaleworm (Polynoidae). Dark, red-purple body; elytra dark with light spot. Commensal with terebellid polychaete, *Amphitrite.* Vi,Co–Sa–Mu,Li,pr,10.1 cm (4 in), (F63,G169,NM415,P19). (FIG. 14.118. Non-elytrum-bearing parapodium; slender, banded dorsal cirrus [dc].)

Lepidonotus squamatus, twelve-scaled worm (Polynoidae). Elytra mottled brown or gray; obvious bumps or tubercles on scale surfaces. Less apt to lose scales than other polynoids. Important fish food. Whole worm illustrates character choices 1 and 2 above. Ac–,Ha–Al–Gr,Es–Li–Su,pr,5.1 cm (2 in), (F63,G169,NM415,P16), ☺. (FIG. 14.119. a, head; b, elytrum; 14.110. Top left, worm.)

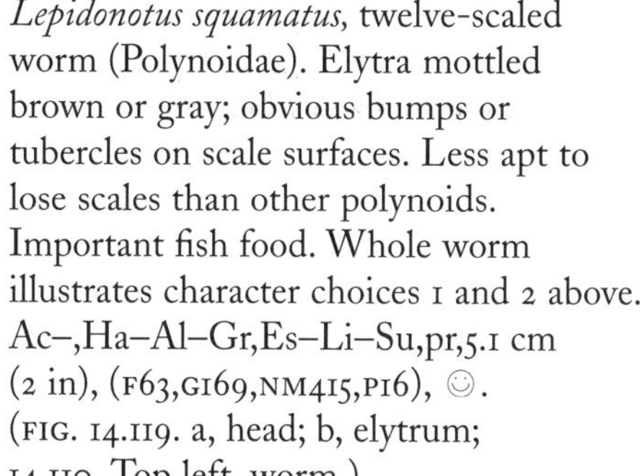
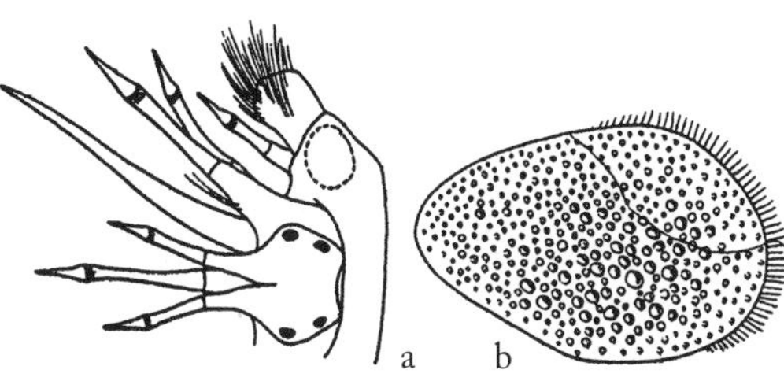

FIG. 14.119

Lepidonotus sublevis, commensal twelve-scaled worm (Polynoidae). Mottled with grayish green or reddish brown; small tubercles on scales. Commensal with hermit crabs; in shell beds. Vi,Co–Ha–Sa,Li–Su,pr,3.3 cm (1.3 in), (F63,G169,P18). (FIG. 14.120. Elytrum.)

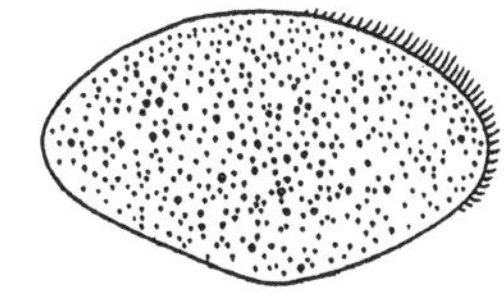
FIG. 14.120

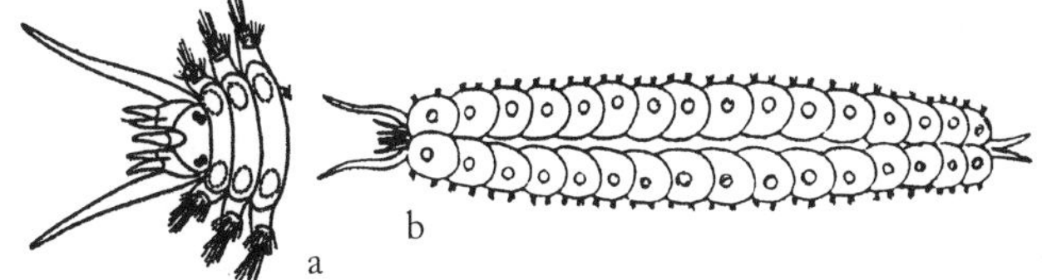

FIG. 14.121

Pholoe minuta (Pholoidae). Scales do not cover middorsal. Yellowish brown to pale pink, mottled with brown. Lack dorsal cirri. Ac–,Ha–Al–Gr–Sa–Mu,Li–Su,pr,25 mm (1 in), (F70,P46). (FIG. 14.121. a, head; b, worm.)

Sthenelais boa, burrowing scaleworm (Sigalionidae). Tan or grayish, opaque scales. Protected beaches and grassbeds. Vi,Gr–Sa–Mu–SG,Li–Si,pr,20.3 cm (8 in), (F40,G171,NM418,P50). (FIG. 14.122. a, anterior end; b, non-elytrum-bearing parapodium.)

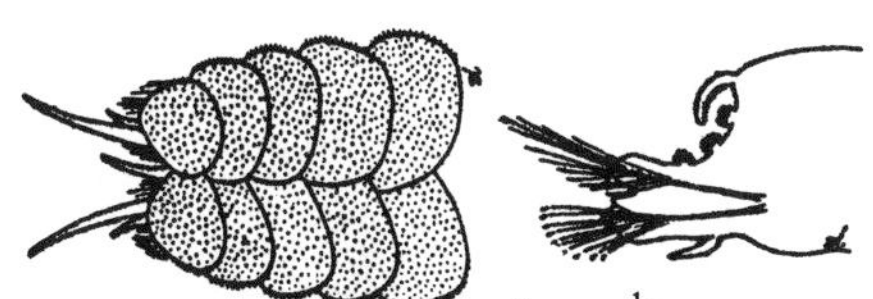

FIG. 14.122

Sthenelais limicola (Sigalionidae). Scales translucent, thin, without tubercles or bumps. Bo,Sa–Mu,Su,pr,10.1 cm (4 in), (F40,P51). (FIG. 14.123. a, elytrum with marginal points; b, parapodium.)

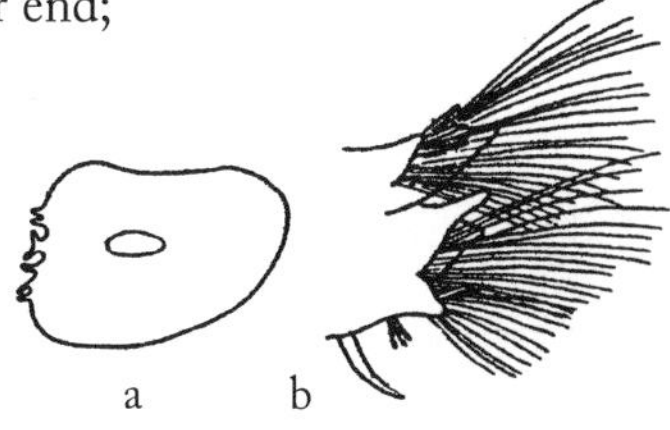

FIG. 14.123

14.60 **Sabellariidae** (from 14.10)

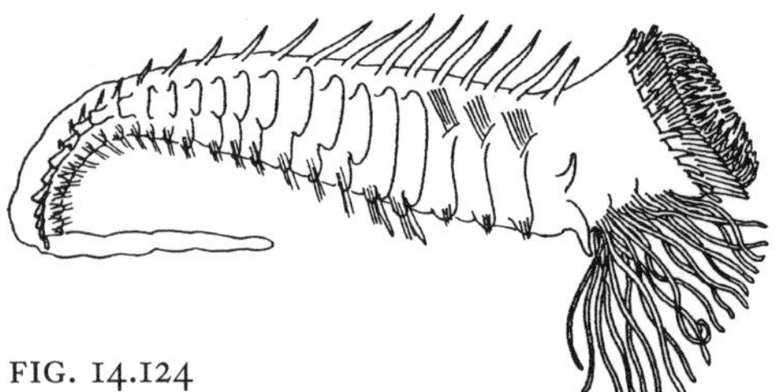

FIG. 14.124

Conical body shape. Head rimmed with stout, golden seta and with a "beard" of oral tentacles. Heavy tubes of sand and gravel cemented together to form blocks or reefs, sometimes over bivalve beds. (F118)

Sabellaria vulgaris, sand-builder worm, reefworm. Pale body with conspicuous golden setae and coiled tentacles. Reefs or mats of sandy tubes crumble easily; often with oyster beds. Vi,Ha–Gr,Su,ff,30 mm (1.2 in), (G189). (FIG. 14.124. Anterior end to right.)

14.61 **Sabellidae, Featherduster Worms** (from 14.8)

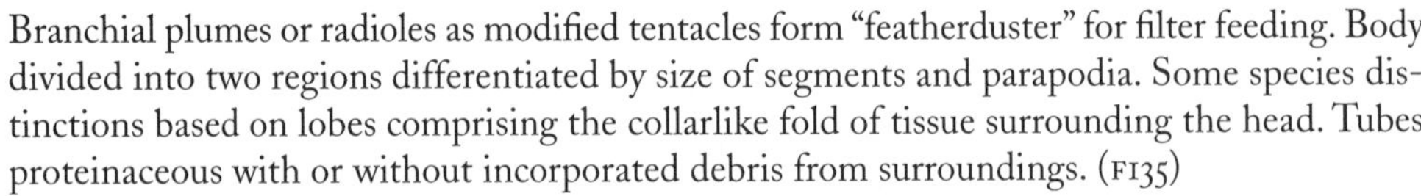

Branchial plumes or radioles as modified tentacles form "featherduster" for filter feeding. Body divided into two regions differentiated by size of segments and parapodia. Some species distinctions based on lobes comprising the collarlike fold of tissue surrounding the head. Tubes proteinaceous with or without incorporated debris from surroundings. (F135)

FIG. 14.125

1. Extent of membranous webbing between elements of branchial plume (FIG. 14.125)
   - A. Membrane extends nearly to *tentacle tips* — tt
   - B. Membrane extends about *one-half* the distance to tips — ½*
   - C. Membrane extends about *one-quarter* the distance to tips — ¼
   - D. Membrane *absent* — x
2. Composition of the worm's tube
   - A. Tube of thick and *gelatinous* — ge
   - B. Tube leathery or muddy, slime *absent* — x*
3. Location of eyes
   - A. Eyes present on *pygidial* (posteriormost) segment — py
   - B. Eyes *paired* along branchial filaments — pa
   - C. Eyes in *single row* — sr
   - D. Eyes *irregularly* scattered along branchial filaments — sc
   - E. Eyes *absent* — x
4. Shape of posterior end of worm
   - A. Posterior end with collarlike *funnel* — fu
   - B. Posterior end with funnel *absent* — x

14.62

Webbing	Tube	Eyes	Funnel	Species
tt	ge	x	x	*Myxicola infundibulum*
½	x	x	fu	*Euchone* Collar at tentacle base without a ventral slit: *E. elegans* OR Collar at tentacle base with a ventral slit: *E. rubrocincta*
½	x	x	x	*Chone infundibuliformis*
¼	x	sr	x	*Pseudopotamilla* (=*Potamilla*) *reniformis*
¼	x	x	x	*Potamilla neglecta*
x	x	py	x	*Fabricia sabella*
x	x	pa	x	*Sabella crassicornis*
x	x	sc	x	*Demonax* (=*Sabella*) *microphthalma*, fanworm

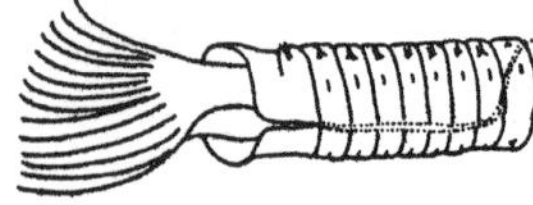

FIG. 14.126

Chone infundibuliformis. Single ventral slit in collar surrounding head. Ac–,Mu,Su,ff,12.2 cm (4.8 in), (F138,G195). (FIG. 14.126. Anterior end.)

Demonax (=*Sabella*) *microphthalma,* fanworm. White, feathery tentacle cluster; small leathery, sandy tubes. Bo,Ha–SG,Li–Su,ff,30 mm (1.2 in), (F140). (FIG. 14.127. Head.)

FIG. 14.127

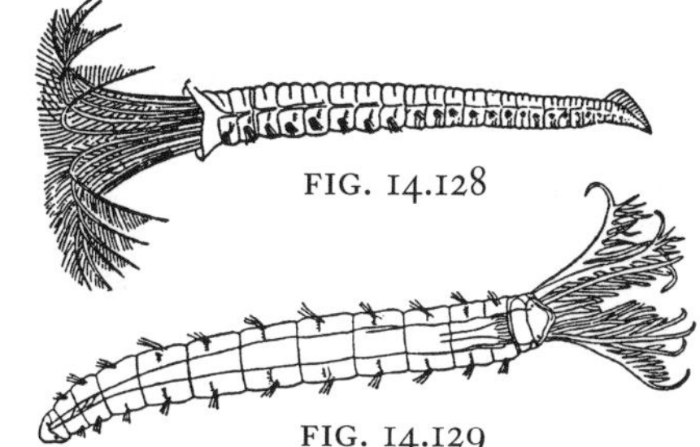
FIG. 14.128

FIG. 14.129

Euchone elegans. Collar surrounding head lacks slit. Ac–,Mu,Su,ff,25 mm (1 in), (F138,G195). (FIG. 14.128)

Euchone rubrocincta. Ac–,Mu,Su,ff,25 mm (1 in), (F138).

Fabricia sabella. Tiny; can leave tube and crawl about backward. Includes a pair of eyes at each end of the body. Bo,Ha–Al,Li–Su,ff,4 mm (0.2 in), (F139,G195), ☺. (FIG. 14.129. Worm with pygidial eyes, anterior to right.)

Myxicola infundibulum. Uniformly dark green or purplish brown. Its giant axon has been used in research. Ac–,Mu,Es–Li–,ff,20.3 cm (8 in), (F139,G195). (FIG. 14.130. Worm in mucous tube.)

FIG. 14.130

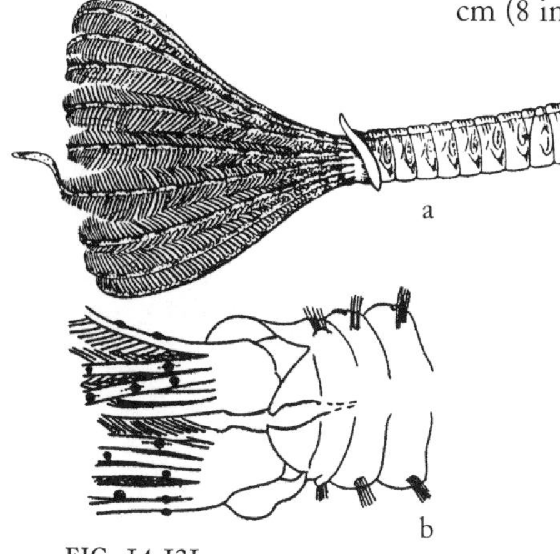

FIG. 14.131

Potamilla neglecta. Reddish orange. Deep, cold waters. Tube with sand grains. Bo,Sa,Su,ff,5.8 cm (2.3 in), (F139,G195).

Pseudopotamilla (=Potamilla) reniformis. Horny, translucent tube with sand and mud. Anterior collar with middorsal and lateral slits. On shelly bottoms. Bo,Ha,Su,ff,11.7 cm (4.6 in), (F139,G195), ☺. (FIG. 14.131. a, anterior end; b, head.)

Sabella crassicornis. Ac,?,Su,ff,5.1 cm (2 in), (F140,G194). (FIG. 14.132. Anterior end.)

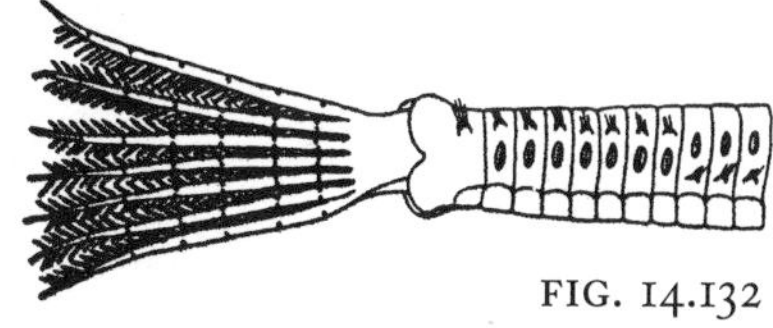
FIG. 14.132

14.63 **Scalibregmidae** (from 14.16)

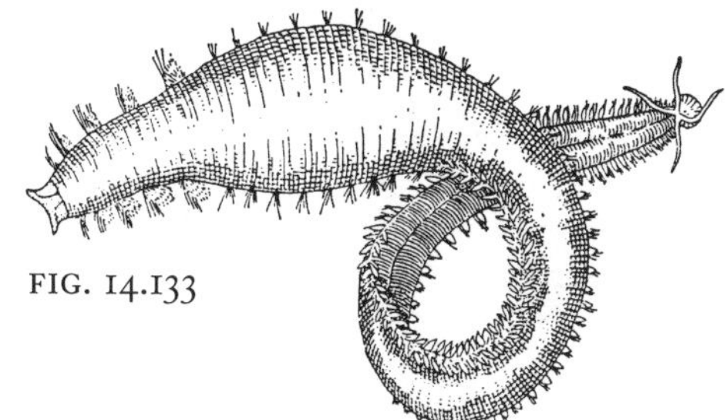
FIG. 14.133

Anterior end enlarged; prostomium with two small anterior tips. Saclike proboscis. Active burrowers. (F43)

Polyphysia crassa. No anal cirri. Short, grublike body. Ac–,Sa–Mu,Su–De,df,30 mm (1.2 in), (F44,G185).

Scalibregma inflatum. With four to seven anal cirri. Yellowish patches on purplish red body. Four pairs of branched gills anteriorly. Bo,Ha–Sa–Mu,Li–,df,10.1 cm (4 in), (F44,G185). (FIG. 14.133)

14.64 **Serpulidae and Spirorbidae, Featherduster or Tubeworms** (from 14.4, 14.8)

Like Sabellidae, a crown of feathery branchiae for filter feeding. Serpulidae (F141) and Spirorbidae (F147) tubes are calcareous, however, and one branchial plume develops into a pluglike operculum that seals off the tube end upon retreat.

1. Calcareous tube shape
   - A. *Coiled* regularly (Spirorbidae) co
   - B. Tubes *irregular* on rocks and shells (Serpulidae) ir
2. Texture of tube
   - A. Distinct *ridges* follow coils ri
   - B. Tube *smooth* or irregularly wrinkled sm
3. Direction of coiling
   - A. Coils toward the *right* or dextral; clockwise ri
   - B. Coils toward the *left* or sinistral; counterclockwise le
   - C. Tube does *not* coil x
4. Relationship among adjacent tubes
   - A. Tubes woven together as tangled *mass* ma
   - B. Tubes may touch, but remain basically *individual* in

14.65

Shape	Texture	Coils	Adjacent Tubes	Species
co	sm	ri	in	*Circeis* (=*Spirorbis*) *spirillum,* dextral spiral tubeworm
co	sm	le	in	*Spirorbis spirorbis* (=*borealis*), sinistral spiral tubeworm
co	ri	ri	in	*S. violaceus*
co	ri	le	in	*S. granulatus*
ir	sm	x	in	*Hydroides dianthus,* limy tubeworm
ir	sm	x	ma	*Filograna implexa,* lacy tubeworm

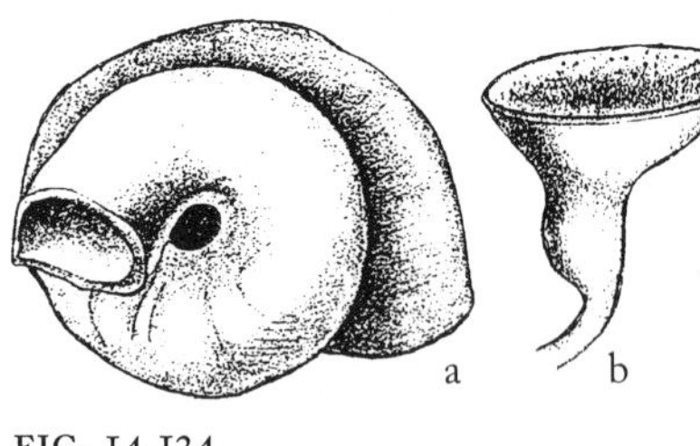

FIG. 14.134

Circeis (=*Spirorbis*) *spirillum,* dextral spiral tubeworm. About 0.64 cm (0.25 in) across coil. On algae, stones, shells, etc. Ac–,Al–Ha,Li–,ff,3 mm (0.1 in), (F151,G196,NM442), ☺. (FIG. 14.134. a, tube; b, operculum.)

Filograna implexa, lacy tubeworm. Can divide by fission. Lacelike webs of calcareous tubes; clusters up to 12 in diameter. Ac,Ha,Su,ff,6 mm (0.25 in), (F144,G196,NM441). (FIG. 14.135. a, worm; b, operculum.)

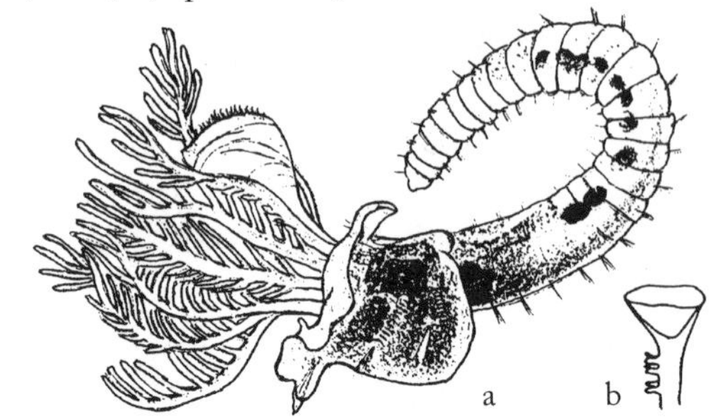

FIG. 14.135

Hydroides dianthus, limy tubeworm. Red or mottled branchial plume. Forms tangles of tubes. Vi+,Ha,Li–Su,ff,7.6 cm (3 in), (F144,G196), ☺.

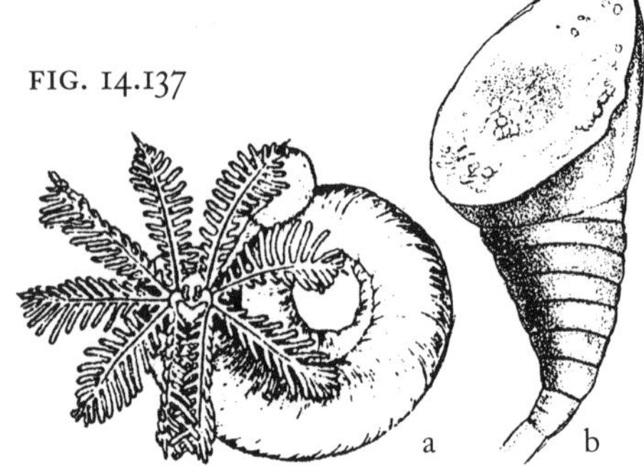

FIG. 14.137

Spirorbis granulatus. Subtidal. Ac–,Al–Ha,Su,ff,3 mm (0.1 in), (F151,G196). (FIG. 14.136. Tube.)

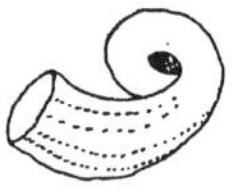

FIG. 14.136

Spirorbis spirorbis (*borealis*), sinistral spiral tubeworm. Size and habitat similar to that of *S. spirillum.* Ac–,Al–Ha,Li–,ff,3 mm (0.1 in), (F151,G196,NM442), ☺. (FIG. 14.137. a, worm in tube; b, operculum.)

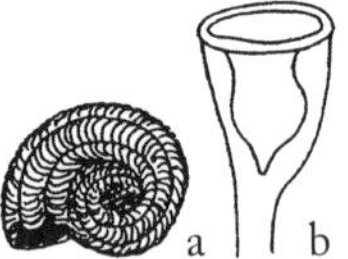

FIG. 14.138

Spirorbis violaceus. Subtidal. Ac–,Ha,Su,ff,3 mm (0.1 in), (F151,G196). (FIG. 14.138. a, tube; b, operculum.)

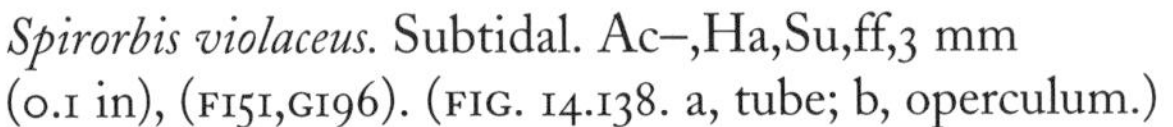

14.66 Sigalionidae (see 14.58)

14.67 Sphaerodoridae (from 14.4)

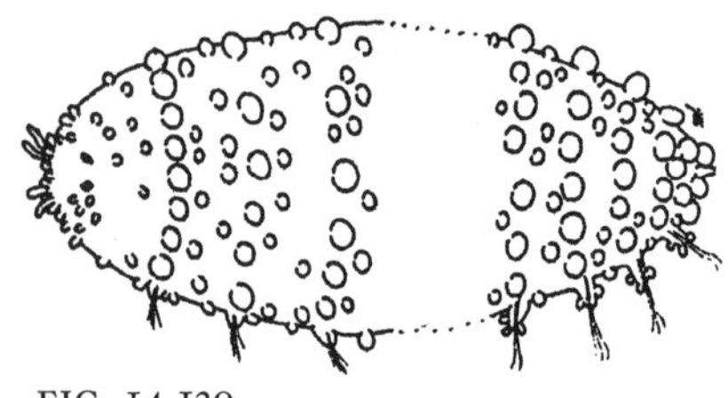
FIG. 14.139

Body short, segments indistinct; parapodia reduced to lobes with setae ventral to spherical bodies. No tentacles or bristles anteriorly. (F97)

Ephesiella minuta. Body covered with rows of spherical papillae with smaller, scattered papillae between. Yellow to white color. In holdfasts of kelp, *Laminaria.* Po–,Al–Gr–Sa,Su,df?,7 mm (0.3 in), (F98,P208). (FIG. 14.139. Composite.)

14.68 Spionidae, Mudworms (from 14.4)

Body not divided into regions. Prostomium is spatulate or pointed; mouth square; long, often coiled, palps for food gathering; palps and gills easily broken off in handling. Parapodia biramous. Some form vertical burrows; others build tubes. (F22)

1. Segment on which gills (branchiae) are visible as small, red, straplike projections above dorsal body surface (FIG. 14.140)
   - A. Provide *number* of segment on which gills begin —
   - B. Gills *absent* x

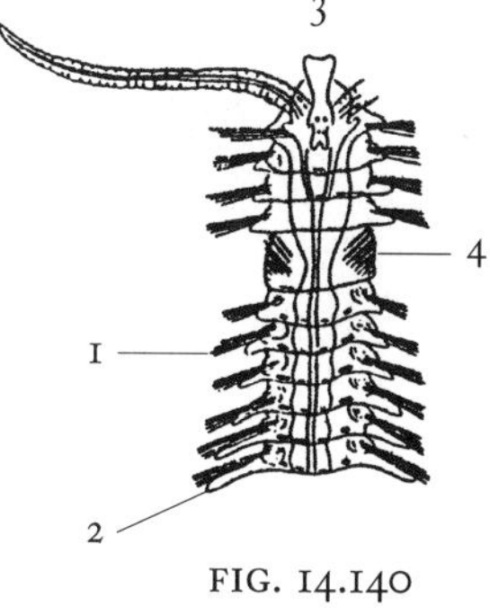

FIG. 14.140

2. Number of pairs of gills (FIG. 14.140)
   - A. *Many* (>6) pairs of gills continue nearly to posterior *end* of the worm — me
   - B. *Many* (>6) pairs, but *absent* from posterior one-third of body — mx
   - C. Few gills (4 or 5 pairs); provide exact *number* of pairs — —
   - D. Gills *absent* — x
3. Shape of prostomium (FIG. 14.140)
   - A. *Anterior* margin distinctly *wider than posterior* margin — a>p
   - B. *Anterior* margin about *equal to posterior* margin — a=p*
   - C. *Anterior* margin distinctly *narrower than posterior* margin — a<p
4. Modifications making individual segments differ from surrounding segments (FIG. 14.140)
   - A. *Fifth* segment without parapodia and with heavy, flattened setae — 5*
   - B. *Fourth* seta-bearing segment with one enlarged seta on each side — 4
   - C. *Second* segment with membranous hood — 2
   - D. Such features *absent* — x
5. Texture of tube (if available)
   - A. Tube *transparent* and clearly segmented or ringed — tr
   - B. Surface primarily of *sand* grains — sa
   - C. Surface primarily *mud* covered — mu
   - D. *No* tube formed — x

14.69

Gills Start	Gill Pairs	Prostomium	Odd Segments	Tube	Species
x	x	a=p	4	tr	*Spiochaetopterus oculatus,* glassy tubeworm
x	x	a>p	x	sa	*Spiophanes bombyx*
1	1	a<p	2	mu	*Streblospio benedicti,* bar-gill mudworm
1	5	a=p	x	?	*Prionospio steenstrupia*
1	4	a=p	x	?	*P. heterobranchia*
1	3	a<p	x	x	*Paraprionospio pinnata,* fringe-gill mudworm
1	me	a>p,a=p	x	sa	*Spio* As many as 16 hooked setae in neuropodium or ventral lobe of parapodia: *S. setosa* OR Six hooked setae in neuropodial lobe: *S. filicornis*
1	mx	a>p	x	mu	*Marenzelleria* (=*Scolecolepides*) *viridis,* red-gill mudworm
2	me	a=p	x	sa	*Scolelepis squamata*
2–3	mx	a>p	x	x	*Laonice cirrata*
6–7	me	a=p	5	mu	*Polydora,* 14.70
13 (11–20)	me	a=p	x	x	*Pygospio elegans*

Laonice cirrata. Prostomium rounded anteriorly. About 12 cirri surround anus. Bo,Mu–Sa–Gr,Su,df,11.9 cm (4.7 in), (F24).

FIG. 14.141

Marenzelleria (=*Scolecolepides*) *viridis,* red-gill mudworm. Builds mud and mucus tube. Vi?,Sa–Mu,Li,df,10.1 cm (4 in), (F25,G188), ☺. (FIG. 14.141. Head.)

Paraprionospio pinnata, fringe-gill mudworm. Large feather-shaped gills; thick, wrinkled palps. Vi,Mu?,Su,df?,?, (F24).

Prionospio heterobranchia. Vi,Sa–Mu,Es–Li–,df,7.6 cm (3 in), (F24,G187).

Prionospio steenstrupia. Tube dweller. Marginally tolerates pollution. Silty sand. Bo,Sa–Mu,Li–De,df,7.6 cm (3 in), (F24,G187).

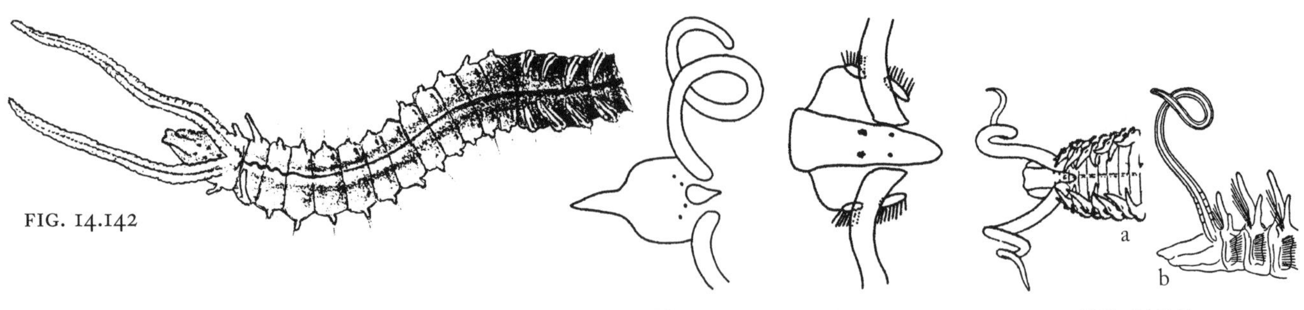

FIG. 14.142

FIG. 14.143 FIG. 14.144 FIG. 14.145

Pygospio elegans. Can use mucous net within its tube to filter feed on plankton. Bo,Ha,Li–,df–ff,18 mm (0.7 in), (F25). (FIG. 14.142. Anterior end.)

Scolelepis squamata. Blue green or whitish with red gills. Palps tougher than most. Crowded sandy tubes. Vi,Sa,Li,ff,3.6 cm (1.4 in), (F25). (FIG. 14.143. Head.)

Spio filicornis. Usually tube-dwelling; dense colonies in sandy beaches. Tentaculate, surface deposit feeders. Bo,Sa,Li–Su,df,30 mm (1.2 in), (F25). (FIG. 14.144. Head.)

Spio setosa. Dark green with red gills; four eyes form square. Body fragile, easily broken with handling. Bo,Sa,Li–,df,7.6 cm (3 in), (F25,G188,NM428), ☺. (FIG. 14.145. a, anterior end, dorsal view; b, anterior end, lateral view.)

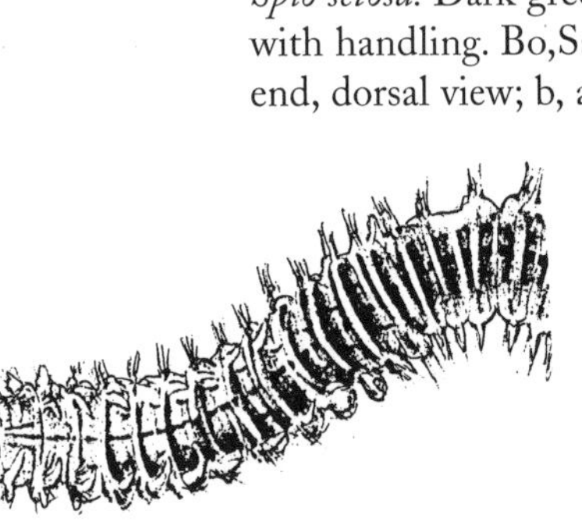

FIG. 14.146

Spiochaetopterus oculatus, glassy tubeworm (Chaetopteridae, see 14.26). Translucent tubes, 6.4 cm (2.5 in) long, with annulations. Feeds like its relative, *Chaetopterus,* by secreting mucous net. Bo,Mu–Sa,Es–Li–Su,ff–df,5.8 cm (2.3 in), (F28,G189).

Spiophanes bombyx. Small worm; sandy tube. Prostomium with frontal horns. Bo,Sa,Es–Li–Su,df,6.1 cm (2.4 in), (F25,G187). (FIG. 14.146. a, prostomium; b, anterior end.)

Streblospio benedicti, bar-gill mudworm. Pair of anteriorly placed gills with greenish bars. Bo,Sa–Mu,Es–Li–,df,7 mm (0.3 in), (F25,G187), ☺.

14.70 **Polydora** (from 14.69)

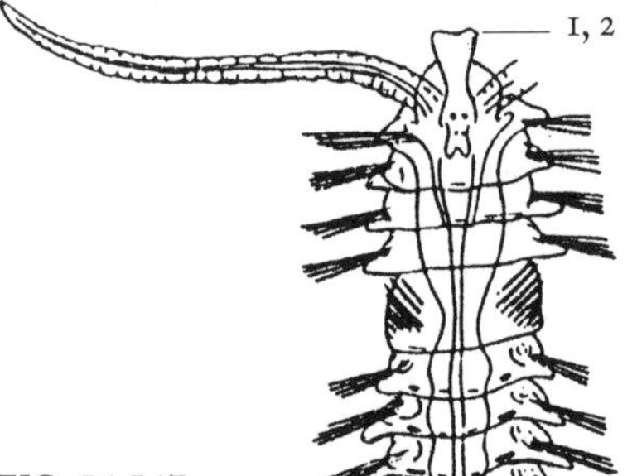

FIG. 14.147

Small spionids that form mud-tubes from which a long pair of threadlike palps extend. Some bore into calcareous shell material. (F24)

1. Presence and extent of caruncle (raised, eye-bearing, prostomial ridge) (FIG. 14.147)
   - A. Provide *number* of seta-bearing setiger at which caruncle ends __*
   - B. Caruncle *absent* x
2. Presence of small tentacle located on caruncle (FIG. 14.147)
   - A. *Tentacle* present te
   - B. Tentacle *absent* x

14.71

Caruncle	Tentacle	Species
x	x	*Polydora commensalis,* hermit-crab worm
2	x	*P. websteri,* oyster mudworm
3	te	*P. cornuta* (=*ligni*), whip mudworm
4	x	Several species, difficult to distinguish; see Blake 1971.
5–9	x	*P. socialis*

Polydora commensalis, hermit-crab worm. Orange to red. Commensal in shells occupied by hermit crabs, especially *Pagurus longicarpus.* Ca+,Sa–Mu–Co,Su,ff–df,3.1 cm (1.2 in), (GI87).

Polydora cornuta (=*ligni*), whip mudworm. Gills on 14 segments. Can form dense, smothering clusters of muddy tubes in shell beds. Bo,Ha–Sa,Es–Li–Su,ff–df,3.3 cm (1.3 in), (GI87,NM428).

FIG. 14.148

Polydora socialis. Orange with faint dark bands, dorsally and ventrally. Bo,Ha–Sa–Mu,Su,ff–df,5.6 cm (2.2 in). (FIG. 14.148. Head.)

Polydora websteri, oyster mudworm. Gills on ca. 100 segments. Often in or on bivalve shells where shell material encases them; mud tubes. Bo,Ha–Sa,Su,ff–df,20 mm (0.8 in), (GI87).

14.72 **Spirorbidae** (see 14.64)

14.73 **Sternaspidae** (from 14.4, 14.12)

Hourglass shape; posterior, coiling gills near pair of brownish red anal shields. Anterior end with setae in semicircular bands; posterior with long stiff setae. Subtidal, muddy bottom. (FI14)

Sternaspis scutata. Light color with dark yellow or reddish brown anal plates. Mud burrower. Ac–,Mu–Sa,Su,df,?, (FI14). (FIG. 14.149)

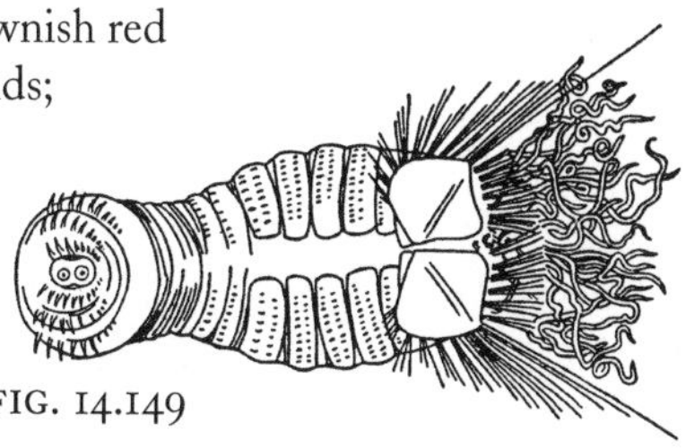

FIG. 14.149

14.74 **Syllidae** (from 14.6)

Small predators of hydroids, sponges, or tunicates. Prostomium with four eyes, two palps, and three antennae. Parapodia uniramous. Distinct sexually dimorphic forms swarm to surface for spawning. Can reproduce asexually by budding. (Members of this family have been the subject of substantial revision in recent years. All identifications to the species level will require verification from sources within the primary literature.) (F79)

1. Body shape
   - A. Body *threadlike* with short prostomial appendages — th
   - B. Body more substantial, *not* threadlike and with long prostomial appendages — x
2. Shape of antennae (FIG. 14.150)
   - A. Antennae *beaded* or moniliform — be*
   - B. Antennae basically *smooth* — sm
3. Degree of fusion along midline between paired, anterior, lobelike palps (FIG. 14.150)
   - A. Palps fully *fused* together — fu
   - B. Palps *partially fused* — pf
   - C. Palps *not* fused, occurs as two lobes — x*
4. Length of antennae compared to palps (FIG. 14.150)
   - A. Antennae *long,* clearly longer in length than palps — lo*
   - B. Antennae *shorter,* or about equal to length of palps — sh
5. Presence of ventral cirri (thin projection from ventral surface of parapodia)
   - A. *Ventral cirri* present — vc
   - B. Ventral cirri *absent* — x
6. Adhesive papillae on body (most easily recognized by adhering debris)
   - A. Body covered with tiny *adhesive papillae* — ap
   - B. Adhesive papillae scarce or *absent* — x

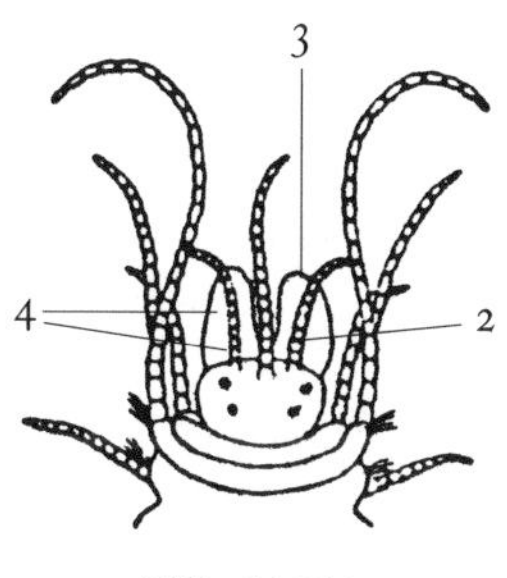

FIG. 14.150

14.75

Shape	Antennae	Palp	Antennae Palps	Ventral Cirri	Adhesive Papillae	Species
x	be	x	lo	vc	x	*Syllis*, 14.78
x	sm	pf	lo	vc	x	Three antennae form row along anterior margin of prostomium: *Eusyllis blomstrandi* OR Medial antenna placed posterior to lateral pair of antennae: *Odontosyllis fulgurans*
x	sm	fu	lo	vc	x	*Eusyllis lamelligera* (=*fragilis*)
x	sm	fu	lo	x	x	*Autolytus* Two darker, longitudinal bands pass just above the parapodia: *A. cornutus* OR Two narrow bands per segment: *A. prolifer* OR Three darker lengthwise dorsal bands: *A. prismaticus*
th	sm	pf	sh	vc	x	*Parapionosyllis longicirrata*
th	sm	fu	lo	vc	x	*Brania clavata*
th	sm	fu	sh	vc	x	*Exogone*, 14.76
th	sm	fu	sh	vc	ap	*Sphaerosyllis erinaceus*

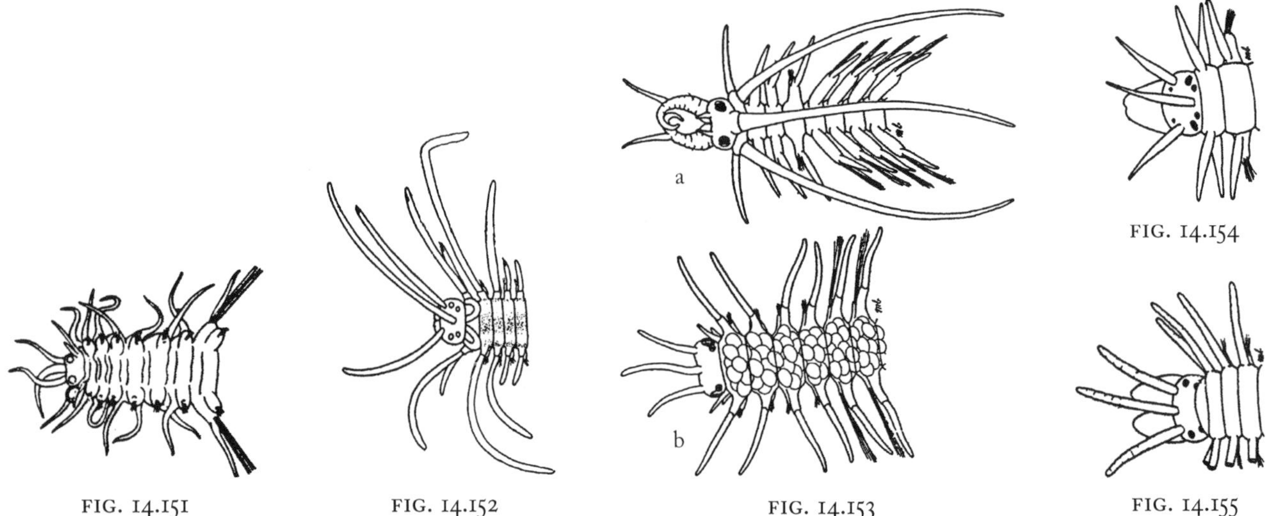

FIG. 14.151 FIG. 14.152 FIG. 14.153 FIG. 14.154 FIG. 14.155

Autolytus cornutus. Cirri and antennae often curled. Long, evertable proboscis; eats hydroids. Tubes on hydroids and algae. Bo,Al–Sa–Mu,Li,pr,20 mm (0.8 in), (F81,P144). (FIG. 14.151. Anterior end.)

Autolytus prismaticus. Probably eats hydroids. Ac,Ha–Al,Li–,pr,25 mm (1 in), (F81,P139). (FIG. 14.152. Anterior end.)

Autolytus prolifer. Bo,Ha–Al,Li,pr,20 mm (0.8 in), (F81,P145). (FIG. 14.153. a, anterior end, male; b, anterior end, female [note eggs].)

Brania clavata. Six eyes, including two small spots toward anterior of prostomium. Bo,Al–Ha–Mu,Li,pr,4 mm (0.2 in), (F82,P133). (FIG. 14.154. Anterior end, fused palps.)

Eusyllus blomstrandi. Pale orange. Ac,Ha–Sa–Mu,Su,pr,3.3 cm (1.3 in), (F82,P119). (FIG. 14.155. Anterior end, partially fused palps.)

Eusyllis lamelligera (=*fragilis*). Vi,Gr–Sa,Su,pr,18 mm (0.7 in), (F82,P120). (FIG. 14.156. Anterior end.)

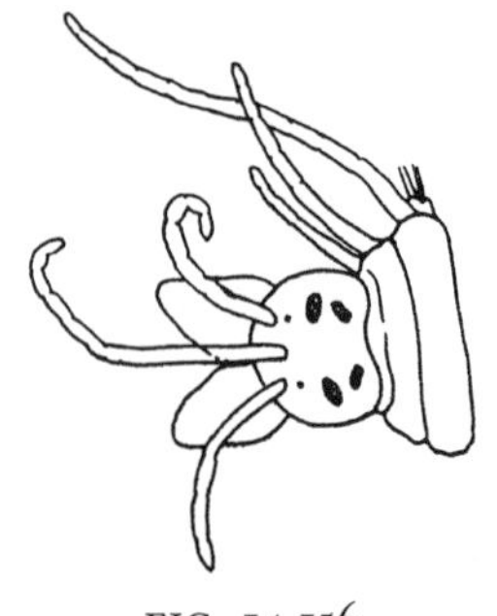

FIG. 14.156

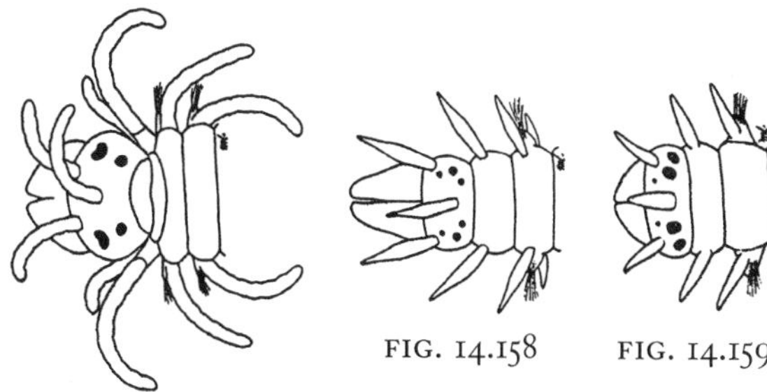

FIG. 14.157 FIG. 14.158 FIG. 14.159

Odontosyllis fulgurans. Often with tunicate *Apylidium.* Vi,Ha–Gr–Sa,Li–Su,pr,20 mm (0.8 in), (F83,P122). (FIG. 14.157. Anterior end, partially fused palps.)

Parapionosyllis longicirrata. Often on tubes of maldanid bambooworms. Vi,Sa–Mu,Li–Su,pr,8 mm (0.3 in), (F83,P132). (FIG. 14.158. Anterior end, long, partially fused palps.)

Sphaerosyllis erinaceus. Six eyes. Bo,Ha–Sa,Li–Su,pr,4 mm (0.2 in), (F84,P135). (FIG. 14.159. Anterior end, stubby, fused palps.)

14.76 **Exogone** (F82) (from 14.75)

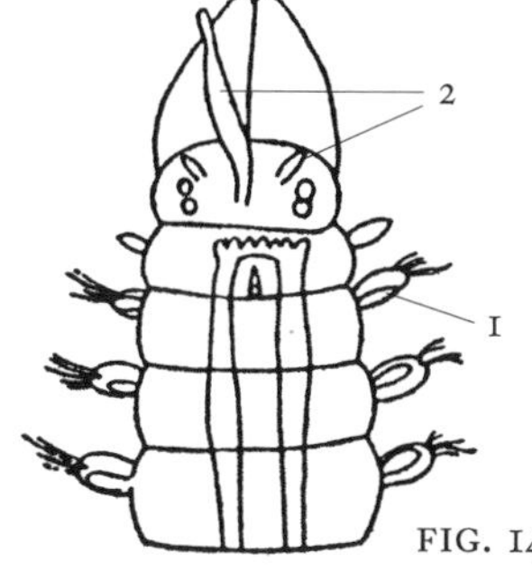

FIG. 14.160

1. Dorsal cirri on second setiger (seta-bearing segment) (FIG. 14.160)
   - A. *Dorsal cirri* present on this segment — dc*
   - B. Dorsal cirri *absent* from this segment — x
2. Relative lengths of three prostomial antennae (FIG. 14.160)
   - A. Antennae approximately *equal* in length — eq
   - B. *Medial* antenna much longer than *lateral* antennae — m/l*

14.77

Dorsal Cirrus	Antenna Lengths	Species
dc	m/l	*Exogone dispar*
dc,x	eq	*E. verugera*
x	m/l	*E. hebes*

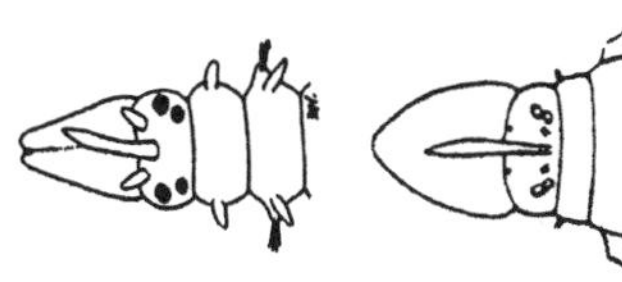

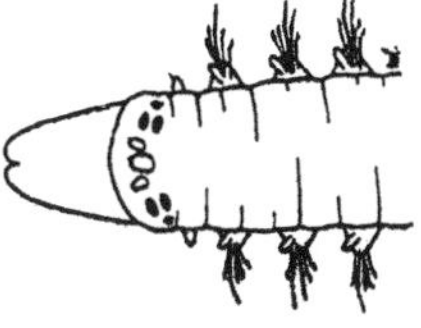

FIG. 14.161 FIG. 14.162 FIG. 14.163

Exogone dispar. Bo,Ha–Al,Li–Su,pr,8 mm (0.3 in), (F82,P130). (FIG. 14.161. Anterior end.)

Exogone hebes. Ac,Sa–Mu–SG,Li–,pr,12 mm (0.4 in), (P131). (FIG. 14.162. Anterior end.)

Exogone verugera. Colorless. Ac,Ha–Al–Sa–Mu,Li–Su,pr,8 mm (0.3 in), (P129). (FIG. 14.163. Anterior end.)

14.78 **Syllis** (F84) (from 14.75)

1. Relative sizes of prostomial antennae (FIG. 14.164)
   - A. *Medial* antenna *longer than lateral* antennae — m>l
   - B. *Medial* antenna *equal to lateral* antennae — m=l*
2. Number of segments comprising beaded dorsal cirri: provide *number* of segments in dorsal cirri (FIG. 14.164) — __

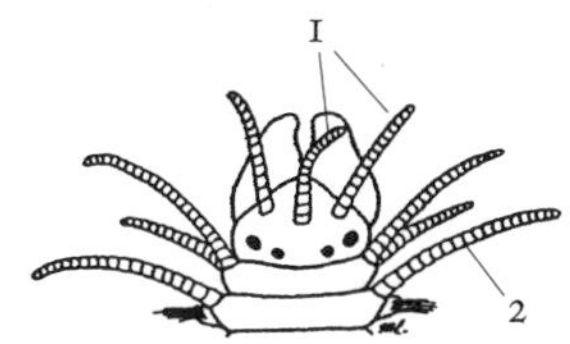

FIG. 14.164

14.79

Antennae	Dorsal Cirri	Species
m>l	50 (40–70)	*Syllis spongiphila*
m=l	10 (7–16)	*S. gracilis*
m=l	22 (11–40)	*S. cornuta*

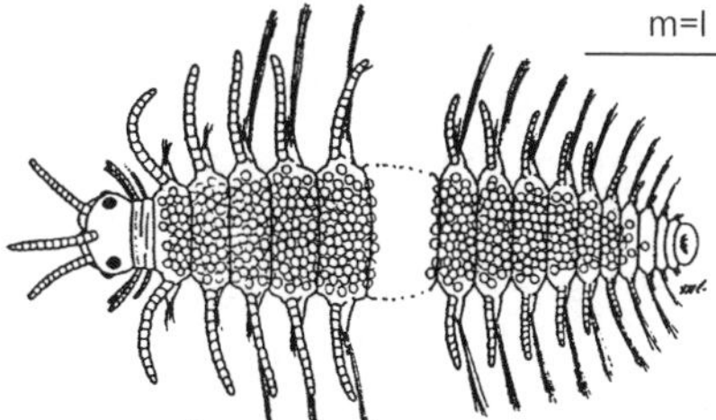

FIG. 14.165

Syllis cornuta. Bo,Sa–Mu,Li–Su,pr,4.6cm (1.8 in), (P118).

Syllis gracilis. Y-shaped setae on parapodia of medial segments. Vi,Ha–Al,Li–Su,pr,5.1 cm (2 in), (P116). (FIG. 14.165. Composite.)

Syllis spongiphila. Vi,Gr–Sa–Mu,Su,pr,25 mm (1 in), (P114).

14.80 **Terebellidae and Trichobranchidae** (from 14.12)

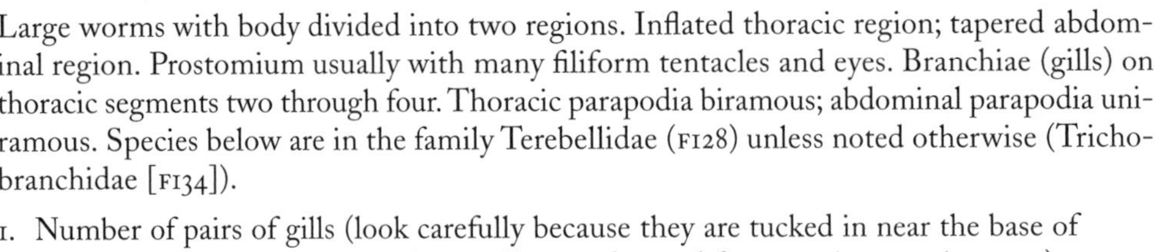

Large worms with body divided into two regions. Inflated thoracic region; tapered abdominal region. Prostomium usually with many filiform tentacles and eyes. Branchiae (gills) on thoracic segments two through four. Thoracic parapodia biramous; abdominal parapodia uniramous. Species below are in the family Terebellidae (FI28) unless noted otherwise (Trichobranchidae [FI34]).

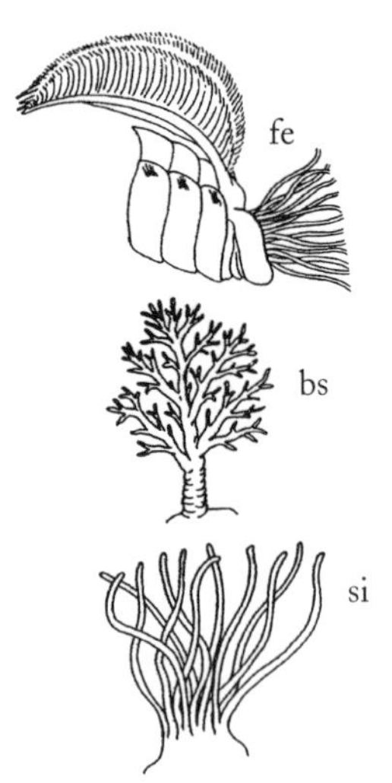

FIG. 14.166

1. Number of pairs of gills (look carefully because they are tucked in near the base of tentacles, which can obscure them; gills are often red from respiratory pigments)
   - A. Provide *number* of gill pairs — __
   - B. Gills *absent* — x
2. Shape of most elaborate gills (some gills may be branched while others are simple) (FIG. 14.166)
   - A. Gills *featherlike* — fe
   - B. Gills *branched* with long support *stem* (i.e., treelike) — bs
   - C. Gills *branched* nearly from base, supporting stem *absent* — bx
   - D. Gills *simple*, unbranched — si
   - E. Gills *absent* — x
3. Number of setigers (segments bearing setae or bristles): provide *number* or "all" — __

14.81

Gill Pairs	Gill Shape	Setigers	Species
1	fe	18	*Terebellides stroemi*
1	bs	16	*Pista maculata*
2	bs	17	*P. cristata*, crested terebellid
2	bx	15–17	*Nicolea venustula*
2	si	all	*Thelepus cincinnatus*
3	si	15	*Trichobranchus glacialis*
3	bx	15–17	*Loimia medusa*, red-spotted terebellid
3	bx	17	*Amphitrite cirrata*
3	bx	23–25	*Neoamphitrite (=Amphitrite) figulus (=johnstoni)*, Johnston ornate terebellid
3	bx	40–50	*Amphitrite ornata*, ornate spaghetti worm
x	x	18–25	*Polycirrus eximius*, red terebellid
x	x	11–13	*P. medusa*
x	x	24–32	*P. phosphoreus*
x	x	all	*Enoplobranchus sanguineus*

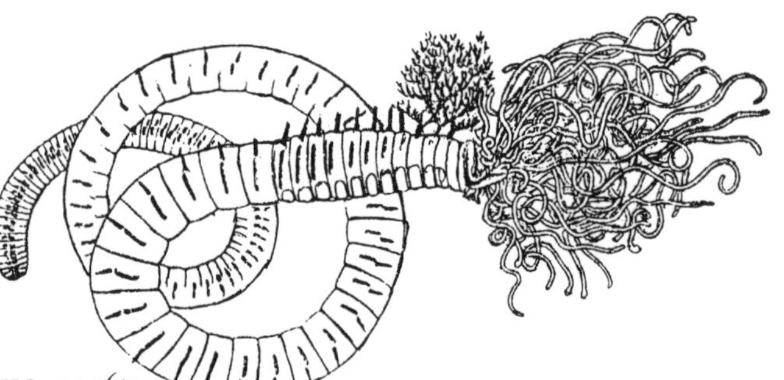

FIG. 14.167

Amphitrite cirrata. Ac,?,Li,df,30.5 cm (12 in), (FI30).

Amphitrite ornata, ornate spaghetti worm. Sand-mud lined, U-shaped burrows. Prefers stiff mud. Look for commensal crab, *Pinnixa* or scale worms. Large worms are 15 cm (6 in). Bo,SM–SG–Mu,Es–Li–,df,38.1 cm (15 in), (FI30,GI92,NM434), ☺. (FIG. 14.167)

Enoplobranchus sanguineus. Blood red; fragile. Midbody parapodia are branched. Vi,Sa–Mu,Li–,df,35 cm (13.8 in), (FI31,GI92). (FIG. 14.166)

Loimia medusa, red-spotted terebellid. Green with red patch near tentacles. Vi,Sa–Mu,Es–Su,df,3.3 cm (1.3 in), (FI31,GI92).

Neoamphitrite (=Amphitrite) figulus (=johnstoni), Johnston ornate terebellid. Rocky. Look for commensal scale worm. Ac–,Sa–Mu,Li–,df,25.4 cm (10 in), (FI30,GI92,NM435).

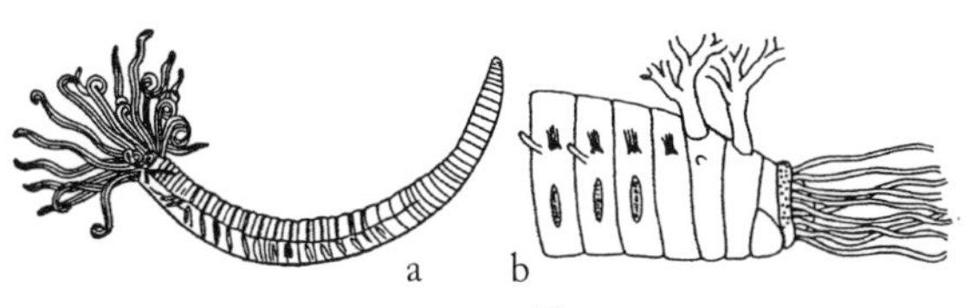

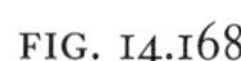
FIG. 14.168

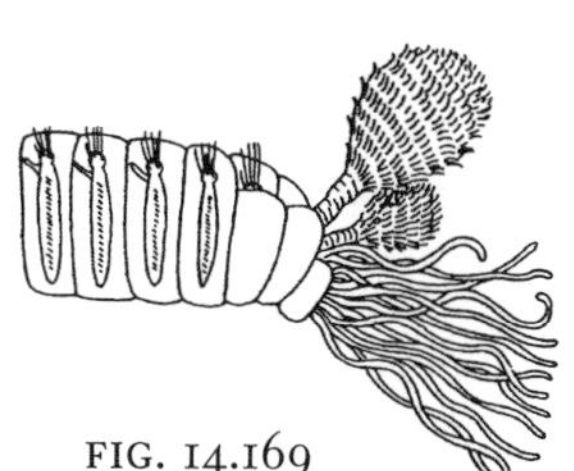
FIG. 14.169

FIG. 14.170

Nicolea venustula. Red with whitespots; bright red gills. Ac–,Al–Sa,Li,df,5.8 cm (2.3 in), (F132,G192), ☺. (FIG. 14.168. a, worm; b, anterior end.)

Pista cristata, crested terebellid. Dark red-orange with brown, pom-pom gills; one gill larger than the rest. Membranous tube with sediments attached. Bo,Sa–Mu,Es–Su,df,8.9 cm (3.5 in), (F132,G192,NM436). (FIG. 14.169. Anterior end.)

Pista maculata. Ac–,Sa–Mu,Es–Su,15.3 cm (6 in), (F132,G192). (FIG. 14.170. Anterior end.)

Polycirrus eximius, red terebellid. Bright red body. Builds burrows, not tubes, in shell beds. Vi,SG–Mu,Es–Li–Su,df,25 mm (1 in), (F132,G192,NM437).

Polycirrus medusa. Tiny, red. Ac,Sa–Mu,Su,df, 7.1 cm (2.8 in), (F132,G192).

Polycirrus phosphoreus. Yellow; luminescent. Ac,Sa–Mu,Su,df,8.1 cm (3.2 in), (F132). (FIG. 14.171)

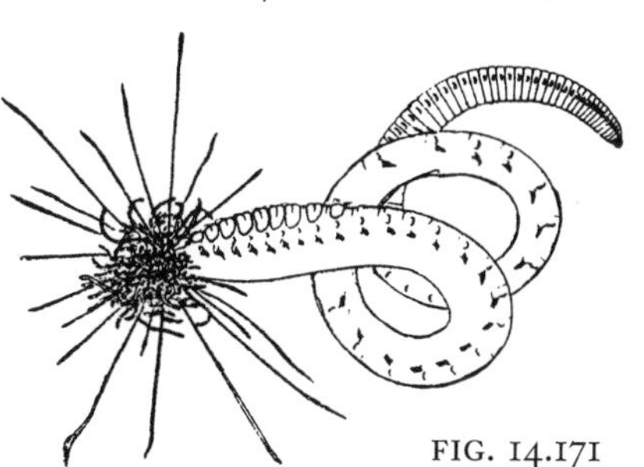
FIG. 14.171

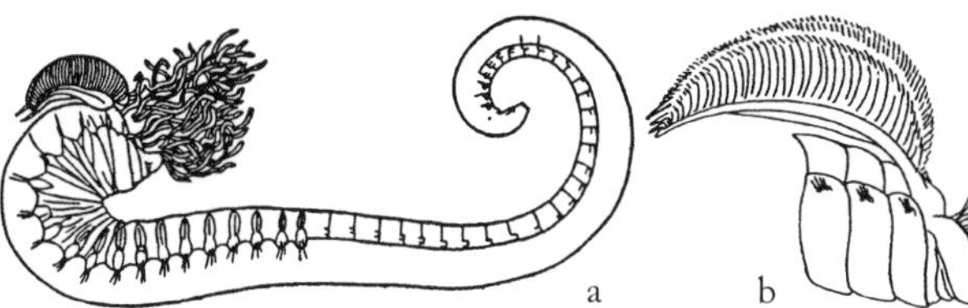

FIG. 14.172

Terebellides stroemi (Trichobranchidae). Pale-pinkish color. Thick based, four-sectioned gill. Long, horizontal, sandy or muddy tubes. Ac,Sa–Mu,Su,df,5.8 cm (2.3 in), (F134). (FIG. 14.172. a, worm; b, anterior end.)

Thelepus cincinnatus. Cream-colored; many eyespots. Dredged; sand-grain tube. Ac,Sa,Li–Su,df,15.3 cm (6 in), (F133). (FIG. 14.173)

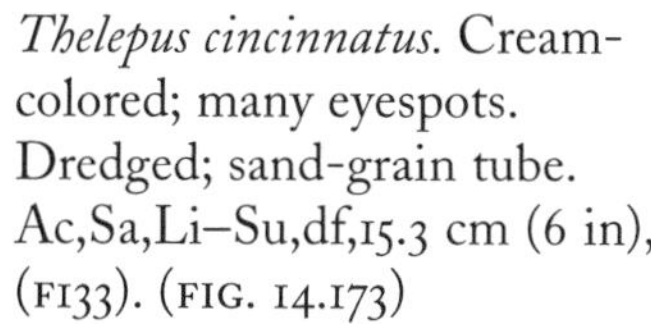
FIG. 14.173

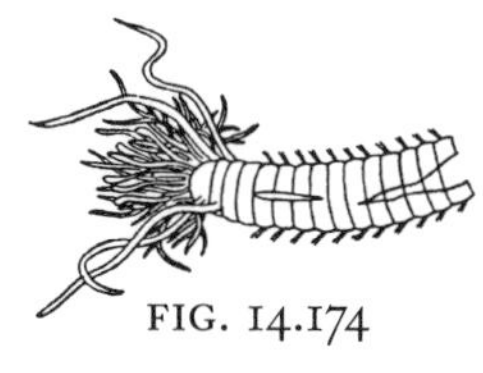
FIG. 14.174

Trichobranchus glacialis (Trichobranchidae). Orange to red with reddish gills; tentacles purplish. Ac–,Al,Li–Su,df,30 mm (1.2 in), (F13). (FIG. 14.174. Anterior end.)

14.82 **Tomopteridae (Plankton Worms)** (from 14.4)

Transparent, leaf-shaped body with large parapodia and one long pair of tentacular cirri (not antennae). Active, voracious, cold-water, planktonic predators. Nocturnal vertical migrators. (F98).

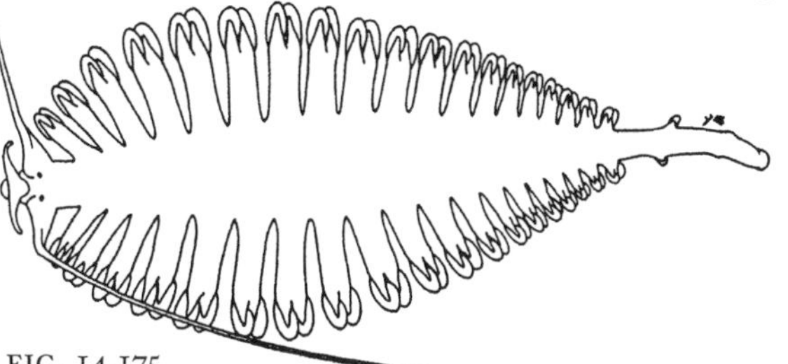
FIG. 14.175

Tomopteris helgolandica. Distinct, cylindrical tail present. Bo,Pe,Ps–Pd,pr,8.9 cm (3.5 in), (G168). (FIG. 14.175)

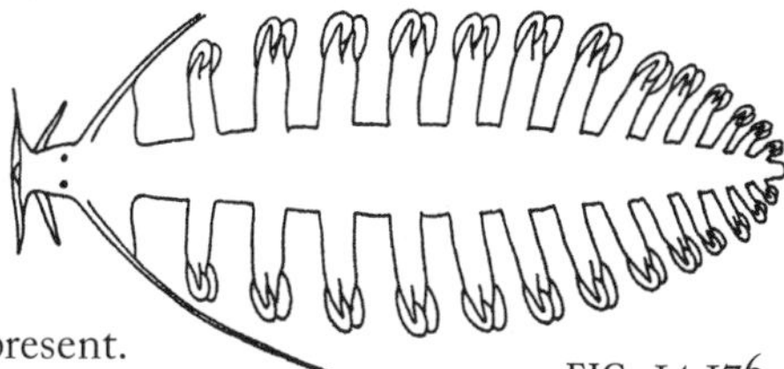
FIG. 14.176

Tomopteris septentrionalis. No tail; young with two pairs of tentacular cirri in addition to antennae. Bo,Pe,Ps–Pd,pr,25 mm (1 in), (G168). (FIG. 14.176)

14.83

Class Oligochaeta

Oligochaetes are small, segmented, earthwormlike animals lacking both anterior appendages and lateral parapodia or flaps. They may be distinguished from the superficially similar family of polychaete worms, the Capitellidae, by the absence of *hooded setae* (bristle-tips cloaked in a thin, clear envelope) in the abdominal region of oligochaetes, which are present in capitellids. Typical segments bear groups of *setae* (bristles) on each side; two groups dorsally and two ventrally. The types of setae and the shapes of their tips are useful features in making identifications. The head comprises an anteriormost, dorsal lobe, the *prostomium,* followed by the first true segment, the *peristomium,* which, as its name suggests, surrounds the mouth opening. For anatomical reference, oligochaete segments can be numbered (with Roman numerals), counting the combined prostomium/peristomium as I. It is useful to look carefully to detect the segment numbers in which openings to the *spermathecae* (seminal receptacles for sperm storage following copulation) and the *male genital pore* may be located. The swollen *clitellum* region, familiar from earthworm construction, is found only in sexually mature marine oligochaetes and, even then, is reduced in size.

This key is designed to distinguish among the three major families of local marine oligochaetes using external features alone. A few of the mostly commonly reported species are listed, but internal details of the reproductive system must be examined for positive specific identifications. Techniques for examining these structures are beyond the scope of this manual. Interested users should refer to Cook and Brinkhurst (1973) or Brinkhurst (1982).

Additional reference: (c) Cook, D. G., and R. O. Brinkhurst, 1973.

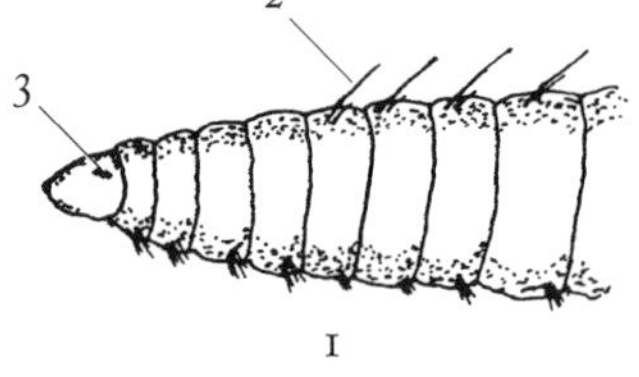

FIG. 14.177

1. Shape of ventral setae (FIG. 14.177)
   - A. All ventral setae bear *simple* (single-pointed) tips — si
   - B. At least some ventral setae are *bifid* (two-pointed) — bi
   - C. Ventral seta *absent* — x
2. Presence of long, thin, "hair" setae (FIG. 14.177)
   - A. Hair setae are found *dorsally* only — do*
   - B. Hair setae *absent* — x
3. Presence of eyes (FIG. 14.177)
   - A. *Eyes* present — ey*
   - B. Eyes *absent* — x
4. Body coloring (a weak but possibly helpful character; exceptions occur)
   - A. *Reddish* — re
   - B. *Whitish* — wh
   - C. *Translucent* — tr
5. Body wall
   - A. Body wall *thin,* worm appears fragile — tn
   - B. Body wall *thicker,* worms appears more rigid — tk
6. Number (in Roman numerals) of the segments in which the male pore and pores to the seminal receptacles are located (pores are tiny and ventrally placed; setae may be modified or absent in these segments): provide segment *numbers* — __

14.84

Ventral Setae	Hair Setae	Eyes	Color	Body Wall	Segments with Pores	Family
bi	do,x	ey,x	tr	tn	V,VI	Naididae, 14.86
bi	do,x	x	re	tn	X,XI	Tubificidae, sludge worms, 14.87
si,x	x	x	wh	tk	V,XII or XVIII	Enchytraeidae, pot worms, 14.85

Families of Oligochaeta (arranged alphabetically)

14.85 **Enchytraeidae, Pot Worms** (from 14.84)
(FIG. 14.178. Generalized anterior end.)

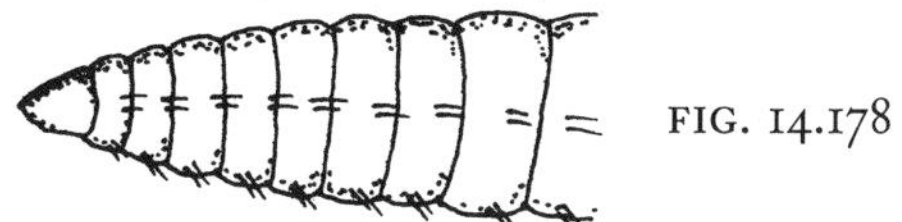

FIG. 14.178

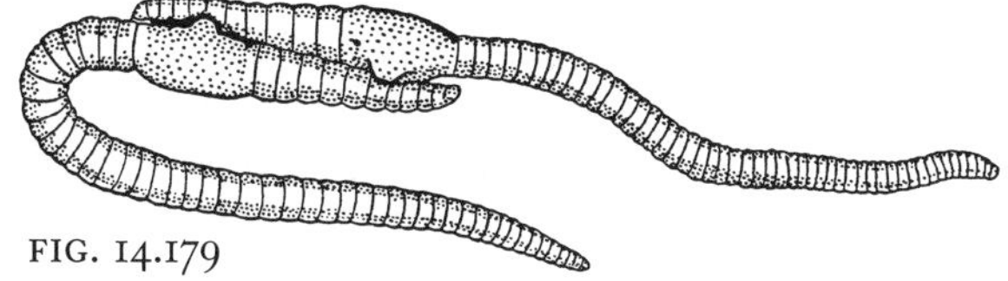

FIG. 14.179

Enchytraeus albidus. Small, fewer than 60 segments; pinkish white color; 2–9 setae per group. High tide line; decaying seaweed. Bo,Al–Mu–SG,Es–Li,df,3.8 cm (1.5 in), (c20). (FIG. 14.179. Pair of worms copulating.)

14.86 **Naididae** (from 14.84)
(FIG. 14.180. Generalized anterior end.)

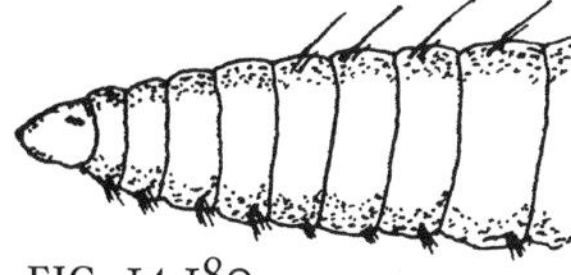

FIG. 14.180

Chaetogaster limnaei. No dorsal setae. Commensal in mantle chamber or kidneys of pulmonate gastropods. Ac,Co,Es–SM–Li–Su,pr,3.5 cm (1.4 in), (c18).

Paranais litoralis. Thirteen to 14 body segments; dorsal setae missing on segments 2 through 4. Ac–,Sa–Mu,Es–Li–Su,df,12 mm (0.5 in), (c18).

14.87 **Tubificidae, Sludge Worms** (from 14.84)
(FIG. 14.181. Generalized anterior end.)

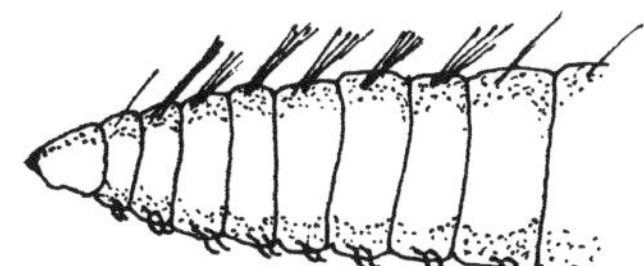

FIG. 14.181

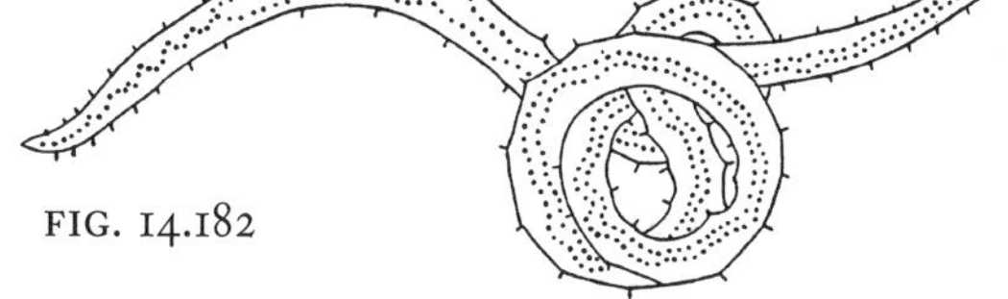

FIG. 14.182

Clitellio arenarius. Reddish color; 2–3 setae per group. Setae with small bifid or trifid tips; ventral setae missing from the eleventh segment. Bo,Sa–Gr,Li,df,6.3 cm (2.5 in), (c19). (FIG. 14.182)

Limnodrilus hoffmeisteri. Smooth body; dorsal setae present on all segments except the peristomium. About seven seta per bundle toward anterior end. Freshwater species that can invade brackish water. Vi,Sa–Mu,Es–Su,df,35 mm (1.4 in), (c19).

Tubifex longipenis. No missing setae; posterior setae single-pointed. Ac,Sa,Su,df,?, (c19).

Tubificoides (=*Peloscolex*) *benedeni.* Body surface papillate. Two setae per bundle; some setae may be trifid. High-tide line. Ac–,Ha–Sa–Gr,Es–Li–Su,df,5.5 cm (2.2 in), (c19).

FIG. 14.183

14.88 **Class Hirudinea, Leeches**

Most frequently encountered marine leeches are blood-sucking ectoparasites of fishes. Their underlying annelid segmentation is often partially masked by secondary annulations (superficial subdivisions of segments). Anterior and posterior suckers are used for attachment usually to site-specific locations on the external surface of particular host species. Some leeches bear ocelli (eyespots) on their suckers, features most easily seen on fresh, live specimens. All temperate marine leeches are in the family Piscicolidae.

Reference: (s) Sawyer, R. T., A. R. Lawler, and R. M. Overstreet, 1975.

1. Body shape in imaginary cross-section (FIG. 14.183)
   - A. Body nearly *cylindrical* cy
   - B. Body clearly *flattened* fl*
2. External gills (FIG. 14.183)
   - A. Pairs of conspicuous, foliose *gills* present gi*
   - B. Gills *absent* x

3. Comparative size of the posterior sucker (FIG. 14.183)
   - A. Posterior sucker *large,* equal to or larger than greatest body width — la*
   - B. Posterior sucker *smaller,* clearly less than greatest body width — sm
4. Presence of eyes on the posterior sucker (FIG. 14.183)
   - A. Provide *number* of pairs of *eyes* — __ey
   - B. Eyes *absent* — x
5. Host type
   - A. *Ray* or skate — ra
   - B. *Sculpins* — sc
   - C. *Other* fish — ot

14.89

Shape	Gills	Posterior Sucker	Eyes	Host	Species
cy	x	sm	x	ra	*Oxytonostoma typica*
cy	x	la	2ey	ot	*Calliobdella vivida*
fl	x	la	>3ey	sc	*Oceanobdella microstoma*
fl	x	sm	x	ot	*Austrobdella (=Pisciola) rapax*
fl	gi	la	x	ra	*Branchellion torpedinis*

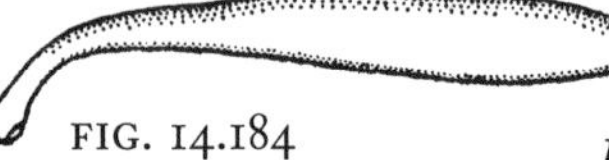
FIG. 14.184

Austrobdella (=*Pisciola*) *rapax.* Dark green with rows of white spots laterally. On upper side of summer flounder. Bo,Co,Ps,pa,?, (s645). (FIG. 14.184)

Branchellion torpedinis. Thirty-three pairs of gills; brownish purple, speckled with white. Six eyes on oral sucker. On tail or between eyes of rays. Vi,Co,Ps,pa,30 mm (1.2 in), (s638). (FIG. 14.185)

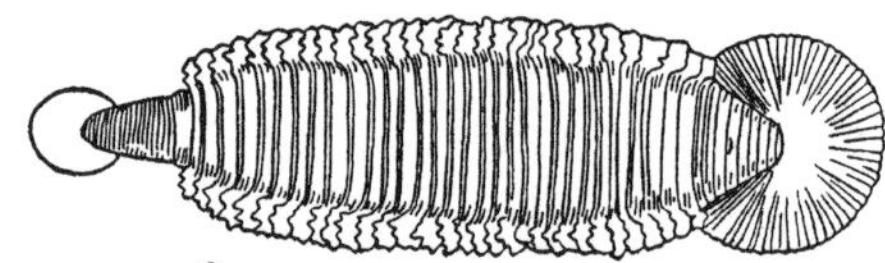
FIG. 14.185

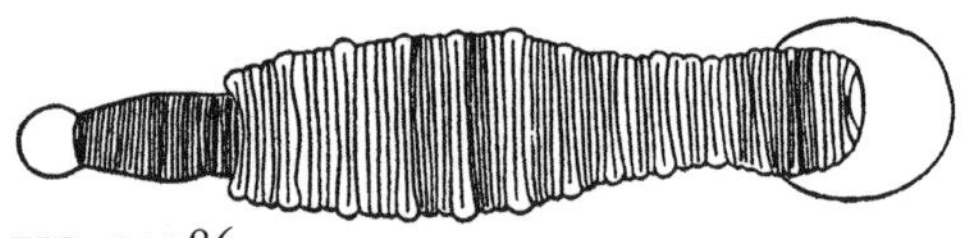
FIG. 14.186

Calliobdella vivida. Smooth. Purplish brown with white spots dorsally. Posterior sucker twice diameter of anterior sucker. Mostly from Chesapeake southward. On blue crabs, fundulus, herring, toadfish, flounder. Ca+,Co,Es–Ps,pa,25 mm (1 in), (s640). (FIG. 14.186)

Oceanobdella microstoma. Translucent, yellowish white with darker bands; blood-filled gut may show through body wall. Circle of 2–12 ocelli around posterior sucker. Found under head of sculpins, ocean pout, cod, cunner. Ac,Co,Ps,pa,25 mm (1 in).

Oxytonostoma typica. Large oral sucker, nearly the size of posterior sucker; small teatlike papillae flank female gonopore. On big skate, *Raja ocellata.* Ac,Co,Ps,pa,36 mm (1.4 in), (s646).

Chapter 15 ■ Phylum Arthropoda

15.1 *Arthropods* are a major animal group as judged by any criterion. The success of their body plan is obvious from their diversity in every habitat type in nature. Their role in energy flow within most ecosystems is often primary. Their economic, human-related significance ranges from the positive, as in aquaculture, fisheries, and agricultural contributions (such as pollination and pest control) to the negative, as agricultural and human health pests.

Following current practice, living members of this phylum are divided into three subphyla. The subphylum Uniramia includes centipedes, millipedes, and insects. The insects most routinely part of the marine world are the tiny, wingless, steel-gray spring-tails.

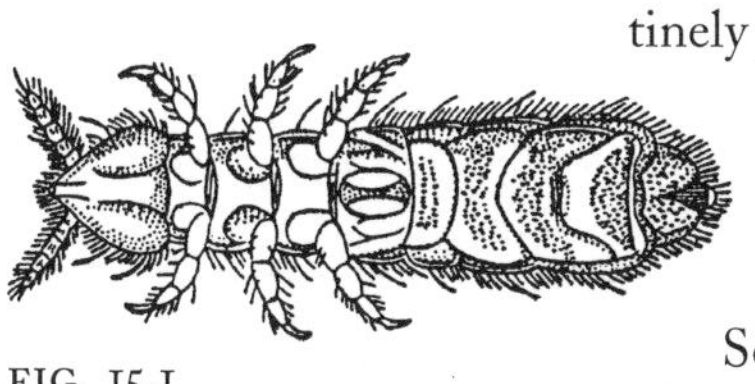

FIG. 15.1

Anurida maritima, springtail (Cl: Insecta, Or: Collembola). Collect in groups on the surface of high intertidal and supralittoral pools along rocky shores. Bo,Ha,Li,pr,6 mm (0.25 in), (G204). (FIG. 15.1)

Some larval stages (usually of the Or: Diptera) along with a few, normally freshwater, adult forms may be found in littoral zone habitats, especially associated with saltmarshes, algal-mussel beds, and tidal pools. Identification of these animals can be approached best through the literature on freshwater invertebrates (e.g., Pennak 1989). Members of this subphylum are not treated further in these keys.

The subphylum Chelicerata includes animals whose body segments are grouped to form an anterior *prosoma,* or head-plus-thorax region, and a posterior *opisthosoma* or abdomen. They also possess a characteristic set of appendages, including pincher-tipped *chelicerae,* followed by a pair of *pedipalps,* and four pairs of *walking legs,* all associated with the prosoma. Spiders, ticks and mites are included in the subphylum but are either too uncommon or too small to fall within the range of this manual. The most conspicuous marine chelicerates from the northeast include the horseshoe crab, *Limulus polyphemus* (for convenience, treated here with the true crabs, 15.104) and the small, slender, long-legged sea spiders or Pycnogonida (15.2).

Finally, the subphylum Crustacea includes the majority of marine arthropods. The distinctive, shell-covered appearance of barnacles separates them from the rest of the Crustacea (see 15.14). For others, an introduction and general key appears on 15.4, followed by more detailed keys for each group.

15.2 Subphylum Chelicerata: Class Pycnogonida, Sea Spiders

Pycnogonids or sea spiders are small and require special observational care both to find them initially (often entangled among stalks of hydroid colonies or filamentous algae) and to identify them. Details of the head and its appendages must be examined (several requiring use of at least low-power microscopy). A bulbous *proboscis* is present in all pycnogonids, and a lateral view should reveal whether it is inserted at the anteriormost extreme of the head or (in one genus) just posterior and ventral to it. In some sea spiders, the proboscis is flanked by a pair of antennalike *chelifores.* Some also have a pair of thin, multisegmented *palps.* In one case, both sets of structures are present. In addition to the four pairs of *walking legs,* there may be an additional pair of ventrally placed, multisegmented legs called *ovigers,* which are used for brooding eggs. These are not easily observed because they are small and tend to be held beneath the body.

Additional reference: (M) McCloskey, L. R., 1973.

A dissecting or compound microscope is necessary for identification of these small animals. Lengths given are for body length. Some or all legs are omitted from illustrations.

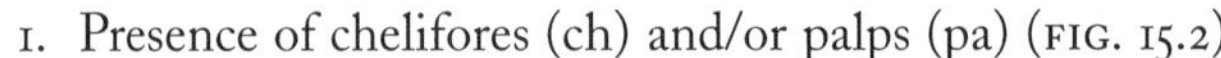

1. Presence of chelifores (ch) and/or palps (pa) (FIG. 15.2)
   - A. *Chelifores* present: *palps* present — ch:pa
   - B. *Chelifores* present: palps *absent* — ch:x
   - C. Chelifores virtually or totally *absent*: *palps* present — x:pa
   - D. Chelifores virtually or totally *absent*: palps *absent* — x:x

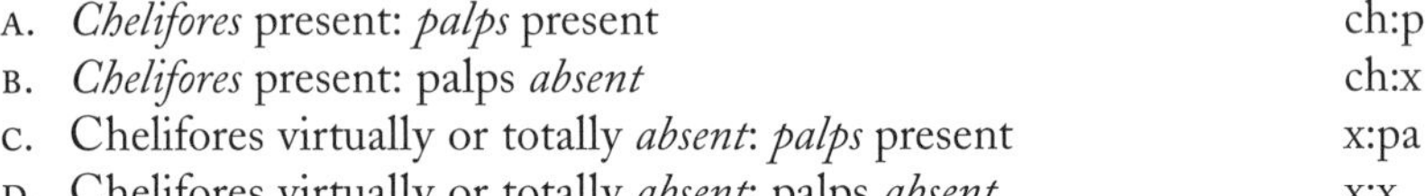

2. Shape of cephalic segment (from base of chelifores to first leg pair) (FIG. 15.2)
   - A. Distinctly *wider than long* — w>l
   - B. Distinctly *longer than wide* — l>w
   - C. *Width* about *equal to length* — w=l
3. Shape of imaginary line drawn around body connecting bases of legs (FIG. 15.2)
   - A. Such a line would be approximately *circular* — ci
   - B. Such a line would be *ovaloid* — ov
4. Length of chelifores compared to proboscis
   - A. *Chelifores longer than proboscis* — ch>pr
   - B. *Chelifores shorter than proboscis* — ch<pr
   - C. Chelifores *absent* — x

FIG. 15.2

15.3

Chelifores/ Palps	Cephalic Segment	Body	Chelifores/ Proboscis	Species
ch:pa	w>l	ci	ch<pr	*Achelia spinosa*
ch:pa	l>w	ov	ch>pr	*Nymphon* spp.
ch:x	l>w	ov	ch>pr	*Callipallene brevirostris,* long-neck sea spider
ch:x	w=l	ov	ch>pr	*Anoplodactylus lentus,* lentil sea spider
ch:x	w>l	ov	ch>pr	*Phoxichilidium femoratum,* clawed sea spider
x:pa	w>l	ci	x	*Tanystylum orbiculare,* ringed sea spider
x:pa	w>l	ov	x	*Endeis spinosa,* sargassum sea spider
x:x	w>l	ov	x	*Pycnogonum littorale,* anemone sea spider

FIG. 15.3

Achelia spinosa. Spiny legs. Ovigerous legs always present on both sexes. On hydroids, ascidians, or under stones. Ac–,Ha–Al,Li–,pr,3 mm (0.1 in), (M9). (FIG. 15.3)

Anoplodactylus lentus, lentil sea spider. Dark reddish purple color; bumpy ovigerous legs only in males. With hydroids, ascidians, or shell-gravel bottoms. Vi,Ha–Al,Li–Su,pr,6 mm (0.25 in), (G201,M9,NM589). (FIG. 15.4)

Callipallene brevirostris, long-neck sea spider. Ovigerous legs always present on both sexes. Vi,Ha–Al,Li,pr,6 mm (0.2 in), (G201,M9), ☺. (FIG. 15.5)

FIG. 15.4

FIG. 15.5

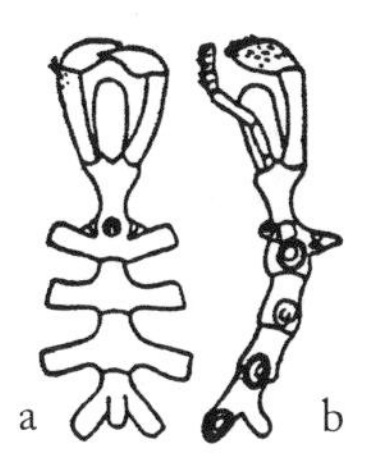

FIG. 15.6

FIG. 15.7

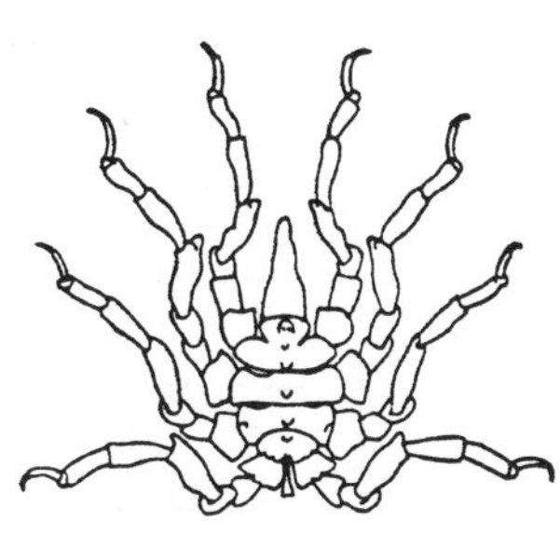

FIG. 15.8

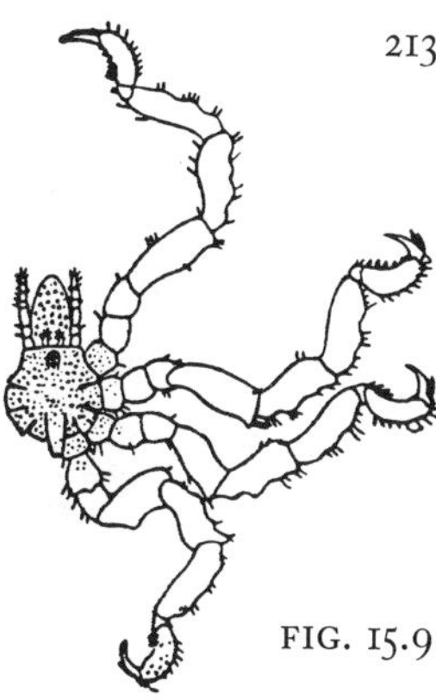

FIG. 15.9

Endeis spinosa, sargassum sea spider. Long legs; long slender proboscis. Found on floating *Sargassum,* gulfweed. Vi,Pe–Al,Ps,pr,6 mm (0.25 in), (G202,M9).

Nymphon spp. Large chelate ends to chelifores. *Nymphon grossipes* is the most common. Ac–,Ha–Al,Li–Su,pr,13 mm (0.5 in), (M9). (FIG. 15.6. a, dorsal view; b, lateral view.)

Phoxichilidium femoratum, clawed sea spider. Proboscis attaches just posterior to anterior margin of head; claws on walking legs bear accessory hooks; ovigerous legs only in males. Eats hydroids. Ac–,Ha–Al,Li–Su,pr,5 mm (0.2 in), (G201,M9,NM590), ☺. (FIG. 15.7)

Pycnogonum littorale, anemone sea spider. Light yellowish brown to dark brown color; ovigerous legs only in males; row of bumps down center-line of body. Attaches to and eats anemones. Ac–,Ha,Li–Su,pr,7 mm (0.3 in), (G202,M10,NM587). (FIG. 15.8)

Tanystylum orbiculare, ringed sea spider. Ovigerous legs always present on both sexes. Small; member of fouling communities, on hydroids, ascidians, etc. Vi,Ha–Al,Li–,pr,6 mm (0.25 in), (G201,M9,NM588). (FIG. 15.9)

15.4 Subphylum Crustacea

Crustacea range widely in size and body form. Overall appearance can be used to separate the Crustacea into subgroups. The body segments tend to be grouped into three major regions or *tagmata*: head, thorax, and abdomen. In some cases, a single-piece *carapace* or shell covers the head and most or all of the thorax. In others, the carapace is lacking, and thoracic segments are clearly visible. The integrity of original segments has been largely obscured in the formation of these groupings. One feature retained by most crustaceans, however, is the presence of a pair of appendages corresponding to each of the original segments. Appendages have become modified to serve a variety of functions, including food capture, food manipulation, sensation, locomotion, courtship, copulation, brooding, respiration, etc. Appendages associated with the head (including a uniquely crustacean feature, two pairs of antennae) are involved with the first three of these functions. Often the abdominal appendages or *pleopods* are modified in form and involve the last three functions. In some groups, these appendages—and for that matter, the abdomen itself—are highly reduced. Thoracic appendages or *pereiopods* are the dominant locomotory appendages or *walking legs* in virtually all groups.

Common names followed by a dagger (†) are designated as standard names by Williams et al. (1989). Additional, unofficial but widely used common names are also listed to facilitate linkage to approved names. Only the standarized names should be used subsequently.

Additional reference: (w) Williams, A.B., 1984.

1. Body length
   - A. Body length *greater than one-half inch* (>12 mm) — >½
   - B. Body length equal to or *less than one-half inch* (<12 mm) but still visible to the unaided eye — <½
   - C. Body *microscopic* in length — mi

15.5

Length	Group
>½	Crustacean Group 1, 15.6
<½	Crustacean Group 2, 15.10
mi	Planktonic Crustacea, 5.3

15.6

Crustacean Group 1
(Larger Crustacea [Class Malacostraca,in part]) (from 15.5)

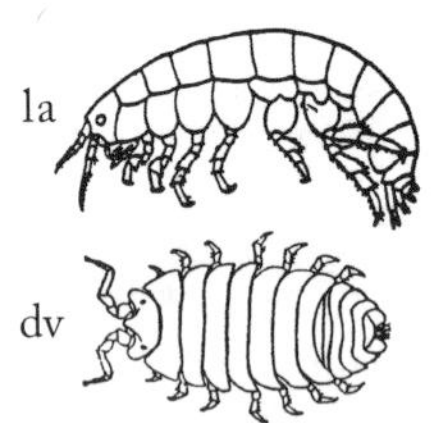

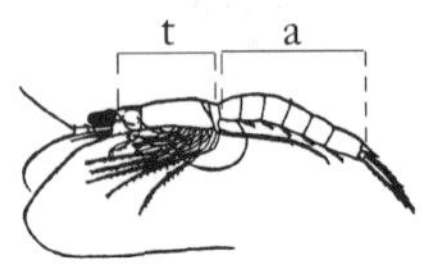

t a

FIG. 15.10

1. Presence of single-piece carapace covering thorax (i.e., those segments bearing the major walking legs)
   A. *Carapace* covers all or nearly all thoracic segments — ca
   B. Carapace *absent,* individual thoracic segments exposed to view — x
2. Body shape in imaginary cross-section through the thorax (FIG. 15.10)
   A. Body compressed *laterally* — la
   B. Body compressed *dorsoventrally* — dv
   C. Body approximately *cylindrical* — cy
3. Length of abdomen compared to length of thorax (walking legs are attached to the thorax; abdomen may be folded underneath the thorax in some [e.g., in crabs] or may appear to be absent in others) (FIG. 15.10)
   A. *Abdomen* clearly *longer than thorax* — a>t*
   B. *Abdomen approximately equal to thorax* — a=t
   C. *Abdomen* clearly *shorter than thorax* — a<t

15.7

Carapace	Cross Section	Abdomen:Thorax	Species or Group
x	cy	a<t	Caprellid amphipoda, 15.86
x	la	a<t,a=t	Gammaridean amphipoda, 15.36
x	dv	a<t	Isopoda, 15.20
ca	cy	a>t	Crustacean Group 1A, 15.8
ca	cy	a=t	*Homarus americanus,* American lobster†
ca	cy	a>t	*Munida iris,* squat lobster†
ca	cy	a>t	Crustacean Group 1A, 15.8
ca	dv	a<t	Crabs, 15.104
ca	la	a>t	Shrimp, 15.92

Homarus americanus, American lobster† (Or: Decapoda, IO: Astacidea, Nephropidae). Typically dark green with orange edges; rarely bluish. Heavy crusher claw and smaller cutter claw. Important commercial species. See Factor (1995) for comprehensive review. Bo,Ha,Su–De,sc,86.4 cm (34 in), (G239,NM619), ☺. (FIG. 15.11)

Munida iris, squat lobster†(Galatheidae). Long, slender rostrum. Pinching chelipeds very long and slim; fourth walking leg much smaller than others. Vi,Ha–Gr–Sa,Su–De,sc,4.8 cm (1.9 in) (carapace length), (W233). (FIG. 15.12)

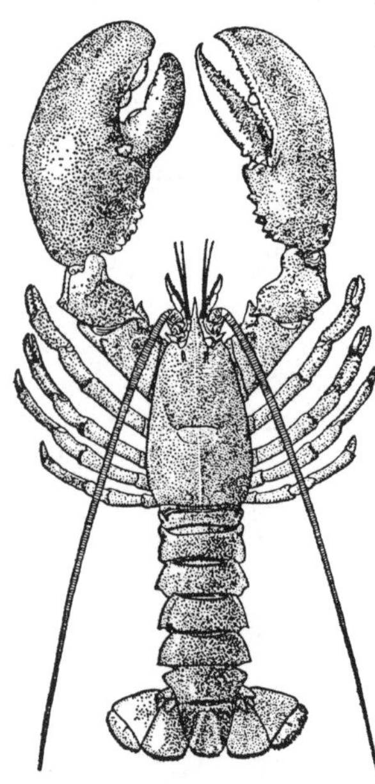

FIG. 15.11

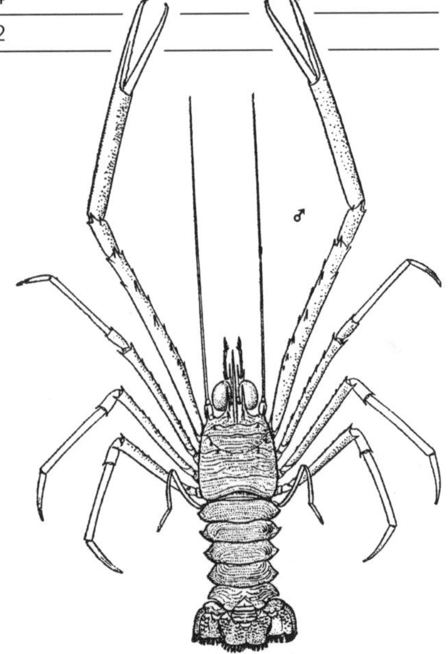

FIG. 15.12

15.8 Crustacean Group 1A

(Crustacea with Large Abdomens [Or: Decapoda and Or: Stomatopoda]) (from 15.7)

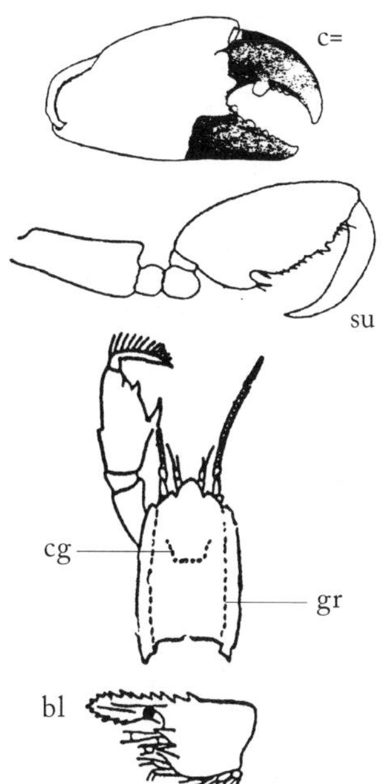

FIG. 15.13

1. Description of ending on first walking legs (FIG. 15.13)
   - A. First leg ending large and praying *mantis*-like — ma
   - B. First legs end in pinching *chela* (which act like a human thumb and forefinger) of approximately *equal* size — c=
   - C. First legs end in pinching *chela* of distinctly *unequal* size — cu
   - D. First leg ending is *subchelate* (like a human forefinger moving against an otherwise closed fist) — su
   - E. First leg ending is simple; chelate or subchelate ending *absent* — x
2. Eyes: size, pigmentation, location
   - A. *Large pigmented* eyes occupying tips of distinct *stalks* — lps
   - B. *Large pigmented* eyes but stalks very short or *absent* — lpx
   - C. *Small pigmented* eyes on short *stalks,* not occupying their tips — sps
   - D. *Small pigmented* eyes; stalks *absent* — spx
   - E. *Small unpigmented* eyes, stalks *absent* — sux
3. Distinct grooves on carapace (FIG. 15.13)
   - A. Semicircular *cervical groove* divides carapace into halves dorsally; distinct anterior to posterior, "shoulder" *grooves* present — cg:gr
   - B. Semicircular *cervical groove* present: grooves *absent* — cg:x
   - C. Cervical groove *absent*: shoulder *groove* present — x:gr
   - D. Cervical groove *absent*: shoulder grooves *absent* — x:x
4. Shape of rostrum or dorsal-anterior projection of carapace (viewed dorsally) (FIG. 15.13)
   - A. Rostrum is a *flat,* horizontally projecting plate with marginal *teeth* — ft
   - B. Rostrum is a *flat,* horizontally projecting plate with marginal teeth *absent* — fx
   - C. Rostrum forms *three* long, sharp, anteriorly directed *points* — 3p
   - D. Rostrum forms a small *point* not extending beyond eyes — po
   - E. Rostrum is a long, vertically oriented, pointed *blade* extending between eyes — bl
5. Habitat
   - A. *Pelagic,* primarily free-swimming in water column — pe
   - B. *Benthic:* inhabit empty *gastropod* snail shells — b:ga
   - C. *Benthic:* crawling on or in *particulate substrates* — b:ps

15.9

Leg 1	Eyes	Carapace	Rostrum	Habitat	Group or Species
ma	lps	x:gr	po	b:ps	*Squilla empusa,* mantis shrimp†
su	spx	cg:gr	fx	b:ps	*Naushonia crangonoides,* Naushon mudshrimp
su,c=	lps	x:x	bl	pe,b:ps	Shrimp, 15.92
c=	lps	cg:x	fx	b:ps	*Upogebia affinis,* coastal mudshrimp†, flat-brow mudshrimp
c=	lps	cg:x	3p	b:ps	*Munida iris,* squat lobster†
c=	sux	cg:x	bl	b:ps	*Calocaris templemani*
cu	lps	cg:x	ft	b:ps	Outermost lobe of tail fan with hinged tip: *Homarus americanus,* American lobster†; see notes on 15.7, ☺. OR Outermost lobe of tail fan not hinged: *Axius serratus,* lobster shrimp† (Axiidae)
cu	lpx	x:x	po	b:ps	*Alpheus heterochaelis,* bigclaw snapping shrimp†
cu	sps	cg:gr	po	b:ps	Callianassidae, mudshrimp: Eyestalks pointed at tips: *Callianassa setimanus* (=*atlantica*) OR Eyestalks rounded at tips: *Biffarius* (=*Callianassa*) *biformis,* biform ghost shrimp†
x	lps	cg:x,x:x	po	pe,b:ps	Mysids and krill, 15.88
cu	lps	cg:x	po	b:ga	Hermit crabs (Superfamily Paguroidea, Family Paguridae), 15.118

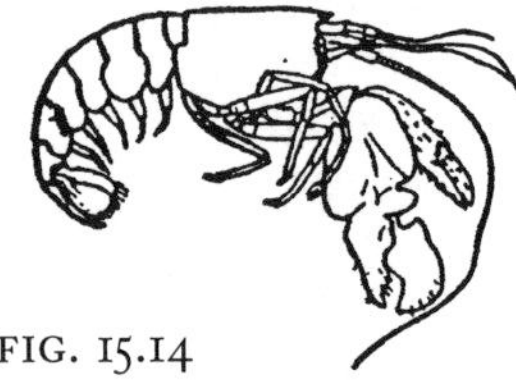
FIG. 15.14

Alpheus heterochaelis, bigclaw snapping shrimp† (IO: Caridae, Alpheidae). Uses large chela to make distinctive snapping sound. Ca+,Mu,Su,ff–df,5.1 cm (2 in). (FIG. 15.14. Lateral view, anterior to right.)

Axius serratus, lobster shrimp† (IO: Astacidea, Axiidae). Small and crayfishlike. Ac–,Sa–Mu,Su,df?,8.4 cm (3.3 in), (W187). (FIG. 15.15)

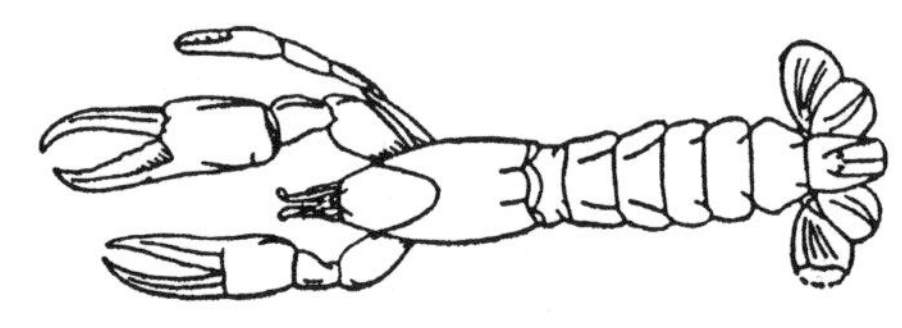
FIG. 15.15

Biffarius (=*Callianassa*) *biformis,* biform ghost shrimp† (IO: Astacidea, Callianassidae). Delicate body. Females have equal claws on first legs. Vi,Mu–Sa,Li–Su,ff,3 cm (1.2 in), (W182).

Calocaris templemanni (IO: Astacidea, Axiidae). Chelipeds are hairy. Bo,Mu,Su–De,om–sc,4.8 cm (1.8 in), (W188). (FIG. 15.16)

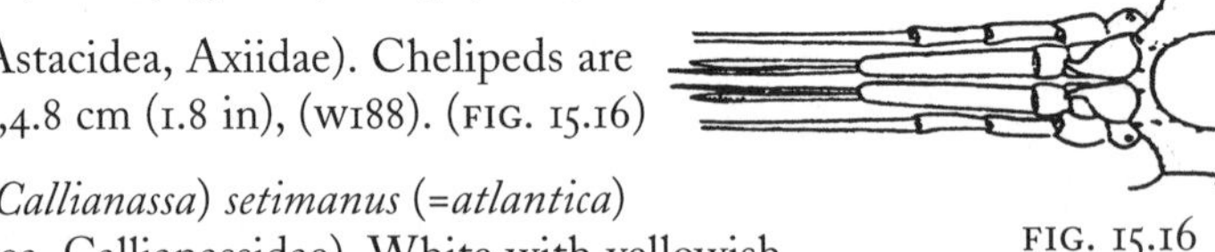
FIG. 15.16

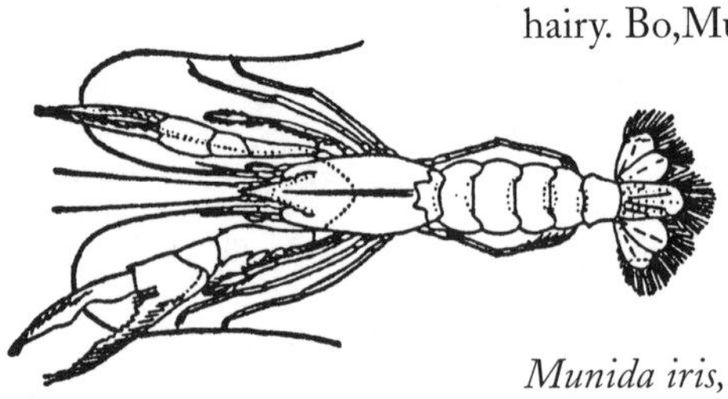

FIG. 15.17

Gilvossius (=*Callianassa*) *setimanus* (=*atlantica*) (IO: Astacidea, Callianassidae). White with yellowish appendages. Often with commensal crab, *Pinnixa cristata.* Bo,Sa–Mu,Li–Su,ff,7.1 cm (2.8 in), (G240,NM623,W180). (FIG. 15.17. Anterior end.)

Munida iris, squat lobster† (IO: Anomura, Galatheidae). Spiny thorax; chelipeds very long and slim. Vi,Sa–Gr,Su–De,sc?,4.8 cm (1.8 in), (carapace length), (W233).

Naushonia crangonoides, Naushon mudshrimp (IO: Astacidea, Laomediidae). Vi,Mu–Sa,Li–Su,df?,3.3 cm (1.3 in), (G240,W189). (FIG. 15.18. Lateral view, anterior to right.)

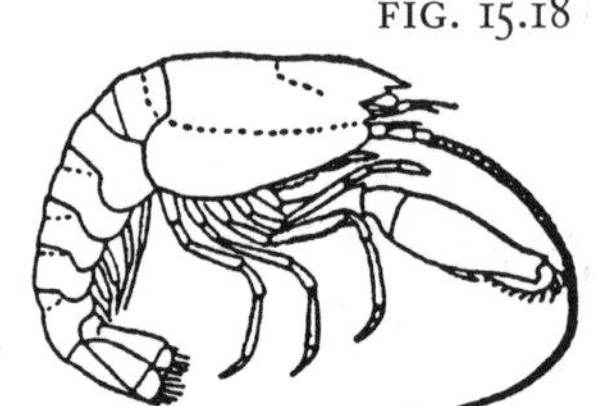
FIG. 15.18

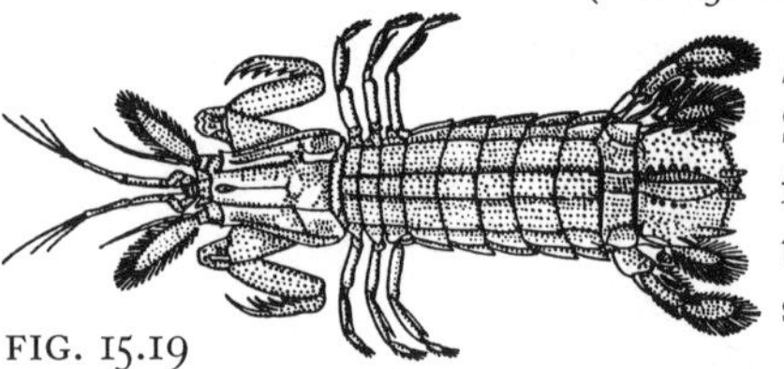
FIG. 15.19

Squilla empusa, mantis shrimp† (Superorder Hoplocarida, Or: Stomatopoda, Squillidae). Yellowish green with pink tinges; bright green eyes; dorsoventrally flattened; very large abdomen. Nocturnal; live in U-shaped burrows but emerge to feed on crustacea and fish. Can pinch! For a detailed account of Stomatopoda, see Manning (1974). Vi,Mu,Li–Su,pr,25.4 cm (10 in), (G218,NM599). (FIG. 15.19)

Upogebia affinis, coastal mudshrimp†, flat-brow mud shrimp (IO: Astacidea, Upogebiidae). Carapace weakly calcified; bluish or yellowish gray; first legs hairy. Lives in pairs in permanent burrows. Vi,Sa–Mu,Su,ff,6.4 cm (2.5 in), (G240,NM622,W191). (FIG. 15.20. Lateral view.)

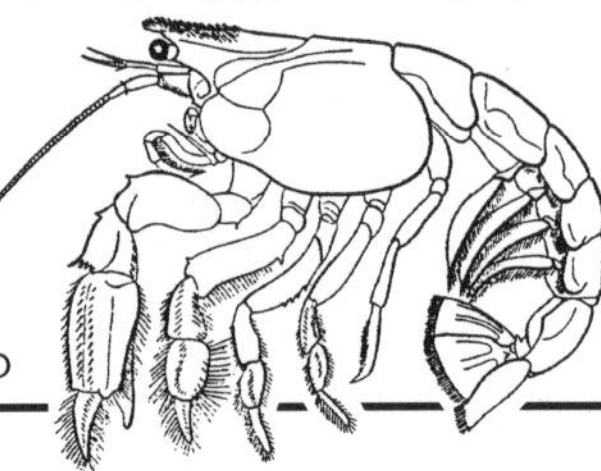
FIG. 15.20

15.10 Crustacean Group 2 (Smaller Crustacea, smaller than 13 mm [0.5 inch]) (from 15.5)

These animals may require immobilization (see Appendix) and observation using a dissecting or compound microscope.

1. Presence of stalked eyes
   - A. Eyes *stalked* — st
   - B. Eyes *sessile* (i.e., set directly on head) or lacking — se
2. Extent of cephalothoracic carapace (FIG. 15.21)
   - A. *Half carapace* present, covers first half of thorax only — hc
   - B. *Carapace* covers nearly all or all of thorax — ca
   - C. Carapace *absent,* thoracic segments clearly visible — x
3. First pair of major thoracic legs (FIG. 15.21)
   - A. First pair of legs *chelate* (move like a human thumb and forefinger) — ch
   - B. First pair of legs *subchelate* (like a human forefinger moving against an otherwise closed fist) — sc
   - C. First pair of legs *not* chelate — x

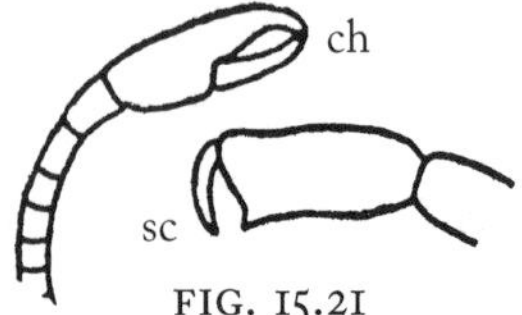

FIG. 15.21

4. Body shape in imaginary cross-section through thorax
   - A. Body compressed *laterally* — la
   - B. Body compressed *dorsoventrally* — dv
   - C. Body approximately *cylindrical* — cy

15.11

Eyes	Carapace	1st Leg	Body	Group
se	hc	x	cy	Cumacea, 15.12
se	x	ch	cy	Or: Tanaidacea
se	x	sc	la	Amphipoda, 15.32
se	x	sc	dv	Isopoda, 15.20
st	ca	x	cy	Thoracic legs arise all along ventral surface of carapace: Mysidacea, 15.90 OR Thoracic legs confined to posterior half of carapace: *Lucifer faxoni* OR If you fail to find a match among the above, try various crustacean larvae (Plankton, 5.3)

Lucifer faxoni (Suborder Dendrobranchiata, Sergestidae). Small peneid shrimp with elongate, cylindrical thorax. Vi,Pe,Es–Ps,ff?,13 mm (0.5 in), (G234,W52). (FIG. 15.22. Lateral view.)

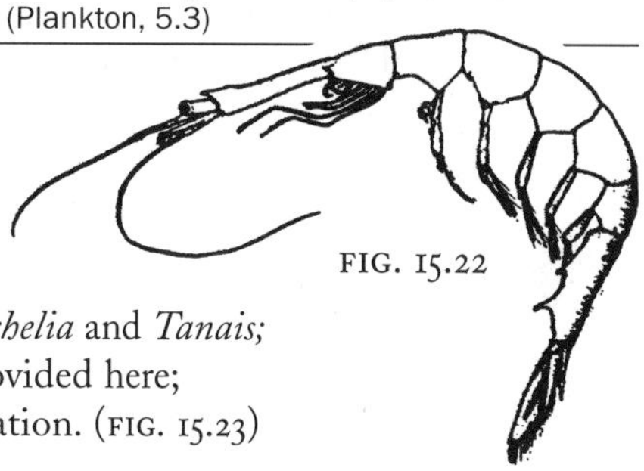

FIG. 15.22

Order Tanaidacea. Most common genera are *Leptochelia* and *Tanais;* light-colored; with seagrasses and algae. No key provided here; see Sieg and Winn (1978) for assistance in identification. (FIG. 15.23)

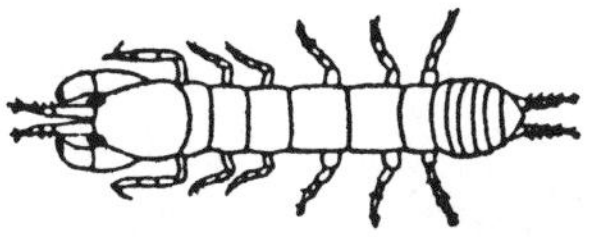

FIG. 15.23

15.12

Cumacean Shrimp (Or: Cumacea) (from 15.11)

The key provided here covers a few representative cumaceans only. Dissecting microscope required. For positive identifications and more inclusive coverage see Watling (1979).

Addtional reference: (W) Watling, L., 1979.

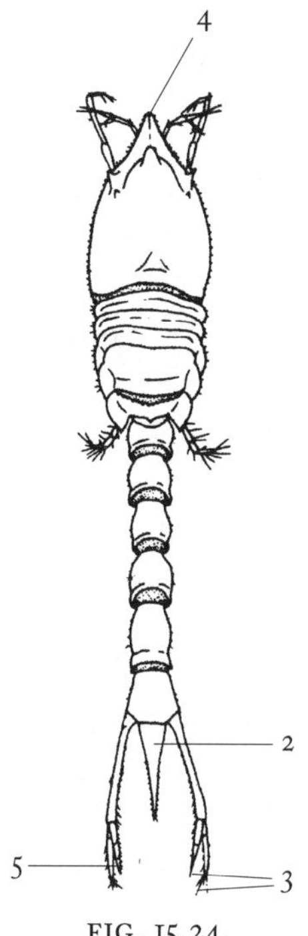

FIG. 15.24

1. Description of the tip of the telson (i.e., the tailmost piece along the center line of the animal) (FIG. 15.24)
   - A. Telson ends in a *point* bearing *five* terminal spines — po+5
   - B. Telson ends in a *point* bearing *two* terminal spines — po+2
   - C. Telson ends in a *point* with lateral but not terminal spines — po*
   - D. Telson short with *rounded* ending — ro
   - E. Telson *absent* — x
2. Number of spines along one side of the telson (FIG. 15.24)
   - A. If spines are present, provide *number* — ___
   - B. Telson *absent* — x
3. Compare inner and outer branches or rami of the uropods that flank the telson or tail piece (FIG. 15.24)
   - A. *Inner* ramus clearly *longer than outer* ramus — i>o
   - B. *Inner* ramus clearly *shorter than outer* ramus — i<o*
   - C. *Inner* ramus about *equal to outer* ramus in length — i=o
4. Shape of anterior end of carapace
   - A. Carapace margin is *pointed* — po†
   - B. Carapace margin is *blunt* — bl
5. Number of segments in inner ramus of uropods (the branch closest to the animal's center line): provide *number* of segments (FIG. 15.24) — ___

15.13

Telson	Telson Spines	Inner vs. Outer Ramus	Anterior	Inner Uropod	Species
po+5	2	i>o	bl	3	*Lamprops quadriplicata*
po+2	14	i<o	po	2	*Diastylis quadrispinosa*
po+2	8	i<o	po	2	*D. sculpta*
po	4–6	i>o,i=o	po	3–4	*Oxyurostylis smithi*
ro	x	i>o	bl	1	*Petalosarsia declivis*
x	x	i<o	po	2	*Leucon americanus*
x	x	i=o	bl	1	*Cyclaspis varians*
x	x	i>o	bl	2	*Mancocuma stellifera* *Eudorella pusilla* Two species, difficult to distinguish, based on relative proportions of appendage segments; see Watling (1979)
x	x	i<o	bl	2	*Pseudoleptocuma (=Leptocuma) minor*
x	x	i<o	po	2	*Almyracuma proximoculi*

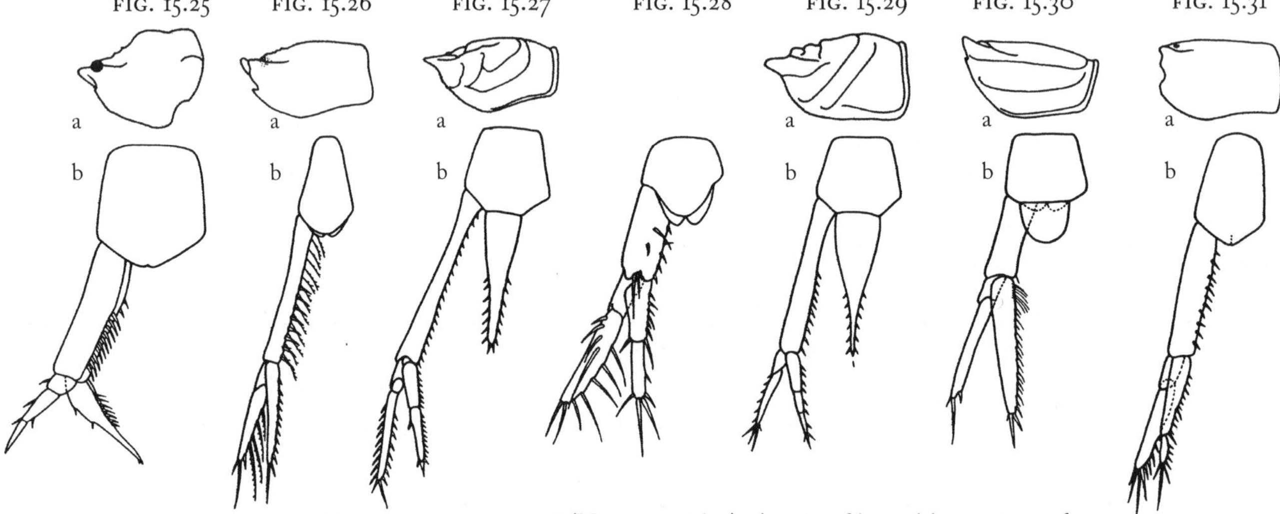

Almyracuma proximoculi (Nannastacidae). A pair of lateral horns toward posterior of carapace. Vi,Mu–Sa–Gr,Es–Su,ff–gr,10 mm (0.4 in), (W19). (FIG. 15.25. a, head, lateral; b, telson and left uropod.)

Cyclaspis varians (Bodotriidae). Bo,Mu,Es–Su,ff–gr,6 mm (0.23 in), (W19). (FIG. 15.26. a, head, lateral; b, telson and left uropod.)

Diastylis quadrispinosa (Diastylidae). Pale with reddish area toward posterior of carapace. Often with four spines toward front of carapace. Soft mud bottoms. Bo,Sa–Mu–Gr,Su–De,ff–gr,12 mm (0.5 in), (G219,W20). (FIG. 15.24)

Diastylis sculpta (Diastylidae). Ac+,Sa–Mu–Gr,Su,ff–gr,10 mm (0.4 in), (W20). (FIG. 15.27. a, head, lateral; b, telson and left uropod.)

Eudorella pusilla (Leuconidae). Bo,Mu–Sa–Gr,Su,ff–gr,10 mm (0.4 in), (W19).

Lamprops quadriplicata (Nannastacidae). Total of nine spines on telson. Ac,Mu–Sa–Gr,Su,ff–gr,12 mm (0.5 in), (W19).

Leucon americanus. Bo,Mu,Es–Su,ff–gr,10 mm (0.4 in). (FIG. 15.28. Telson and left uropod.)

Mancocuma stellifera. Bo,Sa,Li+,ff–gr,5 mm (0.2 in), (W19).

Oxyurostylis smithi. Bo,Sa–Mu,Es–Su,ff–gr,8 mm (0.3 in), (G219). (FIG. 15.29. a, head, lateral; b, telson and left uropod.)

Petalosarsia declivis. Bo,Mu–Sa–Gr,Su,ff–gr,6 mm (0.23 in). (FIG. 15.30. a, head, lateral; b, telson and left uropod.)

Pseudoleptocuma (=Leptocuma) minor. Bo,?,Su,ff–gr,8 mm (0.3 in). (FIG. 15.31. a, head, lateral; b, telson and left uropod.)

15.14 Class Cirripedia, Barnacles

Two groups of barnacles are common throughout the Northeast, acorn and gooseneck. Acorn barnacles attach to hard substrates in both intertidal and subtidal settings. Gooseneck barnacles can be found on floating objects washed in from offshore waters. The arrangement and sculpture of their calcareous protective plates are used in identification.

Additional reference: (z) Zullo, V. A., 1979.

1. Connection to the substrate (FIG. 15.32)
   - A. Barnacle attaches to substrate by prominent, rubbery *stalk* — st
   - B. Barnacle attaches directly to substrate, stalk *absent* — x

15.15

Attachment	Group
st	Gooseneck barnacles, 15.16
x	Acorn barnacles, 15.18

15.16 Gooseneck Barnacles (Or: Thoracica, SO: Lepadomorpha, Lepadidae) (from 15.15)

Gooseneck barnacles are offshore animals that attach themselves to floating objects by means of a muscular stalk and drift with the currents. Naturally, as objects are swept inshore, these barnacles may appear in the shallows or on beaches. Length measurements are for the shelled portion or *capitulum.*

1. Relative thickness of valves
   - A. Valves very *thin,* delicate and papery — th
   - B. Valves more *calcified,* still thin but somewhat resilient — ca
2. Radiating furrows on valves
   - A. Obvious radiating *furrows* — fu
   - B. Minutely striated or scratched, or furrows *absent* entirely — x
3. Number of filamentary (slender, fingerlike) appendages on either side of body (inside valves): provide *number* of pairs — __

15.17

Valves	Furrows	Appendages	Species
th	x	4–5	*Dosima fascicularis,* float gooseneck barnacle†
ca	fu	5–6	*Lepas anserifera*
ca	x	2	*L. anatifera,* common gooseneck barnacle†

Dosima fascicularis, float gooseneck barnacle†. Peduncle yellowish to purplish brown. Attaches to smaller floating objects. Bo,Pe,Ps,ff,3.6 cm (1.4 in), (G217,NM592,Z24). (FIG. 15.32. Lateral view.)

FIG. 15.32

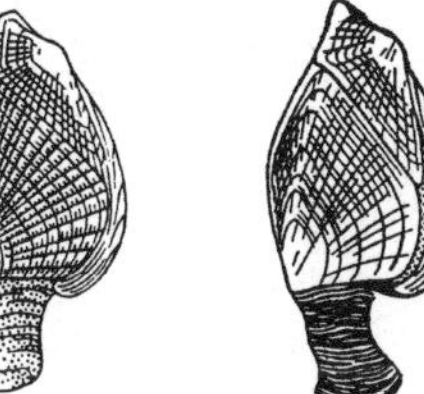

FIG. 15.33 FIG. 15.34

Lepas anatifera, common gooseneck barnacle†. Peduncle is purplish brown. Attaches to objects of all sizes. Bo,Pe,Ps,ff,5.1 cm (2 in), (G217,NM591), ☺. (FIG. 15.33. Lateral view.)

Lepas anserifera. Peduncle is yellow or orange. Tends to attach to large floating objects. Bo,Pe,Ps,ff,3.8 cm (1.5 in), (G217). (FIG. 15.34. Lateral view.)

15.18 Acorn Barnacles (Or: Thoracica, Suborder: Balanomorpha) (from 15.15)

The arrangement and appearance of external, calcareous plates are useful morphological features used in identifications. Pry an individual from the substrate to examine the composition (membranous or calcareous) of the bottom, attachment surface or *basis.* Often, this can be observed more simply by looking for a nearby dead specimen and peering inside. If you can see the rock substrate within, the basis was membranous. Strong forceps may be necessary to remove the *scutum* and *tergum* plates that form the operculum or "doors" covering the opening through which the feeding limbs extend. Try to observe the color of soft tissues surrounding these small, opercular plates.

Species below are members of the family Balanidae unless otherwise noted. Length measurements are for basal diameter.

1. Composition of basis or attachment surface (see note immediately above)
   - A. *Membranous* me
   - B. *Calcareous* ca
2. Relation of endmost rostrum (r) or rostral plates to adjacent (a) plates (FIG. 15.35)
   - A. *Rostrum overlaps adjacent* plates r/a*
   - B. *Adjacent* plates *overlap rostrum* a/r
3. Striations or lengthwise scratches (in addition to growth rings) on outside surface of scutum or small triangular plate that forms part of the operculum (dissecting microscope) (FIG. 15.35)
   - A. Distinct *striations* present on scutum st
   - B. Striations *absent* from scutum x
4. Irregularity of line tracing base of lateral plates where they attach to the substrate (reflects degree of ridge development along plates)
   - A. Basal line *deeply indented* di
   - B. Basal line mildly indented or basically *smooth* sm
5. Presence of grayish or purplish stripes along exterior of lateral plates
   - A. Irregularly arranged *stripes* present st
   - B. Stripes *absent* entirely x
6. Habitat (list all that apply)
   - A. *Upper littoral* or upper intertidal ul
   - B. *Midlittoral* ml
   - C. *Lower littoral* ll
   - D. *Subtidal* su
   - E. *Estuarine*, brackish water es
   - F. Attached to *carapace* of horseshoe crabs, blue crabs, hermit crabs, etc. ca

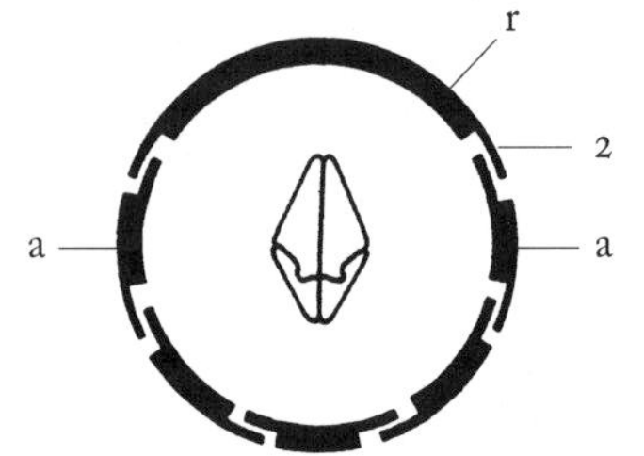

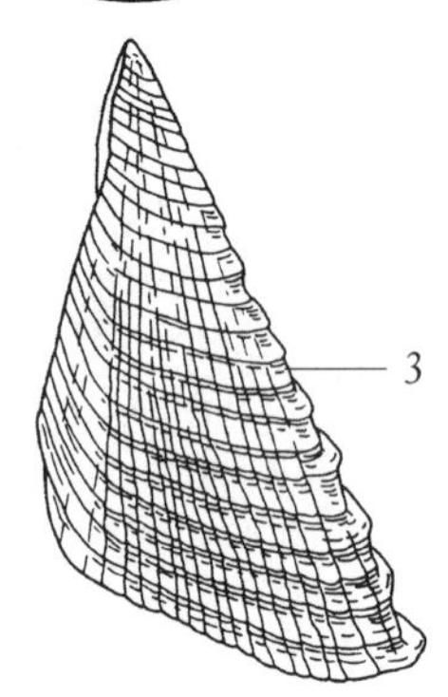

FIG. 15.35

15.19

Basis	Rostrum	Scutum Striations	Basal Line	Stripes	Habitat	Species
me	a/r	x	sm,di	x	ul	*Chthamalus fragilis,* little gray barnacle†
me	r/a	x	sm,di	x	ml,ll	*Semibalanus balanoides,* northern rock barnacle†
me	r/a	x?	sm	x	ca	*Chelonibia patula*
ca	r/a	st	di	x	su	*Balanus balanus,* rough barnacle†
ca	r/a	st	sm	x	su	*Chirona hameri*
ca	r/a	st	sm	x	ll,su,es	*Balanus eburneus,* ivory barnacle†
ca	r/a	x	sm	x	ll,su,es	*B. improvisus,* bay barnacle†
ca	r/a	x	sm,di	x	su	*B. crenatus,* crenate barnacle†
ca	r/a	x	sm	st	ll,su,es,ca	*B. amphitrite,* little striped barnacle†

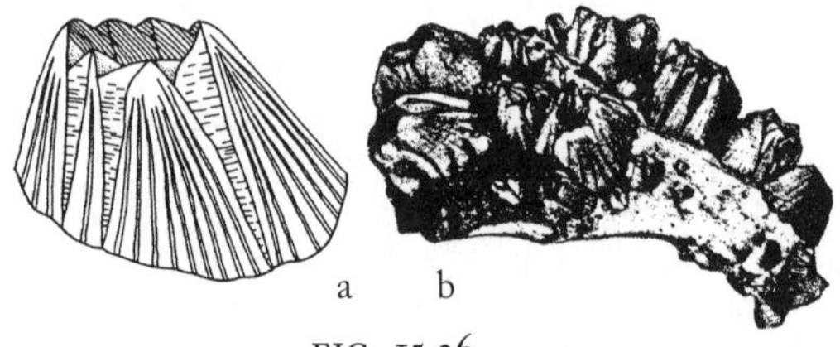

FIG. 15.36

Balanus amphitrite, little striped barnacle†. Small; gray or white with pinkish stripes. Soft parts white with purple or black stripes. Vi,Ha,Es–Li–Su,ff,8 mm (0.3 in), (G215,NM596,Z25). (FIG. 15.36. a, individual, lateral; b, group.)

Balanus balanus, rough barnacle†. Large, dirty white; conical shape. Soft tissues with yellow, brown, and white edge stripes. Ac–,Ha,Li–Su,ff,5.1 cm (2 in), (G216,NM594,Z25), ☺. (FIG. 15.37)

FIG. 15.37

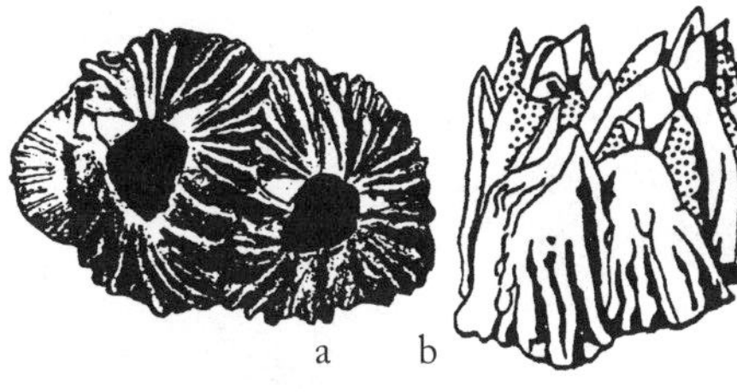

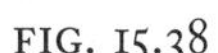

FIG. 15.38

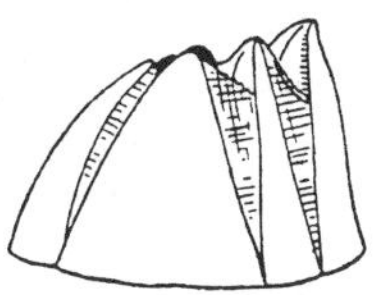

FIG. 15.39

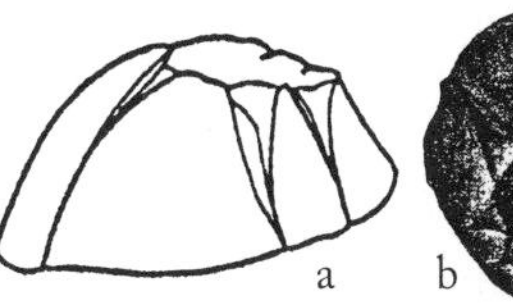

FIG. 15.40

FIG. 15.41

Balanus crenatus, crenate barnacle†. Conical to tubular. Soft parts yellow or dirty white, sometimes with purple stripes. Ac–,Ha,Li–Su,ff,25 mm (1 in), (G216). (FIG. 15.38. a, dorsal view; b, lateral view, group.)

FIG. 15.42

Balanus eburneus, ivory barnacle†. Steeply conical. Soft parts striped in purple, yellow, or dirty white. Dominant form in lower Chesapeake Bay. Vi+,Ha,Es–Li–,ff,25 mm (1 in), (G215,NM594), ☺. (FIG. 15.39)

Balanus improvisus, bay barnacle†. Dirty white; low cone shape. Soft parts white speckled with pink or purple. Dominant form in upper Chesapeake Bay. Bo,Ha,Es–Li–Su,ff,15 mm (0.6 in), (G215,Z26), ☺. (FIG. 15.40. a, individual, lateral; b, individual, dorsal.)

Chelonibia patula (Coronulidae). Opercular plates (tergum and scutum) so small that they do not fill orifice. Ca+,Co,Es–Su,ff,15 mm (0.6 in), (G216,Z25). (FIG. 15.41. Individual, dorsal.)

Chirona hameri (Archaeobalanidae). White or cream; nearly cylindrical shape. Soft parts white. Ac–,Ha,Su–De,ff,7.6 cm (3 in), (Z25). (FIG. 15.42. Individual, lateral.)

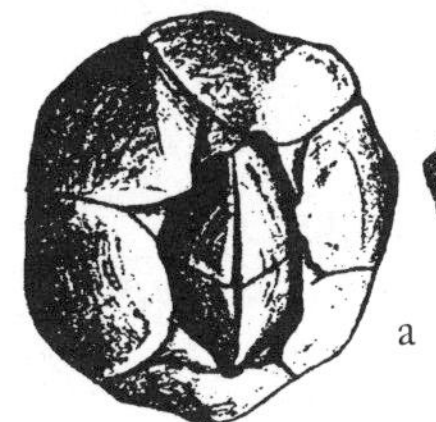

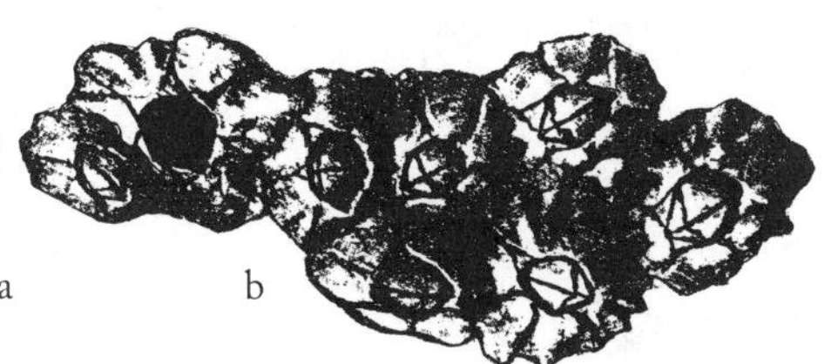

FIG. 15.43

Chthamalus fragilis, little gray barnacle† (Chthamalidae). Small, comparatively fragile. Narrow rostral plate; soft tissues tend to be bluish with orange and black markings. Small groups often higher in intertidal than *Semibalanus balanoides.* Vi,Ha,Li,ff,10 mm (0.4 in), (G215,NM593,Z25). (FIG. 15.43. a, individual, dorsal; b, group.)

Semibalanus balanoides, northern rock barnacle† (Archaeobalanidae). Broad rostral plate. Soft tissues whitish or pinkish. Bo,Ha,Li,ff,25 mm (1 in), (G215,NM593,Z25), ☺. (FIG. 15.44. a, individual, dorsal; b, group showing elongate growth form; c, group.)

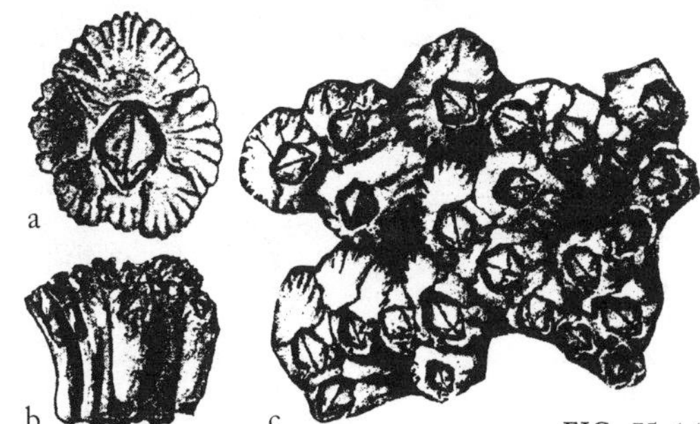

FIG. 15.44

15.20 Class Malacostraca: Order Isopoda, Isopods

The bodies of isopods are generally flattened dorsoventrally with thoracic segments visible beginning just posterior to the head. Seven (rarely five) pairs of large walking legs (*pereiopods*) are associated with the thorax, and fewer and more morphologically varied legs (*pleopods,* including the tailmost *uropods*) are attached to the smaller abdominal segments. The location of the last pair of abdominal legs, the uropods, and the shape of the terminal portion of the body, the *telson,* are used often in identification. Isopods are commonly found among vegetation, especially in rocky intertidal areas and tidal pools.

Species-level identification may require observation of immobilized animals (see Appendix) at the dissecting-microscope level of magnification. Preparing wet mounts may be useful as well (see 0.7).

Additional reference: (R) Richardson, H., 1905.

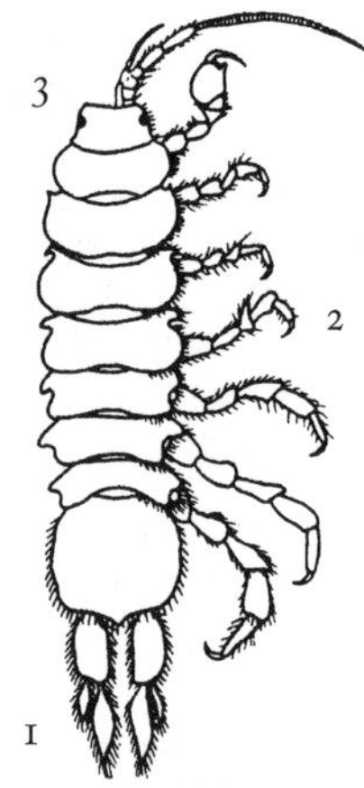

FIG. 15.45

1. Location of uropods (the appendages associated with the tailmost segment) (FIG. 15.45)
   A. Uropods *lateral,* visible dorsally — la
   B. Uropods *terminal,* visible dorsally (these may be tiny) — te*
   C. Uropods not visible dorsally, but are *ventral* and form opercular flaps under abdomen — ve
2. Number of pairs of pereiopods or walking legs: provide *number* of legs (5 or 7) (FIG. 15.45) — __
3. Location of eyes, viewed from dorsal surface (FIG. 15.45)
   A. Eyes located on *dorsal* surface of head — do
   B. Eyes placed on *lateral* surface of head — la*
4. Mode of living
   A. *Parasitic* on fishes or crustacea — pa
   B. *Free-living* — fl

15.21

Uropods	Pereiopods	Eyes	Lifestyle	Group or Species
la	7	la	fl	Anthuridae, Cirolanidae, Limnoriidae, and Sphaeromatidae, 15.22
la	7	la,do	pa	Parasitic isopods: Bopyridae and Cymothoidae, 15.24
te	5	la	pa,fl	*Gnathia cerina*
te	7	la	fl	Ligiidae and Oniscidae, 15.26
te	7	do	fl	*Jaera marina,* little shore isopod
ve	7	la	fl	Idoteidae (in part), 15.28
ve	7	do	fl	Chiridotea spp. (Idoteidae), 15.30

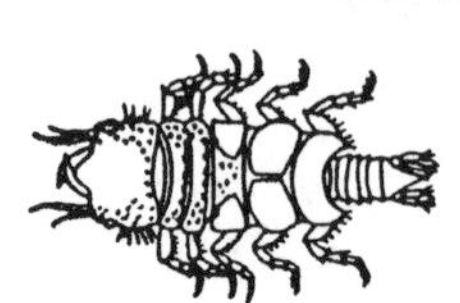
FIG. 15.46

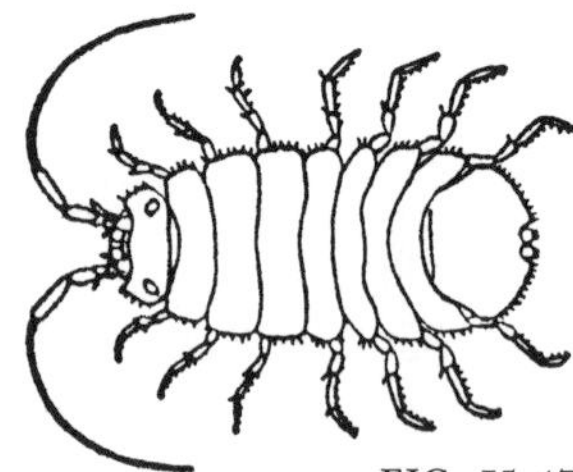
FIG. 15.47

Gnathia cerina (Gnathiidae). Sluggish adults live in benthos; young often ectoparasitic on fish. Ac–,Pa–Ha–Sa–Mu,Su,?,8 mm (0.3 in), (R59). (FIG. 15.46)

Jaera marina, little shore isopod (Janiridae). Usually grayish; small, oval; antenna one, tiny; antenna two reach posterior border of fifth thoracic segment; abdomen as one segment viewed dorsally. Ac–,Ha–Al,Li,om?,4 mm (0.16 in), (G224,R450), ☺. (FIG. 15.47)

15.22 Anthuridae, Cirolanidae, Limnoriidae, and Sphaeromatidae (from 15.21)

Isopods with seven pairs of pereiopods, lateral eyes, and uropods placed laterally.

1. Body proportions: length/width (FIG. 15.48)
   A. Body *slender,* more than seven times longer than wide — sl
   B. Body *stout,* less than two and a half times longer than wide — st
   C. Body *intermediate,* between two and a half and seven times longer than wide — in*
2. Total number of segments in second (outermost) antenna: provide *number* of segments (FIG. 15.48) — __
3. Sections of abdomen (i.e., the region posterior to the last segment bearing walking legs; includes the posteriormost telson, if one is present, in addition to true segments) (FIG. 15.48)
   A. Abdomen in *four or more* distinct sections — >4*
   B. Abdomen in *three* distinct parts, one of which may have scalloped edges that reflect the original segments now fused together — 3
   C. Abdomen in *two* distinct parts (i.e., the telson plus fused abdominal segments) — 2
4. Telson shape (FIG. 15.48)
   A. Telson end *notched* — no
   B. Telson end *rounded* — ro*
   C. Telson end *squared* — sq
   D. Telson end *pointed* — po

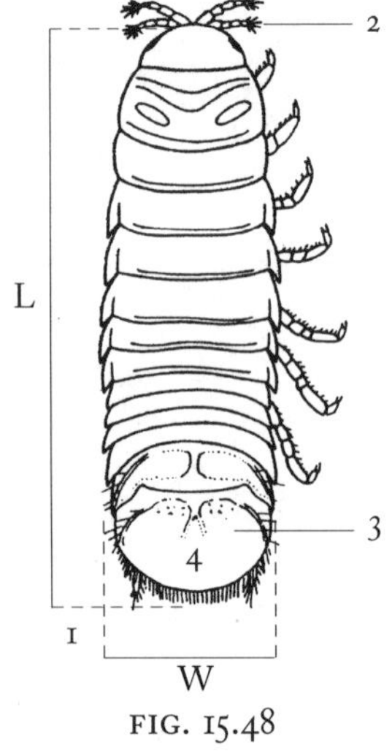

FIG. 15.48

15.23

Body	Antenna 2	Abdominal Segment	Telson	Species
sl	2–4	2	ro,po	*Cyathura polita,* slender isopod
sl	4–5	3	ro,po	*Ptilanthura tenuis*
st	15–20	3	ro	*Sphaeroma quadridentatum,* sea pill bug
st	15–20	3	no,sq	*Paracereis* (=*Cilicaeca*) *caudata,* eelgrass pill bug
in	3–4	>4	ro	*Limnoria lignorum,* gribble
in	15–20	>4	ro	*Politolana* (=*Cirolana*) *polita,* greedy isopod
in	15–20?	>4	no,sq	*P.* (=*Cirolana*) *concharum,* greedy isopod

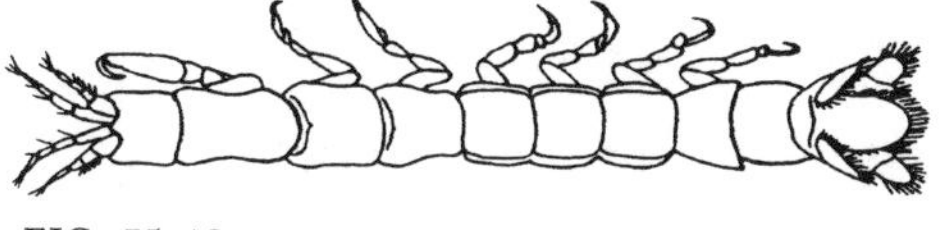
FIG. 15.49

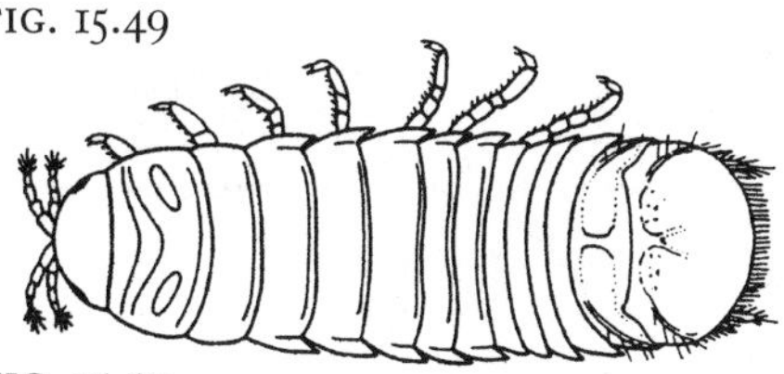

FIG. 15.50

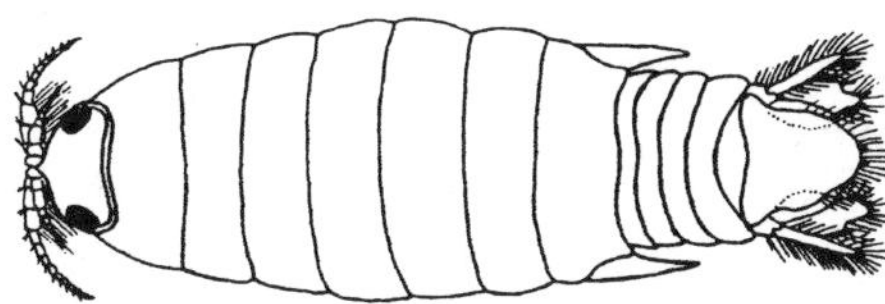
FIG. 15.51

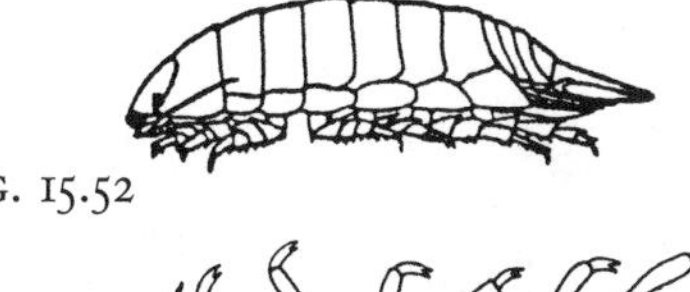
FIG. 15.52

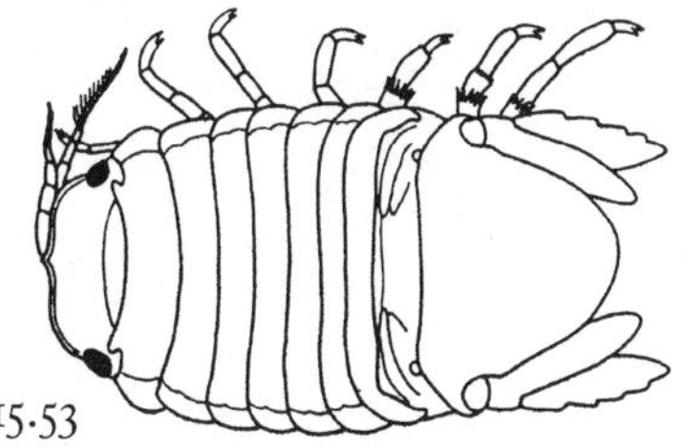

FIG. 15.53

Cyathura polita, slender isopod (Cyathuridae). Most segments longer than wide; uropods arch dorsally over telson; brownish. High intertidal. Bo,SM–Sa–Mu,Es–Li,om,25 mm (1 in), (G221), ☺. (FIG. 15.49)

Limnoria lignorum, gribble (Limnoridae). Eyes lateral; grayish color; uropods small. On or burrowing into submerged wood. Bo,wood,Li–Su,gr,5 mm (0.2 in), (G221,R269), ☺. (FIG. 15.50)

Paracereis (=*Cilicaeca*) *caudata,* eelgrass pill bug (Sphaeromatidae). Females with slightly notched or straight telson margin; smaller males with deep notch and enlarged, crescent-shaped uropods. Vi,SG,Es–Su,pr?,12 mm (0.5 in), (R310).

Politolana (=*Cirolana*) *concharum,* greedy isopod (Cirolanidae). Scavengers on dead material, especially blue crabs. Caution: can bite! Bo,Sa–Mu,Su,sc,3.5 cm (1.4 in), (G221,R95). (FIG. 15.51)

Politolana (=*Cirolana*) *polita,* greedy isopod (Cirolanidae). Uropods large. Scavengers on dead flesh. Bo,Sa–Mu,Su,sc,15 mm (0.6 in), (G221,NM601,R99). (FIG. 15.52. Lateral view.)

Ptilanthura tenuis (Anthuridae). Small eyes; one pair of antennae much longer than the second pair. Bo,Sa–Mu,Es–Su,?,8 mm (0.3 in), (G221).

Sphaeroma quadridentatum, sea pill bug (Sphaeromatidae). Small, dark, mottled; rolls into ball when disturbed. Low intertidal on barnacles or pilings. Vi,Ha–Al,Li,pr?,11 mm (0.43 in), (G222,R281). (FIG. 15.53)

15.24 **Bopyridae and Cymothoidae, Parasitic Isopods** (from 15.21)

Parasites of fishes and crustaceans.

1. Body proportions: length/width
   - A. Body *slender,* more than seven times longer than wide — sl
   - B. Body *stout,* less than seven times longer than wide — st
2. Telson shape
   - A. Telson margin *rounded* — ro
   - B. Telson margin *pointed* — po
3. Host
   - A. *Fish* — fi
   - B. *Crustacean* (shrimp) — cr
4. Location on host
   - A. *Gill* area — gi
   - B. *Mouth* — mo

15.25

Body	Telson	Host	Location	Species
st	ro	fi	gi	*Lironeca (=Livoneca) ovalis,* fish gill isopod
st	ro	cr	gi	*Probopyrus pandalicola,* shrimp parasitic isopod
sl	po	fi	mo	*Olencira praegustator,* fish mouth isopod

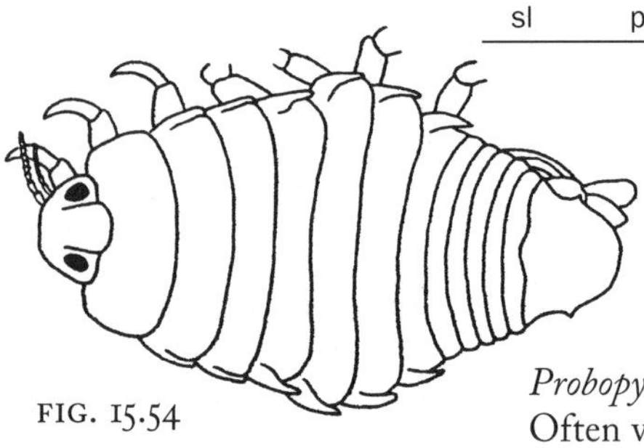

Lironeca (=*Livoneca*) *ovalis,* fish gill isopod (Cymothoidae). Hosts include striped bass, white and silver perch, bluefish, parrotfish. Vi,Pa,Es–Ps,pa,25 mm (1 in), (G225,R263). (FIG. 15.54)

FIG. 15.54

Olencira praegustator, fish mouth isopod (Cymothoidae). Menhaden is the most common host. Vi,Pa,Es–Ps,pa,25 mm (1 in), (G225,R231).

Probopyrus pandalicola, shrimp parasitic isopod (Bopyridae). Often visible through translucent carapace especially in gill area of *Palaemonetes.* Vi,Pa,Es–Ps,pa,5 mm (0.2 in), (G224,R554). (FIG. 15.55. Large, oval female; tiny male; drawn to scale.)

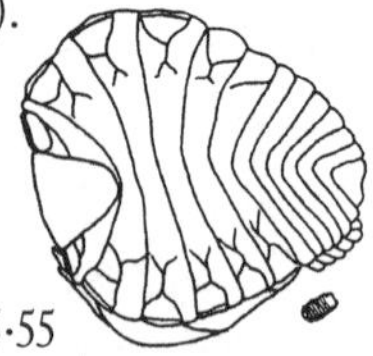

FIG. 15.55

15.26 **Ligiidae and Oniscidae** (from 15.21)

Isopods with seven pairs of pereiopods, lateral eyes, and uropods located terminally. Very active; under and in high intertidal debris and rock crevices; feed on fine, filamentous algae.

1. Length or uropods compared to length of antennae
   - A. Uropods *large,* more than half the length of antennae — la
   - B. Uropods *shorter,* much shorter than antennae — sh
2. Shape of telson (tailmost body piece)
   - A. Telson distinctly *pointed,* triangular in shape — po
   - B. Telson gently *curved,* not pointed — cu

15.27

Uropods vs. Antenna	Telson	Species
la	cu	*Ligia exotica,* sea roach
sh	cu	*L. oceanica,* sea roach
sh	po	*Philoscia vittata,* sow bug

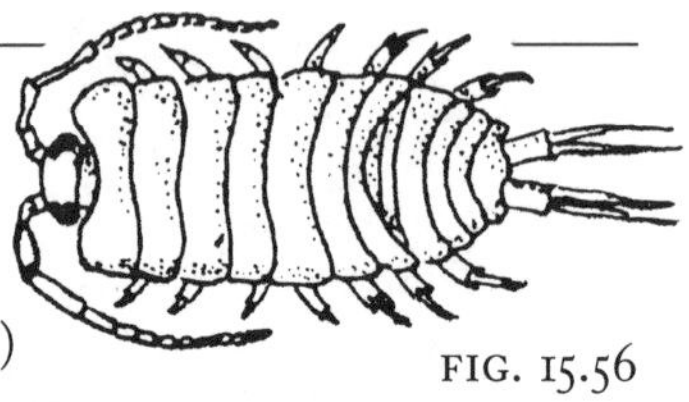

Ligia exotica, sea roach (Ligiidae). South of New Jersey. Ca+,Ha–Sa,Li,sc?,3.3 cm (1.3 in), (G224,NM603,R676), ☺.

Ligia oceanica, sea roach (Ligiidae). North of Cape Cod. Ac,Ha–Sa,Li,sc?,25 mm (1 in), (G224,NM603,R684). (FIG. 15.56)

FIG. 15.56

Philoscia vittata, sow bug (Oniscidae). Brownish with lighter middorsal line. Vi,Ha–Sa,Li,sc?,8 mm (0.3 in), (R605). (FIG. 15.57)

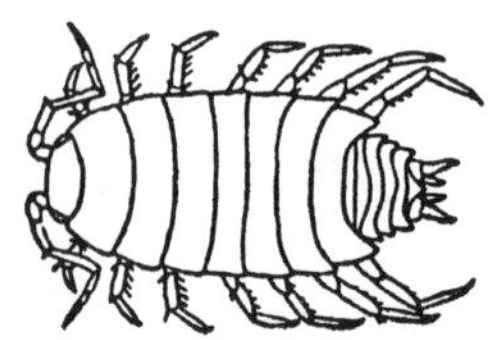

FIG. 15.57

15.28 **Idoteidae (in part)** (from 15.21)

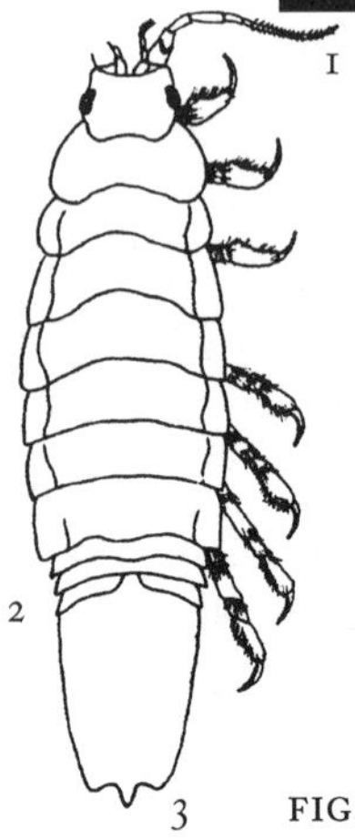

Isopods with seven pairs of similar pereiopods, laterally placed eyes, and uropods forming ventral opercular flaps. See 15.30 also.

1. Relative size of antennal pairs (FIG. 15.58)
   - A. Lateral antennae *long* (at least three times medial antennae) and with *many* segments (six or more) — lm*
   - B. Lateral antennae *long* but *fewer* than six segments — lf
   - C. Lateral antennae *shorter* (less than three times medial antennae) with *fewer* than five segments — sf
2. Presence of distinct lobes along lateral margin of telson (FIG. 15.58)
   - A. Telson with distinct *lobe* along margin — lo
   - B. Telson with *smooth* margin, lobes absent — sm*

FIG. 15.58

3. Shape of posterior margin of telson (FIG. 15.58)
   - A. Telson tapers gradually to a sharp *point* — po
   - B. Telson margin smoothly *rounded* — ro
   - C. Telson margin *square* — sq
   - D. Telson margin *bracket*-shaped, relatively squared but with a terminal point — br*

15.29

Antennae	Telson Sides	Telson End	Species
sf	lo	po	*Edotea triloba* (=*montosa*)
lf	lo	ro	*Erichsonella* Prominent bump or elevation on head between eyes: *E. filiformis* OR Bump between eyes absent: *E. attenuata,* elongated eelgrass isopod
lm	sm	po	*Idotea phosphorea,* sharp-tail isopod
lm	sm	br	*I. balthica,* Baltic isopod
lm	sm	sq	*I. metallica*

Edotea triloba (=*montosa*), mound-back isopod. Telson with raised, midline bumps. Uniramous uropods. Mud-colored with adhering debris. Estuarine mud; pilings, rotting seaweed. Bo,SG–Mu,Es–Su,df–sc,8 mm (0.3 in), (G223,R396). (FIG. 15.59)

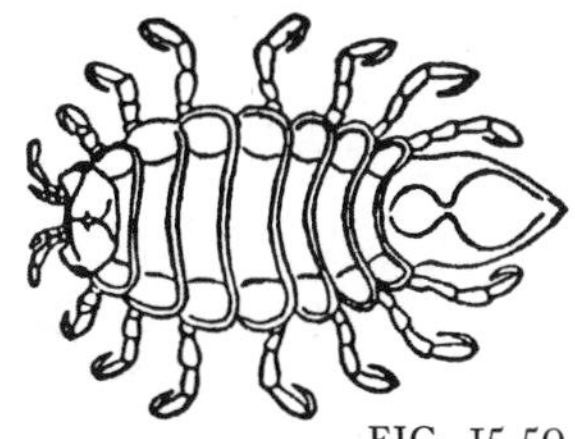
FIG. 15.59

Erichsonella attenuata, elongated eelgrass isopod. Vi,SG,Es–Su,sc?,12 mm (0.5 in), (G223,R400).

Erichsonella filiformis. Slender; thick antennal segments. Outer antennae at least five times longer than medial antennae. Vi,SG,Es–Su,sc?,12 mm (0.5 in), (G223,R401). (FIG. 15.60)

FIG. 15.60

Idotea balthica, Baltic isopod. Color varies widely, often green or brown. Rocky shores and pools; *Chondrus* zone; drift seaweed. Feeds on decaying vegetation, especially *Fucus vesiculosus.* Often seen swarming at night. Bo,Ha–Al,Li–Su,gr–sc?,25 mm (1 in), (G222,NM601,R364), ☺. (FIG. 15.61)

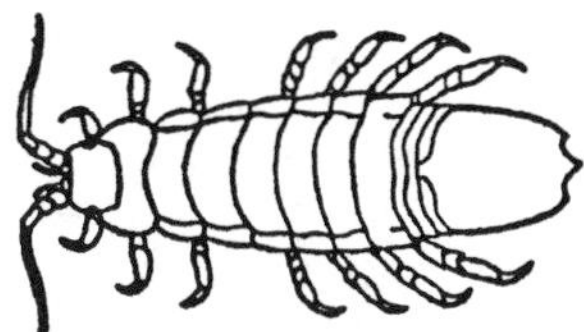
FIG. 15.61

Idotea metallica. Dark body. Bo,Ha–Al–Pe,Li–Su–Ps,sc?,25 mm (1 in), (G222,R362). (FIG. 15.62)

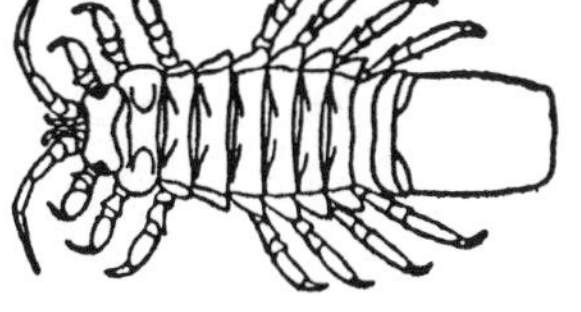
FIG. 15.62

Idotea phosphorea, sharp-tail isopod. Rocky intertidal and pools; gravel bottoms. Highly variable in color and pattern. Ac,Ha–Al–Gr,Li–Sa,sc?,25 mm (1 in), (G222,R367), ☺. (FIG. 15.63)

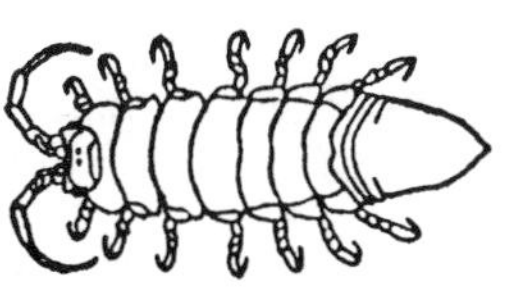
FIG. 15.63

15.30 ***Chiridotea* spp. (Idoteidae, in part)** (from 15.21)

Isopods with seven pairs of pereiopods, first three of which with larger tips than remaining four pairs. Ventrally placed, biramous uropods (i.e., end in two lobes). Dorsally located eyes.

1. Shape of sides of telson
   - A. Sides of telson *straight* to pointed end — st
   - B. sides of telson *curved* to pointed end — cu
2. Shape of anterior margin of head, especially near eyes
   - A. Margin *deeply notched,* divided into four distinct stubby projections — dn
   - B. Margin *barely notched* — bn
3. Comparative lengths of antennae
   - A. Both antennal pairs approximately *equal* in length — =
   - B. One pair of antennae distinctly *longer* — >

15.31

Telson	Head	Antennae	Species
st	bn	>	*Chiridotea tuftsi*
cu	bn	>	*C. almyra*
cu	dn	=	Dark mottled coloring: *C. caeca* OR Black: *C. nigrescens*

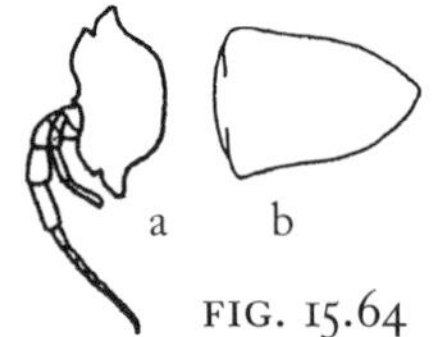

FIG. 15.64

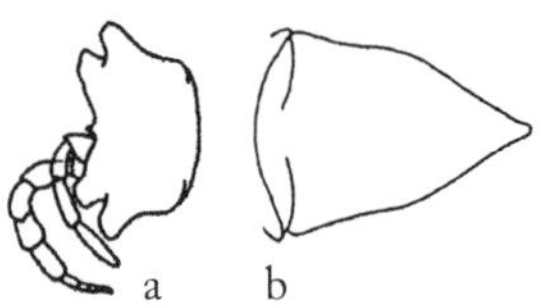

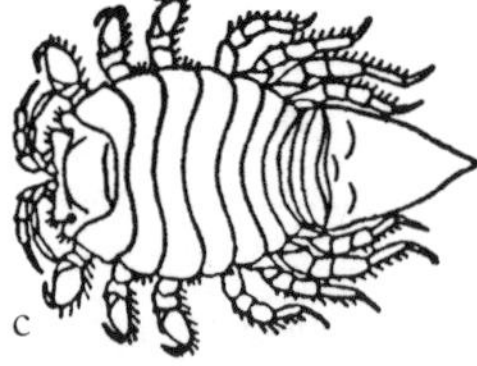

FIG. 15.65

Chiridotea almyra. Low salinity settings. Vi,Es–Mi,Su,sc?,8 mm (0.3 in), (G223). (FIG. 15.64. a, head; b, telson.)

Chiridotea caeca. Intertidal sand flats; builds ridgelike burrows near high-water line. Bo,Sa,Li,sc?,8 mm (0.3 in), (R353). (FIG. 15.65. a, head; b, telson; c, animal.)

Chiridotea nigrescens. Vi,Sa,Es–Su,sc?,8 mm (0.3 in).

Chiridotea tuftsi. Reddish brown mottled; telson with setae at tip. Subtidal, sandy-mud; full marine conditions. Ac,Sa–Mu,Su,sc?,8mm (0.3 in), (R354). (FIG. 15.66. a, head; b, telson; c, animal.)

FIG. 15.66

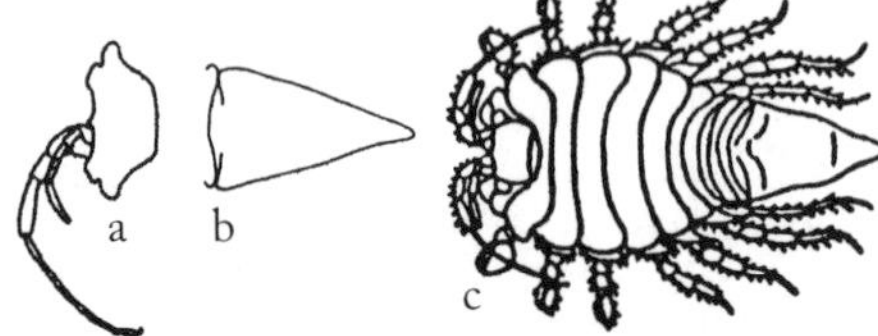

15.32

Class Malacostraca: Order Amphipoda, Amphipods, Gribbles

The classification of amphipod crustaceans is much debated among experts in the field. Four basic types are recognized. The members of the suborder Ingolfiellidea are known only from deep waters off North America and will not be treated here. Abundant and diverse Gammaridea dominate the region's amphipod fauna. Two small, specialized groups, the Hyperiidea and the Caprellidea, are often accorded their own status as suborders, although further review may lead to their inclusion with the Gammaridea. For a more in-depth discussion of these issues, see Bousfield and Shih (1994).

The first key below leads to separate keys for each of the three subgroups to be treated further in this guide. In many instances, requirements for viewing the animals include a dissecting microscope level of magnification and/or wet mounts (for instructions, see 0.7). Immobolization of these active animals will help (see the Appendix).

FIG. 15.67

Amphipod Subgroups

1. Body shape (FIG. 15.67)
   - A. Body *slender,* nearly cylindrical — sl
   - B. Body *laterally* compressed — la
   - C. Body *dorso-ventrally* compressed — dv
2. Eyes
   - A. Eyes very *large,* covering entire sides of the head (almost none of the headsegment can be seen from lateral view) — la
   - B. Eyes moderately large to *small* size, but not covering entire sides of head — sm
3. Location of first pair of walking legs (FIG. 15.67)
   - A. Head and first thoracic segment fused; first leg pair appears on *head* — he[†]
   - B. First leg pair on distinct *thoracic* segment behind the head — th

15.33

Body	Eyes	Legs	Group
la	la	th	Hyperiidea, 15.34
la,dv	sm	th	Gammaridea, 15.36
sl	sm	he	Caprellidea, 15.86

15.34 **Hyperiid Amphipods** (from 15.33)

Laterally compressed amphipods with the first pair of walking legs located on the thorax and with huge, lateral eyes covering the sides of the head. Hyperiid amphipods are offshore, pelagic animals. Adult stages are most frequently encountered in commensalistic association with various gelatinous macroplankton and sometimes drift onshore with their hosts.

1. Appearance of first pair of appendages
   - A. First appendages *chelate* or pincerlike — ch
   - B. First appendages *not* chelate, but are conspicuously *hairy* — xh
   - C. First appendages *not* chelate and *not* especially hairy — xx
2. Macroplanktonic host with which amphipod is associated
   - A. *Cyanea*, lion mane jellyfish (8.37) — cy
   - B. *Aurelia*, the moon jellyfish (8.37) — au
   - C. *Bolinopsis*, a comb jelly (9.2) — bo
   - D. *Salps*, a form of tunicate (17.14) — sa

15.35

Leg 1	Host	Species
ch	bo	*Hyperoche tauriformis,* big-eye amphipod
xh	cy	*Hyperia medusarum,* big-eye amphipod
xx	au	*H. galba,* big-eye amphipod
xx	sa	*Phronima sedentaria,* big-eye amphipod

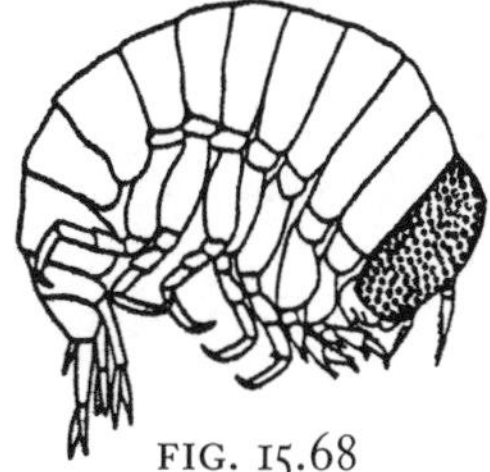
FIG. 15.68

Hyperia galba, big-eye amphipod. Appendages less hairy than the next species. Bo,Pa–Co,Ps,pr–pa,20 mm (0.8 in), (G226). (FIG. 15.68)

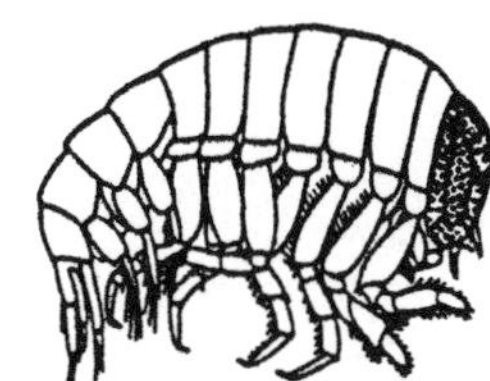
FIG. 15.69

Hyperia medusarum, big-eye amphipod. Anterior appendages very hairy. Bo,Pa–Co,Ps,pr–pa,20 mm (0.8 in), (G226). (FIG. 15.69)

Hyperoche tauriformis, big-eye amphipod. Chelate or pincherlike first appendages. Ac,Pa–Co,Ps,pr–pa,8 mm (0.3 in).

Phronima sedentaria, big-eye amphipod. Found offshore, inside the tests of salps. Bo,Pa–Co,Ps,pr–pa,25 mm (1 in). (FIG. 15.70)

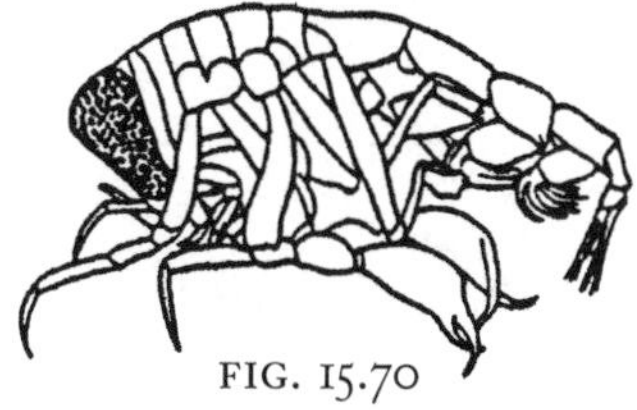
FIG. 15.70

15.36 **Gammaridean Amphipods** (from 15.7, 15.33)

Gammaridean amphipods are abundant and important but difficult to identify. This portion of the key should assist in obtaining an accurate identification, but any tentative identification should be checked carefully against more complete descriptions available in the Bousfield or other references listed below. Good optical equipment and lighting are essential for identifying amphipods. The specimen must be immobile for accurate observation. It may be useful to cut off the abdominal section (the posterior third) of the body in order to orient it so that it can be viewed from above.

Several key characteristics require initial definition. The two pairs of *antennae* comprise a few large basal segments that form the *peduncle* and many smaller segments forming the *flagellum.* Sometimes a second or *accessory flagellum* is present on the first pair of antennae. The seven most conspicuous pairs of legs are attached to thoracic segments. The first two pairs of these are *gnathopods* and often have enlarged ends. Thoracic legs or *pereiopods* have rounded or squared *coxal plates* overlying their attachment to the body. The shape of the fourth coxal plate is especially useful in these keys. The abdomen comprises the posterior third of the amphipod. The last three abdominal legs are *uropods,* and details of their construction may be noted. The third pair of uropods flank a telson or tailpiece whose shape is yet another key character.

If uropods 3 terminate in two lobes or rami, they are *biramous;* termination in a single tip is the *uniramous* condition. As a swimming appendage, uropods 3 tend to be most highly developed in active swimmers and least well developed in tube dwellers.

Recent investigations of lifestyle habits (discussed by Bousfield and Shih 1994) have led to the notion of dividing the group into two large, functionally based categories, the *natantia,* whose activities are primarily associated with the water column, and the *reptantia,* which are more typically benthic. These ties are especially clear with regard to differences in reproductive habits, which, in turn, influence morphological traits. Natant amphipods (swimmers) mate freely in the water column and thus emphasize sensory structures that presumably enable them to locate one another effectively. These features include an enlarged flagellum on antennae 2, as well as a line of *brush setae* along the anterior margin of the peduncle. In addition, bundles of sensory hairs, known as the *callynophore,* are clustered along the posterior margin of antennae 1. Members of this group tend to have well-developed uropods 3 for swimming and often display sexual dimorphism (distinct differences in appearance between the sexes) in swimming and sensory features such as eyes, antennae, and uropods 3. Females are usually larger than males.

Reptant amphipods tend to maintain close association with one another during the reproductive period. In some cases, the male engages in mate-guarding, which includes *mate-carrying,* that is, using the first pair of gnathopods to attach itself to a pre-molt female, allowing the pair to travel together until her molt permits fertilization. In other cases of mate-guarding, males employ *mate-attending,* maintaining close proximity to females awaiting the opportunity to copulate and warding off rivals during this period. Reptant sensory features are modest in comparison with those of natantia, but often they display a form of sexual dimorphism in the first two thoracic limbs, the gnathopods. Typically, gnathopods 2 is conspicuously larger than gnathopods 1, especially in males, where it is used in agonistic (combative) behaviors. Hook or notchlike modifications of certain coxal plates of females may be related to attachment points in mate-carrying species (Borowsky 1984). The sexes are similar with regard to sensory and swimming features. In males both body size and gnathopods 2 are distinctly larger.

Although this linkage between form and function may lead to important revisions of our understanding of natural relationships among the amphipods, comparative information is missing for some groups, exceptions are known in several cases, and some of these features occur only during certain life-history stages. As identification characters for individual specimens collected in the field, such features are valuable when present, but their absence is not as meaningful. Females or immature males may not possess characters that are in fact diagnostic for their species or gender as a whole. Thus, rather than including these characteristics as definitive key choices, the designation natantia or reptantia is given for each family (with modifying comments from Bousfield and Shih 1994; Conlan 1991) as an additional source for confirmation. For natantia, look for brush setae and callynophores on antennal bases of mature males; for reptantia, check for enlarged gnathopods 2 on males. In addition, amphipods found clinging to one another are apt to be reptantia.

It is best to collect several individuals of each type of amphipod to be identified. Because certain appendages (especially the antennae) are easily broken, missing entirely, or in various stages of regeneration following such damage, it is necessary to look for signs of breakage and choose specimens that appear to have all structures intact. It may be necessary to examine several specimens to locate the distinguishing features of sexually mature males.

Additional references: (B) Bousfield, E. L., 1973; (LB) Barnard, J. L., 1969; (BK) Barnard, J. L., and G. S. Karaman, 1991. Note: Barnard (1969) has been updated by Barnard and Karaman (1991). Because the more recent source is less likely to be available to general users of this manual, references are to both sources.

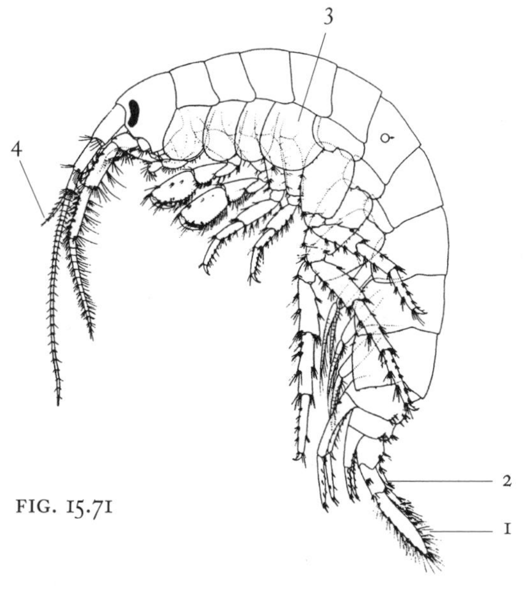

FIG. 15.71

1. Number of branches comprising uropods 3 (the hindmost pair of appendages, flanking the telson); (view them from above) (FIG. 15.71)
   - A. Uropods 3 *uniramous* (end in only 1 tip) — un
   - B. Uropods 3 *biramous* (end in two parts or lobes) — bi*
2. Cleft or split in tailpiece or telson (dorsal view) (FIG. 15.71)
   - A. Telson distinctly *cleft* — cl
   - B. Telson entire, cleft *absent* — x
3. Shape of fourth coxal plate (squared or rounded, lateral, flaplike exoskeletal plate overlying the base of the fourth pair thoracic walking legs) (FIG. 15.71)
   - A. Coxa 4 *excavated* along the posterior edge (has a notch) — ex*
   - B. Coxa 4 *without* excavation (posterior edge is intact) — x
4. Presence of accessory flagellum on antenna one (it may be tiny) (FIG. 15.71)
   - A. *Accessory flagellum* present — af*
   - B. Accessory flagellum *absent* — x

15.37

Uropods 3	Telson	Coxal Plate 4	Accessory Flagellum	Group
bi	cl	ex	af	Amphipod Group 1, 15.38
bi	cl	x	af	Amphipod Group 1, 15.38
bi	cl	ex	x	Amphipod Group 2, 15.40
bi	cl	x	x	Bateidae, 15.54
bi	x	ex	x	Both rami (branches) of uropod three equal in length: Calliopiidae, 15.55 OR Inner ramus distinctly shorter than outer ramus: Pleustidae, 15.70
bi	x	ex	af	Amphipod Group 1, 15.38
bi	x	x	af	Amphipod Group 3, 15.42
bi	x	x	x	Amphipod Group 3, 15.42
un	cl	ex	x	Amphipod Group 4, 15.44
un	cl	x	x	Amphipod Group 4, 15.44
un	x	x	af	Aoridae (in part), 15.50
un	x	x	x	Amphipod Group 4, 15.44

15.38

Amphipod Group 1
(Amphipoda: Gammaridae, Gammarellidae, Haustoriidae, Lysianassidae, Melitidae, and Phoxocephalidae) (from 15.37)

Amphipods with accessory flagellum, coxal plate 4 either excavate or not, and with biramous third uropods flanking a barely cleft telson.

1. Shape of coxal plate 4 (FIG. 15.72)
   - A. Plate distinctly *longer than wide* — l>w
   - B. Plate distinctly *wider than long* — l<w
   - C. Plate about *equal* in *length and width* — l=w*
2. Shape of eyes (FIG. 15.72)
   - A. Eyes *bean*-shaped (anterior edge concave) — be*
   - B. Eyes nearly *round* or smoothly ovaloid — ro
   - C. Eyes *rectangular* — re
   - D. Eyes *absent* — x

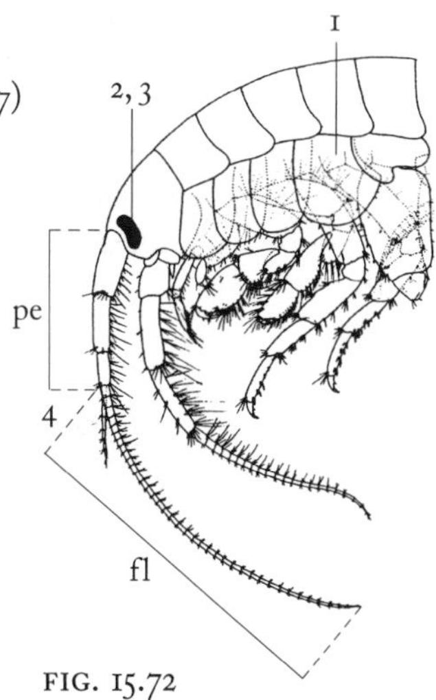

FIG. 15.72

3. Eye pigmentation (FIG. 15.72)
   - A. Eyes well *pigmented* — pi*
   - B. Eye pigmentation *reduced* — re
   - C. Eyes pigmentation *absent* — x
4. Length of first three (large) segments of antenna 1 (peduncle) (FIG. 15.72, pe), compared to the remaining (small) segments of antenna 1 (flagellum) (FIG. 15.72, fl)
   - A. *Flagellum* distinctly *longer than* the *peduncle* — f>p*
   - B. *Flagellum* approximately *equal to* or shorter than the *peduncle* — f=p
5. Distinctive head shape
   - A. Head forms pointed *hood* or rostrum anteriorly overlapping bases of first antennae — ho
   - B. Lower sides of head (beneath the eyes) project anteriorly as a *point*, extending anterior to eyes — po
   - C. Head *rounded* anteriorly or if pointed, does not overlap bases of first antennae — ro*

15.39

Coxal Plate 4	Eyes	Pigmentation	Antennae 1	Head	Family
l<w,l=w	ro	pi	f=p,f>p	ro	Total length of antennae 1 less than first 4 trunk segments: Melitidae, 15.75 OR Total length of antenna 1 greater than first 4 trunk segments: Liljeboriidae, 15.72
l=w	ro	pi,re	f=p,f>p	po	Cheirocratidae, 15.56
l<w,l=w	be,x,ro	re,x	f=p	ho,ro	Haustoriidae, 15.65
l=w	be,re	pi	f>p	ro	Gammaridae and Gammarellidae, 15.61
l>w	be	pi	f=p,f>p	ro	Branches of uropods 3 approximately equal in length: Lysianassidae, 15.73 OR Inner branch of uropods 3 much shorter than (often <½) the length of outer branch: Haustoriidae, 15.65
l>w	ro,x	pi,x	f=p	ho	Phoxocephalidae, 15.79

15.40 Amphipod Group 2

(Amphipods: Ampeliscidae, Liljeborgiidae, and Pontogeneiidae) (from 15.37)

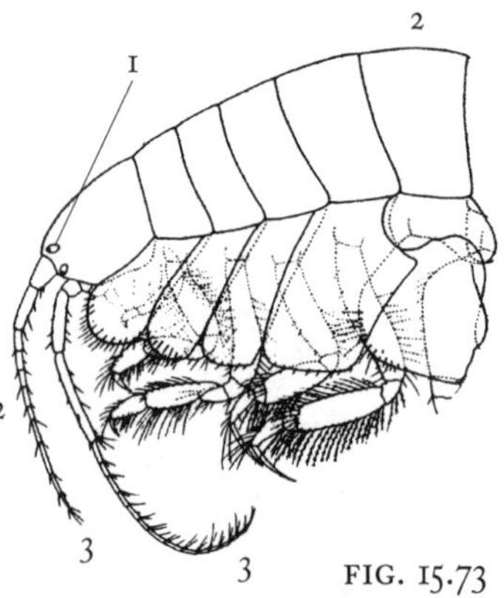

FIG. 15.73

Amphipods lacking accessory flagellum but with coxal plate 4 excavated and with biramous third uropods flanking a deeply cleft telson.

1. Total number of eyes present: provide *number* (usually 2 or 4) (FIG. 15.73) — __
2. Length of antennae 1 compared to body length (antennae are sometimes broken or in the process of regeneration) (FIG. 15.73)
   - A. Antennae *1* distinctly *less than half body* length — 1<½b*
   - B. Antenna *1* distinctly *more than half body* length — 1>½b
   - C. Antenna *1 equal to half body* length — 1=½b
3. Length of antennae 1 compared to antennae 2 (FIG. 15.73)
   - A. Antennae *1* distinctly *shorter than* antennae *2* — 1<2*
   - B. Antennae *1* about *equal to* or greater than antennae *2* — 1=2

15.41

Eyes	Antennae 1 vs. Body	Antennae 1 vs. Antennae 2	Family
4	1<Hb	1<2	Ampeliscidae, 15.46
2	1<½b	1=2	Liljeborgiidae, 15.72
2	1>½b	1=2	Pontogeneiidae, 15.82
2	1=½b	1=2	Dexaminidae, 15.59

15.42 Ampiphod Group 3
(Amphipods: Ampithoidae, Aoridae, Oedicerotidae, Isaeidae [=Photidae], and Ischyroceridae) (from 15.37)

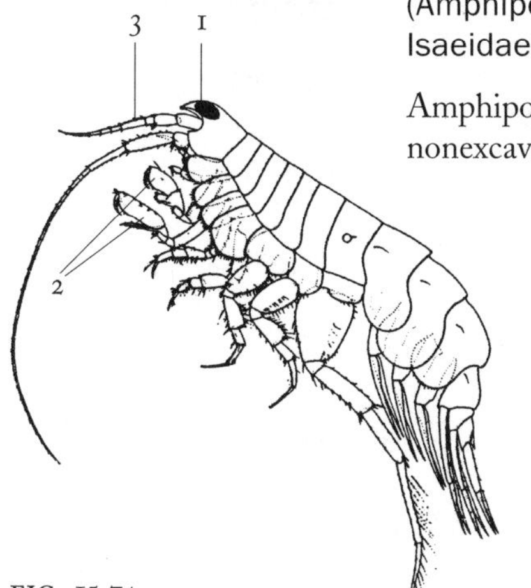

FIG. 15.74

Amphipods with accessory flagellum and biramous third uropods, but with noncleft telson and nonexcavate coxal plate 4.

1. Placement of eyes (FIG. 15.74)
   A. Eyes *fused* middorsally on head — fu*
   B. Eyes *lateral,* not fused middorsally — la
2. Compare size of gnathopods 1 (first obvious leg) to gnathopods 2 (this character is most obvious in males; females tend to show less distinct differences) (FIG. 15.74)
   A. Gnathopods *1 larger than* gnathopods *2* — 1>2
   B. Gnathopods *1 smaller than* or equal to gnathopods *2* — 1<2*
3. Total number of segments in antennae 1 (use an undamaged specimen): provide *number* (FIG. 15.74) — __

15.43

Eye	Gnathopods 1 vs. Gnathopods 2	Antennae 1	Family
fu	1<2	10–11	Oedicerotidae, 15.77
la	1<2	8–12	Ischyroceridae, 15.70
la	1<2	11–15	Isaeidae, 15.69
la	1<2	18–21	Ampithoidae, 15.48
la	1>2	19–25	Aoridae, 15.50

15.44 Amphipod Group 4
(Amphipods: Corophiidae, Hyalidae, Stenothoidae, and Talitridae) (from 15.37)

Amphipods lacking accessory flagellum and lacking excavate coxal plates 4; uniramous third uropods flank a noncleft telson.

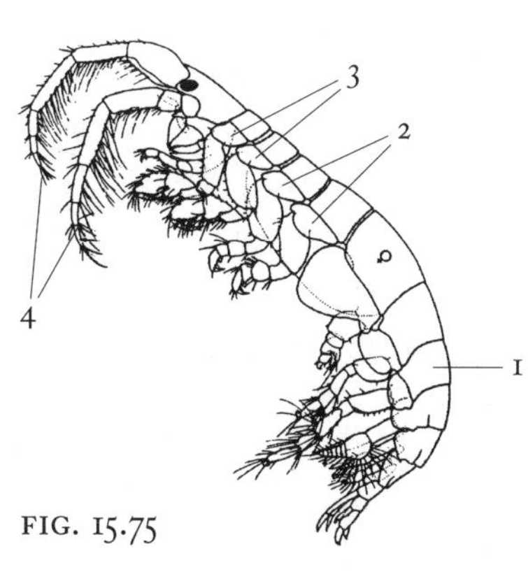

FIG. 15.75

1. Shape of abdomen (FIG. 15.75)
   A. Abdomen flattened *laterally* (like typical amphipod) — la
   B. Abdomen flattened *dorsoventrally* (top to bottom) — dv*
2. Area of coxal plate 4 compared to coxal plate 3 (FIG. 15.75)
   A. Coxal plate *4 much larger* (at least twice) the area of coxal plate *3* — 4>>3
   B. Coxal plate *4* only slightly larger than or *equal to* coxal plate *3* — 4=3*
3. Area of coxal plate 1 compared to coxal plate 2 (FIG. 15.75)
   A. Coxal plate *1 less than half* area of coxal plate *2* — 1<2
   B. Coxal plate *1 about equal to* coxal plate *2*; may be from slightly smaller to slightly larger — 1=2*
4. Length of antennae 1 compared to length of antennae 2 (FIG. 15.75)
   A. Antennae *1 much less* (⅓) than the length of antennae *2* — 1<<2
   B. Antennae *1* at least *half* the length of antennae *2* — 1½2*

15.45

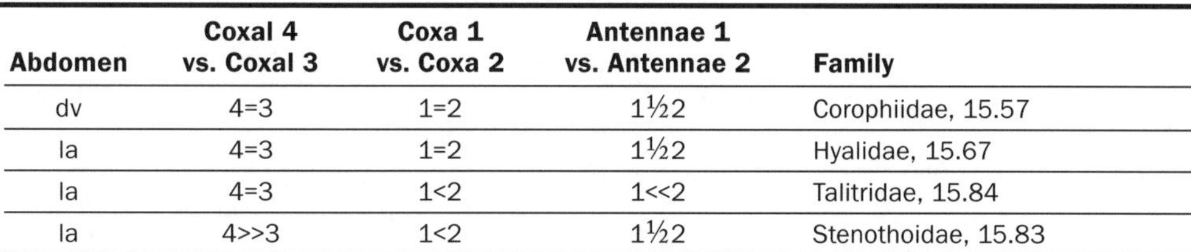

Abdomen	Coxal 4 vs. Coxal 3	Coxa 1 vs. Coxa 2	Antennae 1 vs. Antennae 2	Family
dv	4=3	1=2	1½2	Corophiidae, 15.57
la	4=3	1=2	1½2	Hyalidae, 15.67
la	4=3	1<2	1<<2	Talitridae, 15.84
la	4>>3	1<2	1½2	Stenothoidae, 15.83

Families of Gammaridean Amphipoda (arranged alphabetically)

15.46 Ampeliscidae, Four-Eyed Amphipods (from 15.41)

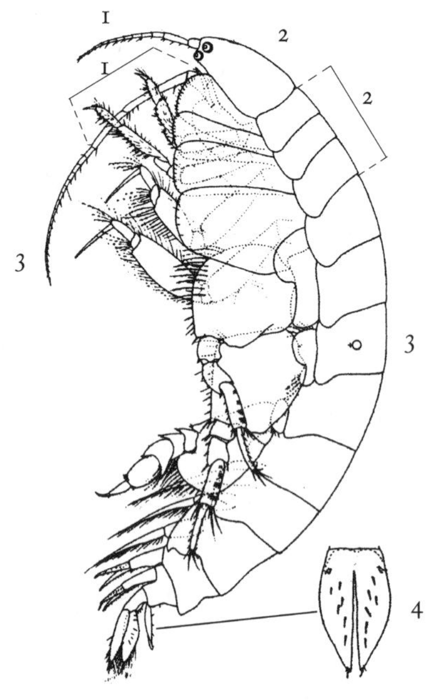

FIG. 15.76

Accessory flagellum absent; uropods 3 biramous; telson usually cleft; antennae 1 unusually short. Suspension feeders and surface detritivores; infaunal; tube dwellers in sand and mud. Natantia, usually with well-developed brush setae. (BK84,G227,LB128)

1. Antennae 1 compared in length to peduncle or basal three segments of antenna 2 (FIG. 15.76)
   - A. Antennae *1 longer than peduncle* of antennae *2* — 1>2p
   - B. Antennae *1 shorter than peduncle* of antennae *2* — 1<2p*
2. Length of head compared to the length of the first three thoracic segments (FIG. 15.76)
   - A. *Head* approximately *equal to* length of first *three* thoracic segments — h=3*
   - B. *Head* distinctly *shorter than* length of first *three* thoracic segments — h<3
3. Length of antennae 2 compared to body length (FIG. 15.76)
   - A. Antennae *2 longer* than *half body* length — 2>½b
   - B. Antennae *2* less than or *equal to half body* length — 2=½b*
4. Proportions of telson (FIG. 15.76)
   - A. *Length* clearly *greater than width* — l>w*
   - B. *Length* approximately *equal to width* — l=w

15.47

Antennae 1 vs. Peduncle of Antennea 2	Head vs. Thorax	Antennae 2 vs. Body	Telson	Species
1<2p	h=3	2=½b	l>w	*Ampelisca verrilli,* narrow-headed four-eyed amphipod
1>2p	h=3	2=½b	l>w	*A. macrocephala*
1>2p	h<3	2=½b	l>w	*A. abdita,* small four-eyed amphipod
1>2p	h<3	2>½b	l>w	*A. vadorum*
1>2p	h<3	2>½b	l=w	*Byblis serrata*

FIG. 15.77

Ampelisca abdita, small four-eyed amphipod. Mud tubes; polyhaline. Uropods 1 slightly longer than uropods 2. Narrow silty tube. Difficult to distinguish from *A. vadorum.* Bo,Sa–Mu,Es–Li–Su,ff–df,8 mm (0.3 in), (B136), ☺. (FIG. 15.77)

Ameplisca macrocephala. Zostera beds; mud. Uropods 1 slightly shorter than uropods 2. Whitish with pink and yellow markings. Ac–,Mu–Ss–SG,Su–De,ff–df,20 mm (0.8 in), (B133).

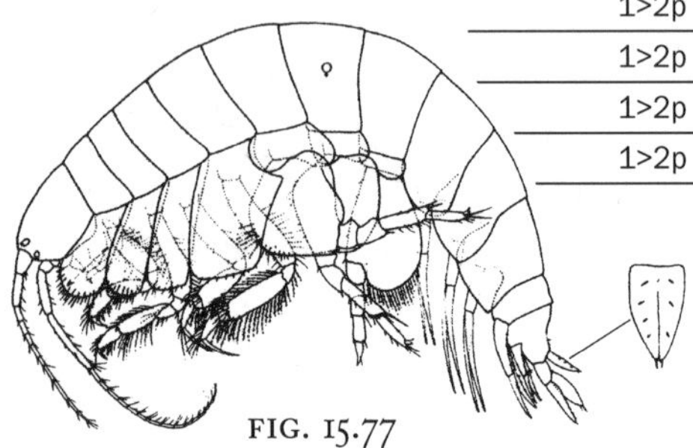
FIG. 15.78

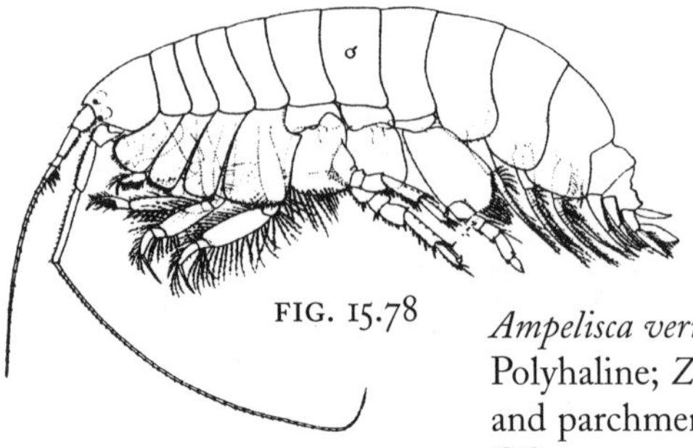

Ampelisca vadorum. Polyhaline; lies on its back within short, wide, fine sandy tubes; coarse sediment. Bo,Sa–SG,Es–Li–Su,ff–df,12 mm (0.5 in), (B135). (FIG. 15.78)

Ampelisca verrilli, narrow-headed four-eyed amphipod. Polyhaline; *Zostera* beds; coarse sand. Narrow sand and parchment tubes with only one open end. Often turns up in "night-light" collections. Vi,Sa–SG,Es–Li–Su,ff–df,12 mm (0.5 in), (B134). (FIG. 15.76)

Byblis serrata. Builds tubes in sand. Vi,Sa–Mu,Su,ff–df?,12 mm (0.5 in), (B137). (FIG. 15.79)

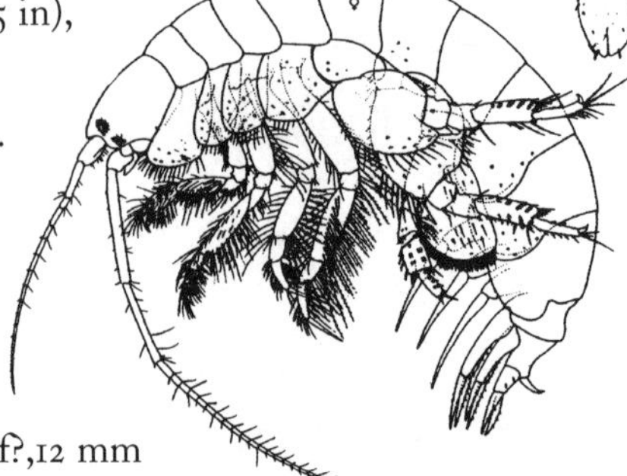
FIG. 15.79

15.48 **Ampithoidae** (from 15.43)

Accessory flagellum present (very small) or absent; gnathopods powerful; uropods 3 biramous, outer ramus with two large hooked spines; telson short, not cleft, fleshy. Grazers; epifaunal tube dwellers. Reptantia. (BK98,LB141)

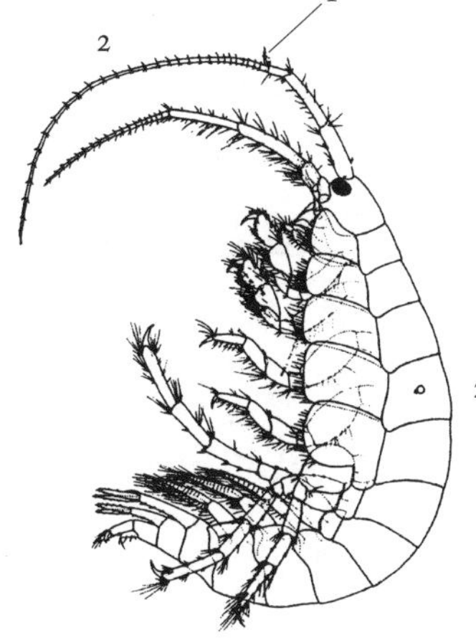

FIG. 15.80

1. Accessory flagellum, attached partway along antennae 1, if present (FIG. 15.80)
   - A. Accessory flagellum *minute* but present — mi*
   - B. Accessory flagellum *absent* — x
2. Length of antennae 1 compared to total body length (FIG. 15.80)
   - A. Antennae *1 equal to body* length — 1=b
   - B. Antennae *1 shorter than body* length — 1<b*
3. Middorsal coloration on abdominal segments
   - A. Row of *white spots* along middorsal abdominal segments — ws
   - B. Color uniform, spots *absent* — x
4. Eye color
   - A. *Red* or dark red — re
   - B. *Black* — bl

15.49

Accessory Flagellum	Antennae 1 vs. Body	Color	Eye	Species
x	1=b	x?	bl	*Ampithoe longimana,* long-antennaed tube-building amphipod
x	1<b	ws,x	re	*A. rubricata,* red-eyed amphipod
x	1<b	x	bl	*A. valida*
mi	1<b	x?	re	*Cymadusa compta,* wave-diver tube-building amphipod

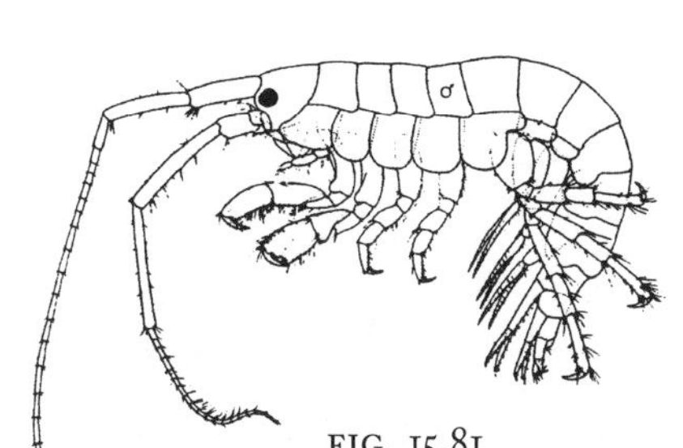
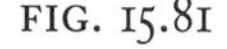
FIG. 15.81

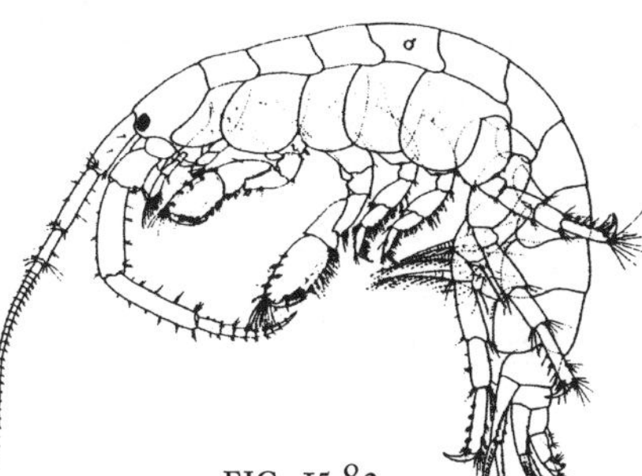
FIG. 15.82

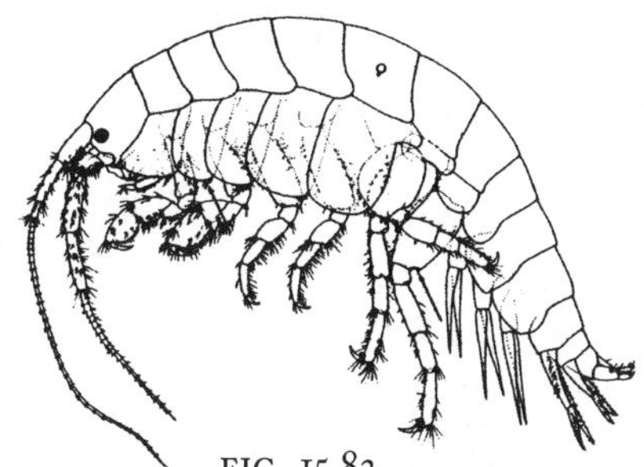
FIG. 15.83

Ampithoe longimana, long-antennaed tube-building amphipod. On algae (especially *Ulva, Ceramium*), or eelgrass, *Zostera;* tubes of tiny bits of algae. Bo,Al–SG,Li–Su,gr,10 mm (0.4 in), (BI81). (FIG. 15.81)

Ampithoe rubricata, red-eyed amphipod. Green-red color. Eats algae and detritus; nests in seaweeds or mussel beds. Ac–,Ha–Al,Li,gr,20 mm (0.8 in), (BI80,NM604), ☺. (FIG. 15.82)

Ampithoe valida. Tidal debris on mud flats; on *Ulva.* Bo,Mu,Es–Li–Su,gr,12 mm (0.5 in), (BI80). (FIG. 15.83)

Cymadusa compta, wave-diver tube-building amphipod. Large eyes. Eelgrass, *Zostera,* beds and shallows. Ac–,SG,Es–Li–Su,gr–df,12 mm (0.5 in), (BI82). (FIG. 15.80)

15.50 **Aoridae** (from 15.37, 15.43)

Accessory flagellum present; uropods 3 short, biramous; telson noncleft but sometimes sculptured. Suspension feeders or surface detritivores; infaunal or epifaunal tube dwellers. Reptantia, but with gnathopods 1 larger than gnathopods 2, the opposite of usual dimorphic patterns. (LB147)

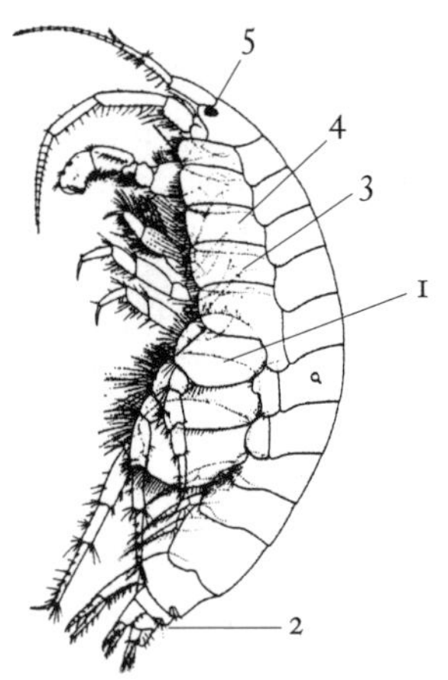

FIG. 15.84

1. Shape of bases of last three pairs of thoracic legs (the largest walking legs) (FIG. 15.84)
   - A. Bases *fat* and *ovaloid* (length/width < 1.7) — fo*
   - B. Bases *slender* (length/width > 1.7) — sl
2. Shape of telson (FIG. 15.84)
   - A. Telson smoothly *rounded* — ro
   - B. Telson *angular* at corners — an
   - C. Posterior corners of telson formed as distinct *points* — po
3. Shape of coxal plate 4 (FIG. 15.84)
   - A. Plate distinctly *longer than wide* — l>w*
   - B. Plate truncated, distinctly *wider than long* — w>l
   - C. Plate with *length* about *equal to width* — l=w
4. Proximity of coxal plates to one another (FIG. 15.84)
   - A. Coxal plates abut or slightly *overlap* one another — ov*
   - B. Coxal plates separated, do *not* touch one another — x
5. Eyes (FIG. 15.84)
   - A. *Eyes* present — ey
   - B. Eyes *absent* — x

15.51

Coxae	Telson	Coxa 4	Overlap	Eyes	Species
fo	ro	l>w	ov	ey	*Leptocheirus plumulosus*, common burrower amphipod
fo	po	l>w	ov	ey	*L. pinguis*
fo	an	l>w	ov	ey	*Microprotopus raneyi*
sl	an	l=w	ov	ey	*Lembos websteri*
sl	po	l<w	x	ey	*L. smithi*
sl	ro	l<w,w=l	ov	ey	*Microdeutopus gryllotalpa*
sl	ro	l<w	x	ey	*Unciola*, 15.52
sl	ro	l<w	x	x	*Pseudunciola obliquua*

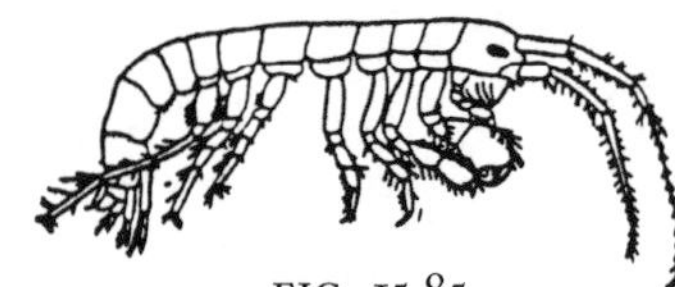
FIG. 15.85

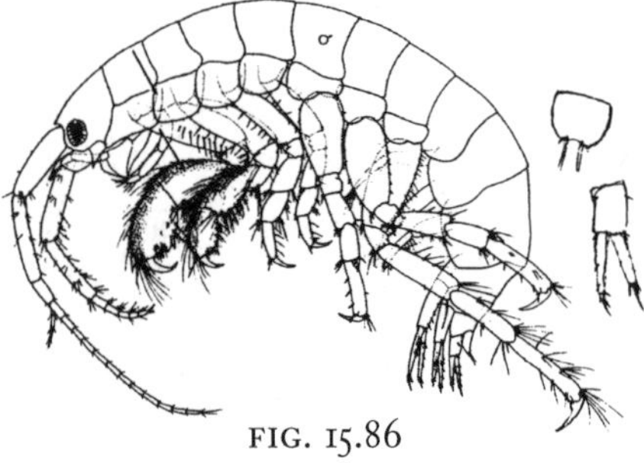
FIG. 15.86

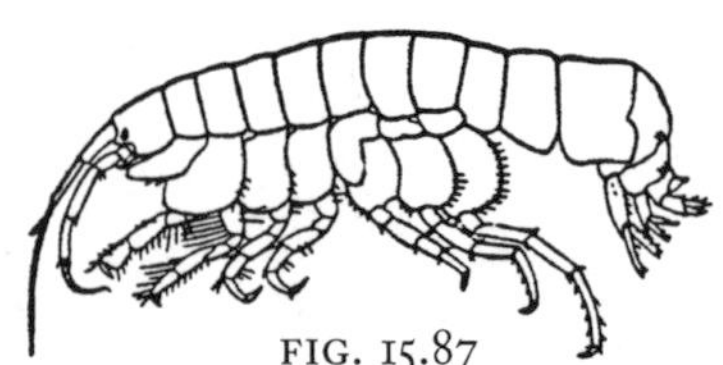
FIG. 15.87

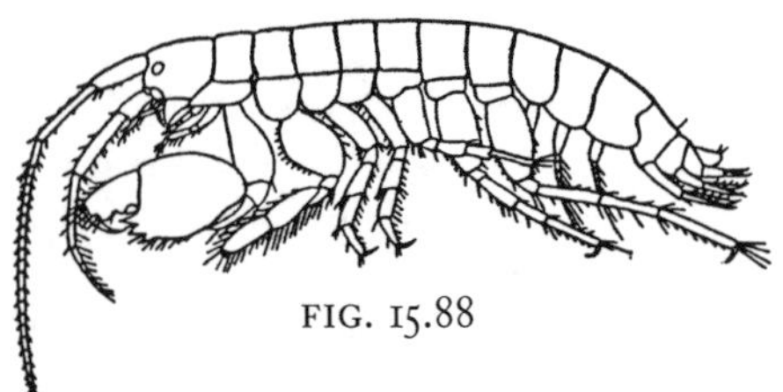
FIG. 15.88

Lembos smithi. Mottled coloring. Tide-pool algae, eelgrass (*Zostera*) beds. Vi,SG–Al,Li–Su,ff–df,6 mm (0.2 in), (B170). (FIG. 15.85)

Lembos websteri. Yellowish brown with brown marks. Subtidal on algae; often on kelp holdfasts. Vi,Al,Su,ff–df,6 mm (0.2 in), (B169). (FIG. 15.86)

Leptocheirus pinguis. Coxal plate 1 drawn forward as point. Epifaunal on sandy-mud; attracted to night lights. Tolerates high organic content. Important as food for fish. Bo,Sa–Mu,Es–Li–De,ff–df,16 mm (0.65 in), (B167). (FIG. 15.87)

Leptocheirus plumulosus, common burrower amphipod. Coxal plate 1 is blunt. Estuarine, burrows into muddy sand; sandy tube. Vi,Mi–Sa,Es–Li–,ff,12 mm (0.5 in), (B168), ☺. (FIG. 15.84)

Microdeutopus gryllotalpa. Tube builder. Vi+,Al–SG–Ha,Li–De,ff–df?,11 mm (0.4 in), (B172,G228). (FIG. 15.88)

Microprotopus raneyi (in family Isaeidae [15.69] but may key out here). Builds sandy tubes. Vi,Sa,Li–,ff–df,5 mm (0.2 in), (B188).

Pseudunciola obliquua. Tube builder. Blind; accessory flagellum only one segment. Bo,Sa,Li–De,ff,6 mm (0.24 in), (B178).

15.52 Unciola (from 15.51)

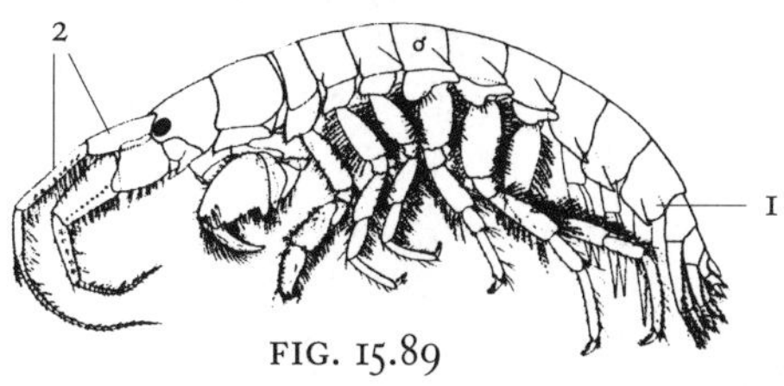

FIG. 15.89

1. Appearance of lateral hook on last abdominal segment (FIG. 15.89)
   - A. Hook top *bends* distinctly upwards — be
   - B. Hook tip *straight* — st*
2. On antennae 1, length of segment 1 compared to segment 2 (FIG. 15.89)
   - A. Segment *2* clearly *longer than* segment *1* — 2>1*
   - B. Segment *2* about *equal to* segment *1* — 2=1

15.53

Hook	Antennae 1: Segment 1 vs. 2	Species
be	2>1	*Unciola irrorata*
st	2>1	*U. dissimilis*
st	2=1	*U. serrata*

Unciola dissimilis. Bo,Sa–Mu,Su–De,ff–df?,16 mm (0.65 in), (B176). (FIG. 15.89)

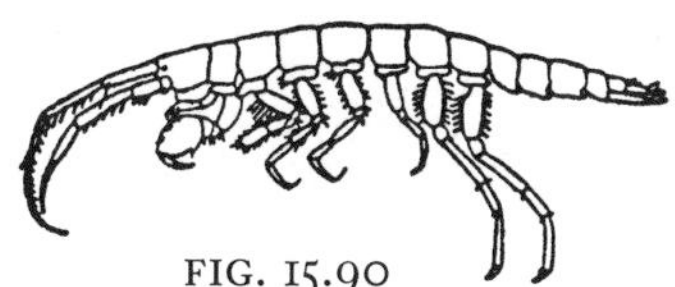
FIG. 15.90

Unciola irrorata. Red middorsal band with red spots on sides; flagella reddish. Medium-coarse sand; estuarine to full marine. Lives in tubes of others; tolerates low salinity and oxygen, including sludge. Ac–,Mu–Sa,Es–Su–De,ff–df?,14 mm (0.55 in), (B174), ☺. (FIG. 15.90)

Unciola serrata. Tube builder. Vi+,Sa–Mu,Su,ff–df?,7 mm (0.3 in), (B176). (FIG. 15.91)

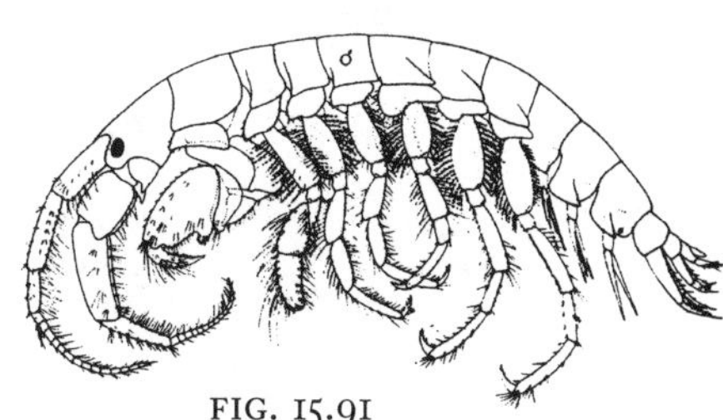
FIG. 15.91

15.54 Bateidae (from 15.37)

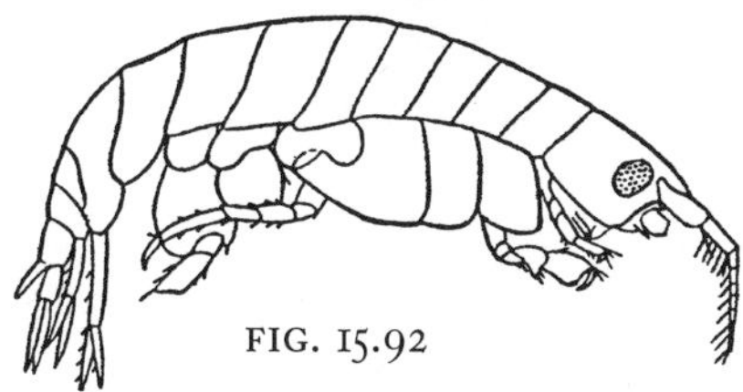
FIG. 15.92

Accessory flagellum absent; uropods 3 biramous; telson deeply cleft; rostrum well developed; coxal plate 1 absent, coxal plates 3 and 4 enlarged. Natantia. (BK114,LB164)

Batea catharinensis, purple-eyed amphipod. Large, squared eyes. Polyhaline; member or "fouling community," on eelgrass, *Zostera,* or hydroids, ectoprocts, sponges, or algae. Attracted to night lights. Surface detritivore and grazer; no tubes. Vi,Gr–Ha–Al–SG,Su,gr–df,10 mm (0.4 in), (B76). (FIG. 15.92)

15.55 Calliopiidae (from 15.37)

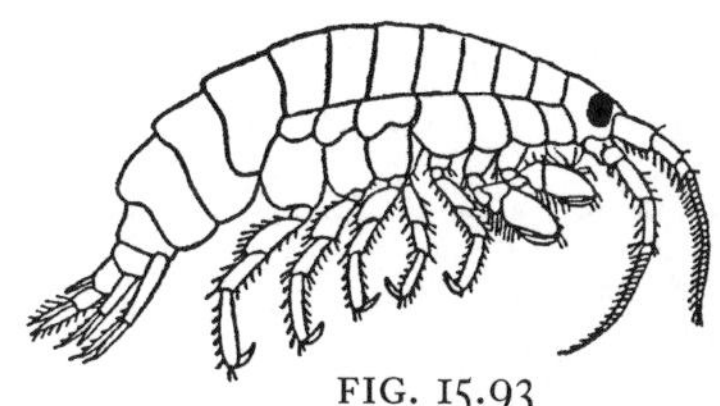
FIG. 15.93

Accessory flagellum absent; stout antennae; uropods 3 biramous with approximately equal rami; telson usually entire or barely cleft. Natantia. (BK117,LB167)

Calliopius laeviusculus. Segment 3 or antennae 1 with small posterior lobe. Pelagic or epibenthic; low, rocky shore, intertidal pools or to deep water. Surface detritivore and grazer; no tubes. Bo,Pe–Ha,Li–Su,df–gr,16 mm (0.6 in), (B80,G226), ☺. (FIG. 15.93)

15.56 Cheirocratidae (from 15.39)

Tiny accessory flagellum; antennae 1 shorter than antennae 2; uropods 3 biramous and with equal rami; telson deeply cleft, with terminal spines. Natantia.

Casco bigelowi. Cold water. Ac–,Mu–Ha–Gr,Su,sc–df?,25 mm (1 in), (B62). (FIG. 15.94)

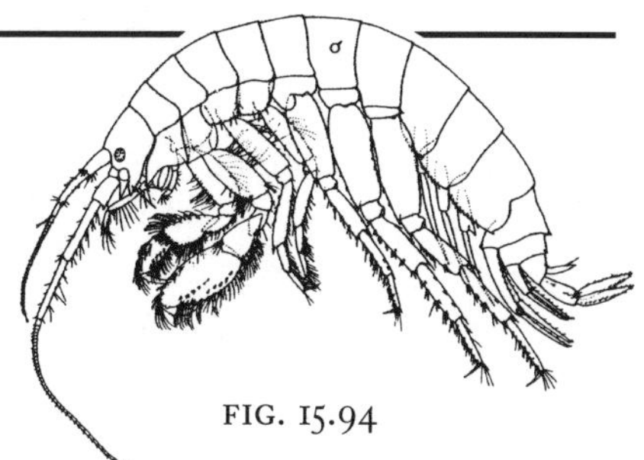
FIG. 15.94

15.57 **Corophiidae** (from 15.45)

Accessory flagellum absent; uropods 3 with one ramus; telson short, fleshy. Most are tube-dwelling and are cylindrical in shape with enlarged antennae for reaching out of tube. Suspension feeders and surface detritivores, either infaunal or epifaunal. Reptantia with great dimorphic distinction between gnathopods 2 (larger) and gnathopods 1 (smaller) in mature males. (BK137,LB184)

1. Diameter of segment 2 of antenna 1 compared to diameter or segment 2 of antenna 2
   - A. Segment 2 of antenna *1 much thinner than* segment 2 of antenna *2* — 1<2
   - B. Segment 2 wo of antenna *1* about *equal* in diameter to segment 2 of antenna *2* — 1=2
2. Shape of telson
   - A. Telson *cleft* — cl
   - B. Telson single piece, cleft *absent* — x
3. Viewed dorsally, the segmentation of urosome (the small abdominal region extending from the telson forward to the last of the major, large body segments; the urosome bears the posteriormost three pairs of appendages, the uropods)
   - A. Urosome is clearly *segmented* in construction — se
   - B. Urosomal segments are *fused* into a single, composite piece — fu

15.58

Antennae 1 vs. Antennae 2	Telson	Urosome	Species
1=2	cl	se	*Cerapus tubularis*
1=2	x	se	*Erichthonius brasiliensis*
1<2	x	se	*Corophium volutator*
1<2	x	fu	*Corophium* spp.

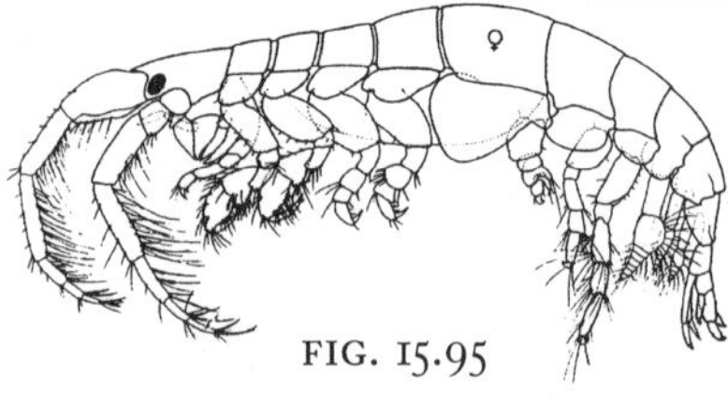

FIG. 15.95

Cerapus tubularis. First segment of antennae 1 inflated. Carries a tube, rectangular in cross-section. Vi,Mu–Sa,Es–Su,ff–gr?,7 mm (0.3 in), (B197). (FIG. 15.95)

Corophium spp. Muddy tubes among hydroids and algae; often estuarine. Several species; males differ in several respects from females. Refer to Bousfield (1973) for identification. (B200–203), ☺.

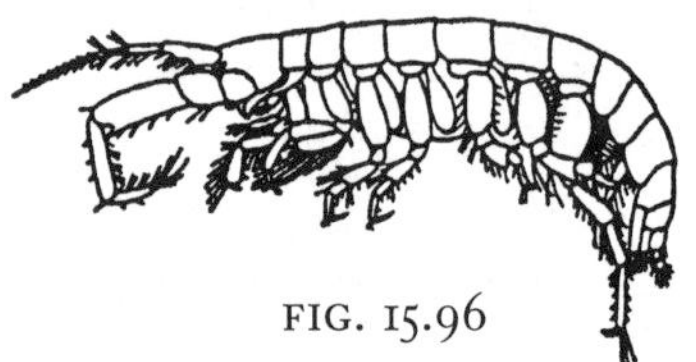

FIG. 15.96

Corophium volutator. Restricted to Bay of Fundy on this coast. Makes tubes in muddy substrates; marine to nearly freshwater. Ac,Mu–SM,Es–Li,df,4 mm (0.15 in), (B201), ☺. (FIG. 15.96)

Erichthonius brasiliensis. Muddy tubes on hydroids (especially *Sertularia*) and ectoprocts. Vi,Al,Es–Su–De,ff–df,6 mm (0.2 in), (B195), ☺. (FIG. 15.97)

Unciola irrorata. In the family Aoridae; may key out here. See notes, 15.52.

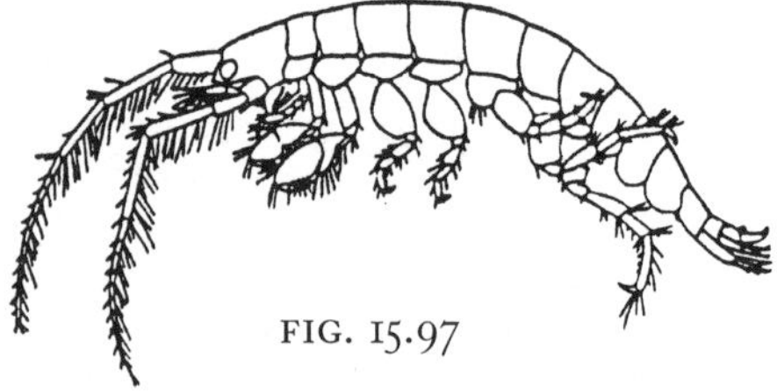

FIG. 15.97

15.59 **Dexaminidae** (from 15.41)

Accessory flagellum absent; coxal plate 4 excavate; gnathopods small; uropods 3 biramous; telson deeply cleft. Natantia with well-developed row of short, brush setae on the anterior surface of the peduncle (basal segments) of antennae 2. (BK260,LB200).

Dexamine thea. Abdominal segments with strong posterior-dorsal tooth. Ac–,Al–Ha–SG,Su,?,7 mm (0.3 in), (B130). (FIG. 15.98)

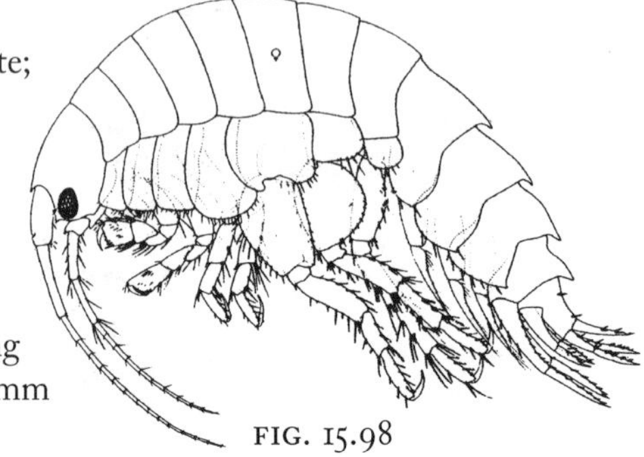

FIG. 15.98

15.60 **Gammarellidae** (see 15.61)

15.61 **Gammaridae and Gammarellidae** (from 15.39)

Accessory flagellum present; uropods 3 biramous; telson usually cleft but can be whole; gnathopods well developed. Species included are Gammaridae unless identified otherwise. Gammaridae are reptantia, employing precopulatory mate-carrying. Gammarellidae are natantia, but with weakly developed brush setae and not engaging in mate-carrying. (BK350,G227,LB231)

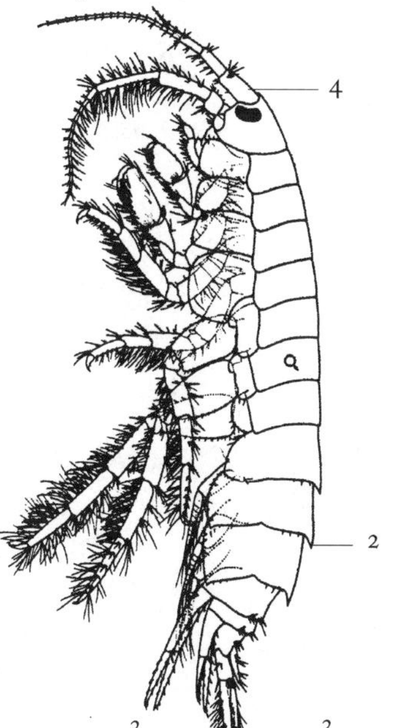

FIG. 15.99

1. Length of antennae 1 compared to antennae 2 (FIG. 15.99)
   - A. Antennae *1* approximately *equal to* or less than total length of antennae *2* — 1=2*
   - B. Antennae *1 longer than* total length of antennae *2* — 1>2
   - C. Antennae *1 shorter than* total length of antennae *2* — 1<2
2. Presence of teeth or keels along middorsal body surface (FIG. 15.99)
   - A. Sharply acute, middorsal *tooth* on posterior edge of each of last three thoracic segments — to*
   - B. Distinct *keel* present along middorsal line of abdominal segments — ke
   - C. Teeth and keels *absent* — x
3. Length of inner ramus (medial branch) of uropods 3 compared to its outer ramus (FIG. 15.99)
   - A. *Inner* ramus equal to or *less than half* the length of the *outer* ramus — i<½o
   - B. *Inner* ramus *more than half* but distinctly less than the length of the *outer* ramus — i>½o*
   - C. *Inner* ramus *equal* in length to *outer* ramus — i=o
4. Number of clusters of setae or bristles along the ventral surface of antennae 1, segment 1: provide *number* of clusters (FIG. 15.99) — —

15.62

Antenna 1 vs. Antenna 2	Tooth	Inner vs. Outer	Clusters	Species
1>2	x	i>½o	1–3	*Gammarus oceanicus*, scud
1>2	x	i>½o	3–5	*G. duebeni*
1>2	x	i<½o	3–5	*Eulimnogammarus (=Marinogammarus) obtusatus*
1>2	x	i<½o	0–1	*Gammarus (=Marinogammarus) finmarchicus*
1=2	x	i<½o	3–5	*Eulimnogammarus (=Marinogammarus) obtusatus*
1=2	x	i>½o	1–2	Gammaridae A, 15.63
1=2	to	i>½o,i=o	1	*Gammarus mucronatus*
1=2,1<2	ke	i=o	0	*Gammarellus angulosus*
1<2	x	i=o	2	*Gammarus annulatus*
1<2	x	i>½o	2	*G. lawrencianus*

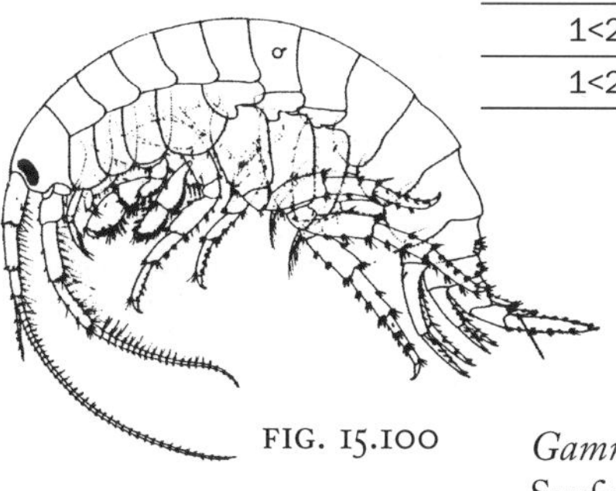
FIG. 15.100

Eulimnogammarus (=*Marinogammarus*) *obtusatus.* Gnathopods 1 larger than gnathopods 2 in males; lots of hairy setae on appendages. Midintertidal: under rocks, tide pools, marshes. Ac–,Ha–SM–Al,Li,gr?,17 mm (0.65 in), (B58). (FIG. 15.100)

Gammarellus angulosus (Gammarellidae). Surfy, rocky shores and tide pools; cling to algae. Often green, orange, or brown with white spots; many with white "saddle." Ac–,Ha–Al,Li–Su,gr–df?,14 mm (0.55 in), (B60), ☺. (FIG. 15.101)

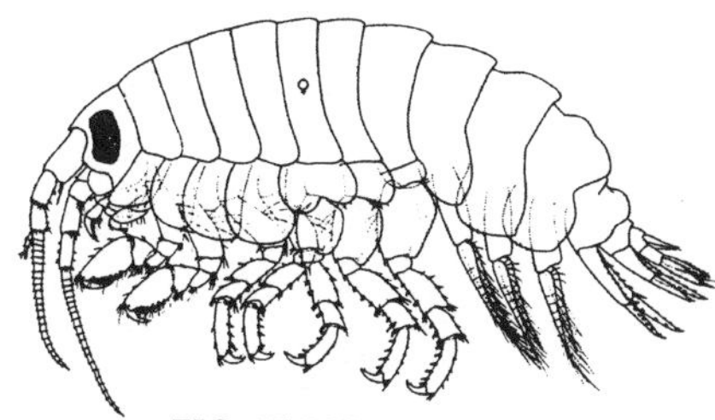
FIG. 15.101

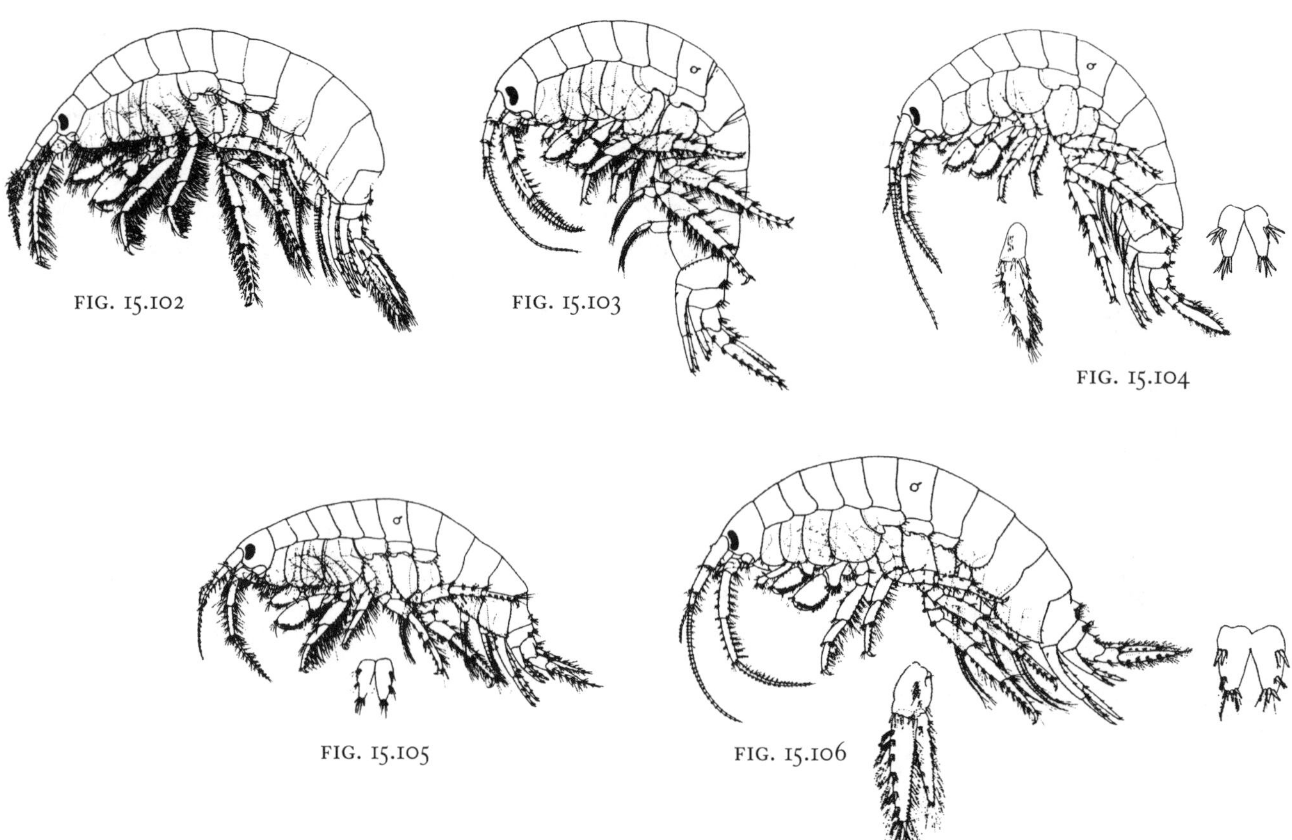
FIG. 15.102 FIG. 15.103 FIG. 15.104 FIG. 15.105 FIG. 15.106

Gammarus annulatus. Dark segmental bands on body. Pelagic species. Ac–,Pe–Sa,Ps,?,20 mm (0.8 in), (B53). (FIG. 15.102)

Gammarus duebeni. Yellowish-green. In estuarine-spray-zone tidal pools; often high intertidal. Ac,Ha,Es–Li–Su,gr–df?,22 mm (0.9 in), (B56). (FIG. 15.103)

Gammarus (=Marinogammarus) finmarchicus. Higher intertidal: under rocks, tidepools, marshes. Gnathopods 1 slightly smaller than gnathopods 2 in males. Ac–,Ha–SM–Al,Li,gr?,25 mm (1 in), (B58), ☺. (FIG. 15.104. With telson and uropods 3.)

Gammarus lawrencianus. Tolerates low salinity; in running waters; under stones and seaweed. Ac–,Sa–Mu,Es–Li–Su,gr–df?,12 mm (0.5 in), (B54). (FIG. 15.105)

Gammarus mucronatus. Hairy appendages; olive green with red eyes. *Spartina* marshes and *Zostera* grassbeds; on algae and debris; estuarine. Also rocky pools. Bo,Ha–SM–SG,Es–Li–Ss,gr–df?,16 mm (0.6 in), (B54,G227), ☺. (FIG. 15.99)

Gammarus oceanicus, scud. Uniform color; green, red, orange, or brown. Under stones; tide pools. Ac–,Ha–Al,Es–Li–Su,gr?,20 mm (0.8 in), (B50,G227,NM604), ☺. (FIG. 15.106. With uropods 3 and telson.)

15.63 **Gammaridae A** (from 15.62)

A difficult group. Consult Bousfield (1973) for confirmation or more help.

1. Seta groups on inside surface of antennae 1, segment 2: provide *number* of tufts of setae on segment 2, antennae 1 —
2. Distinctive coloration on fresh specimens
   - A. Orange *tiger-striping* present along body ts
   - B. Striping *absent* x

15.64

Antennae 1, Segment 2	Color	Species
1	x	*Gammarus fasciatus*
3–5	ts	*G. tigrinus*
3–5	x	*G. daiberi*

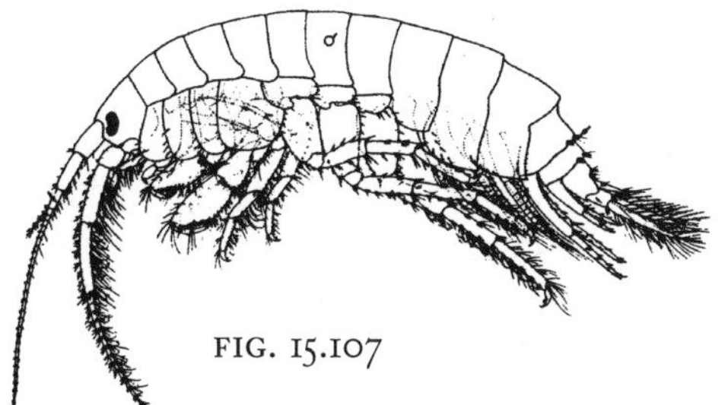

FIG. 15.107

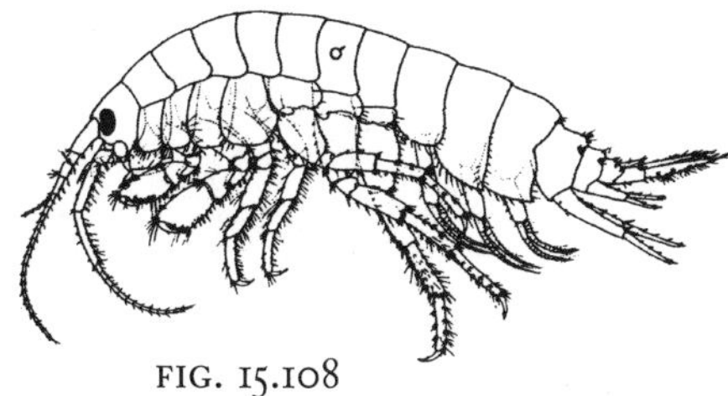

FIG. 15.108

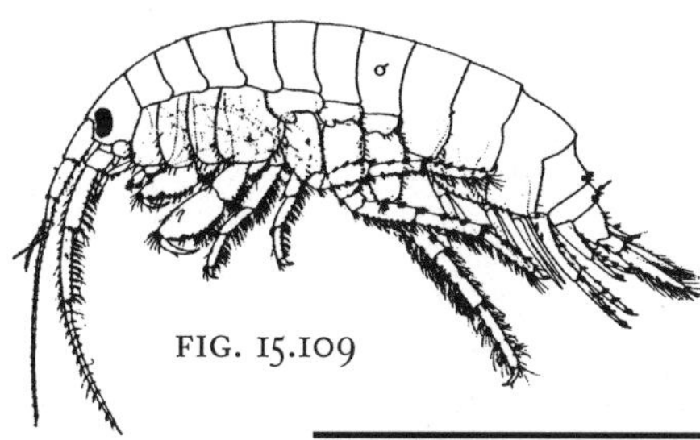

FIG. 15.109

Gammarus daiberi. Upper to midestuaries on hydroids and ectoprocts. Vi,Al,Es–Su,gr–df?,12 mm (0.5 in), (B52). (FIG. 15.107)

Gammarus fasciatus. Upper estuaries, virtually freshwater species. Vi,Al,Es–Su,gr–df?,12 mm (0.5 in), (B53). (FIG. 15.108)

Gammarus tigrinus. Upper to midestuaries in debris, on pilings, etc. Bo,Al,Es–Su,gr–df?,14 mm (0.55 in), (B51). (FIG. 15.109)

15.65 **Haustoriidae, Sand Burrowers** (from 15.39)

Accessory flagellum present; body dorsally smooth; antennae 1 shorter than antennae 2. Broad-backed; distal segments of thoracic legs expanded and dense with setae or spines; uropods 3 distinctly biramous; telson at least partially cleft. Suspension feeders or buried detritivores; burrowers. Natantia, although apparently associated with their benthic life habits, normally emphasized sensory features are less obvious than expected. (BK357,G228,LB248)

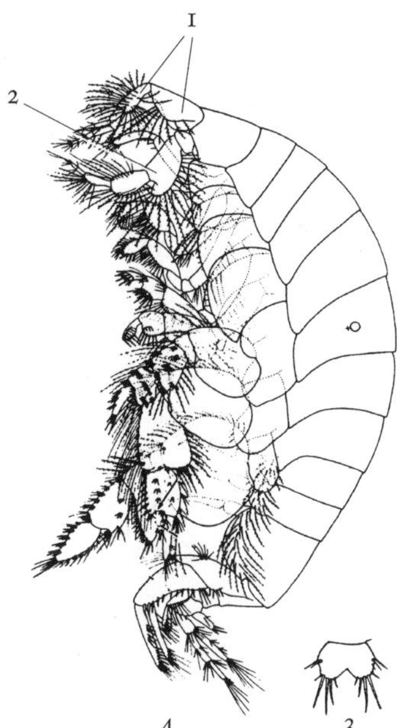

FIG. 15.110

1. Antennas 1, first large segment relative to the second large segment (FIG. 15.110)
   - A. Segment 1 *overhangs* base of segment 2 — ov
   - B. Segment 1 does *not* overhang base of segment 2 — x*
2. Shape of largest segment (actually segment 4) on antennae 2 (FIG. 15.110)
   - A. Segment with conspicuous *lobe* along its posterior margin — lo*
   - B. Such a lobe *absent* — x
3. Shape of telson (FIG. 15.110)
   - A. Telson *deeply* cleft, to or nearly to its base — de
   - B. Telson cleft to about its *midpoint* — mi
   - C. Telson *slightly* but distinctly cleft — sl*
   - D. Telson entire, *not* cleft — x
4. On uropods 1, relative sizes of inner (medial) versus outer (lateral) branches (rami) (FIG. 15.110)
   - A. *Inner* branch clearly *longer than outer* branch — i>o
   - B. *Outer* branch clearly *longer than inner* branch — o>i
   - C. *Inner* branch more or less *equal to outer* branch — i=o*
   - D. Uropods one *uniramous,* only one piece — un
5. Number of segments in accessory flagellum on antennae 1: count *number* of segments — __

15.66

Overhang	Segment 4 Shape	Telson Shape	U1: Inner vs. Outer Rami	Accessory Flagellum	Species
ov	x	de	o>i	2	Subfamily Pontoporeiinae Long spines on telson end: *Bathyporeia quoddyensis* OR Short spines on telson end: *Amphiporeia virginiana*
x	x	de	i=o	2	*Pontoporeia femorata*
x	lo	de	o>i	2	*Acanthohaustorius intermedius*
x	lo	de	i=o	2	*A. millsi*
x	lo	mi	i=o	2	*A. spinosus*
x	lo	mi	o>i,i=o	3–5	*Haustorius canadensis*
x	lo	sl	i>o	2	*Pseudohaustorius caroliniensis*
x	lo	sl	o>i	2	*Parahaustorius* spp.
x	lo	sl	i=o	2	*Protohaustorius:* Coxal plate 4 clearly wider than long: *P. deichmannae* OR Width and length of coxal plate 4 about equal: *P. wigleyi*
x	lo	x	o>i	2	*Neohaustorius biarticulatus*
x	lo	x	un	2	*N. schmitzi*

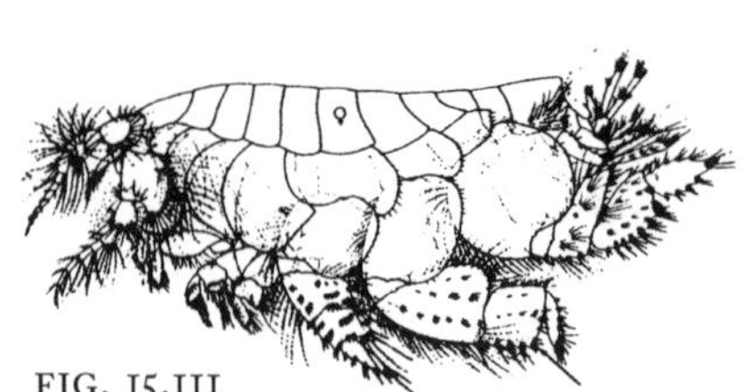
FIG. 15.111

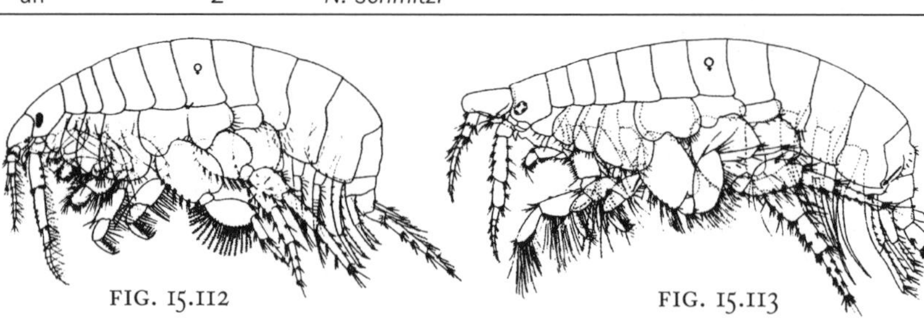
FIG. 15.112

FIG. 15.113

Acanthohaustorius intermedius. Posterior margin of coxal plate 4 rounded. Vi+,Sa,Su,df–ff?,6 mm (0.23 in), (B117).

Acanthohaustorius millsi. Posterior margin of coxal plate 4 squared. Tolerates lower oxygen and salinity. Bo,Sa,Su,df–ff?,11 mm (0.43 in), (B116), ☺. (FIG. 15.111)

Acanthohaustorius spinosus. Ac–,Sa,Li–De,df–ff?,14 mm (0.55 in), (B117).

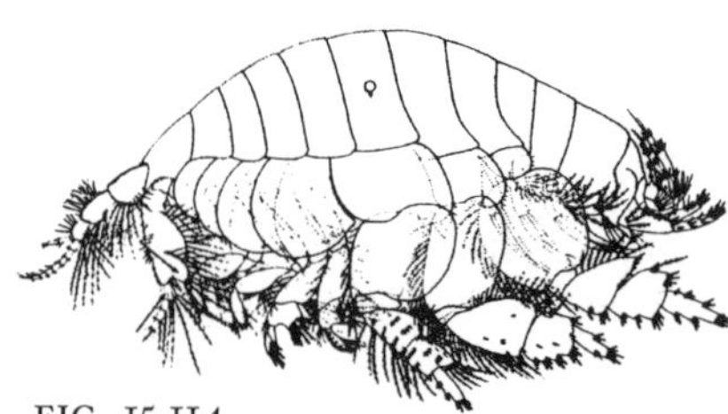
FIG. 15.114

Amphiporeia virginiana. Inner branch of uropods 3 less than half outer branch length. Often near freshwater flows. Bo,Sa,Li,ff–df,7 mm (0.3 in), (B113). (FIG. 15.112)

Bathyporeia quoddyensis. Uropods 3 with outer ramus four to five times length of inner ramus. Bo,Sa,Su,df–ff?,7 mm (0.3 in), (B105), ☺. (FIG. 15.113)

Haustorius canadensis. Unpigmented eyes. Telson with short terminal spines. Bo,Sa,Li,df–ff?,16 mm (0.63 in), (B113), ☺.

Neohaustorius biarticulatus. Vi,Sa,Li,ff–df,7 mm (0.3 in), (B114). (FIG. 15.114)

Neohaustorius schmitzi. Tolerates salinity from full seawater to 6°/oo. Vi,Sa,Es–Li–,ff–df,7 mm (0.3 in), (B114).

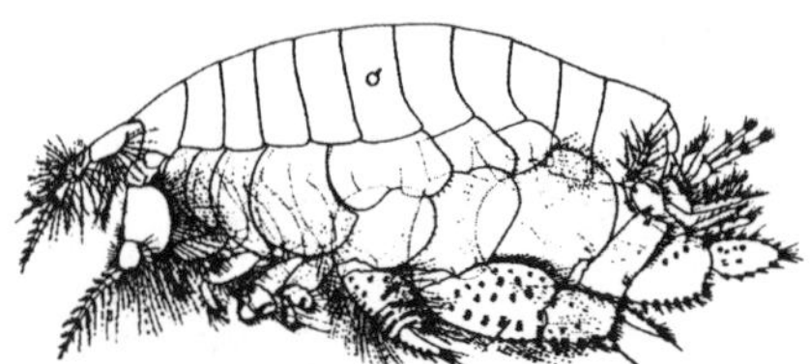
FIG. 15.115

Parahaustorius spp. Difficult species to distinguish. Refer to Bousfield (1973, p. 109) reference. *Parahaustorius attenuatus* and *P. holmesi* are subtidal and often co-occur in fine sand to 50 m depth; *P. longimerus* is intertidal to slightly subtidal in surfy beach sand. (B109). (FIG. 15.115)

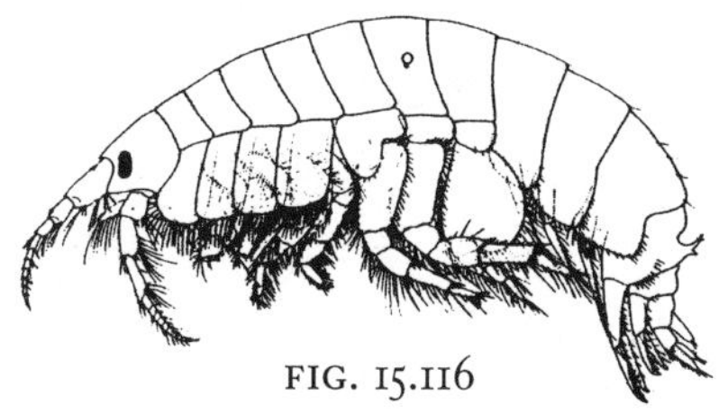
FIG. 15.116

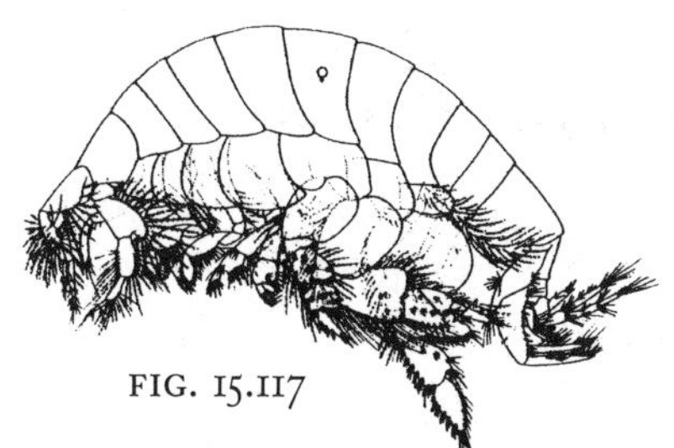
FIG. 15.117

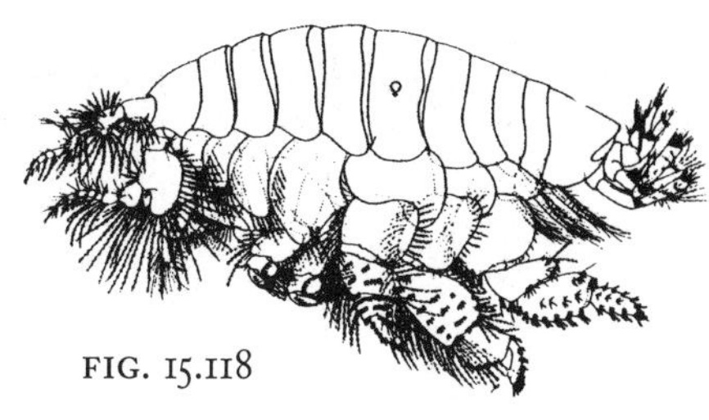
FIG. 15.118

Pontoporeia femorata. Two-toothed projection on the abdominal segment second forward from telson. Po–,Mu–Sa,Su,df?,14 mm (0.6 in), (B101). (FIG. 15.116)

Protohaustorius deichmannae. Free-living burrower. Bo,Sa–Mu,Es–Su,df–ff?,6 mm (0.25 in), (B109).

Protohaustorius wigleyi. Bo,Sa,Su–De,df–ff?,11 mm (0.43 in), (B108). (FIG. 15.117)

Pseudohaustorius caroliniensis. Lateral margin of posteriormost large abdominal segment with sharp, upturned hook. Vi,Mu–Sa,Li–Su,df–ff?,7 mm (0.3 in), (B119). (FIG. 15.118)

15.67 **Hyalidae** (from 15.45)

Accessory flagellum absent, uropods uniramous (in species covered here), telson deeply cleft. In males, gnathopods 2 larger than gnathopods 1. Capable of jumping. Reptantia that engage in mate-carrying. (BK366,LB465)

1. Presence of large spine at base of biramous ending of uropods 1
   - A. Large *spine* present on uropods 1 — sp
   - B. Large spine *absent* — x
2. Bristles along the posterior edge of antennae 2 segment 3
   - A. Third segment of antennae 2 with continuous coverage of *bristles* — br
   - B. Third segment of antennae 2 with a few groups of bristles, continuous coverage *absent* — x

15.68

Spine	Bristles	Species
sp	br	*Hyale plumulosa*
x	x	*H. nilssoni*

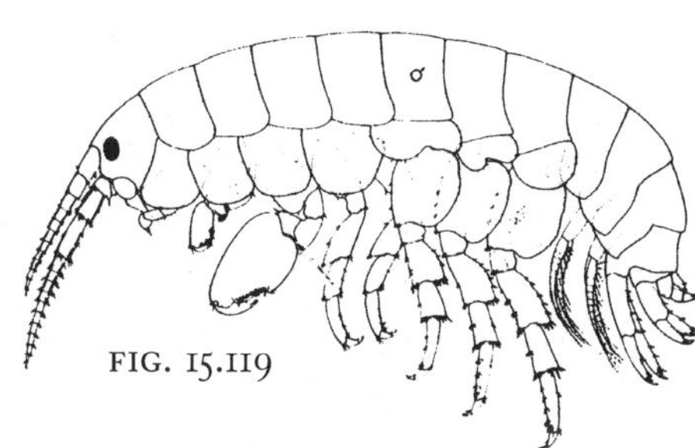
FIG. 15.119

Hyale nilssoni. Often orangish color. Intertidal, subtidal under *Fucus.* Hops when disturbed. Ac–,Ha–Al,Li–,gr,11 mm (0.43 in), (B156), ☺. (FIG. 15.119)

Hyale plumulosa. Protected shores, saltmarshes, tidal pools; may hop when disturbed. Bo,Ha–Al–SM,Li,gr,12 mm (0.5 in), (B155), ☺. (FIG. 15.120)

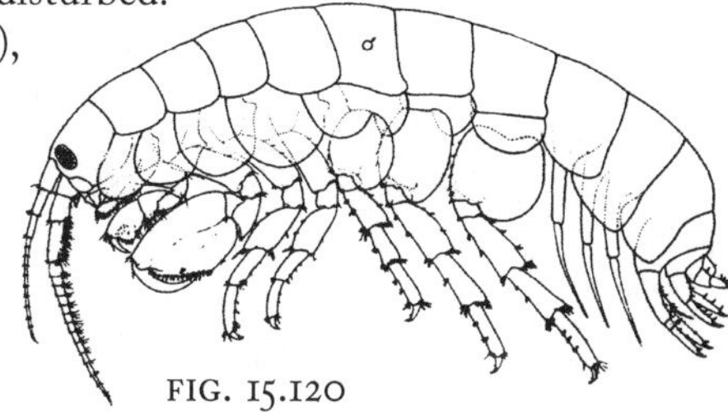
FIG. 15.120

15.69 **Isaeidae (=Photidae)** (from 15.43)

Accessory flagellum present or absent; gnathopods 2 larger than gnathopods 1 ; uropods 3 biramous; telson entire. Reptantia.

Photis spp. Mostly subtidal, coldwater species. Bo,Sa–Mu,Su,df–sc?,5 mm (0.2 in), (B186). (FIG. 15.121)

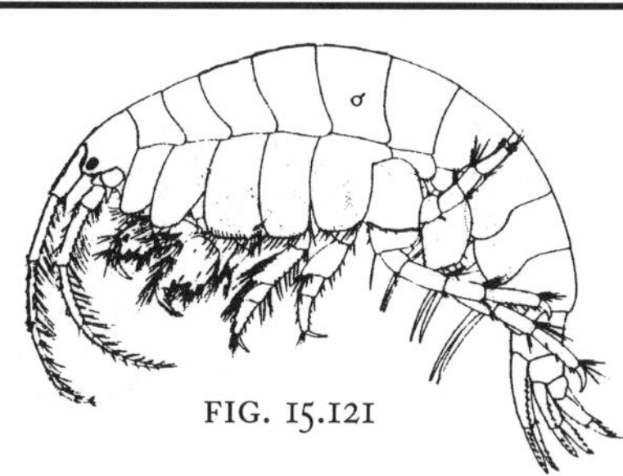
FIG. 15.121

15.70 **Ischyroceridae** (from 15.43)

Gnathopods powerful, with gnathopods 2 much larger than gnathopods 1 (especially in males); peduncle of uropods 3 elongate, one or two short rami; telson short, fleshy, not divided. Walking legs 5–7 often directed dorsally. Normally tube-dwelling. Natantia. (BK403,LB275)

1. Shape of lateral contour of head, anterior to eyes and between antennal bases
   - A. Head contour *pointed* anterior to eyes — pt
   - B. Head contour *rounded* anterior to eyes — ro
2. Shape of gnathopods 2
   - A. Gnathopods 2 is comparatively *huge* (almost grotesque) with large *thumb*like extension — ht
   - B. Gnathopods 2 is comparatively *huge* but *lacks* the thumblike extension — hx
   - C. Gnathopods 2 is larger than gnathopods 1 but is *not* hugely larger — x
3. Teeth on outer ramus of uropods 3
   - A. Outer ramus of uropods 3 with one to three strong *teeth* — te
   - B. Outer ramus of uropods 3 with *more than four* weak *teeth* — >4te

15.71

Head	Gnathopods 2	Uropods 3	Species
pt	x	>4te	*Ischyrocerus anguipes*
ro	ht	te	*Jassa marmorea*—males
ro	hx	te	*J. marmorea*—females

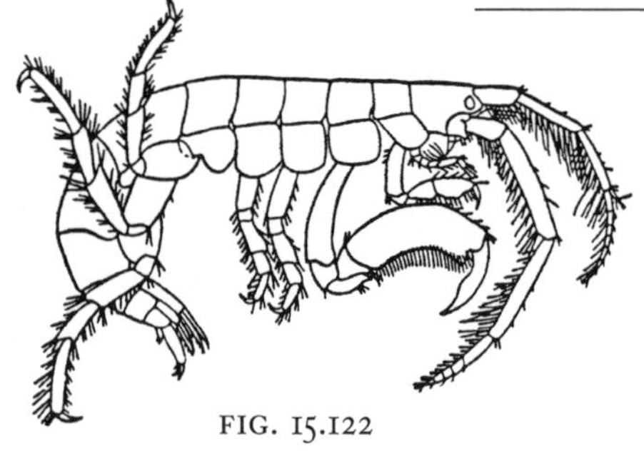

FIG. 15.122

Ischyrocerus anguipes. Olive green to reddish; banded coloring. Often with tubes in kelp holdfasts. Bo,Ha–Al,Li–Su,gr–df?,14 mm (0.55 in), (B192). (FIG. 15.122)

Jassa marmorea. Light, marked with brown. Tube-dwelling member of fouling community; very dense clusters of tubes in strong water currents. Bo,Ha–Al,Li–Ps,df–ff,10 mm (0.4 in). (FIG. 15.123)

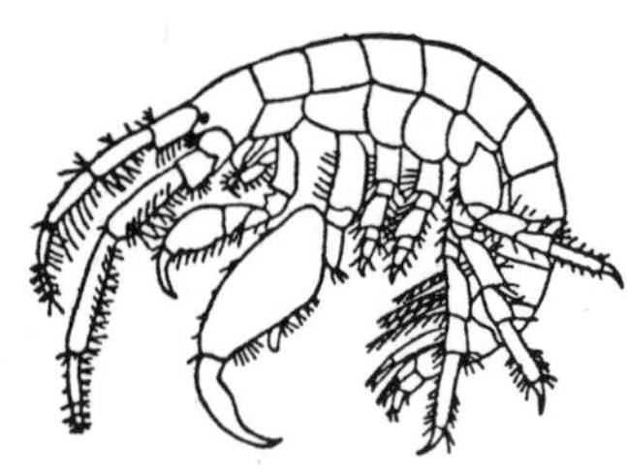

FIG. 15.123

15.72 **Liljeborgiidae** (from 15.39, 15.41)

Accessory flagellum present but small; antennae short; coxa 4 excavate; uropods 3 biramous; telson cleft. Reptantia, displaying dimorphism in gnathopods (2 larger than 1) in mature males. (BK412,LB291)

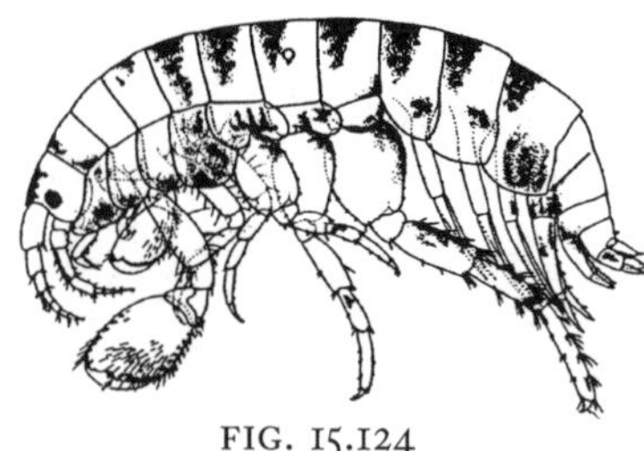

FIG. 15.124

Listriella barnardi. Commensal in tubes of polychaete worm, *Amphitrite ornata.* Vi,Co,Li–Su,ff?,7 mm (0.3 in), (B71). (FIG. 15.124)

Listriella clymenellae. Commensal in tubes of polychaete worm, *Clymenella torquata.* Vi+,Co,Li–Su,df–ff?,7 mm (0.3 in), (B72). (FIG. 15.125)

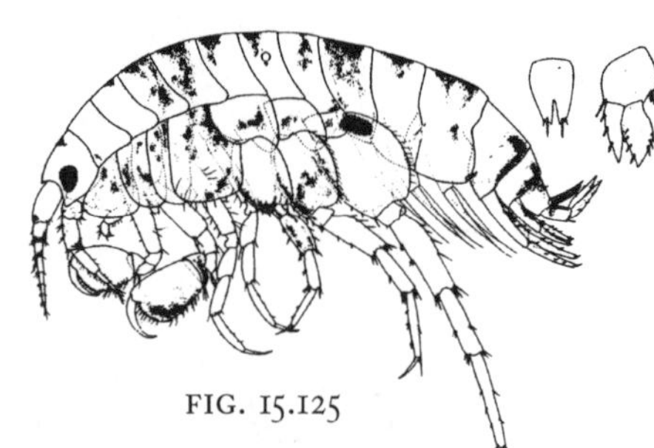

FIG. 15.125

15.73 **Lysianassidae** (from 15.39)

Accessory flagellum present; peduncle of antennae 1 short and stout; gnathopods usually small; uropods 3 biramous; telson cleft, entire or absent. Large, complex family. Mostly scavengers; some as surface detritivores or carnivores. Burrowers or free at surface. Natantia, although benthic-dwelling forms do possess weakly developed sensory structures. (BK420,LB294)

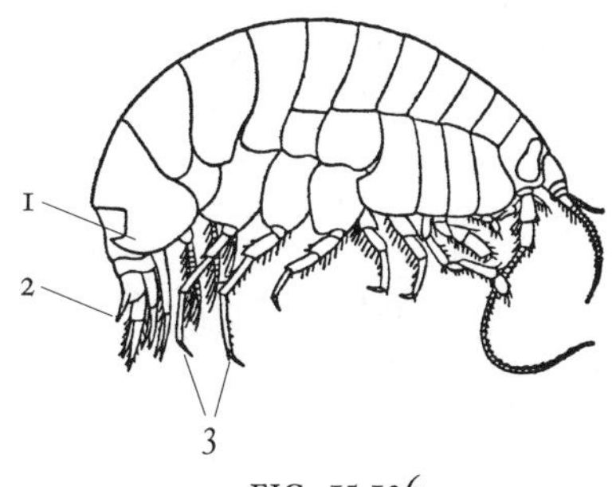

FIG. 15.126

1. Shape of posterior-lateral margin of third (or last, large) abdominal segment (FIG. 15.126)
   - A. Posterior-lateral margin bears posteriorly directed *hook* — ho*
   - B. Posterior-lateral margin not hooked but is *pointed* — po
   - C. Posterior-lateral margin is *rounded* — ro
2. Telson shape (FIG. 15.126)
   - A. Telson deeply *cleft* — cl
   - B. Telson *not* cleft, nearly circular — x
3. Length of thoracic leg 7 (9 = last one) compared to length of thoracic leg 6 (FIG. 15.126)
   - A. Leg 7 clearly *longer than* leg *6* — 7>6
   - B. Leg 7 about *equal* in length to leg *6* — 7=6*

15.74

Abdominal Segment 3	Telson	Leg 7 vs. 6	Species
ho	cl	7=6	*Anonyx* spp.
po	cl	7>6	*Psammonyx nobilis,* noble sand amphipod
ro	cl	7=6	*Orchomenella* spp.
ro	x	7>6	*Lysianopsis alba*

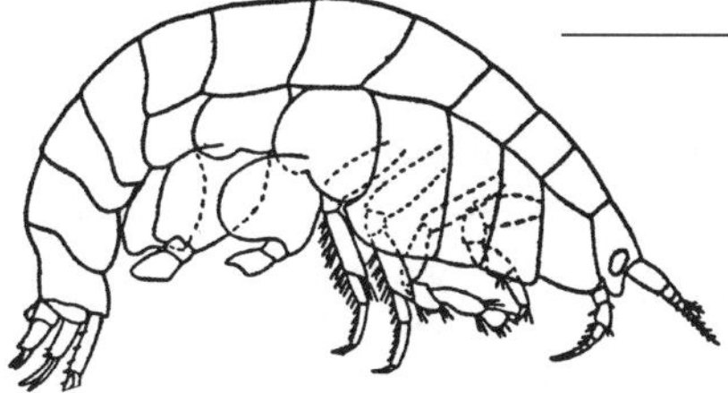
FIG. 15.127

Anonyx spp. Cold water; sandy substrates; subtidal. Ac,Sa,Si,sc,25 mm (1 in), (B149). (FIG. 15.126)

Lysianopsis alba. Whitish to yellowish. Warmer waters; shallow, shelly-sandy-mud. Vi,Gr–Sa–Mu–SG,Su,sc,10 mm (0.4 in), (B146). (FIG. 15.127)

Orchomenella spp. Colder waters; sand; subtidal. Ac–,Sa–SG,Su,sc,12 mm (0.5 in), (B147). (FIG. 15.128)

FIG. 15.128

Psammonyx nobilis, noble sand amphipod. Gray to white, dark eyes. Medium to medium-coarse sandy, coldwater beaches. Ac–,Sa,Li–,sc,20 mm (0.8 in), (B145,NM606). (FIG. 15.129)

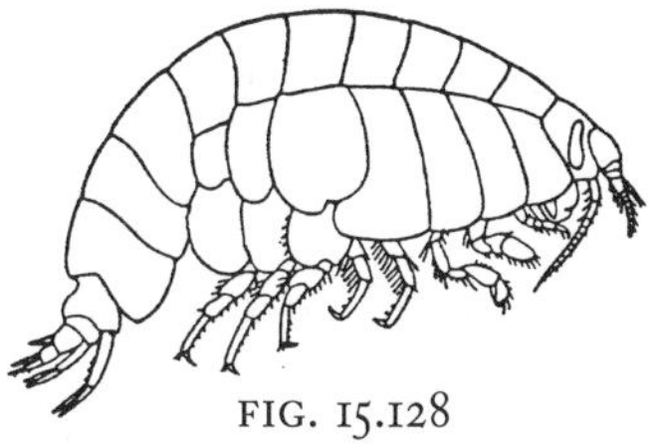

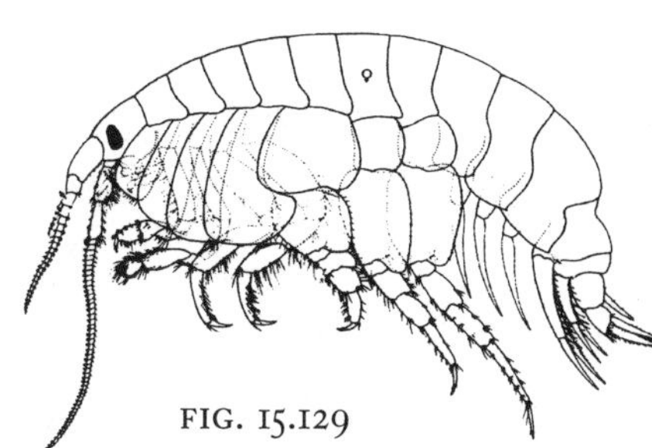
FIG. 15.129

15.75

Melitidae (from 15.39)

Accessory flagellum small or absent; uropods 3 biramous; telson deeply cleft. These reptantia are mate-carriers in which gnathopods 2 are larger than gnathopods 1. (BK545)

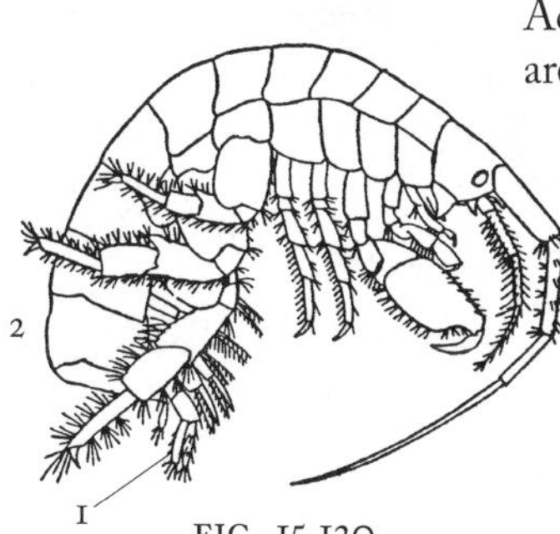

FIG. 15.130

1. Length of inner ramus of uropods 3 compared to outer ramus (FIG. 15.130)
   - A. *Inner* ramus *less than half outer* ramus — i<½o
   - B. *Inner* ramus about *equal to outer* ramus — i=o*
   - C. *Inner* ramus *greater than half outer* ramus, but not equal to it — i>½o
2. Presence of middorsal teeth along posterior margin of abdominal segments (FIG. 15.130)
   - A. Abdominal segments with *teeth* — te
   - B. Such teeth are *absent* — x*

15.76

Inner vs. Outer Ramus	Teeth	Species
i=o	x	*Maera danae*
i<½o	te	*Melita dentata*
i<½o	x	*M. nitida*
i>½o	x	*Elasmopus laevis*

Elasmopus laevis. Vi+,Al–Ha–Gr–SG,Li–Su,df–sc?,12 mm (0.5 in), (B63). (FIG. 15.130)

Maera danae. Mud bottoms; rocky shores. Ac–,Mu–Ha,Su,df–sc?,25 mm (1 in), (B66).

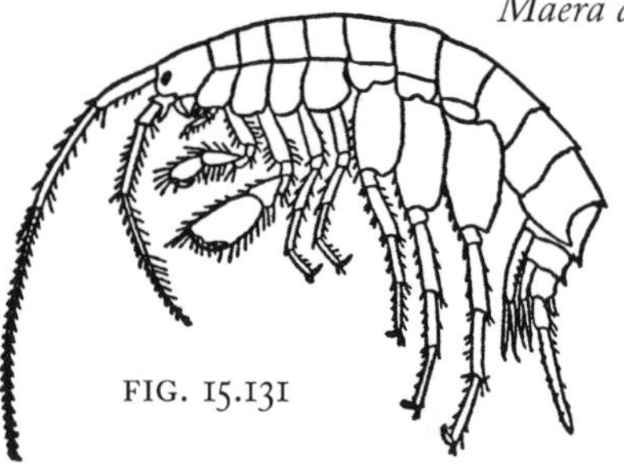

FIG. 15.131

Melita dentata. Yellow with darker bands. Rocky shore; under stones. Ac–,Ha–Gr,Su,df–sc?,25 mm (1 in), (B65), ☺. (FIG. 15.131)

Melita nitida. Dark green. On hydroids and ectoprocts, mud bottoms, saltmarshes, shell fragments. Bo,Mu–SM,Es–Su,sc–df,15 mm (0.6 in), (B65). (FIG. 15.132)

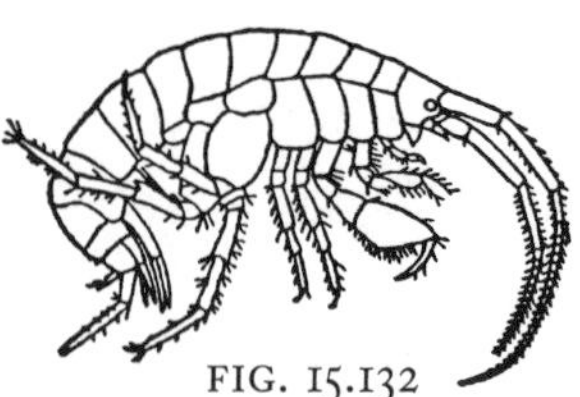

FIG. 15.132

15.77 **Oedicerotidae** (from 15.43)

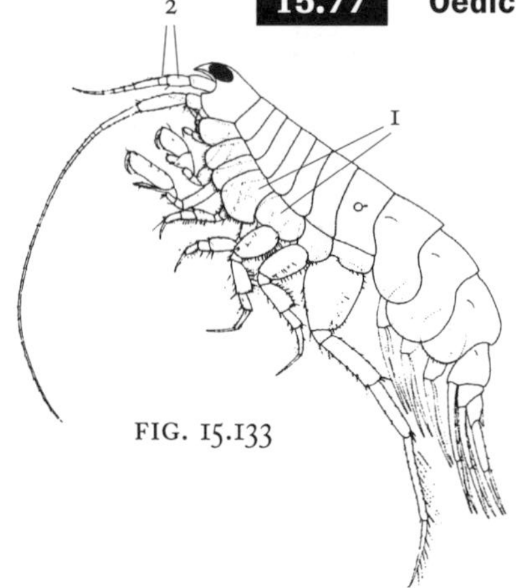

FIG. 15.133

Accessory flagellum absent; head often with conspicuous rostrum; eyes fused dorsally; uropods 3 biramous; telson entire or notched. Shallow seas; burrow in sandy sediments with excursions into plankton at night; buried detritivore. Dorsally fused eyes give one-eyed appearance. Natantia, but sensory features are usually underdeveloped. (BK547,LB373)

1. Size of coxal plate 5 relative to coxal plate 4 (FIG. 15.133)
   - A. Coxal plate *5* nearly *equal to* coxal plate *4* in size — 5=4
   - B. Coxal plate *5* distinctly *shorter than* coxal plate *4* (closer to half its length) — 5<4*
2. Relative size of segment 3, antennae 1 (FIG. 15.133)
   - A. Segment *3* is *less than half* length of segment *2* — 3<½2*
   - B. Segment *3* is approximately *equal to* length of segment *2* — 3=2

15.78

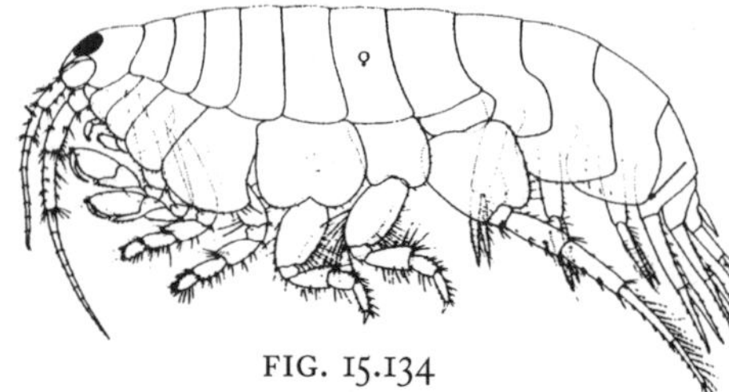

FIG. 15.134

Coxa 5 vs. 4	Antenna 1 Segments	Species
5=4	3=2	*Monoculodes edwardsi*, red-eyed amphipod
5<4	3=2,3<½2	*M. intermedius*

Monoculodes edwardsi, red-eyed amphipod. Euryhaline; southern species. Bo,Es–Li–Su,df–sc?,10 mm (0.4 in), (B97), ☺. (FIG. 15.134)

Monoculodes intermedius. Sandy bottoms. Ac,Sa,Su,df–sc?,9 mm (0.35 in), (B96). (FIG. 15.133)

15.79 **Phoxocephalidae** (from 15.39)

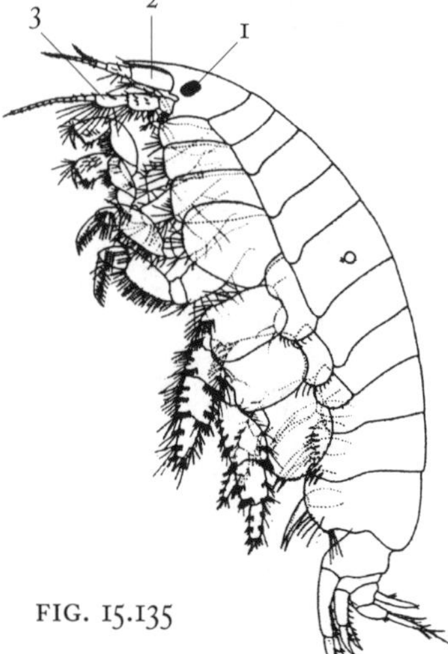

FIG. 15.135

Accessory flagellum present; head with hoodlike cowl; pereiopods with many setae and spines; uropods 3 biramous; telson cleft. Buried detritivores; burrowers. Natantia, although brush setae are not well developed. (BK588,LB412)

1. Eye pigmentation (FIG. 15.135)
   - A. Eyes darkly *pigmented* — pi*
   - B. Eyes present but *unpigmented* — un
   - C. Eyes *absent* — x
2. Shape of lower lateral margin of head (FIG. 15.135)
   - A. Lower margin *straight* or slightly curved, head appears triangular in side view — st
   - B. Lower margin distinctly *curved,* forms slender pointed anterior rostrum — cu*
3. Length of antennae 2 (FIG. 15.135)
   - A. Antennae *2* nearly *equal to body* length — 2=b
   - B. Antennae *2 greater than head* length — 2>h*
   - C. Antennae *2 shorter than head* length — 2<h
   - D. Antennae *2* about *one-half* the *body* length — 2½b

15.80

Eyes	Head	Antennae 2	Organism
pi	cu	2>h	*Trichophoxus epistomus*
pi	st	2½b	*Paraphoxus spinosus*
un	st	2=b	*Phoxocephalus holbolli*
x	cu	2<h	*Harpinia propinqua*

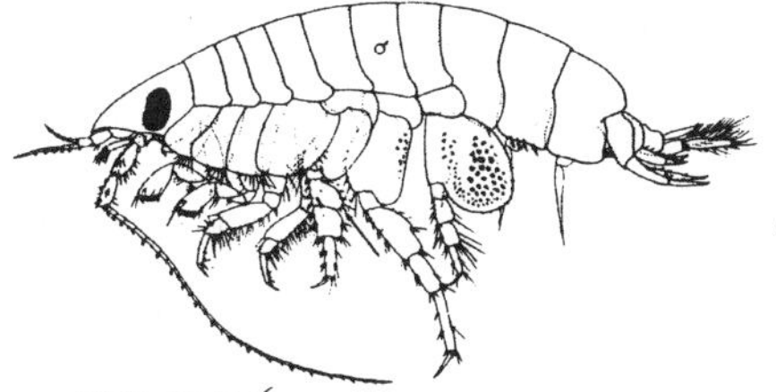

FIG. 15.136

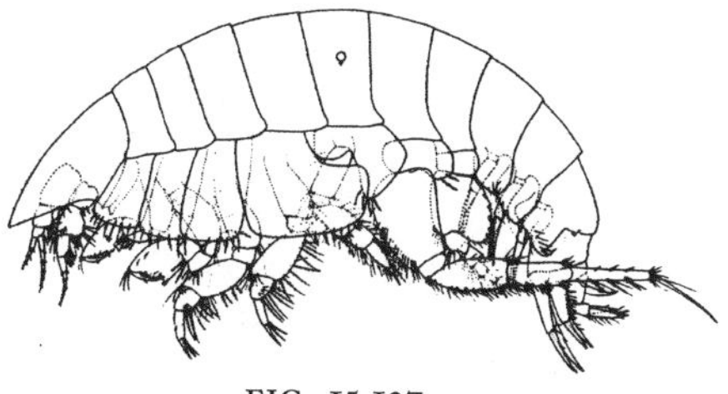

FIG. 15.137

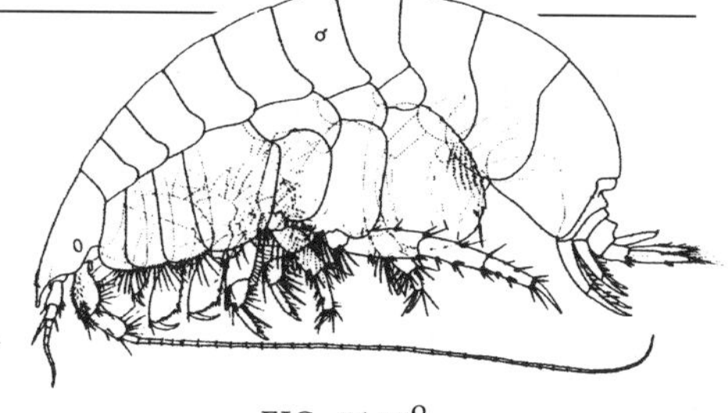

FIG. 15.138

Harpinia propinqua. Ac,Sa–Mu,Su,df–ff?,6 mm (0.25 in), (B128). (FIG. 15.136)

Paraphoxus spinosus. Vi,Sa,Es–Su,ff?,4 mm (0.16 in), (B125). (FIG. 15.137)

Phoxocephalus holbolli. Light brown to orange with white spots. Ac–,Sa–Mu–SG,Li–De,df–ff?,8 mm (0.3 in), (B123), ☺. (FIG. 15.138)

Rhepoxynius (=*Trichophoxus*) *epistomus.* Eyes pigmented in males; unpigmented in females. Burrows in unstable sand. Bo,Sa,Su,df–ff?,10 mm (0.4 in), (B126). (FIG. 15.135)

15.81 **Pleustidae** (from 15.37)

Accessory flagellum absent; uropods 3 biramous but unequal in length; telson entire. Reptantia. (BK644,LB421)

Pleusymtes glaber. Bo,Ha,Li–Su,df–ff?,14 mm (0.55 in), (B84). (FIG. 15.139)

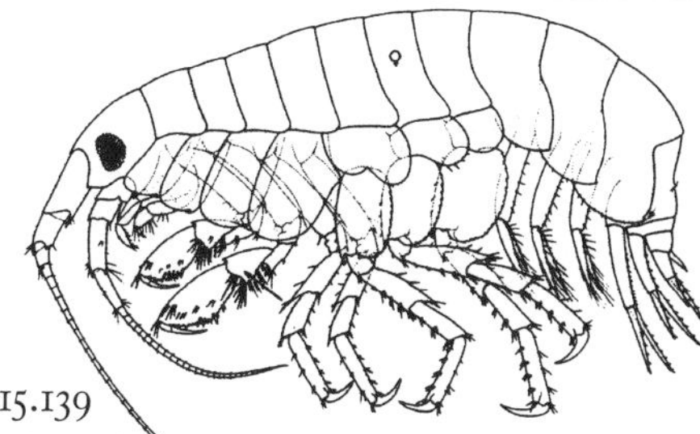

FIG. 15.139

15.82 **Pontogeneiidae** (from 15.41)

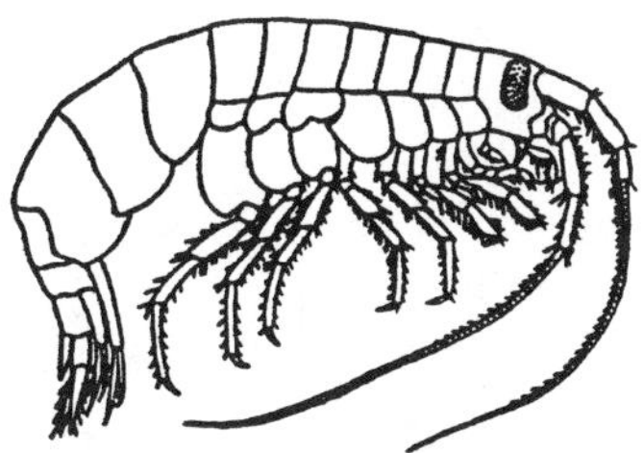

FIG. 15.140

Accessory flagellum absent; uropods 3 biramous; telson deeply cleft; comparatively very large eyes. Natantia.

Pontogeneia inermis. Translucent with some purple markings. Swimming amphipods in water column or on seaweeds; common in the Gulf of Maine. Ac–,Pe–Al,Su–Ps,ff?,12 mm (0.5 in), (B74), ☺. (FIG. 15.140)

15.83 **Stenothoidae** (from 15.45)

Accessory flagellum absent; coxal plate 4 is huge; uropods 3 uniramous; telson entire. Commensals associated with hydroids and ectoprocts. Reptantia with gnathopods 2 larger than gnathopods 1 in mature males. (BK684,LB444).

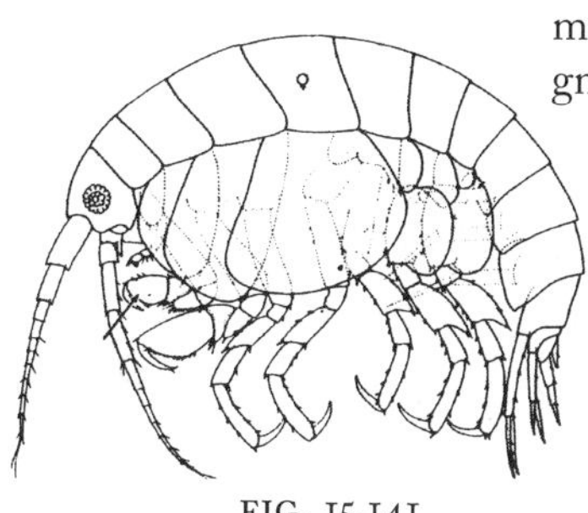

FIG. 15.141

Stenothoe minuta. Coxal plate 4 large, coxal plate 3 well developed and about equal in length to coxal plate 4. Small, deepwater species living on hydroids or ectoprocts. Vi,Ha–Al,Es–Su,?,5 mm (0.2 in), (B87). (FIG. 15.141)

Parametopella cypris. Coxal plate 4 very large, entirely obscures coxal plate 3. Vi,Ha–Al,Es–Su,?,5 mm (0.2 in), (B91,G228). (FIG. 15.142)

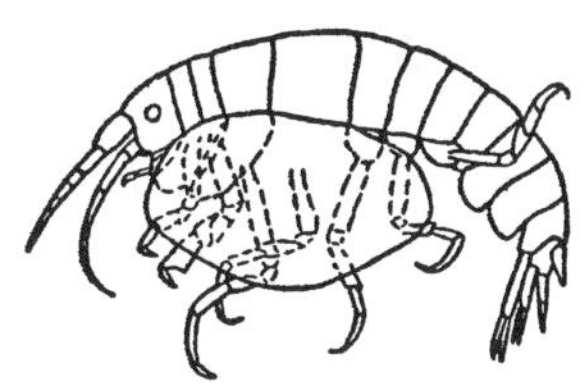

FIG. 15.142

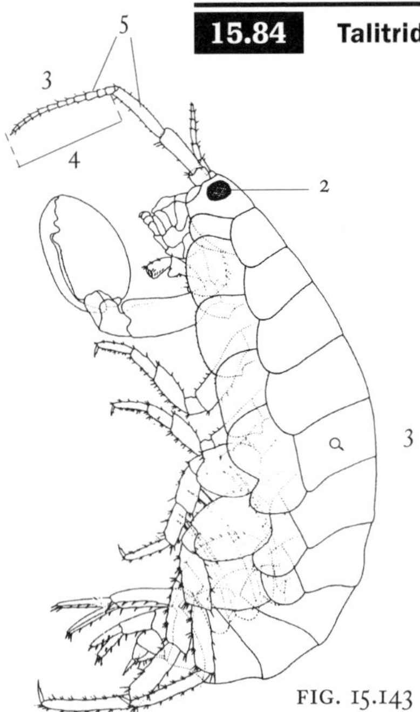

FIG. 15.143

15.84 **Talitridae** (from 15.45)

Accessory flagellum absent; uropods 3 uniramous; telson entire. Burrowers. Reptantia that engage in mate-carrying before copulation. (BK718,G226,LB463)

1. Body color
   - A. *Light,* sandy color — li
   - B. *Darker,* orange-brown color — da
2. Eye (FIG. 15.143)
   - A. Eyes *small,* occupy less than half head width — sm*
   - B. Eyes very *large,* occupy more than half head width — la
3. Length of antennae 2 relative to body length (FIG. 15.143)
   - A. *Antennae 2 less than half body* length — a<½b*
   - B. *Antennae 2 more than or equal to half body* length — a>½b
4. Number of segments in flagellum of antennae 2: provide *number* of segments (FIG. 15.143) — —
5. Diameter of antennae 2 segment 3 compared to diameter of flagellum (FIG. 15.143)
   - A. Antennae 2 segment *3* much *greater* in diameter than *flagellum* — 3>f
   - B. Antennae 2 segment *3* nearly *equal* in diameter to *flagellum* — 3=f*

15.85

		Antennae 2			
Color	**Eye**	**Body Length**	**Segments in Flagellum**	**Flagellum Diameter**	**Organism**
li	la	a<½b	>25	3>f	*Americorchestia* (=*Talorchestia*) *megalophthalma,* beach hopper
li	la	a>½b	>25	3>f	*A.* (=*Talorchestia*) *longicornis,* beach hopper
da	sm	a<½b	12	3>f	*Orchestia platensis,* beach flea
da	sm	a<½b	>20	3=f	*O. grillus*
da	sm	a<½b	12	3=f	*Uhlorchestia uhleri*

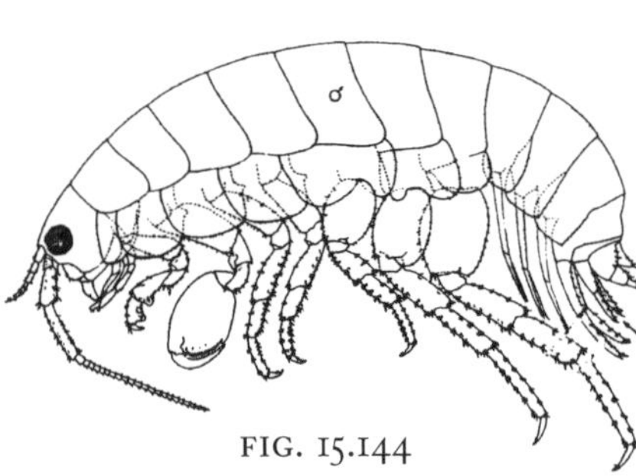
FIG. 15.144

Americorchestia (=*Talorchestia*) *longicornis,* beach hopper. Eyes blue-white in life. High-tide line; under debris; hops when disturbed. Nocturnal. Bo,Sa,Li,df?,3 cm (1.2 in), (B163,NM606), ☺. (FIG. 15.144)

Americorchestia (=*Talorchestia*) *megalophthalma,* beach hopper. Large, bulging eyes. High-tide line; higher salinity (>17°/oo) estuarine; circular burrows in sand. Bo,Sa,Li,df?,25 mm (1 in), (B162,NM606), ☺. (FIG. 15.145)

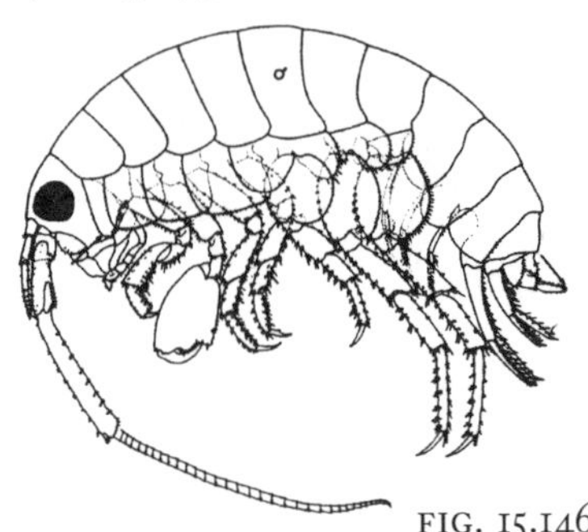
FIG. 15.146

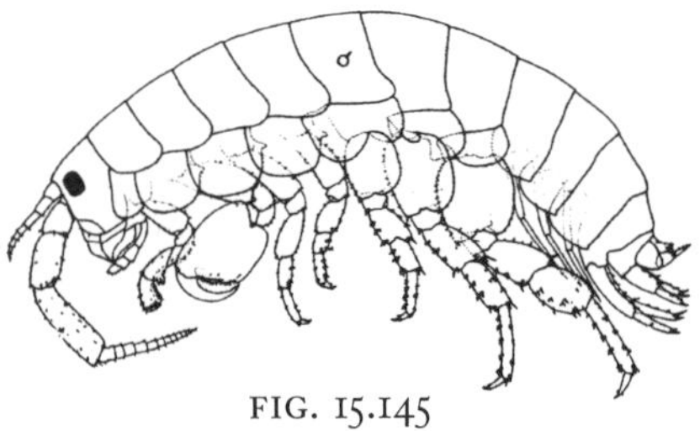
FIG. 15.145

Orchestia grillus. Builds nests at base of *Spartina* marsh grasses; high-tide line and above. Bo,SM–Gr,Li,df?,19 mm (0.75 in), (B150), ☺. (FIG. 15.146)

Orchestia platensis, beach flea. Often under high-tide beach drift; burrows in sand. Nocturnal; jumps when disturbed. Bo,Ha–Gr–Sa–Mu–SM,Li,df?,12 mm (0.5 in), (B160), ☺. (FIG. 15.147)

Uhlorchestia (=*Orchestia*) *uhleri.* Low-salinity saltmarsh debris and roots; high-tidal line and above. Bo,SM,Es–Li,df?,10 mm (0.4 in), (B161). (FIG. 15.143)

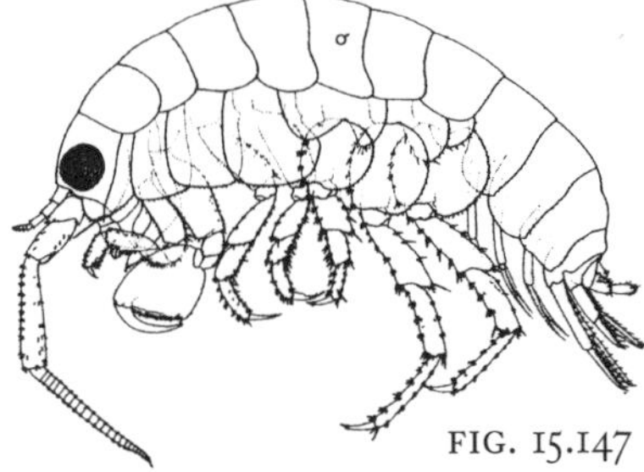
FIG. 15.147

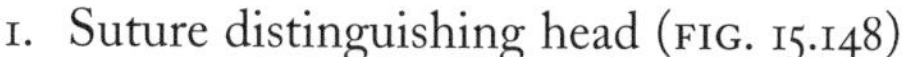

15.86 **Caprellid Amphipods, Skeleton Shrimp** (from 15.7, 15.33)

Slender *caprellids* frequently attach by rear legs to hydroids, ectoprocts, or seastars while waving their upper bodies about in search of food. Females are distinguished by possessing midbody flaps of exoskeleton that form a brood pouch for eggs. Two midbody segments also bear stalked, oval-shaped gills. Reptantia that employ mate-attending behavior.

Additional reference: (M) McCain, J. C., 1968.

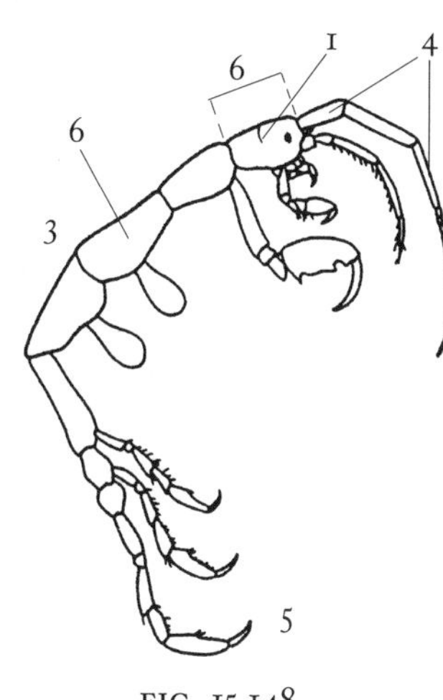

FIG. 15.148

1. Suture distinguishing head (FIG. 15.148)
   - A. Complete *suture* demarcates head (i.e., head appears to be in 2 segments) su
   - B. Incomplete suture does *not* completely demarcate head (i.e., head a single segment) x*
2. Spines and tubercles on head
   - A. Head with *spines* sp
   - B. Head smooth, distinct spines *absent* sm*
3. Spines on the same midbody trunk segments that bear stalked, oval-shaped gills (FIG. 15.148)
   - A. Segments with obvious *spines* sp
   - B. Segments smooth, with spines totally *absent* sm
4. Antennae 1: length of segment 1 compared to length of segment 3 (FIG. 15.148)
   - A. Segment *1* distinctly *longer than* segment *3* 1>3
   - B. Segment *1* about *equal* in length to segment *3* 1=3*
   - C. Segment *1* distinctly *shorter than* segment *3* 1<3
5. Spines, in addition to the large claw, on last segment of terminal legs (FIG. 15.148)
   - A. Distinct *spines* (not just fine bristles or setae) present toward base of terminal feet sp
   - B. Distinct spines *absent* from terminal feet, although setae may be present x*
6. Length of anterior section (from tip of head to start of segment bearing the largest claw) compared to the first segment bearing an oval gill (FIG. 15.148)
   - A. Anterior segment *short,* less than half the length of the first gill segment sh
   - B. Anterior segment *long,* distinctly longer than first gill segment lo
   - C. Anterior segment *intermediate* in length, between one half and equal to first gill segment in*

15.87

Suture	Head Spines	Body Spines	Antennae 1: Segment 1 vs. 3	Leg Spines	Anterior Length	Species
su	sp	sp	1=3	sp	sh	*Aeginella spinosa*
su	sp	sp,sm	1<3	sp	in	*Aeginina longicornis,* long-horned skeleton shrimp
x	sp	sp,sm	1>3	sp	in	*Caprella septentrionalis*
x	sp	sp,sm	1>3	x	in	*C. unica*
x	sp	sm	1=3	sp,x	in	*C. penantis*
x	sm	sp	1=3	sp	lo	*C. linearis,* linear skeleton shrimp
x	sm	sm	1=3,1<3	sp	lo	*C. equilibra*
x	sm	sm	1=3	x	in	*Paracaprella tenuis*

Aeginella spinosa. Strong spines along dorsal surface. Cold, deepwater form; may be associated with red and brown algae, hydroids, or asteroids. Ac,Ha–Al–Co,Su–De, pr?,22 mm (0.85 in), (M8).

Aeginina longicornis, long-horned skeleton shrimp. Very spiny in northern, cold waters; less spiny toward the south. Red, tan, or colorless. Very long antenna. With algae, hydroids, and ectoprocts. Bo,Ha–Al,Su–De,pr?,5.2 cm (2.1 in), (M13,NM608), ☺.

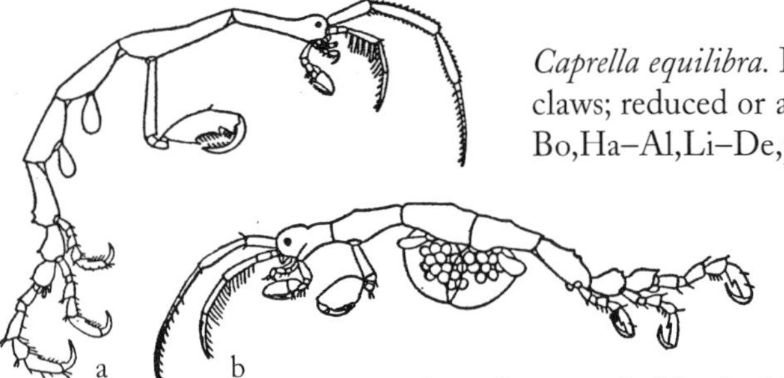

FIG. 15.149

Caprella equilibra. In northern areas, midventral spine between bases of largest claws; reduced or absent to the south. May be found on floating objects. Bo,Ha–Al,Li–De,pr?,23 mm (0.9 in), (M25).

Caprella linearis, linear skeleton shrimp. Pale-colored. Found on algae, sponges, hydroids, and tunicates. Ac–,Ha–Al,Su–De,pr?,22 mm (0.85 in), (M30,NM607). (FIG. 15.149. a, male; b, female.)

Caprella penantis. Head with anterior rostral point. Bo,Ha–Al,Li–Su,pr?,12 mm (0.5 in), (M33), ☺. (FIG. 15.150. a, male; b, female.)

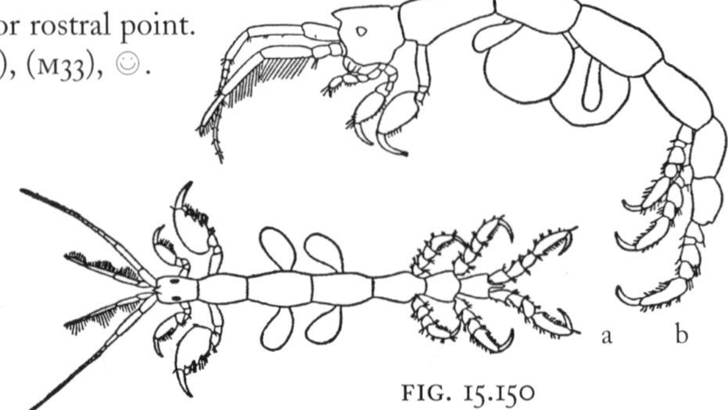

FIG. 15.150

Caprella septentrionalis. Spination varies; head always with at least one spine. Cold water; may be with the calcareous alga, *Corallina,* in tidal pools. Ac,Ha–Al,Li–De,pr?,22 mm (0.85 in), (M44), ☺.

Caprella unica. Larger individuals are less spiny. May be with sea stars, *Asterias.* Ac–,Ha–Co,Su,pr,17 mm (0.67 in), (M49).

Paracaprella tenuis. Found on algae, sea grasses, hydroids, sponges, and Bryozoa. Body curls in preservatives. Bo,Ha–Al,Su,pr?,7 mm (0.3 in), (M86), ☺.

15.88 Class Malacostraca: Order Mysidacea, Mysid Shrimp, and Order Euphausiacea, Krill

Mysid shrimp are small, slender, translucent shrimplike animals with spherical *statocysts* (small, spherical, sensory structures) embedded in the bases of their *uropods* (terminal pair of appendages). They may be found in swarms in shallow waters, often associated with seaweeds or tidal pools. Identifications are difficult, requiring close examination of immobilized specimens (see Appendix for suggested techniques). The thorax is nearly covered by a carapace, although the carapace is only attached to the first four thoracic segments. Females use ventral plates to hold and brood eggs. Important morphological features used in this key include the shape of the bladelike *antennal scale* associated with antennae 2 and the shape and spination of the *telson* (midpiece in the tail fan). Accurate observations may require use of a dissecting microscope. Determinations made from information presented here must be confirmed through reference to the more detailed literature.

Euphausids or *krill* resemble large mysids but lack the uropod statocysts and their gills are not covered by the carapace. Typically oceanic, these coldwater planktonic forms infrequently venture onshore. Offshore they form such dense swarms as to turn the water color pink and to serve as food for baleen whales.

Additional reference: (T) Tattersall, W. M., 1951.

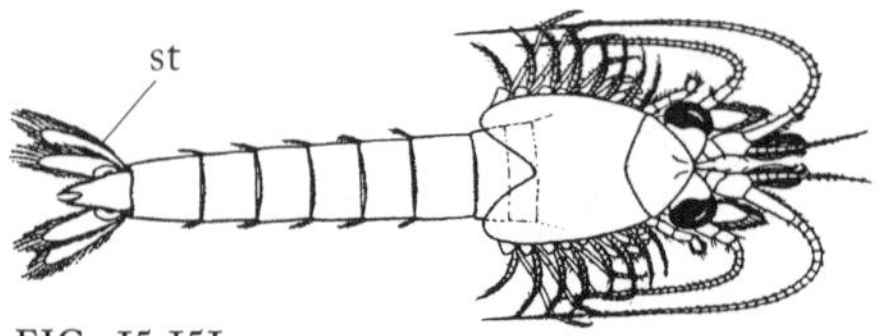

FIG. 15.151

1. Pair of rounded statocysts embedded at bases of uropods (use strong transmitted light [from underneath] to locate these) (FIG. 15.151)
   - A. *Statocysts* present in uropod bases — st
   - B. Statocysts *absent* — x

15.89

Statocysts	Group
st	Mysids, 15.90
x	*Meganyctiphanes norvegica,* horned krill†

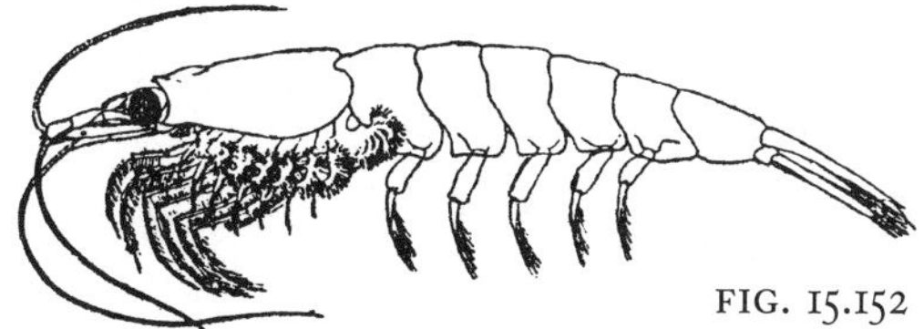

FIG. 15.152

Meganyctiphanes norvegica, horned krill† (Or: Euphausiacea, Euphausiidae). Display bioluminescence. Turns water pinkish in dense swarms. Bo,Pe,Ps–Pd,p?,3.8 cm (1.5 in), (B231,NM611). (FIG. 15.152)

15.90 Mysids (Or: Mysidacea) (from 15.11, 15.89)

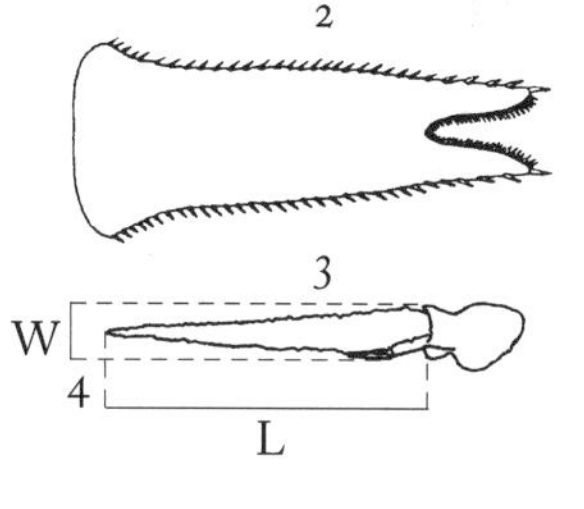

FIG. 15.153

1. Telson shape (FIG. 15.153)
   - A. Telson is *rounded* at its tip — ro
   - B. Telson is *notched* at its tip — no*
   - C. Telson ends with more than three large *spines* — sp
2. Spines along the lateral margins of the telson (FIG. 15.153)
   - A. Marginal spines along *entire* length of telson — en*
   - B. Marginal spines along the *posterior* half of the telson only — po
   - C. Marginal spines end at the level of the terminal *notch* — no
   - D. Marginal spines *absent* — x
3. Shape of ending of antennal scale (FIG. 15.153)
   - A. Antennal scale ends in a *point* — pt*
   - B. Antennal scale is *rounded* at its tip — ro
   - C. Antennal scale includes a prominent *spine* at or near its tip — sp
4. Proportions of antennal scale (FIG. 15.153)
   - A. Antennal scale *long,* length to width ratio greater than 8:1 — lo*
   - B. Antennal scale *short,* length to width ratio less than 3:1 — sh
   - C. Antennal scale *intermediate,* length to width ratio greater than 3:1 but less than 8:1 — in

15.91

Telson Tip	Telson Spines	Scale Tip	Scale L:W	Species
no	no	pt	lo	*Mysis stenolepis*
no	no	sp	lo	*Praunus flexuosus,* bent opossum shrimp
no	en	pt	lo	*Mysis mixta*
no	po	ro	sh	*Heteromysis formosa,* red opossum shrimp
sp	x	ro	in	*Erythrops erythrophthalma*
sp	en	ro	in	*Mysidopsis bigelowi*
ro	en	pt	sh	*Neomysis americana*
ro	po	ro	in	*Metamysidopsis munda*

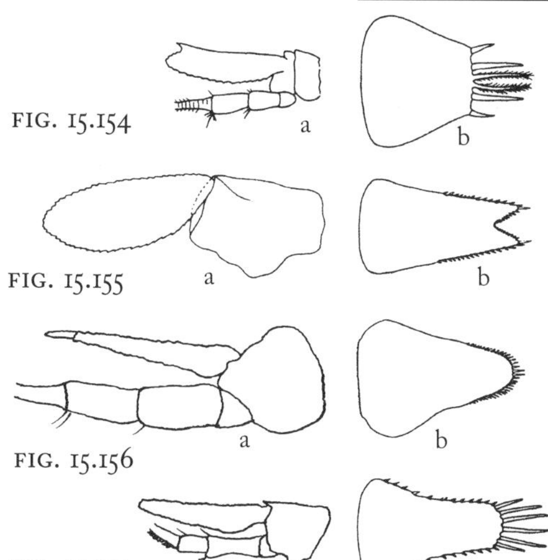

Erythrops erythrophthalma. Brightly colored; eyes red, dorsal carapace red-orange, scattered white and yellow spots. Offshore. Ac–,Sa,Su–De,pr,11 mm (0.43 in), (T110). (FIG. 15.154. a, antennal scale; b, telson.)

Heteromysis formosa, red opossum shrimp. Often associated with valves from dead bivalve mollusks. Females: red or pink; males: translucent. Ac–,Ha–Gr–Sa,Li–De,pr, 8 mm (0.3 in), (G231,NM610,T235). (FIG. 15.155. a, antennal scale; b, telson.)

Metamysidopsis munda. Ca+,?,Su,?,14 mm (0.55 in), (T147). (FIG. 15.156. a, antennal scale; b, telson.)

Mysidopsis bigelowi. Large spines line the telson ending. Small offshore species. Vi,Ha–Al–Pe,Su–Pd,pr,5 mm (0.2 in) (G231,T139). (FIG. 15.157. a, antennal scale; b, telson.)

Mysis mixta. Avoids algae and seagrasses. Ac–,Gr–Sa–Mu,Su,pr,20 mm (0.8 in), (G230,T168). (FIG. 15.153. a, antennal scale; b, telson.)

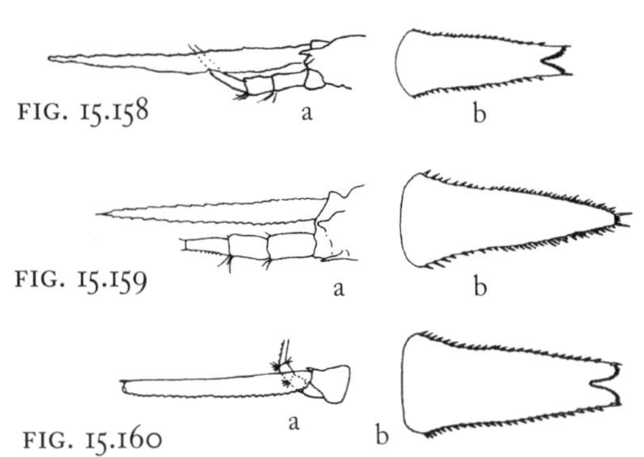

FIG. 15.158

FIG. 15.159

FIG. 15.160

Mysis stenolepis. Large; translucent with a dark, star-shaped spot on each segment. Shallow, seagrass or algal beds. Ac–,SG–Al,Su,pr,25 mm (1 in), (G230,T170). (FIG. 15.158. a, antennal scale; b, telson.)

Neomysis americana. Marine or estuarine shallows. Near bottom by day; toward surface by night. Bo,Ha–Al,Es–Li–De,pr,12 mm (0.5 in), (G231,T195), ☺. (FIG. 15.159. a, antennal scale; b, telson.)

Praunus flexuosus, bent opossum shrimp. Recently introduced from Europe to North America. Colors in life: dark, yellowish, or translucent. Ac,Ha–Al–SG,Li–Su,pr–df,25 mm (1 in), (G230,NM610,T247). (FIG. 15.160. a, antennal scale; b, telson.)

15.92 Class Malacostraca: Order Decapoda: Infraorder Caridea and Family Penaeidae, Shrimp

Shrimp may be either inhabitants of the open water column (pelagic) or members of the benthos (bottom dwellers). Some pelagic species are of economic importance in a fishery that extends along the entire coastline of eastern North America. Others are common members of the shallow-water seagrass community. The head and thorax of shrimp are covered by a single exoskeletal *carapace.* The abdominal region is usually well developed, bearing conspicuous appendage pairs as well as a terminal *tail fan* (the *telson* plus adjacent *uropods*). Many shrimp are brightly and distinctively colored, but this feature is most notable on live, fresh specimens because colors tend to fade quickly following the animal's death.

Characteristics useful in identification include the morphology of key appendages and especially the shape and ornamentation of the *rostrum* (portion of the carapace that extends forward between the eyes). It may be necessary to distinguish specific appendages. Appendages associated with the head region include two pairs of *antennae* (the larger of which often bears a bladelike antennal scale at its base) and a pair of *mandibles* (that serve as crushing jaws and are difficult to see). Eight pairs of thoracic limbs, called *pereiopods,* follow. The first three pairs are *maxillipeds* used in the capture and manipulation of food. The next five pairs of thoracic limbs function primarily as *walking legs.* To locate specific walking legs in order to view identifying features, it is easiest to start with the pair of walking legs located nearest the posterior end of the carapace. Treat them as walking-leg pair 5 and count backward.

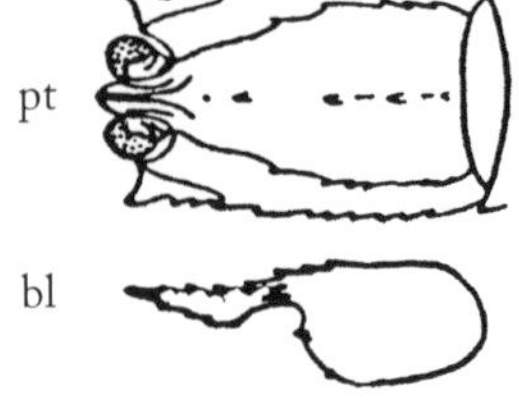

Common names followed by a dagger (†) are designated as standard names according to Williams et al. (1989). Additional, unofficial but widely used common names are also listed to facilitate linkage to approved names. Only the standardized names should be used subsequently.

Additional references: (wa) Williams, A. B., 1974; (wb) Williams, A. B., 1984.

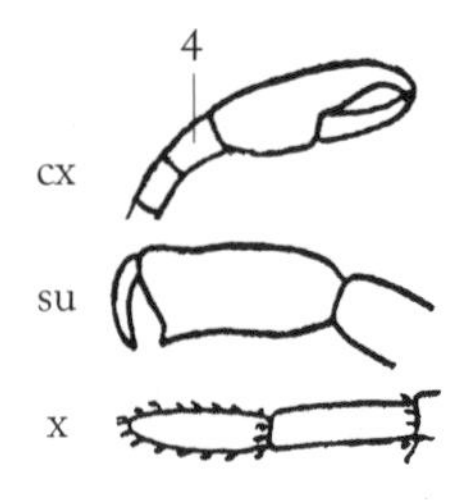

FIG. 15.161

1. Relationship of second abdominal segment compared to the first abdominal segment
   - A. Abdominal segment 2 *overlaps* both segment 1 and segment 3 in lateral view — ov
   - B. Abdominal segment 2 does *not* overlap segment 1, although it does overlap segment 3 — x
2. General shape of the rostrum or anterior projection of the carapace (FIG. 15.161)
   - A. Rostrum forms a simple *point* anteriorly — pt
   - B. Rostrum forms a *blade*like structure — bl

3. Counting the last pair of periopods (nearest the end of the carapace) as walking-legs 5, describe the ending on the first conspicuous walking legs (FIG. 15.161)
   - A. Leg ends in a distinct *chela* or pinching claw (human thumb and forefinger); the inside surface of both "fingers" is completely lined with bristles (i.e., is *comb*like) — cc
   - B. Leg ends in a distinct *chela* but bristles or combs are *not* found on both fingers — cx
   - C. Leg ending is *subchelate* (forefinger against a closed fist) — su
   - D. First walking leg has chelate ending *absent* or microscopically small — x
4. Description of the leg section next closest to the body from the chela on pereiopod 4, the second conspicuous walking legs (FIG. 15.161)
   - A. Section just below the chela is subdivided into several short *segments* — se
   - B. This section is *not* subdivided — x

15.93

Segment 2 vs. Segment 1	Rostrum	Walking Legs 1	Walking Legs 2	Group
ov	pt	cc	x	Pasiphaeidae, 15.102
ov	pt	su	x	Crangonidae, 15.94
ov	bl	cx	x	Palaemonidae, 15.98
ov	bl	cx	se	Hippolytidae, 15.96
ov	bl	x	se	Pandalidae, 15.100
x	bl	cs	x	*Penaeus aztecus*, brown shrimp†

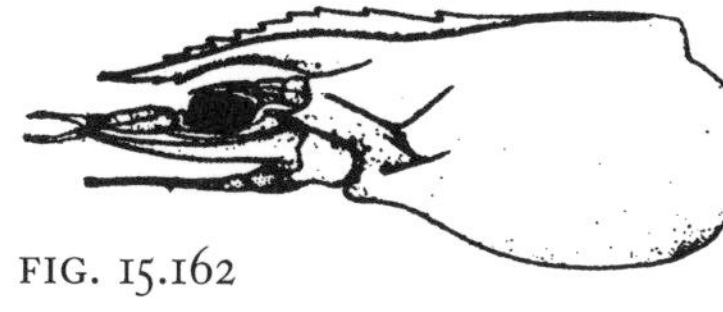
FIG. 15.162

Penaeus aztecus, brown shrimp† (Penaeidae). Thin brownish to olive exoskeleton. The third pair of walking legs is also chelate. Most frequent commercial shrimp along southeastern U.S. coast and Gulf of Mexico. Vi,Mu,Su,om?,24.1 cm (9.5 in), (B233,NM612,wa39,wb24). (FIG. 15.162. Head, lateral view.)

Families of Caridean Shrimp (arranged alphabetically)

15.94

Crangonidae (from 15.93)

1. Number of middorsal spines along the carapace: provide *number* of distinct middorsal spines (FIG. 15.163) — __
2. Projection of the rostrum relative to the eyes (FIG. 15.163)
   - A. *Rostrum* extends *beyond* the *eyes* — r>e*
   - B. *Rostrum is equal* in length to tip of the *eyes* — r=e
   - C. *Rostrum* is *shorter than* the *eye* length — r<e
3. Details of the rostrum (FIG. 15.163)
   - A. With a pair of small *basal spikes* — bs
   - B. With a middorsal raised *keel* — ke*
   - C. Basal spikes and middorsal keel *absent* — x
4. Shape of the terminal tip of the telson (FIG. 15.163)
   - A. *Pointed* ending — po*
   - B. *Blunt* ending — bl

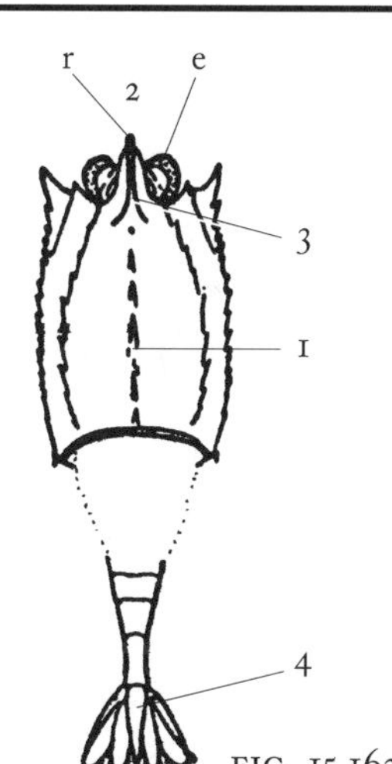

FIG. 15.163

15.95

Spines	Rostrum vs. Eye	Rostrum	Telson	Species
1	r<e	x	po	*Crangon septemspinosa*, sevenspine bay shrimp†, sand shrimp
3	r<e	bs	po	*Pontophilus brevirostris*
3	r=e	bs	po	*P. norvegicus*, Norwegian shrimp†
3	r>e	ke	po	*Sclerocrangon boreas*, sculptured shrimp†
4–5	r>e	ke	po	*Sabinea sarsii*, Sars shrimp†
4–5	r>e	ke	bl	*S. septemcarinata*, sevenline shrimp†

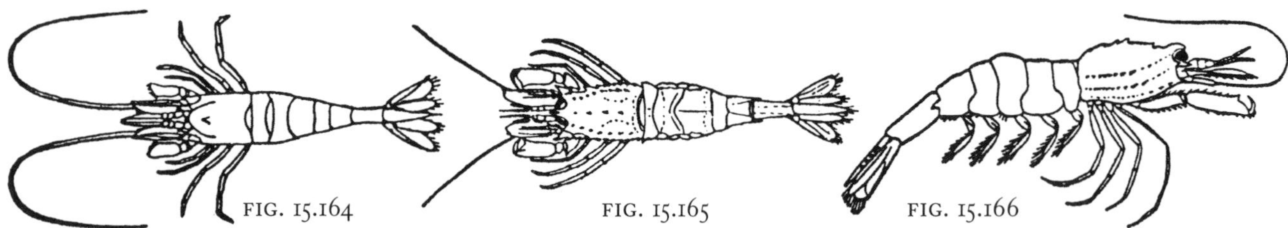

FIG. 15.164 FIG. 15.165 FIG. 15.166

Crangon septemspinosa, sevenspine bay shrimp†, sand shrimp. Carapace: gray with dark speckles, looks like sand. Often motionless in tide pools or on sand flats. Bo,Sa,Li–Su,pr, 7.1 cm (2.8 in), (wa40,wb159), ☺. (FIG. 15.164)

Pontophilus brevirostris. Carapace: mottled, pale, dull, reddish brown. Bo,Sa,Su–De,om?,3.3 cm (1.5 in), (wa40,wb161).

Pontophilus norvegicus, Norwegian shrimp†. Carapace: dull red with two white bands. Bo,Mu–Sa–Gr,Su–De,om,7.8 cm (3.1 in), (wa40,wb162).

Sabinea sarsii, Sars shrimp†. Carapace: gray to brown; telson with four spines. Ac,Sa–Gr,Su–De,om,77.1 cm (2.8 in), (wa40,wb163). (FIG. 15.165)

FIG. 15.167

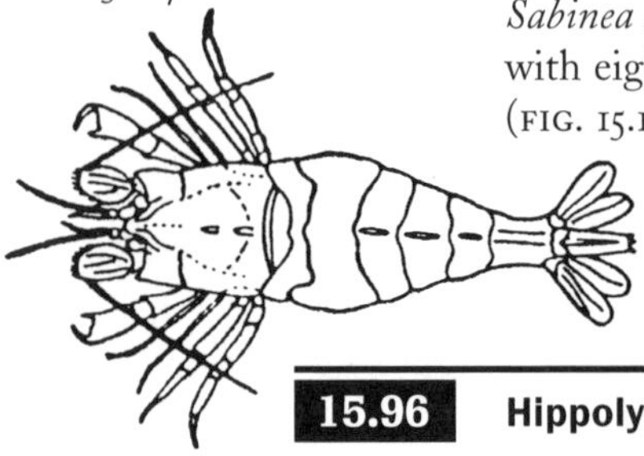

Sabinea septemcarinata, sevenline shrimp†. Carapace: dorsal reddish brown spots; telson with eight or more spines. Ac,Mu–Sa–Gr,Su–De,om,9.1 cm (3.6 in), (wa41,wb164). (FIG. 15.166. Lateral view, anterior to right.)

Sclerocrangon boreas, sculptured shrimp†. Carapace: red with reddish brown spots. Ac,Mu–Sa–Gr,Su–De,om,12.9 cm (5.1 in), (wa41,wb166). (FIG. 15.167. Dorsal view.)

15.96 **Hippolytidae** (from 15.93)

1. Length of rostrum relative to length of bladelike antennal scale
   - A. *Rostrum* is roughly *equal* in length to antennal *scale* — r=s
   - B. *Rostrum* is distinctly *longer than* antennal *scale* — r>s
   - C. *Rostrum* is distinctly *shorter than* antennal *scale* — r<s
2. General shape of rostrum (viewed laterally) (FIG. 15.168)
   - A. *Tapers* gradually toward tip or is approximately uniform in width throughout — ta
   - B. Distinctly *expands* in height between its origin and its tip — ex*
3. Number of teeth along the dorsal edge of the rostrum (FIG. 15.168)
   - A. *Many* (i.e, >10) — ma
   - B. Teeth *absent* along dorsal rostral edge — x
   - C. An *intermediate* number (1–10) present — in*
4. Number of teeth along the ventral edge of the rostrum (FIG. 15.168)
   - A. *Many* (i.e, >10) — ma
   - B. Only *one* tooth present, toward the tip, making it bifid — 1
   - C. Teeth *absent* along ventral edge of rostrum — x
   - D. An *intermediate* number (2–10) present — in*
5. Supraorbital spines (slightly above and posterior to the eye opening or orbit) (does not include any other spines, even those near but below the orbit) (FIG. 15.168)
   - A. Provide *number* of supraorbital spines — __
   - B. Supraorbital spines *absent* — x
6. Basic color-pattern characteristics of the carapace
   - A. *Translucent* — tr
   - B. Colored, but color is *uniform* in color — un
   - C. *Dots* or spots of contrasting color — do
   - D. *Banded* with contrasting color — ba
   - E. Irregularly *mottled* or blotched with a contrasting color — mo

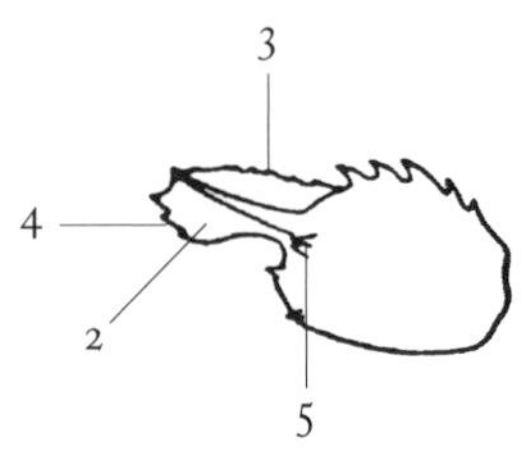

FIG. 15.168

15.97

Rostrum Length	Rostrum	Dorsal	Ventral	Spines	Color	Species
r=s	ex	in	in	x	do	*Eualus fabricii,* Arctic eualid†
r=s	ex	in	in	x	tr	*E. gaimardii,* circumpolar eualid†
r=s	ex	ma	in	2	do	*Spirontocaris liljeborgii,* friendly blade shrimp†
r=s	ex	ma	in	2	mo	*S. spinus,* parrot shrimp†
r>s	ex	x	ma	1	un	*Tozeuma* spp., arrow shrimps†
r<s	ta	in	in	x	un,ba	*Hippolyte zostericola,* zostera shrimp†
r<s	ta	in	1	x	do	*Eualus pusiolus,* doll eualid†
r<s	ta	in	in	1	mo	*Lebbeus groenlandicus,* spiny lebbeid†
r<s	ta	in	x	1	ba	*L. zebra,* zebra lebbeid†
r<s	ex	in	in	1	do	*L. polaris,* polar lebbeid†
r<s	ex	in	in	2	do	*Spirontocaris phippsii,* punctate blade shrimp†
r<s	ex	in	in	x	ba	*Lysmata wurdemanni,* peppermint shrimp†, red cleaning shrimp

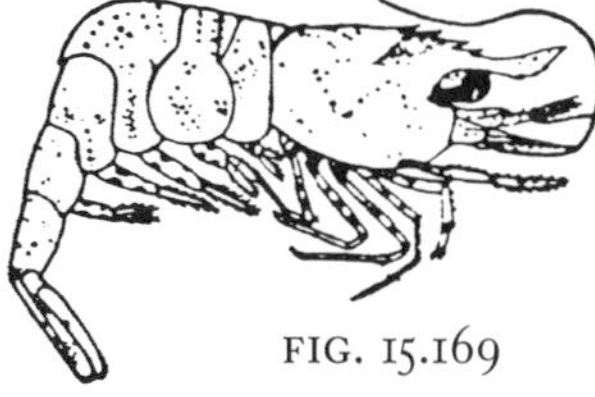

FIG. 15.169

FIG. 15.170

Eualus fabricii, Arctic eualid†. White with red dots. Ac,Mu–Sa–Gr,Su–De,om,5.8 cm (2.3 in), (wa40,wb111). (FIG. 15.169. Anterior to right.)

Eualus gaimardii, circumpolar eualid†. Greenish. Ac,Mu–Sa–Gr,Su–De,om,10.1 cm (4 in), (wa40,wb112).

Eualus pusiolus, doll eualid†. White with reddish orange dots. Bo,Sa–Gr,Li–Su,om,3.3 mm (1.3 in), (wa40,wb113). (FIG. 15.170. Anterior to right.)

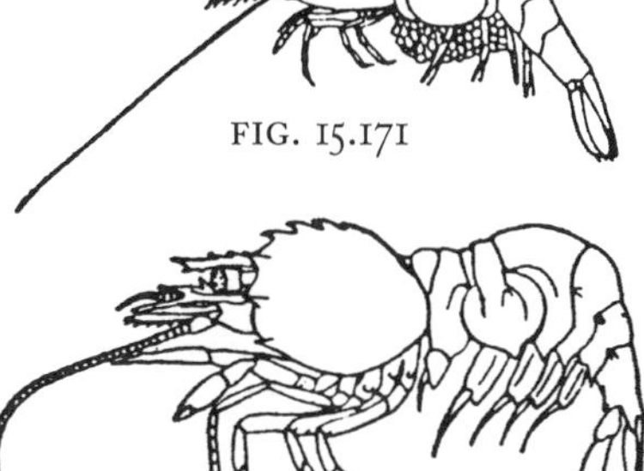

FIG. 15.171

FIG. 15.172

Hippolyte zostericola, zostera shrimp†. Green or brown; some with brown dorsal stripe. Vi,SG–Al,Su,om?,15 mm (0.6 in), (wa40,wb118). (FIG. 15.171. Carrying eggs.)

Lebbeus groenlandicus, spiny lebbeid†. Brown mottled with red or green. Ac–,Ha–Gr,Su,om,10.1 cm (4 in), (wa40,wb122). (FIG. 15.172)

Lebbeus polaris, polar lebbeid†. Light with reddish orange dots. Bo,Mu–Sa–Gr–Ha,Su–De,om,8.9 cm (3.5 in), (wa40,wb123). (FIG. 15.173. Anterior to right.)

FIG. 15.173

Lebbeus zebra, zebra lebbeid†. Light with brownish red mottling. Ac,Ha,Su,om?,5.1 cm (2 in), (wa40,wb125). (FIG. 15.174. Anterior to right.)

Lysmata wurdemanni, peppermint shrimp†, red cleaning shrimp. Light with red bands. Vi,Ha,Li–Su,pr,6 cm (2.4 in), (wb127).

Spirontocaris liljeborgii, friendly blade shrimp†. Reddish with yellow and white dots. Bo,Ha–Mu,Su–De,pr,7 cm (2.8 in), (wa40,wb129). (FIG. 15.175)

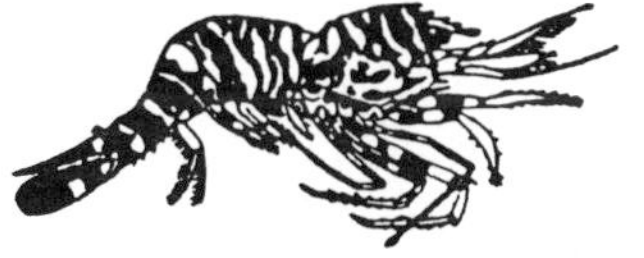

FIG. 15.174

Spirontocaris phippsii, punctate blade shrimp†. Light with reddish brown spots. Ac,Ha–Al,Su–De,om,4.6 cm (1.8 in), (wa40,wb130). (FIG. 15.176. Anterior to right.)

Spirontocaris spinus, parrot shrimp†. Red or brown with green or white blotches. Ac–,Mu–Sa–Gr–Ha,Su–De,om,5.8 cm (2.3 in), (wa40,wb132).

Tozeuma spp. arrow shrimps†. Refer to Williams to distinguish species. (wa9,wb138–140). (FIG. 15.177. Anterior to right.)

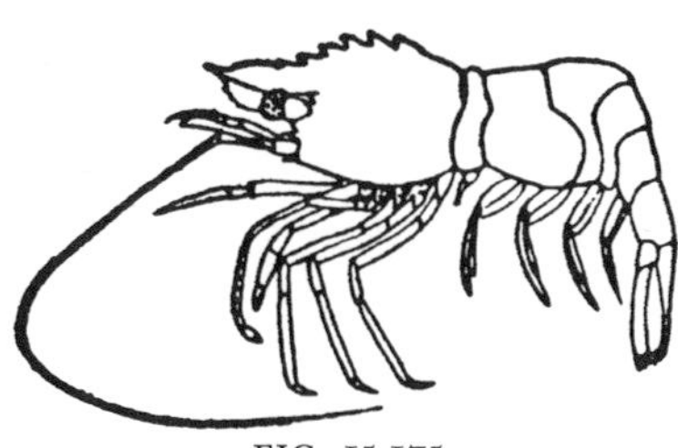

FIG. 15.175

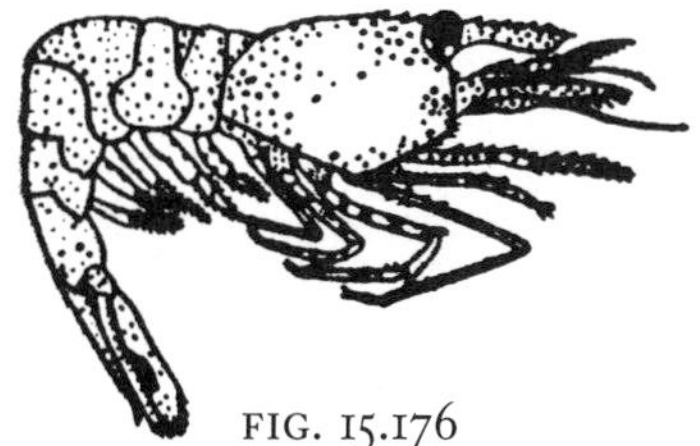

FIG. 15.176

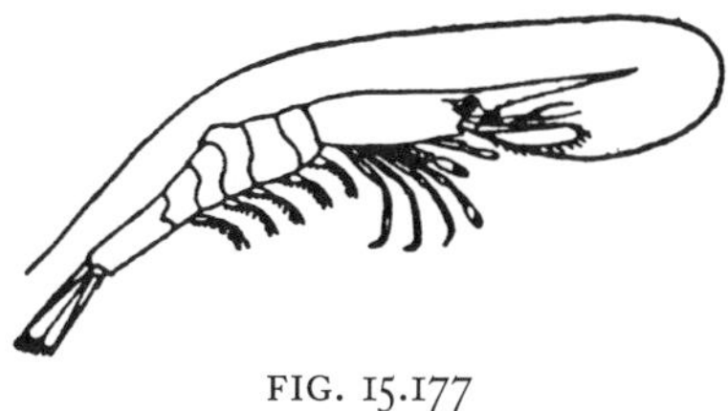

FIG. 15.177

15.98 **Palaemonidae** (from 15.93)

1. Number of teeth along the ventral edge of the rostrum: provide *number* (FIG. 15.178) —
2. Posterior extent of dorsal rostral teeth relative to posterior rim of the orbit (curve in carapace behind the eye) (FIG. 15.178)
   - A. *Two* teeth arise posterior to the orbit — 2
   - B. *One* tooth arises posterior to the orbit — 1*
3. Geographic distribution
   - A. South of Cape Cod only—*Virginian* — vi
   - B. Found *both* north and south of Cape Cod — bo

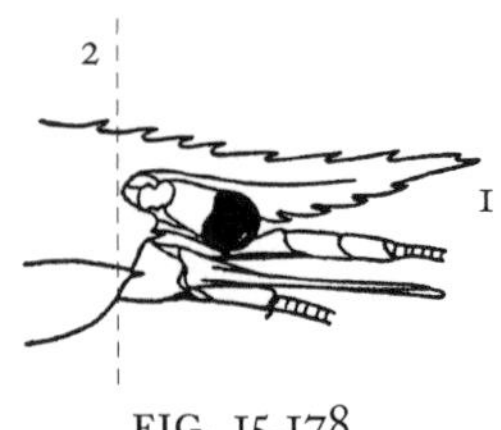

FIG. 15.178

15.99

Teeth	Posterior to Orbit	Distribution	Species
3–5	2	bo	*Palaemonetes vulgaris*, marsh grass shrimp†
4–5	1	vi	*P. intermedius*, brackish grass shrimp†
2–4	1	bo	*P. pugio*, daggerblade grass shrimp†

Palaemonetes intermedius, brackish grass shrimp†. Vi,Al–SG,Es–Su,om,14 mm (0.55 in), (wa39,wb75). (FIG. 15.178. Head, lateral view.)

Palaemonetes pugio, daggerblade grass shrimp†. Tolerates lower salinity; prefers 10–20°/oo. Bo,Al–SE,Es–Su,om,17 mm (0.67 in), (wa39,wb76). (FIG. 15.179. Head, lateral view.)

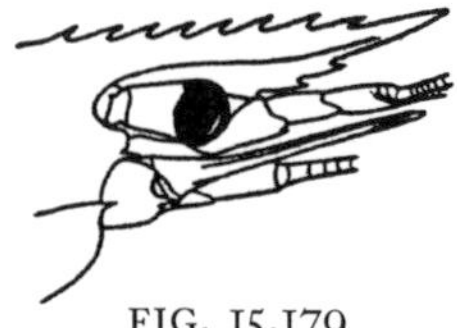
FIG. 15.179

Palaemonetes vulgaris, marsh grass shrimp†. Bo,Al–SG,Es–Su,om,14 mm (0.55 in), (wa40,wb72). (FIG. 15.180. a, head, lateral view; b, whole animal, lateral view.)

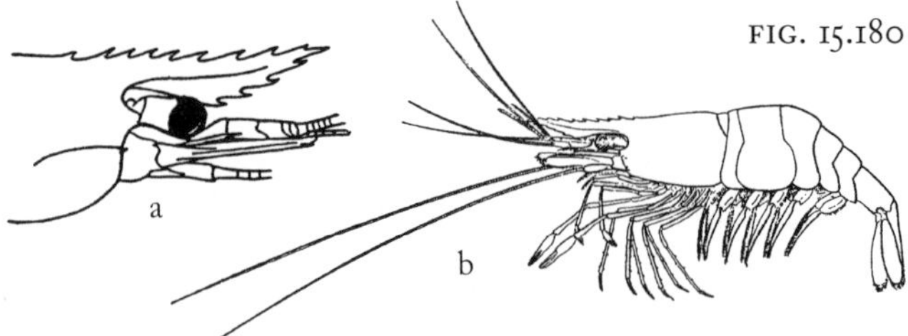

FIG. 15.180

15.100 **Pandalidae** (from 15.93)

1. Length of rostrum (FIG. 15.181)
   - A. Rostrum very *long* (i.e., more than twice the carapace length) — lo
   - B. Rostrum *shorter* (i.e., less than twice the carapace length) — sh*
2. Angle of rostrum (FIG. 15.181)
   - A. Rostrum bends sharply *upward* at a 45° angle — up
   - B. Rostrum bends upward *slightly*, but well less than 45° angle — sl*
3. Extent of teeth along dorsal rostral surface (FIG. 15.181)
   - A. Teeth extend *half* the distance to the rostral tip — ½
   - B. Teeth extend nearly to the rostral *tip* — ti*
4. Texture of the carapace surface (FIG. 15.181)
   - A. Basically *smooth* — sm
   - B. *Rough*, with small ridges and bristles — ro
5. Color pattern on carapace
   - A. Basically *uniform* — un
   - B. *Stripes* of contrasting color — st
   - C. *Speckled* with dots of contrasting color — sp

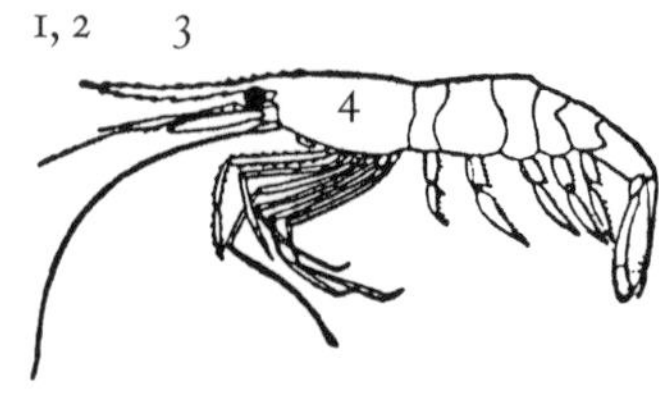

FIG. 15.181

15.101

Rostrum Length	Rostrum Bend	Teeth	Carapace	Color	Species
lo	sl	ti	sm	un	*Stylopandalus richardi*
sh	up	½	sm	un	*Pandalus propinquus*
sh	up,sl	½	sm	st	*P. montagui,* Aesop shrimp†, striped pink shrimp
sh	sl	ti	sm	sp	*P. borealis,* northern shrimp†, pink shrimp
sh	sl	ti	ro	un	*Dichelopandalus leptocerus,* bristled longbeak shrimp†

Dichelopandalus leptocerus, bristled longbeak shrimp†. Brick red. Bo,Sa–Mu,Su–De,pr,10.1 cm (4 in), (wa40,wb150).

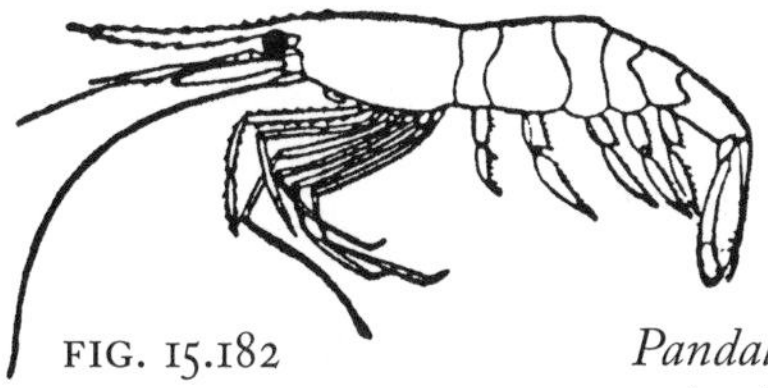

FIG. 15.182

Pandalus borealis, northern shrimp†, pink shrimp. Dark red antennae. Important commercial species in Gulf of Maine. Ac,Mu–Sa–Ha,Su–De,pr,17.2 cm (6.8 in), (wa40,wb151). (FIG. 15.182)

Pandalus montagui, Aesop shrimp†, striped pink shrimp. Pink with red lines. Ac–,Mu–Sa–Gr,Su–De,pr,16.5 cm (6.5 in), (wa40,wb154). (FIG. 15.183)

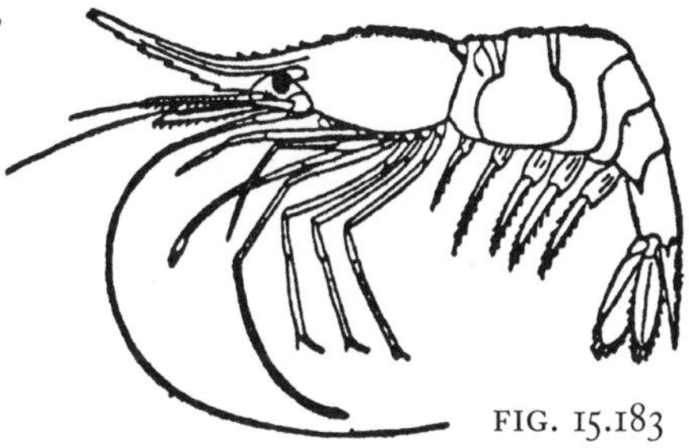

Pandalus propinquus. Abdomen streaked. Bo,Mu,Su–De,pr,10.1 cm (4 in), (wa40,wb156).

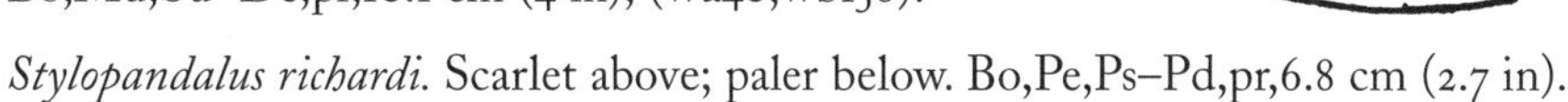

FIG. 15.183

Stylopandalus richardi. Scarlet above; paler below. Bo,Pe,Ps–Pd,pr,6.8 cm (2.7 in).

15.102 **Pasiphaeidae, Glass Shrimps** (from 15.93)

FIG. 15.184

1. Lateral edges of the front of the carapace (FIG. 15.184)
   - A. Edge is *toothed* or notched — to*
   - B. Edge is relatively *smooth* — sm
2. Comparative sizes of the last two pairs of walking legs (walking-legs 4 and 5) (FIG. 15.184)
   - A. Leg *4* is *shorter than* leg *5* — 4<5
   - B. Leg *4* is *longer than* leg *5* — 4>5

15.103

Carapace	Legs 4 vs. 5	Species
sm	4<5	*Leptochela papulata,* light glass shrimp†
to	4>5	*Pasiphaea multidentata,* pink glass shrimp†

Leptochela papulata, light glass shrimp†. Bo,Mu–Sa–Gr,Su,pr,8 mm (0.3 in), (wb57).

Pasiphaea multidentata, pink glass shrimp†. Ac,Pe,Ps–Pd,pr,6.4 cm (2.5 in), (wa39,wb60).

15.104 Class Malacostraca: Order Decapoda: Infraorder Anomura, Anomuran Crabs and Infraorder Brachyura, True Crabs (from 15.7), and Subphylum Chelicerata: Class Merostomata, Horseshoe Crabs

Crabs are active, conspicuous, and important predators and scavengers in marine ecosystems. Beware of their pinching abilities and their voracious appetite that they will demonstrate happily when placed in a confined space (including collecting buckets and aquaria) with other specimens.

A single-piece *carapace* covers their head and thoracic regions. Its shape and marginal *dentition* are useful key characters. The placement of *stalked eyes* and the shape of certain thoracic appendages or *pereiopods* (notably *chelipeds* [largest pinching claws] and hindmost walking legs) are also important to note. The abdominal region is usually a triangular flap folded under the posterior margin of the body. In many cases, sexes can be determined by examining the width of the abdomen and the number of appendages (or *pleopods*) associated with the abdomen. Females have a broad abdomen and several pairs of often enlarged pleopods for carrying eggs. The abdomen of the male is more narrow and often only a pair or two of enlarged and hardened anterior pleopods remain that are used in transfer of sperm packets (spermatophores) during copulation.

Because crustaceans possessing a hard exoskeleton must molt or shed in order to grow in volume, their cast-off exoskeletons frequently are found washed ashore among the high-tidal drift material or caught in supralittoral grasses. The carapace, which once covered the crab's head and thorax, can by raised along its posterior border, revealing the route through which the molting crab exited its old shell.

Reduced size of the fifth (last) pair of walking legs distinguishes anomuran from brachyuran crabs. These legs are often tucked into the gill chamber under the carapace. The second antennae may be attached lateral to the eye stalks rather than between eye stalks as in typical brachyuran crabs. Crabs treated here are in the infraorder Brachyura unless noted otherwise. The horseshoe crab, *Limulus polyphemus,* has also been included here because of its superficial resemblance to true crabs. In actuality, this animal is not even a crustacean but belongs the the subphylum Chelicerata.

When length is called for, use the longest dimension (either width of the carapace or its length).

Common names followed by a dagger (†) are designated as standard names according to Williams et al. (1989). Additional, unofficial but widely used common names are also listed to facilitate linkage to approved names. Only the standardized names should be used subsequently.

Additonal references: (wa) Williams, A. B., 1974; (wb) Williams, A. B., 1984.

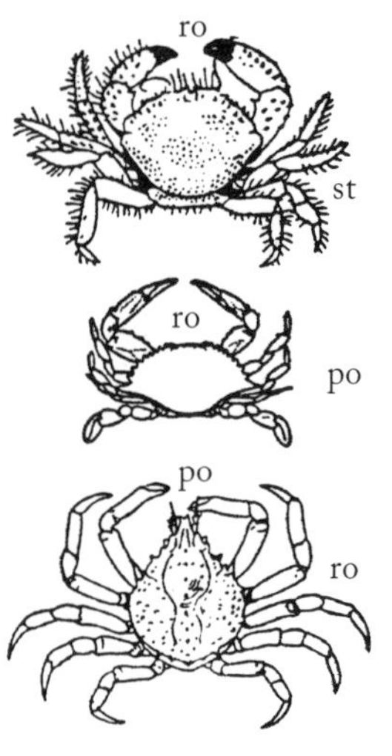

FIG. 15.185

1. Body shape—front: sides (ignore scattered body spines and marginal teeth, but include rostrum ["nose" projection] and large obvious lateral spikes) (FIG. 15.185)
   - A. Front *rounded*: sides *rounded;* smooth, curved contour — ro:ro
   - B. Front *rounded*: sides more or less *straight* or only gently curved — ro:st
   - C. Front *rounded*: sides drawn out into *points* — ro:po
   - D. Front *pointed*: sides *pointed* (i.e., roughly triangular) — po:po
   - E. Front *pointed*: sides *rounded* — po:ro
   - F. Front *straight*: sides *straight* (i.e., square or rectangular) — st:st
   - G. Front *straight*: sides *rounded* — st:ro
   - H. Body *cylindrical,* width narrower than length — cy
   - I. Body hidden from view within a *gastropod* (snail) shell — ga
2. Contour of anterior margin of carapace
   - A. Anterior margin with distinct *teeth* or points — te
   - B. Carapace circular, forming narrow *bottleneck* shaped opening anteriorly, between eyes — bn
   - C. Anterior margin basically smooth, distinct teeth or bottleneck *absent* — x
3. Black tips on pinching claws or chelae
   - A. Distinct *black* tips present on pinching claws — bl
   - B. Black tips *absent* from pinching claws — x
4. Shape of hindmost pair of legs
   - A. Last pair of legs are *paddle*-shaped for swimming — pa
   - B. Last pair of legs are pointed for walking, paddle legs *absent* — x
5. Habitat
   - A. *Saltmarshes* — sm
   - B. *Sandy* or muddy *littoral* habitat — sl
   - C. *Sandy* or muddy *sublittoral* habitat — ss
   - D. *Rocky* substrate — ro
   - E. *Seagrass* beds — sg
   - F. *Commensal,* living on another invertebrate or within its tube or burrow — co
   - G. *Mud* — mu

15.105

Front: Sides	Anterior	Claw Tips	Last Legs	Habitat	Group or Species
st:st	x	x	x	sm,sl,mu	Crab Group 1, 15.106
st:ro	x	x	x	sm,sl,mu	Crab Group 1, 15.106
ro:st	x	x	x	sm,sl,mu	Crab Group 1, 15.106
ro:st	x	x	x	co,ss	Crab Group 2, 15.108
ro:st	te	x	pa	sm,si,sg,ss	*Ovalipes ocellatus,* lady crab†, calico crab
ro:st	te	x	x	sm,sl,ro,sg,ss,mu	Crab Group 3, 15.110
ro:st	te	bl	x	sm,sl,sg,mu	Crab Group 4, 15.112
ro:po	te	x	pa	sm,sl,ro,sg,ss	Crab Group 5, 15.114
po:ro	te	x	x	sm,sl,ro,sg,ss,mu	Crab Group 6, 15.116
po:ro	bn	x	x	ss,mu	*Persephona mediterranea* (=*punctata*), mottled purse crab†
po:po	te	x	x	ss	*Rochinia tanneri,* thorned spiny crab†
po:po	x	x	x	ro,sg	*Heterocrypta granulata,* smooth elbow crab†, chip crab
ro:ro	x	x	x	sm,sl,sq,ss,mu	*Limulus polyphemus,* horseshoe crab†
ro:ro	x	x	x	co,mu	Crab Group 2, 15.108
cy,ga	te	x	x	sm,sl,ro,sg,ss,mu	Occupy empty gastropod shells—Crab Group 7, 15.118
cy	te	x	pa	sl,sm,sg,ss	*Emerita talpoida,* Atlantic sand crab†, mole crab

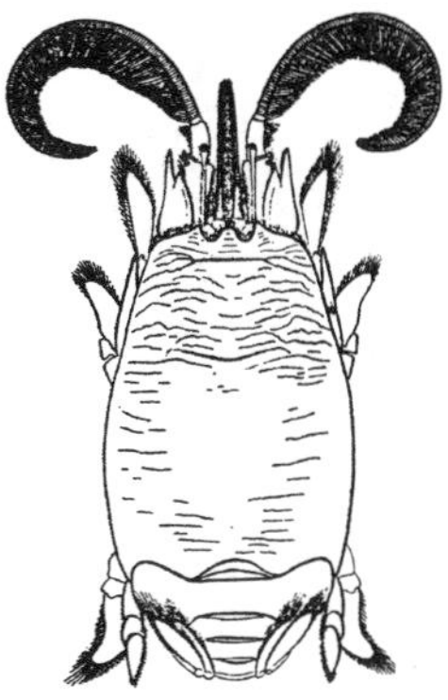

FIG. 15.186

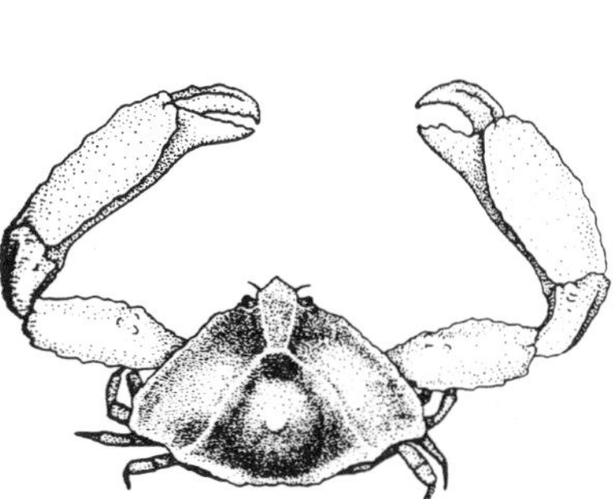

FIG. 15.187

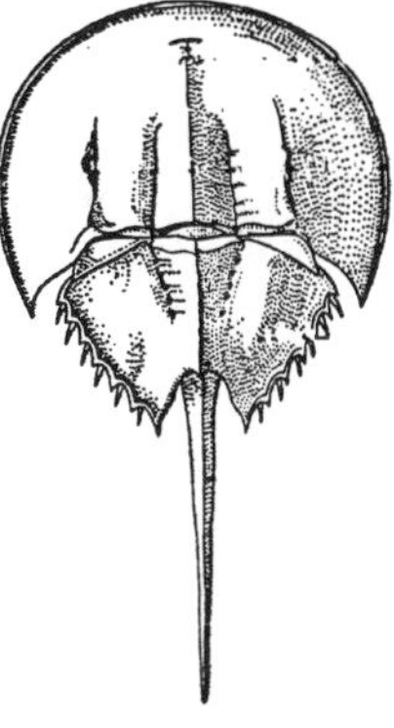

FIG. 15.188

FIG. 15.189

Emerita talpoida, Atlantic sand crab†, mole crab (IO: Anomura, Hippidae). Egg-shaped body; purplish tan color; short, feathery antennae; all legs hairy. Wave zone of surf beaches; fast burrower into sand. Shifts position to follow rise and fall of tides. Bright orange eggs under abdomen of larger females in spring. Vi,Sa,Li–Su,ff,25 mm (1 in), (G244,NM244,Wa42,Wb252), ☺. (FIG. 15.186)

Heterocrypta granulata, smooth elbow crab†, chip crab (Parthenopidae). Chelipeds very large. Light-orange body color. Well-camouflaged on shelly bottom. Vi,Gr–Ha,Su,om,21 mm (0.8 in), (G246,Wa42,Wb347). (FIG. 15.187)

Limulus polyphemus, horseshoe crab†. Dark reddish brown; long caudal spike. These members of the SP: Chelicerata, Cl: Merostomata are not true crabs. Bo,Sa–Mu–SM–SG,Li–Su,pr,61 cm (24 in) (carapace width), (G202,NM202), ☺. (FIG. 15.188)

FIG. 15.190

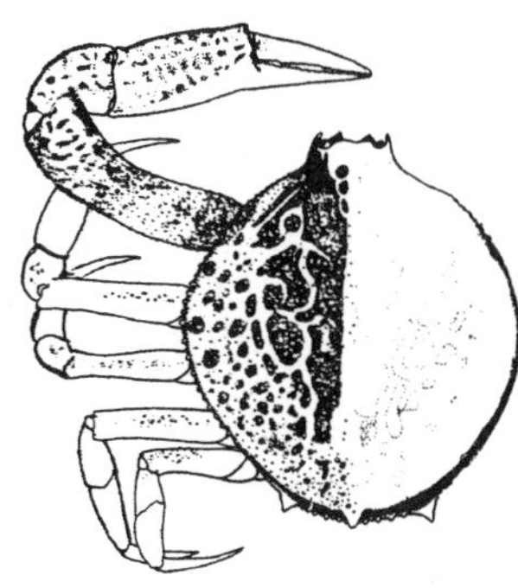

Ovalipes ocellatus, lady crab†, calico crab (Portunidae). Three to five anterolateral teeth between eyes and lateral extremes of carapace. Bo,Sa,Li–Su,pr,8.9 cm (3.5 in), (G248,NM638,Wa43,Wb359), ☺. (FIG. 15.189)

Persephona mediterranea (=*punctata*), mottled purse crab† (Leucosiidae). Globular carapace is roughly granular; color dull with darker spots and blotches. Vi,Mu,Su,om,5.8 cm (2.3 in), (G245,NM636,Wa42,Wb288). (FIG. 15.190)

Rochinia tanneri, thorned spiny crab† (Majidae). Long, sharp body spikes and rostral tips; long thin legs. Vi,Sa,Su–De,om,25 mm (1 in), (Wb323).

15.106

Crab Group 1
(Fiddler and Ghost Crabs [Ocypodidae] and Marsh or Shore Crabs [Grapsidae]) (from 15.105, 15.113)

Fiddler crabs are semiterrestrial and diurnal in activity, foraging for organic detritus on marsh banks and along estuarine creeks when ebb tide exposes their burrows. During flood tides, they use a dirt plug to seal themselves and a small volume of air within their burrows. Males possess one enlarged claw that they wave during courtship to attract females or to warn intruders. In contrast, females have small, equal-sized chelae or pinching claws. Fiddlers live in colonies. Locate groups of 1.25-cm (0.5-in) holes in saltmarsh mud banks at low tide and stand motionless for several minutes to encourage their return to activity to watch their behavioral show. Lengths given are for carapace width.

1. Location of eyes and eyestalks
   - A. Eyes found laterally on *long* stalks that are attached *medially* — lm
   - B. Eyes found laterally on *short* stalks that attach near *lateral* extreme — sl
2. Presence of tubercles arranged in ridge on inner surface of chela (FIG. 15.191)
   - A. *Tuberculate ridge* present — tr*
   - B. Tuberculate ridge *absent* — x

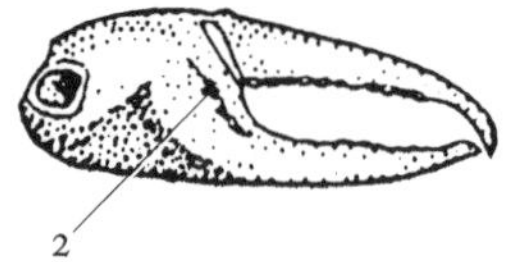

FIG. 15.191

3. Presence of reddish leg joints
   - A. Leg joints *reddish* — re
   - B. Reddish leg joints *absent* — x
4. Shape of carapace
   - A. Carapace *tapers,* posterior margin clearly shorter than anterior margin; sides curved or straight but not toothed — ta
   - B. Carapace with *three* conspicuous *teeth* along lateral margins — 3t
   - C. Carapace roughly *square;* anterior and posterior margins approximately equal; lateral margins straight — sq
5. Basic body coloring
   - A. Uniformly *dark* — dk
   - B. Uniformly *light* — lt
   - C. Light-colored with dark *blotches* — bl

15.107

Eyes	Ridge	Joints	Shape	Color	Species
lm	tr	x	ta	dk	*Uca pugnax,* Atlantic marsh fiddler†, mud fiddler
lm	tr	re	ta	bl	*U. minax,* redjointed fiddler†, brackish fiddler
lm	x	x	ta	bl	*U. pugilator,* Atlantic sand fiddler†, calicoback fiddler
sl	x	x	sq	lt	*Ocypode quadrata,* Atlantic ghost crab†
sl	x	x	sq	dk	Endings of walking legs densely hairy: *Sesarma reticulatum,* heavy marsh crab† OR Endings of walking legs with some but not dense hairs: *Armases (=Sesarma) cinereum,* squareback marsh crab†
sl	x	x	3t	bl	*Hemigrapsus sanguineus*
sl	x	x	ta	bl	*Planes minutus,* gulfweed crab†, Columbus or turtle crab

Armases (=*Sesarma*) *cinereum,* squareback marsh crab† (Grapsidae). Higher marsh, into supralittoral fringe. Ca+,Mu–SM,Li–Su,df,25 mm (1 in), (G251,NM651,wb465). (FIG. 15.192)

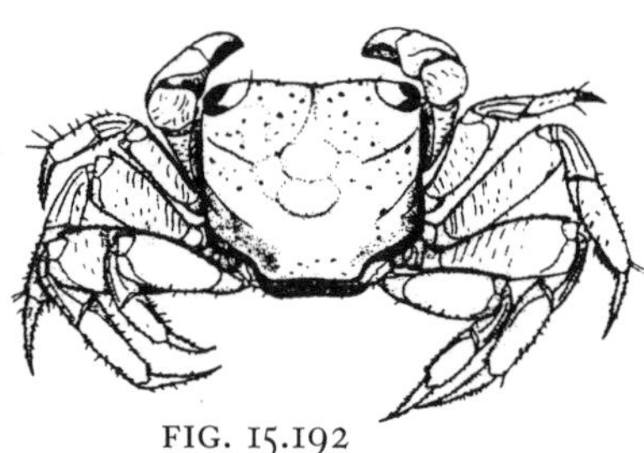
FIG. 15.192

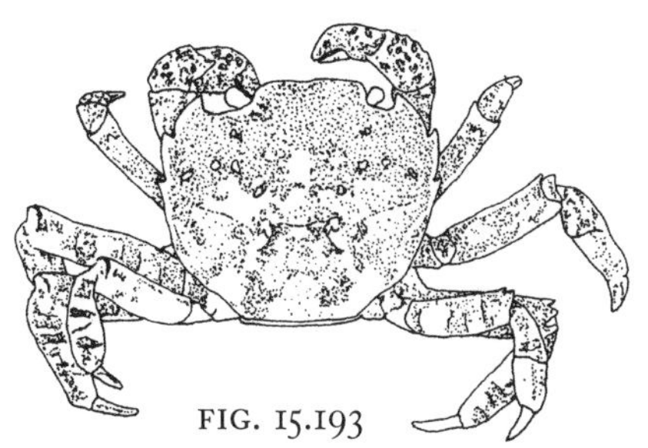
FIG. 15.193

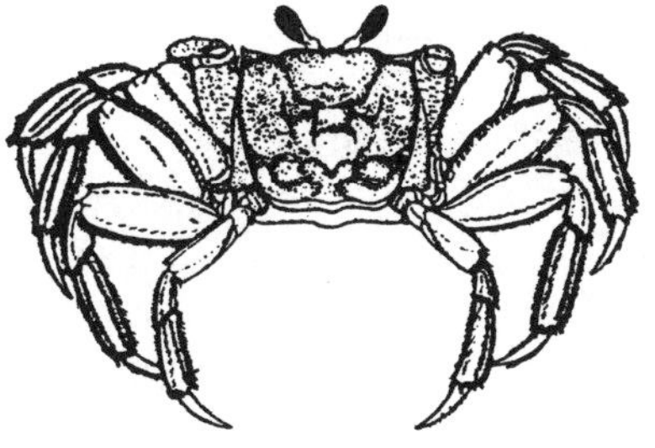
FIG. 15.194

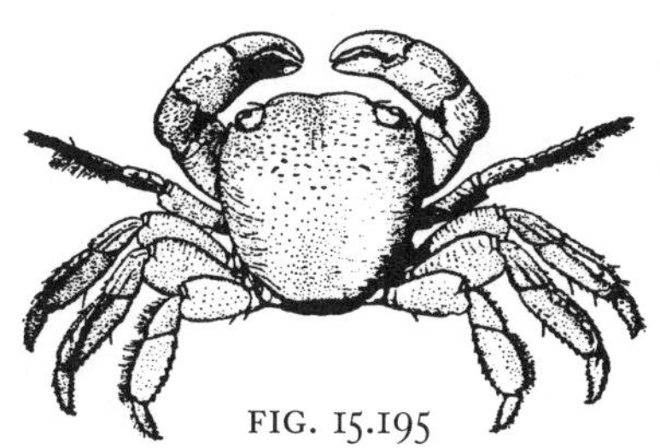
FIG. 15.195

Hemigrapsus sanguineus (Grapsidae). Banded legs; red spots on chelae. Secretive, swift, and abundant crab of upper intertidal rocks in the western Pacific; first found in New Jersey (Williams and McDermott 1990), later in some numbers with egg-bearing females (McDermott 1991). Could become significant competitor with native species and with the previously introduced *Carcinus.* Vi,Ha,Li,sc,25 mm (1 in). (FIG. 15.193)

Ocypode quadrata, Atlantic ghost crab† (Ocypodidae). Chelipeds of nearly equal size in both sexes. Nocturnal; eats mole crabs and small bivalves; fast runner on leg tips. Sand-rimmed burrows with radiating foot paths in high beach areas. Vi+,Sa,Li,sc,5.1 cm (2 in), (G251,NM653,wa44,wb468). (FIG. 15.194)

Planes minutus, gulfweed crab†, Columbus or turtle crab (Grapsidae). Light olive to reddish with white and brown spots. On floating *Sargassum* weed (not on attached species). Vi,Pe–Al,Ps,sc,12 mm (0.5 in), (G251,wa44,wb460). (FIG. 15.195)

Sesarma reticulatum, heavy marsh crab† (Grapsidae). Yellow claws. Muddy, lower marsh. Vi,SM–Mu,Li,df,3.1cm (1.2 in), (G251,NM652,wa44,wb466). (FIG. 15.196)

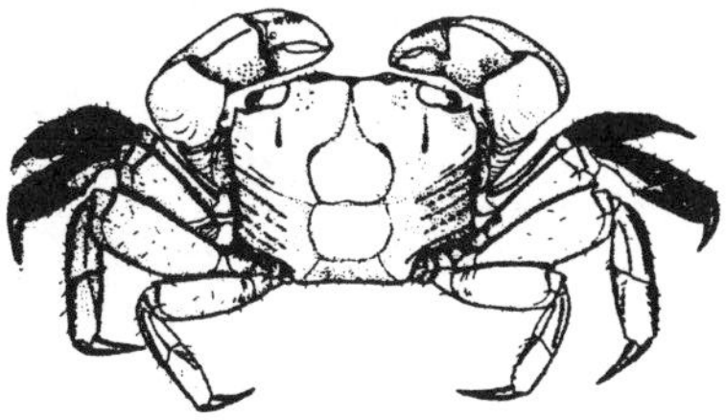

FIG. 15.196

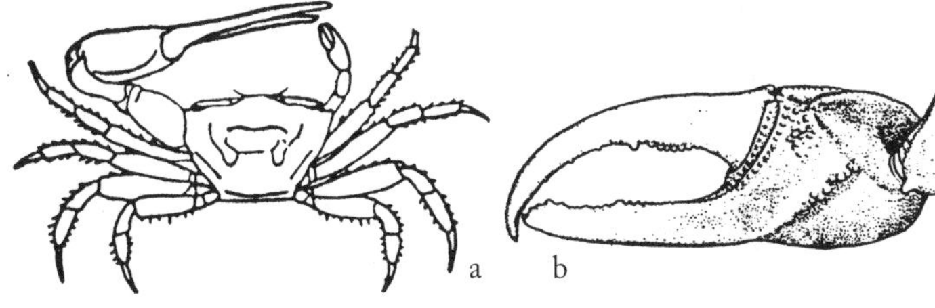
FIG. 15.197

Uca minax, redjointed fiddler†, brackish fiddler (Ocypodidae). Distance between eyestalks greater than one-third carapace width. Lower salinity, 8–20°/oo; mud. Vi,SM–Mu,Li,df,3.8 cm (1.5 in), (G252,wa44,wb473). (FIG. 15.197. a, animal; b, inside surface of chela.)

Uca pugilator, Atlantic sand fiddler†, calicoback fiddler (Ocypodidae). Distance between eyestalks less than one-third carapace width; carapace pinkish-purple. Burrows in sand or mud flats. Vi,SM–Sa,Li,df,25 mm (1 in), (G252,NM654,wa44,wb475), ☺. (FIG. 15.198. a, animal; b, inside surface of chela.)

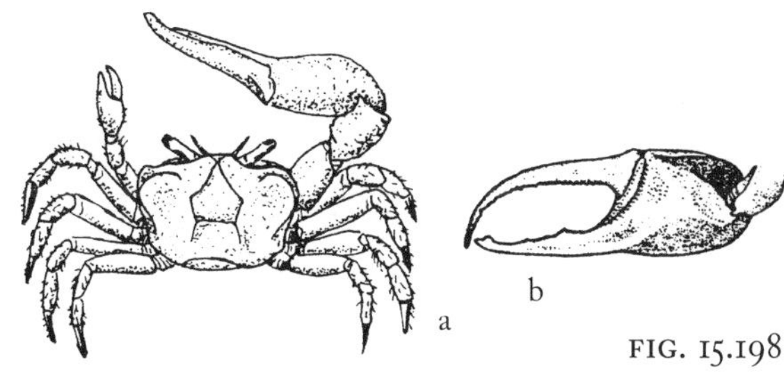
FIG. 15.198

FIG. 15.199

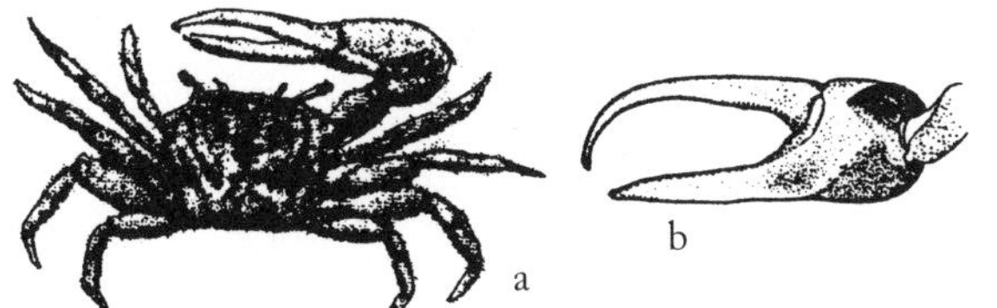

Uca pugnax, Atlantic marsh fiddler†, mud fiddler (Ocypodidae). Distance between bases of eyestalks less than one third carapace width; carapace brown with yellow. Common; burrows in muddy banks; higher salinity, 24–30°/oo. Vi,SM–Mu,Li,df,20 mm (0.8 in), (G252,NM655,wa44,wb478), ☺. (FIG. 15.199. a, animal; b, inside surface of chela.)

15.108 Crab Group 2

(Commensal Crabs [Pinnotheridae and Porcellanidae (1 species)]) (from 15.105)

Pinnotheridae and Porcellanidae are tiny crabs found living in close association with bivalves, polychaete worms, echinoderms, or the tubes of certain crustaceans. In most cases, they live commensally, causing no harm to their host. In other cases, they damage the host and thereby cross the boundary between commensalism and parasitism/predation. These crabs are Pinnotheridae unless noted otherwise.

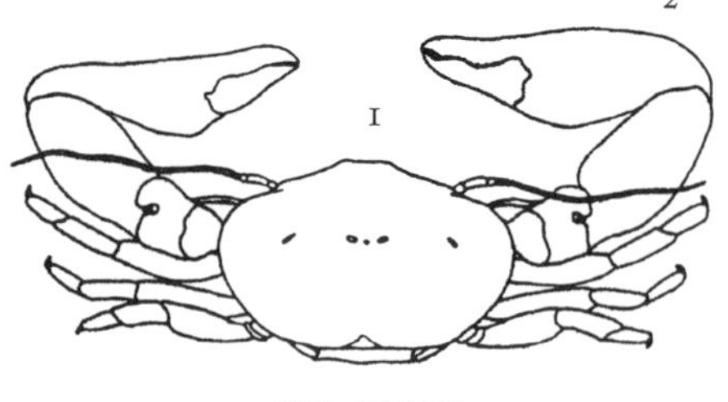

FIG. 15.200

1. Shape of carapace (FIG. 15.200)
   - A. Body more or less *round* — ro
   - B. Body barely wider than long, nearly *square* — sq
   - C. Body clearly much *wider than long* — w>l*
2. Length of cheliped or large claw-bearing arm compared to carapace length (anterior to posterior) (FIG. 15.200)
   - A. *Cheliped more than twice body* length — c2×b*
   - B. *Cheliped about equal to body* length — c=b
3. Field location
   - A. Associated with the mantle cavity of *bivalves* — bi
   - B. Associated with tubes of the polychaete *lugworm, Arenicola* — lu
   - C. Associated with tubes of the polychaete worm, *Chaetopterus* — ch
   - D. Associated with *sand dollars* — sd
   - E. Associated with mudshrimp, *Upogebia affinis* — up
   - F. Free-living in *mud* — mu

15.109

Shape	Chela	Field	Species
ro	c=b	bi	*Pinnotheres* spp., pea crabs†. Tiny crabs commensal in mantle cavity of some bivalves (e.g., *Crassostrea, Mytilus, Anomia, Argopecten*). *P. maculatus*, squatter pea crab†, mussel crab. *P. ostreum*, oyster pea crab†.
w>l	c2xb	ch	*Polyonyx gibbesi*, eastern tube crab†
w>l	c=b	ch	*Pinnixa chaetopterana*, tube pea crab†
w>l	c=b	lu	*P. cylindrica*
w>l	c=b	up	*P. retinens*
w>l	c=b	mu	*P. sayana*
sq	c=b	sd	*Dissodactylus mellitae*, sand-dollar pea crab†

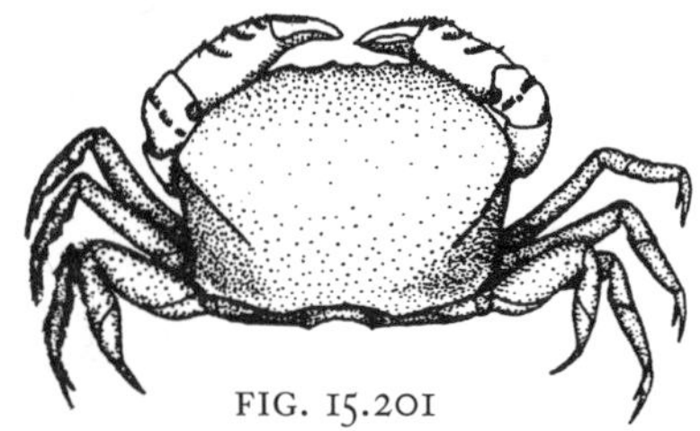
FIG. 15.201

FIG. 15.202

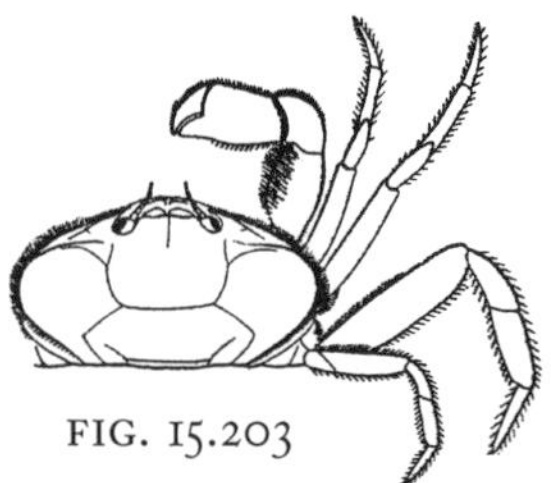
FIG. 15.203

Dissodactylus mellitae, sand-dollar pea crab†. Clings to oral surface of sand dollars, *Mellita* and *Echinarachnius*, with fork-tipped walking legs; up to 88% infestations are reported. Vi,Co,Su,pa,4 mm (0.16 in), (G250,wa44,wb439). (FIG. 15.201)

Pinnixa chaetopterana, tube pea crab†. Also found in tubes of the polychaete, *Amphitrite ornata*. Vi,Co–Mu,Su,ff,12 mm (0.5 in), (G250,wa44,wb451). (FIG. 15.202)

Pinnixa cylindrica. Observed in as many as three-fourths of lugworm tubes in some localities. Vi,Co–Mu,Li–Su,ff,20 mm (0.8 in), (G250,wa44,wb453). (FIG. 15.203)

FIG. 15.204

Pinnixa retinens. All legs and anterior margin of carapace lines with coarse hairs. In adults, basal segment of third walking leg with prominent, stout spine. Vi,Co–Mu,Su,ff,12 mm (0.5 in), (wb456). (FIG. 15.204)

Pinnixa sayana. Light colored; sparse, short hairs on legs; lacks spine on base of third walking leg. Vi,Co–Mu,Su,ff,9 mm (0.36 in), (wa44,wb457). (FIG. 15.205)

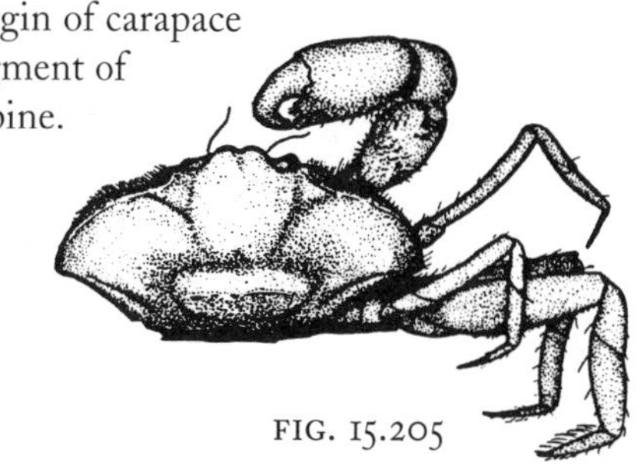
FIG. 15.205

FIG. 15.206

FIG. 15.207

Pinnotheres maculatus, squatter pea crab†, mussel crab. Most frequently with mussels or scallops. Vi,Pa,Li–Su,sc,15 mm (0.6 in), (G250,NM648,wa44,wb441). (FIG. 15.206)

Pinnotheres ostreum, oyster pea crab†. Most frequently in oyster mantle; some damage to oyster's gill. Shell soft and mobility is limited. Vi+,Co,Li–Su,sc,8 mm (0.3 in), (wa44,wb444). (FIG. 15.207)

Polyonyx gibbesi, eastern tube crab† (IO: Anomura, Porcellanidae). Gray with brown mottling. A warm-water species that appears to be spreading northward. Vi,Co,Su,ff,12 mm (0.5 in), (wa41,wb244). (FIG. 15.200)

15.110 Crab Group 3

(*Cancer* and Green Crabs [Cancridae and Portunidae, in part]) (from 15.105)

Included here are the most common intertidal and subtidal crabs of the rocky shore. These abundant crabs are major predators or scavengers in shallow-water communities.

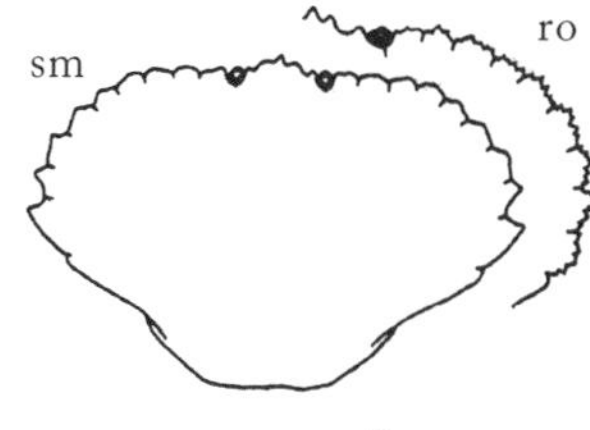

FIG. 15.208

1. Appearance of "teeth" or points between eye and lateral extreme of carapace: the number of such points (FIG. 15.208)
   - A. Each obvious tooth *smooth* edged (or no more than finely granular): provide *number* of teeth sm:__
   - B. Each obvious tooth distinctly *rough* edged: provide *number* of teeth ro:__
2. Shape of carapace; shape index = width/length
   - A. Shape index *less than 1.2* <1.2
   - B. Shape index *greater than 1.2* >1.2

15.111

Teeth	Shape	Species
sm:5	<1.2	*Carcinus maenas,* green crab†
sm:9	>1.2	*Cancer irroratus,* Atlantic rock crab†
ro:9	>1.2	*C. borealis,* Jonah crab†

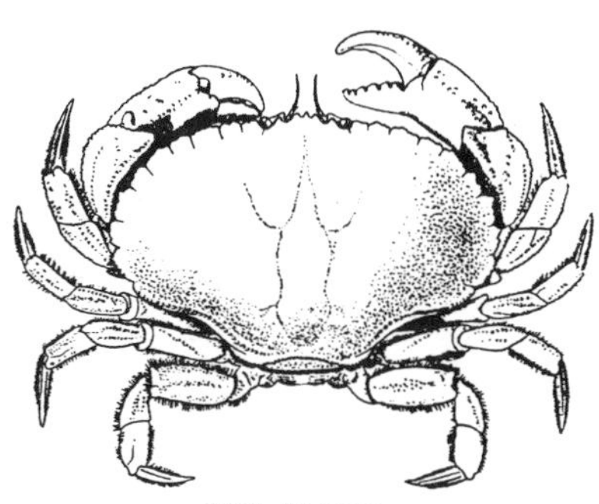
FIG. 15.209

Cancer borealis, Jonah crab† (Cancridae). Carapace granular; purplish brown. Prefers rocky bottom; eats *Mytilus,* ascidians, etc. Bo,Sa–Mu,Li–De,pr,16 cm (6.3 in), (G246,wa43,wb351). (FIG. 15.209)

Cancer irroratus, Atlantic rock crab† (Cancridae). Carapace smooth; reddish. Prefers sandy bottoms; eats bivalves, polychaetes, echinoderms. Smaller nearshore, larger in deeper water. Bo,Sa–Mu–Ha,Li–De,pr,14 cm (5.5 in), (G246,wa43,wb353), ☺. (FIG. 15.210)

FIG. 15.210

Carcinus maenas, green crab† (Portunidae). Green, reddish orange, or tan but with darker mottling. Color varies, but those on algae with more mottled patterns. Very common; under rocks and algae, tide pools, but also marshes, seagrass beds. Introduced species. Eats other invertebrates. A portunid or swimming crab, although it lacks paddle-shaped hindmost feet. Ac–,Ha–Al,Li+,sc,7.6 cm (3 in), (G247,WA42,WB356), ☺. (FIG. 15.211)

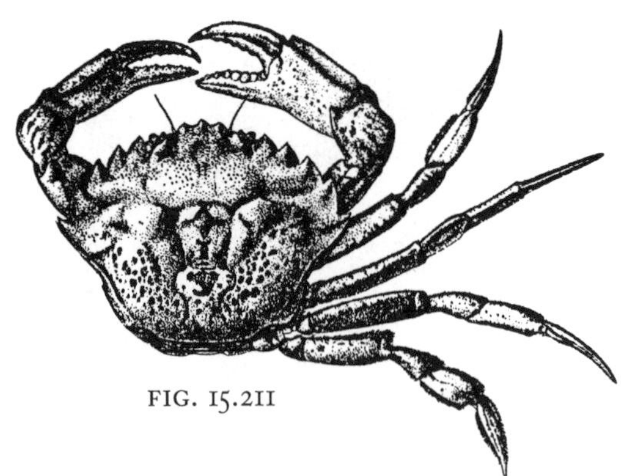

FIG. 15.211

15.112 Crab Group 4

(Mud Crabs [Xanthidae]) (from 15.105)

These small crabs are found in soft substrates. Their typically black-tipped chelae (largest pinching claws) are strong enough to crush small bivalves and barnacles for food.

FIG. 15.212

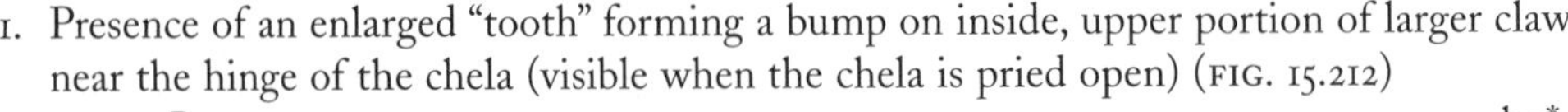

1. Presence of an enlarged "tooth" forming a bump on inside, upper portion of larger claw near the hinge of the chela (visible when the chela is pried open) (FIG. 15.212)
   - A. *Bump* present — bu*
   - B. Bump *absent* — x
2. Comparative sizes of large pinching claws (chelae)
   - A. Large chelae approximately *equal* in size — eq
   - B. Chelae *unequal;* one chela (usually the right) clearly larger than the other (usually the left) — un
3. Presence and size of red spot on inside of third maxilliped (place crab on its back; pry back rectangular shaped mouth appendages and look on inside surface)
   - A. *Large red spot* present, one-quarter the size of the segment it is on — lrs
   - B. *Small red spot* present, occupies less than one-sixth of its segment — srs
   - C. Red spot *absent* — x
4. Extent of dark coloring on lower portion (mandible) of chela (FIG. 15.211)
   - A. Black confined to pointed *tip* only — ti
   - B. Black extends onto *body* of chela also — bo*
   - C. *No* black tip on claws — x

15.113

Bump	Sizes	Spot	Black	Species
bu	un	x	ti	*Hexapanopeus angustifrons,* smooth mud crab†, groove-wrist mud crab
bu	un	srs	ti,bo	*Panopeus herbstii,* Atlantic mud crab†, black-finger or narrow mud crab
x	un	lrs	bo	*Eurypanopeus depressus,* flatbacked mud crab†
x	eq	x,srs	bo	*Dyspanopeus* (=*Neopanope*) *sayi* (=*texana sayi*), Say mud crab†, equal-clawed mud crab
x	eq	x	x	Possibly fiddler crab, Crab Group 1, 15.106
x	un	x	x	*Rhithropanopeus harrisii,* Harris mud crab†, white-finger mud crab

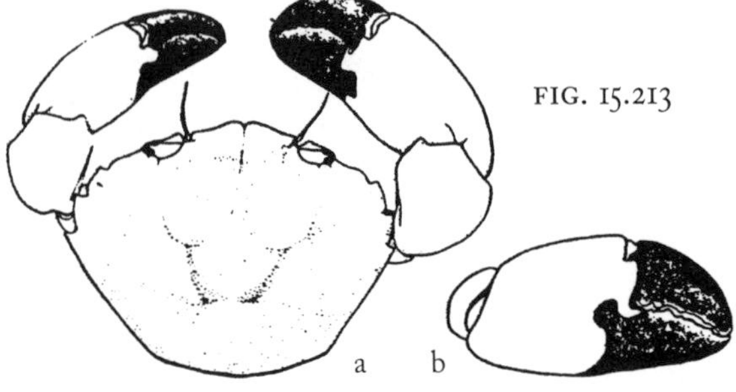

FIG. 15.213

Dyspanopeus (=*Neopanope*) *sayi* (=*texana sayi*), Say mud crab†, equal-clawed mud crab. Few have tiny red spot. Tolerates salinity down to 12°/oo. Preys on clam spat (recently settled larvae) and barnacles. Bo,Mu–SG,Es–Su,pr,24 mm (0.95 in), (G249,NM647,WA43,WB409), ☺. (FIG. 15.213. a, carapace and chelipeds; b, right chela.)

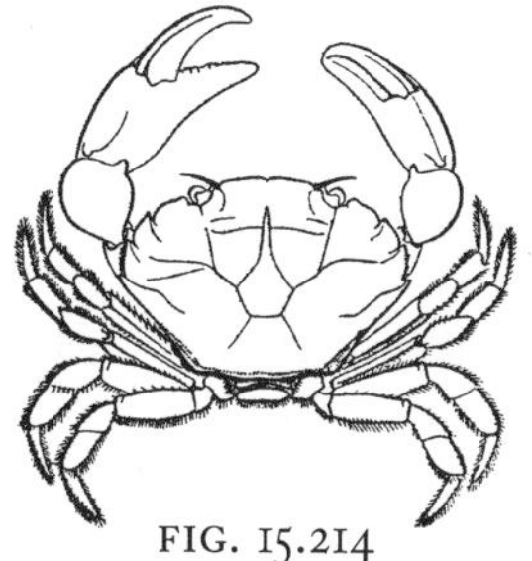
FIG. 15.214

Eurypanopeus depressus, flatbacked mud crab†. Often bluish around mouth. Eats oyster spat (recently settled larvae). Movable portion of smaller chela is scoop-shaped. Vi+,Mu,Es–Su,pr,25 mm (1 in), (G249,wa43,wb408). (FIG. 15.214)

Hexapanopeus angustifrons, smooth mud crab†, groove-wrist mud crab. Has groove on inside of joint to which large claw is attached. Generally dark colored. Prefers salinity 20–33°/oo. Vi,Mu–Sa,Su,pr,3 cm (1.2 in), (G249,wb415). (FIG. 15.215. a, animal; b, right chela.)

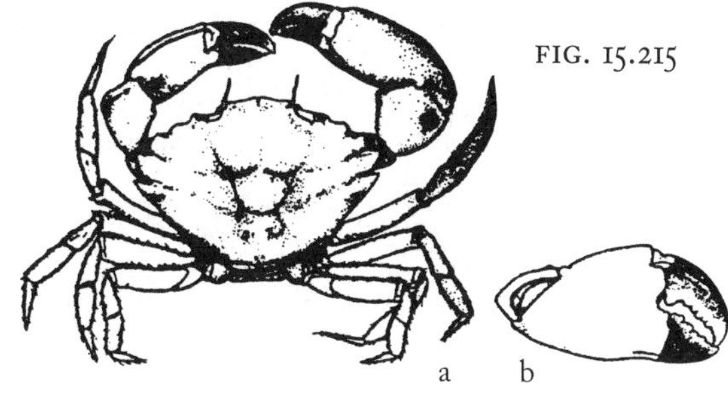

FIG. 15.215

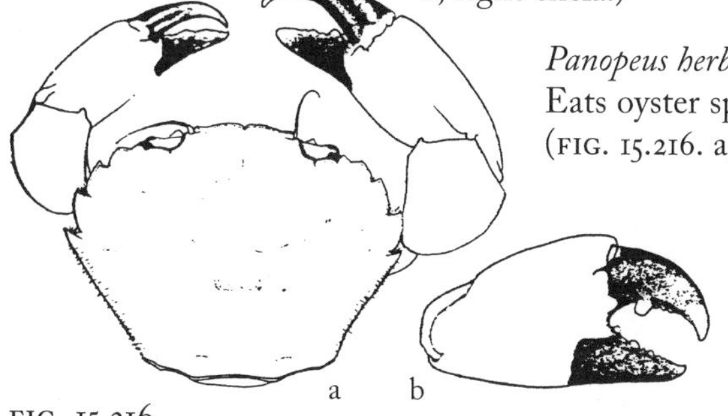

FIG. 15.216

Panopeus herbstii, Atlantic mud crab†, black-finger or narrow mud crab. Eats oyster spat. Vi+,Mu,Es–Li–Su,pr,6.4 cm (2.5 in), (G249,wb412), ☺. (FIG. 15.216. a, animal; b, right chela.)

Rhithropanopeus harrisii, Harris mud crab†, white-finger mud crab. Tolerates estuarine streams, salinity 0–19°/oo. Bo,Mu,Es–Su,om,20 mm (0.8 in), (wa43,wb401). (FIG. 15.217)

FIG. 15.217

15.114 Crab Group 5

(Swimming Crabs [Portunidae, in part]) (from 15.105)

The paddle-shaped hindmost legs of swimming crabs give them great mobility through the water. Be careful; they can give a painful pinch!

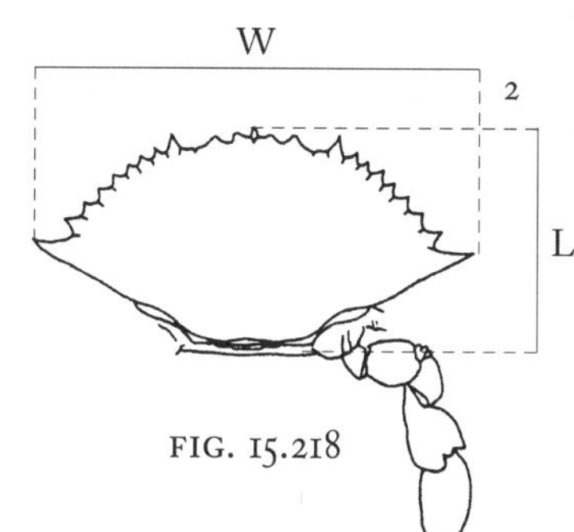

FIG. 15.218

1. Number of teeth along front of carapace between the eyes: provide *number* __
2. Shape of carapace; shape index = width (including lateral points)/length (FIG. 15.218)
   - A. Shape index equal to or *greater than 2* — >2*
   - B. Shape index *less than 2* — <2
3. Bluish tinge to at least some legs
   - A. *Bluish* tinge present — bl
   - B. Bluish tinge *absent* — x
4. Basic body coloring
   - A. *Light* with *darker* spots — lt/dk
   - B. *Dark* with *light* spots — dk/lt
   - C. *Light* with *yellow brown* blotches — lt/yb
   - D. Rather uniformly *dark* — dk

15.115

Teeth	Shape	Blue	Color	Species
3	<2	x	lt/dk	*Ovalipes ocellatus,* lady crab†, calico or oscellated crab
4	>2	bl	dk	*Callinectes sapidus,* blue crab†, hardshell or softshell crab
6	>2	bl	dk	*C. similis,* lesser blue crab†
4–6	>2	x	lt/yb	*Portunus sayi,* sargassum swimming crab†
7–8	>2	x	dk	*P. gibbesii,* iridescent crab†
6–8	>2	x	dk/lt	*Arenaeus cribrarius,* speckled swimming crab†

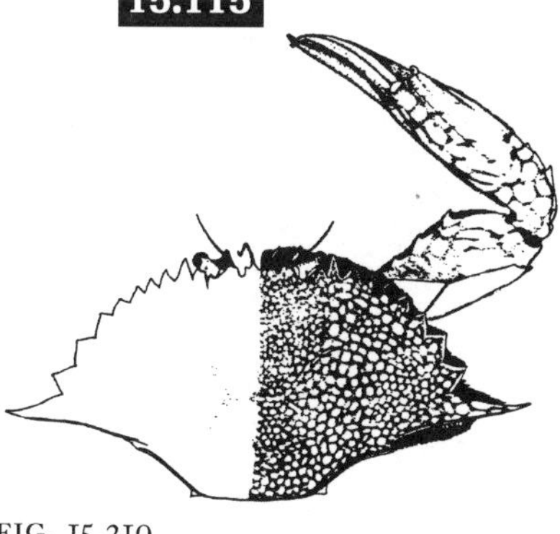
FIG. 15.219

Arenaeus cribrarius, speckled swimming crab†. Outer beaches; buries completely in sand. Vi,Sa,Su,pr,11.4 cm (4.5 in), (G248,wb362). (FIG. 15.219. Carapace and cheliped.)

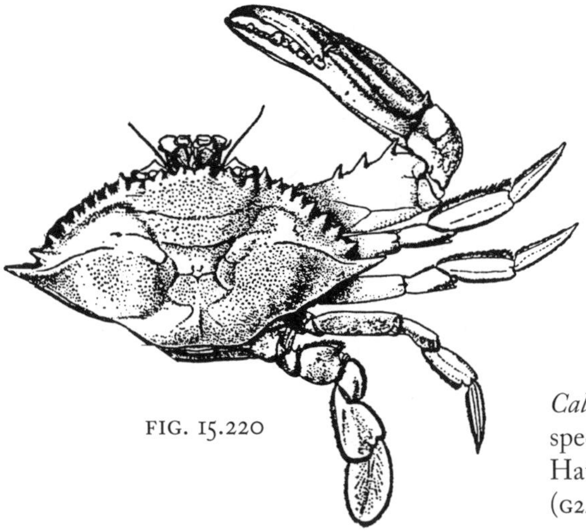

FIG. 15.220

Callinectes sapidus, blue crab†, hardshell or softshell crab. With eight marginal teeth and one large spine. Important commercial soft-shell crab. Aggressive: they pinch hard! Burrow into sand. Sex determination by abdomen shape: triangular shape, immature; slender, male; rounded, female. Peeler, about to molt; buster, in process of molting; soft-shell, post molting, before new shell hardens. Males stay in lower salinities; females migrate toward full seawater for egg laying. Vi+,Sa,Su–Es,sc,22.8 cm (9 in), (G247,NM639,WA43,WB376), ☺. (FIG. 15.220)

Callinectes similis, lesser blue crab†. A southern species; deeper water only north of Cape Hatteras. Ca+,Sa,Su–Es,pr,12.7 cm (5 in), (G248,WA43,WB383).

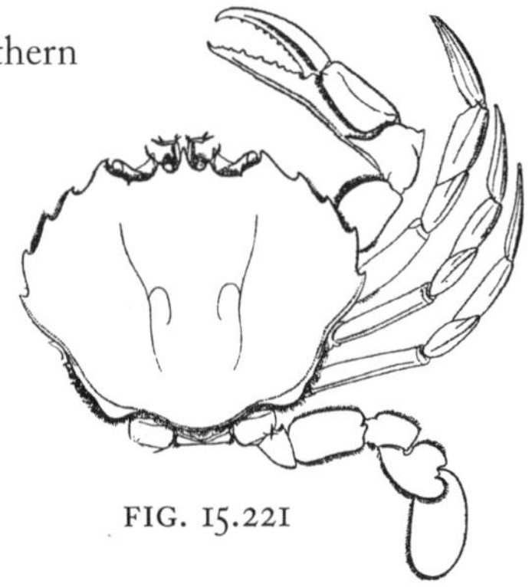

FIG. 15.221

Ovalipes ocellatus, lady crab†, calico or oscellated crab. Light purplish background with darker spots. Three to five teeth between eyes and lateral extremes. Lacks elongated lateral points. Buries completely in sand. Bo,Sa,Li–Sa,pr,8.9 cm (3.5 in), (G248,NM638,WA43,WB359), ☺. (FIG. 15.221)

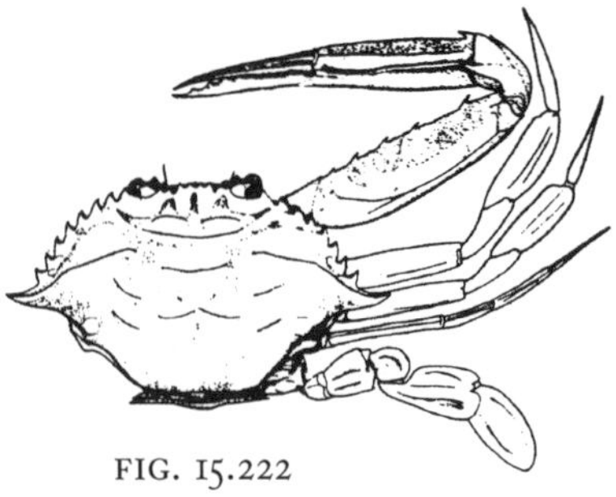

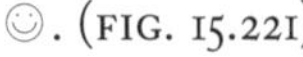

FIG. 15.222

Portunus gibbesii, iridescent crab†. Reddish brown with bright red spines, often with iridescent legs. Four to seven spines along front of first section of chelipeds. Vi,Sa,Su,pr,7.6 cm (3 in), (G248,NM639,WB389). (FIG. 15.222)

Portunus sayi, sargassum swimming crab†. On floating (not attached) alga, *Sargassum.* Bo,Al,Ps,pr,6.4 cm (2.5 in), (G248,NM638,WA43,WB391). (FIG. 15.223)

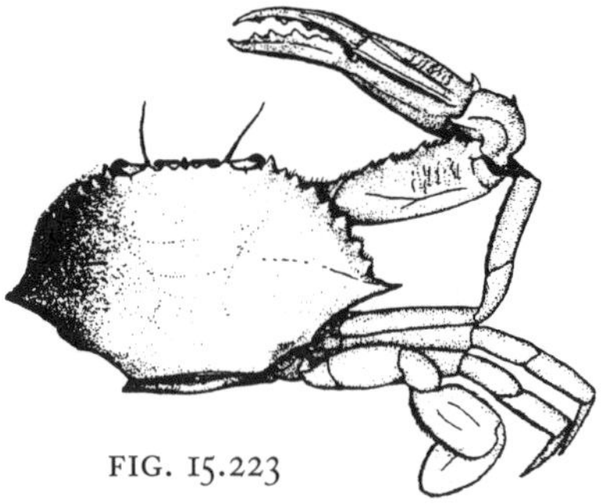

FIG. 15.223

15.116 Crab Group 6

(Spider Crabs [Majidae] and Stone Crabs [IO: Anomura, Lithodidae]) (from 15.105)

Spider crabs and stone crabs are long-legged crabs typical of nearshore and offshore fine particulate sediments. They appear to be a favored food for gulls because their carcasses often litter the shoreline. Some lyre crabs are "decorators," encouraging the settling of a dorsal covering of hydroids, sponges, and algae. This mutualistic interaction provides camouflage for the crab and mobility to the "decorations." All species are included in Majidae unless noted otherwise.

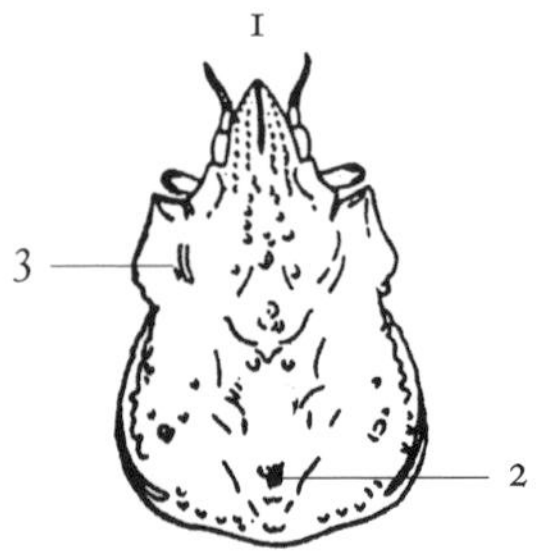

FIG. 15.224

1. Shape of rostrum (FIG. 15.224)
   - A. Rostrum *bifid* (split), with two parts *fused* medially along their entire length — bf
   - B. Rostrum *bifid,* two parts separate but rejoin at *tips* — bt*
   - C. Rostrum *bifid,* two points separated for *less than half* the length of the rostrum — b<½
   - D. Rostrum *bifid,* two points separated for *more than half* the length of the rostrum — b>½
   - E. Rostrum with *bifid* tip plus *two* more midlateral points — b+2
   - F. Rostrum *pointed* with two pairs small lateralteeth — po
2. Number of major spines along the middorsal line of the carapace (FIG. 15.224)
   - A. Provide *number* of major spines or tubercles along the middorsal line — __
   - B. Either spines are *absent* or they fail to form a single recognizable middorsal line — x*

3. Features of lateral margin of carapace (FIG. 15.224)
   - A. Margin with sharp *spines* — sp
   - B. Margin smooth with distinct lateral *shelf* or plate located behind the eye making overall "violin-shaped" appearance to carapace — sh
   - C. Lateral margin smooth and with lateral shelf *absent* — x
4. Number of pairs of visible walking legs, including cheliped: provide *number* or pairs — __

15.117

Rostrum	Spines	Lateral	Legs	Species
bt	x	sh	5	*Hyas coarctatus,* arctic lyre crab†, lesser toadcrab
bf	x	x	5	*H. araneus,* Atlantic lyre crab†, toad crab
po	2	sh	5	*Euprognatha rastellifera,* spider crab
b+2	x	sp	4	*Lithodes maja,* northern stone crab
b>½	5–6	sp	5	*Rochinia crassa,* inflated spiny crab†
b<½	x	sp,x	5	*Chionoecetes opilio,* snow crab†, queen crab
b<½	6	sp	5	*Libinia dubia,* longnose spider crab†
b<½	9	sp	5	*L. emarginata,* portly spider crab†
b<½	x	x	5	*Pelia mutica,* cryptic teardrop crab†, red-spotspider crab

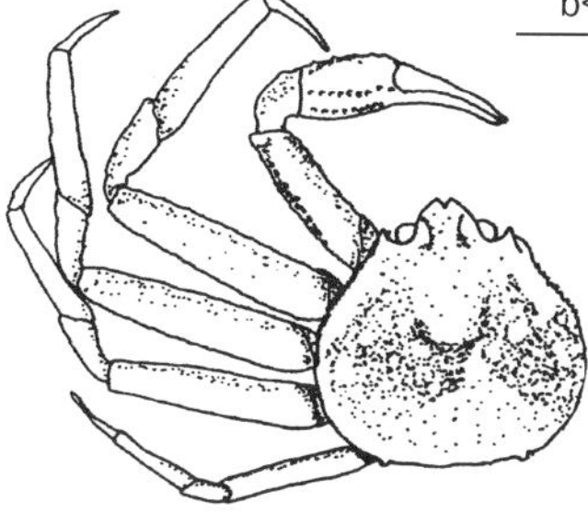

FIG. 15.225

Chionoecetes opilio, snow crab†, queen crab. Brownish to reddish above; yellowish below. Important commercial species in Canada. Ac,Mu–Sa,Su–De,pr,16 cm (6.3 in), (wb307). (FIG. 15.225)

Euprognatha rastellifera, spider crab. Vi,Sa–Mu,Su–De,om,12 mm (0.5 in), (wa42,wb298). (FIG. 15.226)

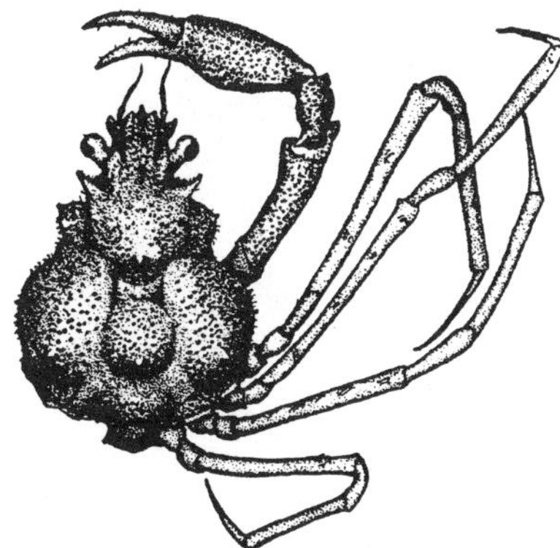

FIG. 15.226

Hyas araneus, Atlantic lyre crab†, toad crab. Dull purplish red. More common on soft substrates. Carapace often camouflaged with algae and sessile invertebrates. Ac–,Sa–Ha–Gr,Li–Su,om,9.6 cm (3.8 in), (G245,NM656,wa42,wb309). (FIG. 15.227)

Hyas coarctatus, arctic lyre crab†, lesser toad crab. Dull purplish red. More common on harder substrates. Eats algae, polychaetes. Bo,Sa–Ha,Su,om,3.3 cm (1.3 in), (G245,NM657,wa42,wb309), ☺. (FIG. 15.224. Ccarapace.)

Libinia dubia, longnose spider crab†. Especially in Chesapeake Bay. Juveniles camouflage themselves with bits of sponge. Tolerates low oxygen. Vi+,Mu–Sa–Gr,Es–Su,pr,10.1 cm (4 in), (G243,NM657,wa42,wb316), ☺. (FIG. 15.228)

Libinia emarginata, portly spider crab†. Dull yellowish brown. Juveniles use sponge pieces for camouflage. Bo,Mu–Sa–Gr,Su,pr,10.1 cm (4 in), (G246,NM657,wa42,wb318). (FIG. 15.229)

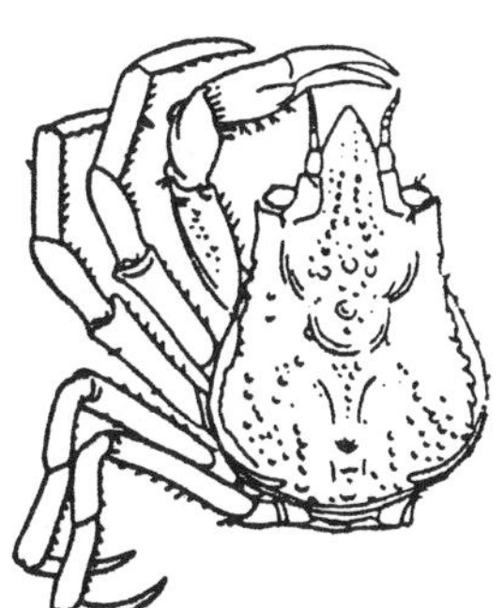

FIG. 15.227

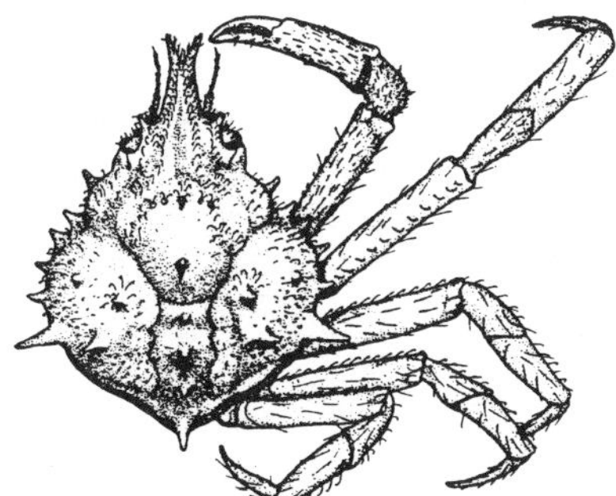

FIG. 15.228

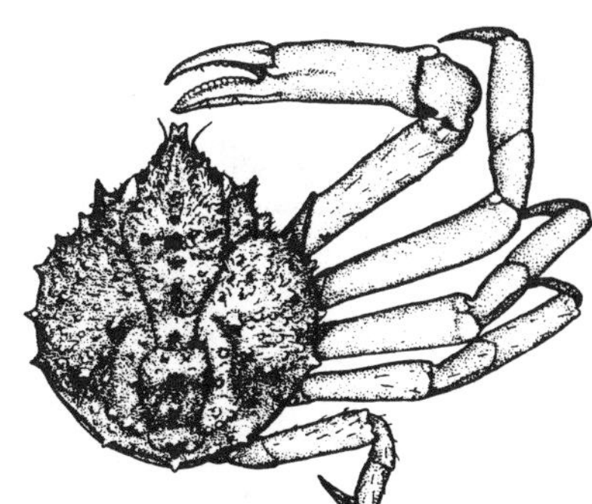

FIG. 15.229

FIG. 15.230

Lithodes maja, northern stone crab (IO: Anomura, Lithodidae). Anomuran crab with fifth walking legs small and tucked into gill chamber. Light brown, purplish to yellowish red. Ac–,Ha–Sa,Su,sc,17.8 cm (7 in), (G243,WA41,WB230). (FIG. 15.230)

Pelia mutica, cryptic teardrop crab†, red-spot spider crab. Often sponge covered; with hydroids. Bright red blotches and bands. Vi,Gr,Su,sc,12 mm (0.5 in), (G245,WA42,WB321). (FIG. 15.231)

FIG. 15.231

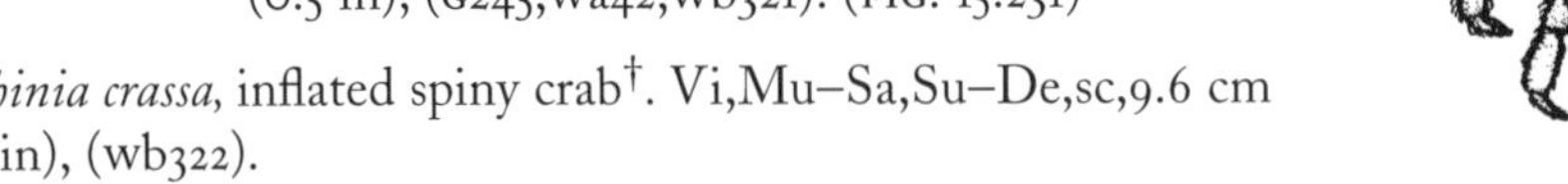

Rochinia crassa, inflated spiny crab†. Vi,Mu–Sa,Su–De,sc,9.6 cm (3.8 in), (WB322).

15.118 Crab Group 7

(Hermit Crabs [IO: Anomura, Paguridae]) (from 15.9, 15.105)

The fifth pair of walking legs curl over the hermit crab's back and hold onto the empty gastropod shells that the crab uses for protection. To remove a clinging hermit crab from its shell, hold the shell so that the crab's legs just contact a dish of seawater. Gently heat the spire of the shell with a match or lighter; the crab will soon voluntarily abandon the heated shell for the cool seawater.

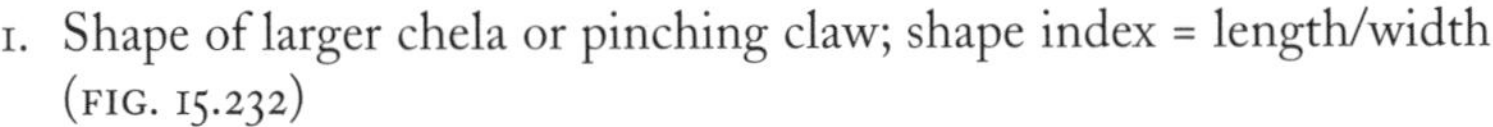

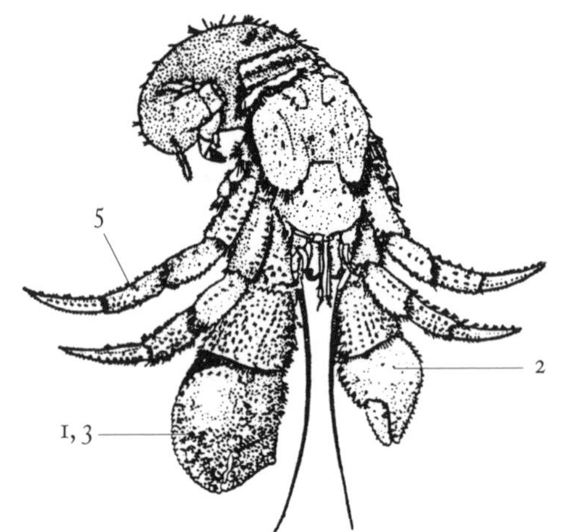

FIG. 15.232

1. Shape of larger chela or pinching claw; shape index = length/width (FIG. 15.232)
   - A. Shape index *greater than 2* — >2
   - B. Shape index *less than 1.6* — <1.6*
   - C. Shape index *intermediate,* between 1.6 and 2 — in
2. Contour of front surface of smaller chela (FIG. 15.232)
   - A. Basically *flat* — fl†
   - B. Distinct mid*ridge* gives triangular appearance — ri
   - C. *Rounded* — ro
3. Distinctive markings of larger chela (FIG. 15.232)
   - A. Chela basically *uniformly* colored (i.e., red-brown, and with *tubercles*) — ut*
   - B. Chela basically *uniformly* colored, tubercles *absent* — ux
   - C. Chela light with *single* darker *stripe* — 1s
   - D. Chela light with *four to five* darker *stripes* — 4–5s
4. Compare lengths of small spine on second antennae (located laterally to eyestalks) and eyestalk
   - A. Antennal *spine* clearly *longer than eyestalk* — s>e
   - B. Antennal *spine* about *equal to eyestalk* — s=e
   - C. Antennal *spine* distinctly *shorter than eyestalk* — s<e
5. Hairiness of legs (FIG. 15.232)
   - A. Legs conspicuously *hairy* — ha
   - B. Legs *not* conspicuously hairy — x*

15.119

Shape	Surface	Marks	Spine:Eyestalk	Hairs	Species
>2	ro	ux,1s	s=e,s>e	x	*Pagurus longicarpus,* longwrist hermit†, longclaw hermit
>2	ro	4–5s	s<e	x	*P. annulipes,* banded hermit
<1.6	fl	ut	s=e	x	*P. pollicaris,* flatclaw hermit†
in	ro	ut,1s	s>e	x	*P. acadianus,* Acadian hermit
in	ri	ut	s=e	ha	*P. arcuatus,* hairy hermit
in	ri	ut	s>e	ha	*P. pubescens*

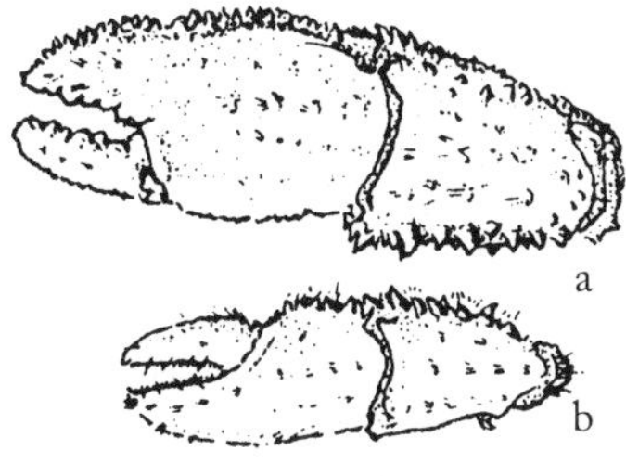

FIG. 15.233

Pagurus acadianus, Acadian hermit. Tide pools; north of Cape Cod. Preferred shells: *Buccinum, Nucella, Littorina,* or *Nassarius.* Ac–,Sa–Ha,Su,om,3.3 cm (1.3 in), (G242,NM629,wa41,wb209). (FIG. 15.233. a, larger chela; b, smaller chela.)

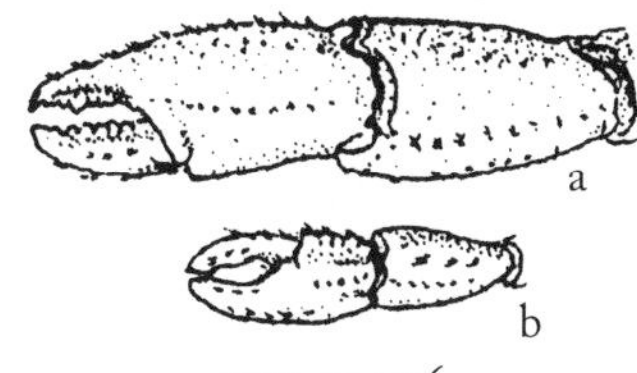

FIG. 15.236

Pagurus annulipes, banded hermit. Small, subtidal species. Shell usually *Nassarius, Littorina,* or *Costoanachis.* Bands on claws and legs. Vi,Sa–Gr,Su,om,7.6 mm (0.3 in), (G241,wa41,wa210), ☺. (FIG. 15.234. a, larger chela; b, smaller chela.)

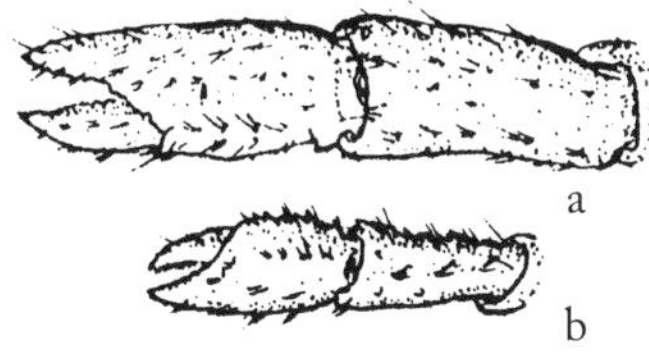

FIG. 15.234

Pagurus arcuatus, hairy hermit. Dense, coarse hairs on chelae. Deep water form. Bo,Ha–Gr–Sa,Su,om,3.3 cm (1.3 in), (G242,NM630,wa41,wb212). (FIG. 15.235. a, larger chela; b, smaller chela.)

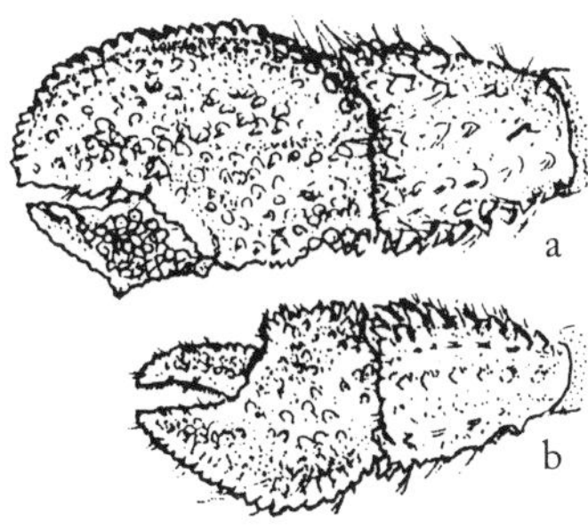

FIG. 15.237

Pagurus longicarpus, longwrist hermit[†], longclaw hermit. Small, very common species. Shell usually *Littorina* spp., *Ilyanassa obsoleta,* or *Urosalpinx cinerea.* Vi+,Mu–Sa,Li–Su,om,8 mm (0.3 in), (G241,wa41,wb216), ☺. (FIG. 15.236. a, larger chela; b, smaller chela.)

Pagurus pollicaris, flatclaw hermit[†]. Short, sharp spines on chelae. Large, common offshore. Shells usually moon snails (Naticidae) or whelks (Buccinidae). Vi+,Sa–Gr,Es–Su,om,3.3 cm (1.3 in), (G242,NM631,wa41,wb220), ☺. (FIG. 15.237. a, larger chela; b, smaller chela.)

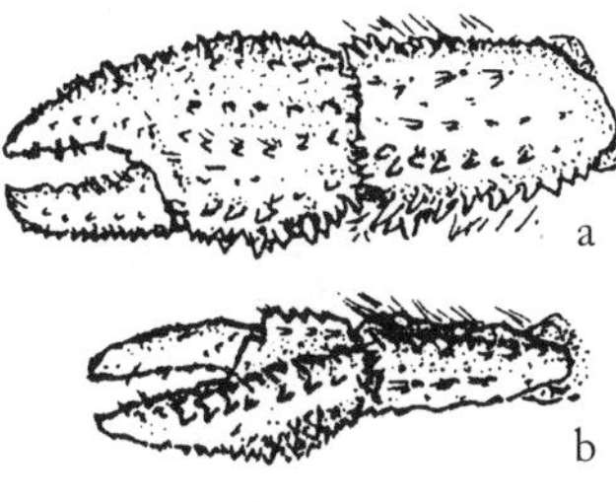

FIG. 15.235

Pagurus pubescens. Fine, yellowish hairs on chelae. Cold-water species; in deeper water toward southern end of range. Bo,Ha–Gr,Su,om,20 mm (0.8 in), (G242,wa41,wb222). (FIG. 15.238. a, larger chela; b, smaller chela.)

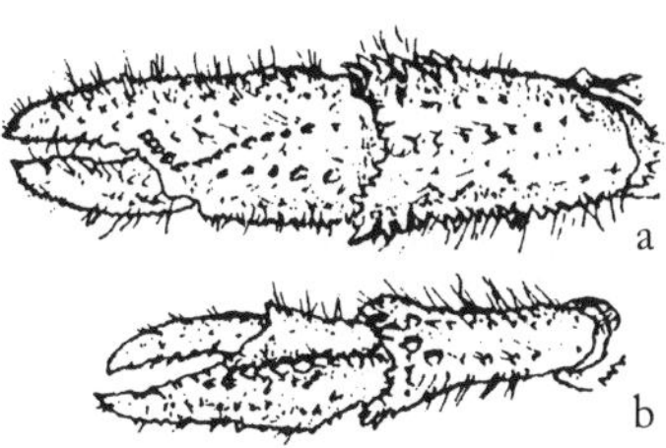

FIG. 15.238

Chapter 16 ■ Phylum Echinodermata

16.1 External body features of most echinoderms are arranged in *pentaradial symmetry* (in five similar fields around a common center). The body surface including the mouth is termed the *oral* surface; the opposite side is *aboral.*

Star-shaped echinoderms may be separated into sea stars of the class Asteroidea (wide, hollow rays abut one another directly around a circular *central body disk;* 16.2) or brittle stars of the class Ophiuroidea (slender, solid rays attached to a central disk leaving obvious spaces between their bases; 16.8). The class Echinoidea (16.14), which includes sea urchins and sand dollars, are characterized by rigid, subspherical, or discoid bodies covered with spines. Sea cucumbers, in the class Holothuroidea (16.12), have elongate, muscular bodies and branched tentacles for feeding.

Echinoderm internal skeletons, comprised of calcareous units, called *ossicles,* range from the fused, rigid, globular *tests* of sea urchins, through the firm but flexible meshwork skeletons of many sea stars, to the vertebral columnlike arms of brittle stars, and finally to the isolated skeletal *plates* imbedded in the flexible, muscular wall of sea cucumbers. Their unique *water vascular system* includes fluid-filled *tube feet* as functional units. Tube feet are the primary locomotory devices for sea stars but are used by the other classes primarily for attachment, food gathering, or sensation. The inability of echinoderms in general to cope with dilute surroundings limits the members of this phylum to full marine conditions, with very few species capable of penetrating into estuarine conditions.

Only the classes mentioned above are treated here. Members of the class Crinoidea (sea lilies or feather stars) and the newly described class Concentricycloidea (sea daisies) are encountered too infrequently to be included.

16.2 Class Asteroidea, Sea Stars

The star-shaped body of the sea star or starfish is familiar to everyone. Using hundreds of tube feet for locomotion, the moving star appears to glide along the substrate when viewed from the aboral (uppermost) surface. *Ambulacral grooves,* which house the tube feet, are conspicuous features of the oral surface (the surface including the mouth), extending like gutters

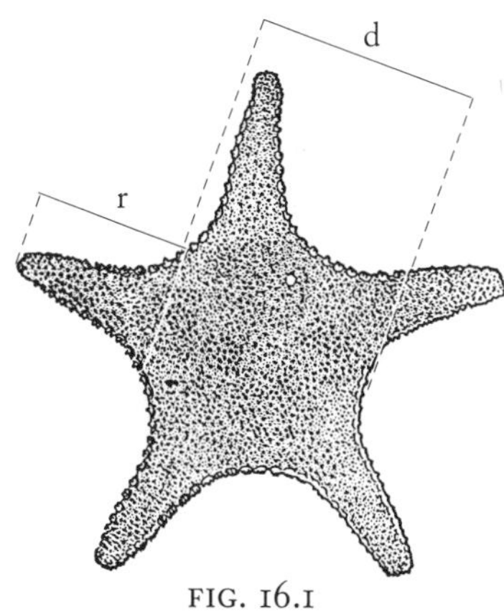

FIG. 16.1

along each ray. The surface of most sea stars is studded with calcareous *spines* and often includes small pinching *pedicellariae* as well. Thin-walled, fingerlike *dermal branchiae* (literally, "skin gills") extend from the aboral surface to assist in respiratory exchange. Species distinctions are based on overall shape, the appearance of *marginal plates* lining each ray, construction of their body spines, and the color of their bodies and of features such as the aboral *madreporite* or sieve-plate entrance into the water vascular system. Length is given as radius, measured from the center of the body to the tip of a ray.

Additional reference: (c) Clark, A. M. and M. E. Downey, 1992.

1. The length of ray or arm (from its junction with the central disc to its tip) compared to the diameter of the central disc (between opposite arm bases) (FIG. 16.1)
   - A. *Ray much longer than disc* diameter — ra>di
   - B. *Ray equal to or shorter than disc* diameter — ra<di*

16.3

Ray to Disc	Group
ra>di	Long-rayed sea stars, 16.4
ra<di	Short-rayed sea stars, 16.6

16.4 **Long-Rayed Sea Stars** (from 16.3)

Rays longer than central disc.

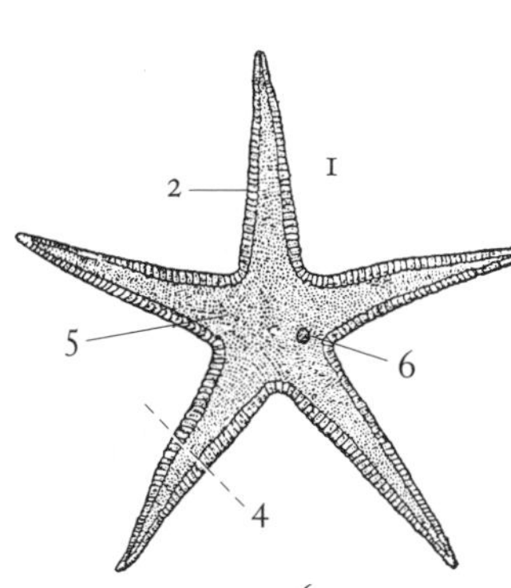

FIG. 16.2

1. Number of rays or arms (FIG. 16.2)
   - A. *Five* or fewer; includes the few instances of six arms — 5*
   - B. *More than six* — >6
2. Distinct marginal or lateral plates bordering rays, clearly visible from above (FIG. 16.2)
   - A. Marginal *plates* present — pl*
   - B. Marginal plates *absent* — x
3. Alignment of rows of tube feet running lengthwise within the ambulacral groove along the undersurface of each ray
   - A. Tube feet form *two* regular rows — 2
   - B. Tube foot placement more complex, does *not* form two distinctly regular rows — x
4. Shape of imaginary cross-section taken midway along a ray or arm (FIG. 16.2)
   - A. Ray nearly *cylindrical* in cross-section — cy
   - B. Ray *ovaloid,* clearly wider than high in cross-section — ov*
5. Spines on aboral surface (observe individual spines with dissecting microscope magnification) (FIG. 16.2)
   - A. Scattered spines, *short* and *multi*tipped (paxillae), aboral surface feels smooth — sm
   - B. Scattered spines, *large* and *single*-tipped, aboral surface feels rough — ls
   - C. Scattered spines, *large* and *multi*tipped (paxillae), aboral surface feels rough — lm
6. Color of madreporite (round spot visible on aboral surface; this is a comparatively weak character because coloring may vary or fall between these categories) (FIG. 16.2)
   - A. *Whitish* — wh
   - B. *Yellowish* — ye
   - C. *Orangish* — or
7. Location of pedicellaria (small, jawed or pinching structures) relative to spines on aboral surface (upper body) (dissecting microscope)
   - A. Pedicellaria form a ring around the *base* of the spines — ba
   - B. Pedicellaria form a ring about *halfway* up the shaft of spines — ha
   - C. Pedicellaria *scattered* on the aboral surface — sc
   - D. Pedicellaria *absent* — x

16.5

No. Rays	Marginal Plates	Tube Foot Rows	Cross-Section	Spine	Madreporite Color	Pedicell	Species
5	x	x	ov	ls	ye	ba	*Asterias rubens* (=*vulgaris*), northern sea star
5	x	x	ov	ls	or	ba	*A. forbesi,* common sea star
5–6	x	x	cy	ls	wh	ha	*Leptasterias* Body color pale; pinkish or purplish: *L. tenera,* slender-armed star OR Body greenish drab: *L. littoralis,* polar or green slender sea star.
5	x	2	cy	sm	wh,ye	x	*Henricia sanguinolenta,* bloodstar
5	x,pl	2	cy	sm	wh	x	*Luidia clathrata,* striped sea star
5	pl	2	ov	sm	ye,wh	x	*Astropecten articulatus,* armored sea star
5	pl	2	ov	lm	ye,wh	sc	*A. americanus*
>6	x,pl	2	ov	lm	ye,wh	x	*Crossaster papposus,* rough, spiny, or spring sun star
>6	x	2	cy,ov	sm	ye,or	x	*Solaster endeca,* smooth or purple sun star

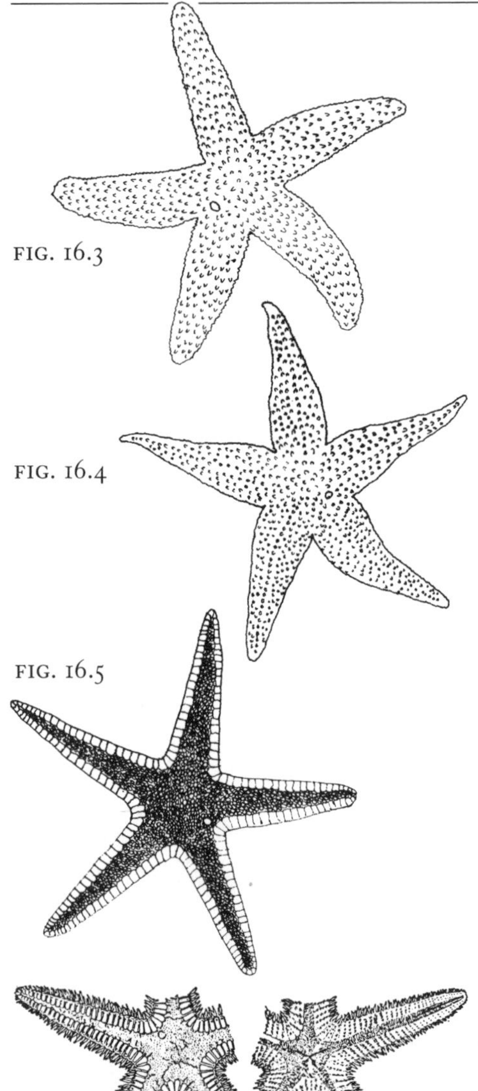

FIG. 16.3

FIG. 16.4

FIG. 16.5

FIG. 16.6

Asterias forbesi, common sea star (Asteriidae). Madreporite reddish orange (but see note for item 6 above). Shades of brown. Five arms blunt-tipped; body rather rigid, even out of water. Row of middorsal spines along arms less distinct than in *A. rubens.* Most common south of Cape Cod. Bo,Ha–Gr–Sa,Li–Su,pr,13.2 cm (5.2 in), (C421,G261,NM679), ☺. (FIG. 16.3. Aboral surface.)

Asterias rubens (=*vulgaris*), northern sea star (Asteriidae). Purple, reddish, brown, orange, or yellowish. Madreporite yellowish (but see note for item 6 above). Five arms pointed at tips; body firm but not rigid but goes limp quickly when removed from water. Spines often form distinct central row along aboral surface or arms. Most common north of Cape Cod. Bo,Ha–Gr,Li–De,pr,20.3 cm (8 in), (C422,G261,NM678), ☺. (FIG. 16.4. Aboral surface.)

Astropecten americanus (Astropectinidae). Reddish brown above, lighter margins, light below. Tall paxillae on aboral surface give shaggy appearance. Vi,Sa–Mu,Su–De,pr–sc,14 cm (5.5 in), (C29), ☺. (FIG. 16.5. Aboral surface.)

Astropecten articulatus, armored sea star (Astropectinidae). Purple or bluish above; large orangish yellow marginal plates; yellowish below. Short paxillae give aboral surface a granular appearance. Swallows food whole. Vi,Sa–Mu,Su–De,pr–sc,12.7 cm (5 in), (C31,G260,NM669). (FIG. 16.6. a, aboral surface; b, oral surface.)

Crossaster papposus, rough, spiny, or spring sun star (Solasteridae). Mottled coloring or solid reddish above; light-colored below. Prominent multitipped paxillae in clusters. Typically with 12 arms. Eats stars, *Mytilus* and other mussels. Ac,Ha,Su–De,pr,17.8 cm (7 in), (C297,G260,NM673). (FIG. 16.7. Aboral surface.)

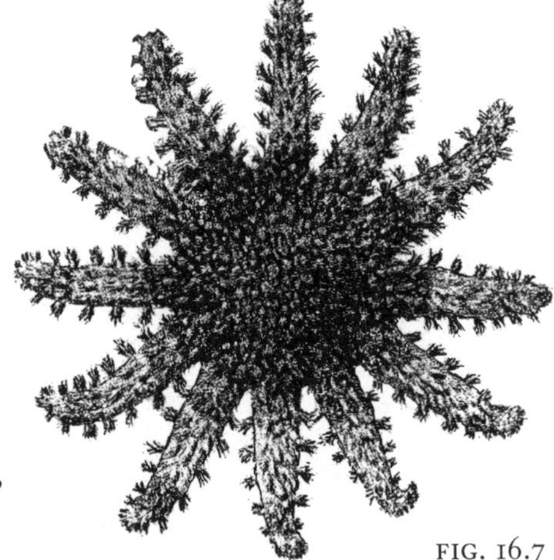

FIG. 16.7

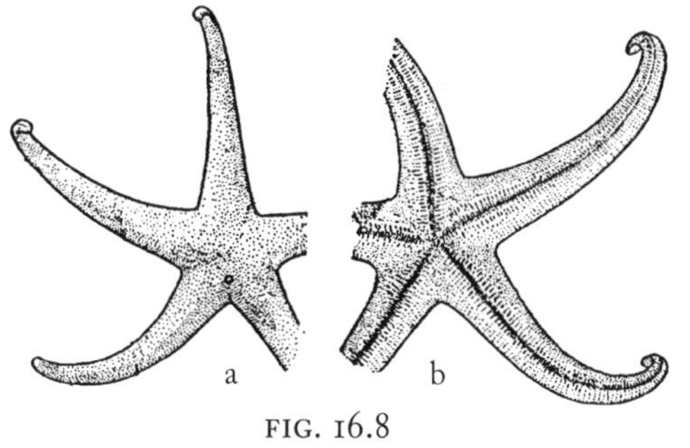

FIG. 16.8

Henricia sanguinolenta, blood star (Echinasteridae). Reddish purple, black, and yellow color phases. Arms taper evenly to tips. Eats particulate matter or sponges. Broods eggs. (Actually several species, difficult to distinguish. Refer to Clark and Downey 1992.) Bo,Ha,Su–De,pr–df,10.1 cm (4 in) [5.1 cm (2 in)], (C396,G261,NM675), ☺. (FIG. 16.8. a, aboral surface; b, oral surface.)

Leptasterias littoralis, polar or green slender sea star (Asteriidae). Typically with six arms; arms wider near disc, taper to narrow tips. Ac,Ha–Al,Su,pr,12.7 cm (5 in), (C438,NM680). (FIG. 16.9. Aboral view.)

FIG. 16.9

Leptasterias tenera, slender-armed star (Asteriidae). Five arms, slender throughout length. Broods eggs externally near mouth. Bo,Ha–Gr,Su–De,pr,3.8 cm (1.5 in), (C441,G261,NM680).

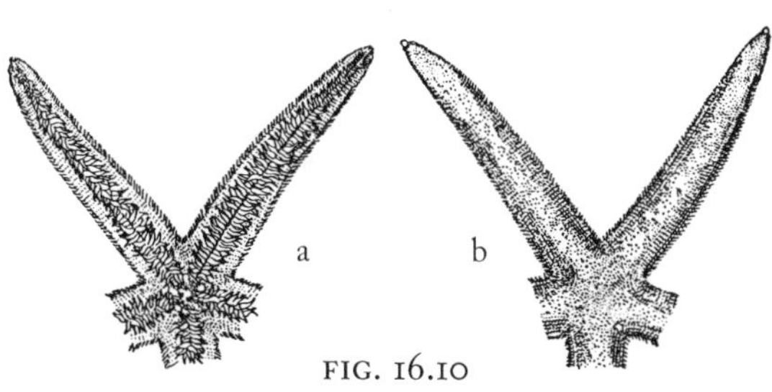

FIG. 16.10

Luidia clathrata, striped sea star (Luidiidae). Grayish coloring. Madreporite may be partially or totally obscured by surrounding ossicles. Spiny marginal plates clearly visible from oral or under surface, but not from above. Often with dark middorsal stripe along rays. Eats brittle stars. Vi,Sa,Li–Su,pr,17.8 cm (7 in) [10.1 cm (4 in)], (C13,G259,NM669). (FIG. 16.10. a, oral surface; b, aboral surface.)

Solaster endeca, smooth or purple sun star (Solasteridae). Purple, orange, and yellow color phases. Typically with 10 arms. Eats sea stars, molluscs, sea cucumbers, small sea urchins. Ac,Ha–Gr,Li–De,pr,20.3 cm (8 in), (C303,G250,NM672). (FIG. 16.11. Aboral surface.)

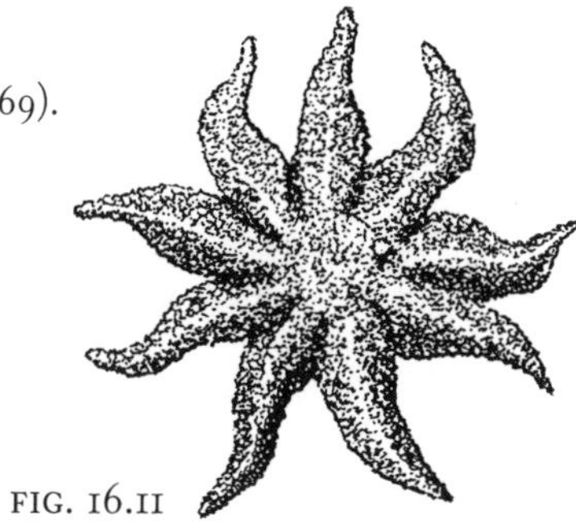
FIG. 16.11

16.6 **Short-Rayed Sea Stars** (from 16.3)

Short arms compared to central disc; includes "bat stars."

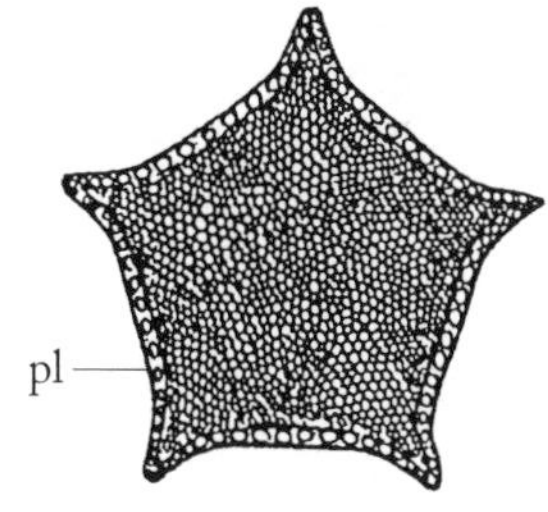

FIG. 16.12

1. Distinct lateral plates bordering rays visible from above (FIG. 16.12)
   - A. Lateral *plates* present — pl
   - B. Lateral plates *absent* — x
2. Distinctive features of aboral or uppermost surface
   - A. Aboral surface with scattered, prominent, soft, white, fingerlike projections or *dermal branchiae* or "skin gills" — db
   - B. Aboral surface with distinct, central *pointed* projection — po
   - C. Aboral surface overlain by *membranous* covering with central hole or opening — me
   - D. Such distinctive features *absent* — x
3. Dominant color, aboral or upper surface
   - A. *Reddish* — re
   - B. *Yellowish* — ye

16.7

Plates	Aboral Features	Aboral Color	Species
pl	po,x	ye	*Ctenodiscus crispatus,* mud star
pl	x	re	*Hippasteria phrygiana,* horse sea star
x	db	re	*Porania pulvillus insignis* (=*insignis*), badge sea star
x	me	ye,re	*Pteraster militaris,* winged sea star

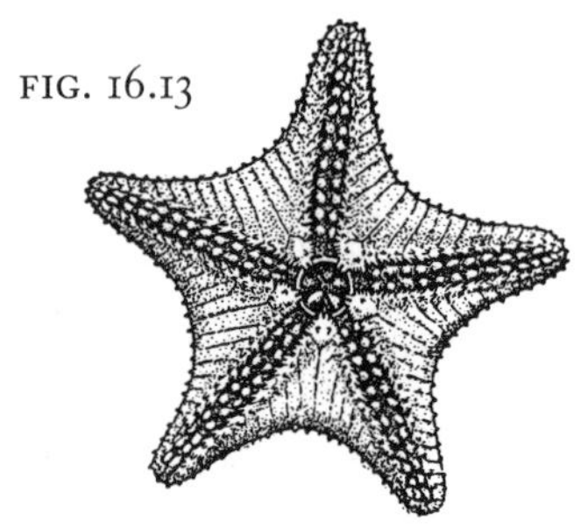

FIG. 16.13

Ctenodiscus crispatus, mud star (Goniopectinidae). Madreporite yellow; pale or brownish color. Pointed projection may be raised or lowered. Ac–,Mu,Su–De,df,5.1 cm (2 in), (C109,G260,NM670), ☺. (FIG. 16.13. Oral surface.)

Hippasteria phrygiana, horse sea star (Goniasteridae). Scarlet above; white below. Marginal plates orange and beaded. Clam-shell or valvelike pedicellarial flaps. Ac,Sa–Mu,Su–De,pr–sc,12.7 cm (5 in), (C247,G260,NM669). (FIG. 16.14. Aboral surface.)

Porania pulvillus insignis (=*insignis*), badge sea star (Poraniidae). Plump star with red above with prominent, white dermal branchiae; whitish below. Slimy, smooth feel to aboral surface. Eats sponges. Bo,Ha–Al,Su–De,df–pr,7.1 cm (2.8 in), (C210). (FIG. 16.15. a, aboral surface; b, oral surface.)

Pteraster militaris, winged sea star (Pterasteridae). Plump star with yellow to reddish above; tan below. Broods eggs under aboral membrane. Burrows into substrate. Ac,Ha–Al,Su–De,df–pr,15.2 cm (6 in), (C332,NM673). (FIG. 16.16. Oral surface.)

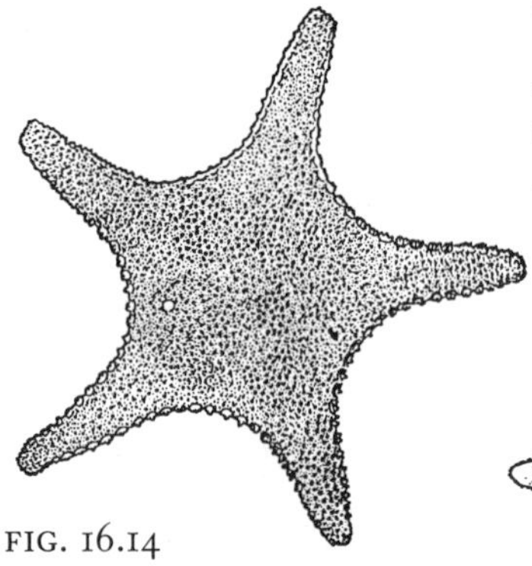

FIG. 16.14

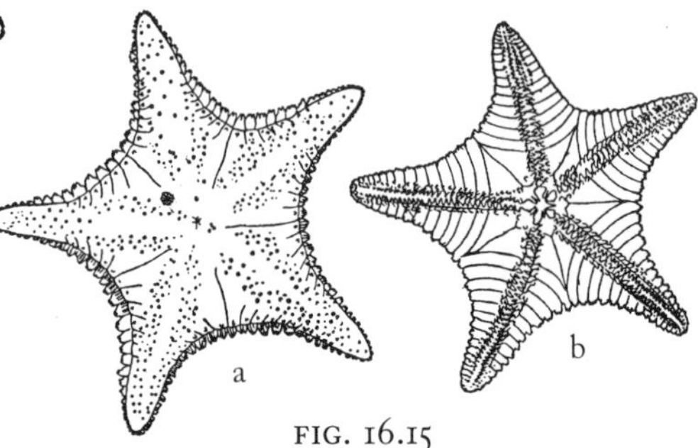

FIG. 16.15

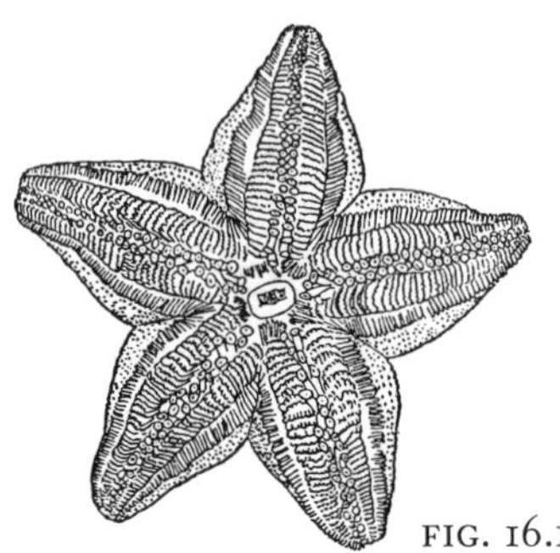

FIG. 16.16

16.8 Class Ophiuroidea, Brittle Stars

As their common name suggests, arms of brittle stars break easily; handle them carefully. Note that primary locomotion in these animals is by arm movements and not by tube feet. Species are difficult to distinguish, requiring close review of spines bordering each arm joint and the distribution or shape of scales on the *aboral* (uppermost) and *oral* (lowermost) surfaces of their body disc. These scales include *radial shields* (paired plates often located on aboral or upper surface near the point of attachment of arms to the body disc), and *oral shields* (plates located between points of star-shaped mouth opening on oral or under-surface of body disc). Dimensions of these plates include width (parallel to circumference or oral disc) and length (parallel to axis of the arm). Sizes given in species descriptions refer to diameter of body disc.

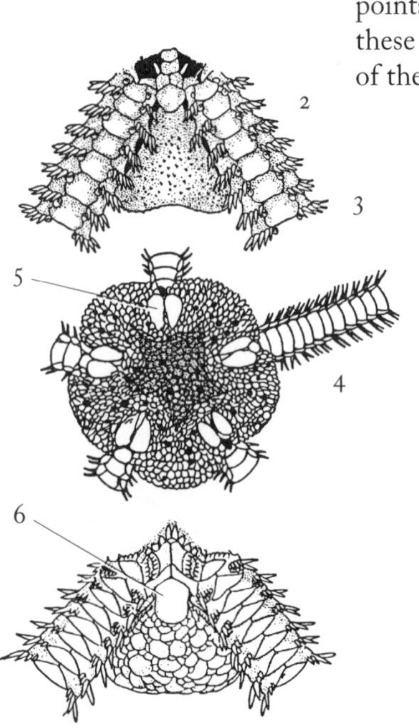

FIG. 16.17

1. Relative length of arms compared to disc diameter
   - A. Arms very *long,* more than six times the disc diameter — lo
   - B. Arms *moderate* in length, less than six times the disc diameter — mo
2. Orientation of arm spines relative to arm segments (FIG. 16.17)
   - A. Spines *erect* (stick outward at nearly right angle to arm segments) — er*
   - B. Spines lie *flat* along arm segments — fl
   - C. Spines *not* conspicuous, present as tiny hooks — x
3. A vertically oriented row of lateral spines is visible between adjacent arm segments (intersegmental spines); determine the number of spines in a typical row (FIG. 16.17)
   - A. Provide *number* of spines per row — ___
   - B. Lateral spines *absent* between arm segments — x
4. Length of intersegmental arm spines from proximal region (portion near disc) (FIG. 16.17)
   - A. Spines *long,* equal to or greater than the width of adjacent dorsal arm plates — lo
   - B. Spines *short,* clearly shorter than width of adjacent dorsal arm plates — sh*
   - C. Intersegmental arm spines *absent* — x

5. Position of radial shields (large plates near arm attachment sites; aboral surface) (FIG. 16.17)
   - A. Radial shields are platelike and *touch* one another — to*
   - B. Radial shields are platelike but *separated* from one another for at least part of their length — se
   - C. Radial shields *absent* or not visible because of granular body covering — x
6. Shape of oral shields (single plates on aboral surface between arm bases) (FIG. 16.17)
   - A. Oral shield *width* approximately *equal to length* — w=l
   - B. Oral shield *width* clearly *greater than length* — w>l
   - C. Oral shield *width* clearly *less than length* — w<l*
   - D. Oral shields *absent* — x

16.9

Arm Length	Spine Orientation	Spine No.	Spine Length	Radial Shield	Oral Shield	Species
lo	x	x	x	x	x	*Gorgonocephalus arcticus,* northern basket star
lo	er	3	sh	to	?	*Micropholis (=Amphiodia) atra,* burrowing brittle star
lo	er	3–4	sh	se	w<l	*Amphioplus abditus,* burrowing brittle star
mo	fl	2–3	sh,lo	se	w<l,w=l	Ophiuridae (in part), 16.10
mo	fl	7–8	sh	se,x	w=l	*Ophioderma brevispina,* green, short-spined, smooth, or mud brittle star
mo	er	2–6	lo	to	w>l	*Ophiothrix angulata,* Atlantic spiny brittle star
mo	er	3–4	sh	to	w=l	*Axiognathus (=Amphipholis) squamatus,* dwarf or brooding snake star
mo	er	5–6	lo	x	w>l	*Ophiopholis aculeata,* daisy brittle star
mo	er	4–6	lo	se,to	w=l	*Ophiomitrella clavigera*
mo	er	6–8	lo	x	w>l	*Ophiacantha bidentata*

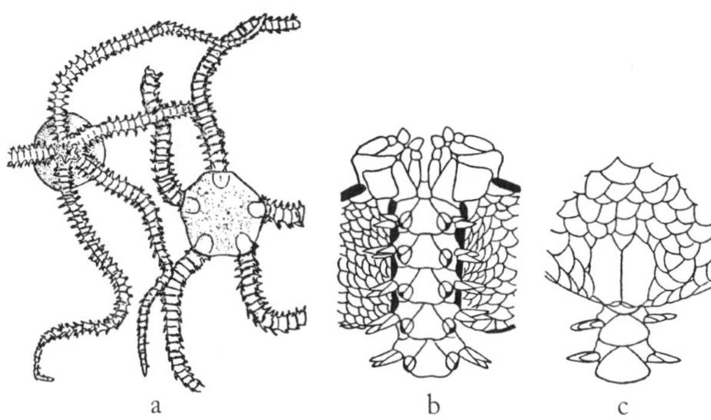

FIG. 16.18

Amphioplus abditus, burrowing brittle star (Amphiuridae). Extremely long arms, more than 10 times disc diameter; dull grayish brown. Deep water, mud. Vi,Mu,Su,df,12 mm (0.5 in), (G264).

Axiognathus (=Amphipholis) squamatus, dwarf or brooding snake star (Amphiuridae). Disc scale-covered; small. Light spot at base of each arm. Bioluminescent; broods young in bursal slits near arm attachment sites. Often with tunicates. Ac–,Ha–Al–Gr,Li–Su,df–ff,10 mm (0.4 in), (G264,NM686). (FIG. 16.18. a, animal; b, portion of oral surface; c, portion of aboral surface.)

Gorgonocephalus arcticus, northern basket star (Gorgonocephalidae). With branching arms. Yellowish brown coloring. Thick skin covers over disc features. Ac,Sa–Gr–Ha,Su–De,ff,3.8 cm (1.5 in), (G263,NM683). (FIG. 16.19. Oral surface.)

FIG. 16.19

Micropholis (=Amphiodia) atra, burrowing brittle star (Amphiuridae). Radial shields are short and wide; gray color. Ca+,Mu,Su,df,12 mm (0.5 in).

Ophiacantha bidentata (Ophiacanthidae). Bo,Mu–Ot,Su–De,ff,12 mm (0.5 in). (FIG. 16.20. a, portion of arm and disc, oral surface; b, portion of arm, from above.)

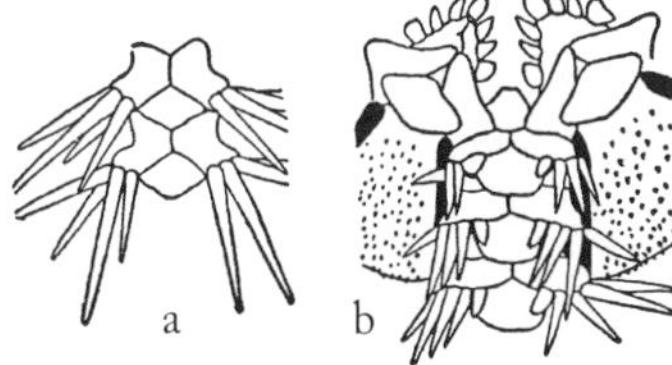

FIG. 16.20

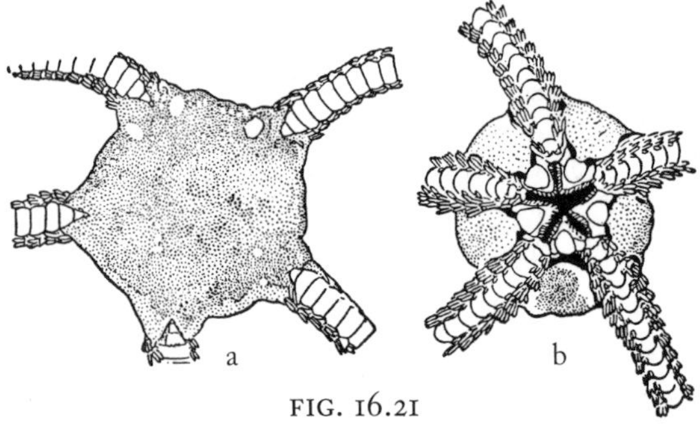

FIG. 16.21

Ophioderma brevispina, green, short-spined, smooth, or mud brittle star (Ophiodermatidae). Gray, green, or brown. Disc granular; not scale covered. Arm spines pressed against arms. Vi,Mu–Sa–SG,Su,om,15 mm (0.6 in), (G263,NM684), ☺. (FIG. 16.21. a, disc and arm bases, aboral view; b, oral view.)

Ophiomitrella clavigera (Ophiacanthidae). Bo,Mu,De,df?,7 mm (0.3 in).

Ophiopholis aculeata, daisy brittle star (Ophiactidae). Mottled; colorful. Major disc plates and arm plates ringed with small scales. Strong negative response to light. Ac–,Ha–Al,Li–De,df–pr,20 mm (0.8 in), (G263,NM685), ☺. (FIG. 16.22. a, aboral view; b, oral view.)

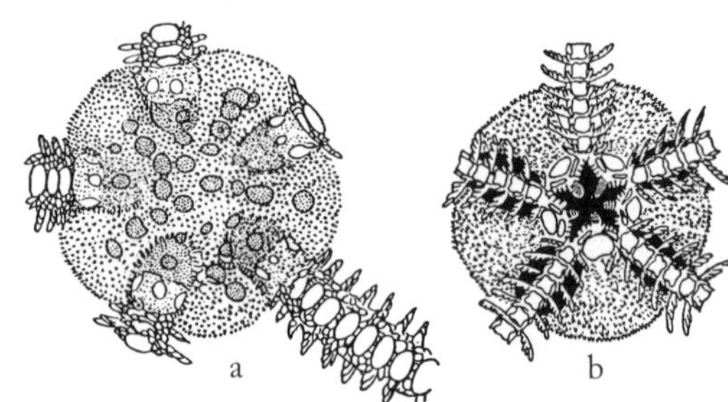

FIG. 16.22

Ophiothrix angulata, Atlantic spiny brittle star (Ophiothricidae). Long, glassy spines from greenish gray to bright orangish red body; distinctive stripe along dorsal surface of each ray. In fouling community. Ca+,Ha–Al,Su,sc,12 mm (0.5 in), (G263). (FIG. 16.23. Aboral view of disc.)

FIG. 16.23

16.10 **Ophiuridae (in part)** (from 16.9)

Brittle stars with comparatively short arms; arm spines lie flat against arm; radial shields are separated from one another. Sizes given in species descriptions refer to diameter of body disc.

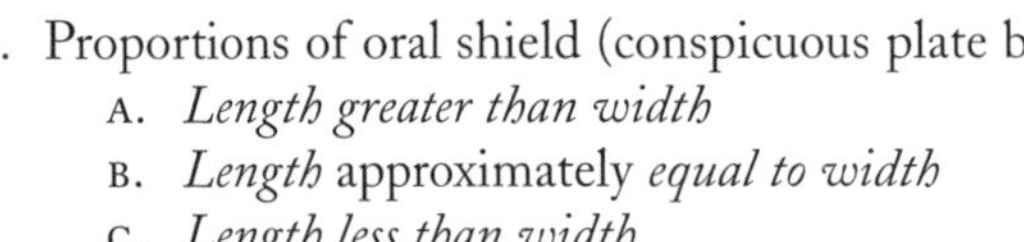

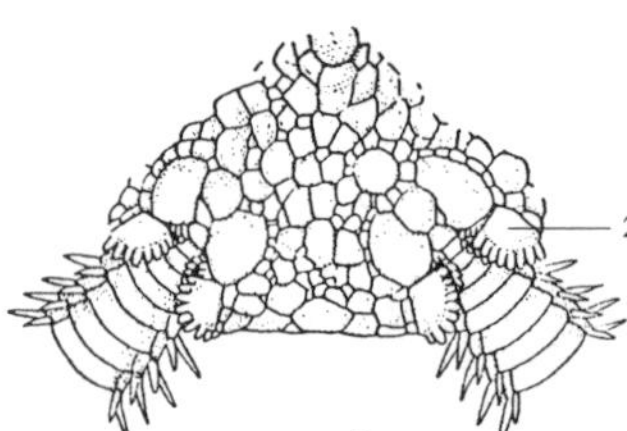

FIG. 16.24

1. Proportions of oral shield (conspicuous plate between arm bases; oral surface)
   - A. *Length greater than width* — l>w
   - B. *Length* approximately *equal to width* — l=w
   - C. *Length less than width* — l<w
2. Appearance of "arm combs" (rows of small spines bordering points of attachment of rays to the body disc)
   - A. Arm combs form *single*, continuous row — si
   - B. Arm combs formed as *two* separated rows — 2
3. Dominant body color (variability makes this a weaker character)
   - A. *Purplish red* — pr
   - B. *Gray mottled* with darker brown or red — gm
   - C. *Bluish black* — bb

16.11

Oral Shield	Arm Combs	Color	Species
l=w,l<w	2	gm	*Ophiura robusta*
l=w,l>w	2	pr	*O. sarsi*
l>w	si	bb	*Ophiocten scutatum*

Ophiocten scutatum. Ac,?,Su–De,?,17 mm (0.7 in). (FIG. 16.25. Portion of oral surface.)

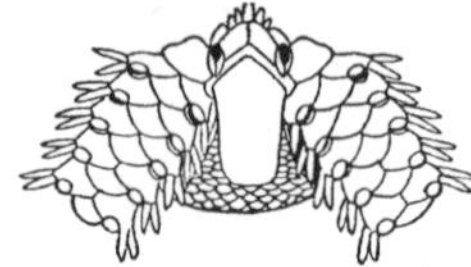
FIG. 16.25

Ophiura robusta. Radial shields small; arms banded. Ac,Mu–Sa,Li–De,df–pr,10 mm (0.4 in), (G264). (FIG. 16.26. a, portion of aboral surface; b, oral surface.)

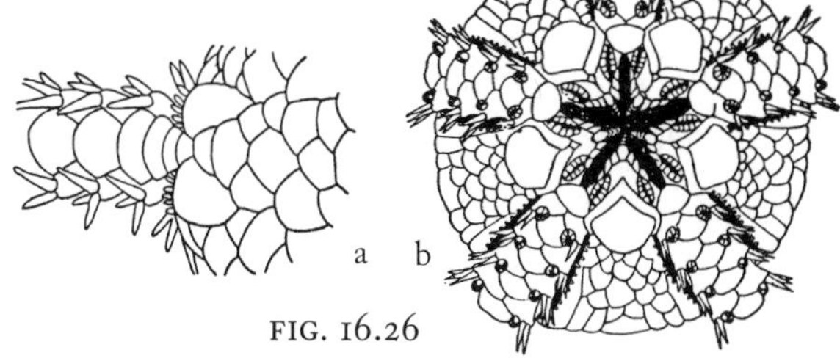

FIG. 16.26

Ophiura sarsi. Radial shields larger than preceding species. Feeds on small invertebrates. Ac,Mu,Su–De,pr,4.1 cm (1.6 in), (G264). (FIG. 16.27. a, portion of aboral surface; b, portion of oral surface.)

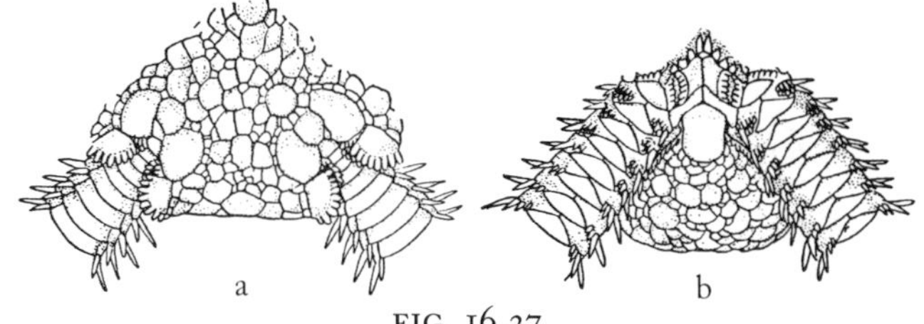

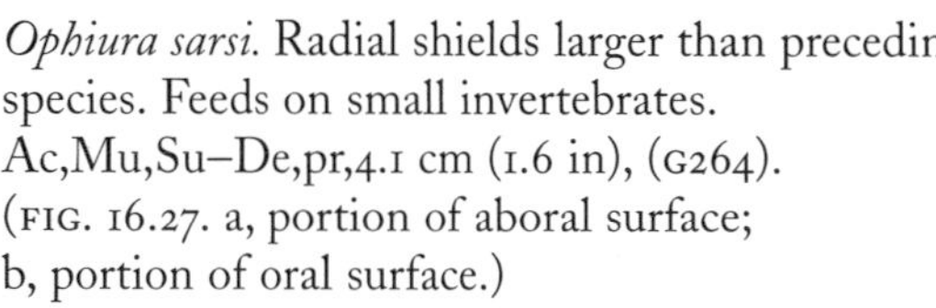

FIG. 16.27

16.12 Class Holothuroidea, Sea Cucumbers

A closer look reveals that, despite their elongate, muscular body construction, sea cucumbers possess echinoderm characteristics such as tube feet, which in many cases are arranged in *pentaradial* or five-patterned rows. Anterior tube feet are modified to form *tentacles* used for *suspension feeding* by some and for *deposit feeding* by others. Sea cucumbers must be left in a dish of cold seawater for some time following disturbance before they relax enough to re-extend their tentacles fully. The number and shape of tentacles and the distribution of tube feet on the body are used in identifications. Skeletal elements or *ossicles,* which are typically embedded like islands within the muscular body wall, may be revealed by strong transmitted light if the cucumbers are thin-walled. Otherwise, a piece of body wall may be placed in bleach to dissolve away the organic components, leaving behind these tiny, bony plates. Vermiform burrowing anemones (Phylum Cnidaria, Class Anthozoa, 8.48) look like vermiform burrowing holothurians. Tentacle shape (simple in anemones, branched in holothurians) and ossicles embedded in the body wall (present in holothurians, absent in anemones) are distinguishing features.

Additional reference: (P) Pawson, D. L., 1977.

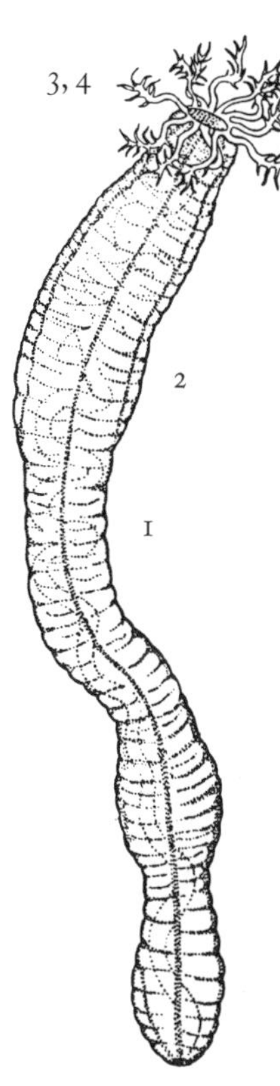

FIG. 16.28

1. Body shape
   - A. *Sack* or elongate football-shaped (FIG. 16.28) sa
   - B. Sausage or pear-shaped with distinctive tapered *tail* ta
   - C. *Domed* dorsal surface (usually with scales) but flattened ventral side for attachment to substrate do
   - D. *Cylindrical* or worm-shaped, more than four times longer than wide cy*
2. Placement of tube feet along body surface (FIG. 16.28)
   - A. Tube feet *scattered* irregularly over body surface sc
   - B. Tube feet arranged only in clear *rows* lengthwise along body ro
   - C. Most tube feet in clear *rows* but some *scattered,* especially dorsally ro:sc
   - D. Tube feet confined to *sole* or flattened ventral attachment surface so
   - E. Tube feet absent; but rows of small adhesive *papillae* present on body surface pa*
   - F. Both tube feet and papillae *absent* x
3. Shape of tentacles (FIG. 16.28)
   - A. Stalk leads to fingerlike branching forming a terminal *star* st
   - B. Stalk with only simple branching or singly *pinnate* pi*
   - C. Stalk with branched branches, more complex and *tree*like tr
4. Number of large tentacles + number of small tentacles: provide *number* of large tentacles + provide *number* of distinctly smaller tentacles (FIG. 16.28) __+__

16.13

Shape	Tube Feet	Tentacle Shape	Tentacle Number	Species
sa	sc	tr	10+0	*Sclerodactyla* (=*Thyone*) *briareus,* hairy cucumber
sa	ro:sc	tr	8+2	*Cucumaria frondosa,* orange-footed cucumber
sa	ro:sc	tr	10+0	*Duasmodactyla commune*
sa	ro	tr	10+0	*Pentamera* (=*Cucumaria*) Ossicle plates usually with four holes: *P. pulcherrima,* pale sea cucumber OR Ossicle plates usually with many holes: *P. calcigera*
sa	ro	tr	8+2	*Thyonella gemmata,* green sea cucumber
cy	pa	pi	12–15+0	Rose color; 4–6 branches per tentacle: *Epitomapta* (=*Leptosynapta*) *roseola,* pink synapta OR White; 10–16 branches per tentacle: *Leptosynapta tenuis* (=*inhaerens?*), white synapta
cy	x	st	12–15+0	*Chiridota laevis,* silky cucumber
do	so	tr	10+0	*Psolus* Sole more than twice as long as wide; well-developed tail: *P. phantapus* OR Sole about twice as long as wide; tail a bump to absent: *P. fabricii,* scarlet psolus
ta	x	st	15+0	Red to black; deep water: *Molpadia oolitica* OR Rose to white; shallow to deep water: *Caudina arenata,* rat-tailed cucumber

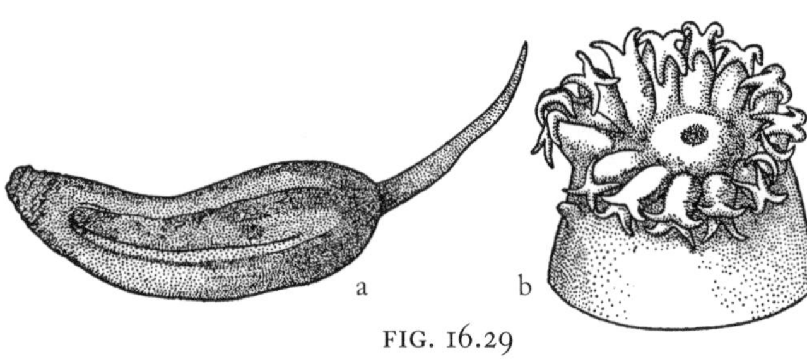

FIG. 16.29

Caudina arenata, rat-tailed cucumber (Caudinidae). May be found beached following storms. Ac–,Sa–Mu,Li–Su,df,17.8 cm (7 in), (G256,P13). (FIG. 16.29. a, animal; b, tentacles.)

Chiridota laevis, silky cucumber (Chiridotidae). Rose or white. Skeletal elements like spoked wheel. Ac,Mu–Al,Su,df,15.2 cm (6 in), (G256,NM705,P13).

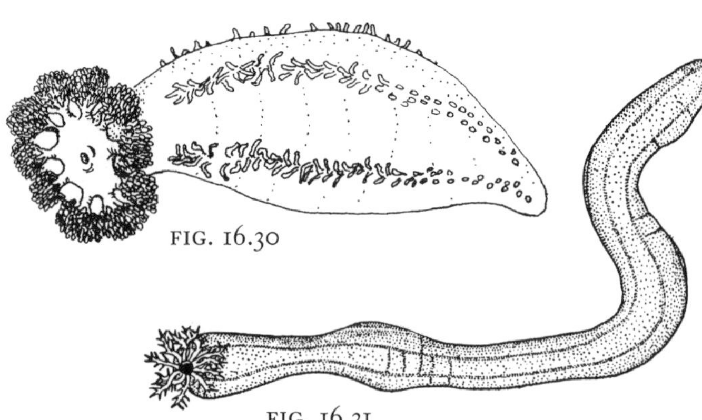

FIG. 16.30

FIG. 16.31

Cucumaria frondosa, orange-footed cucumber (Cucumariidae). Large, cylindrical tube feet. Red, brown, or cream colored; darker above, lighter below. Can constitute more than half of the shallow subtidal biomass in some areas of eastern Maine. Modest fishery for their muscles, sold as "trepang" or "bêche-de-mer" in Asia. Ac,Ha–Sa,Li–Su,ff,48.3 cm (19 in) [15.2 cm (6 in)], (G254,NM701,P12), ☺. (FIG. 16.30)

Duasmodactyla commune (Cucumariidae). Yellow to white. Ac,Ha–Gr,Li+,15.2 cm (6 in) (P12)

Epitomapta (=*Leptosynapta*) *roseola,* pink synapta (Synaptidae). Ac–,Mu–Sa–Ha,Su, ff,10.1 cm (4 in), (G255,NM705,P13). (FIG. 16.31)

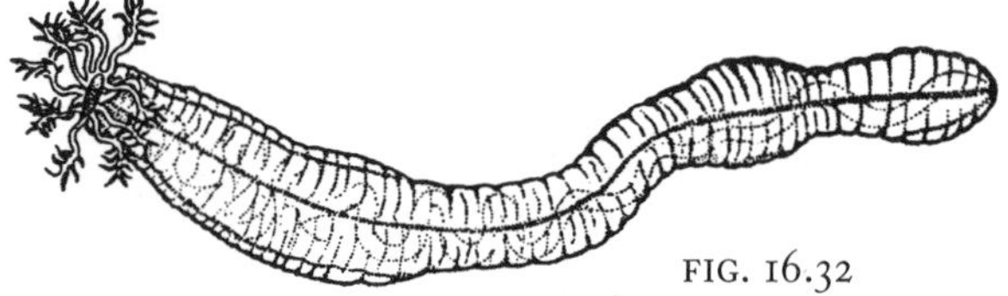

FIG. 16.32

Leptosynapta tenuis (=*inhaerens*), white synapta (Synaptidae). Translucent body wall. U-shaped burrow with conical depression for head end. Bo, Sa–Mu,Li–De, df,15.2 cm (6 in), (G255,NM705,P13), ☺. (FIG. 16.32)

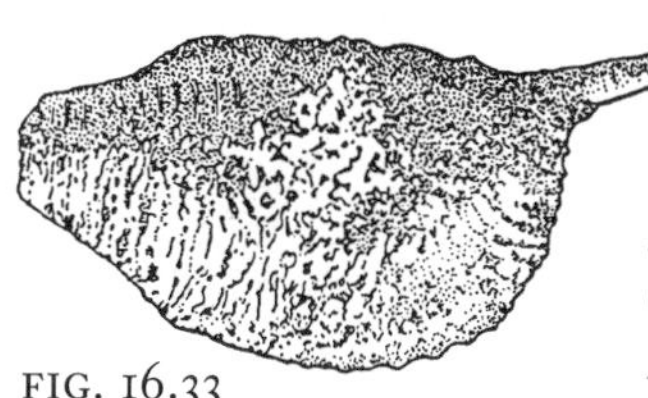
FIG. 16.33

Molpadia oolitica (Molpadiidae). Becomes blacker with age; subtidal. Ac,Mu,Su–De,df, 25.4cm (10 in), (G256,PI3). (FIG. 16.33)

Pentamera (=*Cucumaria*) *calcigera* (Cucumariidae). White. Cape Cod north. Ac,Sa,Su,df,8.1 cm (3.5 in), (PI2).

Pentamera (=*Cucumaria*) *pulcherrima*, pale sea cucumber (Cucumariidae). White to mottled brown color. Cape Cod south. Vi,Mu,Su,df,5.1 cm (2 in), (PI2). (FIG. 16.34)

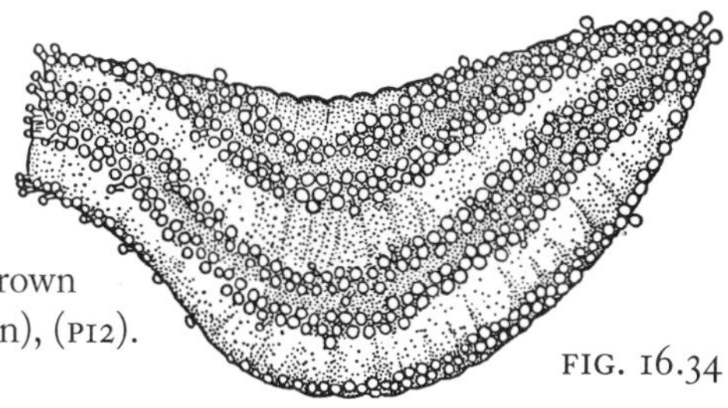
FIG. 16.34

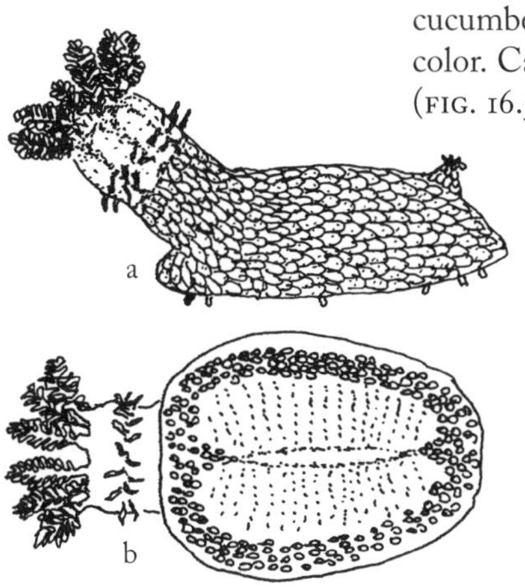

Psolus fabricii, scarlet psolus (Psolidae). Bright red or reddish orange. Tentacles densely branched. Ac,Ha,Li–De,ff,20.3 cm (8 in), (G254,NM703,PI2). (FIG. 16.35. a, lateral view; b, ventral view.)

FIG. 16.35

Psolus phantapus (Psolidae). Dull reddish orange (young) to black (older). Tentacles orange color; sparsely branched. Ac,Ha–Sa,Su–De,ff,15.2 cm (6 in), (G254,PI2). (FIG. 16.36. a, lateral view; b, ventral view.)

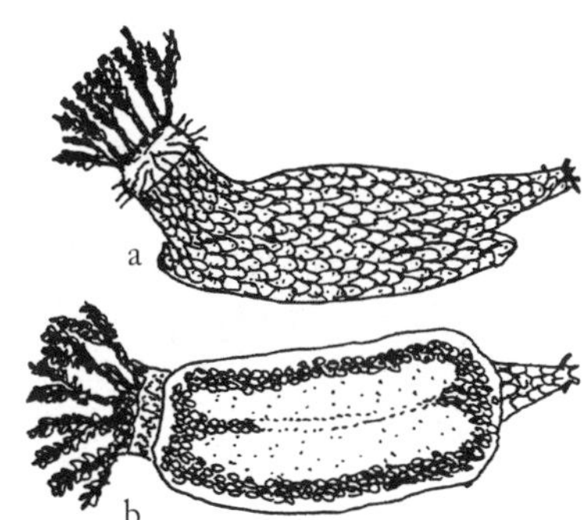
FIG. 16.36

Sclerodactyla (=*Thyone*) *briareus*, hairy cucumber (Sclerodactylidae). Slender, hairlike tube feet. Green, brown, or black color; soft-bodied. Bo,Mu–Sa–SG,Su,ff,15.2 cm (6 in), (G254,NM702,PI2), ☺. (FIG. 16.37)

FIG. 16.37

Thyonella gemmata, green sea cucumber. Brownish to greenish. May be found beached following storms. Vi,Mu–Sa,Su,ff,25.4 cm (10 in), (PI3).

16.14 Class Echinoidea, Sea Urchins and Sand Dollars

Sea urchins and sand dollars are unmistakable as a group. Their characteristic *spines* tend to be lost upon death, leaving behind a delicate white skeleton or "test." In the case of sea urchins, the test is perforated by patterns of pores to accomodate tube feet and rows of knobs that form part of the ball-and-socket bases for moveable spines. On the *oral* surface, the firmer, flattened test of sand dollars includes a central mouth opening, whereas the *aboral* surface features an inflated, star pattern of slitlike openings, known as the *petaloid*, which accomodates a group of flattened, modified *tube feet* used to contribute to respiratory exchange in these animals. Urchins use a complex, toothed apparatus (known from its shape as *Aristotle's lantern*) to graze on vegetation or scrape up detrital material. Sand dollars use tiny tube feet arranged along detritic grooves on the oral surface to collect and transport organic detritus to the mouth. Overall body shape and the relative size of spines are sufficient to differentiate among regional echinoids.

Sizes given are for diameter of the test, excluding spines.

Additional reference: (s) Serafy, D. K., and F. J. Fell, 1985.

FIG. 16.38

1. Body shape (FIG. 16.38)
   - A. *Hemispheric* or subglobular — he*
   - B. Flattened *disc* perforated by *five* slits or "lunules" — d5
   - C. Flattened *disc* with slits *absent* — dx
   - D. Ovaloid or *egg* shaped — eg
2. Size and color of spines (FIG. 16.38)
   - A. Prominent, short (<¼ diameter of test) but rigid and sharp *green*-brown *spines* — gs*
   - B. Prominent, long (>¼ diameter of test) sharp *purplish*-brown *spines* — ps
   - C. Slender, short spines (green, brown, or purple) appear fuzzy or *hairy* — ha

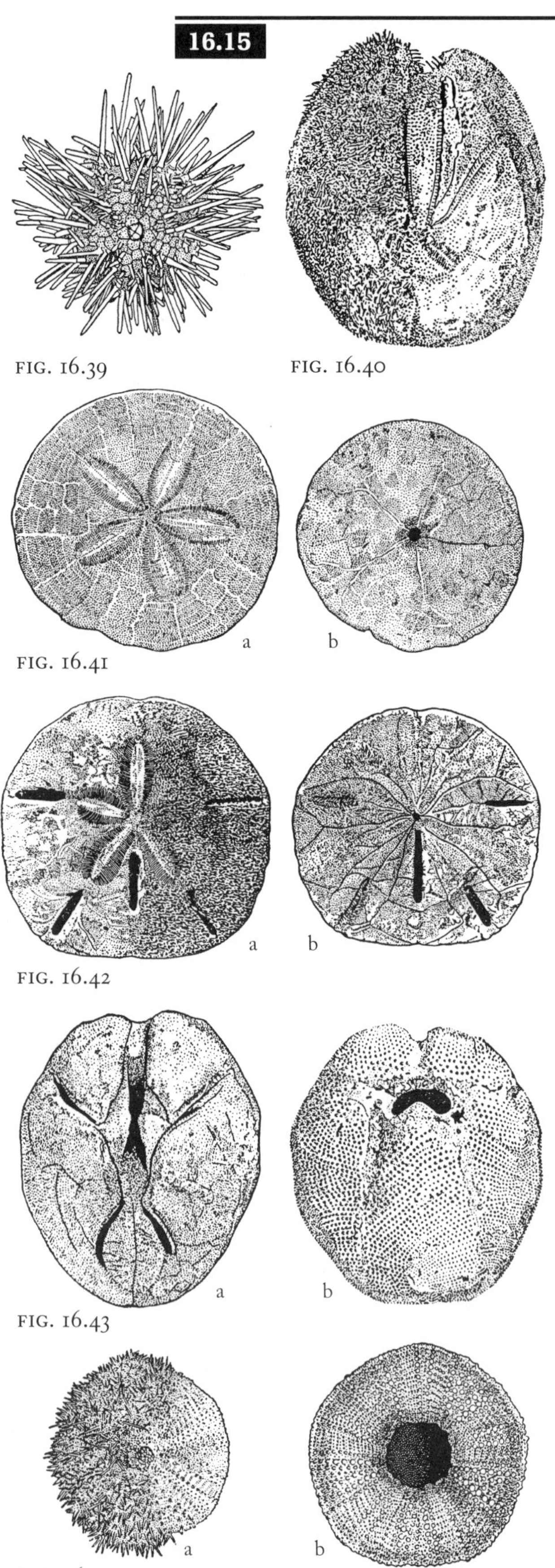

FIG. 16.39

FIG. 16.40

FIG. 16.41

FIG. 16.42

FIG. 16.43

FIG. 16.44

Body	Spines	Species
he	gs	*Strongylocentrotus droebachiensis,* green sea urchin, sea egg
he	ps	*Arbacia punctulata,* Atlantic purple urchin, million-dollar urchin
dx	ha	*Echinarachnius parma,* sand dollar
d5	ha	*Mellita quinquiesperforata,* keyhole dollar
eg	ha	All five rays of starlike, aboral pattern deeply grooved: *Moira atropos,* mud heart urchin OR One ray of starlike aboral pattern deeply grooved, the other four are shallow *Brisaster fragilis*

Arbacia punctulata, Atlantic purple urchin, million-dollar urchin (derived from its use in embryological research) (Arbaciidae). Longer, thicker, and darker spines than green sea urchin. Vi,Sa–Gr–Ha,Li–De,om,5.6 cm (2.2 in), (G257,NM689,S21), ☺. (FIG. 16.39. Aboral surface.)

Brisaster fragilis (Schizasteridae). Deep water form. Ac,Mu,Su–De,df,8.8 cm (3.5 in), (S23). (FIG. 16.40. Aboral surface with spines partially removed.)

Echinarachnius parma, sand dollar (Echinarachniidae). Purplish brown fades to dark green in air. Shallow to north of range, deeper toward the south. Tests frequently found among beached shell deposits. Senstive to sewage sludge. Ac–,Sa,Li–De,df,7.6 cm (3 in), (G258,NM694,S22), ☺. (FIG. 16.41. a, aboral surface; b, oral surface—both with spines removed.)

Mellita quinquiesperforata, keyhole dollar (Mellitidae). Barely penetrates the southern margin of our region. Shallows with strong water currents. Ca+,Sa,Li–De,df,7.6 cm (3 in), (G258,NM695,S22). (FIG. 16.42. a, aboral surface; b, oral surface—both with spines removed.)

Moira atropos, mud heart urchin (Schizasteridae). Ovaloid and bilaterially symmetrical. Scatterred black spots on brownish surface are pinching pedicellariae. Forms burrow with ventilation shaft to muddy-sand surface. Ca+,Mu–Sa,Su,df,6.4cm (2.5 in), (G258,NM696). (FIG. 16.43. a, aboral surface; b, oral surface—both with spines removed.)

Strongylocentrotus droebachiensis, green sea urchin, sea egg (Strongylocentrotidae). Basically herbivorous but also ingests encrusting invertebrates including young *Mytilus.* Supports a modest fishery. Overpopulation gives rise to "urchin fronts," in which urchins eat their way across subtidal sea floor leaving it denuded of algae, creating "urchin barrens." Ac–,Ha–Al,Li–Su,gr–pr,8.8 cm (3.5 in), (G257,NM523,S22), ☺. (FIG. 16.44. a, aboral surface, spines removed from half; b, oral surface of test.)

Chapter 17 ■ Phylum Chordata

17.1 Basic chordate construction includes a unique set of anatomical features, including a *notochord* for support; a dorsal, hollow *nerve cord* for coordination; a pharyngeal chamber with multiple *gill slits;* and a postanal *tail* for locomotion. In the simplest chordates, these features are used for effective swimming and/or burrowing by filter-feeding animals that strain food material from water flow otherwise used for respiration. Subsequently, the chordate plan became elaborated to produce the diverse, sophisticated, and highly successful vertebrates.

The Chordata comprises three subphyla. The Urochordata (or Tunicata) and the Cephalochordata are considered the lower chordates. The urochordate tunicates (17.2) include sessile sea squirts and pelagic salps. The name "urochordate" emphasizes the fact that the supportive notochord, found only during larval stages, is restricted to the tail of the animal. In *cephalochordates,* the notochord extends into the head. These small, sand-burrowing animals, the lancelets, barely extend into the southern limits of our area (1.2).

Marine members of the subphylum Vertebrata (17.16) include fishes (17.17), reptiles (17.69), and mammals (17.71). Although there are many marine birds as well, they are not strictly aquatic animals and are not treated here.

17.2 Subphylum Urochordata (Tunicata), Sea Squirts, Salps, and Larvaceans

Either as adults or as larvae, urochordates display such chordate anatomical features as a hollow *nerve cord,* which lies dorsal to a semirigid *notochord* for support, and a pharynx with *gill slits* through which water flows and against which food particles are screened. Representatives of three classes are found in our region. Ascidiacea (sea squirts) are frequently encountered in shallow waters. Gelatinous Thaliacea (salps) are either solitary or colonial planktonic animals, preferring warmer, offshore waters in general. The planktonic Appendicularia (larvacea) are tiny tadpole-shaped animals that live in association with "houses" made from secreted mucus.

1. Habitat
   A. *Benthic* or bottom-dwelling, either attached to firm objects or burrowing in sediments — be
   B. *Pelagic,* within the water column — pe
2. Size of individuals or colonies
   A. *Less than* ¼ inch in length — <¼
   B. *More than* ¼ inch in length — >¼

17.3

Habitat	Size	Group
be	>¼	Class Ascidiacea (sea squirts), 17.4
pe	>¼	Class Thaliacea (salps), 17.14
pe	<¼	Class Appendicularia; larvaceans, included in zooplankton, 5.16

17.4 Class Ascidiacea, Sea Squirts (from 17.3)

Sea squirts or tunicates reveal their chordate affiliation through their "tadpole" larvae. As adults, they are sessile, attached to solid substrates or, in a few cases, superficially burrow into soft sediments. They occur either individually or organically integrated into colonies. Regardless of growth format, individual zooids use ciliary currents to draw surrounding water through an *incurrent* or *branchial siphon,* into a *pharyngeal basket* perforated with rows of openings or *stigmata,* and then out through an *excurrent* or *atrial siphon.* Food items, trapped in mucus upon their passage through stigmata, are swept into the digestive system, which eventually terminates in an anus near the atrial siphon. Sea squirts secrete an outer, protective *tunic* that may be distinctively colored or elaborated with surface features. Identifications can be difficult in this group because shapes and colors can vary. Siphons also can be extended or contracted. Still, overall body plan (solitary or colonial), basic shape, and distinctive coloration can provide a start.

Additional references: (P) Plough, H. H., 1978; (V) Van Name, W. G., 1945.

1. Growth form
   A. *Solitary,* even if growing closely together in groups — so
   B. *Colonial,* individual zooids imbedded in a common *gelatinous* matrix — cg
   C. *Colonial* or social, with individual zooids linked by a *stolon* or runner — cs

17.5

Form	Group
so	Solitary tunicates, 17.6
cg	Colonial tunicates, 17.10
cs	*Perophora viridis,* creeping ascidian

FIG. 17.1

Perophora viridis, creeping ascidian (Perophoridae). Tiny, translucent or greenish individuals can form colonies covering patches of substrate several inches long. Vi,Ha,Es–Li–Su,ff,5 mm (0.2 in) (zooids), (G269,P67,V165). (FIG. 17.1)

17.6 Solitary Tunicates (from 17.5)

Gentle squeezing pressure can force out water through siphons, demonstrating the origin of the name "sea squirt." Provide specimens with an opportunity to relax in seawater before attempting identifications. In some cases, accurate determinations require examination of internal features such as the pattern of stigmatal openings in the pharygeal basket, the number of *folds* on the stomach and basket, and the number and shape of *gonads.* This requires that the superficial tunic of the specimen be removed. Fox and Ruppert (1985) suggest starting at the outside base of one siphon, cutting around the margin of the tunic up to the outside base of the other siphon, and then around the two siphons. Remove the front piece of the tunic leaving the back piece, the siphons, and a piece of tunic connecting them intact.

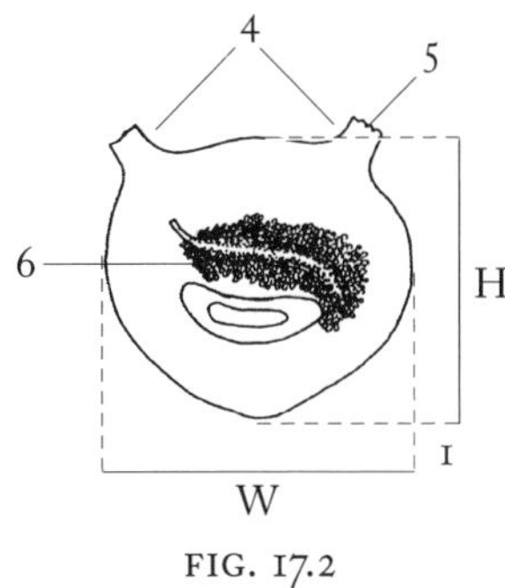

FIG. 17.2

1. Overall shape; height = line extending from between siphons to opposite side, width = line perpendicular to midpoint of height line (FIG. 17.2)
   A. Markedly *wider than high* — w>h
   B. Markedly *higher than wide* — h>w
   C. *Height* and *width* approximately *equal* — h=w*
2. Attachment to substrate
   A. Attaches by means of a distinct, slender *stalk* — st
   B. Attaches directly to substrate; stalk *absent* — x
3. Tunic and/or siphons show reddish coloring
   A. At least some distinctly *reddish* coloring present — re
   B. Reddish coloring entirely *absent* — x
4. Orientation of siphons (FIG. 17.2)
   A. Both siphons open *dorsally* — do*
   B. Both siphons open *laterally* — la
   C. Siphons open at *right angles* to one another — ra
5. Number of lobes defining orifice of incurrent or branchial siphon: provide *number* of lobes (8, 6, or 4) (FIG. 17.2) — __
6. Number of dark (sometimes reddish, sometimes fuzzy) gonads on the left side (the side including the gut) versus the right side (requires removal of the tunic; see introduction to the section): provide *number* of gonads on the left side: provide *number* of gonads on the right side (FIG. 17.2) — __:__

17.7

Height: Width	Attachment	Red Coloring	Siphons	Lobe No.	Gonads	Species
h>w	st	x	do	4	5-6:2	*Styela clava,* club tunicate
h>w	x	re	do	4	4-6:4-6	*Halocynthia pyriformis,* sea peach
h>w	x	x	do	8	1:0	*Ciona intestinalis,* sea vase
w>h	st	re	la	4	1:1	*Boltenia ovifera,* stalked sea squirt, sea potato
w>h	x	re	do	4	0:1	*Dendrodoa carnea,* blood-drop sea squirt
w>h	x	x	do	4	2:2	*Styela partita,* rough sea squirt
h=w	x	re	do	4	1:1	*Boltenia echinata,* cactus sea squirt
h=w	x	x	do	6	1:0	*Bostrichobranchus pilularis*
h=w	x	x	do	6	1:1	Molgulidae, 17.8
h=w	x	x	ra	8	1:0	*Ascidia* (requires dissection to determine pattern of openings or stigmata in branchial basket): 8–10 openings between adjacent longitudinal vessels: *A. callosa,* callused sea squirt OR 4–5 openings between adjacent longitudinal vessels: *A. prunum*

FIG. 17.3

Ascidia callosa, callused sea squirt (Ascidiidae). Firmer than next species. Tan to brownish. Ac,Ha,Li–Su,ff,8.9 cm (3.5 in), (G270,NM735,P71,V178). (FIG. 17.3. Lateral view, left side, tunic removed to show gut and gonad.)

Ascidia prunum (Ascidiidae). Grayish or opaque. Ac,Ha,Su–De,ff,5.1 cm (2 in), (G270,P71,V175).

Boltenia echinata, cactus sea squirt (Pyuridae). Flesh-colored or brownish red; globular body with four to eight branched, flexible spines. Ac–,Ha,Li–De,ff,25 mm (1 in), (G271,NM741,P75,V354). (FIG. 17.4. Lateral views to show gut and gonads. a, right side; b, left side; c, detail of surface spines.)

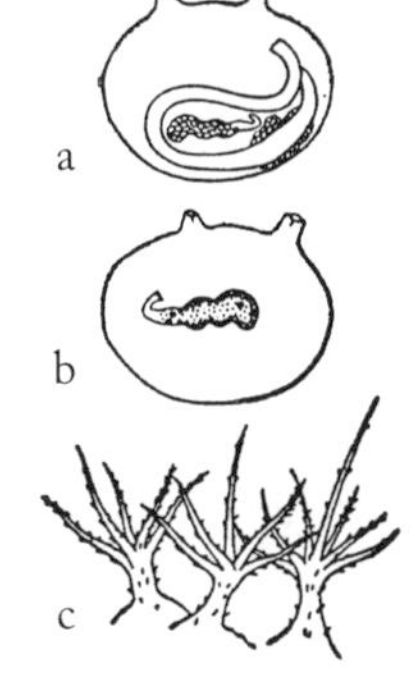

FIG. 17.4

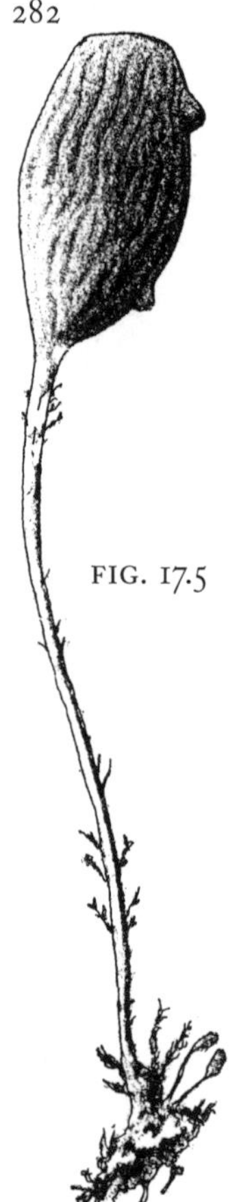
FIG. 17.5

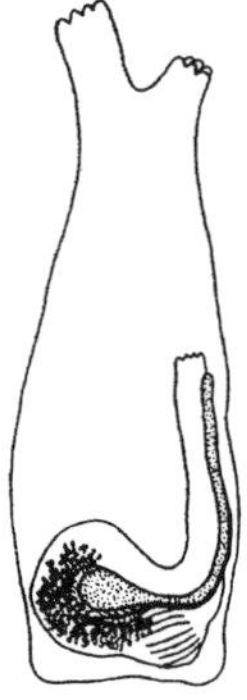
FIG. 17.6

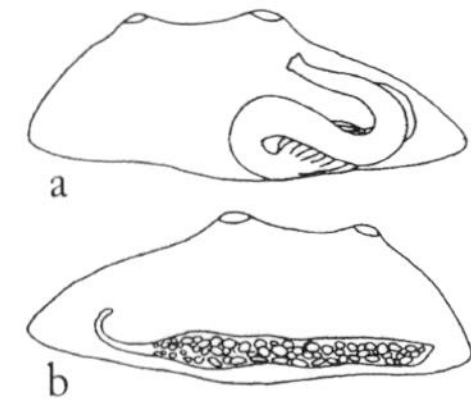

FIG. 17.7

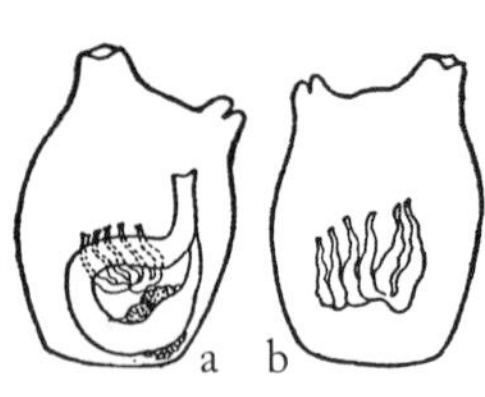

FIG. 17.8

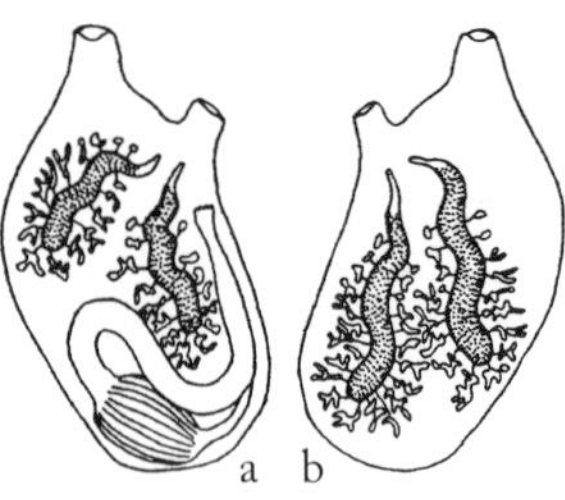

FIG. 17.9

Boltenia ovifera, stalked sea squirt, sea potato (Pyuridae). Body reddish or orange; siphons reddish. Stiff stalk with attached algae and invertebrates. Ac–,Ha–Gr,Su–De,ff,8.9 cm (3.5 in) (body), 30.5 cm (12 in) (stalk), (G271,NM741,P73,V351). (FIG. 17.5)

Bostrichobranchus pilularis (Molgulidae). Often covered with mud. Bo,Sa–Mu,Su–De,ff,25 mm (1 in), (G272,P84,V438). (See FIG. 17.11)

Ciona intestinalis, sea vase (Cionidae). Elongate, translucent, pale yellow; yellow rim to siphons. Eight light-sensitive spots around oral siphon; six around atrial siphon. Ac,Ha–Gr,Li–De,ff,15.2 cm (6 in), (G270,NM733,P57,V160), ☺. (FIG. 17.6. Right side, tunic removed; shows gut and gonad.)

Dendrodoa carnea, blood-drop sea squirt (Styelidae). Often encrusted with debris with only red siphons showing. Ac–,Ha,Su,ff,12 mm (0.5 in), (width), (G270,NM736,V281). (FIG. 17.7. a, right side, shows gut; b, left side, shows gonad; both with tunic removed.)

Halocynthia pyriformis, sea peach (Pyuridae). Fuzzy, velvety look; clean but sandpaper-like touch. Body reddish, orange, or yellowish: all are red toward top. Ac,Ha,Su–De,ff,12.7 cm (5 in), (G271,NM742,P75,V359). (FIG. 17.8. a, right side; b, left side; both with tunic removed; shows groups of flamelike gonads and gut.)

Styela clava, club tunicate (Styelidae). Narrow-bodied; tough, bumpy exterior. Introduced in 1960s; originally from Japan (see Berman et al. 1992). Vi,Ha,Su,ff,15.2 cm (6 in), (NM736,P92,V317).

Styela partita, rough sea squirt (Styelidae). Rough, leathery, reddish tunic. South shore of New England only. Vi,Ha,Su,ff,4 cm (1.6 in), (P91,V290). (FIG. 17.9. a, right side; b, left side; both with tunic removed; shows two gonads/side.)

17.8 **Molgulidae** (from 17.7)

Solitary sea squirts, approximately equal in height and width, with six-lobed branchial siphon opening. Examination of internal features by dissection may be necessary for identification to species. Size given in species descriptions represents maximal dimension.

1. Attachment to substrate
   - A. *Attached* firmly to solid substrate — at
   - B. *Not* attached; freely moving, burrows in soft substrates — x
2. Relative spacing of siphons (FIG. 17.10)
   - A. Siphons *closely* placed, basically immediately adjacent — cl
   - B. Siphons *separated* from one another by a space equal to at least one siphon diameter — se

FIG. 17.10

3. Number (per side) of outwardly bowing, longitudinal folds in the branchial basket (requires removal of outer tunic)
   - A. Provide *number* of folds per side (6 or 7) — 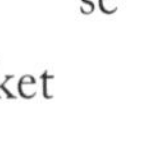
   - B. Folds in branchial basket absent, surface dotted with *papillae* — pa

4. Life-history stage released from adult following reproduction. This character may be visible upon dissection; mature viviparous species will have tadpole larvae within atrial chamber. Warning: larval production is seasonal; absence of larvae does not prove oviparity.
   A. *Oviparous,* eggs released; never with tadpole larva upon dissection — ov
   B. *Viviparous;* tadpole larvae develop in atrial chamber before release — vi

17.9

Attachment to Substrate	Siphon Spacing	Folds	Eggs or Larva	Species
at	cl	pa	ov	*Bostrichobranchis pilularis*
at	cl	6	vi	*Molgula complanata*
at	cl	7	ov	*M. siphonalis*
at	cl	6	ov	*M. manhattensis,* common sea grape
at	se	7	ov	*M. retortiformis*
at	se	7	vi	*M. citrina,* orange sea grape
at	se	6	ov	*M. provisionalis*
x	se	6	ov	*M. arenata*

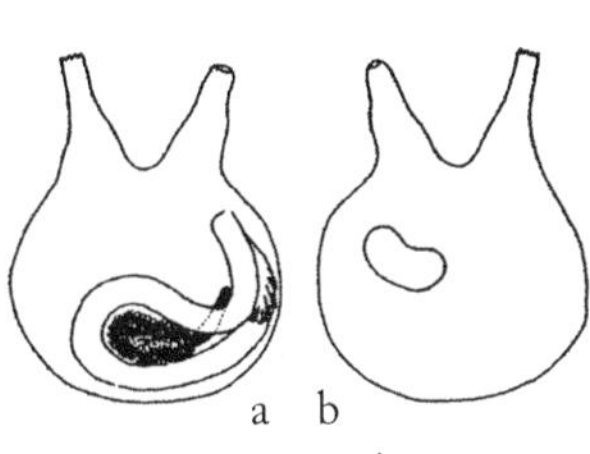

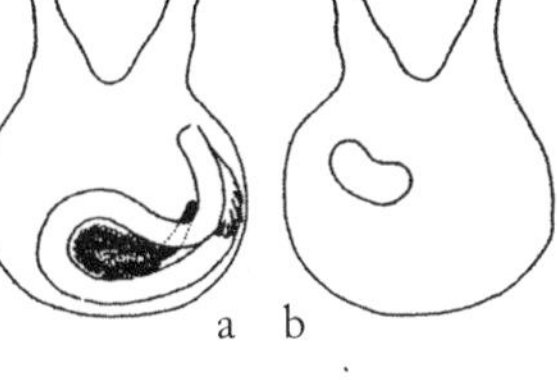

FIG. 17.11

Bostrichobranchis pilularis. Often covered with mud. Bo,Sa–Mu,Su–De,ff,25 mm (1 in), (G272,P84,V438). (FIG. 17.11. a, right side; b, left side; both with tunic removed; shows gut and gonad on left side.)

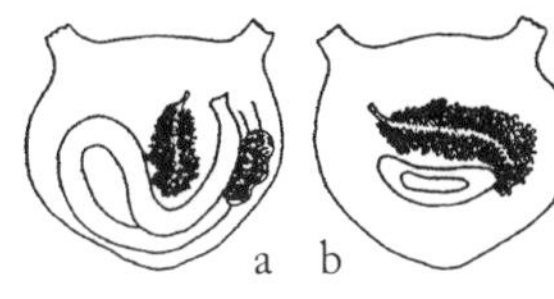

FIG. 17.12

Molgula arenata. Buries in sand; sand encrusts body. Vi+,Sa–Mu,Su,ff,22 mm (0.8 in), (G272,P84,V393), ☺. (FIG. 17.12. a, right side; b, left side;both with tunic removed to show gonads and gut.)

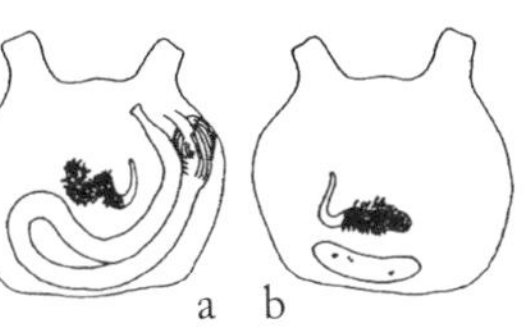

FIG. 17.13

Molgula citrina, orange sea grape. Gonads orange. Usually clean of debris. Ac–,Ha,Li–Su,ff,18 mm (0.7 in), (G272,NM272,P79,V379), ☺. (FIG. 17.13. a, right side; b, left side; both with tunic removed to show gonads and gut.)

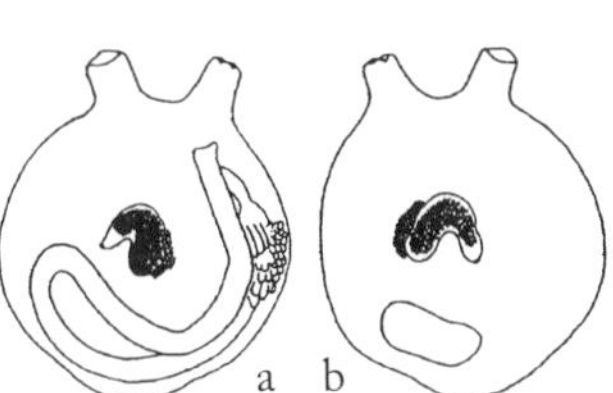

FIG. 17.14

Molgula complanata. Siphons can be pinkish; attach using slender fibers. Ac–,Ha–Sa–Gr,Li–De,ff,15 mm (0.6 in), (P79,V382). (FIG. 17.14. a, right side; b, left side; both with tunic removed to show gonads and gut.)

Molgula manhattensis, common sea grape. Grayish green; tunic often hairy and muddy. Tolerates pollution. Sometimes in large groups. Bo,Ha–SG,Es–Li–Su,ff,5.1 cm (2 in), (G272,NM743,P81,V385), ☺. (FIG. 17.15. a, right side; b, left side; both with tunic removed to show gonads and gut.)

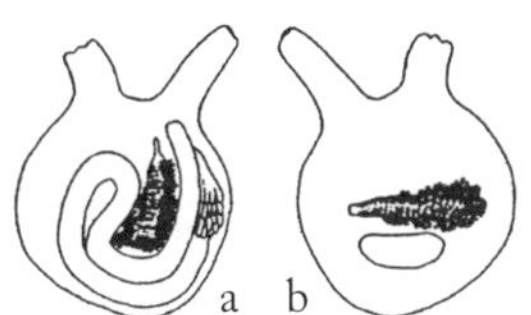

FIG. 17.15

Molgula provisionalis. Maine northward. Po–,Ha–SG,Li–,ff,15 mm (0.6 in), (G272,P82,V389).

Molgula retortiformis. Large arctic species. Po–,Ha,Li–Su,ff,7.6 cm (3 in), (G272,P83,V422). (FIG. 17.16. a, right side; b, left side; both with tunic removed to show gonads and gut.)

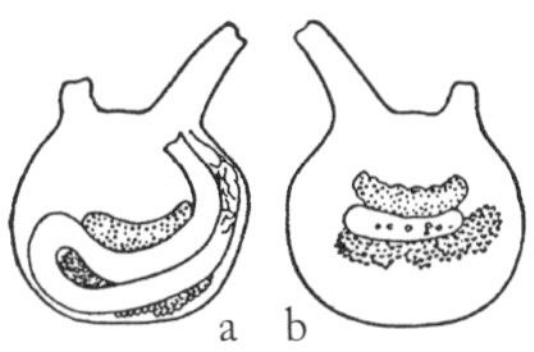

FIG. 17.16

Molgula siphonalis. Attaches by threads. Often covered with coarse sand. Arctic species. Po–,Ha,Li–Su,ff,25 mm (1 in), (G272,P82,V377).

17.10 **Colonial Tunicates** (from 17.5)

Tiny individual zooids are embedded in a common matrix. Identification may require dissection to view internal features.

1. Growth form
   - A. Colony *thin* sheet (<⅛ in thick); basically two-dimensional growth — tn
   - B. Colony *thick* sheet or lobes (>⅛ in thick): basically three-dimensional growth — tk
2. Colony surface with obvious, openings (often with small incurrent openings and fewer but larger collective outflow openings) visible to the unaided eye (FIG. 17.17)
   - A. Surface with many clear *circular* openings — ci*
   - B. Surface with scattered *cross-shaped* openings — cr
   - C. Surface *lacks* distinct openings visible to the unaided eye — x

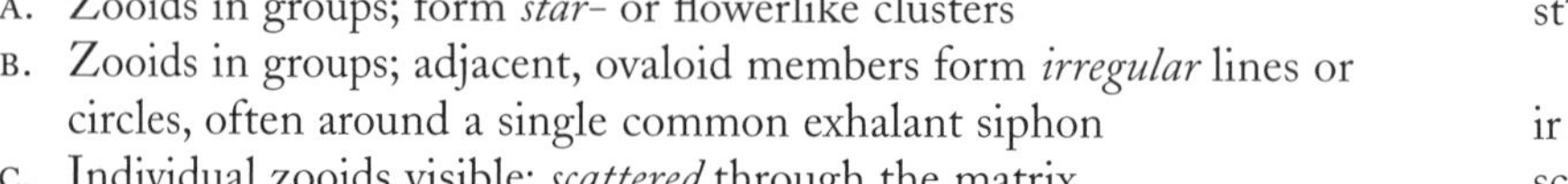

3. Arrangement of individuals within the common matrix (scrape away surface debris if present) (FIG. 17.17)
   - A. Zooids in groups; form *star*- or flowerlike clusters — st*
   - B. Zooids in groups; adjacent, ovaloid members form *irregular* lines or circles, often around a single common exhalant siphon — ir
   - C. Individual zooids visible; *scattered* through the matrix — sc
   - D. Individual zooids *not* visible — x

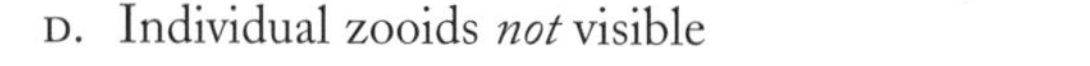

FIG. 17.17

4. Dominant colors
   - A. Whole colony *bright reddish* or purple — br
   - B. Whole colony *bright orange* — bo
   - C. Individuals are *pale*, yellowish to orangish within translucent matrix — pa
   - D. Clustered members *contrast* with background; light zooids in dark matrix or the opposite — co*
   - E. Colony appears *grayish* or white — gr

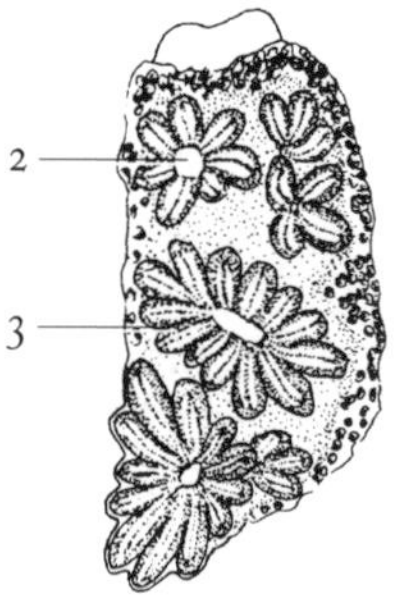

17.11

Growth Form	Openings	Individuals	Color	Species
tk	ci	ir	br,bo	*Botrylloides diegense,* orange sheath tunicate
tk	ci	st	co	*Botryllus schlosseri,* golden star tunicate
tn	cr,x	x	gr	*Didemnum* spp., northern white crust tunicate. Sometimes slippery feeling, but firm from embedded calcareous spicules: Spicules with points about equal in width and height: *D. albidum* OR Spicules with points longer than wide: *D. candidum*
tk	x	sc,st,x	pa	*Aplidium* (=*Amaroucium*), sea pork, 17.12

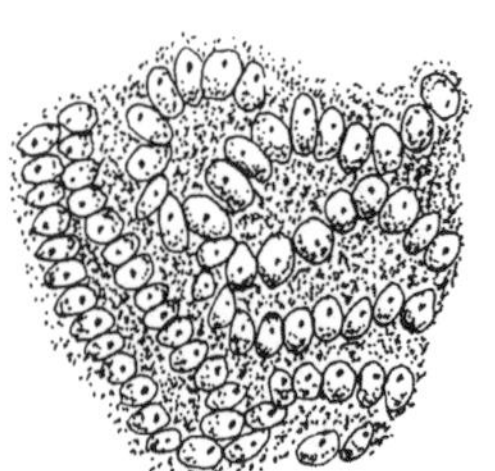

FIG. 17.18

Botrylloides diegense, orange sheath tunicate (Styelidae). Zooids (2.5 mm [0.1 in] diameter) arranged within matrix in rows, loose circles, or dense clusters around occasional, raised excurrent pores. Bright, uniform coloration. Apparently introduced on east coast at Woods Hole, Massachusetts in early 1980s (see Berman et al. 1992). Rapidly increasing in abundance and distribution north and south of Cape Cod. Ac–,Ha–Al,Su,ff,12.7 cm (5 in), (NM738,P89,V226). (FIG. 17.18. Portion of colony.)

Botryllus schlosseri, golden star tunicate (Styelidae). Zooids with white or yellow margins. Wide range of striking color patterns. Bo,Ha–Al–SG,Es–Li–Su,ff,10.1 cm (4 in), (G267,NM738,P88,V220), ☺. (FIG. 17.17. Colony.)

Didemnum albidum (Didemnidae). Cream to white; feels gritty. Ac,Ha–Al,Su,ff,10.1 cm (4 in), (G267,NM731,P65,V80). (FIG. 17.19. a, skeletal spicules; b, colony.)

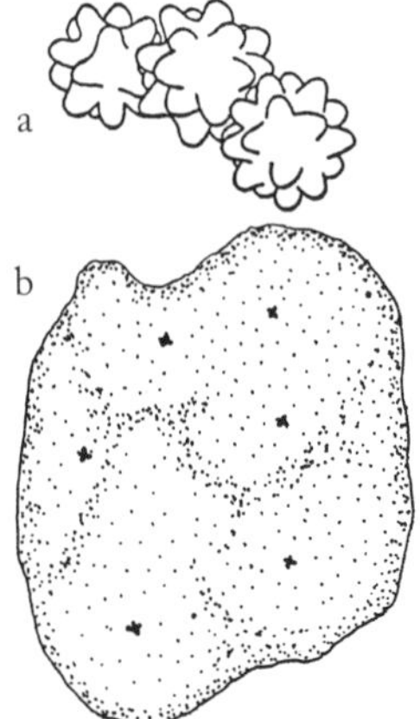

FIG. 17.19

Didemnum candidum (Didemnidae). Vi+,Sa–Gr,Su–De,ff,10.1 cm (4 in), (G267,NM731,P66,V83), ☺. (FIG. 17.20. Sharp skeletal spicules.)

FIG. 17.20

17.12 *Aplidium* (=*Amaroucium*) (Polyclinidae) (from 17.11)

Identification of *Aplidium* requires cutting some individuals out of the common matrix to view ridges on the stomach (orange-colored in fresh specimens). Sizes given are colony dimensions.

FIG. 17.21

1. Count the number of longitudinal ridges in the stomach of individuals dissected out of the common gelatinous matrix: provide *number* of ridges (FIG. 17.21) —
2. Sand coating
   - A. *Sand* coating present, embedded in gelatinous matrix sa
   - B. Sand coating *absent* x
3. Colony shape
   - A. Overall *dome*-shaped do
   - B. Irregular *lobes* lo

17.13

Ridges	Sand	Shape	Species
18–25	sa	do	*Aplidium constellatum* (=*pellucidum forma constellatum*), northern sea pork
10–12	x	lo	*A. stellatum*, common sea pork

Aplidium constellatum (=*pellucidum forma constellatum*), northern sea pork. Zooids reddish orange. Forms stalked lobes, upright in sand. Bo,Ha–Gr,Li–,ff,7.6 cm (3 in) wide by 25 mm (1 in) high. (G268,NM730,P64,V38), ☺. (FIG. 17.22. a, individual zooid; b, colony.)

Aplidium stellatum (=*palladum*), common sea pork. Upright, irregular lobes; zooids red-orange, arranged in star groupings. Thick, salmon-colored masses, common on beaches following storms. Bo,Ha–Al–Sa,Li–De,ff,30.5 cm (12 in) wide by 25 mm (1 in) high. (P64,V34), ☺. (FIG. 17.23. a, individual zooid; b, colony.)

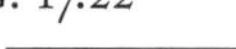

FIG. 17.22

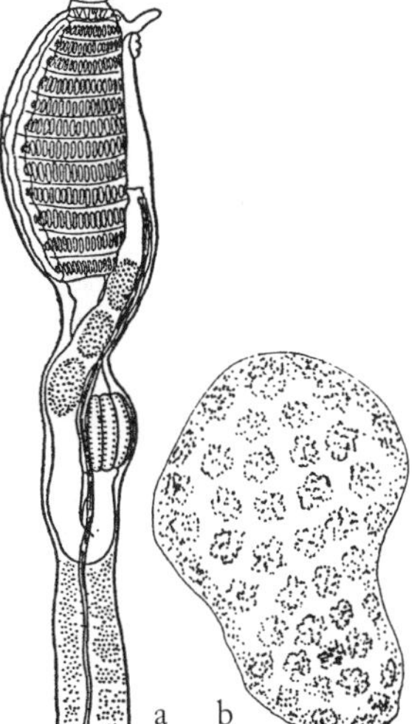

FIG. 17.23

17.14 Class Thaliacea, Salps

Salps are found individually or in colonial chains of individuals. In general, their bodies are transparent cylinders of cellulose material, normally ringed or nearly ringed by *muscle bands*. Action of these bands draws water into the anterior end, through a perforated pharynx, and out the posterior end, "jet propelling" the animal forward. Individuals of the same species differ markedly depending on whether they occur singly or in a colonial group. As planktonic drifters, several species could drift into inshore waters along the northeastern coastline. Only the most common ones have been included here.

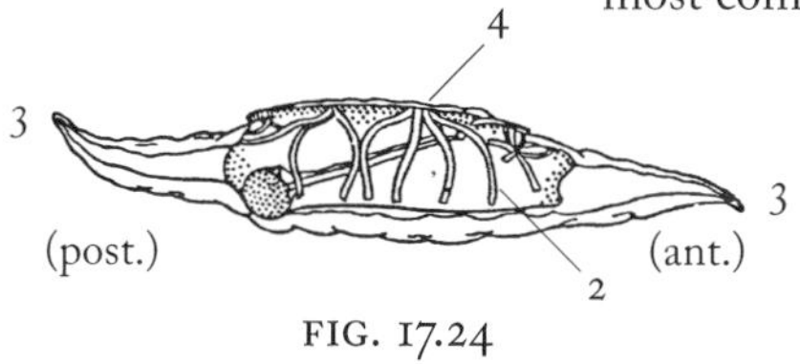

FIG. 17.24

1. Form
   - A. *Solitary* so
   - B. *Colonial*, forming chains co
2. Number of distinct muscle bands fully or partially ringing the animal's body: provide *number* of bands (FIG. 17.24) —
3. Presence of a distinctly elongate point (not just a small pointed prominence) at the anterior end and at the posterior end (FIG. 17.24. Anterior to right.)
   - A. Anterior end with *point*: posterior end with *point* po:po*
   - B. Anterior end *lacks* a pointed part: posterior end with *point* x:po
   - C. Anterior end *lacks* a pointed part: posterior end *lacks* a pointed part x:x
4. For colonial or chain forms only, number of muscles among the anteriormost four bands that fuse dorsally (FIG. 17.24)
   - A. Provide *number* of bands that fuse dorsally —
   - B. Anteriormost bands converge dorsally but do *not* actually fuse x

17.15

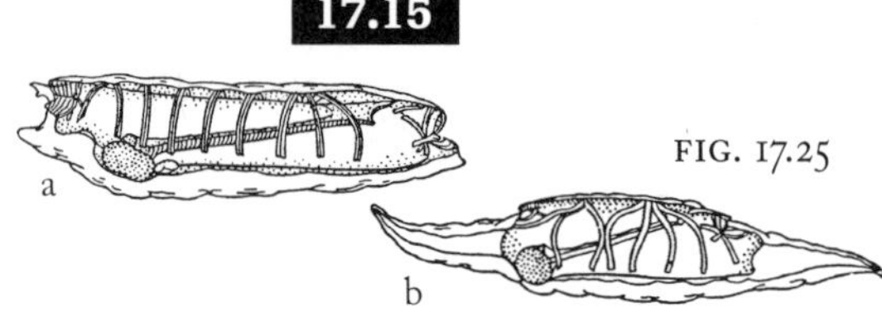
FIG. 17.25

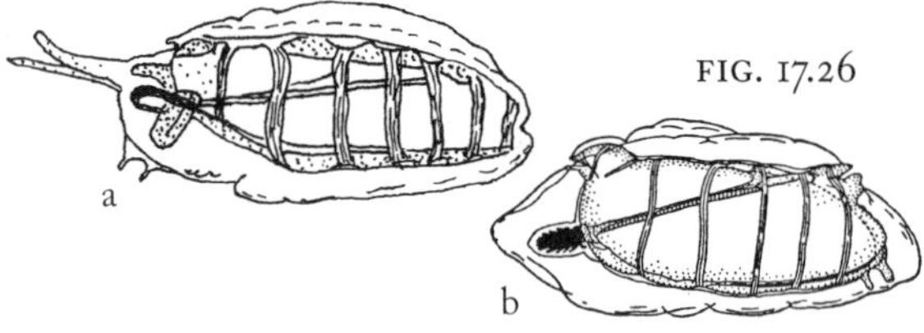
FIG. 17.26

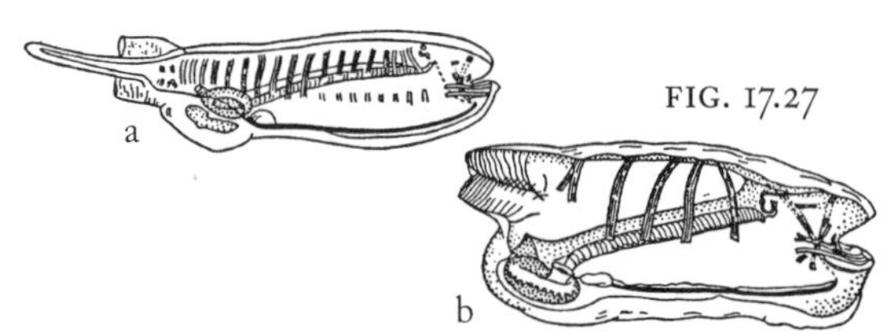
FIG. 17.27

Form	Bands	Points	Fusion	Species
so	18–22	x:po	–	*Thetys vagina*
so	9–10	x:x	–	*Salpa fusiformis,* common salp
so	6–7	x:po	–	*Thalia (=Salpa) democratica,* horned salp
co	6	po:po	4	*Salpa fusiformis*
co	5	x:x,po	3	*Thalia (=Salpa) democratica*
co	5	x:x	x	*Thetys vagina*

Salpa fusiformis, common salp. Solitary form: smoothly cylindrical body. Gulf Stream drifter. Colonial form: elongated, with points. Vi,Pe,Ps,ff,8.1 cm (3.2 in), (G273,NM745). (FIG. 17.25. a, solitary form; b, colonial form.)

Thalia (=Salpa) democratica, horned salp. Solitary form: compact, with two posterior points. Colonial form: smooth, cylindrical surface. Vi,Pe,Ps,ff,25 mm (1 in), (G273). (FIG. 17.26. a, solitary form; b, colonial form.)

Thetys vagina. Solitary form: very large; prefers warmer waters. Vi,Pe,Ps,ff,21.6 cm (8.5 in). (FIG. 17.27. a, solitary form; b, colonial form.)

17.16 Subphylum Vertebrata

Annotation coding used for habitat and lifestyle characters in the preceding sections on the invertebrates has been modified for the fishes, reptiles, and marine mammals. The notation for this section is provided below.

A. Geographic range: same designations as for the invertebrates.

B. Habitat type is indicated by a two-letter code.

Code	Description	Code	Description
Es	Estuarine or brackish water	Li	Littoral or intertidal
Co	Coastal or shallow water	SM	Salt marsh
Oc	Oceanic or offshore		

When appropriate, distributions are also characterized by depth ranges (in feet). Also where benthic or demersal or bottom-dwelling fishes are associated with particular substrates, these are indicated by a two-letter code.

Code	Description	Code	Description
ha	hard substrate such as rock or cobble	mu	mud
sa	sandy substrate, including gravel		

C. Trophic type
Because the majority of fishes are carnivorous, their dietary preferences, where known, are symbolized by a two- or three-letter code.

Code	Description	Code	Description	Code	Description
An	annelid worms	He	herbivore, plants	Ot	other
As	Ascidians or tunicates	In	invertebrates in general	Ph	phytoplankton
Cr	Crustacea	Ma	mammals	Pl	plankton in general
Ec	echinoderms	Mo	shelled molluscs	Sc	scavengers—detritus
Fi	fishes	Mos	molluscs, squid	Zo	zooplankton

D. Length

Unless otherwise noted, measurements given are for maximal length in northeastern North American waters. In some cases, the largest recorded size is given but is followed, in brackets, by more-typical large-size limits. For example, for a species listed at "6.1 m (20 ft) [4.9 m (16 ft)]," a record length of 6.1 meters (20 feet) is recorded, but 4.9 meters (16 feet) is a more typical upper limit for length.

17.17 Fishes: Superclass Agnatha (Jawless Fishes) and Superclass Gnathostomata (Jawed Fishes)

Basic understanding of the external morphology of fishes is prerequisite to their identification. Multiple openings from the gill area are typical of the cartilaginous or nonbony fishes, whereas a single bony flap, the *operculum,* covers the gill area of bony fishes, creating a single, crescent-shaped, external gill opening. Fins fall into two broad categories. *Paired fins,* found in all but the jawless fishes, include *pectoral fins,* usually located close to the gill operculum, and *pelvic fins,* that usually occur somewhere along the ventral surface. Unpaired *medial fins* are found vertically oriented along the fish's midline. The midback *dorsal fin* takes on a variety of shapes used here for preliminary sorting of bony fishes. *Caudal fins* provide a conspicuous identification character as well. Finally, the *anal fin* is found along the midventral line usually just posterior to the anus or vent. Fins of bony fishes are supported either by rays or by a combination of rays and spines. *Rays* are segmented in construction, typically Y-shaped with bifurcate ends, and are flexible. In contrast, *spines* are solid, rigid, and sharply pointed.

A careful review of a fish's construction can also reveal much about its lifestyle and relationships. For example, its overall body shape, types of sensory structures, size of its fins and mouth, its dentition, the condition of its gill rakers and gill filaments, and its coloration are often linked to survival needs. Similarly, the deployment of fins, complexity of jaw construction, composition of skeleton, and number of gill openings reflect ancestry. All of these features are used in making identifications.

Fishes are the most diverse vertebrates. Cryptic coloration, rapid movement, and secretive habits make field identifications especially difficult. Keys provided here work primarily for fishes "in the hand." In all cases, characters listed here apply to intact, adult animals. Juveniles may differ in both appearance and habits. Even among adults, however, the condition of the specimen along with individual color variations may confound identification. Be flexible in your interpretations and be prepared to try alternative routes should you wind up at an obvious mismatch. In most cases, separations used here have been based on useful morphology and not on systematic criteria. Family designations for each species have been provided, either by group or individually.

Scientific and common names follow recommendations of the Committee on Names of Fishes of the American Fisheries Society (Robins et al. 1991). Common names followed by a dagger (†) are designated standard names, and users are urged refer to fishes by these names. In some cases, widely used, alternative names are also given to facilitate linkage to scientific names and to preferred, standard names.

References for this section: (B) Bigelow, H. B., and W. C. Schroeder, 1953; (H) Hildebrand, S. F., and W. C. Schroeder, 1928; (M) McClane, A. J., 1978; (S) Scott, W. B., and M. G. Scott, 1988.

1. Presence of distinct lower jaw (FIG. 17.28)
   - A. *Jaw* present (Superclass Gnathostomata) ja*
   - B. Jaw *absent* (Superclass Agnatha) x

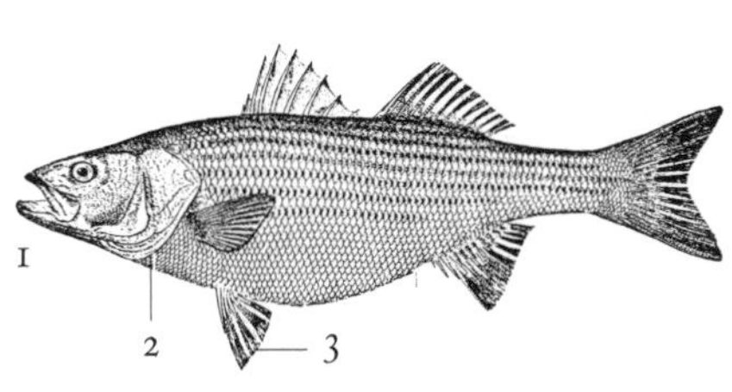

FIG. 17.28

2. Number of gill openings visible on one side of head (FIG. 17.28)
   - A. *Single* slit or hole on each side of head 1*
   - B. *Five* vertical gill slits on each side of head 5
   - C. *Five* gill slits on *ventral* surface of head 5v
   - D. *Seven* round openings on either side of head 7
3. Presence of at least one set of paired fins (pectoral and/or pelvic fins) (FIG. 17.28)
   - A. At least one set of *paired fins* present pf*
   - B. Paired fins *absent* x

17.18

Jaw	Gill Opening	Paired Fins	Group or Species
ja	1	pf	Bony fishes, 17.23
ja	5	pf	Sharks, 17.19
ja	5v	pf	Skates and rays, 17.21
x	7	x	*Petromyzon marinus,* sea lamprey†
x	1	x	*Myxine limosa,* Atlantic hagfish†

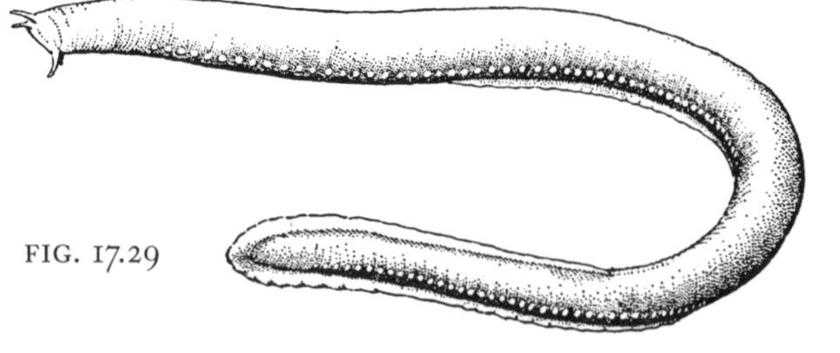
FIG. 17.29

Myxine limosa, Atlantic hagfish† (Cl: Agnatha, Myxinidae). Cartilaginous skeleton; scaleless; six tentacles around face. Deep water, soft-mud bottoms. Produces copious mucus when disturbed. Often misidentified as *M. glutinosa;* western Atlantic hagfish are darker, often with a pale middorsal streak (see Wisner and McMillan 1995). Bo,90–120,Fi–In–Sc,78.7 cm (31 in), (B10,R15,S3), ☺. (FIG. 17.29)

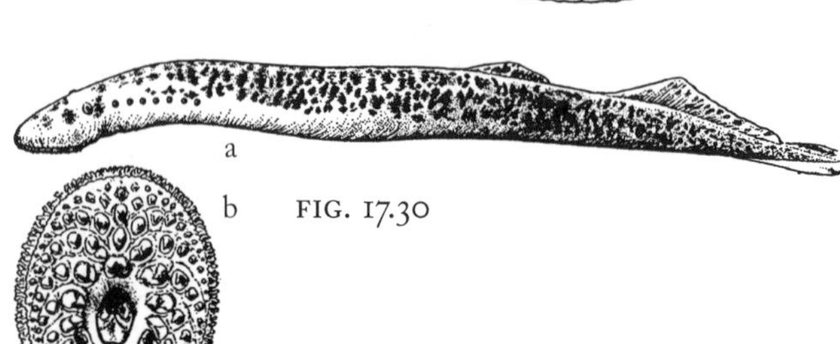

FIG. 17.30

Petromyzon marinus, sea lamprey† (Cl: Agnatha, Petromyzontidae). Cartilaginous skeleton; two dorsal fins. Anadromous, breeds in fresh water, lives much of its life in saltwater. Feeds by attaching to various fishes and eating their tissues. Bo,Oc–Es,Fi,91.4 cm (36 in), (B12,H43,R14,S6). (FIG. 17.30. a, animal; b, mouth disk.)

Class Chondrichthyes, Cartilagenous Fishes

17.19 **Sharks** (from 17.18)

Sharks are an easily distinguished group of fish. Their large dorsal, pectoral, and caudal fins are distinctive, as is the pointed snout with subterminal (ventral, back from tip of snout), well-toothed mouth. Species distinctions emphasize relative size and placement of fins. Sharks have been the object of considerable press, much of it negative. Documented shark attacks on humans in our region are few, but these widely ranging predatory fish always warrant caution.

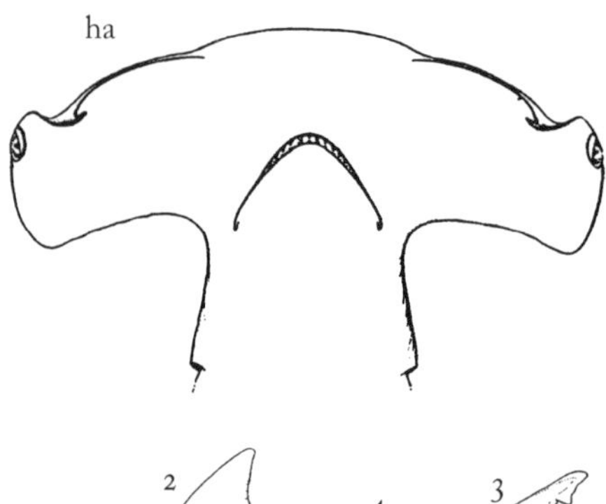

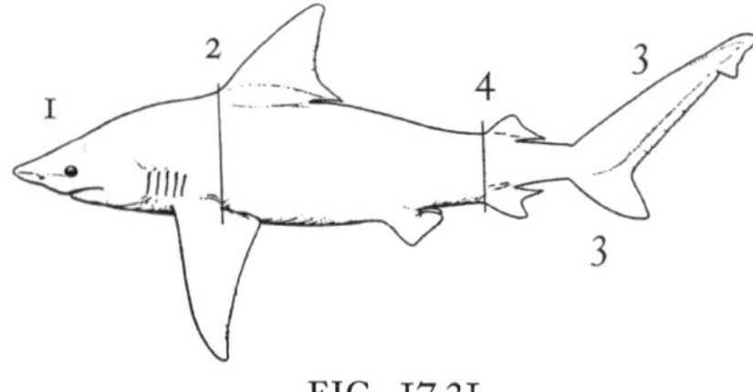

FIG. 17.31

1. Distinctive hammerhead shape to head, eyes at edges of squared lateral flanges (FIG. 17.31)
   - A. *Hammerhead* present — ha
   - B. Hammerhead *absent* — x
2. Position of first dorsal fin relative to paired fins (pectoral and pelvic fins) (FIG. 17.31)
   - A. Dorsal fin begins at a point *over* the posterior part of the *pectoral* fins — op*
   - B. Dorsal fin begins posterior to the pectoral fins, but distinctly closer to *pectoral* fins than to pelvic fins — pc
   - C. Dorsal fin distinctly closer to *pelvic* fins — pl
   - D. Dorsal fin *midway* between pectoral and pelvic fins — mi
   - E. Dorsal fin begins *posterior* to the *pelvic* fins — p/d
3. Shape of tail; dorsal lobe compared to ventral lobe (measure along outer edges of respective lobes) (FIG. 17.31)
   - A. *Dorsal* and *ventral* lobes about *equal* in length, or ventral lobe longer than dorsal lobe — d=v
   - B. Dorsal lobe from *1.3 to 2* times longer than ventral lobe — 1.3–2
   - C. Dorsal lobe from *two to four* times longer than ventral lobe — 2–4*
   - D. Dorsal lobe *more than four* times longer than ventral lobe — >4

4. Relative position of anterior edge of second dorsal fin relative to anterior edge of the anal fin (i.e., the posteriormost ventral fin) (FIG. 17.31)
   A. Second *dorsal* fin starts distinctly *anteriorly* to *anal* fin — d/a
   B. Second *dorsal* fin starts *above* or even slightly posterior to anal fin — d=a*
   C. Anal fin *absent* — x

17.20

Hammerhead	Dorsal vs. Paired	Tail Lobes	Dorsal vs. Anal	Species
ha	pc	2–4	d=a	*Sphyrna zygaena,* smooth hammerhead†
x	op	2–4	d=a	*Carcharhinus* (in part) Most fins with distinct black tips: *C. limbatus,* blacktip shark† OR No black tip to fins: *C. plumbeus* (=*milberti?*), sandbar shark†, brown shark
x	pc	2–4	d=a	*Carcharhinus* (in part) First dorsal fin begins above midpoint of pectoral fins: *C. leucas,* bull shark† OR Dorsal fin begins above posterior margin of pectoral fins: *C. obscurus,* dusky shark†
x	op,pc	2–4	d/a	*Mustelus canis,* smooth dogfish†
x	op	1.3–2	d/a	*Carcharodon carcharias,* white shark†
x	op	1.3–2	d=a	*Lamna nasus,* porbeagle†, mackerel shark
x	pc	1.3–2	d=a	*Isurus oxyrinchus,* shortfin mako†
x	mi	1.3–2	d/a	*Cetorhinus maximus,* basking shark†
x	mi	2–4	x	*Squalus acanthias,* spiny dogfish†
x	mi,pl	>4	d/a	*Alopias vulpinus,* thresher shark†
x	pl	2–4	d=a	*Prionace glauca,* blue shark†
x	pl	2–4	d/a	*Odontaspis taurus,* sand tiger†
x	p/d	1.3–2	d=a	*Scyliorhinus retifer,* chain dogfish†
x	p/d	d=v	x	*Squatina dumeril,* Atlantic angel shark†

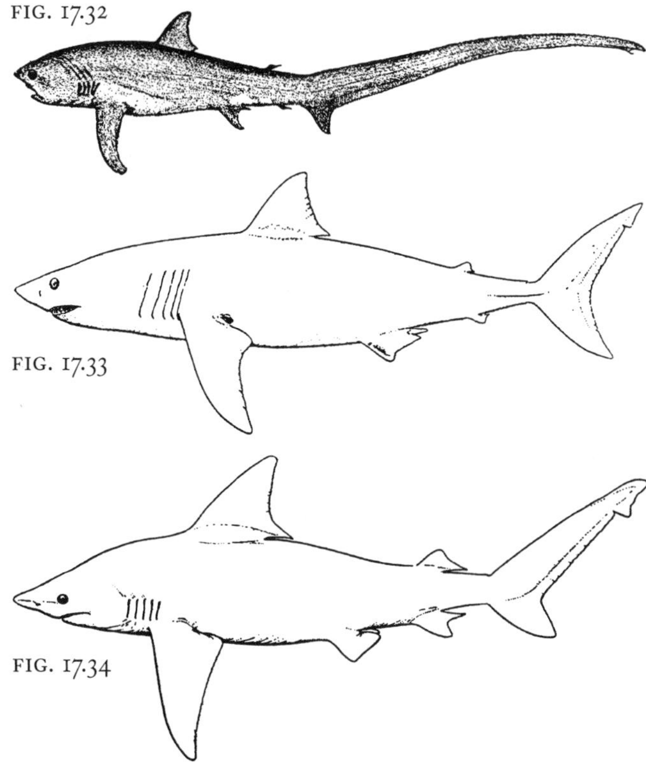

FIG. 17.32

FIG. 17.33

FIG. 17.34

Alopias vulpinus, thresher shark† (Alopiidae). Enormous caudal fin; very small second dorsal and anal fins. Brown or black above, white below. Feeds on surface schooling fishes. Bo,Oc,Fi,6.1 m (20 ft) [4.9m (16 ft)], (B32,M6,R18,S13). (FIG. 17.32)

Carcharodon carcharias, white shark† (Lamnidae). Second dorsal and anal fins very small; large triangular, serrated teeth; dangerous human-eater. Bo,Oc,Ma–Fi,11 m (36 ft) [6.4 m (21 ft)], (B25,H46,M12,R19,S14). (FIG. 17.33)

Carcharhinus limbatus, blacktip shark† (Carcharhinidae). Light stripe along sides of body. Vi,Co–Oc,Fi,2.5 m (8.3 ft), (M18,R26).

Carcharhinus plumbeus (=*milberti?*), sandbar shark†, brown shark (Carcharhinidae). Ridge between dorsal fins. Short snout; blue gray above, white below. Numerous off New York, New Jersey, and into Chesapeake Bay, where young ones hunt blue crabs in grassbeds. Vi,Co,Fi–Cr,2.4 m (8 ft), (B43,H48,M26,R27). (FIG. 17.34)

Carcharhinus leucas, bull shark† (Carcharhinidae). Very short snout; gray above, white below. Tolerates brackish or even fresh water. Dangerous. Vi,Co–Es,Fi,2.5 m (8.3 ft), (M22,R26).

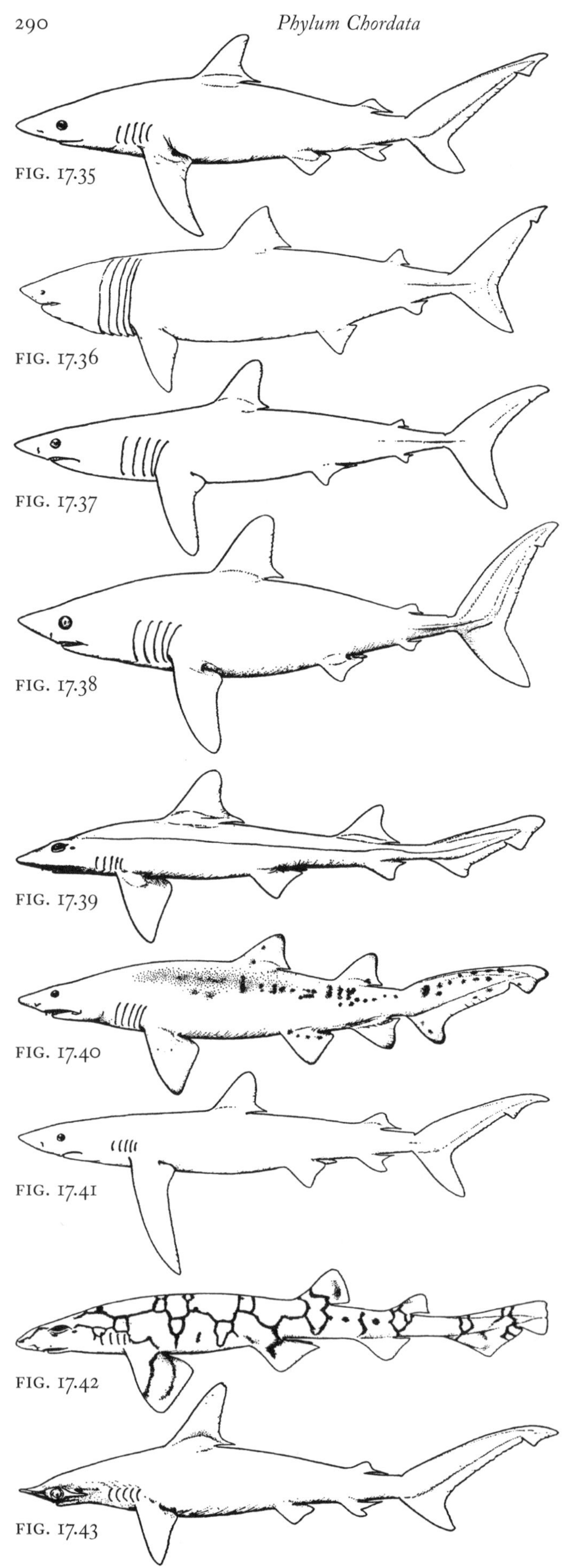

FIG. 17.35

FIG. 17.36

FIG. 17.37

FIG. 17.38

FIG. 17.39

FIG. 17.40

FIG. 17.41

FIG. 17.42

FIG. 17.43

Carcharhinus obscurus, dusky shark† (Carcharhinidae). Ridge between two dorsal fins. Blue gray above, white below. Dangerous. Bo,Oc,Fi,3.3 m (10.7 ft), (M21,R27,S25). (FIG. 17.35)

Cetorhinus maximus, basking shark† (Cetorhinidae). Huge gill slits; tiny teeth in large mouth and long gill rakers for plankton feeding. Grayish brown all over. Found at surface, slow and lazy, easy to approach. Bo,Oc,Pl,9.8 m (32 ft), (B28,M8,R19,S17), ☺. (FIG. 17.36)

Isurus oxyrinchus, shortfin mako† (Lamnidae). Slender tail base or caudal peduncle with single ridge or keel. Conical snout; short pectoral fins. Deep blue above, white below. Active, feeds on surface schooling fish; habit of leaping clear of water. Vi+,Oc,Fi,4 m (13 ft) [2.4 m (8 ft)], (B23,M9,R20,S19). (FIG. 17.37)

Lamna nasus, porbeagle†, mackerel shark (Lamnidae). Very slender tail base or caudal peduncle with two ridges or keels. Oval eyes; often with white patch toward base of dorsal fin. Oceanic. Ac–,Co–Oc,Fi,2.4 m (8 ft), (B20,M9,R20). (FIG. 17.38)

Mustelus canis, smooth dogfish† (Carcharhinidae). Low, flat teeth line jaw. Slender; large spiracle behind eyes. Olive above, yellowish gray below. Prefer larger crustacea and squid. Schools. Vi+,Co–Oc,Cr–Mo–Fi,1.5 m (5 ft) [0.9 m (3 ft)], (B34,H47,M27,R28,S26). (FIG. 17.39)

Odontaspis taurus, sand tiger† (Odontaspididae). Second dorsal about equal to first dorsal and anal fins; smallish pectoral fins. Greenish gray above, gray to white below. Summers in coastal New England. Bo,Co–Oc,Fi–Cr,2.7 m (9 ft) [1.8 m (6 ft)], (B18,M4,R18,S11), ☺. (FIG. 17.40)

Prionace glauca, blue shark† (Carcharhinidae). Pectoral fins very long and slender; dark, bright blue back; long snout. Prefers open water. Dangerous. Bo,Oc,Fi,3.4 m (11 ft) [2.4 m (8 ft)], (B38,M18,R29,S26), ☺. (FIG. 17.41)

Scyliorhinus retifer, chain dogfish† (Scyliorhinidae). Pale with darker netlike markings. Vi,120–750,?,43.2 cm (17 in), (M21,R22,S25). (FIG. 17.42)

Sphyrna zygaena, smooth hammerhead† (Sphyrnidae). Unmistakable head morphology. Olive brown above, white below. Dangerous. Vi+,Co–Oc,Fi–Cr,2.1 m (7 ft), (B45,H50,M17,R30). (FIG. 17.43)

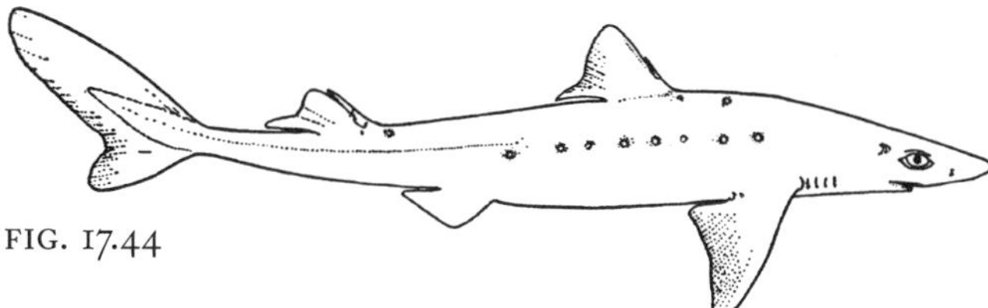

FIG. 17.44

Squalus acanthias, spiny dogfish† (Squalidae). Large, venomous spine in front of each dorsal fin; no anal fin. Small white spots scattered on body. Voracious and gregarious; seasonal migration up the coast in spring, return during mid-fall. Bo,Co,Fi–Cr,91.4 cm (36 in), (B47,H52,M32,R33,S35), ☺. (FIG. 17.44)

Squatina dumeril, Atlantic angel shark (Squatinidae). Blend of shark and ray characters. Very large, yet distinct, pectoral and pelvic fins; partially flattened dorsoventrally; swims like a shark, not a ray. Eyes on top of head; large spiracles; two small dorsal fins near tail. Vi,?–4260,?,1.5 m (5 ft), (H54,R34).

17.21 Skates (Rajidae) and Rays (Dasyatidae, Myliobatidae, and Torpedinidae)

(from 17.18)

Skates have thick tails with fins. They lay individual eggs within purse-shaped capsules known by beach combers as "mermaid's purses" (see 6.4). Rays have thin tails devoid of fins but with bony stingers. Rays bear their young live.

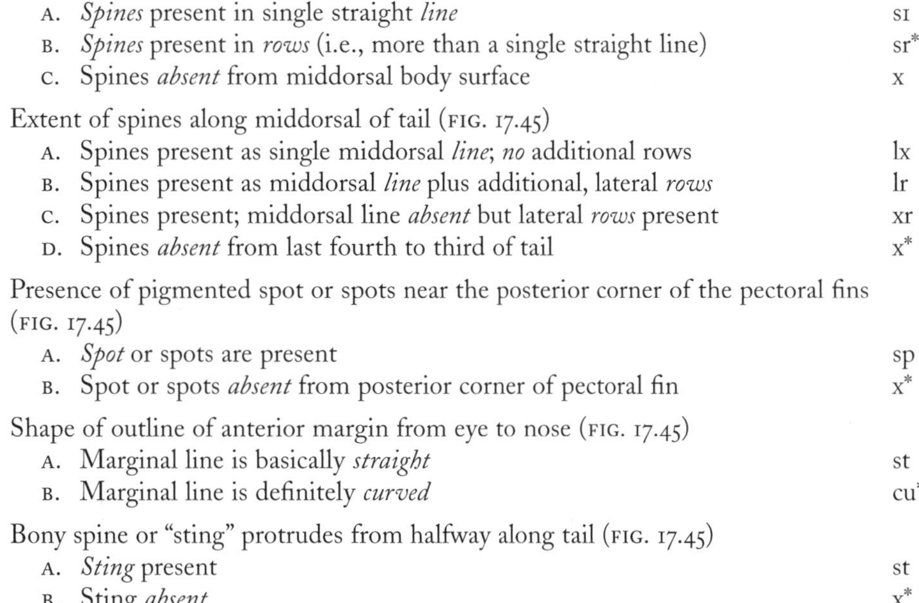

FIG. 17.45

1. Presence of distinct thorny spines along middorsal body between level of eyes and start of pelvic fins (FIG. 17.45)
   - A. *Spines* present in single straight *line* — sl
   - B. *Spines* present in *rows* (i.e., more than a single straight line) — sr*
   - C. Spines *absent* from middorsal body surface — x
2. Extent of spines along middorsal of tail (FIG. 17.45)
   - A. Spines present as single middorsal *line*; *no* additional rows — lx
   - B. Spines present as middorsal *line* plus additional, lateral *rows* — lr
   - C. Spines present; middorsal line *absent* but lateral *rows* present — xr
   - D. Spines *absent* from last fourth to third of tail — x*
3. Presence of pigmented spot or spots near the posterior corner of the pectoral fins (FIG. 17.45)
   - A. *Spot* or spots are present — sp
   - B. Spot or spots *absent* from posterior corner of pectoral fin — x*
4. Shape of outline of anterior margin from eye to nose (FIG. 17.45)
   - A. Marginal line is basically *straight* — st
   - B. Marginal line is definitely *curved* — cu*
5. Bony spine or "sting" protrudes from halfway along tail (FIG. 17.45)
   - A. *Sting* present — st
   - B. Sting *absent* — x*

17.22

Body Spines	Tail Spines	Spot	Margin	Sting	Species
x	x	x	cu	x	*Torpedo nobiliana,* Atlantic torpedo†
x	x	x	cu	st	*Rhinoptera bonasus,* cownose ray†
x	lr	x	st	x	*Raja laevis,* barndoor skate†
sr	x	x	cu	x	*R. senta,* smooth skate†
sr	x	x	cu	st	*Dasyatis say,* bluntnose stingray†
sr	xr	x	cu	x	*Raja erinacea,* little skate†
sr	xr	sp	cu	x	*R. ocellata,* winter skate†, big skate
sr	xr	x	st	st	*Dasyatis centroura,* roughtail stingray†
sr	lx	x	cu	x	*Raja radiata,* thorny skate†
sr	lr	sp	cu	x	*R. garmani,* rosette skate†
s1	lr	x	cu	x	*R. eglanteria,* clearnose skate†, brier skate

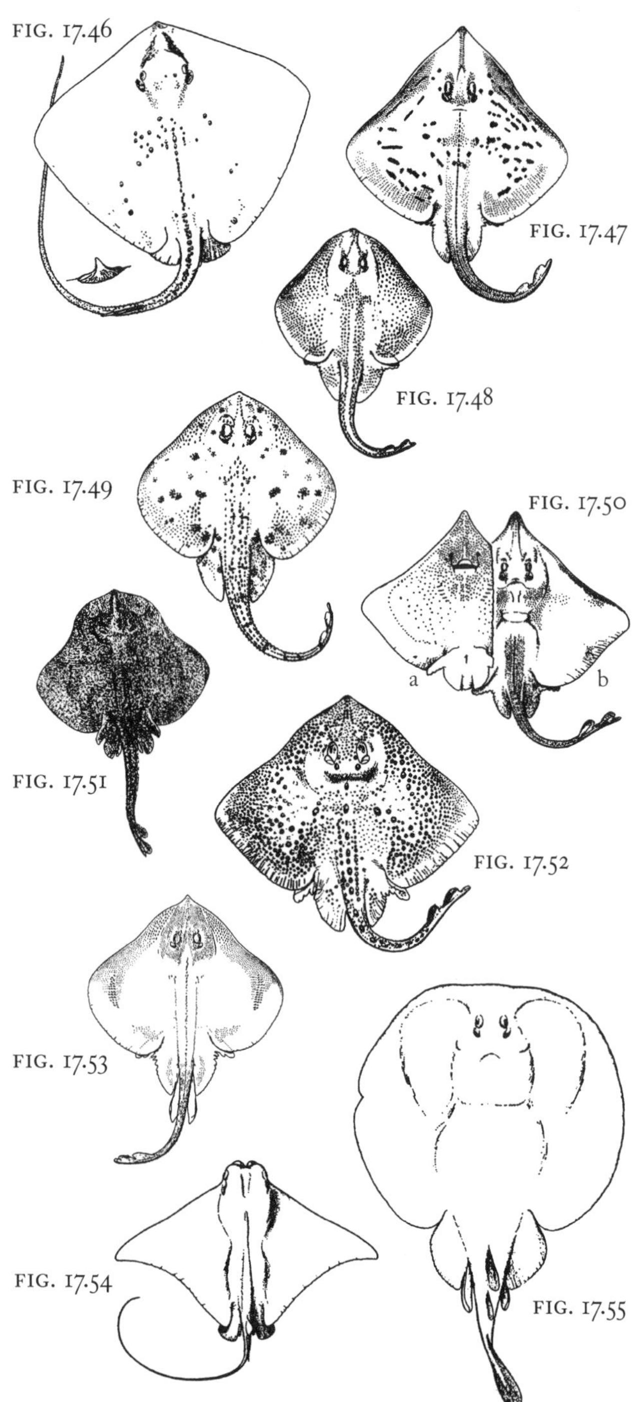

Dasyatis centroura, roughtail stingray† (Dasyatidae). Scattered spines over body surface. Vi,Co,Cr,4.3 m (14 ft), (H64,R41,S57). (FIG. 17.46)

Dasyatis say, bluntnose stingray† (Dasyatidae). Finlike skin folds along tail. Can inflict painful sting if provoked. Vi,Co,Cr–Fi,0.9 m (3 ft), (H66,R41).

Raja eglanteria, clearnose skate†, brier skate (Rajidae). Sharp nose. Irregular lines and bars among darker spots on dorsal surface. Two clear patches toward snout. Vi+,Co,Cr–Fi,83.8 cm (33 in), (B65,H58,M37,R38), ☺. (FIG. 17.47)

Raja erinacea, little skate† (Rajidae). Blunt nose. Sand or mud bottoms. Bo,Co,Cr–Mo–Fi,50.8 cm (20 in), (B67,H60,M38,R38), ☺. (FIG. 17.48)

Raja garmani, rosette skate† (Rajidae). Deep-water species. Vi,180–1740,?,40.6 cm (16 in), (B66,R39). (FIG. 17.49)

Raja laevis, barndoor skate† (Rajidae). Very pointed snout. Prefers shallow mud or sand bottoms. Bo,Co–360,Mo–An–Cr–Fi,1.5 m (5 ft), (B61,H59,M35,R39,S46). (FIG. 17.50. a, ventral; b, dorsal.)

Raja ocellata, winter skate†, big skate (Rajidae). Blunt nose. Eats rock crabs, squid, and polychaetes. Bo,Co,Cr–Mos–An–Fi,0.9 m (3 ft), (B63,M39,R38,S49). (FIG. 17.51)

Raja radiata, thorny skate† (Rajidae). Bo,60–2760,An–Cr–Fi,1 m (3.3 ft), (B38,S51), ☺. (FIG. 17.52)

Raja senta, smooth skate† (Rajidae). Ac–,150–3000,In?,61 m (24 in), (B70,R40,S53), ☺. (FIG. 17.53)

Rhinoptera bonasus, cownose ray† (Myliobatidae). Flattened nose in two distinct lobes; pointed pectoral "wings;" small fin at base of tail. Dig into sediments with pectoral fins to expose molluscs. Will inflict painful sting if irritated. Vi,Co,Cr–Mo,2.1 m (7 ft), (M48,R44). (FIG. 17.54)

Torpedo nobiliana, Atlantic torpedo† (Torpedinidae). Large caudal fin; dark brown coloring. Capable of generating electrical discharge of up to 220 volts. Bo,Oc–Co,Fi,1.8 m (6 ft), (B58,H62,R36,S39). (FIG. 17.55)

17.23 Class Osteichthyes, Bony Fishes

Identification of bony fishes involves examination of details beyond the number, shape, and deployment of fins. In some cases, species distinctions are made by counting the number of rays or spines supporting specific fins. Variability in feeding options is reflected in mouth construction. In closely related species, the overall size of the mouth can be gauged in comparison to the eye; for example, does the mouth end before or after the pupil of the eye? Surface feeders tend to have upturned mouths. Many fish capture food in front of them and thus have terminal mouths, whereas the mouths of benthic feeders are often subterminal. This construction can be evaluated by comparing the relative lengths of the upper and lower jaws. Finally, although body coloration is often highly variable, certain striking patterns or features can be useful.

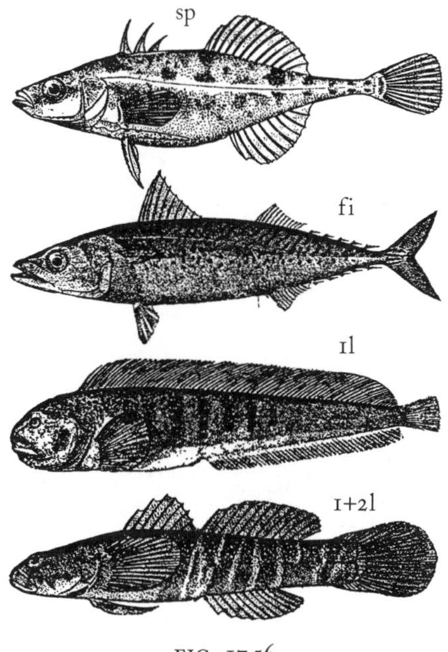

FIG. 17.56

1. Number and appearance of medial fins and spines along the dorsal surface between the head and caudal fin or tail (FIG. 17.56)
   - A. Distinct, isolated *spine*, spines, rays or other non-finlike prominences preceding fins along middorsal line — sp
   - B. Several small *finlets* occur just posterior to typical dorsal fin or fins — fi
   - C. *One long* fin, four or more times as long as it is high — 1l
   - D. *One short* fin, less than three times as long as it is high (often length = width) — 1s
   - E. *Two* distinct fins, each one about *equal* in length and height — 2e
   - F. Two distinct fins, the *first* much *longer* than it is high, the *second* about equal in height and width — 1l+2
   - G. Two distinct fins, the *first* about equal in height and length, the *second* much *longer* than high — 1+2l
   - H. *Three* fins along dorsal surface — 3
2. Distinct covering of scales present on lower half of body, even if minute (does not include fish with a few isolated scales, isolated row(s) of scales, or those with scales restricted to the head or back)
   - A. Many, overlapping or abutting *scales* present — sc
   - B. Scales *absent* — x

17.24

Dorsal Fins	Scales	Group
1s	sc	Fish Group 1, 17.25
1s	x	Fish Group 2, 17.31
1l	sc,x	Fish Group 3, 17.33
1+2l	sc	Fish Group 4, 17.49
1+2l,1l+2	x	Without sword: Fish Group 5, 17.57 OR With sword: Fish Group 6, 17.59
2e	sc	Fish Group 7, 17.61
2e	x	Fish Group 5, 17.57
3	sc,x	Fish Group 8, 17.63
fi	sc	Fish Group 9, 17.65
sp	sc,x	Fish Group 10, 17.67

17.25 **Fish Group 1** (from 17.24)

Scaled fish with one dorsal fin of about equal height and length.

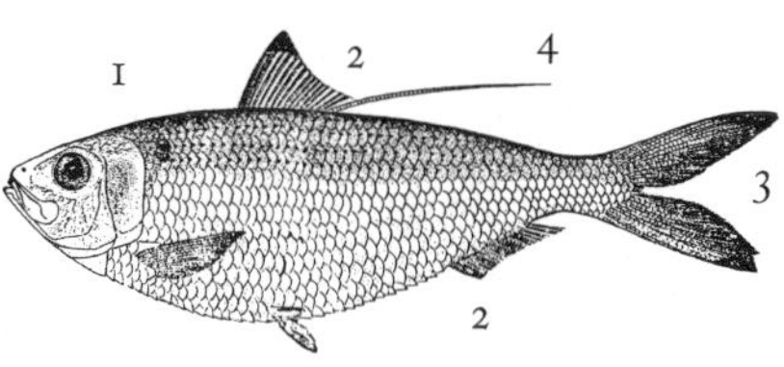

FIG. 17.57

1. Body shape: length (tip of snout to base of tail), compared to height at tallest point (use a ruler; not estimate) (FIG. 17.57)
   - A. Very slender, length at least *six* times height — >6
   - B. Body length is *four to six* times height — 4–6
   - C. Body length is *three to four* times height — 3–4
   - D. Body length is *two to three* times height — 2–3*
2. Relationship of dorsal fin to anal fin (FIG. 17.57)
   - A. *Dorsal* fin ends *anterior* to point where *anal* fin begins — d/a*
   - B. *Dorsal* fin *overlaps* start of *anal* fin — d=a
3. Shape of caudal fin (FIG. 17.57)
   - A. Top lobe of caudal fin distinctly longer than bottom lobe (*heterocercal* tail) — he
   - B. Homocercal tail (symmetrical), *deeply crescent* — dc*
   - C. Homocercal tail, *squared* — sq
   - D. Homocercal tail, *rounded* — ro

4. Long, threadlike projection from posterior edge of dorsal fin (FIG. 17.57)
   A. *Thread* projection present — th*
   B. Thread projection *absent* — x

17.26

Length vs. Height	Dorsal vs. Anal	Caudal Fin Shape	Thread Projection	Species
>6	d=a	ro	x	*Aspidophoroides monopterygius,* alligatorfish†
>6	d=a	dc	x	*Hyporhamphus unifasciatus,* silverstripe halfbeak†
4–5	d=a	dc	x	*Anchoa* With 24–27 rays in anal fin: *A. mitchilli,* bay anchovy† OR With 20–24 rays in anal fin: *A. hepsetus,* striped anchovy†
4–5	d/a	dc	x	*Elops saurus,* ladyfish†, ten pounder
4–5	d/a	he	x	*Acipenser oxyrhynchus,* Atlantic sturgeon†
3–4,4–5	d=a	ro,sq	x	Killifishes, 17.27
3–4	d/a	dc	x,th	Herrings (in part), 17.29
2–3	d/a	dc	th	Posterior thread reaches nearly to tail base: *Opisthonema oglinum,* Atlantic thread herring† OR Posterior thread extends halfway to tail base: *Dorosoma cepedianum,* gizzard shad†
2–3	d=a	sq	x	*Cyprinodon variegatus,* sheepshead minnow†

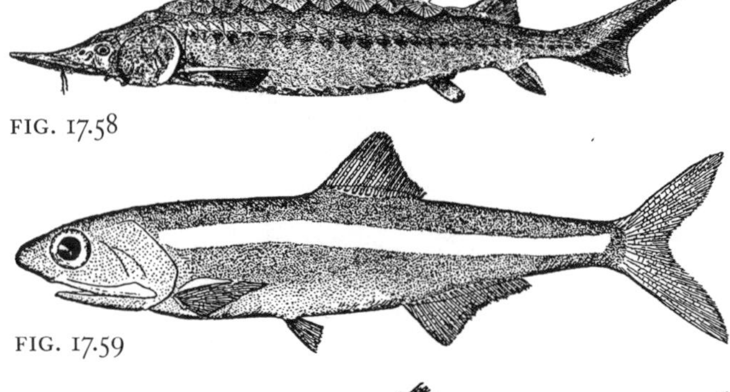

FIG. 17.58

FIG. 17.59

Acipenser oxyrhynchus, Atlantic sturgeon† (Acipenseridae). Rows of bony shields dorsally and laterally; long narrow snout; four barbels in front of mouth. Anadromous; 7 years in river, goes to sea, returns at 9 or 10 years for first spawn. At least 7000 in Hudson River. Bo,Es–Co(sa–mu),Cr–Mo–An–Fi,3.7 m (12 ft) [3 m (10 ft)], (B81,H72,R46,S68), ☺. (FIG. 17.58)

Anchoa hepsetus, striped anchovy† (Engraulidae). Ca+,Co,Pl,15.1 cm (6 in) [12.6 cm (5 in)], (B119,R73,S119). (FIG. 17.59)

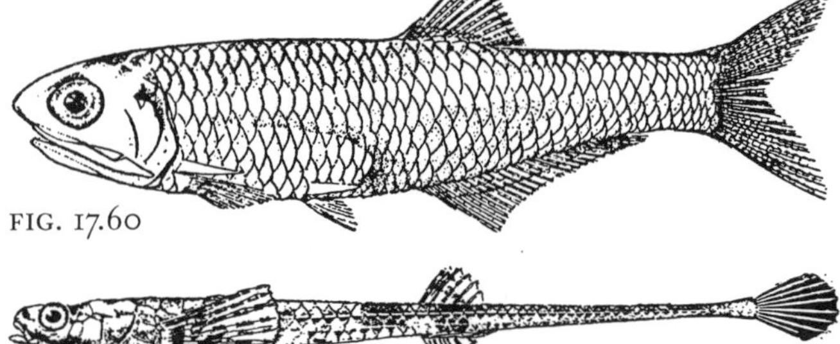

FIG. 17.60

FIG. 17.61

Anchoa mitchilli, bay anchovy† (Engraulidae). Large, thin, easily detached scales. Whitish or silvery with dark dots. Bo,Co–Es(sa),Pl,8.9 cm (3.5 in), (H109,R74), ☺. (FIG. 17.60)

Aspidophoroides monopterygius, alligatorfish† (Agonidae). Head and body encased in "armor;" very slender. Ac,58–630,In?,17.8 cm (7 in), (B457,R284,S514). (FIG. 17.61)

Cyprinodon variegatus, sheepshead minnow† (Cyprinodontidae). Large pectoral fins. Olive above, yellow below. Females with black posterior edge to dorsal and anal fins. Vi,SM–Es–Co,He–Fi,7.6 cm (3 in), (B165,H135,M81,R108), ☺. (FIG. 17.62)

FIG. 17.62

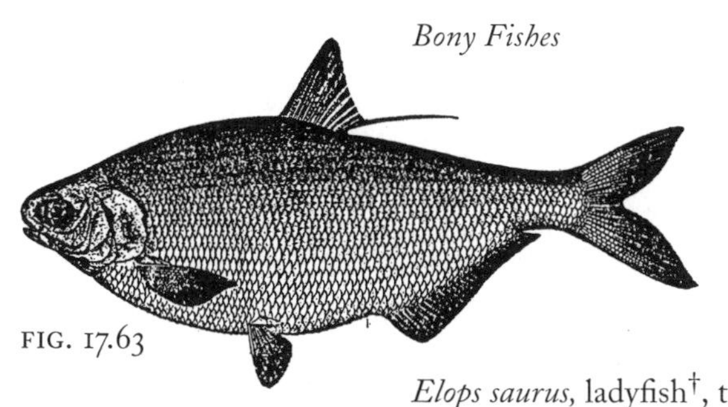

FIG. 17.63

Dorosoma cepedianum, gizzard shad† (Clupeidae). Metallic blue above, silver below. Small or absent dark spot behind operculum. Mouth slightly subterminal below bulbous snout. Vi,Es–Co,In–He–Sc,38.1 cm (15 in), (H106,R70), ☺. (FIG. 17.63)

Elops saurus, ladyfish†, ten pounder (Elopidae). Large mouth. Vi,Es?–Co,Cr,91.4 cm (36 in) [50.8 cm (20 in)], (B86,H78,R47). (FIG. 17.64)

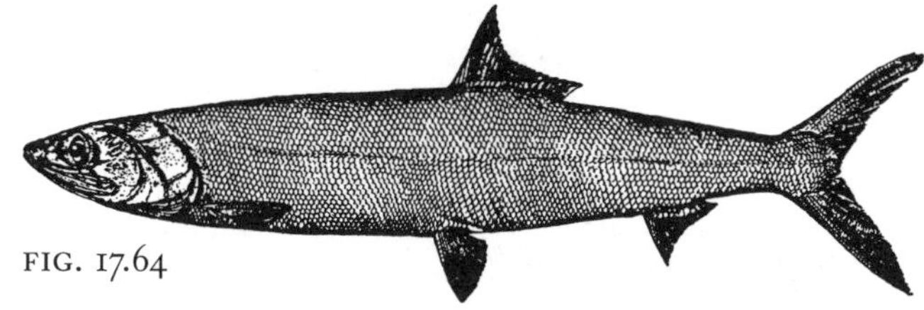

FIG. 17.64

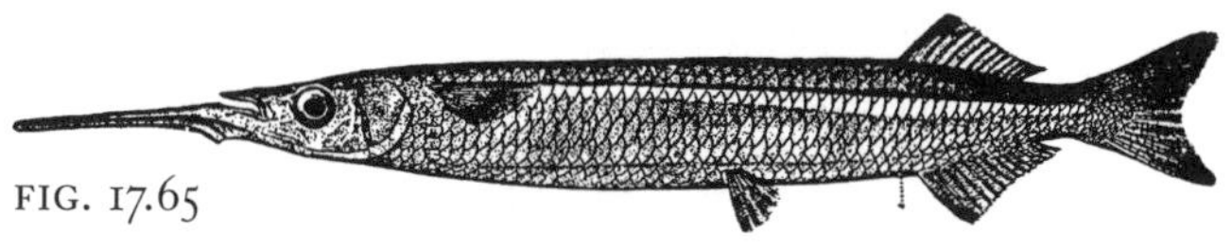

FIG. 17.65

Hyporhamphus unifasciatus, silverstripe halfbeak† (Exocoetidae). Long lower jaw, very short upper jaw. Greenish above, silvery below. Can jump clear of water. Vi,Co,Cr–Mo,30.5 cm (12 in), (B169,H153,R102,S310). (FIG. 17.65)

Opisthonema oglinum, Atlantic thread herring† (Clupeidae). Dark spot just behind opercular plate. Vi,Co,Cr–Pl,30.5 cm (12 in), (B112,H101,R70). (FIG. 17.66)

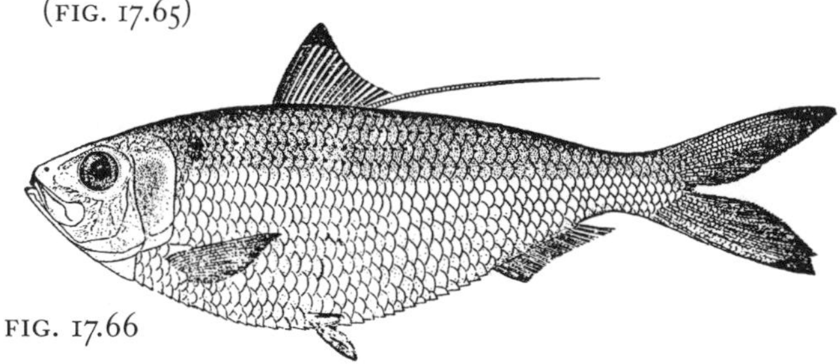

FIG. 17.66

17.27 Killifishes or Bull Minnows (Cyprinodontidae) (from 17.26)

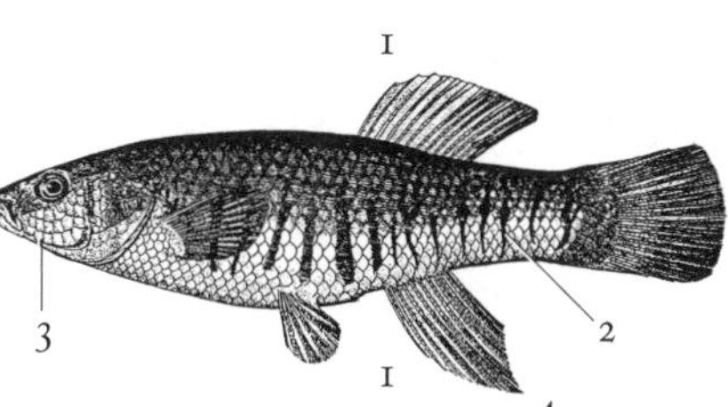

FIG. 17.67

1. Location of dorsal fin relative to anal fin (FIG. 17.67)
   - A. *Dorsal* fin begins directly *above* the *anal* fin — d=a
   - B. *Dorsal* fin begins distinctly *anteriorly* to *anal* fin — d/a*
2. Presence of dark stripes (horizontal) or bars (vertical) on body (FIG. 17.67)
   - A. Provide *number* of vertical *bars* — __b*
   - B. Provide *number* of horizontal *stripes* — __s
3. Appearance of "cheek" scales, between eye and ventral margin of head (FIG. 17.67)
   - A. *Two* rows of *rectangular* scales — 2re
   - B. *Three* rows of *square* scales — 3sq*
   - C. Cheek scales *absent* — x
4. Proportions of anal fin (FIG. 17.67)
   - A. Anal fin length nearly *twice* its width at its line of insertion — 2x*
   - B. Anal fin clearly *less than twice* its width — <2
5. Number of rakers or bony points along anterior surface of each gill (this feature is useful if the specimen is already dead; you may need to snip away one opercular cover to see the gills): provide *number* of gill rakers per gill — __

17.28

Dorsal vs. Anal	Stripes or Bars	Cheek Scales	Anal Fin	Gill Raker No.	Species
d=a	11–14b	x	<2	x	*Fundulus luciae,* spotfin killifish†
d/a	14–22b	x	<2	4–6	*F. diaphanus,* banded killifish†
d/a	10–18b	2re	<2	8–12	*F. heteroclitus,* mummichog†, killifish, chub
d/a	15–20b	3sq	2x	?	*F. majalis,* striped killifish†–male
d/a	2–4s	3sq	2x	?	*F. majalis,* striped killifish†–female

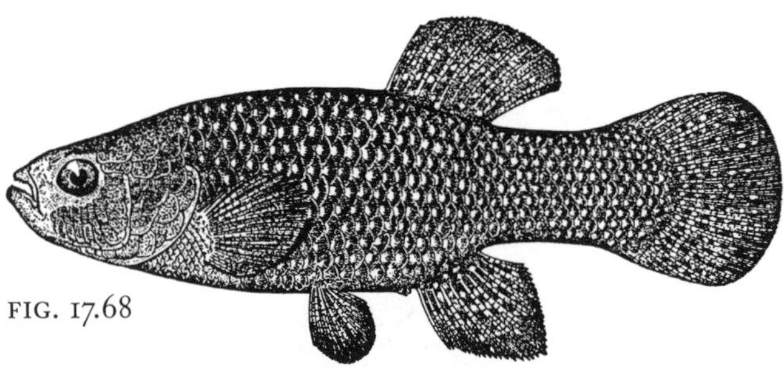

FIG. 17.68

Fundulus diaphanus, banded killifish†. Slender. Fresh water to brackish water. Bo,Es–SM,Cr–Mo–An–He,10.9 cm (4.3 in), (H143,R109), ☺.

Fundulus heteroclitus, mummichog†, killifish, chub. Schools along sheltered shore, eelgrass, and salt marshes. Bo,Es–SM–Co,In–Fi–He,15.2 cm (6 in) [10.1 cm (4 in)], (B162,H138,R109,S314), ☺. (FIG. 17.68)

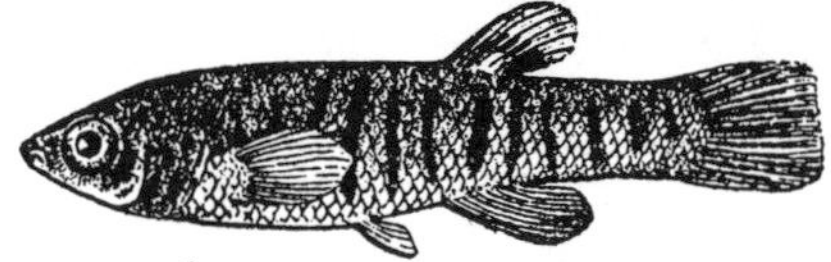

FIG. 17.69

Fundulus luciae, spotfin killifish†. Vi,Es–Co,Cr–Mo–An,3.8 cm (1.5 in), (H144,R109). (FIG. 17.69)

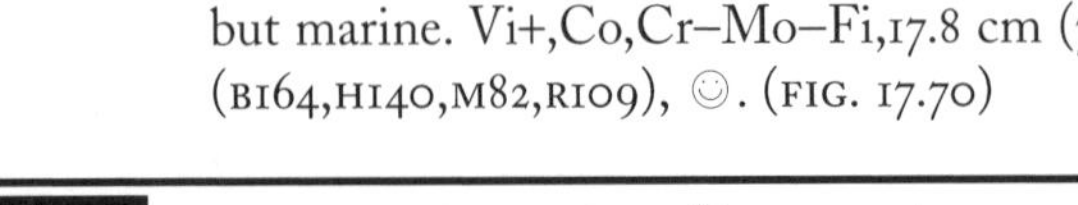

FIG. 17.70

Fundulus majalis, striped killifish†. Paler in color with darker bars than *F. heteroclitus.* Shallow but marine. Vi+,Co,Cr–Mo–Fi,17.8 cm (7 in), (B164,H140,M82,R109), ☺. (FIG. 17.70)

17.29 Herrings (Clupeidae) (from 17.26)

Many members of the herring family are described as *anadromous,* which means that they spend much of their lives in offshore schools, frequently feeding on plankton, but after 3 to 5 years of maturation, they return in spring to the freshwater streams of their origins to spawn. Most adults have returned to the sea by summer, whereas their offspring are apt to be stalled in brackish waters for a while before migrating to the sea during the fall.

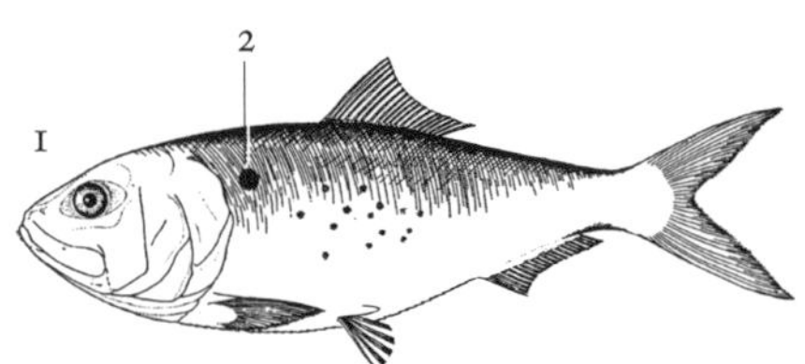

FIG. 17.71

1. Contour of upper jaw bone (FIG. 17.71)
   - A. Anterior margin of upper jaw sharply *angled* upward so that lower jaw projects anteriorly — an
   - B. Upper jaw *not* angled sharply upward — x*
2. Dark spot or spots behind operculum or gill cover (FIG. 17.71)
   - A. *Single* dark spot behind operculum — 1
   - B. Spots for a single *row* — ro
   - C. *Scattered* dark spots — sc
   - D. *Single* dark spot and *scattered* smaller spots — 1:sc*
   - E. Dark spot or spots *absent* from behind operculum — x
3. Coloration of anal fin
   - A. *Translucent* — tr
   - B. *Greenish* — gr
   - C. *Brown* — br
   - D. *White* — wh
   - E. *Yellow* — ye

17.30

Jaw	Spot(s)	Anal Fin	Species
x	ro	gr	*Alosa sapidissima,* American shad†, white shad
x	1:sc,sc	ye	*Brevoortia tyrannus,* Atlantic menhaden†, pogy, fatback, bunker
an	1,x	gr,ye	*Alosa* Interior lining of belly is dark: *A. aestivalis,* blueback herring†, glut herring OR Interior lining of belly is light or pink: *A. pseudoharengus,* alewife†, gaspereau, sawbelly, big-eye or branch herring
an	ro	tr	*A. mediocris,* hickory shad†, greenback
an	x	tr,wh	*Clupea harengus,* Atlantic herring†

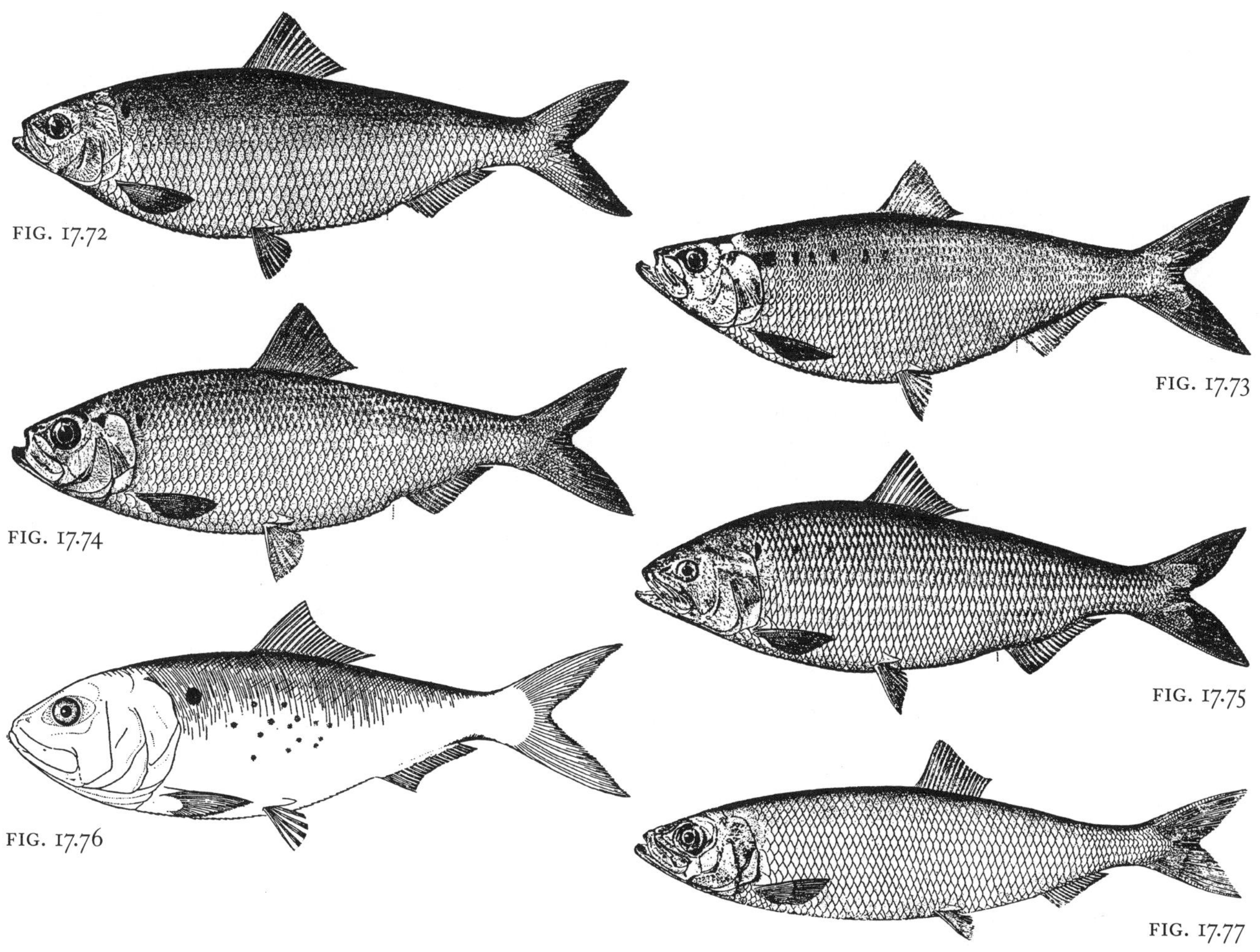

FIG. 17.72

FIG. 17.73

FIG. 17.74

FIG. 17.75

FIG. 17.76

FIG. 17.77

Alosa aestivalis, blueback herring†, glut herring. Bo,Es–Co,Zo,37.7 cm (15 in) [27.6 cm (11 in)], (B106,R67,S102). (FIG. 17.72)

Alosa mediocris, hickory shad†, greenback. More of a fish-eater than other herring. Bo,Es–Co,Cr–Fi,61 cm (24 in), (B100,R68). (FIG. 17.73)

Alosa pseudoharengus, alewife†, gaspereau, sawbelly, big-eye or branch herring. A little higher (deeper) body than blueback. Bo,Es–Co,Zo,38.1 cm (15 in) [27.9 cm (11 in)], (B101,H89,R67,S105). (FIG. 17.74)

Alosa sapidissima, American shad†, white shad. Mouth extends to posterior edge of eye. Dark blue to green above, light and silvery below. Surface schools. Preferred eating fish; roe is a delicacy. Anadromous. Bo,Es–Co,Zo,76.2 cm (30 in), (B108,H93,R68,S109), ☺. (FIG. 17.75)

Brevoortia tyrannus, Atlantic menhaden†, pogy, fatback, bunker. Large scaleless head; dark blue back. Large schools at surface on warm days. Oily; used for chicken food and fertilizer. Not anadromous; enters estuaries to feed. Bo,Es–Co,Ph–Zo,55 cm (20 in) [38.1 cm (15 in)], (B113,H102,M76,R69,S112), ☺. (FIG. 17.76)

Clupea harengus, Atlantic herring†. Dorsal fin begins relatively farther back on body than other herring. Schooling fish. Bo,Co–Oc,Zo,43.2 cm (17 in), (B88,H81,M75,R69,S114), ☺. (FIG. 17.77)

17.31 Fish Group 2 (from 17.24)

Scaleless fish with one dorsal fin of about equal height and length, or at least less than three times as long as it is tall.

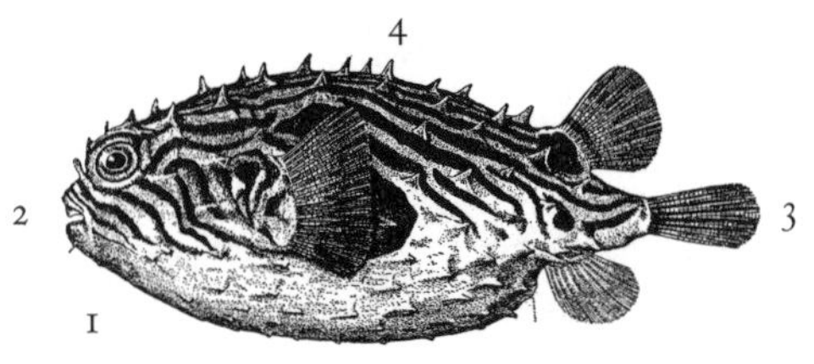

FIG. 17.78

1. Distinctive shapes (FIG. 17.78)
   - A. *Seahorse* shape with long, pointed prehensile tail — se
   - B. Dorsoventrally flattened, frying pan or *skillet* shape — sk
   - C. Body *inflates* like balloon when disturbed — in*
   - D. *Slender* body, more than four times as long as it is high — sl
   - E. Large, *laterally* flattened body with long pointed dorsal and anal fins — la
   - F. Body otherwise, *not* matching these distinctive shapes — x
2. Shape of mouth (FIG. 17.78)
   - A. *Tiny* mouth at end of tapered *snout* — ts
   - B. No tapered snout, but *small* mouth ending anterior to eyes — sm*
   - C. *Larger* mouth, extends at least to eye level — la
3. Shape of caudal fin (FIG. 17.78)
   - A. Caudal fin *slightly crescent* — sc
   - B. Caudal fin *rounded* — ro*
   - C. Caudal fin *squared* — sq
   - D. Caudal fin with *scalloped margin* — sm
   - E. Caudal fin *absent* — x
4. Body spines (FIG. 17.78)
   - A. *Large,* sharp *spines* on back, sides and abdomen — ls*
   - B. *Small,* sharp spines on *back* and *abdomen* — s/ba
   - C. Skin with *small,* sharp spines on *abdomen* only — s/a
   - D. Bony plates form horizontal *ridges* along sides — ri
   - E. Several longitudinal rows of heavy, triangular *studs* — st
   - F. Such body spines *absent* — x

17.32

Body Shape	Mouth	Caudal Fin	Spines	Species
in,x	sm	sc	s/a	*Lagocephalus laevigatus,* smooth puffer†, rabbitfish
in,x	sm	ro	s/ba	*Sphaeroides maculatus,* northern puffer†
in,x	sm	ro	ls	*Chilomycterus schoepfi,* striped burrfish†, porcupinefish
la	sm	sm	x	*Mola mola,* ocean sunfish†
x	sm	sq	st	*Cyclopterus lumpus,* lumpfish†
sl	ts	ro	ri	*Syngnathus* Dorsal fin long, with 35–41 rays: *S. fuscus,* northern pipefish† OR Dorsal fin short, with 28–30 rays: *S. floridae,* dusky pipefish
se	ts	x	ri	*Hippocampus erectus* (=*hudsonius*), lined seahorse†
sk	la	ro	x	*Gobiesox strumosus,* skilletfish†, oyster clingfish

FIG. 17.79

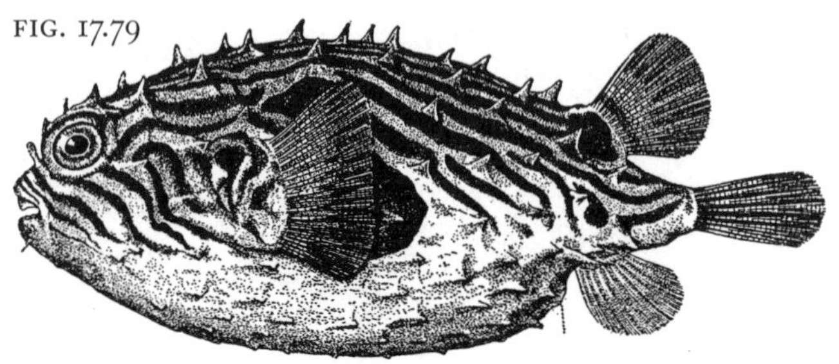

Chilomycterus schoepfi, striped burrfish†, porcupinefish (Diodontidae). Stout, triangular spines. Contoured, roughly parallel dark lines over dorsal half of body. Eats hermit crabs in particular. Vi,Co?,Cr, 25.4 cm (10 in), (B527,H350,R309). (FIG. 17.79)

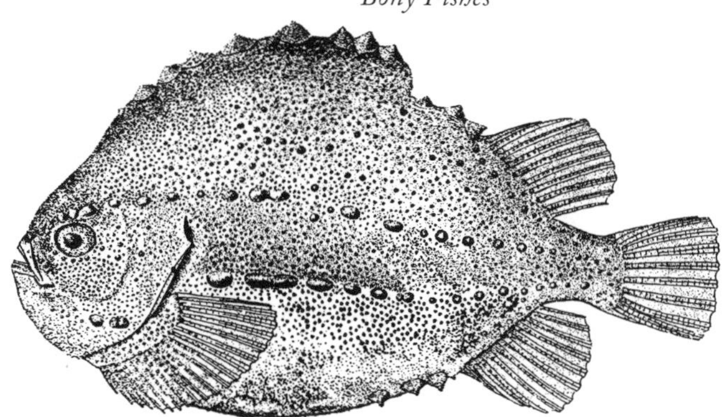

FIG. 17.80

Cyclopterus lumpus, lumpfish† (Cyclopteridae). Cartilaginous skeleton. Female larger than male. Attaches to objects by anteroventral sucker. Bo,Co–329 (ha),In–Fi,58.4 cm (23 in) [40.6 cm (16 in)], (B459,H311,R285,S518), ☺. (FIG. 17.80)

Gobiesox strumosus, skilletfish†, oyster clingfish (Gobiesocidae). Olive brown with darker mesh pattern. Pelvic fins form ventral sucker for attachment. Vi,Es–Co,Cr–An,7.6 cm (3 in), (H339,R86).

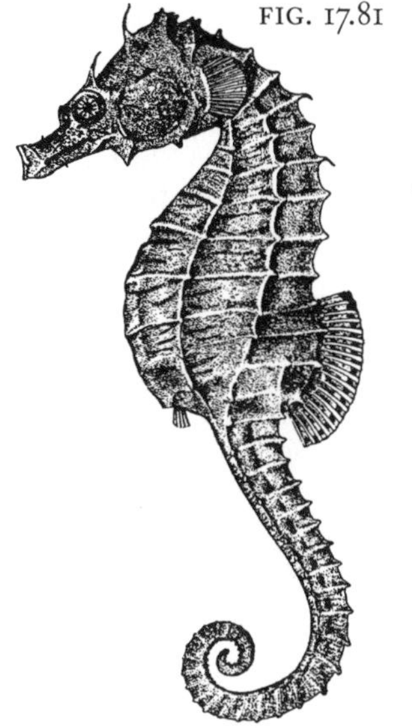

FIG. 17.81

Hippocampus erectus (=*hudsonius*), lined seahorse† (Syngnathidae). Male with brood pouch. Eelgrass or algal beds. (See 17.48 for description.) (FIG. 17.81)

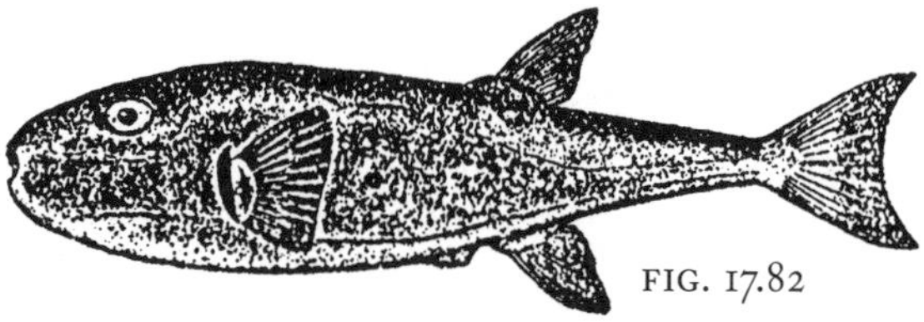

FIG. 17.82

Lagocephalus laevigatus, smooth puffer†, rabbitfish (Tetraodontidae). Vi,Co,?,61 cm (24 in), (H347,M94,R305). (FIG. 17.82)

Mola mola, ocean sunfish† (Molidae). Thick slime layer; tough skin; cartilaginous skeleton. Often heavily parasitized. Surface drifter. Bo,Oc,In–Fi,2.4 m (8 ft) [1.5 m (5 ft)], (B529,R310), ☺. (FIG. 17.83)

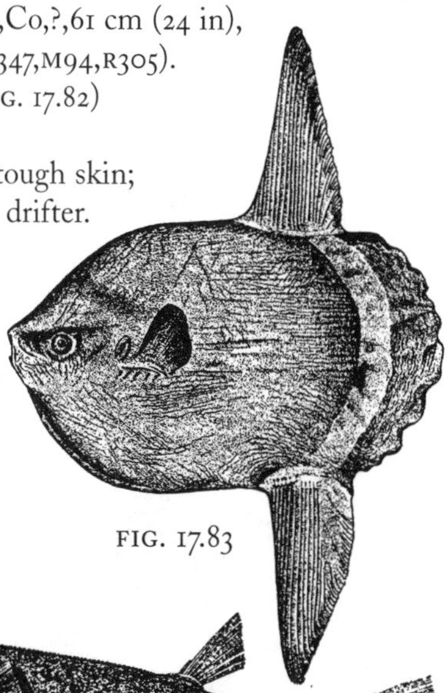

FIG. 17.83

Sphaeroides maculatus, northern puffer†, balloonfish (Tetraodontidae). Small mouth at tip of snout; gill openings small and slanted. Eyes set high on head. Inflates with water or air when disturbed. Greenish above, yellowish below with dark bars. Vi+,Es–Co,Cr–Mo–An,35.6 cm (14 in) [25.4 cm (10 in)], (B526,H365,M94,R307). (FIG. 17.84)

Syngnathus floridae, dusky pipefish, (Syngnathidae). Mottled coloring. Subtidal grassbeds. Ca+,Es–SG,Cr,22.9 cm (9 in), (H183,R126).

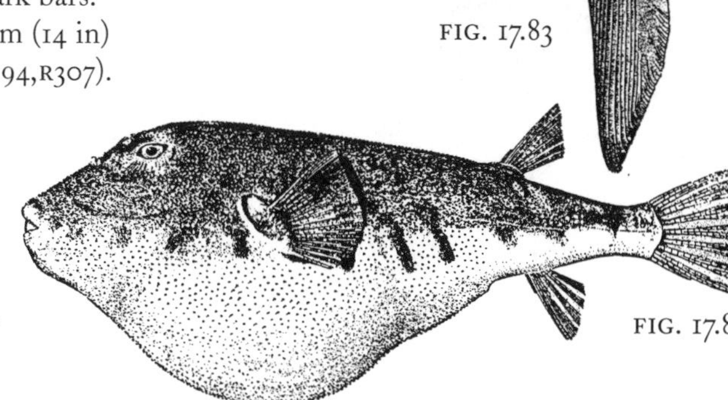

FIG. 17.84

FIG. 17.85

Syngnathus fuscus, northern pipefish†, (Syngnathidae). Dark greenish to brownish. Female with ventral folds. Often orient vertically for camouflage. Bo,Es–SG,Cr,30.5 cm (12 in), (B312,H182,R126), ☺. (FIG. 17.85)

17.33 **Fish Group 3** (from 17.24)

Scaled or scaleless fish with one long dorsal fin.

1. Location of eyes on head
   - A. Both eyes on *same* side of body — sa
   - B. Eyes more *typical,* one per side — ty
2. Compare body length (nose to base of tail) to body height at highest point: body is (provide *number*) times as long as it is high — __x

17.34

Eyes	Length vs. Height	Subgroup
sa	1–3X	Flatfish, 17.35
ty	<4	Stocky Group 3 fish, 17.41
ty	>4	Slender Group 3 fish, 17.43

17.35 Flatfish (from 17.34)

Flatfish or flounders display extreme adaptation to their benthic (bottom dwelling) lifestyle. They are highly compressed laterally and swim or rest on one side. The eye normally associated with the functional undersurface migrates to the upper side during development. Coloration of the upper surface is closely matched (and is even adjustable using chromatophores) with the surrounding substrate. Flatfish are distinguished by determining which side is uppermost, the relative size and shape of the mouth and fins, and the contour of the "lateral line" or row of pressure-sensitive receptors forming a distinct line midway along the body.

1. Right-eyed or left-eyed fish (i.e., holding the side with eyes upward and the tail toward you, does the mouth face to the right or to the left?)
   - A. Mouth faces *right* — ri
   - B. Mouth faces *left* — le

17.36

Right or Left	Subgroup
le	Left-eyed flatfish, 17.37
ri	Right-eyed flatfish, 17.39

17.37 *Left-Eyed Flatfish (Bothidae) and Tonguefish (Cynoglossidae, 1 species listed)* (from 17.36)

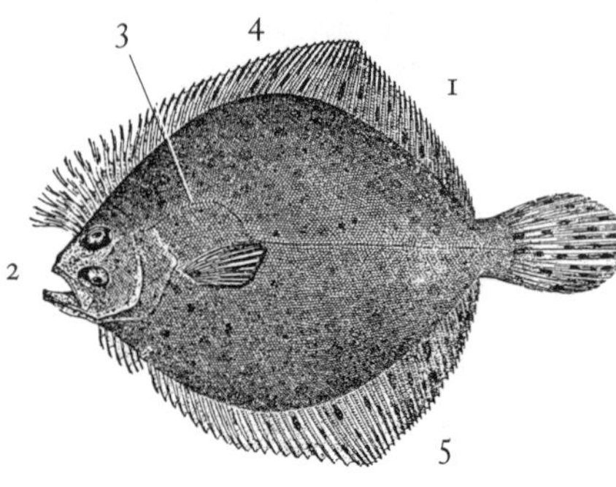

FIG. 17.86

1. Body proportion: length (to tail base) compared to maximal height (not including fins) (FIG. 17.86)
   - A. Body length *more than two* times its greatest height — >2
   - B. Body length *less than two* times its greatest height — <2*
2. Length of mouth relative to eye (FIG. 17.86)
   - A. Mouth *large,* jaw extends at least half-way past the eye — la*
   - B. Mouth *small,* jaw just reaches or falls short of eye — sm
3. Contour of lateral line (pressure sensors forming midbody line along sides) (FIG. 17.86)
   - A. Lateral line with *abrupt rise* above the pectoral fin — ar*
   - B. Lateral line straight or with *gradual rise* above the pectoral fin — gr
4. Number of rays in dorsal fin (FIG. 17.86)
   - A. Provide *number* of rays — __*
   - B. Dorsal fin continuous with and indistinguishable from *caudal* fin — ca
5. Number of rays in anal fin (FIG. 17.86)
   - A. Provide *number* of rays — __*
   - B. Anal fin continuous with and indistinguishable from *caudal* fin — ca

17.38

Body	Mouth	Lateral Line	Dorsal Fin	Anal Fin	Species
<2	la	ar	63–70	40–52	*Scophthalmus aquosus,* windowpane†, sand flounder
>2	la	ar	72–81	60–67	*Paralichthys oblongus,* fourspot flounder†
>2	la	ar	85–94	60–73	*P. dentatus,* summer flounder
>2	sm	gr	71–75	53–57	*Etropus microstomus,* smallmouth flounder†
>2	sm	gr	85	60	*E. crossotus,* fringed flounder†
>2	sm	gr	84–86	70	*Citharichthys arctifrons,* gulf stream flounder
>2	sm	gr?	ca	ca	*Symphurus plagiusa,* blackcheek tonguefish

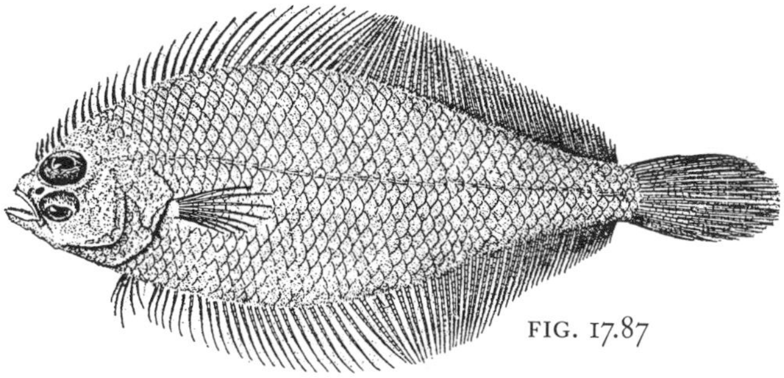

FIG. 17.87

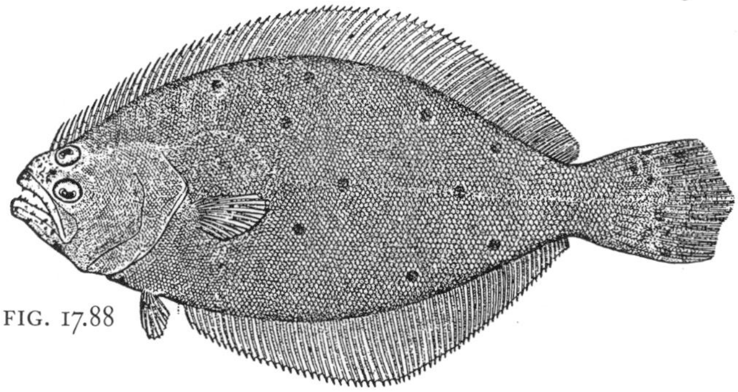

FIG. 17.88

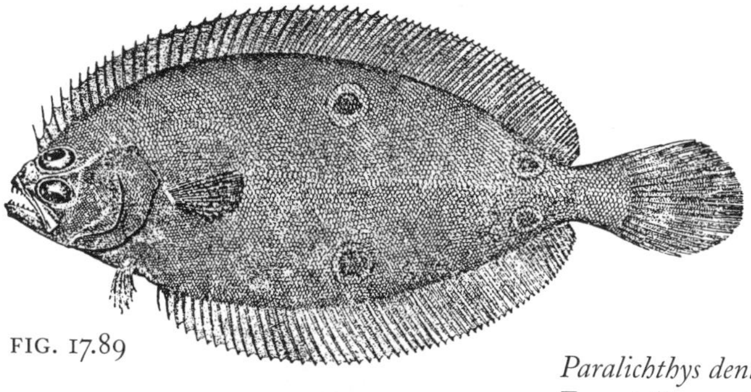

FIG. 17.89

Citharichthys arctifrons, Gulf Stream flounder. Pointed projection above mouth. Bo,150–1186,?,17.8 cm (7 in), (B294,R291). (FIG. 17.87)

Etropus crossotus, fringed flounder†. Ca+,Co?,In,11.4 cm (4.5 in), (H173,R292).

Etropus microstomus, smallmouth flounder†. Vi,?,In?,12.7 cm (5 in), (H172,R292,S537).

Paralichthys dentatus, summer flounder. Narrow-bodied; variable coloration. Especially south of Cape Cod. Young move into estuaries to feed. Bo,48–150,Fi–In,91.4 cm (36 in), (B267,H165,M157,R290,S588), ☺. (FIG. 17.88)

Paralichthys oblongus, fourspot flounder†. Bo,42–900,In,40.6 cm (16 in), (B270,R290,S538), ☺. (FIG. 17.89)

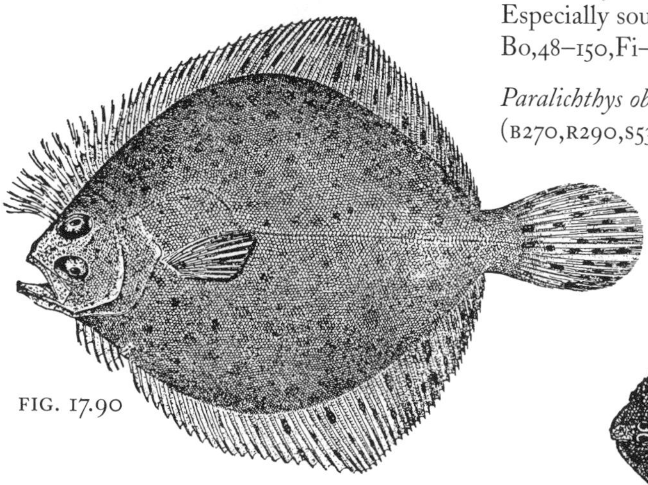

Scophthalmus aquosus, windowpane†, sand flounder. Note shape of dorsal and anal fins; pale greenish. Bo,Li–150,In,45.7 cm (18 in), (B290,M160,R294,S539), ☺. (FIG. 17.90)

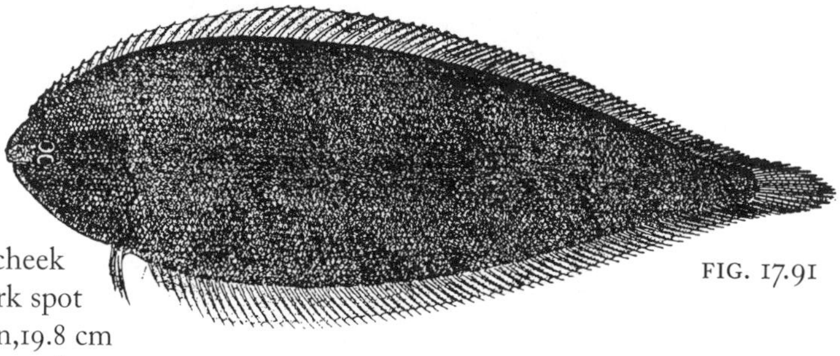

FIG. 17.90

FIG. 17.91

Symphurus plagiusa, blackcheek tonguefish (Soleidae). Dark spot on operculum. Ca+,Oc?,In,19.8 cm (7.8 in), (H177,R297). (FIG. 17.91)

17.39 *Right-Eyed Flatfish (Pleuronectidae) and Soles (Solidae, 1 species listed)* (from 17.36)

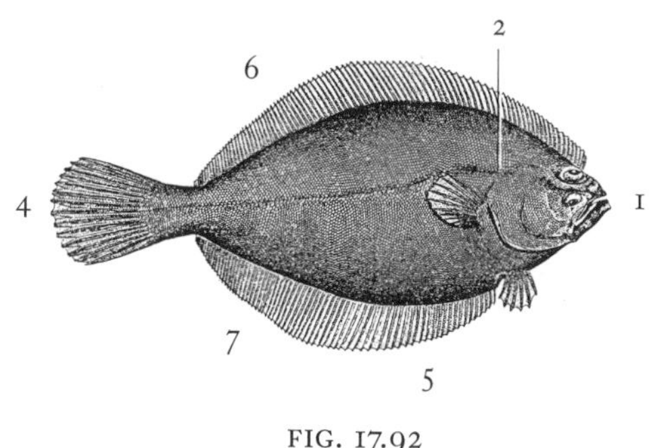

FIG. 17.92

1. Length of mouth relative to eye (FIG. 17.92)
   - A. Mouth *large,* jaw extends at least half-way past the eye — la*
   - B. Mouth *small,* jaw just reaches or falls short of eye — sm
2. Contour of lateral line (pressure sensors forming midbody line along sides) (FIG. 17.92)
   - A. Lateral line with *abrupt rise* above the pectoral fin — ar
   - B. Lateral line straight or with *gradual rise* above the pectoral fin — gr*
3. Body proportion: length (to tail base) compared to maximal height (not including fins)
   - A. Body length *more than two* times its greatest height — >2
   - B. Body length *less than two* times its greatest height — <2*

4. Shape of caudal fin (FIG. 17.92)
   - A. Caudal fin *rounded* — ro*
   - B. Caudal fin *slightly crescent* — sc
5. Number of separate fins along ventral edge (includes pelvic and anal fins, if present): provide *number* (1 or 2) (FIG. 17.92) —
6. Number of rays in dorsal fin: provide *number* of rays (FIG. 17.92) —
7. Number of rays in anal fin: provide *number* of rays (FIG. 17.92) —

17.40

Mouth Size	Lateral Line	Body Proportion	Caudal Fin	No. of Fins	Dorsal Rays	Anal Rays	Species
la	gr	<2	ro	1	50–56	36–42	*Trinectes maculatus,* hogchoker†, American sole
la	gr	>2	ro	2	76–96	64–77	*Hippoglossoides platessoides,* American plaice†, dab
la	ar	>2	sc	2	98–105	73–79	*Hippoglossus hippoglossus,* Atlantic halibut
sm	ar	>2	ro	2	76–85	56–63	*Pleuronectes* (=*Limanda*) *ferrugineus,* yellowtail flounder†, rusty flounder
sm	gr	>2	ro	2	55	35–40	*P.* (=*Liopsetta*) *putnami,* smooth flounder†
sm	gr	>2	ro	2	60–76	45–58	*P.* (=*Pseudopleuronectes*) *americanus,* winter flounder†, blackback flounder, lemon sole
sm	gr	>2	ro	2	100–105	87–100	*Glyptocephalus cynoglossus,* witch flounder†, gray sole

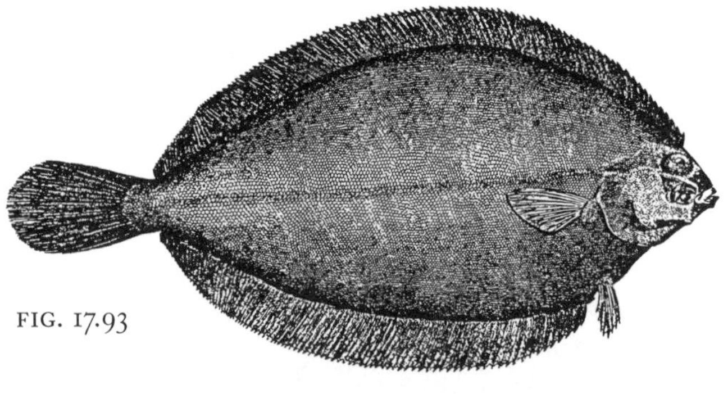

FIG. 17.93

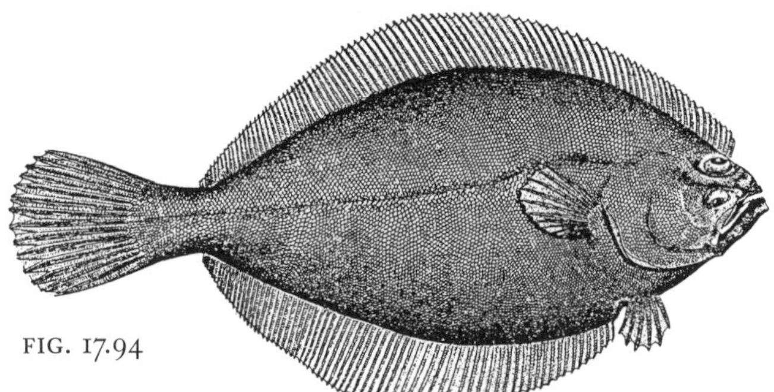

FIG. 17.94

FIG. 17.95

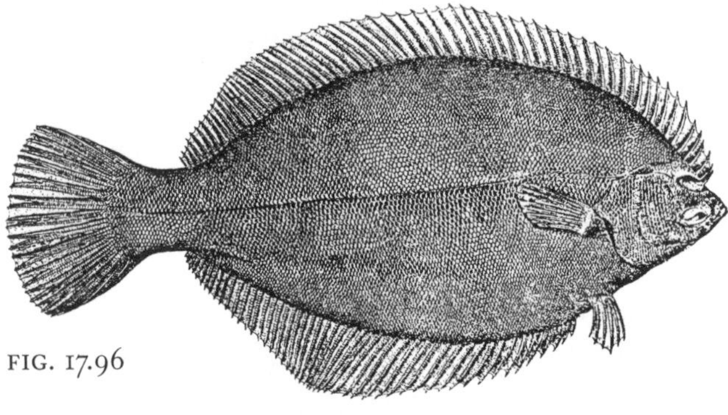

FIG. 17.96

Glyptocephalus cynoglossus, witch flounder†, gray sole. Brownish or russet gray. Fine sand to mud bottoms. Bo,60–4700,In,63.5 cm (25 in), (B285,R294,S542). (FIG. 17.93)

Hippoglossoides platessoides, American plaice†, dab. Reddish to grayish brown. Ac,30–350,Ec–An–Mo,61 cm (24 in), (B259,M160,R294,S544), ☺. (FIG. 17.94)

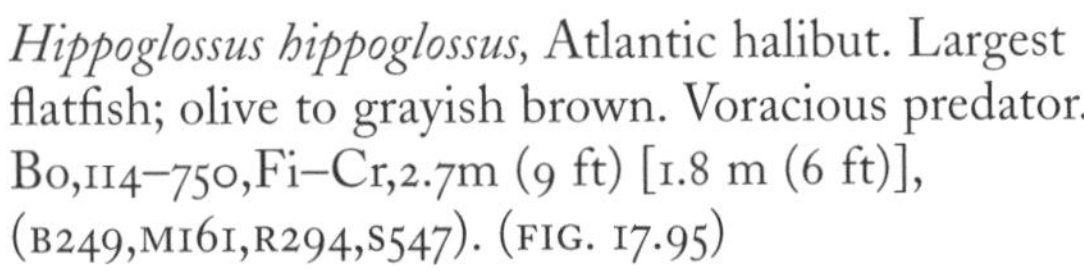

Hippoglossus hippoglossus, Atlantic halibut. Largest flatfish; olive to grayish brown. Voracious predator. Bo,114–750,Fi–Cr,2.7m (9 ft) [1.8 m (6 ft)], (B249,M161,R294,S547). (FIG. 17.95)

Pleuronectes (=*Pseudopleuronectes*) *americanus,* winter flounder†, blackback flounder, lemon sole. Variable color. Offshore in summer, onshore in eelgrass in winter. Bo,Li–270,In–Fi,55.9 cm (22 in), (B276,H168,M167,R295,S554), ☺. (FIG. 17.96)

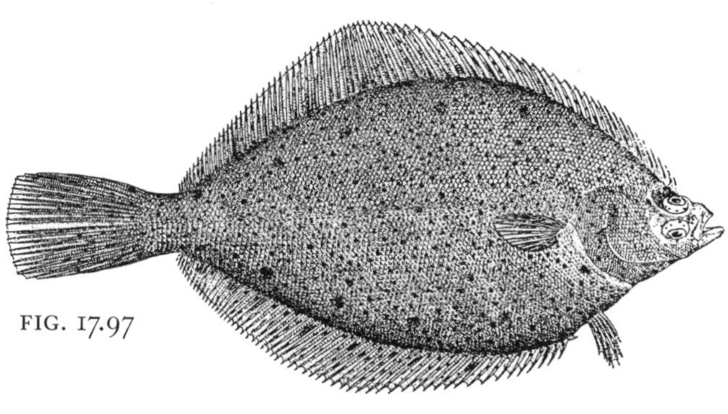

FIG. 17.97

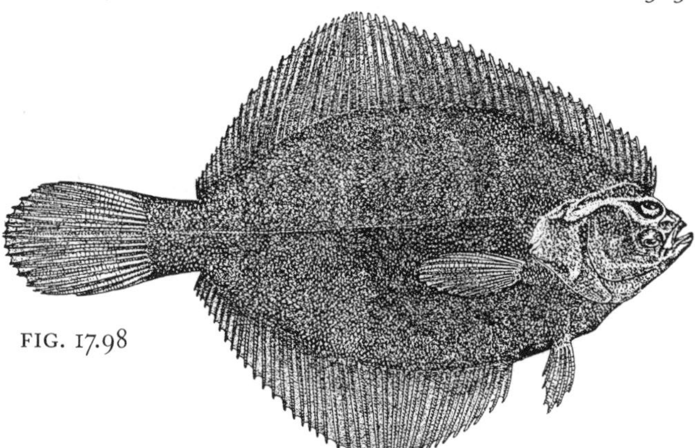

FIG. 17.98

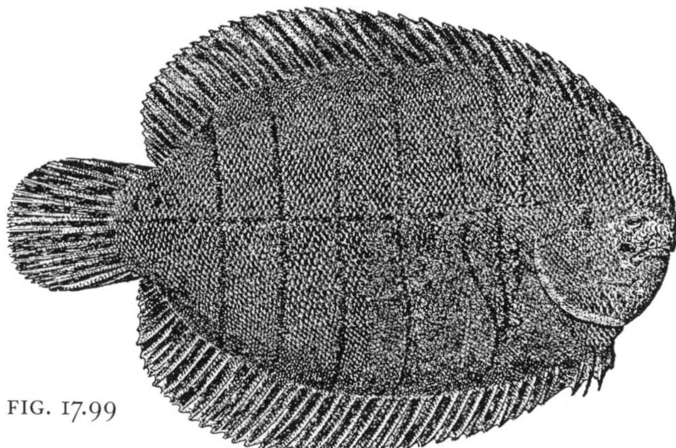

FIG. 17.99

Pleuronectes (=*Limanda*) *ferrugineus,* yellowtail flounder†, rusty flounder. Very pointed snout; brownish to olive with red spots. Bo,30–360,Cr–Mo–An,45.7 cm (18 in), (B271,H168,M169,R294,S550), ☺. (FIG. 17.97)

Pleuronectes (=*Liopsetta*) *putnami,* smooth flounder†. Scaleless between eyes; grayish to brown. Ac,Li–90,Cr–An,30.5 cm (12 in), (B283,R295,S553). (FIG. 17.98)

Trinectes maculatus, hogchoker†, American sole (Solidae). No pectoral fin on either side; rounded head; dark. Vi+,Es–Co,An–Cr,20.3 cm (8 in), (B296,H175,M153,R296), ☺. (FIG. 17.99)

17.41 Stocky Group 3 Fish (from 17.34)

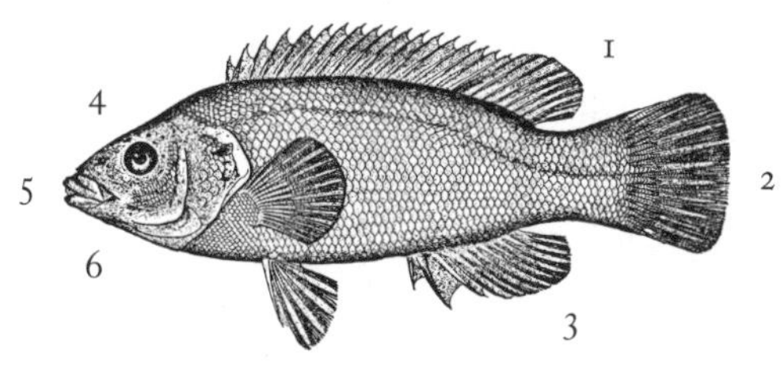

FIG. 17.100

1. Shape of dorsal fin (FIG. 17.100)
   - A. *Tapers* from anterior toward posterior ta
   - B. Ends posteriorly in a distinctly *rounded* edge ro*
2. Shape of caudal fin (FIG. 17.100)
   - A. *Rounded* ro*
   - B. *Slightly crescent* sc
   - C. *Deeply crescent* dc
3. Shape of anal fin (posteriormost ventral fin) (FIG. 17.100)
   - A. Length equal to or *more than twice* fin height >2*
   - B. Length clearly *less than twice* fin height <2
4. Head shape, maximal height compared to width from snout to posterior extreme of opercular plate (FIG. 17.100)
   - A. *Height* approximately *equal to width* h=w
   - B. *Height* distinctly *less than width,* face somewhat elongate h<w*
   - C. *Height* distinctly *greater than width,* face comparatively flattened h>w
5. Mouth size (FIG. 17.100)
   - A. Mouth *large,* extends at least to level of eye la
   - B. Mouth *small,* ends anterior to eye sm*
6. Mouth with large, obvious teeth (FIG. 17.100)
   - A. Both jaws lined with large obvious *teeth* te*
   - B. Such teeth *absent* x

17.42

Dorsal Fin	Caudal Fin	Anal Fin	Head Shape	Mouth	Teeth	Species
ta	dc	>2	h>w	sm	x	Anterior slope of head straight: *Selene (=Vomer) setapinnis,* Atlantic moonfish† OR Anterior slope of head curved: *Peprilus alepidotus,* harvestfish†, silver dollar
ta	dc	>2	h=w	sm	x	*Peprilus triacanthus,* butterfish†
ta	dc	>2	h=w	la	x	*Trachinotus carolinus,* Florida pompano†
ro	sc	<2	h=w,h<w	la	x	Dark blotch toward middle of dorsal fin: *Helicolenus dactylopterus,* blackbelly rosefish† OR No dark blotch on dorsal fin: *Sebastes norvegicus (=marinus),* golden redfish†, rosefish, ocean perch
ro	sc,dc	>2	h=w	sm	te	*Archosargus probatocephalus,* sheepshead†
ro	dc	>2	h=w,h>w	sm	x	*Stenotomus chrysops (=versicolor),* scup†, porgy
ro	dc	>2	h<w	sm	x	*Orthopristis chrysoptera,* pigfish†
ro	ro	>2	h<w	sm	x	*Liparis atlanticus,* Atlantic seasnail†
ro	ro	>2	h=w	la	x	Family: Blenniidae Branched tentacle above eye: *Hypsoblennius hentzi,* feather blenny† OR No tentacle above eye: *Chasmodes bosquianus,* striped blenny†, banded blenny
ro	ro	<2	h=w	sm	te	*Tautoga onitis,* tautog†, blackfish
ro	ro	>2	h<w	sm	te	*Tautogolabrus adspersus,* cunner†, blue perch
ro	ro	<2	h<w	la	x	*Centropristis striata,* black seabass†, blackfish

FIG. 17.101

FIG. 17.102

Archosargus probatocephalus, sheepshead† (Sparidae). Greenish yellow with seven dark bars. Solitary. Vi,Co?,Mo–Cr,76.2 cm (30 in), (B416,H267,R181). (FIG. 17.101)

Centropristis striata, black seabass†, blackfish (Serranidae). Large, long pectoral and pelvic fins. Top ray of caudal fin extended. Mottled gray to black. Solitary. Vi,Li–460,Cr–Mo,61 cm (24 in), (B407,H251,R137,S363), ☺. (FIG. 17.102)

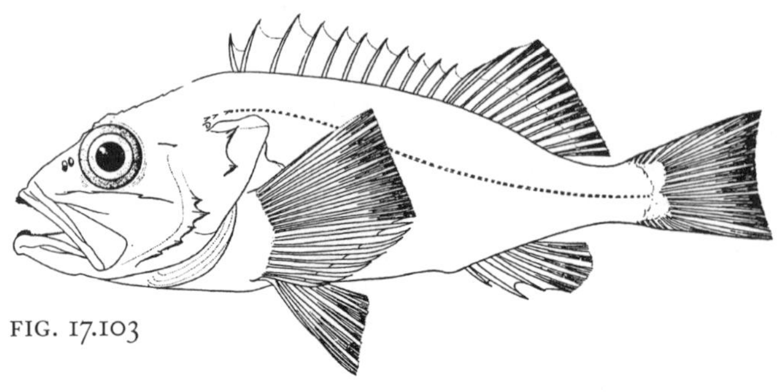

FIG. 17.103

Chasmodes bosquianus, striped blenny†, banded blenny (Blenniidae). Bright blue longitudinal lines. Dorsal fin with bright blue spot. Vi,Co(ha),Cr–Mo,10.2 cm (4 in), (H332,R230).

Helicolenus dactylopterus, blackbelly rosefish† (Scorpaenidae). Reddish body. Bo,360–2400,?,38 cm (15 in), (B437,R271,S480). (FIG. 17.103)

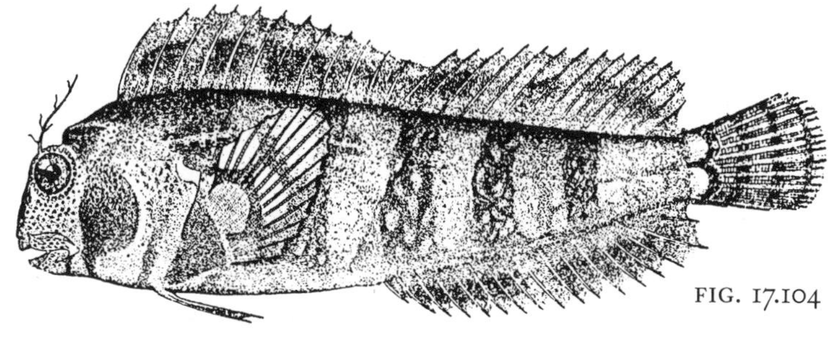

FIG. 17.104

Hypsoblennius hentzi, feather blenny† (Blenniidae). Vertical frontal margin. Bright blue spot on dorsal fin. Deeper flats and oyster beds. Ca+,Li–150,Cr–Mo–As,10.2 cm (4 in), (H334,R231). (FIG. 17.104)

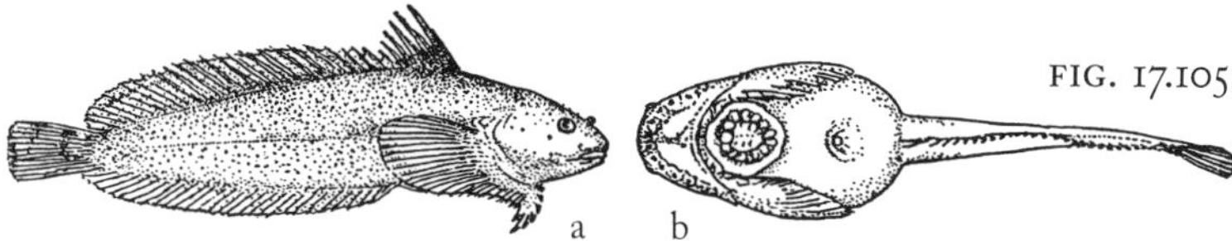

FIG. 17.105

Liparis atlanticus, Atlantic seasnail† (Cyclopteridae). Shaped like a tadpole; sucking disc is ventrally located. Inconspicuous, often coiled under stones. Ac–,Li–,Cr–Mo,12.7 cm (5 in), (B464,R286,S522), ☺.
(FIG. 17.105. a, lateral view; b, ventral view with sucker.)

Orthopristis chrysoptera, pigfish† (Haemulidae). Sloping head, small mouth, straight ventral surface. Bluish purple above, silver below; gold spots form lengthwise lines along body. Makes grunting noise with teeth when disturbed. Vi,Co?,In–Fi,38 cm (15 in), (H258,R177). (FIG. 17.106)

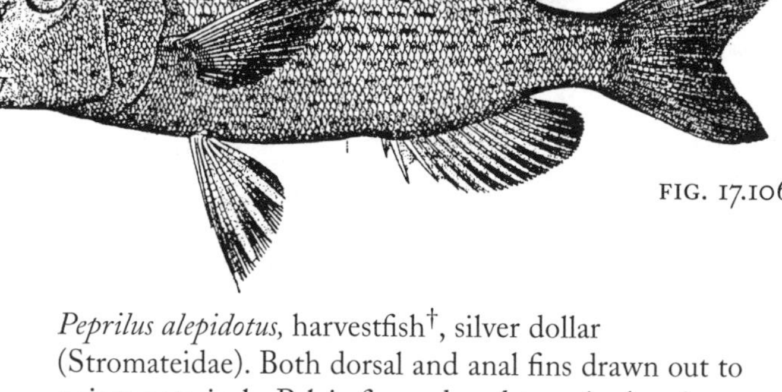

FIG. 17.106

Peprilus alepidotus, harvestfish†, silver dollar (Stromateidae). Both dorsal and anal fins drawn out to points anteriorly. Pelvic fins reduced to a single spine. Ca+,Co,Fi–Co–Mos,27.9 cm (11 in), (B368,H210,R267). (FIG. 17.107)

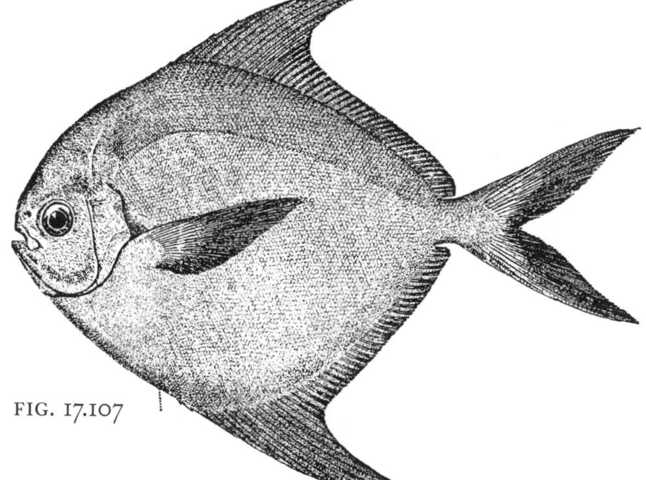

FIG. 17.107

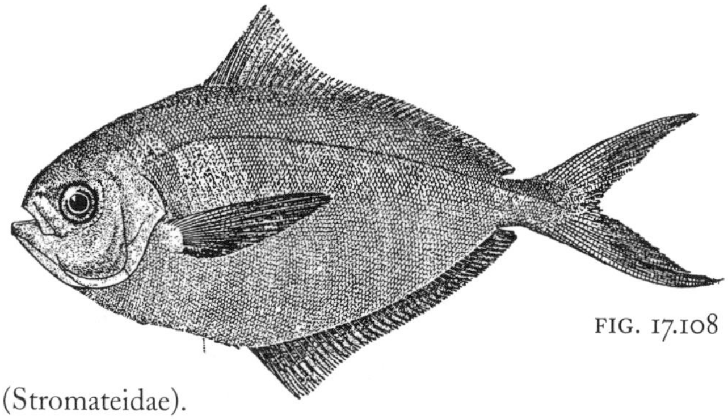

FIG. 17.108

Peprilus triacanthus, butterfish† (Stromateidae).
No pelvic fins. Very thin body; bluish color. Forms loose schools on sandy bottoms. Bo,Es–Co,Fi–An–Mo–Cr,30.5 cm (12 in), (B363,H213,R267,S475). (FIG. 17.108)

Sebastes norvegicus (=*marinus*), golden redfish†, rosefish, ocean perch (Scorpaenidae). Large bony head, large eyes. Orange-red color. Ac–,Li–1050,Cr–Mo,61 cm (24 in), (B430,R277,S483).

Selene (=*Vomer*) *setapinnis,* Atlantic moonfish† (Carangidae). First dorsal fin reduced to spines. Second dorsal fin drawn out to a point anteriorly. Pelvic fins small. Anal fin longer than second dorsal fin. Vi,Oc?,Fi,30.5 cm (12 in), (B378,H226,R163).

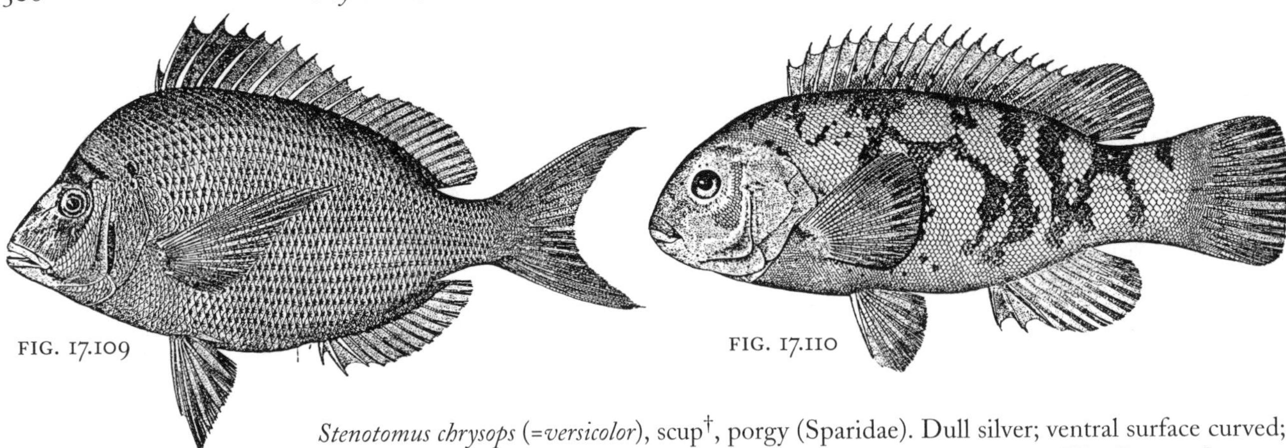

FIG. 17.109

FIG. 17.110

Stenotomus chrysops (=*versicolor*), scup†, porgy (Sparidae). Dull silver; ventral surface curved; 12–15 indistinct longitudinal stripes. Small schools inshore in summer. Vi+,Co–420,In,45.7 cm (18 in), (B411,H261,M130,R184,S390), ☺. (FIG. 17.109)

Tautoga onitis, tautog†, blackfish (Labridae). Blunt snout. Dark mottled to black. Rocky shores, pilings, piers, jetties, wrecks. Bo,Li–78,In,91.4 cm (36 in), (B478,H318,R207,S396), ☺. (FIG. 17.110)

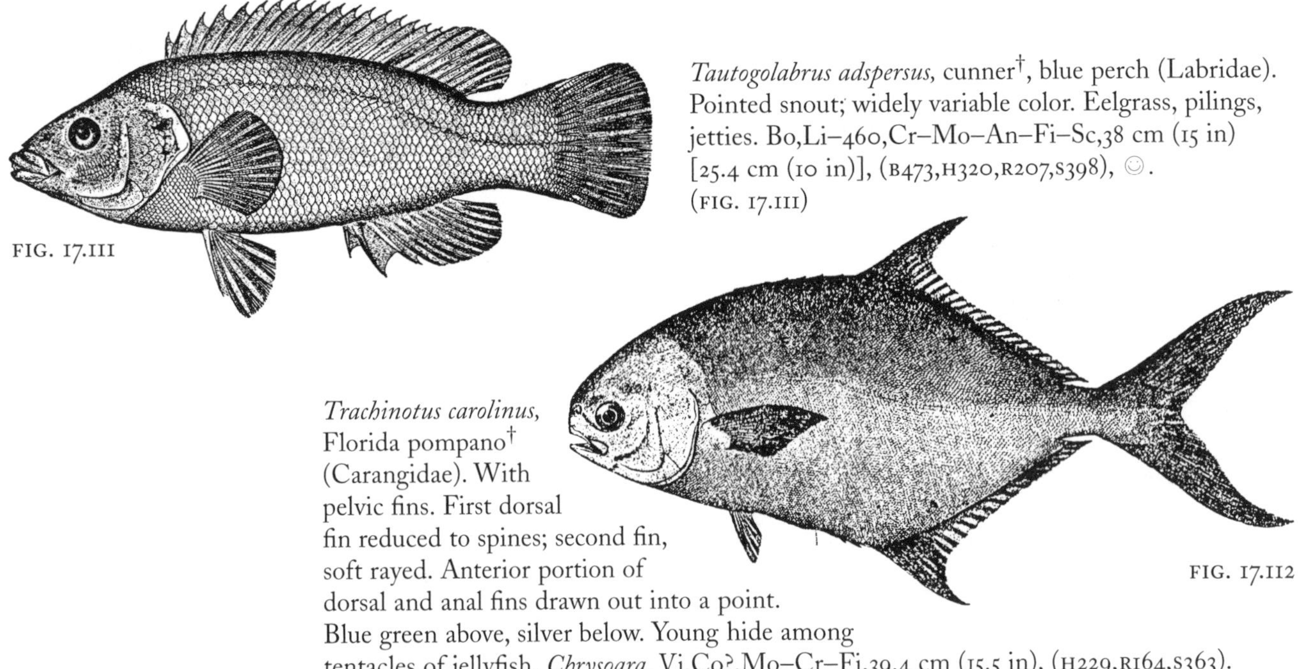

FIG. 17.111

Tautogolabrus adspersus, cunner†, blue perch (Labridae). Pointed snout; widely variable color. Eelgrass, pilings, jetties. Bo,Li–460,Cr–Mo–An–Fi–Sc,38 cm (15 in) [25.4 cm (10 in)], (B473,H320,R207,S398), ☺. (FIG. 17.111)

FIG. 17.112

Trachinotus carolinus, Florida pompano† (Carangidae). With pelvic fins. First dorsal fin reduced to spines; second fin, soft rayed. Anterior portion of dorsal and anal fins drawn out into a point. Blue green above, silver below. Young hide among tentacles of jellyfish, *Chrysoara.* Vi,Co?,Mo–Cr–Fi,39.4 cm (15.5 in), (H229,R164,S363). (FIG. 17.112)

17.43 Slender Group 3 Fish (from 17.34)

I. Some spines in obvious, webbed portion of dorsal fin (check anterior and posterior ends especially)

   A. At least some *spines* present in dorsal fin — sp

   B. Spines *absent,* soft rays only — x

17.44

Spines	Subgroup
sp	Spiny-finned, slender Group 3 fish, 17.45
x	Spineless, slender Group 3 fish, 17.47

17.45 *Spiny-Finned, Slender Group 3 Fish* (from 17.44)

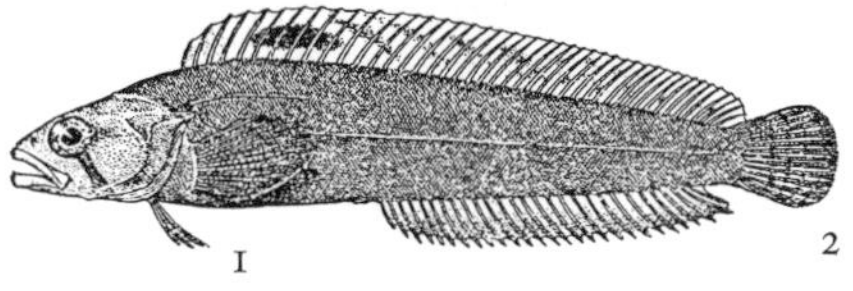

FIG. 17.113

1. Condition of pelvic or ventralmost paired fins (FIG. 17.113)
   - A. Pelvic fins *tiny,* at level of or anterior to pectoral fins — ti
   - B. Obvious, paired, *pelvic* fins of typical type — pe*
   - C. Paired pelvic fins *absent* — x
2. Shape of caudal fin (FIG. 17.113)
   - A. Caudal fin *rounded* — ro*
   - B. Caudal fin *squared* — sq
   - C. Caudal end *pointed* — po
   - D. Caudal fin *absent* — x

17.46

Pelvic Fins	Caudal Fin	Species
ti	ro	*Pholis gunnellus,* rock gunnel†
ti	sq	*Lumpenus (=Leptoclinus) maculatus,* daubed shanny†
ti	po	*L. lumpretaeformis,* snakeblenny†
pe	ro	*Ulvaria subbifurcata,* radiated shanny†
x	sq	*Anarhichas lupus,* Atlantic wolffish†

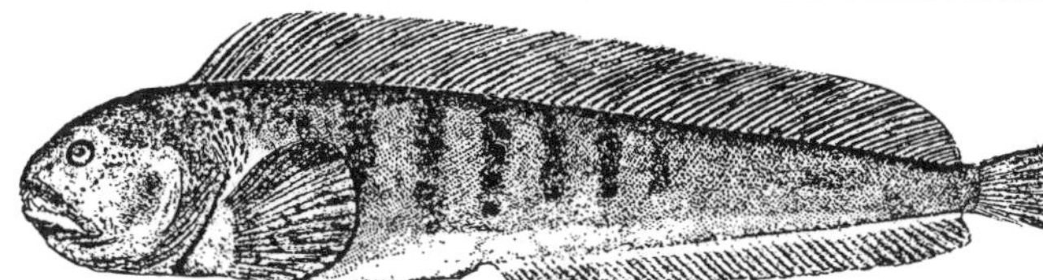
FIG. 17.114

Anarhichas lupus, Atlantic wolffish† (Anarhichadidae). No pelvic fins. Prominent protruding teeth; small eyes; dull color, bars along sides. Solitary. Can give bad bite. Ac–,6–510(ha),Cr–Mo–Ec,1.5 m (5 ft) [0.9 m (3 ft)], (B503,M67,R235,S432). (FIG. 17.114)

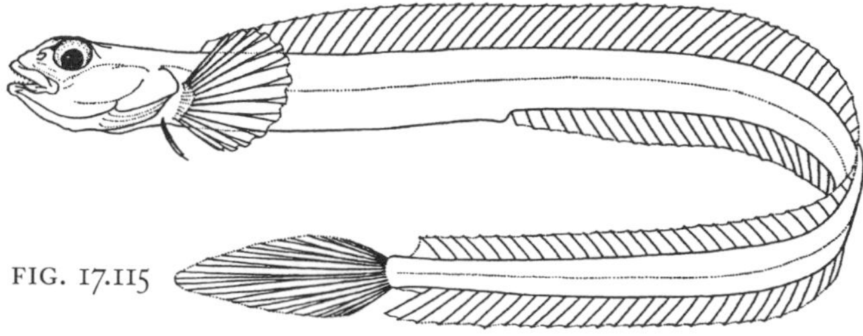
FIG. 17.115

Lumpenus lumpretaeformis, snakeblenny† (Stichaeidae). Very thin; length/height ratio = ca. 20. Large eyes; large pectoral fins; pelvic fins posterior to eyes. Pale with brown blotches; dorsal fin with darker bars. Ac–,1–300,Cr–Mo–Ec,48.3 cm (19 in), (B494,R233,S419). (FIG. 17.115)

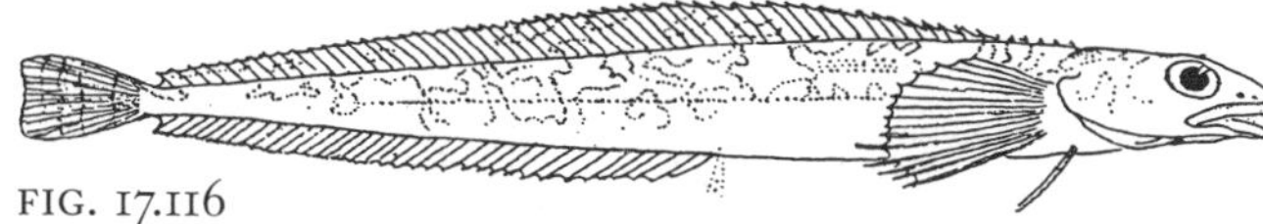
FIG. 17.116

Lumpenus (=Leptoclinus) maculatus, daubed shanny† (Stichaeidae). Length/height ratio = 10–12. Lower pectoral spines exposed at tips. Yellowish with brown blotches on body and dorsal fin. Ac,Co–Oc,An–Cr,17.8 cm (7 in), (B497,R234,S420). (FIG. 17.116)

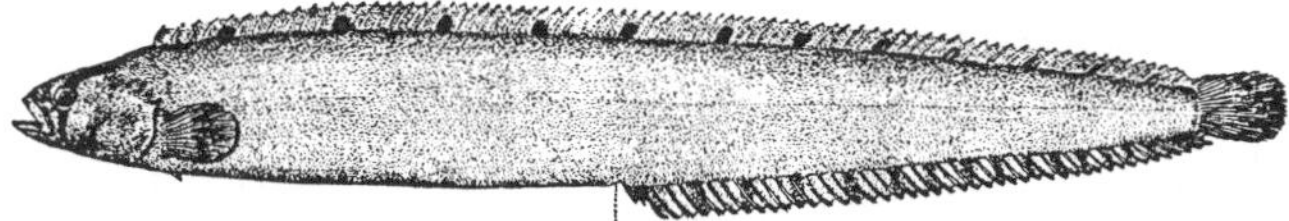
FIG. 17.117

Pholis gunnellus, rock gunnel† (Pholidae). Small pectoral fins; brown to red above, pale below; dark spots along back and onto dorsal fin. Low-tidal level in pools and under stones; thick slime layer. Bo,Co,Mo–Cr,30.5 cm (12 in) [20.3 cm (8 in)], (B492,R234,S427), ☺. (FIG. 17.117)

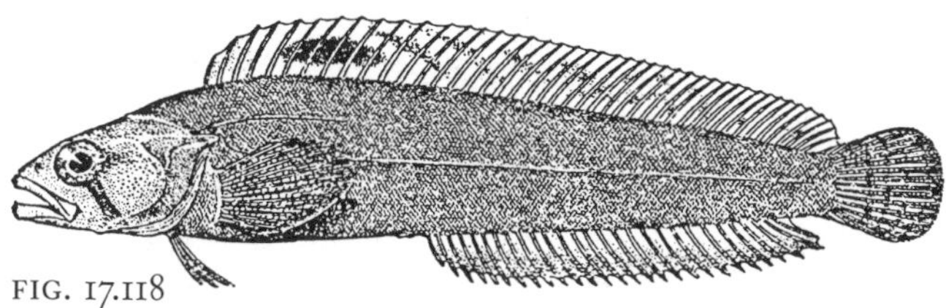
FIG. 17.118

Ulvaria subbifurcata, radiated shanny† (Stichaeidae). No barbel. Pelvic fins anterior to pectoral fins. Dorsal fin with large, dark blotch anteriorly. Blotched brownish above, paler below. Bars on caudal fin. In cold-water rock pools and littoral algae. Ac,Li–325,?,16.5 cm (6.5 in), (B498,R233,S424). (FIG. 17.118)

17.47 *Spineless, Slender Group 3 Fish* (from 17.44)

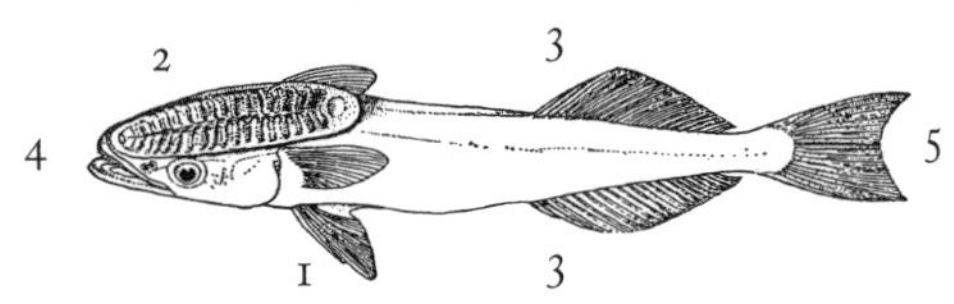

FIG. 17.119

1. Condition of pelvic or ventralmost paired fins (FIG. 17.119)
   - A. Pelvic fins *tiny,* at level of or anterior to pectoral fins — ti
   - B. Obvious, paired, *pelvic* fins of typical type — pe*
   - C. Paired pelvic fins *absent* — x
2. Distinctive features (FIG. 17.119)
   - A. *Sucker* located on head — su*
   - B. Distinct *caudal* fin *absent;* dorsal and caudal fins are blended with one another — cx
   - C. These distinctive features *absent* — x
3. Relationship of beginning of dorsal fin to beginning of anal fin (FIG. 17.119)
   - A. *Dorsal* fin starts *posterior to* start of *anal* fin — a/d
   - B. *Dorsal* fin starts *anterior to* start of *anal* fin — d/a
   - C. *Dorsal* fin starts immediately *above* start of *anal* fin — d=a*
   - D. Anal fin *absent* — x
4. Shape of mouth and relationship of lower jaw to upper jaw (FIG. 17.119)
   - A. Mouth a tiny opening at the tip of a long, narrow *snout* — sn
   - B. *Lower* jaw projects *beyond upper* jaw — l/u*
   - C. *Lower and upper* jaws about *equal* in length — l=u
5. Shape of caudal fin (FIG. 17.119)
   - A. *Rounded* — ro
   - B. *Squared* — sq
   - C. *Slightly crescent* — sc*
   - D. *Deeply crescent* — dc
   - E. *Pointed* — po
   - F. *Absent* — x

17.48

Pelvic Fins	Features	Dorsal vs. Anal	Jaw	Caudal Fin	Species
ti	cx	d/a	l=u	po	*Macrozoarces americanus,* ocean pout†, eel pout
pe	su	d=a	l/u	sc	*Remora remora,* remora†
pe	su	d=a	l/u	ro,sq	*Echeneis naucrates,* sharksucker†
pe	x	a/d	l/u	sc	*Strongylura (=Tylosurus) marina,* Atlantic needlefish†, silver gar
pe	x	d/a	l=u	ro	*Brosme brosme,* cusk†
pe	x	d/a	l=u	dc	*Coryphaena hippurus,* dolphin†
x	x	d/a	l/u	dc	*Ammodytes americanus,* American sand lance†
x	x	d/a	sn	ro	*Syngnathus fuscus,* northern pipefish†
x	cx	d/a	sn	po	*Hippocampus erectus (=hudsonius),* lined seahorse†
x	cx	d/a	l/u	po	Dorsal fin starts midway down body: *Anguilla rostrata,* American eel† OR Dorsal fin starts just behind level of pectoral fins: *Conger oceanicus,* conger eel†
x	cx	x	l/u	po	*Trichiurus lepturus,* Atlantic cutlassfish†

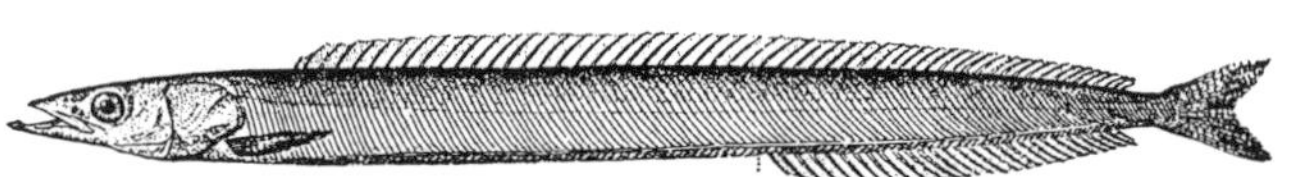

FIG. 17.120

Ammodytes americanus, American sand lance† (Ammodytidae). Irridescent luster. Dense schools; important food for whales and other marine mammals. Bury themselves in sand when pursued in shallow water. Difficult to distinguish from *A. dubius,* which occurs offshore and more northerly (see S438–442). Bo,Co,Cr–Fi,17.8 cm (7 in), (B488,H157,M65,S438), ☺. (FIG. 17.120)

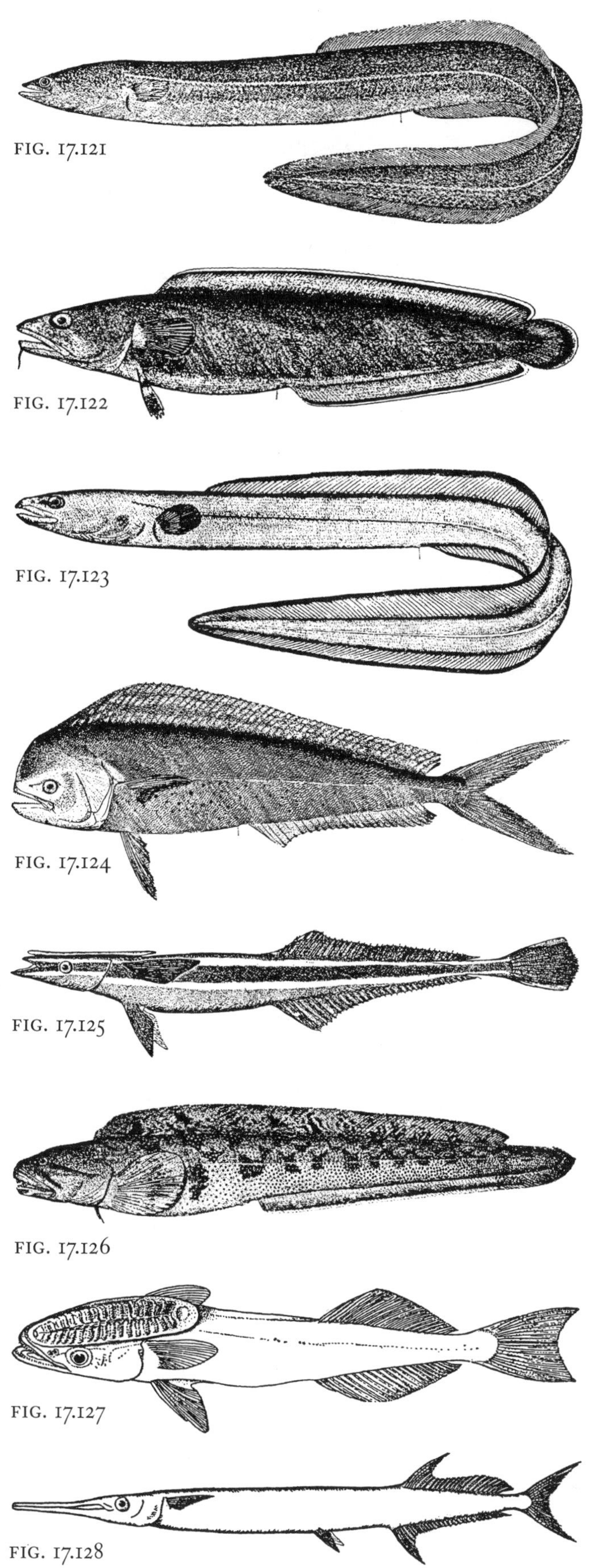

FIG. 17.121

FIG. 17.122

FIG. 17.123

FIG. 17.124

FIG. 17.125

FIG. 17.126

FIG. 17.127

FIG. 17.128

Anguilla rostrata, American eel† (Anguiliidae). Muddy to olive brown color. Anadromous; leptocephalus larvae in brackish water; elver stage in estuaries. Nocturnal feeder. Males remain in lower reaches of rivers; female ascends to streams and lakes. Coastal and estuarine. Bo,SM–Es–Co,Sc,1.2 m (4 ft) [0.9 m (3 ft)], (B151,H112,R49,S75), ☺. (FIG. 17.121)

Brosme brosme, cusk† (Gadidae). Barbel on chin; large gaping mouth; dorsal, pelvic and caudal fins usually with black margins. Solitary. Ac–,60–3180(ha),Cr–Mo,106.7 cm (42 in), (B238,R96,S262). (FIG. 17.122)

Conger oceanicus, conger eel† (Congridae). Ac,Co–852,Fi,2.1 m (7 ft) [1.2 m (4 ft)], (B154,H116,R57,S82). (FIG. 17.123)

Coryphaena hippurus, dolphin† (Coryphaenidae). Massive, blunt head; tapering body. Long dorsal fin; brilliant colors. Fast swimmer. Vi,Oc,Fi,1.8 m (6 ft), (B360,R165,S383). (FIG. 17.124)

Echeneis naucrates, sharksucker† (Echeneidae). Uniform drab color. Attaches to sharks, fishes, turtles, etc. Vi,Co–Oc,Sc,96.5 cm (38 in), (B485,H329,R157,S373). (FIG. 17.125)

Hippocampus erectus (=*hudsonius*), lined seahorse† (Syngnathidae). Male broods eggs; eelgrass and algal beds. Those living in *Sargassum* weed have scattered tabs over their bodies. Vi,Co–Oc,Pl–In,18.3 cm (7.2 in) [15.2 cm (6 in)], (B315,H185,R124,S349). (See FIG. 17.80)

Macrozoarces americanus, ocean pout†, eel pout (Zoarcidae). Very soft-bodied; slimy skin; yellowish brown. Commercial food fish. Bo,48–270,Mo–Cr–Ec,0.9 m (3 ft) [76.2 cm (30 in)], (B510,R237,S412), ☺. (FIG. 17.126)

Remora remora, remora† (Echeneidae). Pectoral and pelvic fins pointed; distinctive stripes. Dorsal fin modified for attachment to sharks, large fish, turtles, ships, etc. Vi,Co–Oc,Sc,45.7 cm (18 in), (B487,R158,S374). (FIG. 17.127)

Strongylura (=*Tylosurus*) *marina,* Atlantic needlefish†, silver gar (Belonidae). Upper and lower jaws greatly extended and well armed; pelvic, anal, and dorsal fins found well back on body. Greenish or silvery. More common toward southern portion of range. Bo,Es–Co,Fi,122 cm (48 in), (B167,H148,R106), ☺. (FIG. 17.128)

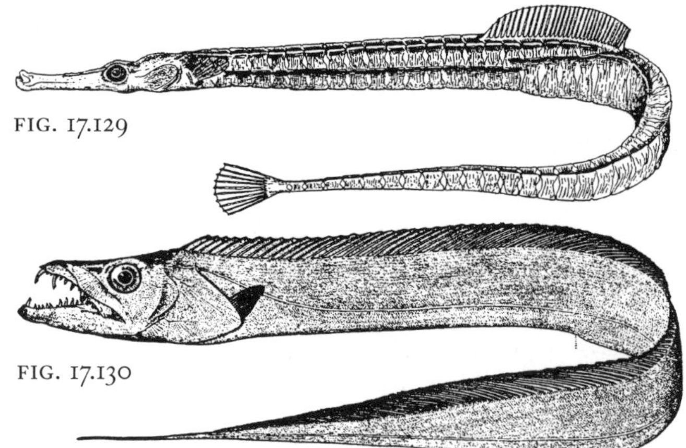

FIG. 17.129

FIG. 17.130

Syngnathus fuscus, northern pipefish† (Syngnathidae). Body hexagonal in cross-section; greenish to brownish. (If fewer than 30 rays in dorsal fin, see *S. floridae* in fish Group 2, 17.32). Bo,Es–Co,Cr–Fi,30.5 cm (12 in) [20.3 cm (8 in)], (B312,H182,R126,S350), ☺. (FIG. 17.129)

Trichiurus lepturus, Atlantic cutlassfish† (Trichiuridae). Long, thin; pointed at both ends; silvery color. Pelvic, anal, and caudal fins lacking. Large teeth and eyes. Vi,Co–Oc,Fi,96.2 cm (38 in), (B350,M65,R258). (FIG. 17.130)

17.49 **Fish Group 4** (from 17.24)

Scaled fish with two dorsal fins, the second one much longer than tall.

1. Shape of caudal fin
   - A. Caudal fin slightly or deeply *crescent* — cr
   - B. Caudal fin *squared* — sq
   - C. Caudal fin *rounded* (even if true only for bottom half of the fin) — ro

17.50

Caudal Fin	Subgroup
cr	Group 4 fish with crescent caudal fin, 17.51
sq	Group 4 fish with squared caudal fin, 17.53
ro	Group 4 fish with rounded caudal fin, 17.55

17.51 Group 4 Fish with Crescent Caudal Fin (from 17.50)

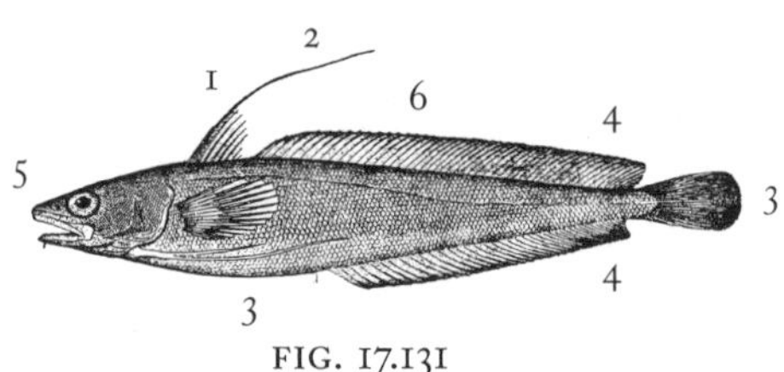

FIG. 17.131

1. Supports in first dorsal fin (FIG. 17.131)
   - A. At least some *spines* present, interconnected by webbed part of fin — sp
   - B. *Isolated* spines present but not interconnected by webbing — is
   - C. Spines entirely *absent,* soft rays only — x
2. Elongated rays or spines (at least twice the length of the midfin supports) at the anterior end of the first or second dorsal fin or from both dorsal fins (FIG. 17.131)
   - A. *Elongate* ray(s) present: *number* of the dorsal fin involved (1 or 2) — el:__*
   - B. Elongate ray(s) *absent* — x
3. Length of pectoral fin compared to length of top lobe of caudal fin (FIG. 17.131)
   - A. *Pectoral* fin *equal* in length to top lobe of *caudal* fin — p=c*
   - B. *Pectoral* fin *shorter* than top lobe of *caudal* fin — p<c
   - C. *Pectoral* fin *longer* than top lobe of *caudal* fin — p>c
4. Shape of anal fin compared to second dorsal fin (FIG. 17.131)
   - A. *Anal* fin *less than* half the length of the second *dorsal* fin — a<d
   - B. *Anal* fin about *equal* to or only slightly shorter than second *dorsal* fin — a=d*
5. Extent of upper jaw (maxillary bone) relative to eye (FIG. 17.131)
   - A. Jaw extends to or beyond *posterior* edge of *pupil* — pp*
   - B. Jaw ends *anterior* to *eye* — ae
   - C. Jaw ends at a point below the *eye,* between its anterior margin and the posterior margin of the pupil — ey

6. Highest part of the second dorsal fin (FIG. 17.131)
   A. Second dorsal fin approximately *uniform* in height throughout — un*
   B. Maximal height toward *anterior* third of second dorsal fin — an
   C. Maximal height toward *posterior* third of second dorsal fin — po

17.52

Spine	Long Rays	Pectoral vs. Caudal	Anal vs. Dorsal	Jaw vs. Eye	2nd Dorsal	Species
is	el:2	p=c,p>c	a=d	ae	an	*Selene vomer,* lookdown
is	el:2	p<c	a=d	ey	an	*Trachinotus carolinus,* Florida pompano†
sp	el:2	p=c	a=d	pp	an	*Caranx hippos,* crevalle jack
sp	el:1,2	p<c	a=d	ae	an	*Chaetodipterus faber,* Atlantic spadefish†
sp	x	p<c	a=d	pp	an	*Pomatomus saltatrix,* bluefish†
sp	x	p=c,p<c	a<d	pp	un	*Cynoscion* Some irregular, dark blotches above lateral line: *C. regalis,* weakfish†, gray sea trout, squeteague OR Many round, black spots above lateral line: *C. nebulosus,* spotted sea trout†
x	x	p=c	a=d	ey	po	*Merluccius bilinearis,* silver hake†, whiting
sp	x	p=c	a=d	ey	an	*Selar crumenophthalmus,* bigeye scad†
sp	x	p=c	a<d	ey	un	*Leiostomus xanthurus,* spot†, lafayette
sp	x	p>c	a=d	ae	un	*Prionotus carolinus,* northern searobin†

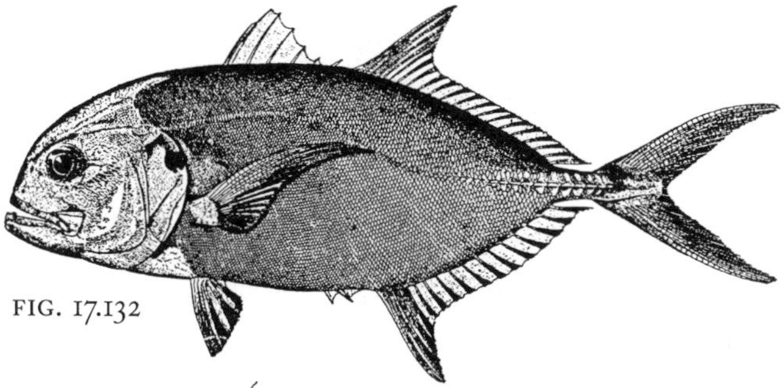

FIG. 17.132

Caranx hippos, crevalle jack (Carangidae). Long pointed pectoral fins; keeled scales along sides of tail base. Greenish gold or silver coloring; black blotch on gill cover. Vi+,Co,Fi,19.8 cm (7.8 in), (B375,H221,R160,S376). (FIG. 17.132)

Chaetodipterus faber, Atlantic spadefish† (Ephippidae). Very compressed body. Anal and both dorsal fins with extended edges; small mouth. Grayish green with four to five darker bars. Ca+,Co,Cr–He,91.4 cm (36 in) [30.5 cm (12 in)], (H306,R192). (FIG. 17.133)

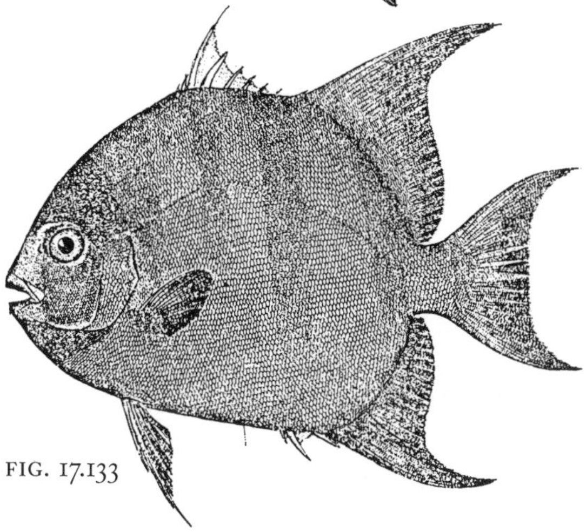

FIG. 17.133

Cynoscion nebulosus, spotted sea trout†, weakfish, speckled sea trout (Sciaenidae). Gray above, silvery below. Vi,Co,Fi–Cr,68.6 cm (27 in), (H296,R189).

Cynoscion regalis, weakfish†, gray sea trout, squeteague (Sciaenidae). Olive green above, white to silvery below. Vi+,Co,Cr–Mo–Mos–An–Fi,81.3 cm (32 in), (B417,H300,R189,S391). (FIG. 17.134)

Leiostomus xanthurus, spot†, lafayette (Sciaenidae). Deep body, blunt snout. Dark spot above operculum. Blue gray with vertical stripes. Vi,Co,An–Cr,35.6 cm (14 in), (B423,H271,R188), ☺. (FIG. 17.135)

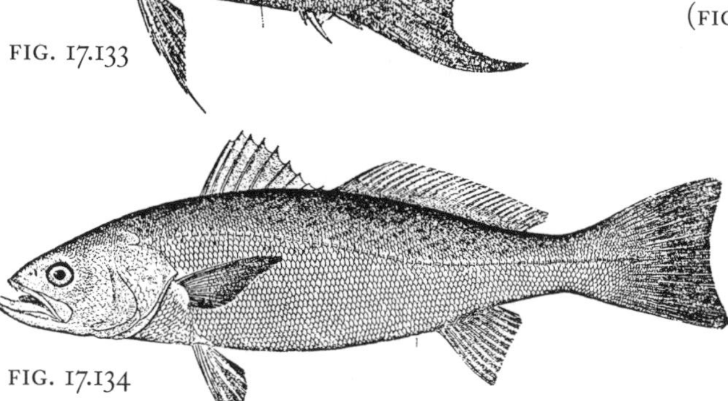

FIG. 17.134

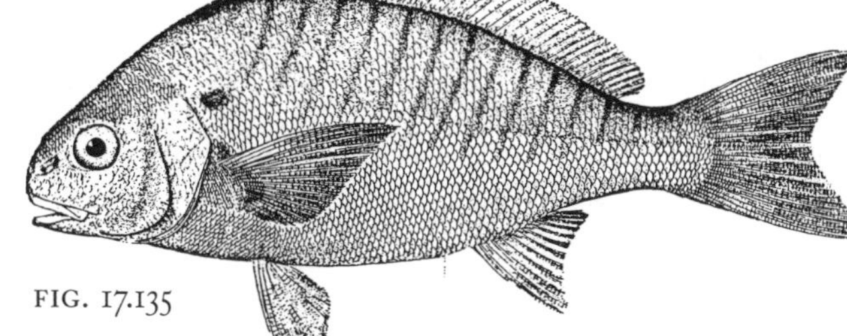

FIG. 17.135

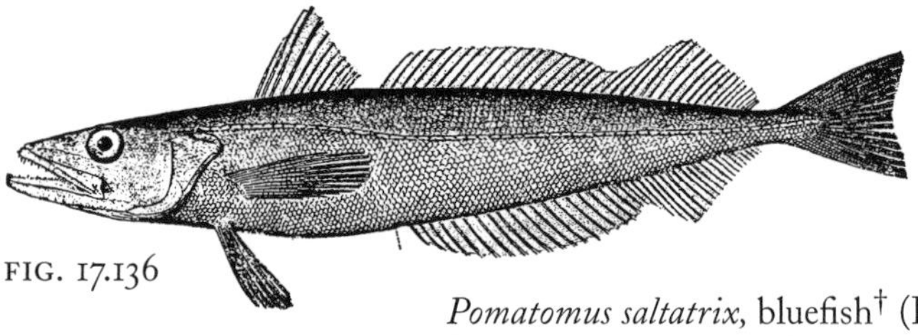

FIG. 17.136

Merluccius bilinearis, silver hake†, whiting (Gadidae). Normal pelvic fins. Eats schooling fishes. Bo,Co–3000,Fi–Cr,76.2 cm (30 in), (B174,H162,R94,S278), ☺. (FIG. 17.136)

Pomatomus saltatrix, bluefish† (Pomatomidae). Both jaws with sharp conical teeth. Greenish above, silvery below. Large schools; voracious predator; can be dangerous to swimmers. Small ones called "snappers." Vi+,Es–Co,Fi,106.7 cm (42 in), (B383,H231,R155,S370), ☺. (FIG. 17.137)

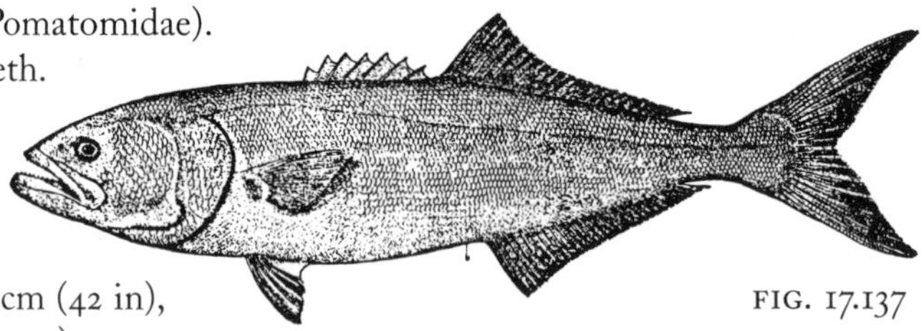

FIG. 17.137

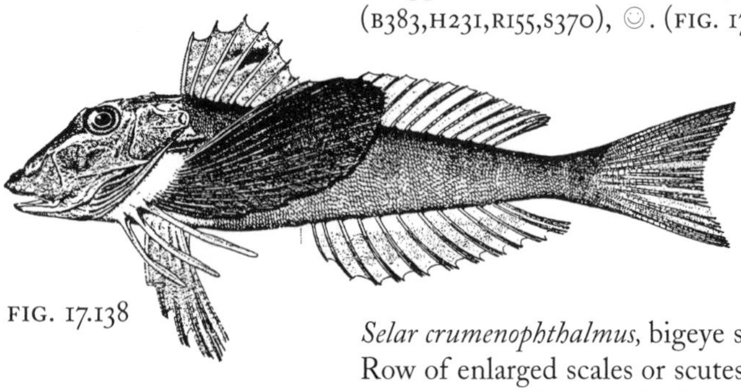

FIG. 17.138

Prionotus carolinus, northern searobin† (Triglidae). Large bony head; one spine at corners of operculum. Grayish or reddish brown; dark spot on first dorsal fin. Bright blue eyes. Can bury themselves in sand. Vi,Co–558(sa),Cr–Mo–Mos–An–Fi,40.6 cm (16 in) [30.5 cm (12 in)], (B467,H314,R279,S488), ☺. (FIG. 17.138)

Selar crumenophthalmus, bigeye scad† (Carangidae). Row of enlarged scales or scutes laterally beneath second dorsal fin. Large eye. Bluish above, silver below. Vi,Co,?,38 cm (15 in), (B377,H217,R161). (FIG. 17.139)

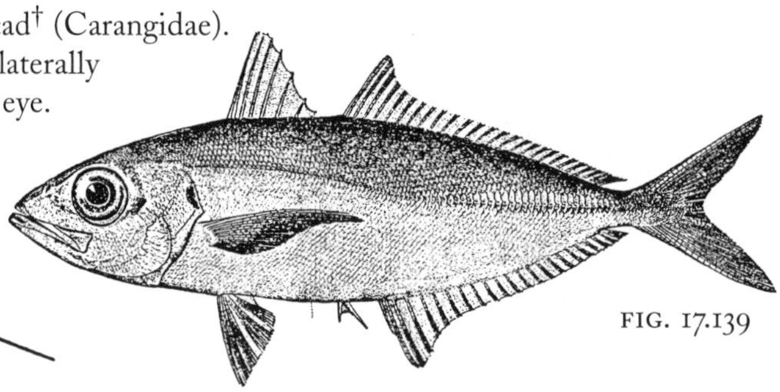

FIG. 17.139

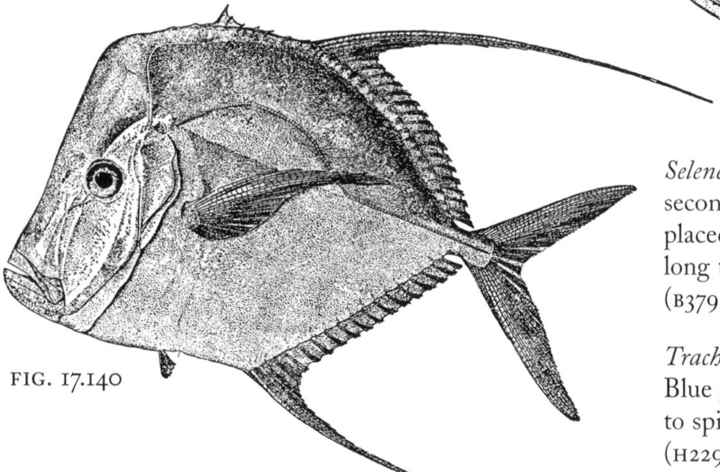

FIG. 17.140

Selene vomer, lookdown (Carangidae). Very long rays to second dorsal fin; strongly laterally compressed. Low-placed mouth; high-placed eyes. Silvery body. Young with long trailing fin filaments. Vi+,Co,Fi–Cr,23.6 cm (9.3 in), (B379,H225,R163,S380). (FIG.17.140)

Trachinotus carolinus, Florida pompano† (Carangidae). Blue green above, silvery below. First dorsal fin reduced to spines. Vi,Co?,Mo–Cr–Fi,39.4 cm (15.5 in), (H229,R164,S363). (FIG. 17.112)

17.53 Group 4 Fish with Squared Caudal Fin (from 17.50)

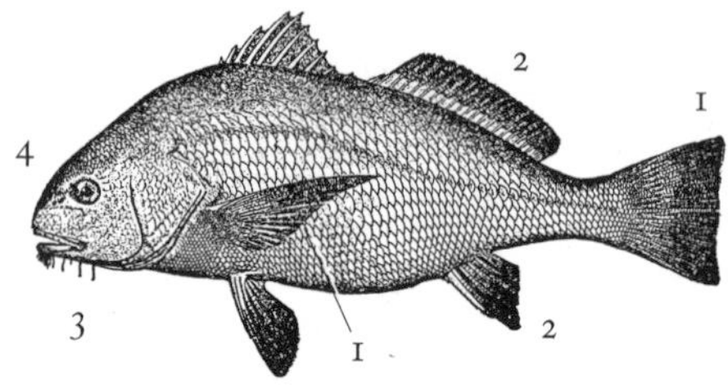

FIG. 17.141

1. Length of pectoral fin compared to length of top lobe of caudal fin (FIG.17.141)
   - A. *Pectoral* fin *equal* in length to top lobe of *caudal* fin — p=c
   - B. *Pectoral* fin *shorter* than top lobe of *caudal* fin — p<c
   - C. *Pectoral* fin *longer* than top lobe of *caudal* fin — p>c*
2. Shape of anal fin compared to second dorsal fin (FIG.17.141)
   - A. *Anal* fin *less than* half the length of the second *dorsal* fin — a<d*
   - B. *Anal* fin about *equal* to or only slightly shorter than second *dorsal* fin — a=d

3. Extent of upper jaw (maxillary bone) relative to eye (FIG.17.141)
   - A. Jaw extends to or beyond *posterior* edge of *pupil* — pp
   - B. Jaw ends *anterior* to *eye* — ae
   - C. Jaw ends at a point below the *eye,* between its anterior margin and the posterior margin of the pupil — ey*
   - D. Jaw ends *posterior* to *eye* — pe
4. Relationship of upper and lower jaws (FIG.17.141)
   - A. *Lower* jaw extends more *anteriorly* than *upper* jaw — l/u
   - B. *Upper* jaw extends more *anteriorly* than *lower* jaw — u/l*

17.54

Pectoral vs. Caudal	Anal vs. Dorsal	Jaw vs. Eye	Upper vs. Lower	Species
p=c	a<d	ey,pp	l/u	*Bairdiella chrysoura,* silver perch†, sand perch, mademoiselle
p=c	a<d	pe	u/l	*Sciaenops ocellatus,* red drum†, channel bass
p=c,p>c	a<d	ey	u/l	*Pogonias cromis,* black drum†
p>c	a=d	ae	u/l	*Prionotus evolans,* striped searobin†
p=c,p<c	a<d	pp	l/u	*Cynoscion* Some irregular, dark blotches above lateral line: *C. regalis,* weakfish†, gray sea trout, squeteague OR Many round, black spots above lateral line: *C. nebulosus,* spotted sea trout†, weakfish, speckled sea trout

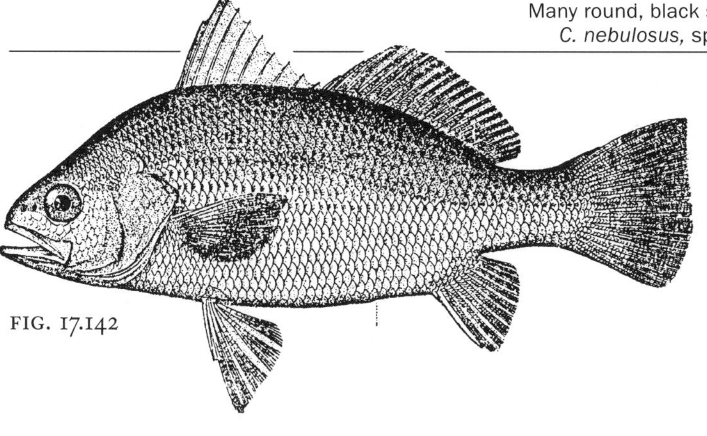

FIG. 17.142

FIG. 17.143

Bairdiella chrysoura, silver perch†, sand perch, mademoiselle (Sciaenidae). Greenish or blue gray above, silver below; fins yellowish. Often in rivers. Vi,Co,Cr–Fi,24.1 cm (9.5 in) [20.3 cm (8 in)], (H279,R185), ☺. (FIG. 17.142)

Cynoscion nebulosus, spotted sea trout†, weakfish, speckled sea trout (Sciaenidae). Gray above, silvery below. Vi,Co,Fi–Cr,68.6 cm (27 in), (H296,R189). (FIG. 17.143)

Cynoscion regalis, weakfish†, gray sea trout, squeteague (Sciaenidae). Olive green above, white to silvery below. Vi+,Co,Cr–Mo–Mos–An–Fi,81.3 cm (32 in), (B417,H300,R189,S391).

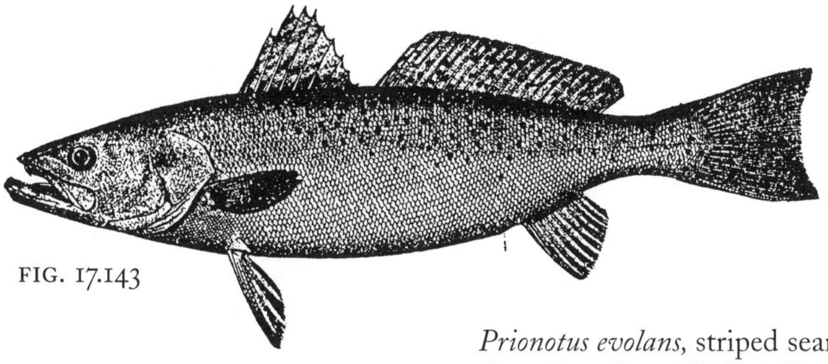

Pogonias cromis, black drum† (Sciaenidae). Brassy to darker coloring. Crushing teeth. Numerous small chin barbels. Ventral surface straight. Juveniles or "puppy drums" with vertical bars. Vi,Co,Mo–Cr,1.3 m (4.3 ft), (B425,H287,R188). (FIG. 17.144)

Prionotus evolans, striped searobin† (Triglidae). Larger mouth, longer pectoral fins than *P. carolinus.* Vi,Co?,Cr,45.7 cm (18 in), (H312,M90,R280,S489). (FIG. 17.145)

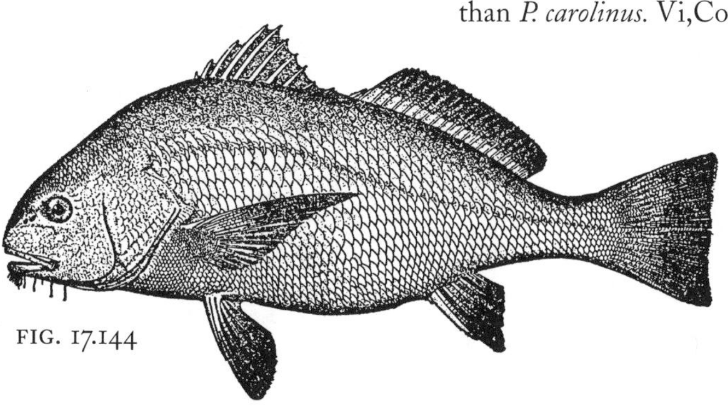

FIG. 17.144

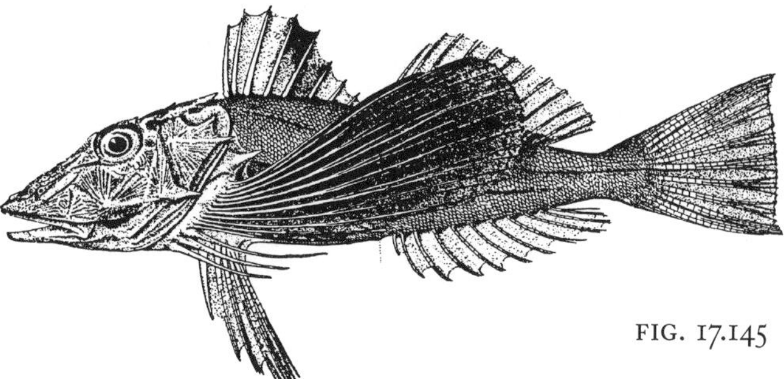

FIG. 17.145

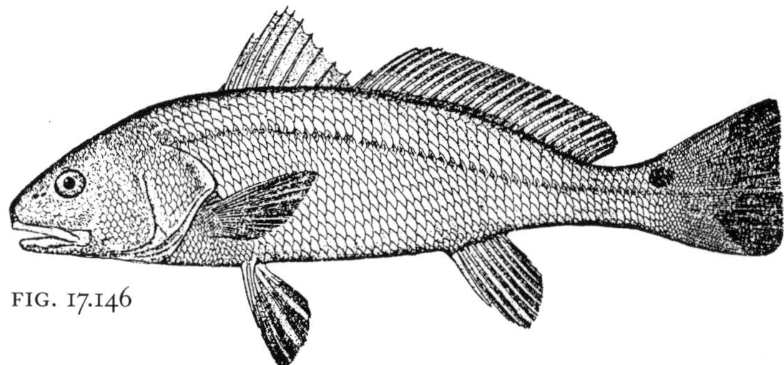

FIG. 17.146

Sciaenops ocellatus, red drum[†], channel bass, (Sciaenidae). One large round, black spot near base of caudal fin. Silvery to copper when alive; some turn reddish after death. Breeds in shallow reed beds. Up to 22.7 kg (50 lb). Vi,Co,Cr,1.5 m (4.8 ft), (H276,R188). (FIG. 17.146)

17.55 Group 4 Fish with Rounded Caudal Fin (from 17.50)

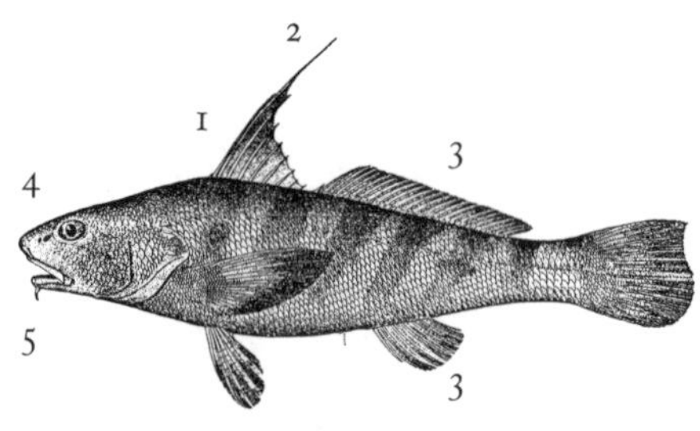

FIG. 17.147

1. Supports for dorsal fin (FIG. 17.147)
   - A. At least some *spines* present in webbed portion of dorsal fin — sp
   - B. Spines *absent,* soft rays only — x
2. Elongate ray or rays at the anterior end of the first dorsal fin (FIG. 17.147)
   - A. *Elongate* ray(s) present — el*
   - B. Elongate ray(s) *absent* — x
3. Shape of anal fin compared to second dorsal fin (FIG. 17.147)
   - A. *Anal* fin *less than* half the length of the second *dorsal* fin — a<d*
   - B. *Anal* fin clearly *more than* half the length of the second *dorsal* fin — a>d
4. Extent of upper jaw (maxillary bone) relative to eye (FIG. 17.147)
   - A. Jaw extends to or beyond *posterior* edge of *pupil* — pp
   - B. Jaw ends *anterior* to *eye* — ae
   - C. Jaw ends at a point below the *eye,* between its anterior margin and the posterior margin of the pupil — ey*
5. Presence of slender processes beneath head (FIG. 17.147)
   - A. Chin *barbel(s)* present — ba*
   - B. *Several* small *barbels* under chin — sb
   - C. Threadlike *pelvic* fins beneath gill operculum — pe
   - D. Both *barbels* and threadlike *pelvic* fins — bp

17.56

Spines	Elongate Rays	Anal vs. Dorsal	Jaw vs. Eye	Barbel	Species
sp	x	a<d	ey	ba	*Menticirrhus americanus,* southern kingfish[†], whiting
sp	x	a<d	ae	sb	*Micropogon undulatus,* Atlantic croaker[†], hardhead
sp	el	a<d	ey	ba	*Menticirrhus saxatilis,* northern kingfish[†], whiting
x	el	a>d	ey,pp	bp	*Urophycis* Threadlike pelvic fins shorter barely reach anal fin: *U. chuss,* red hake[†], squirrel hake, ling OR
x	el	a>d	ey	bp	Threadlike pelvic fins long, nearly to posterior end of anal fin: *U. chesteri,* longfin hake[†] OR
x	el	a>d	pp	bp	Threadlike pelvic fins short, do not reach anal fin: *U. tenuis,* white hake[†]
x	x	a>d	pp	bp	*U. regia,* spotted hake[†]

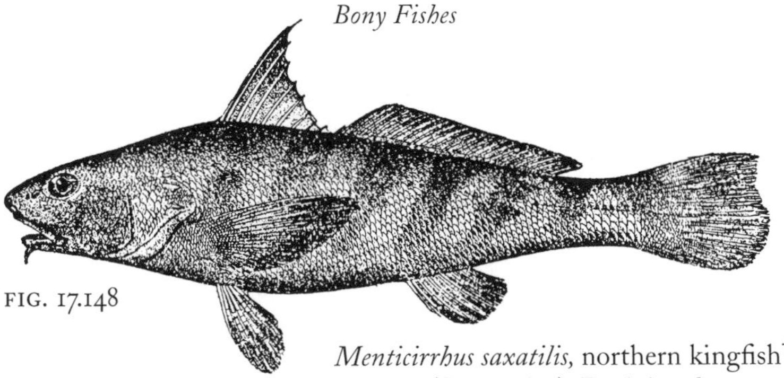

FIG. 17.148

Menticirrhus americanus, southern kingfish†, whiting (Sciaenidae). Bands do not form V-shape near pectoral fins. Sandy bottoms. Vi,Co,Cr–Fi,42 cm (16.5 in), (H291,R187). (FIG. 17.148)

Menticirrhus saxatilis, northern kingfish†, whiting (Sciaenidae). Dark bands near shoulder form V-pattern just in front of pectoral fins. Gray to black color. Schools near sandy bottoms. Vi,Co,Mo–Cr–An–Fi,43.2 cm (17 in) [25.4 cm (10 in)], (B423,H290,R187). (FIG. 17.149)

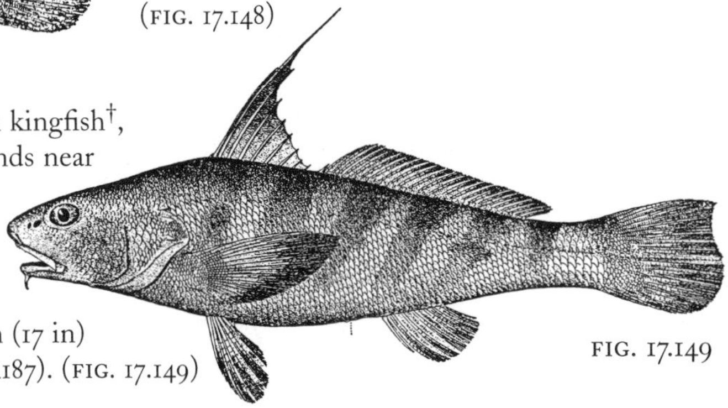

FIG. 17.149

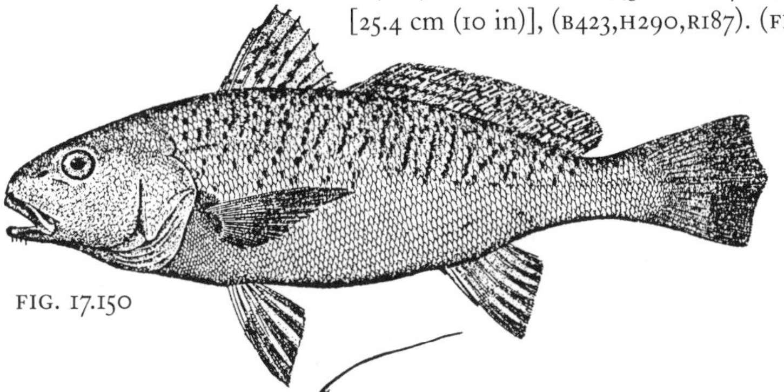

Micropogon undulatus, Atlantic croaker†, hardhead (Sciaenidae). Bulbous nose, with mouth beneath. Several short chin barbels. Both sexes make loud croaking sound. Ventral surface curved. Vi,Co,In,50.8 cm (20 in), (H283,R188), ☺. (FIG. 17.150)

FIG. 17.150

Urophycis chesteri, longfin hake† (Gadidae). Dorsal and anal fins with dark edges. Bo,600–3228,Cr–Fi,38 cm (15 in), (B232,R94).

Urophycis chuss, red hake†, squirrel hake, ling (Gadidae). Reddish to olive brown. Bo,Co–Oc(ha–sa),Cr–Mo–Fi,76.2 cm (30 in), (B223,H159,R95,S290), ☺. (FIG. 17.151)

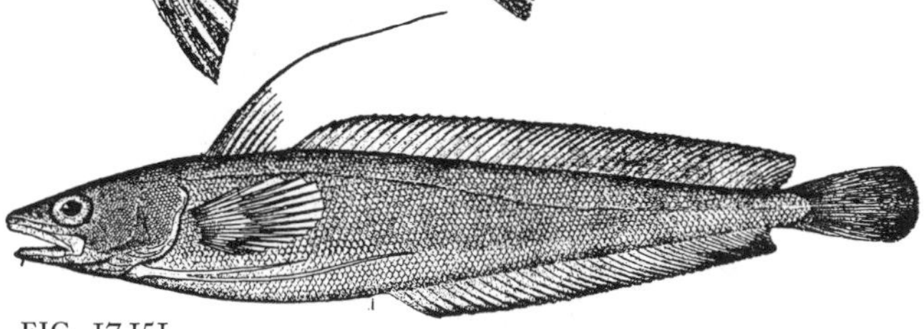

FIG. 17.151

Urophycis regia, spotted hake† (Gadidae). Threadlike pelvic fin about equal to pectoral fin, extends posteriorly to start of anal fin. Rear margin of first dorsal fin light; row of white spots along sides. Bo,Co?,Fi–Mo–Cr,40.6 cm (16 in), (B230,H160,R96,S293), ☺. (FIG. 17.152)

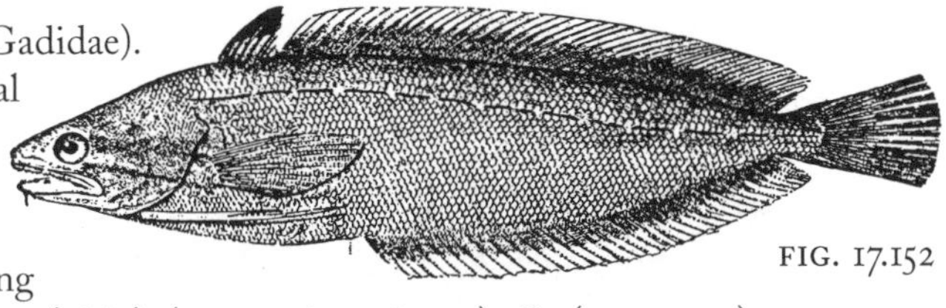

FIG. 17.152

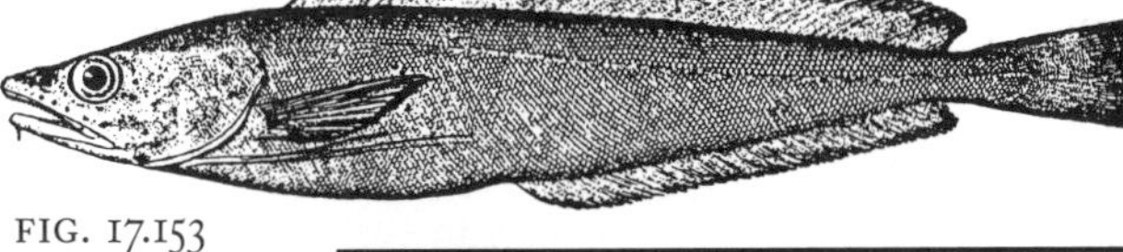

FIG. 17.153

Urophycis tenuis, white hake† (Gadidae). Grayish olive. Bo,1200–6000(m),Fi–Cr,132 cm (52 in), (B221,R95,S293), ☺. (FIG. 17.153)

17.57 Fish Group 5 (from 17.24)

Virtually scaleless fish, lacking swordlike upper jaw but with two dorsal fins, one of which is much longer than it is tall.

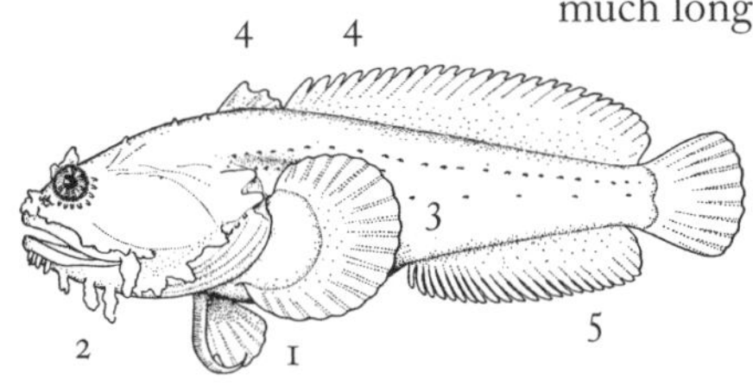

FIG. 17.154

1. Pelvic fins held horizontally along body and form an adhesive disc (FIG. 17.154)
   - A. Pelvic fins form *adhesive disc* — ad
   - B. Pelvic fins more typical, do *not* form adhesive disc — x*
2. Fleshy tabs under chin (FIG. 17.154)
   - A. Fleshy *tabs* present — ta*
   - B. Fleshy tabs *absent* — x

3. Number of rows of large scales associated with lateral line (midbody line of pressure sensors formed along sides) (FIG. 17.154)
   A. Provide *number* of rows of large scales along the sides of the body —
   B. Lateral line scales *absent* x*
4. Number of spines or rays in first dorsal fin: in second dorsal fin: provide *number* of spines or rays in first:second dorsal fins (FIG. 17.154) —:—
5. Number of spines or rays in the anal fin: provide *number* of spines or rays (FIG. 17.154) —

17.58

Adhesive Disc	Tabs	Scales	Dorsal 1 vs. Dorsal 2	Anal	Species
ad	x	x	7:12-13	11	*Gobiosoma,* close-set dorsal eyes No scales: *G. bosci,* naked goby†, clinging goby OR Two scales per side near tail base: *G. ginsburgi,* seaboard goby†
ad	x	x	7:14	14	*Microgobius thalassinus,* green goby†
x	ta	x	3:26-28	21-22	*Opsanus tau,* oyster toadfish†, dowdie
x	ta	x	16:13	13-14	*Hemitripterus americanus,* sea raven†
x	x	2	9-11:16-17	13-14	*Myoxocephalus scorpius,* shorthorn sculpin†
x	x	1	8-9:15-16	14	*M. octodecemspinosus,* longhorn sculpin†
x	x	x	9:13-14	10-11	*M. aenaeus,* grubby†
x	x	x	10-12:20-25	20-22	*Triglops murrayi* (=*ommatistius*), moustache sculpin†

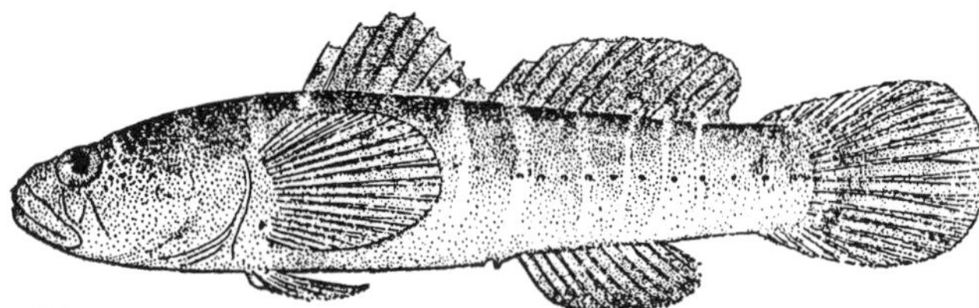
FIG. 17.155

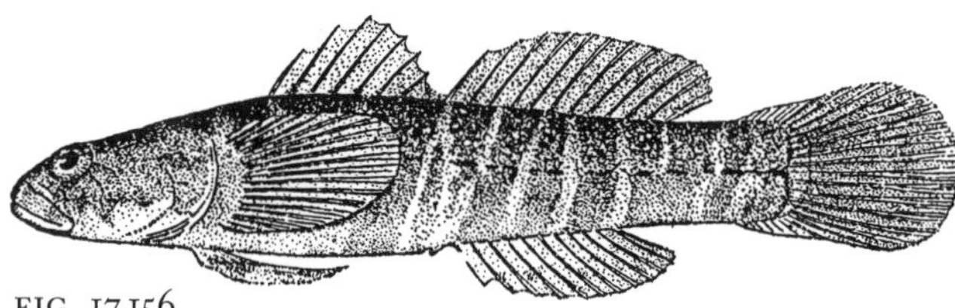
FIG. 17.156

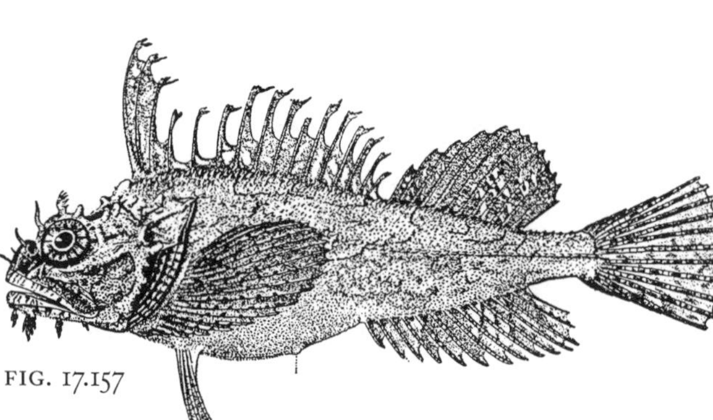
FIG. 17.157

Gobiosoma bosci, naked goby†, clinging goby (Gobiidae). Broad brown bars. Vi,Co–Es,Cr–An,6.4 cm (2.5 in), (H323,R249), ☺. (FIG. 17.155)

Gobiosoma ginsburgi, seaboard goby† (Gobiidae). Irregular darker bars. Vi,Co–Es,Cr,6.4 cm (2.5 in), (H324,R249). (FIG. 17.156)

Hemitripterus americanus, sea raven† (Cottidae). Red or yellow color forms. Can inflate belly with water. Swallows crabs whole. Bo,100–350(ha),In–Fi,62.8 cm (25 in) [50.8 cm (20 in)], (B454,H310,M93,R282,S496), ☺. (FIG. 17.157)

Microgobius thalassinus, green goby† (Gobiidae). Scales on posterior body. Greenish blue; males: with dark spots on first dorsal fin. Ca+,Co–Es(mu),?,6.4 cm (2.5 in), (R252).

Myoxocephalus aenaeus, grubby† (Cottidae). Shallows, especially in eelgrass beds. Ac–,Es–150(mu–sa–ha),In–Fi,20.3 cm (8 in), (B443,R282,S500), ☺. (FIG. 17.158)

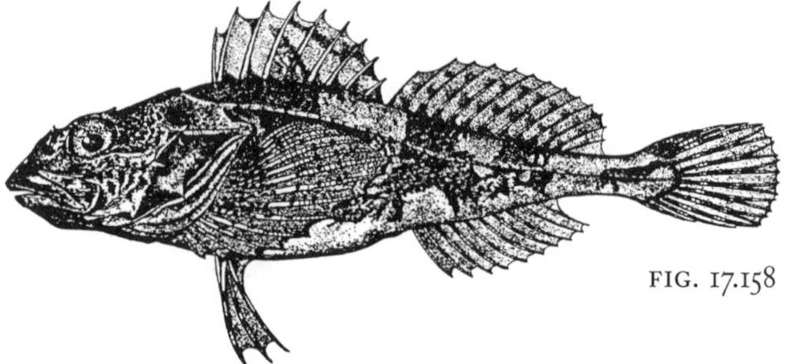
FIG. 17.158

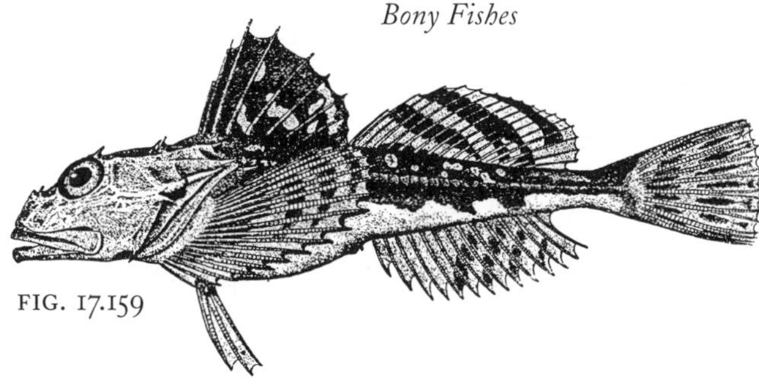
FIG. 17.159

Myoxocephalus octodecemspinosus, longhorn sculpin† (Cottidae). Very sharp spines on each operculum. Bo,Es–650(mu–sa),In–Fi,45.7 cm (18 in) [35.6 cm (14 in)], (B449,R282,S502), ☺. (FIG. 17.159)

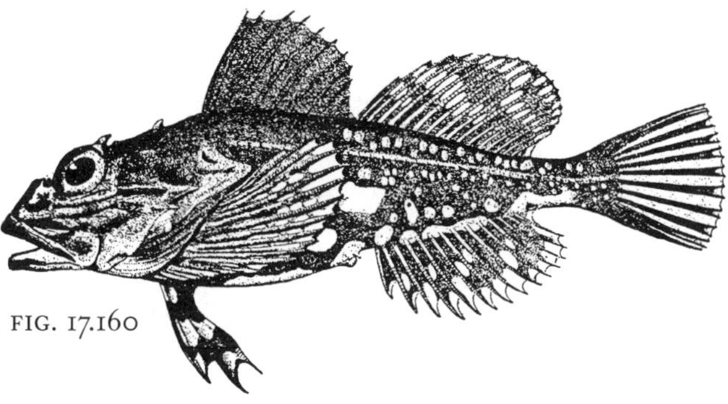
FIG. 17.160

Myoxocephalus scorpius, shorthorn sculpin† (Cottidae). Brownish with lighter blotches. Ac,Co–350(mu–sa–ha),In–Fi,0.9 m (3 ft) [35.6 cm (14 in)], (B445,R283,S507). (FIG. 17.160)

Opsanus tau, oyster toadfish†, dowdie (Batrachoididae). Large head, huge mouth, tapering body. Has only three gill bars. Pelvic fins under throat. Thick slimy mucus. Brownish green or yellow. Foghorn call during spawning; also grunts when disturbed. Can give nasty bite. Vi,Co,In–Fi,38 cm (15 in) [30.5 cm (12 in)], (B518,H337,M84,R85), ☺. (FIG. 17.161)

Triglops murrayi (=*ommatistius*), moustache sculpin† (Cottidae). Dark patch below eye; olive above to yellowish white below. Slender; very long anal fin. Ac,Co–Oc?,In?,20.3 cm (8 in), (B441,R283,S510). (FIG. 17.162)

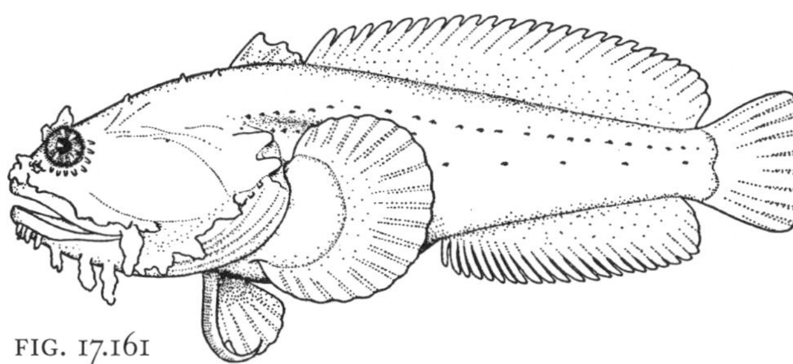
FIG. 17.161

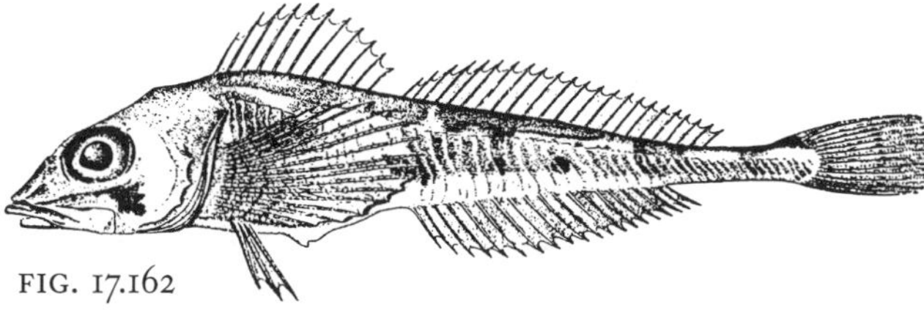
FIG. 17.162

17.59 Fish Group 6, Billfish (Istiophoridae) or Swordfish (Xiphiidae, 1 species included)

(from 17.24)

Scaled or scaleless fish with swordlike upper jaw and with two dorsal fins, the first long and the second shorter.

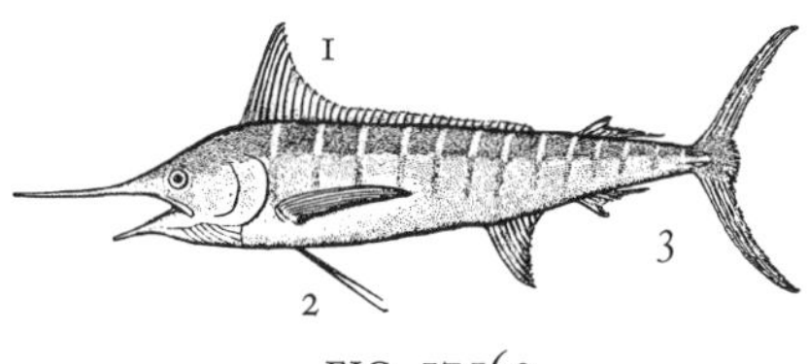

FIG. 17.163

1. Shape of first dorsal fin (FIG. 17.163)
   - A. Large, *sail*like — sa
   - B. *Pointed* anteriorly, then *tapering* — pt*
   - C. *Pointed*, ends abruptly (i.e., *short*) — ps
   - D. *Rounded* anteriorly, then *tapering* — rt
2. Appearance of pelvic fins (FIG. 17.163)
   - A. Pelvic fins as long *spines* below pectoral fins — sp*
   - B. Pelvic fins *absent* — x
3. Hard, ridgelike keels along caudal peduncle (slender portion of body just preceding tail) (FIG. 17.163)
   - A. *Two* ridgelike *keels* along peduncle — 2k*
   - B. *One* ridgelike *keel* along peduncle — 1k

17.60

Dorsal Fin	Pelvic Fin	Keel	Species
sa	sp	2k	*Istiophorus platypterus*, sailfish†
pt	sp	2k	*Makaira nigricans* (=*ampla*), blue marlin†
rt	sp	2k	*Tetrapturus albidus*, white marlin†
ps	x	1k	*Xiphias gladius*, swordfish†

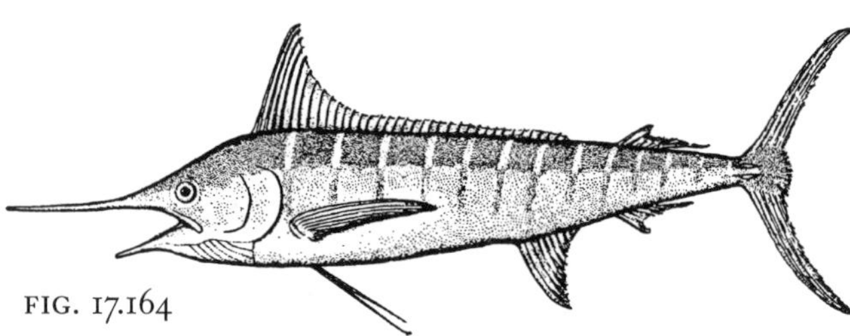

FIG. 17.164

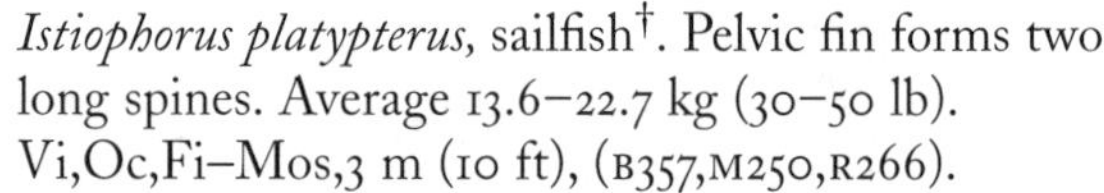

Istiophorus platypterus, sailfish†. Pelvic fin forms two long spines. Average 13.6–22.7 kg (30–50 lb). Vi,Oc,Fi–Mos,3 m (10 ft), (B357,M250,R266).

Makaira nigricans (=*ampla*), blue marlin†. Bill rounded on top. Dark body. Vi+,Oc,Fi–Mos,4.5 m (15 ft), (B358,M241,R264). (FIG. 17.164)

Tetrapturus albidus, white marlin†. Greener body color. Bill rounded on top. Especially acrobatic when hooked. Ca+,Oc,Fi–Mos,2.6 m (8.5 ft), (B360,M250,R265,S469). (FIG. 17.165)

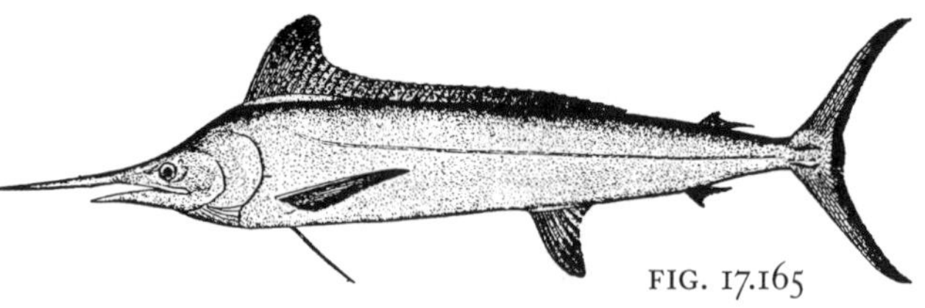

FIG. 17.165

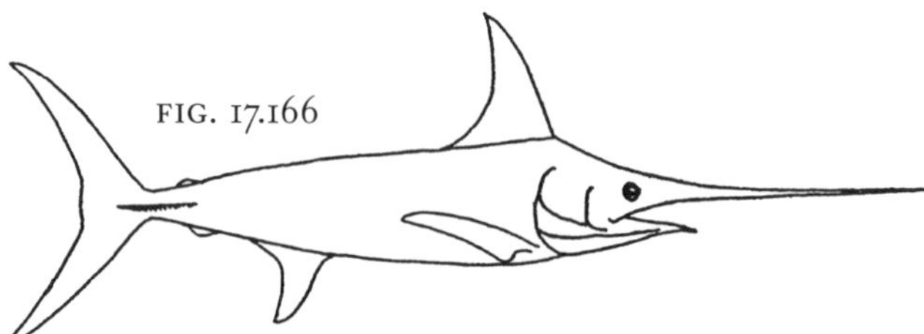

FIG. 17.166

Xiphias gladius, swordfish† (Xiphiidae). Bill is a third of total length; bill flattened on top. Second dorsal fin tiny. Shows both dorsal and caudal fins above water while surface swimming. Aggressive game fish. Bo,Oc,Fi,4.6 m (15 ft), (B351,H209,R266). (FIG. 17.166)

17.61 **Fish Group 7** (from 17.24)

Scaled fish with two dorsal fins, neither of which is especially long. Caudal fin slightly or deeply crescent.

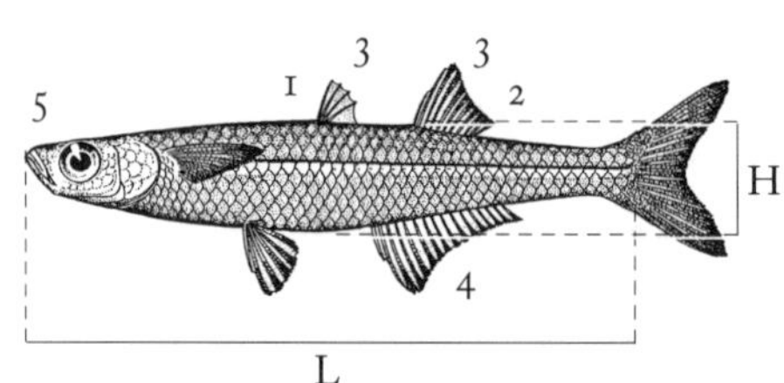

FIG. 17.167

1. Supports for first dorsal fins (FIG. 17.167)
   - A. At least some *spines* present — sp
   - B. Spines *absent,* soft rays only — x
2. Construction of the second dorsal fin (FIG. 17.167)
   - A. Second dorsal fin a small, *fleshy* (adipose) tab — fl
   - B. Fleshy tab *absent* — x*
3. Relative length of first dorsal fin compared to second dorsal fin (FIG. 17.167)
   - A. *First* dorsal fin distinctly *longer than second* dorsal fin — 1>2
   - B. *First* dorsal fin distinctly *shorter than second* dorsal fin — 1<2*
   - C. *First* dorsal fin approximately *equal to second* dorsal fin — 1=2
4. Location of ventral, paired pelvic fins relative to the first dorsal fin (FIG. 17.167)
   - A. *Pelvic* fins begin distinctly *more anteriorly* than *dorsal* fin — p/d*
   - B. *Dorsal* fin begins distinctly *more anteriorly* than *pelvic* fins — d/p
   - C. *Pelvic* fins begin at a line about *equal to* the start of the *dorsal* fin — p=d
5. Relationship of upper and lower jaws (FIG. 17.167)
   - A. *Lower* jaw extends more *anteriorly* than *upper* jaw — l/u*
   - B. *Upper* jaw extends more *anteriorly* than *lower* jaw — u/l
   - C. *Upper* and *lower* jaws are about *equal* — u=l
6. Body proportions—length/height ratio (does not count sword where appropriate) (FIG. 17.167)
   - A. *Slender*—length:height ratio greater than 5.5 — sl*
   - B. *Stout*—length:height ratio less than 4 — st
   - C. *Intermediate*—length:height ratio between 4 and 5.5 — in
7. Stripes or streaks along body
   - A. Provide *number* of horizontal stripes along body — __*
   - B. Stripes along body *absent* — x

17.62

Spines	2nd Dorsal	Dorsal 1 vs. Dorsal 2	Pelvic vs. Dorsal	Upper vs. Lower Jaw	Body Proportions	Stripes	Species
x	fl	1>2	d/p	u=l	in	x	*Salmo salar,* Atlantic salmon†
x	fl	1>2	p=d	u=l	in	x,1	*Osmerus mordax,* rainbow smelt†
x	fl	1>2	p/d	u=l	sl	x	*Synodus foetens,* inshore lizardfish†
x	x	1>2	d/p	u/l	in	x	*Xiphias gladius,* swordfish†, roadbill
sp	x	1<2	p/d	u=l,l/u	sl	1	*Menidia,* silversides Anal fin with 20–26 rays: *M. menidia,* Atlantic silversides† OR Anal fin with 14–20 rays: *M. beryllina,* inland silversides†, tidewater or waxen silversides
x	x	1<2	p=d	l/u	sl	x	*Mallotus villosus,* capelin†
sp	x	1=2	p/d	u=l	st	7–8	*Morone* (=*Roccus*) *saxatilis* (=*lineatus*), striped bass†, rockfish
sp	x	1=2	p/d	u=l	in	7–9	*Mugil cephalus,* striped mullet†, black mullet
sp	x	1=2	p/d	u=l	in	x	*M. curema,* white mullet†
sp	x	1=2	p/d	u=l,l/u	st	x	*Morone americana,* white perch†

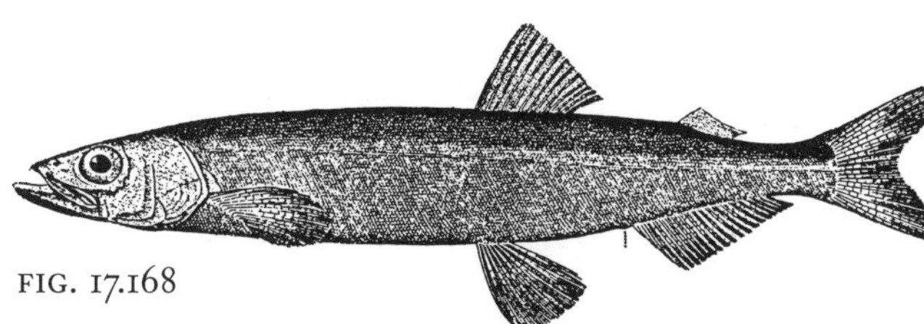
FIG. 17.168

Mallotus villosus, capelin†(Osmeridae). Slender body; minute scales. Greenish above, whitish below. Spawn in inshore gravel; mature offshore in huge schools. Ac,Co–Oc,Zo,18.8 cm (7.5 in), (B134,R77,S145), ☺. (FIG. 17.168)

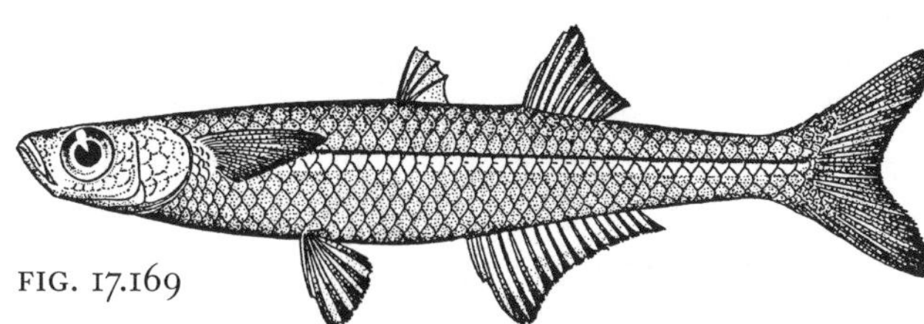
FIG. 17.169

Menidia beryllina, inland silversides†, tidewater or waxen silversides (Antherinidae). Paler coloring than *M. menidia.* Vi,Es–Co,Cr–Mo–An,7.6 cm (3 in), (B304,H189,R112), ☺. (FIG. 17.169)

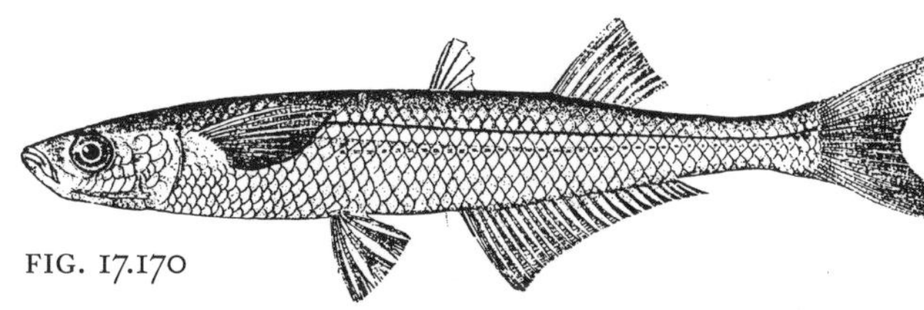
FIG. 17.170

Menidia menidia, Atlantic silversides† (Antherinidae). Schools over sand or gravel. Bo,Es–Co,Cr–An–Mo,14 cm (5.5 in), (B302,H187,M80,R113,S316), ☺. (FIG. 17.170)

Morone americana, white perch† (Percichthyidae). Deeper-bodied than striped bass. Greenish above, silvery below. Restricted to estuarine conditions; common in Hudson River as far up as Albany, N.Y. Bo,Es–Co,Cr–Fi,38 cm (15 in), (B405,H244,R129,S354), ☺. (FIG. 17.171)

Morone (=*Roccus*) *saxatilis* (=*lineatus*), striped bass†, rockfish (Percichthyidae). Stout body, thick caudal peduncle. Dark olive green above, silvery below. Larger than 11.2 kg (30 lb) are female. Strong swimmers in surf. Leave estuarine waters in winter and spring to migrate along coasts. Bo,Es–Co,In–Fi,1.8 m (6 ft) [1.2 m (4 ft)], (B389,H247,M111,R130,S356), ☺. (FIG. 17.172)

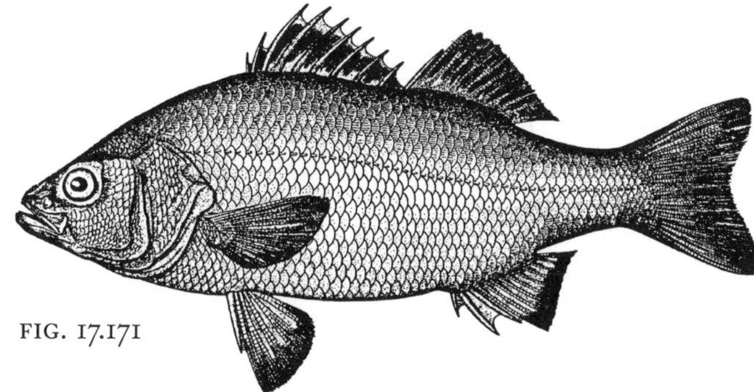
FIG. 17.171

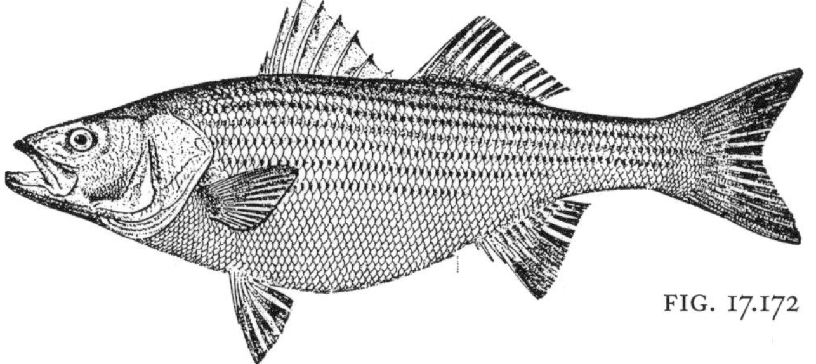
FIG. 17.172

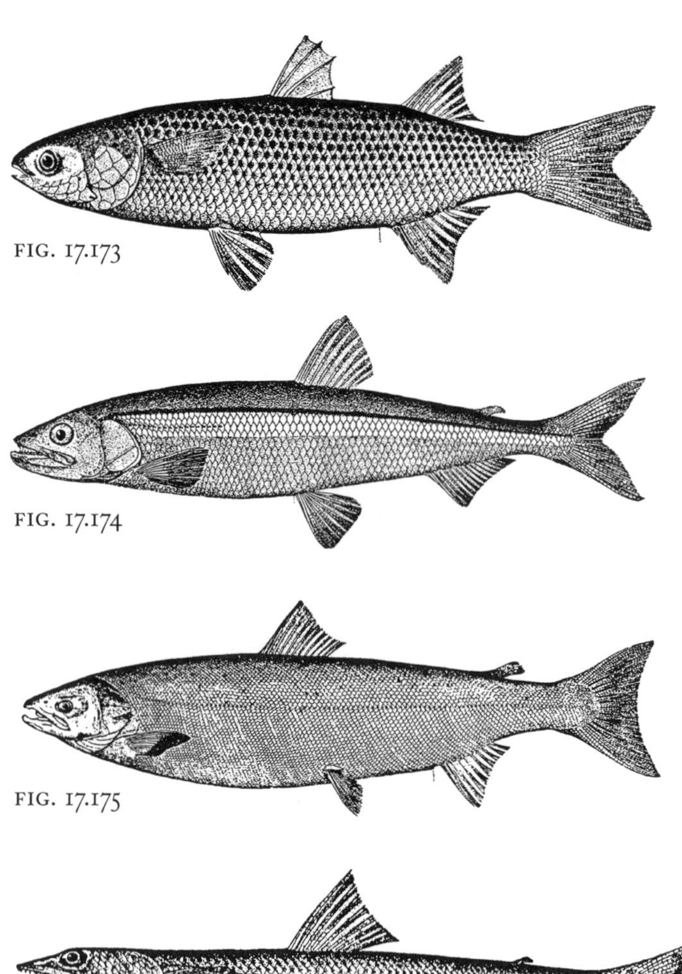

FIG. 17.173

FIG. 17.174

FIG. 17.175

FIG. 17.176

Mugil cephalus, striped mullet†, black mullet (Mugilidae). Somewhat indistinct lines formed by dark-centered scales; dorsal and anal fins lack scales. Forms large schools; forages in mud. Can jump over seine nets. Vi,Co–Es,He,91.4 cm (36 in), (B305,H193,R210,S400). (FIG. 17.173)

Mugil curema, white mullet† (Mugilidae). Slender-bodied. Dark blotch on pectoral fin. Vi,Co,In–He,76.2 cm (30 in), (H196,R213,S400), ☺.

Osmerus mordax, rainbow smelt† (Osmeridae). Large, toothed mouth and tongue. Green above; silvery below. Anadromous; clusters in bays and river mouths in summer and fall. Many lakes have land-locked populations. Bo,Es–Co,Cr–An–Fi,35.6 cm (14 in) [30.5 cm (12 in)], (B135,R78,S150). (FIG. 17.174)

Salmo salar, Atlantic salmon† (Salmonidae). Small head and eye; silvery, speckled dorsally. Second dorsal fin is adipose or fleshy. Anadromous (to freshwater to breed); lives to breed again. *Alevins* while in stream gravel; *parr* before returning to sea; *smolt* going to the sea; *grilse* returners after first winter; *salmon* thereafter. Ac,Es–Co,Fi,1.4 m (4.5 ft), (B121,R76,S129). (FIG. 17.175)

Synodus foetens, inshore lizardfish† (Synodontidae). Elongate, cylindrical body; large mouth. Olive or gray brown. Bury themselves in sand or prop on pectoral fins to ambush prey. Vi,Co,Fi–Cr–An,45.7 cm (18 in), (H130,R80). (FIG. 17.176)

Xiphias gladius, swordfish†, broadbill (Xiphiidae). Upper jaw extends as obvious sword. See description under Group 6, 17.60.

17.63 **Fish Group 8, Gadidae** (from 17.24)

With three, soft-rayed, dorsal fins.

1. Color of lateral line (pressure sensors forming midbody line along sides) (FIG. 17.177)
   - A. *White* or light-colored, contrasts with background color — wh*
   - B. *Black* or dark-colored, contrasts with background color — bl
   - C. *Intermediate* tone, blends with background; lateral line difficult to discern — in
2. Presence of distinctive coloration
   - A. Single *black* patch present above pectoral fin — bl
   - B. Body with scattered, but distinctive *spots* — sp
   - C. Back and sides of body with dark, marbled *mottling* — mo*
   - D. Such distinctive patches and spots *absent* — x
3. Length of longest spines on pelvic fins (FIG. 17.177)
   - A. Pelvic spines *more than twice* the length of the webbed part of the fin — >2*
   - B. Pelvic spines *less than twice* the length of the webbed part of the fin — <2
4. Relationship of upper and lower jaws (FIG. 17.177)
   - A. *Lower* jaw extends more *anteriorly* than *upper* jaw — l/u
   - B. *Upper* jaw extends more *anteriorly* than *lower* jaw — u/l*
   - C. *Upper* and *lower* jaws are about *equal* — u=l

FIG. 17.177

5. Shape of caudal fin (FIG. 17.177)
   A. Caudal fin *slightly crescent* sc
   B. Caudal fin *squared* sq
   C. Caudal fin *rounded* ro*

17.64

Lateral Line	Patch or Spot	Pelvic vs. Spine	Upper vs. Lower Jaws	Caudal Fin	Species
wh	sp	<2	u=l,u/l	sq	*Gadus morhua,* Atlantic cod†
wh	mo	>2	u/l	sq,ro	*Microgadus tomcod,* Atlantic tomcod†, frostfish
wh	x	<2	l/u	sc	*Pollachius virens,* pollock†
bl	bl	<2	u/l	sc	*Melanogrammus aeglefinus,* haddock†
in	x	<2	u/l	sq	*Gadus ogac,* Greenland cod†

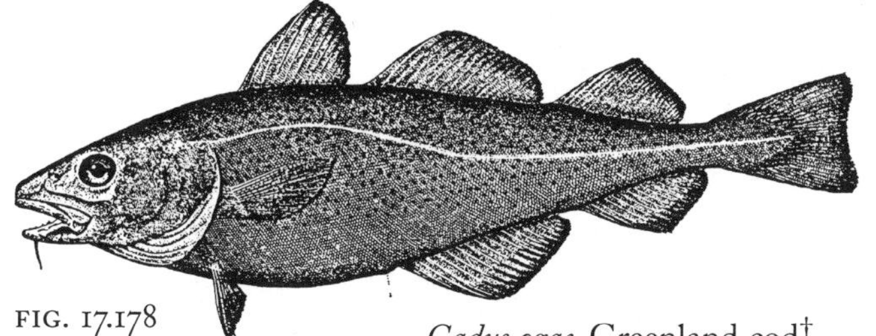

FIG. 17.178

Gadus morhua, Atlantic cod†. Two color phases: dark to olive brown or reddish orange; both spotted. Younger ones nearshore. Scrod are young cod or haddock. Seriously depleted stocks in the Northeast. Bo,Co–250(ha,sa),Fi–Mo,1.8 m (6 ft), (B182,H156,M129,R93,S266), ☺. (FIG. 17.178)

Gadus ogac, Greenland cod†. Po–,Co,Fi–In,71 cm (28 in), (R93,S270).

Melanogrammus aeglefinus, haddock†. Dark gray above, white below. First dorsal fin sharply pointed. Often sold as cod. Seriously depleted stocks in the Northeast. Bo,150–1000(sa),In–Fi,111.8 cm (44 in), (B199,M132,R93,S174). (FIG. 17.179)

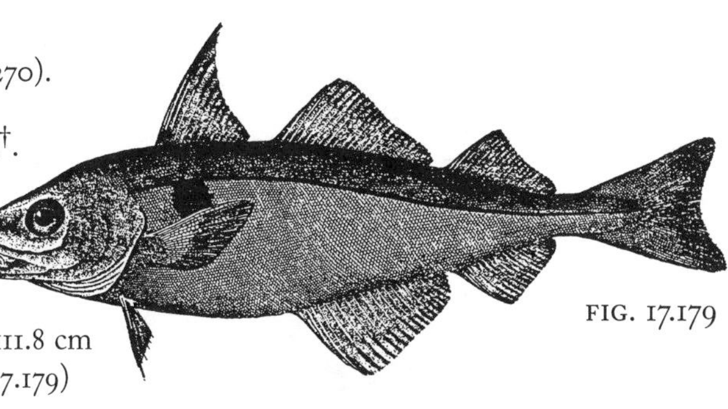

FIG. 17.179

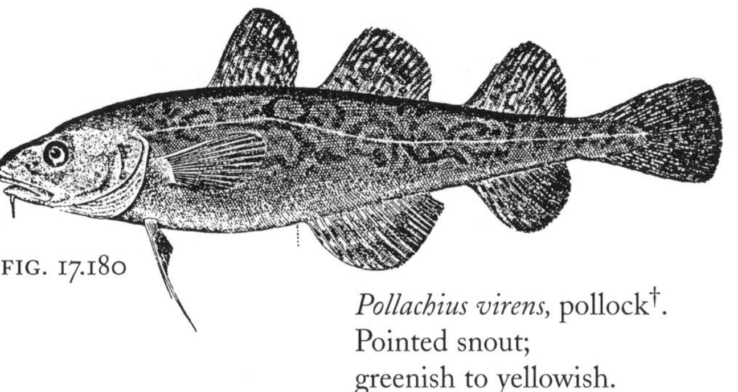

FIG. 17.180

Microgadus tomcod, Atlantic tomcod†, frostfish. Olive to brownish green, with blotches. Brackish water north of Cape Cod. Bo,Es–Co(mu),Cr–An–Mo–Mos–Fi,38 cm (15 in), (B196,M131,R93,S281). (FIG. 17.180)

Pollachius virens, pollock†. Pointed snout; greenish to yellowish. Bo,Co–600,Zo–Cr–Fi,1.1 m (3.5 ft), (B213,H155,M131,R93,S286), ☺. (FIG. 17.181)

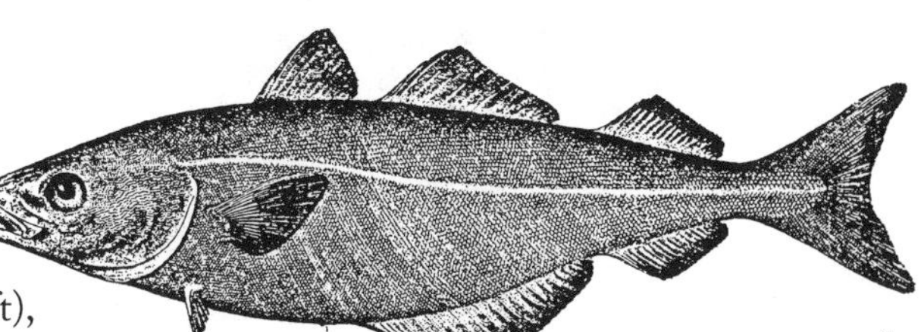

FIG. 17.181

17.65 **Fish Group 9, Scombridae** (from 17.24)

With several small finlets between second dorsal and caudal fins.

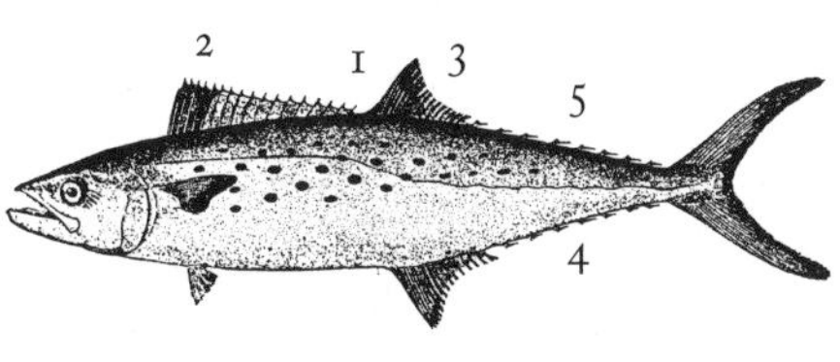

FIG. 17.182

1. Relationship of first dorsal fin to second dorsal fin (FIG. 17.182)
   A. A distinct *gap* or space exists between first and second dorsal fins ga*
   B. First dorsal fin extends nearly to second dorsal fin, gap or space *absent* x
2. Number of spines in first dorsal fin: provide *number* of spines (FIG. 17.182) —

3. Shape of second dorsal fin (FIG. 17.182)
   - A. *Height* of second dorsal fin clearly *less than* its *length* — h<l
   - B. *Height* of second dorsal fin clearly *greater than* its *length* — h>l
   - C. Second dorsal fin about *equal* in *height* and *length* — h=l*
4. Body color pattern below (ventral to) lateral line (midbody row of pressure sensors forming line along sides) (FIG. 17.182)
   - A. Scattered, distinct *spots* present — sp*
   - B. *Anterior* to *posterior* streaks present — ap
   - C. Body silvery: spots and streaks *absent* — x
5. Body color pattern above (dorsal to) lateral line (FIG. 17.182)
   - A. *Anterior* to *posterior* streaks present — ap
   - B. *Dorsal* to *ventral* streaks present — dv
   - C. Scattered dark *spots* present — sp*
   - D. Coloration uniform; streaks and spots *absent* — x

17.66

Dorsal 1 vs. Dorsal 2	Dorsal Spines	Dorsal 2 Shape	Ventral Color	Dorsal Color	Species
x	9–10	h=l	sp	dv	*Scomber japonicus* (=*colias*), chub mackerel†, hardhead, bull's-eye
x	10–14	h=l	x	dv	*S. scombrus* (=*scombrusa*), Atlantic mackerel†
ga	ca. 21	h<l	ap	ap,dv	*Sarda sarda,* Atlantic bonito†, horse mackerel, skipjack
ga	18	h=l	sp	sp	*Scomberomorus maculatus,* Spanish mackerel†
ga	16–18	h>l	sp	ap	*Euthynnus alletteratus,* little tunny†, false albacore
ga	15	h>l	ap	x	*Katsuwonus* (=*Euthynnus*) *pelamis,* skipjack tuna†, striped bonito, oceanic bonito
ga	13–14	h>l	sp,x	x	*Thunnus thynnus,* bluefin tuna†

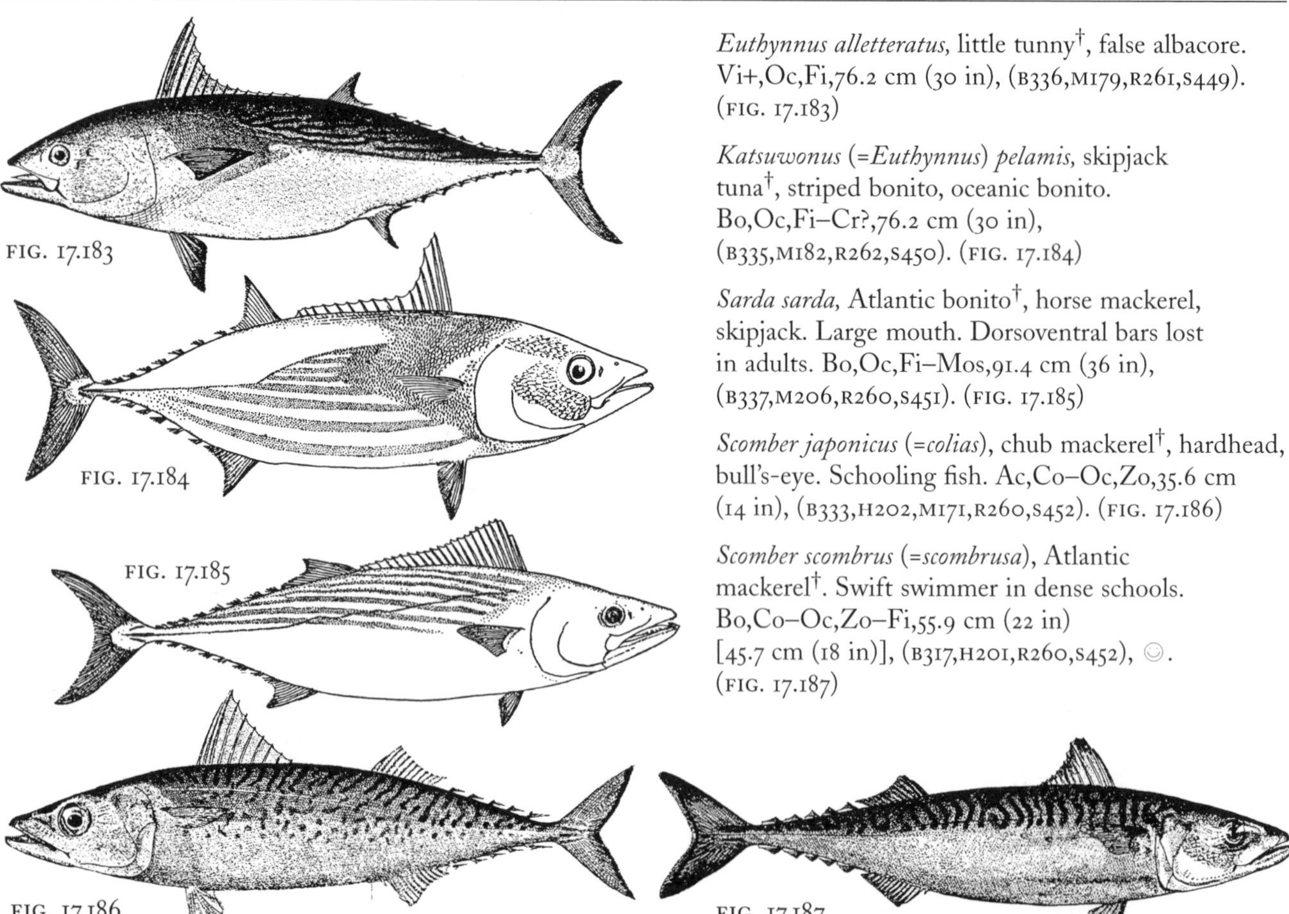

FIG. 17.183

FIG. 17.184

FIG. 17.185

FIG. 17.186

FIG. 17.187

Euthynnus alletteratus, little tunny†, false albacore. Vi+,Oc,Fi,76.2 cm (30 in), (B336,M179,R261,S449). (FIG. 17.183)

Katsuwonus (=*Euthynnus*) *pelamis,* skipjack tuna†, striped bonito, oceanic bonito. Bo,Oc,Fi–Cr?,76.2 cm (30 in), (B335,M182,R262,S450). (FIG. 17.184)

Sarda sarda, Atlantic bonito†, horse mackerel, skipjack. Large mouth. Dorsoventral bars lost in adults. Bo,Oc,Fi–Mos,91.4 cm (36 in), (B337,M206,R260,S451). (FIG. 17.185)

Scomber japonicus (=*colias*), chub mackerel†, hardhead, bull's-eye. Schooling fish. Ac,Co–Oc,Zo,35.6 cm (14 in), (B333,H202,M171,R260,S452). (FIG. 17.186)

Scomber scombrus (=*scombrusa*), Atlantic mackerel†. Swift swimmer in dense schools. Bo,Co–Oc,Zo–Fi,55.9 cm (22 in) [45.7 cm (18 in)], (B317,H201,R260,S452), ☺. (FIG. 17.187)

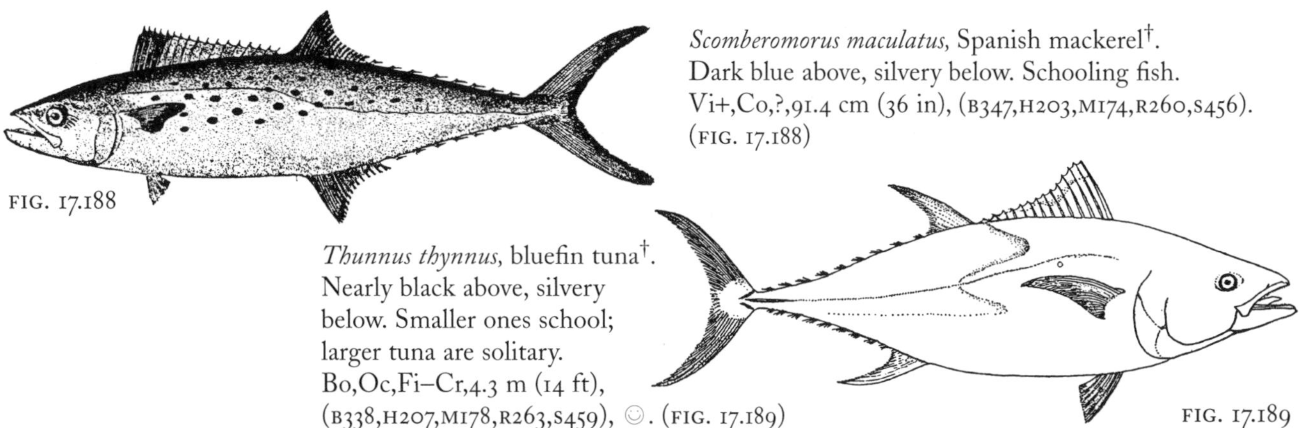

Scomberomorus maculatus, Spanish mackerel†. Dark blue above, silvery below. Schooling fish. Vi+,Co,?,91.4 cm (36 in), (B347,H203,M174,R260,S456). (FIG. 17.188)

FIG. 17.188

Thunnus thynnus, bluefin tuna†. Nearly black above, silvery below. Smaller ones school; larger tuna are solitary. Bo,Oc,Fi–Cr,4.3 m (14 ft), (B338,H207,M178,R263,S459), ☺. (FIG. 17.189)

FIG. 17.189

17.67 **Fish Group 10** (from 17.24)

With distinct spines or other protuberances preceding dorsal fin.

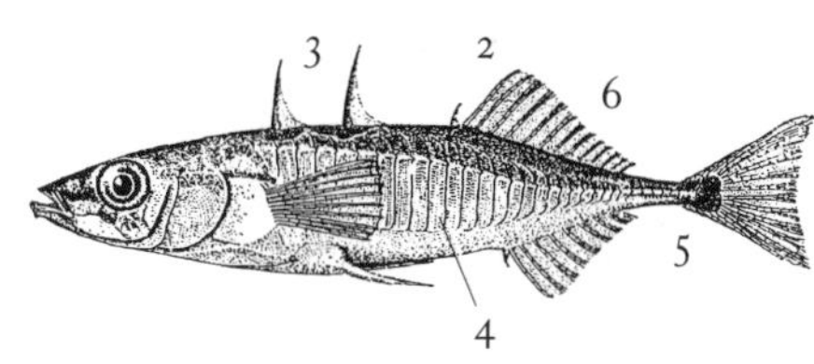

FIG. 17.190

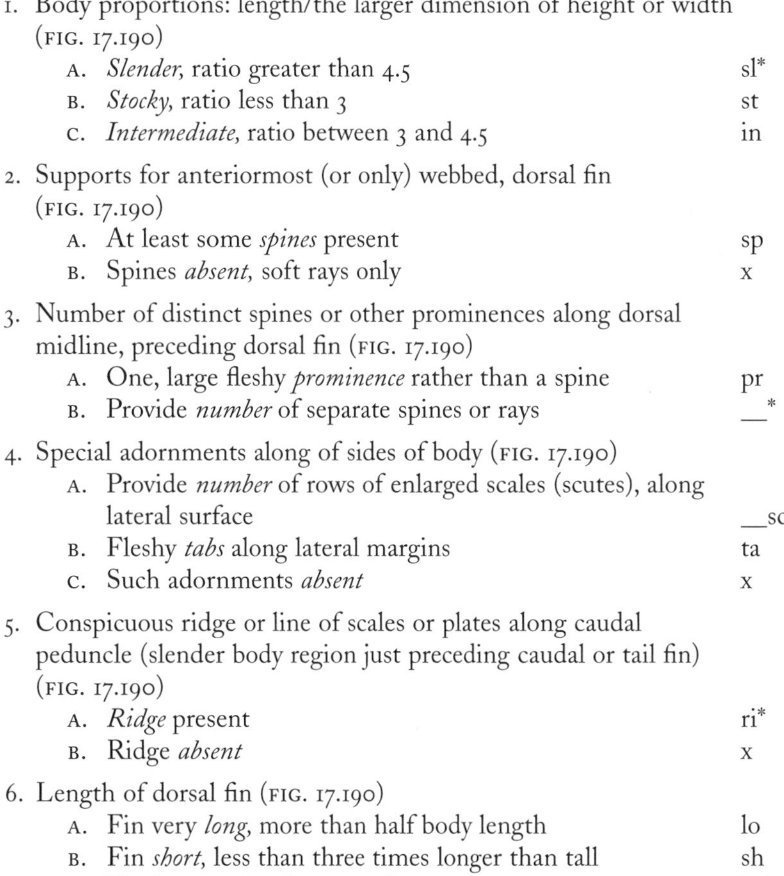

1. Body proportions: length/the larger dimension of height or width (FIG. 17.190)
   - A. *Slender,* ratio greater than 4.5 — sl*
   - B. *Stocky,* ratio less than 3 — st
   - C. *Intermediate,* ratio between 3 and 4.5 — in
2. Supports for anteriormost (or only) webbed, dorsal fin (FIG. 17.190)
   - A. At least some *spines* present — sp
   - B. Spines *absent,* soft rays only — x
3. Number of distinct spines or other prominences along dorsal midline, preceding dorsal fin (FIG. 17.190)
   - A. One, large fleshy *prominence* rather than a spine — pr
   - B. Provide *number* of separate spines or rays — __*
4. Special adornments along of sides of body (FIG. 17.190)
   - A. Provide *number* of rows of enlarged scales (scutes), along lateral surface — __sc*
   - B. Fleshy *tabs* along lateral margins — ta
   - C. Such adornments *absent* — x
5. Conspicuous ridge or line of scales or plates along caudal peduncle (slender body region just preceding caudal or tail fin) (FIG. 17.190)
   - A. *Ridge* present — ri*
   - B. Ridge *absent* — x
6. Length of dorsal fin (FIG. 17.190)
   - A. Fin very *long,* more than half body length — lo
   - B. Fin *short,* less than three times longer than tall — sh
   - C. Fin *intermediate* between these extremes — in*

17.68

Body	Spines	No. Free Spines	Side Feature	Ridge	Dorsal Fin	Species
st	x	9	2sc	x	sh	*Cyclopterus lumpus,* lumpfish†. See description under 17.32.
st	x	6–8	x	x	in	*Selene setapinnis,* Atlantic moonfish† Compare with first 3 species in 17.52; see description under 17.42.
st	x	5–7(6)	x	x	in	*Trachinotus carolinus,* Florida pompano†. See description under 17.42.
st	x	1	x	x	in	*Aluterus schoepfi,* orange filefish†
st	sp	3	ta	x	sh	*Lophius americanus,* goosefish†, monkfish, allmouth
in	sp	pr	x	x	lo	*Lopholatilus chamaeleonticeps,* tilefish†
in	sp	2–4	x	x	in	*Apeltes quadracus,* fourspine stickleback†, bloody stickleback
in	x	2–3	1sc	x	in	*Gasterosteus wheatlandi,* twospine stickleback† , blackspot stickleback
in	x	8–10	x	x	lo	*Rachycentron canadum,* cobia†, crabeater
sl	x	7–12(9)	1sc	ri	in	*Pungitius pungitius,* ninespine stickleback†
sl	x	9–11	1sc	ri	sh	*Acipenser oxyrhynchus,* Atlantic sturgeon†. See description under 17.26.
sl	x	3–4(3)	1sc	ri	in	*Gasterosteus aculeatus,* threespine stickleback†
sl	x	1	x	x	lo	*Enchelyopus cimbrius,* fourbeard rockling†

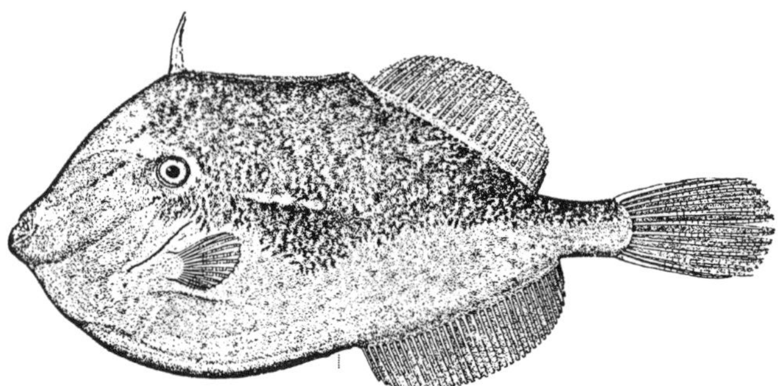
FIG. 17.191

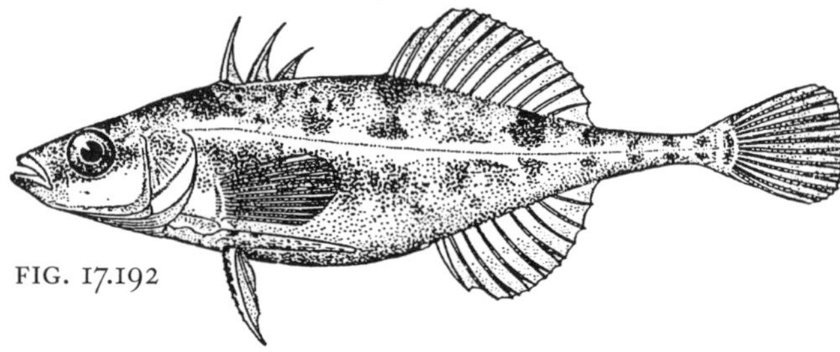
FIG. 17.192

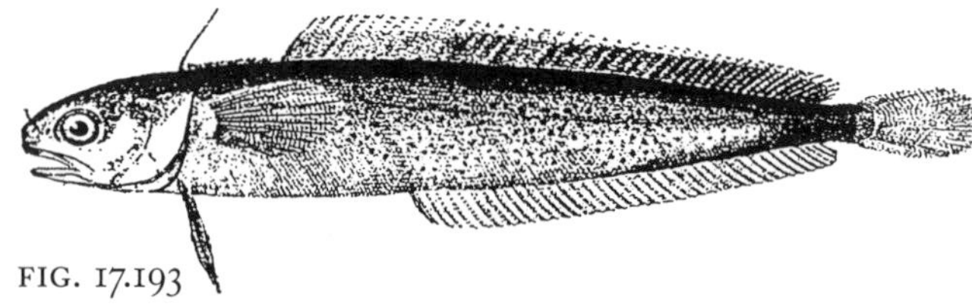
FIG. 17.193

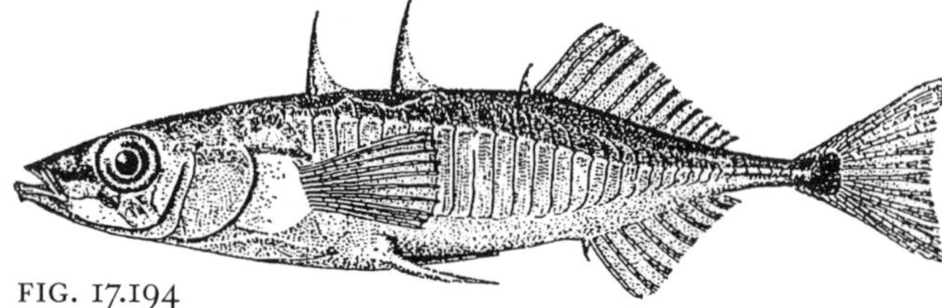
FIG. 17.194

Aluterus schoepfi, orange filefish† (Monacanthidae). Single, large middorsal spine. Small mouth. Dark or orange above with darker blotches. Frequently with jellyfish. Vi+,Co,In–He,61 cm (24 in), (B524,H344,R300). (FIG. 17.191)

Apeltes quadracus, fourspine stickleback†, bloody stickleback (Gasterosteidae). Small; greenish brown with dark mottling above, silvery white below. Male with red pelvic fins. Three freestanding spines; fourth incorporated into the webbed dorsal fin. Salt marshes into fresh water. Male builds cone-shaped nest. Bo,Es–SM–Co,Cr,6.4 cm (2.5 in), (B311,H180,R121,S335). (FIG. 17.192)

Enchelyopus cimbrius, fourbeard rockling† (Gadidae). Four barbels on top of snout. Dark, dusky brown above, light dotted with brown below. Bo,175–1790(mu–sa),Cr–An,30.5 cm (12 in), (B234,R94,S264). (FIG. 17.193)

Gasterosteus aculeatus, threespine stickleback† (Gasterosteidae). Variable coloring, gray to brown above, silvery below. Third spine just anterior to dorsal fin. Pelvic fins reddish in males. Estuarine or full marine. Builds ball-shaped nests. Bo,Es–Co,In–Fi,10.2 cm (4 in), (B308,H178,R122,S338), ☺. (FIG. 17.194)

Gasterosteus wheatlandi, twospine stickleback†, blackspot stickleback (Gasterosteidae). Green with blotches above, sides golden with blotches, silvery below. Reddish pelvic fins. Bo,Es–Co,In,7.7 cm (3 in), (B310,R122,S341).

FIG. 17.195

Lophius americanus, goosefish†, monkfish, allmouth (Lophiidae). Dorsoventrally flattened. Soft texture, no scales; enormous mouth with many sharp, hinged teeth. First dorsal fin with "lure;" dark brown and blotchy. Produces large gelatinous sheet of eggs. Known to capture seabirds along with more usual fish and invertebrates. Bo,Co–2180(sa–mu),Fi–In,122 cm (48 in), (B532,H352,M85,R86,S235), ☺. (FIG. 17.195)

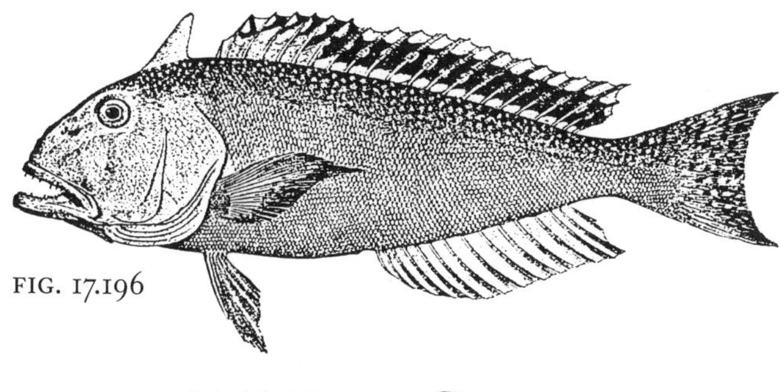

FIG. 17.196

Lopholatilus chamaeleonticeps, tilefish† (Malacanthidae). Large head; eye high on body. Bluish green with white spots above, rose on sides, white below. Prefers crabs. Constructs burrows toward edge of continental shelf. Bo,480–1750(sa–mu),Cr–An–Ec,106.7 cm (42 in), (B426,H305,R154,S368), ☺. (FIG. 17.196)

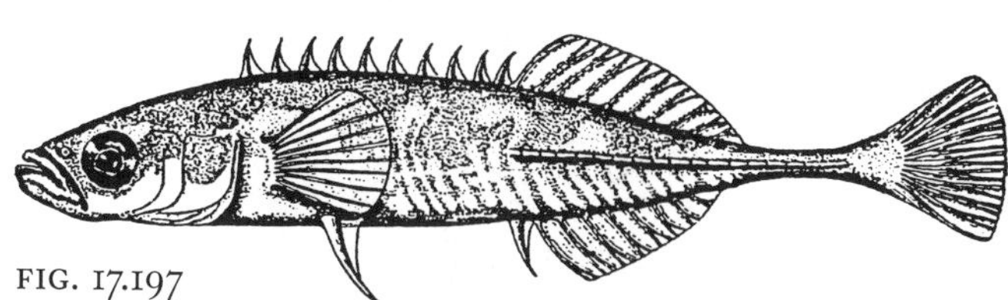

FIG. 17.197

Pungitius pungitius, ninespine stickleback†(Gasterosteidae). Slender; pelvic fins as two curved spines. Olive brown above, sides blotchy, silvery below. Estuaries and marsh creeks especially. Bo,Es–Co,Cr,7.6 cm (3 in), (B307,R122), ☺. (FIG. 17.197)

Rachycentron canadum, cobia†, crabeater (Rachycentridae). Long anal fin. Lower jaw projects beyond upper jaw. Dark brown above, pale below. Good food fish. Young with darker, lateral stripe. Often with floating objects. Migrates into our region in summer. Vi,Co,Cr–Fi,1.8 m (5 ft 10 in), (H235,R156).

17.69 Class Reptilia: Subclass Anapsida: Order Testudines, Marine Turtles

Marine turtles are not common in northeastern North America. The pattern of plates on the *carapace* (dorsal surface of the shell) is useful in distinguishing species. *Vertebral plates* (FIG. 17.198, vp) are found along the midline of the carapace, while *marginal plates* (FIG. 17.198, mp) form the border around it. *Pleural plates* (FIG. 17.198, pp) are located between vertebral and marginal plates. Coloration and size of the carapace may also help in identifications.

Nearshore species prefer warmer waters and seldom stray this far northward. Leatherback turtles are offshore pelagic (open-water) species. For years, all species of marine turtles have been subjected to serious overharvesting by humans for both meat and eggs. Their reproductive habits add to risk factors. Egg laying is often restricted to a limited number of remote beaches. Mass hatchings force their young to run a gauntlet of potential predators. Finally, their slow growth and late onset of sexual maturity increase the period of their vulnerability.

All species are in the family Cheloniidae unless otherwise noted.

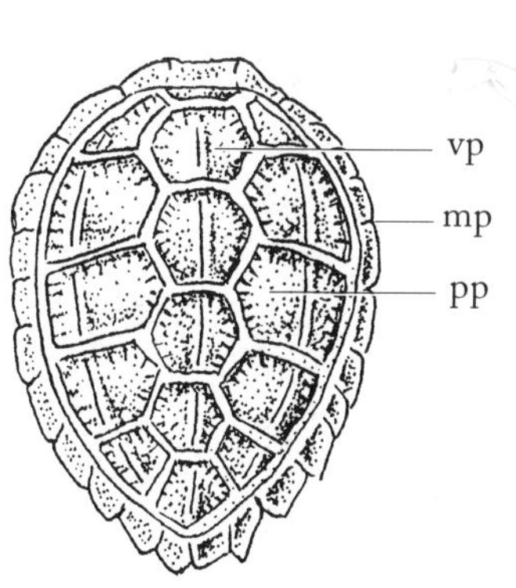

FIG. 17.198

1. Description of carapace (FIG. 17.198)
   - A. Carapace covered with distinct *plates* — pl*
   - B. Carapace lacks plates but possesses *longitudinal ridges* — lr
2. Vertebral plates (vp) (FIG. 17.198)
   - A. Posterior edge of each plate clearly *overlaps* the leading edge of the next plate — ov
   - B. Adjacent vertebral plates *abut* one another without overlapping — ab*
   - C. Vertebral plates *absent* — x
3. Number of pairs of pleural plates (pp) (FIG. 17.198)
   - A. Provide *number* of pairs — __
   - B. Pleural plates *absent* — x
4. Number of pairs of plates between the eyes
   - A. Provide *number* (1 or 2) — __
   - B. Plates *absent* from head — x

17.70

Carapace	Vertebral Plates	Pleural Plates	Eyes	Species
lr	x	x	x	*Dermochelys coriacea,* leatherback or coffinback turtle
pl	ov	4	2	*Eretmochelys imbricata,* hawksbill turtle
pl	ab	4	1	*Chelonia mydas,* green turtle
pl	ab	5	2	Reddish brown color; >0.6 m (2 ft) in length: *Caretta caretta,* loggerhead turtle OR Olive-brown color, up to 0.6 m (2 ft) in length (rare): *Lepidochelys kempi,* Atlantic or Kemp's ridley turtle

FIG. 17.199

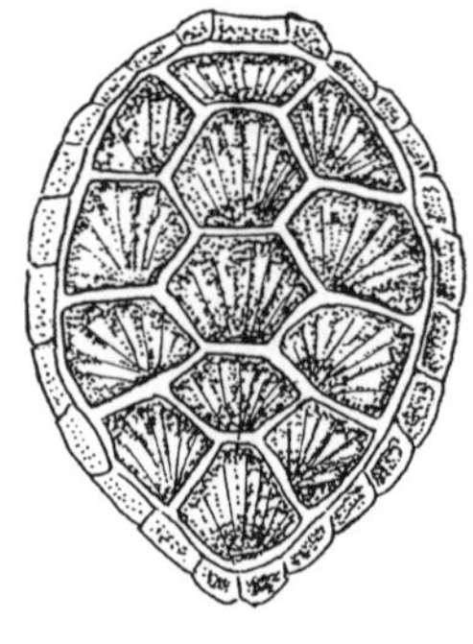

FIG. 17.200

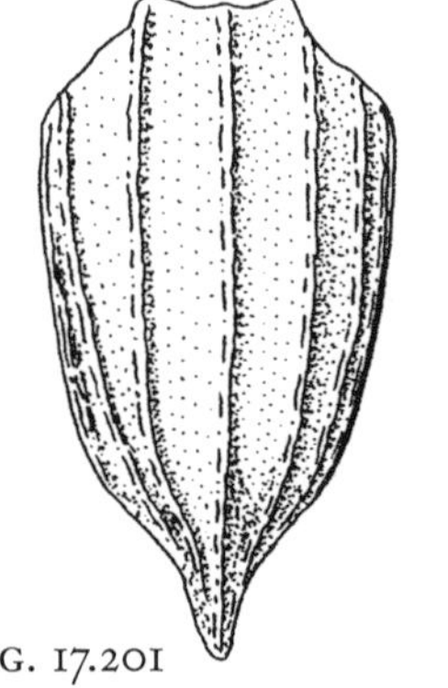

FIG. 17.201

FIG. 17.202

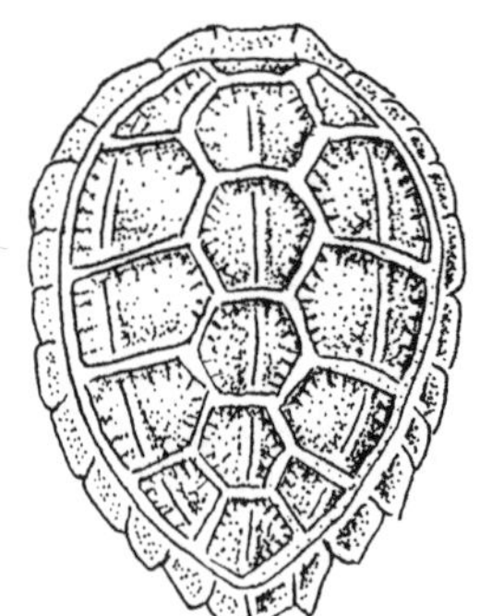

FIG. 17.203

Caretta caretta, loggerhead turtle. Reddish-brown above; yellow brown below. Large head; to 204 kg (450 lb). Eats slower moving crabs, jellyfishes, mollusks. Nests on Florida beaches. Vi+,Co,Cr–Mo,1.2 m (4 ft), ☺ (not common, but most likely of the sea turtles in temperate waters). (FIG. 17.199)

Chelonia mydas, green turtle. Dark brownish to olive with streaks or spots above; yellowish white below. To 226.8 kg (500 lb). Active by day feeding on *Thalassia,* turtle grass. Ca+,Co,He–Ot,1.2 m (4 ft). (FIG. 17.200)

Dermochelys coriacea, leatherback or coffinback turtle (Dermochelyidae). Dark greenish black or brown. Largest marine reptile. Very long forelimbs lack claws. Eats jellyfishes and other planktonic organisms (making them vulnerable to plastic debris in ocean waters). Typically far offshore. Weight to 907 kg (2000 lb). Bo,Oc,In,1.8 m (6 ft). (FIG. 17.201)

Eretmochelys imbricata, hawksbill turtle. Greenish brown with radiating dark streaks. Thick plates; posterior margin strongly toothed. Small, narrow head; some weigh 113.4 kg (250 lb); most are 18–22.7 kg (40–50 lb). Prefers warm waters, especially coral reefs. Eats sponges. Ca+,Co,In,91.4 cm (36 in). (FIG. 17.202)

Lepidochelys kempi, Atlantic or Kemp's ridley turtle. Olive gray. Small. Feeds on fast-moving crabs, clams, and seagrasses in shallow water. Most endangered sea turtle. Adults in Gulf of Mexico; young can stray into our region. Ca+,Co,Cr–Mo–He,91.4 cm (36 in). (FIG. 17.203)

17.71 Class Mammalia

Marine mammals of the northeastern United States have received much attention in recent years. Antics of captive seals and dolphins have endeared these animals to many. Whale watching and other ecologically oriented boat trips are widely available along our coastline and provide excellent opportunities for appreciating the diversity and abundance of these impressive animals.

Marine mammals treated here may be placed within two orders. Seals, sea lions, and walruses are included in the Order Pinnipedia (17.78); and whales, dolphins, and porpoises are found in the Order Cetacea (17.72).

17.72 Order Cetacea: Suborder Mysticeti (Baleen Whales) and Suborder Odontoceti (Toothed Whales, Dolphins, and Porpoises)

(from 17.73)

Cetacea are divided into two suborders on the basis of feeding structures. In the Mysticeti ("moustached whales"), whalebone plates (baleen) consisting of horny (proteinaceous) material hang downward from the roof of the mouth. The fibrous inner edges of these plates form a dense mesh that these whales use to filter plankton or small fish from huge gulps of ingested seawater. Odontoceti ("toothed whales," but also includes all the dolphins and porpoises) use jaws lined with teeth to capture individual, larger prey, such large invertebrates, fish, and other vertebrates. Several are capable of using echolocation to find food.

References: (K) Katona, S. K., V. Rough, and D. T. Richardson, 1993; (L) Leatherwood, S., D. K. Caldwell, and H. E. Winn, 1976.

1. Body length
   - A. *Larger* cetaceans, body length in excess of 4.5 m (15 ft) — la
   - B. *Smaller* cetaceans, body length less than 4.5 m (15 ft) — sm

17.73

Length	Group
la	Larger cetacea, whales and dolphins, 17.74
sm	Smaller cetacea, dolphins and porpoises, 17.76

17.74 Larger Cetacea, Whales and Dolphins (in part) (from 17.73)

Whales are observed alive at sea or stranded on the shoreline. Detailed features often used for the identification of stranded individuals are often useless at sea. The following key emphasizes field characters. More detailed features are included in species descriptions and especially in references cited.

Despite their enormous size, whales at sea often provide only fleeting glimpses of body features, making field determination much more difficult than it may seem. Here again, the robin theory (see introduction) is a useful strategy. The task of identification is made much easier by keeping in mind which species are most likely to be encountered, given circumstances of time and place. Features used are those that are often visible when the dorsal portion of the whale breaks the water surface as the animal empties and then fills its lungs several times in preparation for a dive. They also include features often visible from above, a perspective available to boat-based observers who can view nearby whales just beneath the water surface.

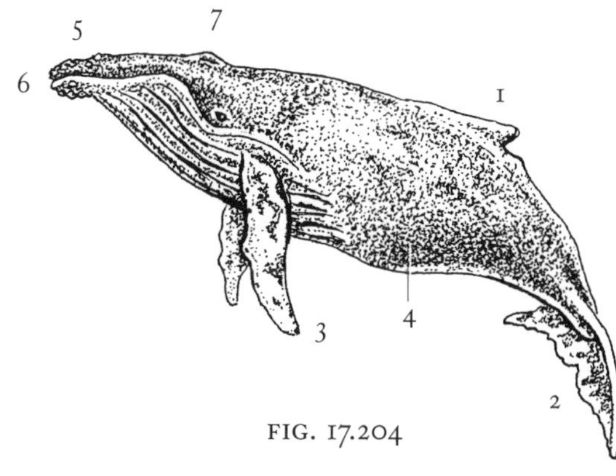

FIG. 17.204

1. Dorsal fin (FIG. 17.204)
   - A. Dorsal fin *tall*, much higher than long — ta
   - B. Distinct *dorsal fin* present, no more than two times taller than long — df*
   - C. Dorsal fin *absent* — x
2. Appearance of underneath surface of flukes or tail (add an asterisk [*] to this character if undersurface of tail is shown as the whale dives; otherwise, you may need to leave this character blank) (FIG. 17.204)
   - A. Undersurface all *dark* — da
   - B. At least some of undersurface *white* — wh*
3. Coloration of pectoral fin or flippers (FIG. 17.204)
   - A. Flippers all *white* and *large*, with rough edges — wl*
   - B. Flippers all *white* but *small* — wm
   - C. Flippers dark but with *white band* — wb
   - D. Flippers all *dark* — da

4. Distinctive white or lighter coloration on body (other than light underbelly that most whales have) (FIG. 17.204)
   - A. Body all *white* — wh
   - B. Irregular, bumpy patches of white on head, called *callosities* — ca
   - C. White *oval* just behind eye — ov
   - D. *Right* upper *jaw* white — rj
   - E. Light patch just behind *dorsal fin* — df
   - F. Light patch just behind *pectoral fins* or flippers — pf
   - G. Light, mottled *spots* all over dorsal body surface — sp
   - H. Additional light markings *absent* — x*

5. Appearance of rostrum or nose (FIG. 17.204)
   - A. *Knobby*, all black — kn*
   - B. Irregular white concretions or *callosities* — ca
   - C. Head blunt ended or *squared* — sq
   - D. Rostrum *smooth* — sm

6. Location of mouth (FIG. 17.204)
   - A. Mouth located towards *dorsal* surface of head — do
   - B. Mouth located on *ventral* surface of head — ve
   - C. Mouth approximately *terminal*, located at front of head — te*

7. Shape of "blow" or spout (this character is placed last because it is easily modified by surrounding conditions and is more subject to individual interpretation) (FIG. 17.204)
   - A. Blow consists of two distinct parts in *V*-shape — v
   - B. Blow directed distinctly *forward* — fo
   - C. Blow *tall* and narrow — ta
   - D. Blow *short* and bushy — sh
   - E. *No* blow visible upon repeated surfacing — x

17.75

Dorsal Fin	Tail Color	Pectoral Fin	Body Light	Rostrum	Mouth	Blow	Species
df	wh*	wl	x	kn	do	sh	*Megaptera novaeangliae*, humpback whale
df	wh	da	df,rj	sm	do	ta	*Balaenoptera physalus*, finback whale
df	da	wb	df	sm	do	x	*B. acutorostrata*, minke whale
df	da*	da	sp	sm	do	ta	*B. musculus*, blue whale
df	da	da	sp	sm	do	ta,sh	*B. borealis*, sei whale
df	da	da	x	sm	te	x	*Globicephala melas* (=*melaena*), pilot whale, pothead, blackfish
ta	da	da	ov	sm	te	x	*Orcinus orca*, killer whale
x	da*	da	ca	ca	do	v	*Eubalaena glacialis*, right whale
x	da*	da	x	sq	ve	fo	*Physeter catodon*, sperm whale
x	wh	wm	wh	sm	te	x	*Delphinapterus leucas*, beluga whale

*shows tail on deep dives

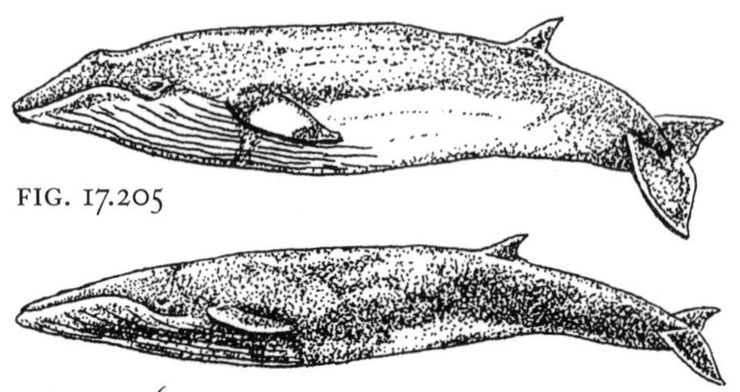

FIG. 17.205

FIG. 17.206

Balaenoptera acutorostrata, minke whale (Balaenopteridae). Rostrum pointed. Baleen is light-colored. On some, light crescent patches from midback to flippers. Relatively fearless of boats. Bo,Co,Fi,9.1 m (30 ft), (K40,L63), ☺. (FIG. 17.205)

Balaenoptera borealis, sei whale (Balaenopteridae). Speckling more sparse than on blue whales; usually limited to sides. Head uniformly dark. Surface feeder on copepods. Bo,Oc,Cr–Fi,15.3 m (50 ft), (K73,L32). (FIG. 17.206)

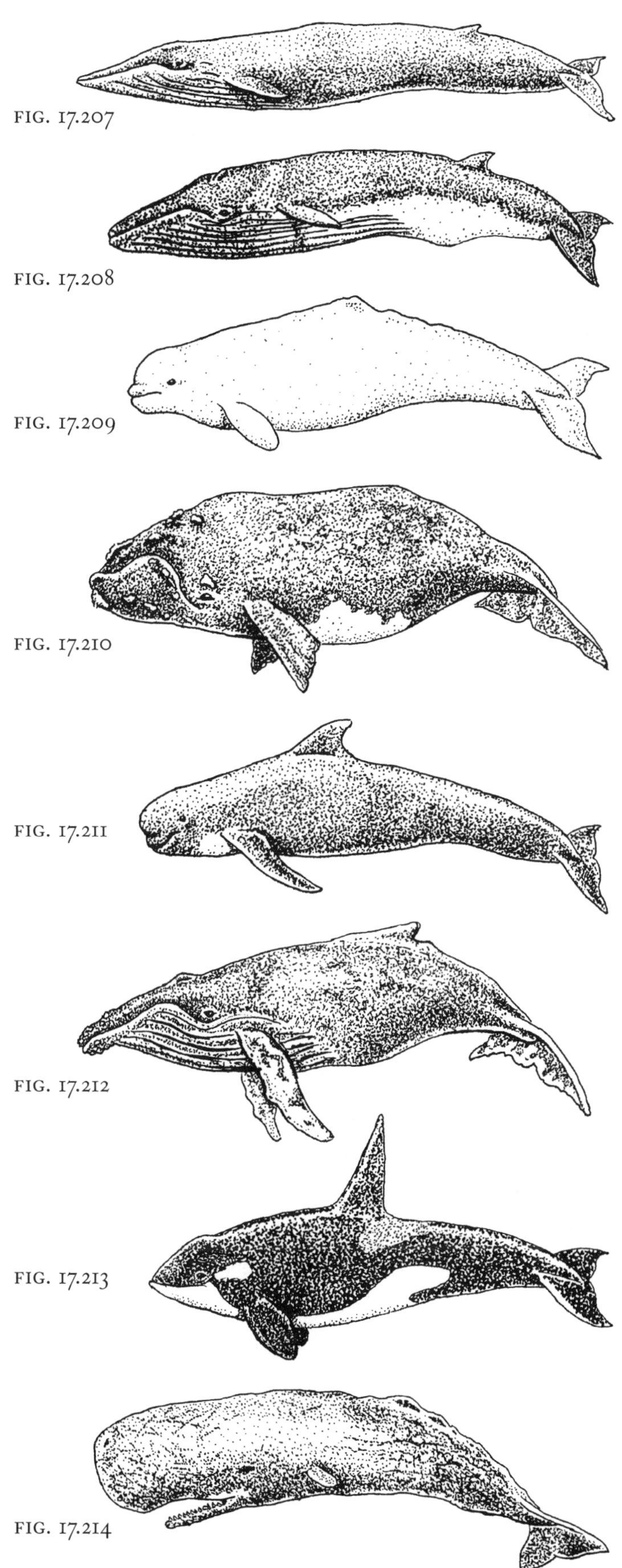

Balaenoptera musculus, blue whale (Balaenopteridae). Speckling extends across back. Baleen all black. Largest living animal. Feeds on krill (Crustacea: Euphausiacea) in polar waters. Populations decimated globally by whaling. Bo,Oc,Cr,25.9 m (85 ft), (K66,L19). (FIG. 17.207)

Balaenoptera physalus, finback whale (Balaenopteridae). Back ridged toward tail. Front third of baleen and lower lip on right side is yellowish white. Right side of head lighter than left side. Bo,Co–Oc,Cr–Mos–Fi,24 m (79 ft), (K31,L26), ☺. (FIG. 17.208)

Delphinapterus leucas, beluga whale (Monodontidae). Distinct neck constriction. Young are gray. Were especially abundant in Gulf of St. Lawrence and northward; numbers sinking in recent years. Ac–,Co,Mos–Cr–Fi,4.9 m (16 ft), (K135,L99). (FIG. 17.209)

Eubalaena glacialis, right whale (Balaenidae). Chunky body; large head; jawline strongly arched. Whitish callosities form individual specific patterns. Slow, not wary of boats, carcass floats upon death; all these factors explain why it was the "right" whale for near-extermination by whaling. Feeds on copepods. Bo,Oc,Cr,16.1 m (53 ft), (K57,L52). (FIG. 17.210)

Globicephala melas (=*malaena*), pilot whale, pothead, blackfish (Delphinidae). Rounded head. Long-based dorsal fin set well forward on body. Gray patch under chin and on belly. Forms large herds; occasionally known to strand themselves on beaches. Bo,Co–Oc,Mos–Fi,6 m (20 ft), (K87,L90), ☺. (FIG. 17.211)

Megaptera novaeangliae, humpback whale (Balaenopteridae). Chunky body; pectoral fin and fluke margins rough. Wonderful behaviors: spy hopping, tail lobbing, flipper slapping, breaching. Bo,Co–Oc,Cr–Fi,16.2 m (53 ft), (K47,L40), ☺. (FIG. 17.212)

Orcinus orca, killer whale (Delphinidae). Extremely tall dorsal fin; white belly extends high on sides behind dorsal fin as grayish saddle; oval white patch just behind eye. Fast swimmer; prefers colder waters; travels in small groups. Feeds primarily on seals. Bo,Co,Ma–Mos–Fi,9.2 m (30 ft), (K125,L84). (FIG. 17.213)

Physeter catodon, sperm whale (Physeteridae). Huge, squared head; narrow lower jaw. Blowhole is forward and toward left side. Ridged back; wrinkled skin on posterior half of body; brownish coloring. Solitary or in groups to 40. Bo,Oc,Mos–Fi,21 m (69 ft), (K140,L57). (FIG. 17.214)

17.76 Smaller Cetacea (Dolphins and Porpoises) (from 17.73)

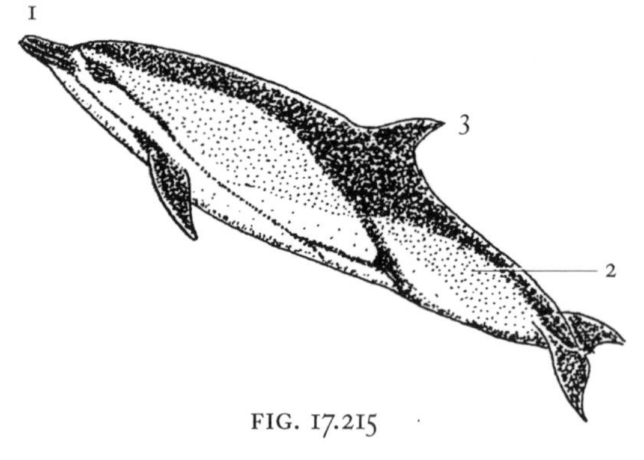

FIG. 17.215

Many species of dolphins and porpoises are gregarious and, to the delight of many, seem to be attracted to boats. Some ride the bow waves of larger vessels, using the pressure wave present as the vessel pushes its way through the water. Acrobatic maneuvers, including jumping clear of the water surface, are among their more spectacular behavioral characteristics. Identifying these fast-moving animals requires a sharp eye and the good fortune of their close approach. Usually, the shape of their head and dorsal fin, along with distinctive, light-colored body areas, are easiest features to spot.

Common names used here follow the convention that members of the family Delphinidae are referred to as dolphins and members of the family Phocoenidae, porpoises.

1. Shape of head (FIG. 17.215)
   - A. Head with narrow *rostrum* (beak) — ro*
   - B. Rostrum *absent,* head gradually tapered or squared — x
2. Distinctive coloration to sides of body (FIG. 17.215)
   - A. Isolated *white patch* in darker background below and behind dorsal fin — wp
   - B. Hour-glass shaped cream-tan patches *criss-cross* one another along sides below dorsal fin — cc*
   - C. Distinctly paler patches found high on sides *anterior* to and especially *posterior* to the dorsal fin — ap
   - D. Light coloring forms *flame*like lateral pattern; one branch upward from side toward dorsal fin, others found on posterior portion of body — fl
   - E. Dark grayish coloring with many light *scratches* and lines all over body — sc
   - F. Light coloring from belly extends high on sides, especially *anterior* of dorsal fin — an
   - G. Such distinctive light coloration *absent*; dark upper body fades to whitish belly coloring — x
3. Shape of dorsal fin (FIG. 17.215)
   - A. Dorsal fin *triangular* in shape (i.e., both anterior and posterior margins approximately straight) — tr
   - B. Dorsal fin *falcate* in shape (i.e., posterior margin concave or curves inward distinctly) — fa*

17.77

Rostrum	Light Color	Shape of Dorsal Fin	Species
x	an	tr	*Phocoena phocoena,* harbor porpoise
x	sc	fa	*Grampus griseus,* Risso's dolphin, gray grampus
ro	wp	fa	*Lagenorhynchus acutus,* Atlantic white-sided dolphin
ro	cc	tr,fa	*Delphinus delphis,* common or saddleback dolphin
ro	ap	fa	*Lagenorhynchus albirostris,* white-beaked dolphin
ro	fl	fa	*Stenella coeruleoalba,* striped dolphin
ro	x	fa	*Tursiops truncatus,* Atlantic bottlenosed dolphin

FIG. 17.216

Delphinus delphis, common or saddleback dolphin (Delphinidae). Pointed snout often white-tipped; black line connects tip of snout to flipper. Eyes with white framed, dark mask. Dorsal fin grayish toward its interior. Large herds; active bow wave riders. Bo,Co–Oc,Mos–Fi,2.6 m (8.5 ft), (K105,L116). (FIG. 17.216)

FIG. 17.217

FIG. 17.218

FIG. 17.219

FIG. 17.220

FIG. 17.221

FIG. 17.222

Grampus griseus, Risso's dolphin or gray grampus (Delphinidae). Dorsal fin darker than body color. Vertical crease in center of forehead. No teeth in upper jaw; less than seven in each side of lower jaw. Forms large groups. Sometimes rides bow waves. Bo,Oc,Mos–Fi,4 m (13 ft), (K120,L96). (FIG. 17.217)

Lagenorhynchus acutus, Atlantic white-sided dolphin (Delphinidae). Between lateral white patch and white belly is distinctive band of yellowish tan. Dark line connects eye to flipper. Travels in small to large herds. Wary of boats; tends not to ride bow waves. Ac–,Oc,Fi?,2.7 m (9 ft), (K94,L123). (FIG. 17.218)

Lagenorhynchus albirostris, white-beaked dolphin (Delphinidae). Tall, dark dorsal fin. Rostrum can be grayish or dark. Light-colored rump area is distinctive. Forms large herds; not usually bow wave rider. Ac,Oc,Mos–Cr–Fi,3.1 m (10 ft), (K102,L126). (FIG. 17.219)

Phocoena phocoena, harbor porpoise (Phocoenidae). Small, stocky body. Travels in small groups. Wary of boats; does not ride bow waves. Coastal; prefers colder waters. Bo,Co,Fi–In,1.8 m (6 ft), (K81,L150), ☺. (FIG. 17.220)

Stenella coeruleoalba, striped dolphin (Delphinidae). Also with curved dark line(s) from eye to anus. Forms large herds; low jumps; sometimes bow wave rider. Bo,Co–Oc,Mos–Cr–Fi,2.4 m (8 ft), (K109,L113). (FIG. 17.221)

Tursiops truncatus, Atlantic bottlenosed dolphin (Delphinidae). Dark dorsal surface blends to pinkish white belly. Warm-water species found increasingly offshore toward its northern limit. Forms large and small groups; often accompanies whales. The most common aquarium species. Vi+,Oc,Cr–Fi,3.7 m (12 ft), (K113,L128). (FIG. 17.222)

17.78 **Order Pinnipedia, Seals** (from 17.00)

Because there are no fur seals or sea lions in the North Atlantic, and because walruses only rarely stray into the northernmost extremes of our region, this treatment will be limited to the true seals of the family Phocidae. Phocids have chubby but streamlined bodies. They lack external ear flaps (i.e., pinnae) and a distinct neck but possess anterior and posterior pairs of flippers. Head shape and body coloration form the primary means of distinction among regional species.

1. Distinctive features of the head profile
   - A. Inflatable mass of tissue forms a *hood* covering this surface — ho
   - B. Profile line is *straight* or slightly convex (i.e., creates horse-head look) — st
   - C. Profile line is slightly *concave* (i.e., creates dog-head look) — cc
2. Relative position of eye
   - A. Front margin of eye located approximately *half* the distance from the snout to the ear — ½
   - B. Front margin of eye distinctly more *posterior,* located closer to ear than snout — po

3. Dark to black coloration of face area
   - A. *Dark* to black facial area — da
   - B. Facial area *not* especially darker than rest of head — x
4. Body color pattern
   - A. *Blue-gray* background with large, irregularly placed black patches — bg
   - B. *Light* body with dark blotches blended along sides to form dark *bands* — lb
   - C. *Light* body with scattered, irregular *darker* spots — ld
   - D. *Light* body color, *no* spots or blotches — lx
   - E. *Dark* body with scattered, irregular *lighter* spots — dl
   - F. *Intermediate* body color with scattered, irregular *darker* spots — id
   - G. *Intermediate* body color with scattered, irregular darker spots *ringed* with lighter color — ir

17.79

Profile	Eye	Face	Color	Species
st	po	da,x	dl	*Halichoerus grypus,* gray seal, horsehead seal—male
st	po	da,x	ld	*H. grypus,* gray seal, horsehead seal—female
cc	½	x	id,ld	*Phoca vitulina,* harbor seal
cc	½	x	ir	*P. hispida,* ringed seal
cc	½	da	lb	*P. groenlandica,* harp seal
cc	½	da	bg	*Cystophora cristata,* hooded seal—pup
cc	½	da	ld	*C. cristata,* hooded seal—female
ho	½	da	ld	*C. cristata,* hooded seal—male
st,cc	po,½	da,x	lx	Pups of all species except *C. cristata,* the hooded seal, are born light-colored (whitecoat) and change to adult colors, usually within a few weeks. Young seals are often lighter in coloration than adults.

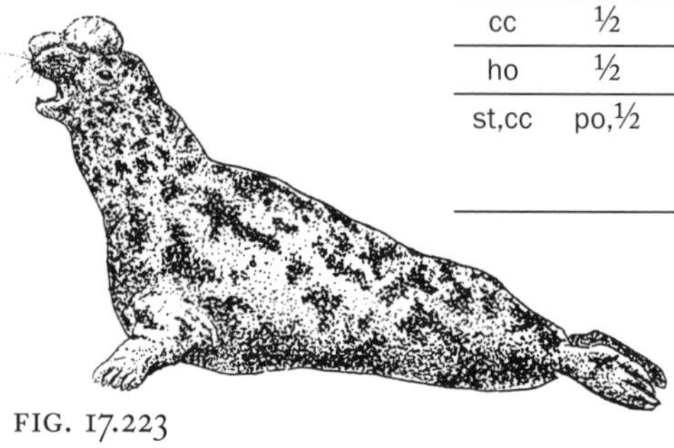

Cystophora cristata, hooded seal. Both hood and nasal sac inflatable when animal is alarmed. Females smaller than males. Po–,Co,Fi–Mo–Mos–Cr,2.6 m (8.5 ft), (K224). (FIG. 17.223)

FIG. 17.223

Halichoerus grypus, gray seal, horsehead seal. Gulf of St. Lawrence northward; strays to the south. Po–,Co,Fi–Mos,2.4 m (8 ft), (K212), ☺. (FIG. 17.224)

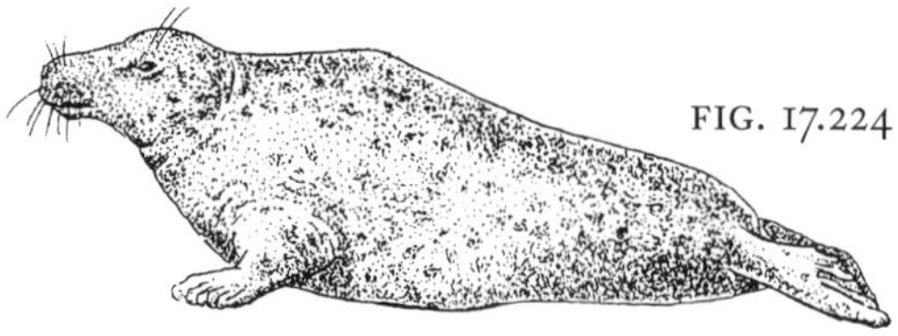

FIG. 17.224

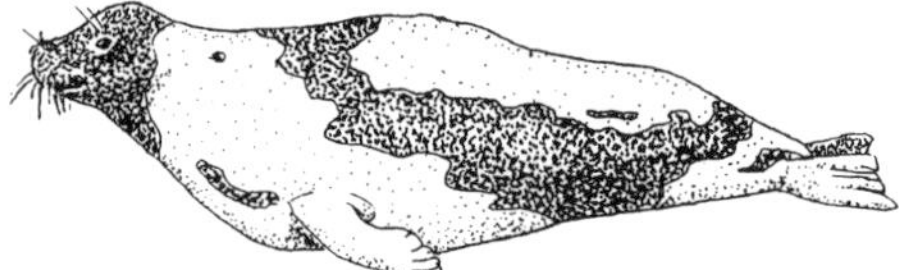

Phoca groenlandica, harp seal. Most abundant seal of eastern Canada. Po–,Co,Fi–Cr,1.7 m (5.7 ft), (K219). (FIG. 17.225)

FIG. 17.225

Phoca hispida, ringed seal. Large eyes and long facial bristles. Uncommon strandings south to New Jersey. Po–,Co,Fi,1.7 m (5.5 ft), (K229). (FIG. 17.226)

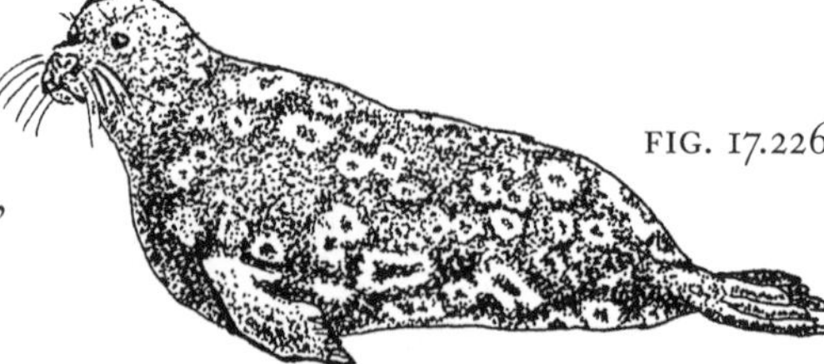

FIG. 17.226

Phoca vitulina, harbor seal. In northern New England and Canada common to see in groups on intertidal rocks during low tide. Winters farther south. Ac,Co,Fi–Mos,1.7 m (5.6 ft), (K203), ☺. (FIG. 17.227)

FIG. 17.227

Appendix ■ Recommendations for Anesthetization, Fixation, and Preservation of Specimens

Occasionally it is useful or necessary to immobilize active organisms for examination and perhaps to preserve them for later study. Conservation and humane concerns should discourage this practice when possible. The following procedures have been suggested by various sources as being most effective if this approach is required.

1. Anesthetizing agents are required to relax the muscles of animals that otherwise would react to fixatives and preservatives by contracting strongly. Frequently, anesthetization helps calm specimens so that accurate observations can be made. If exposure to these agents is kept at suggested levels for brief periods, most specimens will revive fully when returned to fresh seawater following examination. It is important to expose animals to such agents slowly, often by adding the agent to seawater containing the specimen.
   - Chloral hydrate (CH): crystals added to seawater medium; or animal may be immersed in 2% solution.
   - Ethanol (Eth): drop-by-drop addition of 10% ethyl alcohol solution serves as an anesthetizing agent for some animals.
   - Magnesium chloride ($MgCl_2.6H_2O$): used in 7.5% solution, approximately isotonic with ambient seawater. Fox and Ruppert (1985) suggested using a refractometer to adjust concentration to match local conditions.
   - Magnesium sulfate ($MgSO_4$) (MS): epsom salts often used by adding crystals to seawater; frequently requires long time periods to be effective. This may be, however, the most effective pre-preservation method for such groups as Ectoprocta or Holthuroidea.
   - MS-222 (ethyl m-aminobenzoate, sold also as Finquel™): used for relaxing cold-blooded animals (especially fishes), but also works for some invertebrates. Used in 0.01–0.02% solutions (0.5% for polychaetes); requires from seconds to 15 min to work. Must be made up fresh each time. Ayerst Laboratories, Inc., New York.
   - Propylene phenoxytol (PP): add to seawater medium to form less than 1% total volume; may take 15 min or more to be effective. Can be reused repeatedly. PPO Goldschmidt Chemical Corporation, New York.

2. Fixation is required to harden tissues so that they retain their original shape. For serious histological work, a strong fixative such a Bouin's solution, glutaraldehyde solution, or osmium tetroxide are often required. Preparation and use of these materials lie beyond the scope of this presentation. Each of these fixatives is extremely dangerous to work with and costly to dispose of. They are toxins by their very nature. Be extremely careful using them and dispose of them safely. When working with fixatives or preservatives, always wear protective equipment, especially for eyes and hands, and work in a well-ventilated area. Disposal of these materials must follow procedures required by federal and state regulations. For additional information and precautions, refer to the OSHA Material Safety Data Sheet, which should be supplied with such material.

   For routine fixation, formalin and alcohol are most commonly employed.

   - Ethanol: for some specimens, ethyl alcohol is adequate as a fixative (see Table A).
   - Formalin: a 40% solution of formaldehyde. Percentage dilutions listed in Table A are for formalin, not formaldehyde. A 5% seawater solution is made by adding 0.5 parts formalin to 9.5 parts seawater. Formalin is acidic and is not a good long-term storage medium for animals with calcareous parts, which tend to dissolve. These problems can

TABLE A. **Recommended procedures for anesthetization, fixation, and preservation of marine invertebrates. Numbers refer to percent solutions.**

	Anesthetization					**Fixation**			**Preservation**	
Animal Group	Menth	MS222	PP	MgCl	Other	Formalin	Ethanol	Other	Formalin	Ethanol
Porifera						10				70
Hydrozoa	x	x	x	x	CO	20			5–10	70
Scyphozoa	x	x			EU	10			5–10[§]	70
Anthozoa	x	x		x		20			10	70
Ctenophora					CH		70*			70
Platyhelminthes					CH, CT		3–5	HW	5–10	70–90
Nemertea					CH, CT,MS	10	30–50		3–5[§]	70–90
Entoprocta	x									70
Ectoprocta	x				MS, CO	5			5	70–90
Brachiopoda		x	x		Eth, MS		70–90			70–90
Sipuncula			x	x	Eth	4	70–90		4	70–90
Echiuroidea	x		x	x	Eth*	5	70		5	70
Mollusca	x		x	x	MS, temp[‡]	5			5[§]	70
Polychaeta		x	x		MS	5–10				70–90
Oligochaeta			x		Eth	4				70–90
Hirudinea			x		Eth	4			4	
Crustacea, small					F	5		SS	5[§]	70–90
Crustacea, large	x	x			CH, temp	5[†]				70
Chaetognatha						5	70–90		6	70–90
Echinodermata	x		x		MS	10–12	90			50–90
Hemichordata				x	Eth*	5			5	70
Urochordata	x	x	x		MS, CH	5			5	70

SOURCES: Brusca, 1980, Fox and Ruppert, 1985, Lincoln and Sheals, 1979. CT, chloretone; CO, clove oil; EU, ethyl urethane; Menth, menthol; MS, magnesium sulfate; pp, propylene phenoxytol; SS, Steedman's Solution; temp, low temperature.

* Add gradually, e.g., drop by drop.

[†] Inject formalin into tissues of large specimens; add glycerol to prevent specimen from becoming brittle.

[‡] Relax nudibranchs at low temperature; then add frozen formalin and allow it to thaw for fixation.

[§] Calcareous structures (e.g., shells, coral skeleton, nemertean stylets, etc.) must be preserved in alcohol; they dissolve in formalin.

be eliminated in part by neutralizing the formalin solution by buffering it with base. Formalin is now suspected of being toxic. Because of the dangers in handling formalin and the expense of disposing of it as hazardous waste, most laboratories rely on substitutes for preservation. Initial fixation of specimens (usually for 48 hours) may still be necessary.

Standard 10% solution of buffered formalin: 100 ml concentrated formalin; 900 ml distilled water; 4 g monobasic sodium phosphate ($NaH_2PO_4.H_2O$); 6.5 g dibasic sodium phosphate (Na_2HPO_4). After 48 hours, remove specimen from formalin and place in ethanol.

- Steedman's Solution (SS): combines a fixative with a long-term preservative for plankton: 0.5 ml propylene phenoxytol; 4.5 ml propylene glycol; 5 ml formalin; 90 ml seawater (or distilled water).

3. Preservation is required for long-term storage of specimens. The same precautions described for fixatives apply to preservatives.
   - Carosafe™: a preservative used by the Carolina Biological Supply Company. Contains ethylene glycol as a deterrent to mold and tissue deterioration and is applied to specimens initially fixed with formalin. This substance is used to minimize exposure to the unpleasant fumes of formalin.
   - Ethanol: the most widely used long-term preservative. A high concentration (70–90%) is used. Its disadvantages include the fact that it evaporates even from apparently well-sealed storage jars. It also tends to remove color from specimens. A small quantity of glycerine will help with color retention.
   - Formalin: a 4–10% solution of formalin (see notes above) has been widely used as a preservative. Its drawbacks include its suspected toxicity, its unpleasant odor, and its tendency to dissolve calcareous body parts.
   - Isopropanol: another alcohol sometimes employed as a preservative.
   - Propylene phenoxytol: a 1–2% aqueous solution is a good long-term preservative. Although it is expensive, it preserves color and does not leave the specimen rigid and brittle. This material must be used on previously fixed specimens, however.

■ Bibliography

General

Berman, J., L. Harris, W. Lambert, M. Buttrick, and M. Dufresne. 1992. "Recent Invasions of the Gulf of Maine: Three Contrasting Ecological Histories." *Conservation Biology* 6:435–441.

Beston, H. 1928. The Outermost House. New York: Viking Press.

Brusca, R. C. 1980. *Common Intertidal Invertebrates of the Gulf of California.* Tempe: University of Arizona Press.

Carriker, M. R. 1996. "History of a Systematic Odyssey: The Marine Flora and Fauna of the Eastern United States." *Marine Fisheries Review* 58:1–23.

Coyer, J., and J. Witman. 1990. *The Underwater Catalog. A Guide to Methods in Underwater Research.* Ithaca, N.Y.: Shoals Marine Laboratory, Cornell University.

Dowds, R. E. 1979. "References for the Identification of Marine Invertebrates on the Southern Atlantic Coast of the United States." *NOAA Technical Report SSRF* 729:1–37.

Fox, R. S., and E. E. Ruppert. 1985. *Shallow-Water Marine Benthic Macroinvertebrates of South Carolina. Species Identification, Community Composition and Symbiotic Associations.* Columbia: University of South Carolina Press.

Gosner, K. L. 1971. *Guide to Identification of Marine and Estuarine Invertebrates.* New York: Wiley-Interscience.

———. 1979. *A Field Guide to the Atlantic Seashore. Invertebrates and Seaweeds of the Atlantic Coast from the Bay of Fundy to Cape Hatteras.* The Peterson Field Guide Series, No. 24. Boston: Houghton Mifflin.

Gurney, R. 1942. *Larvae of Decapod Crustaceans.* London: Ray Society.

Hayward, P. J., and J. S. Ryland, eds. 1990. *The Marine Fauna of the British Isles and North-West Europe.* Vols. 1, 2. Oxford: Clarendon Press.

Kozloff, E. N. 1990. *Invertebrates.* Philadelphia: Saunders College.

Lincoln, R. J., and J. G. Sheals. 1979. *Invertebrate Animals: Collection and Preservation.* London: British Museum (Natural History), Cambridge University Press.

Lippson, A. J., and R. L. Lippson. 1997. *Life in the Chesapeake Bay,* 2d ed. Baltimore: The Johns Hopkins Press.

Martinez, A. J. 1994. *Marine Life of the North Atlantic. Canada to New England.* Wenham, Mass.: Marine Life.

Maurer, D. and L. Watling. 1973. "The Biology of the Oyster Community and Its Associated Fauna in Delaware Bay." *Delaware Bay Report Series* 6:1–97.

Meinkoth, N. A. 1981. *The Audubon Society Field Guide to North American Seashore Creatures.* New York: Alfred A. Knopf.

Miner, R. W. 1950. *Field Book of Seashore Life.* New York: G. P. Putnam's Sons.

Nybakken, J. W. 1988. *Marine Biology, an Ecological Approach,* 2d ed. New York: Harper & Row.

Olmstead, N. C., and P. E. Fell. 1974. "Tidal Marsh Invertebrates of Connecticut." *Connecticut Arboretum, Bulletin* 20:1–36.

Omori, M., and T. Ikeda. 1984. *Methods in Marine Zooplankton Ecology.* New York: John Wiley & Sons.

Parker, S. P., ed. 1982. *Synopsis and Classification of Living Organisms,* Vols. 1, 2. New York: McGraw-Hill.

Pennak, R. W. 1989. *Freshwater Invertebrates of the United States.* 3d ed. New York: John Wiley and Sons.

Perry, B. 1985. *A Sierra Club Naturalist's Guide to the Middle Atlantic Coast, Cape Hatteras to Cape Cod.* San Francisco: Sierra Club Books.

Ruppert, E. E., and R. D. Barnes. 1994. *Invertebrate Zoology,* 6th ed. Orlando: Harcourt Brace.

Ruppert, E. E., and R. Fox. 1988. *Seashore Animals of the Southeast. A Guide to Common Shallow-Water Invertebrates of the Southeastern Atlantic Coast.* Columbia: University of South Carolina Press.

Smith, D. L., and K. B. Johnson. 1996. *A Guide to Marine Coastal Plankton and Marine Invertebrate Larvae.* Dubuque, Ia: Kendall/Hunt.

Smith, R. I., ed. 1964. *Keys to Marine Invertebrates of the Woods Hole Region.* Woods Hole, Mass.: Marine Biological Laboratory.

Sterrer, W. 1986. *Marine Fauna and Flora of Bermuda.* New York: John Wiley and Sons.

Ward, N. 1995. *Stellwagen Bank. A Guide to the Whales, Sea Birds, and Marine Life of the Stellwagen Bank National Marine Sanctuary.* Camden, Maine: Down East Books.

Watling, L., and D. Maurer. 1973. *Guide to the Macroscopic Estuarine and Marine Invertebrates of the Delaware Bay Region.* Newark: College of Marine Studies, University of Delaware.

Weiss, H. M. 1995. "Marine Animals of Southern New England and New York. Identification Keys to Common Nearshore and Shallow Water Macrofauna." *Bulletin of the State Geological and Natural History Survey of Connecticut* 115.

Whitlach, R. 1982. "The Ecology of New England Tidal Flats: a Community Profile." *Fish and Wildlife Service/OBS* 81/01:1–125.

White, C. P. 1989. *Chesapeake Bay. Nature of the Estuary. A Field Guide.* Centerville, Md.: Tidewater Publishers.

Zinn, D. J. 1975. *The Beach Strollers Handbook from Maine to Cape Hatteras.* Chester, Conn.: The Pequot Press.

Checklists and Faunal Surveys

Borror, A. C. 1995. *Checklist of Flora and Fauna of the Isles of Shoals.* Ithaca, N.Y.: Shoals Marine Laboratory, Cornell University.

Brinkhurst, R. O., L. E. Linkletter, E. I. Lord, S. A. Connors, and M. J. Dadswell. 1975. *A Preliminary Guide to the Littoral and Sublittoral Marine Invertebrates of Passamaquoddy Bay.* St. Andrews, N.B.: Huntsman Marine Science Centre.

Caracciolo, J. V., and F. W. Steimle, Jr. 1983. "An Atlas of the Distribution and Abundance of Dominant Benthic Invertebrates in the New York Bight Apex with Reviews of Their Life Histories." *NOAA Technical Report NMFS-SSRF* 766:1–58.

Center for Natural Areas. 1976. *A Preliminary Listing of Noteworthy Natural Features in Maine.* Augusta: Maine Critical Areas Program, Maine State Planning Office.

Hulbert, A. W., K. J. Pecci, J. D. Witman, L. G. Harris, J. R. Sears, and R. A. Cooper. 1982. "Ecosystem Definition and Community Structure of the Macrobenthos of the NEMP Monitoring Station at Pigeon Hill in the Gulf of Maine." *NOAA Technical Memorandum NMFS-F/NEC* 14:1–144.

Jury, S. H., J. D. Field, S.L. Stone, D. M. Nelson, and M.E. Monaco. 1994. "Distribution and Abundance of Fishes and Invertebrates in North American Estuaries." *NOAA/NOS Strategic Environmental Assessments Division, ELMR Report* 13:1–221.

Kinner, P., D. Maurer, and W. Leathem. 1974. "Benthic Invertebrates in Delaware Bay: Animal-Sediment Associations of the Dominant Species." *International Revue Gesamten Hydrobiologie* 59:685–701.

Knowlton, R. E. 1971. "Preliminary Checklist of Maine Marine Invertebrates." Publications of TRIGOM, The Research Institute of the Gulf of Maine 1:1–11.

Larsen, P. F., L. D. Doggett, and V. M. Berounsky. 1977. "Data Report on Intertidal Invertebrates on the Coast of Maine." *Bigelow Laboratory for Ocean Sciences, Technical Report* 12–77.

Larsen, P. F., A. C. Johnson, and L. F. Doggett. 1983. "Environmental Benchmark Studies in Casco Bay-Portland Harbor, Maine, April, 1980." *NOAA Technical Memorandum, NMFS/NEC* 19:1–174.

McErlean, A. J., C. Kerby, and M. L. Wass, eds. 1972. "Biota of Chesapeake Bay." *Chesapeake Science* 13 Suppl. S1–S197.

Rasmussen, E. 1973. "Systematics and Ecology of the Isefjord Marine Fauna (Denmark)." *Ophelia* 11:1–507.

Reid, R. N., D. J. Radosh, A. B. Frame, and S. A. Fromm. 1991. "Benthic Macrofauna of the New York Bight, 1979–89." *NOAA Technical Report NMFS Circular* 103:1–50.

Smith, D. E., and J. W. Jossi. 1984. "Net Phytoplankton and Zooplankton in the New York Bight, January 1976 to February 1978, with Comments on the Effects of Wind, Gulf Stream Eddies, and Slope Water Intrusions." *NOAA Technical Report NMFS Circular* 5:1–41.

Stone, S. L., T. A. Lowrey, J. D. Field, C. D. Williams, D. M. Nelson, S. H.Jury, M. E. Monaco, and L. Andreasen. 1994. "Distribution and Abundance of Fishes and Invertebrates in Mid-Atlantic Estuaries." *NOAA/NOS Strategic Environmental Assessment Division, ELMR Report* 12:1–280.

Turner, J. T. 1984. "The Feeding Ecology of Some Zooplankters That Are Important Prey Items of Larval Fish." *NOAA Technical Report NMFS Circular* 7:1–28.

Wass, M. L., ed. 1972. "A Check List of the Biota of Lower Chesapeake Bay." *Virginia Institute of Marine Science, Special Publ.* 65:1–290.

Porifera

Hartman, W. D. 1958. "Natural History of the Marine Sponges of Southern New England." *Bulletin of the Peabody Museum of Natural History* 12:1–144.

Hyman, L. H. 1940. *The Invertebrates: Protozoa through Ctenophora.* New York: McGraw-Hill.

Laubenfels, M. W. de. 1953. "A Guide to the Sponges of Eastern North America." *Special Publication, The Marine Laboratory, University of Miami.*

Wells, H. W., M. J. Wells, and I.E. Gray. 1960. "Marine Sponges of North Carolina." *Journal of the Elisha Mitchell Scientific Society* 76:200–245.

Cnidaria and Ctenophora

Berrill, M. 1962. "The Biology of Three New England Stauromedusae, with a Description of a New Species." *Canadian Journal of Zoology* 40:1249–1262.

Cairns, S. D. 1981. "Marine Flora and Fauna of the Northeastern United States. Scleractinia." *NOAA Technical Report NMFS Circular* 438:1–15.

Cairns, S. D., D. R. Calder, A. Brinckmann-Voss, C. B. Castro, P. R. Pugh, C. E. Cutress, W. C. Jaap, D. G. Fautin, R. J. Larson, G. R. Harbison, M. N. Arai, and D. M. Opresko. 1991. "Common and Scientific Names of Aquatic Invertebrates from the United States and Canada: Cnidaria and Ctenophora." *American Fisheries Society Special Publication* 22:1–75.

Calder, D. R. 1971. "Hydroids and Hydromedusae of Southern Chesapeake Bay." Virginia Institute of Marine Science, Special Papers in Marine Science 1:1–125.

———. 1975. "Biotic Census of Cape Cod Bay: Hydroids." *Biological Bulletin* 149:287–315.

Deichmann, E. 1936. "The Alcyonaria of the western part of the Atlantic Ocean." Harvard University, Museum of Comparative Zoology Memoirs 53:1–317.

Fraser, C. M. 1944. *Hydroids of the Atlantic Coast of North America.* Toronto: University of Toronto Press.

———. 1946. *Distribution and Relationship in American Hydroids.* Toronto: University of Toronto Press.

Hyman, L. H. 1940. *The Invertebrates: Protozoa through Ctenophora.* New York: McGraw-Hill.

Kramp, P. L. 1961. "Synopsis of the Medusae of the World." *Journal of the Marine Biological Association of the United Kingdom* 40:7–469.

Larson, R. J. 1976. "Marine Flora and Fauna of the Northeastern United States. Cnidaria: Scyphozoa." *NOAA Technical Report NMFS Circular* 397:1–18.

Naumov, D. V. 1960. "Hydroids and Hydromedusae of the USSR." *Fauna of the USSR* 70:1–600 (Israel Program for Scientific Translations, Catalog No. 5108).

Shih, C. T. 1977. "A Guide to the Jellyfish of Canadian Atlantic Waters." *National Museum of Natural Sciences, Natural History Series* 5:1–90.

Stephenson, T. A. 1928–1935. *The British Sea Anemones,* Vol. 1 (1928), Vol. 2 (1935). London: Ray Society.

Platyhelminthes

Hyman, L. H. 1939. "Some Polyclads of the New England Coast, Especially of the Woods Hole Region." *Biological Bulletin* 76:127–152.

———. 1941. "The Polyclad Flatworms of the Atlantic Coast of the United States and Canada." *Proceedings of the U.S. National Museum* 89:449–495.

———. 1944. "Marine Turbellaria from the Atlantic Coast of North America." *American Museum of Natural History Novitiates* 1266:1–15.

———. 1951a. *The Invertebrates: Platyhelminthes and Rhynchocoela.* New York: McGraw-Hill.

———. 1952. "Further Notes on the Turbellarian Fauna of the Atlantic Coast of North America." *Biological Bulletin* 103:195–200.

Pearse, A. S. 1938. "Polyclads of the East Coast of North America." *Proceedings of the U.S. National Museum* 86:67–98.

Nemertea (Rhynchocoela)

Coe, W. R. 1943. "Biology of the Nemerteans of the Atlantic Coast of North America." *Transactions of the Connecticut Academy of Arts and Sciences* 35:129–328.

Hyman, L. H. 1951a. *The Invertebrates: Platyhelminthes and Rhynchocoela.* New York: McGraw-Hill.

McCaul, W. E. 1963. "Rhynchocoela: Nemerteans from Marine and Estuarine Waters of Virginia." *Journal of the Elisha Mitchell Scientific Society* 79: 111–124.

Riser, N. W. 1993. "Observations on the Morphology of Some North American Nemertines with Consequent Taxonomic Changes and a Reassessment of the Architectonics of the Phylum." *Hydrobiologia* 266:141–157.

Entoprocta

Hyman, L. H. 1951b. *The Invertebrates: Acanthocephala, Aschelminthes and Entoprocta. The Pseudocoelomate Bilateria.* New York: McGraw-Hill.

Nielsen, C. 1966. "Some Loxosomatidae (Entoprocta) from the Atlantic Coast of the United States." *Ophelia* 3:249–275.

Ectoprocta

Hutchins, L. W. 1945. "An Annotated Check-List of the Salt Water Bryozoa of Long Island Sound." *Transactions of the Connecticut Academy of Arts and Sciences* 36:533–551.

Hyman, L. H. 1959. *The Invertebrates: Smaller Coelomate Groups.* New York: McGraw-Hill.

Kluge, G. A. 1975. "Bryozoa of the Northern Seas of the USSR." *Keys on the Fauna of the USSR* 76:1–711.

Maturo, F. J. S. 1957. "A Study of the Bryozoa of Beaufort, NC and Vicinity." *Journal of the Elisha Mitchell Scientific Society* 73:11–68.

———. 1968. "The Distributional Pattern of the Bryozoa of the East Coast of the United States Exclusive of New England." *Atti della Societa Italiana di Scienze Naturali e del Museo Civico di Storia Naturale di Milano* 108:261–284.

Osburn, R. C. 1912. "The Bryozoa of the Woods Hole Region." *Bulletin of the Bureau of Fisheries, Washington* (1910) 30:205–266.

Rogick, M. S, and H. Croasdale. 1949. "Studies on Marine Bryozoa. III. Woods Hole Region Bryozoa Associated With Algae." *Biological Bulletin* 96:32–69.

Ryland, J. S., and P. J. Hayward. 1991. "Marine Flora and Fauna of the Northeastern United States. Erect Bryozoa." *NOAA Technical Report NMFS* 99:1–47.

Sipuncula

Cutler, E. B. 1977. "Marine Flora and Fauna of the Northeastern United States. Sipuncula." *NOAA Technical Report NMFS Circular* 403:1–7.

Mollusca

Abbott, R. T. 1974. *American Seashells. The Marine Mollusca of Atlantic and Pacific Coasts of North America,* 2d ed. New York: Van Nostrand Reinhold.

———. 1993. *Seashells of the Northern Hemisphere.* Stamford, Conn.: Longmeadow Press.

Abbott, R. T., and P. A. Morris. 1995. *A Field Guide to Shells. Atlantic and Gulf Coasts and the West Indies.* The Peterson Field Guide Series, No. 3, 4th ed. Boston: Houghton Mifflin.

Boss, K. J. 1966a. "The Subfamily Tellininae in the Western Atlantic. The Genus *Tellina* (Part I)." *Johnsonia* 4:217–272, plates 127–142.

———. 1966b. "The Subfamily Tellininae in the Western Atlantic. The Genus *Tellina* (Part II)." *Johnsonia* 4:273–344, plates 143–163.

Boss, K. J., and A. S. Merrill. 1965. "The Family Pandoridae in the Western Atlantic." *Johnsonia* 4:181–216.

Clench, W. J. 1951. "The Genus *Epitonium* in the Western Atlantic. Part I." *Johnsonia* 2:249–288.

Clench, W. J., and L. C. Smith. 1944. "The Family Cardiidae in the Western Atlantic." *Johnsonia* 1:1–37.

Franz, D. R. 1968. "Occurrence and Distribution of New Jersey Opisthobranchia." *Nautilus* 82:7–12.

———. 1970. "Zoogeography of Northwest Atlantic Opisthobranch Molluscs." *Marine Biology* 7:171–180.

Henderson, J. B. 1920. "A Monograph of the East American Scaphopod Mollusks." *Bulletin of the U.S. National Museum* 111:1–177.

Hyman, L. H. 1967. *The Invertebrates: Mollusca I.* New York: McGraw-Hill.

Loveland, R. E., G. Hendler, and G. Newkirk. 1969. "New Records of Nudibranchs from New Jersey." *Veliger* 11:418–420.

Lowden, R. D. 1965. "The Marine Mollusca of New Jersey and Delaware Bay, an Annotated Checklist." *Proceedings of the Philadelphia Shell Club* 1(8–9):5–61.

Marcus, E. 1958. "On Western Atlantic Opisthobranchiate Gastropods." *American Museum of Natural Hististory Novitiates* 1906:1–82.

Nesis, K. N. 1987. *Cephalopods of the World: Squids, Cuttlefishes, Octopuses, and Allies.* Neptune City, N.J.: T. F. H. Publications.

Rehder, H. A. 1981. *The Audubon Society Field Guide to North American Seashells.* New York: Alfred A. Knopf.

Stanley, S. M. 1970. "Relations of Shell Form to Life Habits of the Bivalvia (Mollusca)." *Geological Soceity of America, Memoirs* 125:1–296.

Turgeon, D. D., A. E. Bogan, E. V. Coan, W. K. Emerson, W. G. Lyons, W. L. Pratt, C. F. E. Roper, A. Scheltema, F. G. Thompson, and J. D. Williams. 1988. "Common and Scientific Names of Aquatic Invertebrates from the United States and Canada: Mollusks." *American Fisheries Society Special Publication* 16:1–277.

Turner, R. D. 1966. *A Survey and Illustrated Catalogue of the Teredinidae.* Cambridge, Mass.: Museum of Comparative Zoology, Harvard University.

Vecchione, M., C. F. E. Roper, and M. J. Sweeney. 1989. "Marine Flora and Fauna of the Eastern United States. Mollusca: Cephalopoda." *NOAA Technical Report NMFS Circular* 73:1–23.

Annelida

Blake, J. A. 1971. "Revision of the Genus *Polydora* from the East Coast of North America (Polychaeta: Spionidae)." *Smithsonian Contributions to Zoology* 75:1–32.

Brinkhurst, R. O. 1982. "British and Other Marine and Estuarine Oligochaetes." *Cambridge Synopsis of the British Fauna* 21:1–127.

Brinkhurst, R. O., and B. G. M. Jamieson. 1971. *Aquatic Oligochaeta of the World.* Edinburgh: Oliver & Boyd.

Cook, D. G., and R. O. Brinkhurst. 1973. " Marine Flora and Fauna of the Northeastern United States. Annelida: Oligochaeta." *NOAA Technical Report NMFS Circular* 374:1–22.

Day, J. H. 1967. *A Monograph of the Polychaeta of South Africa, Part I: Errantia, Part II: Sedentaria.* London: British Museum of Natural History.

———. 1973. "New Polychaeta from Beaufort, with a Key to All Species Recorded From North Carolina." *NOAA Technical Report NMFS Circular* 375:1–140.

Fauchald, K. 1977. "The Polychaete Worms. Definitions and Keys to the Orders, Families, and Genera." Natural History Museum of Los Angeles County, Science Series 28:1–188.

Fauchald, K., and P. A. Jumars. 1979. "The Diet of Worms: a Study of Polychaete Feeding Guilds." *Annual Review of Oceanography and Marine Biology* 17:193–284.

Frame, A. B. 1992. "The Lumbrinerids (Annelida: Polychaeta) Collected in Two Northwestern Atlantic Surveys with Descriptions of a New Genus and Two New Species." *Proceedings of the Biological Society, Washington* 105:185–218.

Gardiner, S. L. 1975. "Errant Polychaete Annelids from North Carolina." *Journal of the Elisha Mitchell Scientific Society* 91:77–220.

Hartman, O. 1944. "New England Polychaeta. Part 2. Including the Unpublished Plates by Verrill with Reconstructed Captions." *Bulletin of the American Museum of Natural History* 82:327–344.

———. 1945. *The Marine Annelids of North Carolina.* Durham, N.C.: Duke University Press.

———. 1959. "Catalog of the Polychaetous Annelids of the World." *Allan Hancock Foundation Publications Occasional Paper* 23:1–628.

Kahn, R. A., and M. C. Meyer. 1976. "Taxonomy and Biology of Some Newfoundland Marine Leeches (Rhynchobdellae: Piscicolidae)." *Journal of the Fisheries Research Board of Canada* 33:1699–1714.

Kinner, P. and D. Maurer. 1978. "Polychaetous Annelids of the Delaware Bay Region." *Fishery Bulletin* 76:209–224.

Pettibone, M. H. 1963. "Marine Polychaete Worms of the New England Region. 1. Aphroditidae through Trochochaetidae." *U.S. National Museum Bulletin* 227, Part 1:1–356.

Sawyer, R. T., A. R. Lawler, and R. M. Overstreet. 1975. "Marine Leeches of the Eastern United States and the Gulf of Mexico with a Key to the Species." *Journal of Natural History* 9:633–667.

Wells, H. W., and I. E. Gray. 1964. "Polychaetous Annelids of the Cape Hatteras Area." *Journal of the Elisha Mitchell Scientific Society* 80:70–78.

Arthropoda

Banner, A. H. 1954. "A Supplement to W. M. Tattersall's Review of the Mysidacea of the United States National Museum." *Proceedings of the U.S. National Museum* 103 (3334):575–583.

Barnard, J. L. 1969. "The Families and Genera of Marine Gammaridean Amphipoda." *U.S. National Museum Bulletin* 271:1–535.

Barnard, J. L., and G.S. Karaman. 1991. "The Families and Genera of Marine Gammaridean Amphipoda (except Marine Gammaroids). Parts 1 and 2." *Records of the Australian Museum,* Suppl. 13, Parts 1 and 2:1–866.

Borowsky, B. 1984. "The Use of the Males' Gnathopods during Precopulation in Some Gammaridean Amphipods." *Crustaceana* 47:245–250.

Bousfield, E. L. 1958. "Littoral Marine Arthropods and Molluscs Collected in Western Nova Scotia, 1956." *Nova Scotian Institutue of Science* 24(3):303–325.

———. 1965. "Haustoriidae (Crustacea: Amphipoda) of New England." *Proceedings of the U.S. National Museum* 117:159–240.

———. 1970. "Adaptive Radiation in Sand-burrowing Amphipod Crustaceans." *Chesapeake Science* 11:143–154.

———. 1973. *Shallow-Water Gammaridean Amphipoda of New England.* Ithaca, NY: Cornell University Press.

Bousfield, E. L., and C.-T. Shih. 1994. "The Phyletic Classification of Amphipod Crustaceans: Problems in Resolution." *Amphipacifica* 1:76–134.

Calman, W. T. 1912. "The Crustacea of the Order Cumacea in the Collection of the United States National Museum." *Proceedings of the U.S. National Museum* 41:603–676.

Conlan, K. E. 1991. "Precopulatory Mating Behavior and Sexual Dimorphism in the Amphipod Crustacea." *Hydrobiologia* 223:255–282.

Cressey, R. F. 1978. "Marine Flora and Fauna of the Northeastern United States. Crustacea: Branchiura." *NOAA Technical Report NMFS Circular* 413:1–9.

Croker, R. A. 1967. "Niche Diversity in Five Sympatric Species of Intertidal Amphipods (Crustacea: Haustoriidae)." *Ecological Monographs* 37:173–200.

Dexter, D. M. 1967. "Distribution and Niche Diversity of Haustoriid Amphipods in North Carolina." *Chesapeake Science* 8:187–192.

Dickinson, J. J., and R. L. Wigley. 1981. "Distribution of Gammaridean Amphipoda (Crustacea) on Georges Bank." *NOAA Technical Report NMFS-SSRF* 746:1–25.

Dickinson, J. J., R. L. Wigley, R. D. Brodeur, and S. Brown–Leger. 1980. "Distribution of Gammaridean Amphipoda (Crustacea) in the Middle Atlantic Bight Region." *NOAA Technical Report NMFS-SSRF* 741:1–46.

Factor, R. J., ed. 1995. *Biology of the Lobster, Homarus americanus.* New York: Academic Press.

Farfane, I. P. 1988. "Illustrated Key to Penaeoid Shrimps of Commerce in the Americas." *NOAA Technical Report NMFS Circular* 64:1–32.

Gotto, R. V. 1979. "The Association of Copepods with Marine Invertebrates." *Advances in Marine Biology* 16:1–109.

Hedgpeth, J. W. 1948. "The Pycnogonida of the Western North Atlantic and the Caribbean." *Proceedings of the U.S. National Museum* 97:157–343.

Ho, J.-S. 1971. "Parasitic Copepods of the Family Chondracanthidae from fishes of Eastern North America." *Smithsonian Contributions to Zoology* 87:1–39.

———. 1977. "Marine Flora and Fauna of the Northeastern United States. Copepoda: Lernaeopodidae and Sphyriidae." *NOAA Technical Report NMFS Circular* 406:1–12.

———. 1978. "Marine Flora and Fauna of the Northeastern United States. Copepoda: Cyclopoids Parasitic on Fishes." *NOAA Technical Report NMFS Circular* 409:1–11.

Hopkins, T. L. 1965. "Mysid Shrimp Abundance in Surface Waters of Indian River Inlet, Delaware." *Chesapeake Science* 62:86–91.

Manning, R. B. 1969. "Stomatopod Crustacea of the Western Atlantic." *Studies in Tropical Oceanography* 8:1–380.

———. 1974. "Marine Flora and Fauna of the Northeastern United States. Crustacea: Stomatopoda." *NOAA Technical Report NMFS Circular* 387:1–6.

McCain, J. C. 1968. "The Caprellidae (Crustacea: Amphipoda) of the Western North Atlantic." *U.S. National Museum Bulletin* 278:1–147.

McCloskey, L. R. 1973. "Marine Flora and Fauna of the Northeastern United States. Pycnogonida." *NOAA Technical Report NMFS Circular* 386:1–12.

McDermott, J. J. 1991. "A Breeding Population of the Western Pacific Crab, *Hemigrapsus sanguineus* (Crustacea: Decapoda: Grapsidae) Established on the Atlantic Coast of North America." *Biological Bulletin* 181:195–198.

Menzies, R. J., and D. Frankenberg. 1966. *Handbook on the Common Marine Isopod Crustacea of Georgia.* Athens: University of Georgia Press.

Mills, E. L. 1963. "A New Species of *Ampelisca* (Crustacea: Amphipoda) from Eastern North America, with Notes on Other Species of the Genus." *Canadian Journal of Zoology* 41:971–989.

———. 1965. "The Zoogeography of North Atlantic and North Pacific Ampeliscid Amphipod Crustaceans." *Systematic Zoology* 14:119–130.

———. 1967. "A Re-examination of Some Species of *Ampelisca* (Crustacea: Amphipoda) from the East Coast of North America." *Canadian Journal of Zoology* 45:635–652.

Pilsbry, H. A. 1916. "The Sessile Barnacles (Cirripedia) Contained in the Collections of the U.S. National Museum, Including a Monograph of the American Species." *U.S. National Museum Bulletin* 93:1–366.

Pohle, G. 1988. "A Guide to the Deep-Sea Shrimp and Shrimp-Like Decapod Crustacea of Atlantic Canada." *Canadian Technical Report of Fisheries and Aquatic Sciences* 1657:1–29.

———. 1990. "A Guide to Decapod Crustacea from the Canadian Atlantic: Anomura and Brachyura." *Canadian Technical Report of Fisheries and Aquatic Sciences* No.1771:1–30.

Rathbun, M. J. 1918. "The Grapsoid Crabs of America." *U.S. National Museum Bulletin* 97:1–461.

———. 1925. "The Spider Crabs of America." *U.S. National Museum Bulletin* 129:1–598.

———. 1930. "The Cancroid Crabs of America of the Families Euryalidae, Portunidae, Atelecyclidae, Cancridae, and Xanthidae." *U.S. National Museum Bulletin* 152:1–609.

Richardson, H. 1905. "Monograph on the Isopods of North America." *U.S. National Museum Bulletin* 54:1–727.

Ryan, E. P. 1956. "Observations on the Life Histories and the Distribution of the Xanthidae (Mud Crabs) of Chesapeake Bay." *American Midland Naturalist* 56:138–162.

Sameoto, D. D. 1969. "Comparative Ecology, Life Histories, and Behavior of Intertidal Sand-Burrowing Amphipods (Crustacea: Haustoriidae) at Cape Cod." *Journal of the Fisheries Research Board of Canada* 26:361–388.

———. 1969. "Some Aspects of the Ecology and Life Cycle of Three Species of Subtidal Sand-burrowing Amphipods (Crustacea: Haustoriidae)." *Journal of the Fisheries Research Board of Canada* 26:1321–1345.

Schmitt, W. L. 1935. "Mud Shrimps of the Atlantic Coast of North America." *Smithsonian Miscellaneous Collections* 93(2):1–21.

Sieg, J., and R. Winn. 1978. "Keys to Suborders and Families of Tanaidacea (Crustacea)." *Proceedings of the Biological Society, Washington* 91:840–846.

Tattersall, W. M. 1951. "A Review of the Mysidacea of the United States National Museum." *U.S. National Museum Bulletin* 201:1–292.

Watling, L. 1979. "Marine Flora and Fauna of the Northeastern United States. Crustacea: Cumacea." *NOAA Technical Report NMFS Circular* 423:1–23.

Watling, L., and D. Maurer. 1972. "Shallow Water Amphipods of the Delaware Bay Region." *Crustaceana,* Suppl. 3:251–266.

Wigley, R. W., and B. R. Burns. 1971. "Distribution and Biology of Mysids (Crustacea: Mysidacea) from the Atlantic Coast of the U.S. in the NMFS Woods Hole Collections." *Fishery Bulletin* 69:717–745.

Williams, A. B. 1965. "Marine Decapod Crustaceans of the Carolinas." *Fishery Bulletin* 65:1–298.

———. 1974. "Marine Flora and Fauna of the Northeastern United States. Crustacea: Decapoda." *NOAA Technical Report NMFS Circular* 389:1–50.

———. 1984. *Shrimps, Lobsters, and Crabs of the Atlantic Coast of the Eastern United States, Maine to Florida.* Washington: Smithsonian Institution Press.

Williams, A. B., and J. J. McDermott. 1990. "An Eastern United States Record for the Western Indo-Pacific Crab, *Hemigrapsus sanguineus* (Crustacea: Decapoda: Grapsidae)." *Proceedings of the Biological Society, Washington* 103:108–109.

Williams, A. B., L. G. Abele, D. L. Felder, H. H. Hobbs, Jr., R. B. Manning, P. A. McLaughlin, and I. P. Farfante. 1989. "Common and Scientific Names of Aquatic Invertebrates from the United States and Canada: Decapod Crustaceans." *American Fisheries Society Special Publication* 17:1–77.

Zimmer, C. 1980. "Cumaceans of the American Atlantic Boreal Coast Region (Crustacea, Peracarida)." *Smithsonian Contribributions to Zoology* 302:1–29.

Zullo, V. A. 1979. "Marine Flora and Fauna of the Northeastern United States. Arthropoda: Cirripedia." *NOAA Technical Report NMFS Circular* 425:1–29.

Echinodermata

Clark, A. M., and M. E. Downey. 1992. *Starfishes of the Atlantic.* New York: Chapman and Hall.

Clark, H. L. 1904. "The Echinoderms of the Woods Hole Region." *Bulletin of the U.S. Fisheries Commission* (1902) 22:545–576.

———. 1915. "Catalogue of Recent Ophiurans Based on the Collections of the Museum of Comparative Zoology." Harvard University, Museum of Comparative Zoology Memoirs 25:163–376.

Coe, W. R. 1912. "Echinoderms of Connecticut." *Bulletin of the State Geological and Natural History Survey of Connecticut* 19:1–52.

Fell, H. B. 1960. "Synoptic Keys to the Genera of Ophiuroidea." *Zoology Publications from Victoria University of Wellington* 26:1–44.

Gray, I. E., M. E. Downey, and M. J. Cerame-Vivas. 1968. "Sea-Stars of North Carolina." *Fishery Bulletin* 67:127–163.

Hyman, L. H. 1955. *The Invertebrates: Echinoderms.* New York: McGraw-Hill.

Pawson, D. L. 1977. "Marine Flora and Fauna of the Northeastern United States. Echinodermata: Holothuroidea." *NOAA Technical Report NMFS Circular* 405:1–15.

Serafy, D. K., and F. J. Fell. 1985. "Marine Flora and Fauna of the Northeastern United States. Echinodermata: Echinoidea." *NOAA Technical Report NMFS Circular* 33:1–27.

Chordata

Bigelow, H. B., and W. C. Schroeder. 1953. "Fishes of the Gulf of Maine." *Fishery Bulletin* 53 (74):1–577. (Reprinted 1964 by Woods Hole Oceanographic Institution and the Museum of Comparative Zoology, Harvard University.)

Castro, J. I. 1993. "A Field Guide to Sharks Commonly Caught in Commercial Fisheries of the Southeastern United States." *NOAA Technical Memorandum, NMFS-SEFSC* 338:1–47.

Elliott, E. M., and D. Jimenez. 1981. *Laboratory Manual for the Identification of Ichthyoplankton from the Beverly-Salem Harbor Area.* Boston: Commonwealth of Massachusetts, Dept. of Fisheries, Wildlife and Recreational Vehicles, Division of Marine Fisheries.

Flescher, D. D. 1980. "Guide to Some Trawl-Caught Marine Fishes from Maine to Cape Hatteras, NC." *NOAA Technical Report NMFS Circular* 431:1–34.

Hildebrand, S. F., and W. C. Schroeder. 1927. "Fishes of Chesapeake Bay." *Fishery Bulletin* 43, Part 1:1–388. (Reprinted 1972 for the Smithsonian Institution by T.F.H. Publications, Neptune, N.J.)

Katona, S. K., V. Rough, and D. Richardson. 1993. *A Field Guide to the Whales and Seals from Cape Cod to Newfoundland,* 4th ed. Washington: Smithsonian Institution Press.

Leatherwood, S., D. K. Caldwell, and H. E. Winn. 1976. "Whales, Dophins, and Porpoises of the Western North Atlantic. A Guide to Their Identification." *NOAA Technical Report NMFS Circular* 396:1–176.

McClane, A. J. 1978. *McClane's Field Guide to Saltwater Fishes of North America.* New York: Henry Holt.

Plough, H. H. 1978. *Sea Squirts of the Atlantic Continental Shelf from Maine to Texas.* Baltimore: The Johns Hopkins University Press.

Robins, C. R., and G. C. Ray. 1986. *A Field Guide to Atlantic Coast Fishes of North America. The Peterson Field Guide Series, No. 32.* Boston: Houghton Mifflin.

Robins, C. R., R. M. Bailey, C. E. Bond, J. R. Brooker, E. A. Lachner, R. N. Lea and W. B. Scott. 1991. "Common and Scientific Names of Fishes from the United States and Canada." *American Fisheries Society Special Publication* 20:1–183.

Russo, J. L. 1981. "Field Guide to Fishes Commonly Taken on Longline Operations in the Western North Atlantic." *NOAA Technical Report NMFS Circular* 435:1–51.

Scott, W. B., and M. G. Scott. 1988. *Atlantic Fishes of Canada.* Toronto: University of Toronto Press.

Thomson, K.S., W. H. Weed III, and A. G. Taruski. 1971. "Saltwater Fishes of Connecticut." *Bulletin of the State Geological and Natural History Survey of Connecticut* 105:1–165.

Van Name, W. G. 1945. "The North and South American Ascidians." *Bulletin of the American Museum of Natural History* 84:1–476.

Wisner, R. L., and C. B. McMillan. 1995. "Review of New World Hagfishes of the Genus *Myxine* (Agnatha, Myxinidae) with Descriptions of Nine New Species." *Fishery Bulletin* 93:530–550.

■ Illustration Credits and Sources

Illustration Credits

This volume has been enriched immeasurably by figures taken from a variety of sources. Where required, permission to use illustrations from copyright-protected sources has been procured (as indicated in the listing below). Older literature (pre-1940), as well as publications by official agencies of the United States government, are part of the public domain and therefore are available for use. Some original figures were skillfully drawn by Eric Brothers. Other original figures were drawn by the author either from life, from general sources, or modified from literature sources noted. For convenience, references to sources are noted by chapter and figure number. A list of full citations for this literature may be found at the end of this section.

Chapter 1. Groups of Marine Invertebrates. McIntosh 1922 (1.38); Miner 1950 (1.1, 1.3, 1.4, 1.7, 1.8, 1.10, 1.11, 1.12, 1.14, 1.18, 1.19, 1.20, 1.21, 1.22, 1.23, 1.26, 1.28, 1.29, 1.30, 1.32, 1.35, 1.40, 1.42, 1.45, 1.46, 1.47, 1.48, 1.50, 1.53, 1.54, 1.55, 1.56, 1.57, 1.58); Sars 1878 (1.59); Smith 1964 (1.2, 1.6, 1.15, 1.24, 1.27, 1.33, 1.37, 1.44, 1.49, 1.51); Verrill 1885 (1.39); Verrill 1892a (1.5,1.16, 1.51); Verrill 1892b (1.9); Williams 1965 (1.43); original (1.13, 1.17, 1.25 after Miner 1950; 1.31, 1.34, 1.36, 1.41, 1.52).

Chapter 2. Gelatinous Organisms. All figures from Miner 1950.

Chapter 3. Miscellaneous Worm-Shaped Organisms. All figures from Miner 1950 except for those from Smith 1964 (3.15, 3.16, 3.17, 3.18, 3.21), and an original drawing (3.4).

Chapter 5. Zooplankton. Bigelow and Schroeder 1953 (5.33); Brooks 1890 (5.10, 5.16); Gurney 1942 (5.7, 5.8, 5.9, 5.11, 5.13); Hoek 1909 (5.4); Lohmann 1901a (5.53a); Miner 1950 (5.3, 5.28, 5.29, 5.30, 5.31, 5.41); Mortensen 1901 (5.54, 5.56); Sars 1878 (5.37, 5.38, 5.47); Williamson 1915 (5.12, 5.15); Verrill and Smith 1874 (5.2, 5.14); Verrill 1892b (5.46); original (5.1, 5.32, 5.36, 5.40), (5.34, 5.35, 5.39 after Lohmann 1901b), (5.5, 5.6 after Sanders 1963), (5.17, 5.18, 5.19, 5.20, 5.21, 5.22, 5.23, 5.26, 5.27 after Sars 1903), (5.24, 5.25 after Sars, 1918), (5.42, 5.43, 5.44, 5.45, 5.48, 5.49, 5.50, 5.51, 5.52, 5.53b, 5.55, 5.57 after Vannucci 1959).

Chapter 6. Eggs, and Egg Masses. Original (6.2, 6.3, 6.4, 6.5, 6.6, 6.7, 6.8, 6.16, 6.17, 6.20, 6.21, 6.22), (6.9, 6.10, 6.11, 6.12, 6.13, 6.14, 6.15 after D'Asaro 1993), (6.1, 6.18, 6.19 after photographs in Martinez 1994).

Chapter 7. Porifera. Miner 1950 (7.1-pa,pr,ir, 7.3a, 7.4, 7.6a, 7.8, 7.10, 7.13b, 7.15a, 7.16, 7.18b); Smith 1964 (7.1-tu,va,ei,bo, 7.2, 7.3b, 7.5, 7.6b, 7.7, 7.9, 7.11, 7.12, 7.13a, 7.14, 7.15b, 7.17, 7.18a).

Chapter 8. Cnidaria. Brothers—original (8.110a); Calder 1975 (8.12, 8.35, 8.38, 8.51); Fraser 1944 (8.3, 8.4, 8.5, .8.7b, 8.10, 8.11, 8.14, 8.17, 8.18, 8.20, 8.21, 8.22, 8.23, 8.24, 8.25, 8.26, 8.27, 8.28, 8.29, 8.30, 8.31, 8.32, 8.34, 8.37, 8.39, 8.40, 8.41, 8.43, 8.44, 8.45, 8.46, 8.47, 8.49, 8.50, 8.52, 8.53); Hartlaub 1907 (8.82, 8.84); Kramp 1933 (8.70); Maturo 1957 (8.56); Miner 1950 (8.1, 8.2, 8.6, 8.7a, 8.8, 8.13, 8.15, 8.16b, 8.19, 8.42, 8.48, 8.54, 8.57, 8.58, 8.59, 8.60, 8.61, 8.62, 8.63, 8.64, 8.65, 8.66, 8.67, 8.68, 8.69, 8.71, 8.72, 8.73, 8.74, 8.75, 8.76, 8.77, 8.78, 8.79, 8.80, 8.81, 8.83, 8.87, 8.88, 8.92, 8.93, 8.94, 8.95, 8.96, 8.97, 8.98, 8.99, 8.100, 8.101, 8.102, 8.103, 8.104, 8.105, 8.106, 8.107, 8.108, 8.109, 8.111, 8.112, 8.113, 8.114, 8.116, 8.117, 8.119, 8.121); Rogick and Croasdale 1949 (8.55); Smith 1964 (8.9, 8.16a, 8.33, 8.36, 8.90, 8.91, 8.115, 8.118, 8.120); Verrill 1901–02 (8.89); original (8.85 after Miner 1950; 8.86 after Hargitt 1904, 8.110b).

Chapter 9. Ctenophora. All figures from Miner 1950.

Chapter 10. Platyhelminthes. Hyman 1939 (10.15, 10.16); Miner 1950 (10.1, 10.2, 10.6, 10.7, 10.10, 10.14, 10.18); Smith 1964 (10.3, 10.4, 10.5, 10.9, 10.11, 10.13, 10.19); Verrill 1892a (10.8, 10.12, 10.17)

Chapter 11. Nemertea. Miner 1950 (11.1, 11.2a, 11.5, 11.9, 11.11a); Smith 1964 (11.2b, 11.11b, 11.14b); Verrill 1892b (11.3, 11.4, 11.7, 11.8, 11.10, 11.12, 11.13, 11.14a); original (11.6).

Chapter 12. Ectoprocta. Maturo 1957 (12.2, 12.13, 12.20, 12.33); Miner 1950 (12.18, 12.19, 12.22, 12.24a, 12.27); Osburn 1912 (12.29); Rogick and Croasdale 1949 (12.32, 12.38); Smith 1964 (12.1, 12.3, 12.4, 12.5, 12.6, 12.7, 12.8, 12.9, 12.10, 12.11, 12.12, 12.14, 12.15, 12.16, 12.17,

12.21, 12.23, 12.24b, 12.25, 12.26, 12.28, 12.29, 12.30, 12.31, 12.34, 12.35, 12.36, 12.37).

Chapter 13. Mollusca. Brothers—orginal (13.13, 13.54, 13.75, 13.76a, 13.77, 13.183); Miner 1950 (13.1, 13.2, 13.4, 13.5, 13.6, 13.7—top, 13.8, 13.9, 13.10, 13.11, 13.12, 13.15, 13.16, 13.17, 13.25, 13.26, 13.27, 13.28, 13.35, 13.36, 13.37, 13.38, 13.40, 13.41, 13.42, 13.43, 13.44, 13.47, 13.50, 13.52, 13.55, 13.56, 13.58, 13.65, 13.66, 13.67, 13.68, 13.69, 13.71, 13.74, 13.79, 13.81, 13.83, 13.84, 13.87, 13.88, 13.89, 13.90, 13.91, 13.92, 13.93, 13.96, 13.98, 13.101, 13.102, 13.103, 13.104, 13.105, 13.107, 13.108, 13.109, 13.110, 13.111, 13.112, 13.114—center, bottom, 13.115, 13.116, 13.120, 13.122, 13.133, 13.136, 13.140, 13.141, 13.144, 13.146, 13.147, 13.148, 13.149, 13.150, 13.151, 13.152, 13.153, 13.154, 13.155, 13.156, 13.157, 13.158, 13.159, 13.162, 13.163, 13.166, 13.167, 13.168, 13.172, 13.175, 13.176, 13.177, 13.180, 13.181, 13.182, 13.186a, 13.187, 13.188, 13.189, 13.191, 13.193, 13.194, 13.195, 13.196, 13.197, 13.199, 13.200, 13.201, 13.202, 13. 203, 13.204, 13.205, 13.206, 13.207, 13.209, 13.210, 13.211, 13.212, 13.214, 13.216, 13.217, 13.218, 13.220, 13.223, 13.224, 13.226a, 13.227, 13.228, 13.229, 13.230, 13.231, 13.232, 13.233, 13.234, 13.235, 13.236); Sars 1878 (13.14, 13.22, 13.29, 13.53, 13.80, 13.85, 13.99, 13.100, 13.107, 13.114—top, 13.128, 13.165, 13.170, 13.171, 13.225); Smith 1964 (13.39, 13.72, 13.121, 13.123, 13.124, 13.125, 13.126, 13.127, 13.129, 13.130, 13.131, 13.132, 13.135, 13.137, 13.138, 13.139, 13.142, 13.149, 13.226b); Verrill andSmith 1874 (13.3, 13.24, 13.30, 13.31, 13.32, 13.33, 13.34, 13.48, 13.49, 13.62, 13.64, 13.73, 13:76b, 13.78, 13.82, 13.86, 13.95, 13.117, 13.118, 13.119, 13.134, 13.135, 13.145, 13.160, 13.161, 13.178, 13.179, 13.184, 13.186b, 13.190, 13.192, 13.198, 13.208, 13.213, 13.215, 13.219, 13.221, 13.222, 13.226b; Verrill 1884 (13.7—bottom, 13.51); Verrill 1885 (13.113, 13.237); original (13.18, 13.19, 13.20, 13.21, 13.23, 13.45, 13.46, 13.57, 13.59, 13.60, 13.61, 13.63, 13.70, 13.94, 13.97, 13.106, 13.173, 13.174, 13.185, 13.206 after photographs in Abbott 1993), (13.143, 13.164, 13.169).

Chapter 14. Annelida. Miner 1950 (14.1, 14.2—ml,fc,el,as, 14.6, 14.7—po,ro,sq, 14.8, 14.9, 14.10, 14.11, 14.13, 14.15, 14.16, 14.19, 14.22, 14.30, 14.32, 14.34, 14.35—bottom, 14.44, 14.47, 14,58, 14.59a, 14.60-top, 14.80, 14.81, 14.82, 14.83, 14.84-right, 14.91, 14.111a, 14.121b, 14.128, 14.130, 14.133, 14.140, 14.145b, 14.147, 14.150, 14.151, 14.160, 14.168a, 14.172a, 14.173, 14.174, 14.179, 14.182, 14.183, 14.184, 14.185, 14.186); McIntosh 1922 (14.4, 14.56a, 14.57a, 14.134, 14.135, 14.137, 14.138, 14.142, 14.146b); Pettibone 1963 (14.2—pl,bu,sl,sp,do, 14.7—an, 14.14, 14.26, 14.27, 14.28, 14.29, 14.35-top, 14.36, 14.37, 14.38. 14.39, 14.40, 14.41, 14.42, 14.43, 14.45, 14.46, 14.48, 14.49, 14.50, 14.51, 14.52, 14.53, 14.54, 14.60-bottom, 14.61, 14.62, 14.63, 14.64, 14.65, 14.66, 14.67, 14.68, 14.69, 14.70, 14.71, 14.72, 14.73, 14.74, 14.75, 14.76, 14.77, 14.78, 14.79, 14.84—left, 14.85, 14.86, 14.88, 14.89, 14.92, 14.93, 14.94, 14.96, 14.97, 14.98, 14.99, 14.100, 14.101, 14.102, 14.103, 14.104, 14.105, 14.106, 14.107, 14.108, 14.109, 14.110, 14.111b, 14.112, 14.113, 14.114, 14.115, 14.116. 14.117, 14.118, 14.119, 14.120, 14.122, 14.123, 14.139, 14.152, 14.153, 14.154, 14.155, 14.156, 14.157, 14.158, 14.159, 14.161, 14.162, 14.163, 14.164, 14.165, 14.175, 14.176); Smith 1964 (14.3, 14.5, 14.12, 14.17, 14.20, 14.21a, 14.24, 14.25, 14.33, 14.55, 14.56b, 14.57b, 14.59b, 14.95b, 14.121a, 14.124, 14.126, 14.127, 14.129, 14.131b, 14.132, 14.136, 14.141, 14.143, 14.144, 14.146b, 14.148, 14.149, 14.166, 14.168b, 14.169, 14.170, 14.172b); Verrill 1881 (14.21b, 14.23, 14.31); Verrill in Hartman 1944 (14.2—tr, 14.87, 14.90, 14.95—a and c, 14.125, 14.131a, 14.145a, 14.167, 14.171); orginal (14.18, 14.177, 14.178, 14.180, 14.181).

Chapter 15. Arthropoda. Brothers—orginal (15.56, 15.189, 15.198a, 15.233, 15.234, 15.235, 15.236, 15.237, 15.238); Miner 1950 (15.1, 15.2, 15.3, 15.4, 15.5, 15.6, 15.7, 15.8, 15.9, 15.10, 15.11, 15.13—bottom three, 15.14, 15.15, 15.17, 15.18, 15.19, 15.20, 15.21, 15.32, 15.33, 15.34, 15.38b, 15.39, 15.40a, 15.46, 15.47, 15.52, 15.57, 15.59, 15.60, 15.61, 15.62, 15.63, 15.65c, 15.66c, 15.67, 15.68, 15.69, 15.70, 15.85, 15.87, 15.88, 15.90, 15.92, 15.93, 15.95, 15.96, 15.97, 15.122, 15.123, 15.124, 15.125, 15.126, 15.127, 15.128, 15.130, 15.131, 15.132, 15.140, 15.142, 15.149, 15.150, 15.160, 15.161, 15.162, 15.163, 15.164, 15.165, 16.166, 15.167, 15.168, 15.169, 15.170, 15.171, 15.172, 15.173, 15.174, 15.175, 15.176, 15.177, 15.181, 15.182, 15.183, 15.184, 15.185, 15.188, 15.191, 15.192, 15.197a, 15.199a, 15.202, 15.215, 15.217, 15.224, 15.227, 15.230, 15.232); Pilsbry 1916 (15.36b, 15.37, 15.38a, 15.40b, 15.41, 15.43, 15.44); Smith 1964 (15.23, 15.24, 15.25, 15.26, 15.27, 15.28, 15.29, 15.30, 15.31, 15.35, 15.36a, 15.45, 15.48, 15.49, 15.50, 15.51, 15.52, 15.53, 15.54, 15.55, 15.58, 15.64a and b, 15.65a and b, 15.66a and b, 15.148, 15.151, 15.152, 15.153, 15.154, 15.156, 15.157, 15.158, 15.159, 15.200, 15.218); Verrill and Smith 1874 (15.180b, 15.186, 15.203, 15.214, 15.221); Williams 1965 (15.12, 15.13—top, 15.16, 15.22, 15.161, 15.162, 15.178, 15.179, 15.180a, 15.187, 15.190, 15.194, 15.195, 15.196, 15.197b, 15.198b, 15.199b, 15.201, 15.204, 15.205, 15.206, 15.207, 15.208, 15.209, 15.210, 15.211, 15.213, 15.216, 15.219, 15.220, 15.222, 15.223, 15.226, 15.228, 15.229, 15.231); original (15.42 after Pilsbry 1916), (15.155 after Tattersall 1951), (15.193 after photograph in Williams and McDermott 1990), (15.212, 15.225 after photographs in Williams 1984).

The following illustrations are taken from *Shallow Water Gammaridean Amphipods of New England,* by E. L. Bousfield (1973). Courtesy of the Canadian Museum of Nature, Ottawa, Canada. The plate number in the original work is followed by the figure number or numbers in this book.

I (15.106) *Gammarus oceanicus;* II (15.102, 15.105) *G. annulatus* and *G. lawrencianus;* III (15.99) *G. mucronatus;* IV (15.109, 15.107) *G. tigrinus* and *G. daiberi;* V (15.103, 15.108) *G. duebeni* and *G. fasciatus;* VI (15.72, 15.100) *Marinogammarus obstustatus;* VII (15.71, 15.104, 15.101) *M. finmarchicus* and *Gammarellus angulosus;* XI (15.94, 15.129) *Casco bigelowi;* XII (15.124, 15.123) *Listriella clymenellae* and *L. barnardi;* XIII (15.140) *Pontogeneia inermis; xiv* (15.93,15.139) *Calliopius laeviusculus* and *Pleusymetes glaber; xvi* (15.141) *Stenothoe minuta;* XIX (15.134, 15.74, 15.133) *Monoculodes edwardsi* and *M. intermedius;* XXI (15.116) *Pontoporeia femorata;* XXII (15.112) *Amphiporeia virginiana;* XXIII (15.113) *Bathyporeia quoddyensis;* XXIV (15.110, 15.117) *Protohaustorius wigleyi;* XXV (15.115) *Parahaustorius longimerus;* XXVII (15.111) *Acanthohaustorius millisi,* XXIX (15.114) *Neohaustorius (sic) biarticulatus;* XXX (15.118) *Pseudohaustorius caroliniensis;* XXXIV (15.137, 15.135) *Paraphoxus spinosus* and *Trichophoxus epistomus;* XXXV (15.138, 15.136) *Phoxocephalus holbolli* and *Harpinia propinqua;* XXXVI (15.76) *Ampelisca verrilli;* XXXVII (15.78, 15.73, 15.77) *Ampelisca vadorum* and *A. abdita;* XXXVIII (15.79) *Byblis serrata;* XXXIX (15.98) *Dexamine thea;* XL (15.129) *Psammonyx nobilis;* XLIV (15.119, 15.120) *Hyale nilssoni* and *H. plumulosa;* XLV (15.146) *Orchestia grillus;* XLVI (15.143, 15.147) *O. uhleri* and *O. platensis;* XLVII (15.145, 15.144) *Talorchestia megalophthalma* and *T. longicornis;* XLVIII (15.84, 15.87) *Leptocheirus plumulosus* and *L. pinguis;* L (15.86) *Lembos websteri;* LI (15.85, 15.90) *L. smithi* and *Unciola irrorata;* LII (15.89, 15. 91) *U. dissimilis* and *U. serrata;* LIV (15.82, 15.81) *Ampithoe rubricata* and *A. longimana;* LV (15.83, 15.80) *A. valida* and *Cymadusa compta;* LVI (15.121) *Photis reinhardi;* LX (15.75, 15.95) *Cerapus tubularis.*

Chapter 16. Echinodermata. Brothers—original (16.11); Miner 1950 (16.1, 16.2, 16.6, 16.7, 16.8, 16.9, 16.10, 16.12, 16.13, 16.14, 16.15, 16.16, 16.17, 16.18, 16.19, 16.20, 16.21, 16.22, 16.23, 16.24, 16.25, 16.26, 16.27, 16.28, 16.29, 16.31, 16.32, 16.33, 16.34, 16.38, 16.39, 16.40, 16.41, 16.42, 16.43, 16.44); original (16.3, 16.4, 16.5 after Gray et al. 1968), (16.30, 16.35, 16.36, 16.37 after Miner, 1950).

Chapter 17. Chordata. Bigelow and Schroeder, 1953 (all fish, Figs. 17.28 through 17.197, except for those listed under Hildebrand and Schroeder below); Brothers—original (17.198, 17.199, 17.200, 17.201, 17.202, 17.203, 17.204, 17.205, 17.206, 17.207, 17.208, 17.209, 17.210, 17.211, 17.212, 17.213, 17.214, 17.215, 17.216, 17.217, 17.218, 17.219, 17.220, 17.221, 17.222, 17.223, 17.224, 17.225, 17.226, 17.227); Hildebrand and Schroeder, 1927 (17.56—1+2l, 17.63, 17.64, 17.82, 17.91, 17.104, 17.106, 17.112, 17.133, 17.142, 17.143, 17.148, 17.149, 17.155, 17.156,17.176, 17.191); Miner, 1950 (17.1, 17.2, 17.3, 17.4, 17.5, 17.6, 17.7, 17.8, 17.16, 17.19a, 17.20, 17.21, 17.24, 17.25, 17.26, 17.27); Smith, 1964 (17.9, 17.10, 17.11, 17.12, 17.13, 17.14, 17.15, 17.16. 17.17, 17.22, 17.23). Original (17.18, 17.19b).

Illustration Sources

Abbott, R. T. 1993. *Seashells of the Northern Hemisphere.* Stamford, Conn.: Longmeadow Press. Permission to redraw figures from photographs in this volume was granted from American Malacologists, Inc., Melbourne, Fla.

Bigelow, H. B., and W. C. Schroeder. 1953. "Fishes of the Gulf of Maine." *Fisheries Bulletin* 74:1–577.

Bousfield, E. L. 1973. *Shallow-Water Gammaridean Amphipoda of New England.* Ithaca, N.Y.: Cornell University Press. Figures used with permission of the Canadian Museum of Nature, Ottawa, Canada.

Brooks, W. K. 1890. *Handbook of Invertebrate Zoology for Laboratories and Seaside Work.* Boston: Bradlee Whidden.

Calder, D. R., 1975. "Biotic Census of Cape Cod Bay: Hydroids." *Biological Bulletin* 149:287–315. Figures used with permission of the *Biological Bulletin.*

D'Asaro, C. N. 1993. "Gunnar Thorson's World-Wide Collection of Prosobranch Egg Capsules: Nassariidae." *Ophelia* 38:149–215.

Fraser, C. M. 1944. *Hydroids of the Atlantic Coast of North America.* Toronto: University of Toronto Press. Figures used with permission of the University of Toronto Press.

Gray, I. E., M. E. Downey, and M. J. Cerame-Vivas. 1968. "Sea-Stars of North Carolina." *Fishery Bulletin* 67:127–163.

Gurney, R. 1942. *Larvae of Decapod Crustacea.* London: Ray Society.

Hargitt, C. W. 1904. "The Medusae of the Woods Hole Region." *Bulletin of the Bureau of Fisheries* 24:21–79 + 7 plates.

Hartlaub, C. 1907. "Anthomedusen." *Nordisches Plankton: Zoologischer Teil: Sechster Band: Coelenterata: XII. Craspedote Medusen* 1:1–136.

Hartman, O. 1944. "New England Annelida. Part 2. Including the Unpublished Plates by Verrill with Reconstructed Captions." *Bulletin of the American Museum of Natural History* 82:327–344 + 23 plates.

Hildebrand, S. F., and W.C. Schroeder. 1927. "Fishes of Chesapeake Bay." *Fisheries Bulletin* 43, Part 1:1–388. (Reprinted 1972 for the Smithsonian Institution by T.F.H. Publications, Neptune, N. J.)

Hoek, P. P. C. 1909. "Entomostraca: Cirripedien und Cirripedienlarven." *Nordisches Plankton: Zoologischer Teil* 4:265–332.

Hyman, L. H. 1939. "Some Polyclads of the New England Coast, Especially of the Woods Hole Region." *Biological Bulletin* 76:127–152.

Kramp, P. L. 1933. "Coelenterata: XII. Craspedote Medusen, 3. Leptomedusan." *Nordisches Plankton: Zoologischer Teil* 6:541–602.

Lohmann, H. 1901a. "Echinoderma, Vermes: IX. Cyphonautes." *Nordisches Plankton: Zoologischer Teil* 5:31–40.

———. 1901b. "Tunicata and Mollusca: Ascidienlarven." *Nordisches Plankton: Zoologischer Teil* 2:31–47.

Martinez, A. J. 1994. *Marine Life of the North Atlantic, Canada to New England.* Wenham, Mass.: Marine Life.

Maturo, F. J. S. 1957. "A Study of the Bryozoa of Beaufort, NC and Vicinity." *Journal of the Elisha Mitchell Scientific Society* 73:11–68. Figures used with permission of the North Carolina School of Science and Mathematics.

McIntosh, W. C. 1922. *A Monograph of the British Marine Annelids. Vol. IV. Part 1. Polychaeta—Hermellidae to Sabellidae.* London: Ray Society.

Miner, R. W. 1950. *Field Book of Seashore Life.* New York: G. P. Putnam's Sons. Figures used with permission of the Putnam Publishing Group, New York.

Mortensen, T. 1901. "Echinoderma, Vermes: IX. Echinodermen-Larven." *Nordisches Plankton: Zoologischer Teil* 5:1–30.

Osburn, R.C. 1912. "The Bryozoa of the Woods Hole Region." *Bulletin of the U.S. Bureau of Fisheries* 30:18–31.

Pettibone, M. H. 1963. "Marine Polychaete Worms of the New England Region. 1. Aphroditidae through Trochochaetidae." *Bulletin of the U.S. National Museum* 227:1–356. Figures used with permission of the author.

Pilsbry, H. A. 1916. "The Sessile Barnacles (Cirripedia) Contained in the Collections of the U.S. National Museum, Including a Monograph of the American Species." *Bulletin of the U.S. National Museum* 93:1–366.

Rogick, M.D., and H. Croasdale. 1949. "Studies on Marine Bryozoa. III. Woods Hole Region Bryozoa Associated with Algae." *Biological Bulletin* 96:32–69. Figures used with permission of the *Biological Bulletin.*

Sanders, H. L. 1963. "The Cephalocarida: Functional Morphology, Larval Development, Comparative External Anatomy." *Memoirs of the Connecticut Academy of Arts and Sciences* 15:1–80.

Sars, G. O. 1878. "Mollusca Regionis Arcticae Norvegiae." *Bidrag til Kundskaben om Norges Arktiske Fauna* 1:1–466 pages + 52 plates.

———. 1903. "Copepoda Calanoida." *An Account of the Crustacea of Norway* 4:1–171.

———. 1918. "Copepoda Cyclopoida." *An Account of the Crustacea of Norway* 6:1–225.

Smith, R. I. 1964. *Keys to Marine Invertebrates of the Woods Hole Region.* Woods Hole, Mass.: Marine Biological Laboratory. Figures used with permission of the Director of Communications, Marine Biological Laboratory.

Tattersall, W. M. 1951. "A Review of the Mysidacea of the United States National Museum." *Bulletin of the U.S. National Museum* 201:1–292.

Vannucci, M. 1959. *Catalogue of Marine Larvae.* Sao Paolo: Universidade de Sao Paolo, Instituto Oceanografico.

Verrill, A. E. 1881. "New England Annelida. Part 1—Historical Sketch with Annotated Lists of the Species Hitherto Recorded." *Transactions of the Connecticut Academy of Arts and Sciences* 4:285–324 + 10 plates.

———. 1884. "Second Catalogue of Mollusca Recently Added to the Fauna of the New England Coast and the Adjacent Parts of the Atlantic, Consisting Mostly of Deep-Sea Species, with Notes on Others Previously Recorded." *Transactions of the Connecticut Academy of Arts and Sciences* 6:139–294 + 5 plates.

———. 1885. "Third Catalogue of Mollusca Recently Added to the Fauna of the New England Coast and the Adjacent Parts of the Atlantic, Consisting Mostly of Deep-Sea Species, with Notes on Others Previously Recorded." *Transactions of the Connecticut Academy of Arts and Sciences* 6:395–452 + 3 plates.

———. 1892a. "Marine Nemerteans of New England and Adjacent Waters." *Transactions of the Connecticut Academy of Arts and Sciences* 8:328–456.

———. 1892b. "Marine Planarians of New England." *Transactions of the Connecticut Academy of Arts and Sciences* 8:459–520.

———. 1901–02. "The Bermuda Islands: Their Scenery, Climate, Products, Natural History, and Geology; with Sketches of Their Early History and the Changes Due to Man." *Transactions of the Connecticut Academy of Arts and Sciences* 11:413–956 + plates 65–104.

Verrill, A. E., and S. I. Smith. 1874. "Report upon the Invertebrate Animals of Vineyard Sound and Adjacent Waters, with an Account of the Physical Characters of the Region." *Report of the U.S. Fisheries Commission* 1871 and 1872:295–771 + 38 plates.

Williams, A. B. 1965. "Marine Decapod Crustaceans of the Carolinas." *Fisheries Bulletin* 65:1–298.

———. 1984. *Shrimps, Lobsters, and Crabs of the Atlantic Coast of the Eastern United States, Maine to Florida.* Washington: Smithsonian Institution Press.

Williams, A. B., and J. J. McDermott. 1990. "An Eastern United States Record for the Western Indo-Pacific Crab, *Hemigrapsus sanguineus* (Crustacea: Decapoda: Grapsidae)." *Proceedings of the Biological Society of Washington* 103:108–109.

Williamson, C. 1915. "Crustacea. Decapoda-Larven." *Nordisches Plankton: Zoologischer Teil* 3:315–588.

■ Index

Two notations appear with each species listing in this index. They are included here to save space in the text. First, the author and date of the species's initial description are given. Following taxonomic convention, the author and date appear in parentheses in cases in which the species was originally described in a genus other than that of its current placement. Second, a species number is given, which follows the National Oceanographic Data Center's Taxonomic Serial Numbers (NODC Taxonomic Code, version 8.0; National Oceanic and Atmospheric Administration, National Environmental Satellite, Data, and Information Service, NODC, Silver Spring, MD 20910). Numeric codes are intended to improve management efficiency in computerized data sets. As an economy in this index, the official "SPE" (species) or "SSPE" (subspecies) parts of serial numbers have been shortened to "S" or "SS" respectively.

■ About the Author

Leland W. Pollock is a professor of biology at Drew University in Madison, New Jersey. A native of greater Boston, his formal interest in the natural history of the northeastern United States began through employment as a naturalist at the Blue Hills Trailside Museum (Canton, Mass.). A marine emphasis developed during his undergraduate years at Bates College (Lewiston, Maine) and was nurtured at the University of New Hampshire, where he earned a Ph.D. in 1969. His research interests in the microscopic members of the invertebrate phylum Tardigrada, or "water bears," led to predoctoral research with the Systematics Ecology Program of the Marine Biological Laboratory (Woods Hole, Mass.), followed by postdoctoral work there and at the Wellcome Marine Laboratory (Robin Hood's Bay, England). These studies have resulted in the publication of 17 articles in the research literature. Work at the George M. Gray museum (formerly located at Woods Hole) and teaching field courses in marine science to undergraduate students for 25 years at Drew University and for 10 years at the Shoals Marine Laboratory (Isles of Shoals, Maine) stimulated the development of this identification guide.